College Algebra and Trigonometry

College Algebra and Trigonometry

RICHARD N. AUFMANN

VERNON C. BARKER

RICHARD D. NATION, JR.

Palomar College

HOUGHTON MIFFLIN COMPANY   BOSTON
Dallas   Geneva, Illinois   Palo Alto   Princeton, New Jersey

Cover photograph by Edward Slaman

Library of Congress Number 89-080910

ISBN: Examination copy 0-395-52622-1
        Text 0-395-38096-0

Printed in the U.S.A.

ABCDEFGHIJ-VH-9543210/89

Contents

Preface

This text provides a comprehensive and mathematically sound treatment of the topics considered essential for a college algebra and trigonometry course. It is intended for the student who has successfully completed an intermediate algebra course.

To help the student master the concepts in this text, we have tried to maintain a balance among theory, application, and drill. Each definition and theorem is precisely stated and many theorems are proved. Carefully developed mathematics is complemented by abundant, creative applications that are both contemporary and representative of a wide range of disciplines.

Extensive exercise sets ranging from routine exercises to thought-provoking problems are provided at the end of each section. An ample selection of review exercises can be found at the end of each chapter.

Features

Interactive Presentation   *College Algebra and Trigonometry* is written in a style that encourages the student to interact with the textbook. At various places throughout the text, a question in the form of (Why?) is asked of the reader. This question encourages the reader to pause and think about the current discussion and to answer the question. To make sure the student does not miss important information, the answer to the question is provided as a footnote on the same page.

Each section contains a variety of worked examples. Each example is given a name so that a student can see at a glance the type of problem being illustrated. Each example is accompanied by annotations that assist the student in moving from step to step. Following the worked example is a suggested exercise from the exercise set for the student to work. The exercises are color coded by number in the exercise set and the complete solution of that exercise can be found in an appendix in the text.

Extensive Exercise Sets   The exercise sets of *College Algebra and Trigonometry* were carefully developed to provide the student with a variety of exercises. The exercises range from drill and practice to interesting challenges and were chosen to illustrate the many facets of topics discussed in the text. Besides the regular exercise sets, there is a set of supplemental problems that includes material from previous chapters, present extensions of topics, or are of the form "prove or disprove."

Applications   One way to motivate a student to an interest in mathematics is through applications. The applications in *College Algebra and Trigonometry* have been taken from agriculture, architecture, biology, business, chemistry, earth science, economics, engineering, medicine, and physics. Besides providing motivation to study mathematics, the applications provide an avenue to problem solving. The applications problems require the student to organize and implement a problem solving scheme.

Supplements for the Student

In addition to the student Study Guide, two computerized study aids, the Computer Tutor and the Math Assistant, accompany this text.

Study Guide   The study guide contains complete solutions to all odd-numbered problems in the text, plus practice chapter tests with answers.

The Computer Tutor   The Computer Tutor is an interactive instructional microcomputer program for student use. Each section in the text is supported by a lesson on the Computer Tutor. Lessons on the tutor provide additional instruction and practice and can be used in several ways: (1) to cover material the student missed because of absence from class; (2) to reinforce instruction on a concept that the student has not yet mastered; (3) to review material in preparation for examinations. This tutorial is available for the IBM PC and compatible microcomputers.

The Math Assistant   The Math Assistant is a collection of programs that can be used by both the instructor and the student. Some programs are instructional and allow the student to practice a skill like finding the inverse of a matrix. Other programs are computational routines that perform numerical calculations. In addition, there is a function grapher that graphs elementary functions and polar equations. The Math Assistant is available for the IBM PC and compatible microcomputers.

Supplements for the Instructor

College Algebra and Trigonometry has an unusually complete set of teaching aids for the instructor.

Solutions Manual   The Solutions Manual contains worked-out solutions for all end-of-section, supplemental, challenge and review exercises.

Instructor's Manual with Testing Program   The Instructor's Manual contains the printed testing program, which is the first of three sources of testing material available to the user. Four printed tests (in two formats — free response and multiple choice) are provided for each chapter. In addition, the Instructor's Manual includes documentation of all the software ancillaries — the Math Assistant, the Computer Tutor, and the Computerized Test Generator. Finally, it contains answers to all the even-numbered exercises in the text.

Computerized Test Generator The Computerized Test Generator is the second source of testing material. The data base contains more than 1500 test items. These questions are unique to the test generator and do not repeat items provided in the Instructor's Manual testing program. The Test Generator is designed to produce an unlimited number of tests for each chapter of the text, including cumulative tests and final exams. It is available for the IBM PC and compatible microcomputers, the Macintosh microcomputer and the Apple II family of microcomputers.

Printed Test Bank The Printed Test Bank, the third component of the testing material, is a printout of all items in the Computerized Test Generator. Instructors using the Test Generator can use the test bank to select specific items from the data base. Instructors who do not have access to a computer can use the test bank to select items to be included on a test being prepared by hand.

Acknowledgments

We would like to express our gratitude to Linda Murphy of Northern Essex Community College who read the entire book for accuracy. In addition, we sincerely wish to thank the following reviewers who reviewed the manuscript at various stages of development for their valuable contributions.

Frank Battles, Massachusetts Maritime Academy
Diane Johnson, University of Rhode Island
Alberto L. Delgado, Kansas State University
Judith L. Willoughby, Minneapolis Community College
Mel Hamburger, Laramie County Community College
Janice McFatter, Gulf Coast Community College
John W. Burns, Mt. San Antonio College
Cynthia Moody, City College of San Francisco
Archie L. Ritchie, Peace College
W. R. Wilson, Delta State University
Pauline Lowman, Western Kentucky University
F. P. Mathur, California State Polytechnic University
James Younglove, University of Houston
Jim Delany, California Polytechnic State University
David C. Kay, University of North Carolina
Beth D. Layton, Appalachian State University
Gillian C. Raw, University of Missouri–St. Louis
Burdette C. Wheaton, Mankato State University
Ann Thorne, College of DuPage
Gerald Schrag, Central Missouri State University

1

Algebra Review

The infinite! No other question has ever moved so profoundly the spirit of man.

DAVID HILBERT (1921)

HOW LARGE IS INFINITY?

$\aleph_0$ $\aleph_1$ $\aleph_2$ $\aleph_3$ $\aleph_4$ $\aleph_5$ $\aleph_6$ $\aleph_7$ $\aleph_8$ $\aleph_9$ $\aleph_{10}$ $\aleph_{11}$ $\aleph_{12}$ ...

The **cardinality** of a finite set is the number of elements in the set. For example, the set $\{6, 28, 496, 8128\}$ has a cardinality of four, because it contains four elements. In general, the cardinality of $\{a_1, a_2, a_3, \ldots, a_n\}$ is n.

The German mathematician Greg Cantor (1845–1918) devoted much of his life to the study of the cardinality of infinite sets. He made use of the concept of a one-to-one correspondence between the elements of two sets. For example, the sets

$$\{1, 2, 3\} \quad \text{and} \quad \{5, 7, 11\}$$

each have three elements. One method of establishing a one-to-one correspondence between the elements of these sets is as follows:

$$\{1, \quad 2, \quad 3\}$$
$$\updownarrow \quad \updownarrow \quad \updownarrow$$
$$\{5, \quad 7, \quad 11\}$$

The two sets are said to be in a one-to-one correspondence because each element of the first set corresponds with exactly one element of the second set, and each element of the second set corresponds with exactly one element of the first set.

Cantor reasoned that if a one-to-one correspondence could be established between the elements of two sets, then the two sets have the same cardinality. The set of natural numbers

$$\{1, 2, 3, 4, 5, \ldots\}$$

is an infinite set. Its cardinality is denoted $\aleph_0$, which is read as aleph null. But what is the cardinality of each of the following sets?

Integers $\{\ldots, -3, -2, -1, 0, 1, 2, 3, \ldots\}$
Irrational numbers $\{$all nonterminating, nonrepeating decimals$\}$

Because Cantor was able to establish a one-to-one correspondence between the set of natural numbers and the set of integers, he concluded that the cardinality of the set of integers is also $\aleph_0$. You might be tempted to think that all infinite sets have a cardinality of $\aleph_0$; however, Cantor was able to show that the set of irrational numbers has a cardinality that is larger than $\aleph_0$.

1.1

The Real Number System

Human beings share the desire to organize and classify. Ancient astronomers classified stars into groups called constellations. Modern astronomers continue to classify stars by such characteristics as color, mass, size, temperature, and distance from earth. In mathematics it is useful to classify numbers into groups called **sets**. The following sets of numbers are used extensively in the study of algebra:

Integers	$\{\ldots, -3, -2, -1, 0, 1, 2, 3, \ldots\}$
Rational numbers	{all terminating or repeating decimals}
Irrational numbers	{all nonterminating and nonrepeating decimals}
Real numbers	{all rational or irrational numbers}

The real numbers are denoted by the symbol R. Every real number can be written as a (1) terminating decimal, (2) repeating decimal, or (3) nonterminating and nonrepeating decimal.

If a decimal terminates or repeats a block of digits, then the number is a rational number. Rational numbers can also be defined as all numbers that can be written in the form p/q, where p and q are integers and $q \neq 0$. It is important to be able to recognize a rational number, whether it is written in its **fractional form** p/q or as a decimal that either terminates or repeats. For example, the rational number $\frac{3}{4}$ can be written as the terminating decimal 0.75 and the rational number $\frac{5}{11}$ can be written as the repeating decimal $0.\overline{45}$. The bar over the 45 means that the block of digits 45 repeats without end; that is, $0.\overline{45} = 0.454545\ldots$.

In its decimal form, an irrational number neither terminates nor repeats. For example, $0.272272227\ldots$ is a nonterminating, nonrepeating decimal and thus is an irrational number.

One of the best known irrational numbers is pi, denoted by the Greek symbol π. The number π is defined as the ratio of the circumference of a circle to its diameter. Often in applications the rational numbers 3.14 or $\frac{22}{7}$ are used as approximations of the irrational number π.

Additional examples of irrational numbers include $\sqrt{2}$, $-\sqrt{5}$, and $\sqrt{10}$. Each of these numbers has a decimal form that does not terminate and does not repeat.

Remark Many calculators display $\sqrt{2}$ as 1.414213562. Do *not* interpret this as a terminating decimal; it is merely an approximation of $\sqrt{2}$, which in decimal form is a nonterminating, nonrepeating decimal.

Each number in a set is called an **element** of the set. A set A is a **subset** of a set B if every element of set A is also an element of set B. The set of **negative integers** $\{-1, -2, -3, -4, -5, \ldots\}$ is a subset of the set of integers. The set of **positive integers** $\{1, 2, 3, 4, 5, \ldots\}$ (also known as the set of **natural numbers**) is also a subset of the integers.

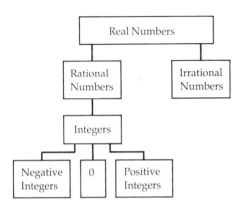

Figure 1.1

Figure 1.1 illustrates the subset relationships among the sets defined above. Notice that every integer is a rational number, every rational number is a real number, every irrational number is a real number, and the real numbers consist of both rational and irrational numbers.

EXAMPLE 1   Classify Real Numbers

Determine which of the numbers -5, -0.2, 0, $0.\overline{3}$, 7, $0.40440444044440\ldots$ are:

a. integers
b. rational numbers
c. irrational numbers
d. real numbers

Solution

a. integers: -5, 0, 7
b. rational numbers: -5, -0.2, 0, $0.\overline{3}$, 7
c. irrational number: $0.40440444044440\ldots$
d. real numbers: -5, -0.2, 0, $0.\overline{3}$, 7, $0.40440444044440\ldots$

■ *Try Exercise* **2**, *page 9.*

Addition, multiplication, subtraction, and **division** are the operations of arithmetic. Addition of the two real numbers a and b is designated by $a + b$. If $a + b = c$, then c is the **sum** and the real numbers a and b are called the **terms.**

Multiplication of the real numbers a and b is designated by ab or $a \cdot b$. If $ab = c$, then c is the **product** and the real numbers a and b are called **factors** of c.

The number $-b$ is referred to as the **additive inverse** of b. The subtraction designated by $a - b$ can be performed by adding a and the additive inverse of b. That is,

$$a - b = a + (-b).$$

If $a - b = c$, then c is called the **difference** of a and b.

Two real numbers whose product is 1 are called **multiplicative inverses** or **reciprocals** of each other. The reciprocal of the nonzero number b is $1/b$. The division of a and b designated by $a \div b$ with $b \neq 0$ can be performed by multiplying a and the reciprocal of b. That is,

$$a \div b = a\left(\frac{1}{b}\right) \quad \text{provided } b \neq 0.$$

If $a \div b = c$, then c is called the **quotient** of a and b.

The notation $a \div b$ is often represented by the fractional notation a/b or $\frac{a}{b}$. The real number a is the **numerator,** and the nonzero real number b is the **denominator** of the fraction. The following **properties of fractions** will be used throughout this text.

Properties of Fractions

For all fractions a/b and c/d where $b \neq 0$ and $d \neq 0$:

Equality $\qquad\qquad \dfrac{a}{b} = \dfrac{c}{d}$ if and only if $ad = bc$

Equivalent fractions $\quad \dfrac{a}{b} = \dfrac{ac}{bc}, \qquad c \neq 0$

Addition $\qquad\qquad \dfrac{a}{b} + \dfrac{c}{b} = \dfrac{a+c}{b}$

Subtraction $\qquad\quad \dfrac{a}{b} - \dfrac{c}{b} = \dfrac{a-c}{b}$

Multiplication $\qquad \dfrac{a}{b} \cdot \dfrac{c}{d} = \dfrac{ac}{bd}$

Division $\qquad\qquad \dfrac{a}{b} \div \dfrac{c}{d} = \dfrac{a}{b} \cdot \dfrac{d}{c} = \dfrac{ad}{bc}, \qquad c \neq 0$

Sign properties $\qquad -\dfrac{a}{b} = \dfrac{-a}{b} = \dfrac{a}{-b}$

Remark The equality property of fractions contains the terminology "if and only if," which implies each of the following:

$$\text{If } \frac{a}{b} = \frac{c}{d}, \qquad \text{then } ad = bc.$$

$$\text{If } ad = bc, \qquad \text{then } \frac{a}{b} = \frac{c}{d}.$$

The number zero has many special properties. The following division properties of zero will play an important role in this text.

Division Properties of Zero

For $a \neq 0$, $\dfrac{0}{a} = 0$.   Zero divided by any nonzero number is zero.

$\dfrac{a}{0}$ is undefined.   Division by zero is undefined.

The properties of fractions and the division properties of zero can be used to find the sum, difference, product, or quotient of fractions.

EXAMPLE 2   **Compute with Fractions**

Use the properties of fractions or the division properties of zero to find the following indicated sums, differences, products, or quotients. Assume that a is a nonzero real number.

a. $\dfrac{2a}{3} - \dfrac{a}{5}$ b. $\dfrac{2a}{5} \cdot \dfrac{3a}{4}$ c. $\dfrac{5a}{6} \div \dfrac{3a}{4}$ d. $\dfrac{0}{3a}$

Solution

a. Because the fractions do not have a common denominator, we first rewrite each fraction as an equivalent fraction with a common denominator of 15. This is accomplished by multiplying both the numerator and the denominator of $2a/3$ by 5 and by multiplying both the numerator and the denominator of $a/5$ by 3.

$$\frac{2a}{3} - \frac{a}{5} = \frac{2a(5)}{3(5)} - \frac{a(3)}{5(3)} = \frac{10a}{15} - \frac{3a}{15}$$

$$= \frac{10a - 3a}{15} = \frac{7a}{15}$$

b. $\dfrac{2a}{5} \cdot \dfrac{3a}{4} = \dfrac{(2a)(3a)}{(5)(4)} = \dfrac{6a^2}{20} = \dfrac{2 \cdot 3a^2}{2 \cdot 10} = \dfrac{3a^2}{10}$

c. $\dfrac{5a}{6} \div \dfrac{3a}{4} = \dfrac{5a}{6} \cdot \dfrac{4}{3a} = \dfrac{20a}{18a} = \dfrac{2 \cdot 10a}{2 \cdot 9a} = \dfrac{10}{9}$

d. $\dfrac{0}{3a} = 0$ Zero divided by any nonzero number is zero.

■ *Try Exercise* **12,** *page 10.*

An **equation** is a statement of equality between two numbers or two expressions. An equation consists of two parts called **sides**, which are separated by an **equal sign**. For example, the equation $a = b$ has a as its left-hand side and b as its right-hand side. There are four basic **properties of equality** that relate to equations.

Properties of Equality

Let a, b, and c be real numbers	
Reflexive property	$a = a$
Symmetric property	If $a = b$, then $b = a$.
Transitive property	If $a = b$ and $b = c$, then $a = c$.
Substitution property	If $a = b$, then a may be replaced by b in any expression that involves a.

EXAMPLE 3 **Identify Properties of Equality**

Identify the property of equality illustrated by each of the following:

a. If $3a + b = c$, then $c = 3a + b$.

b. $5(x + y) = 5(x + y)$

c. If $4a - 1 = 7b$ and $7b = 5c + 2$, then $4a - 1 = 5c + 2$.

d. If $a = 5$ and $b(a + c) = 72$, then $b(5 + c) = 72$.

Solution

a. Symmetric   b. Reflexive   c. Transitive   d. Substitution

■ *Try Exercise* **24,** *page 10.*

A set is **closed** under addition if the sum of any two elements of the set is also an element of the set. The set of positive integers is closed under addition since it is possible to show that if a and b are both positive integers, $a + b$ is also a positive integer. The set of positive integers is not closed under subtraction, however, as shown by the example $3 - 5 = -2$. Notice that the difference of the positive integers 3 and 5 is -2, which is not a positive integer.

The real numbers satisfy the following properties.

Properties of the Set of Real Numbers

Let a, b, and c be real numbers.

	Addition Properties	**Multiplication Properties**
Closure	$a + b$ is a unique real number.	ab is a unique real number.
Commutative	$a + b = b + a$	$ab = ba$
Associative	$(a + b) + c = a + (b + c)$	$(ab)c = a(bc)$
Identity	There exists a unique real number 0 such that $a + 0 = 0 + a = a$.	There exists a unique real number 1 such that $a \cdot 1 = 1 \cdot a = a$.
Inverse	For each real number a there is a unique real number $-a$ such that $a + (-a) = (-a) + a = 0$.	For each *nonzero* real number a there is a unique real number $1/a$ such that $a(1/a) = (1/a)a = 1$.
Distributive	$a(b + c) = ab + ac$	

Notice that $\qquad 5(x + 3) = 5(3 + x)$

is an example of the commutative property of addition. It is not an example of the associative property of addition because the same numbers occur inside each of the parentheses. In brief, the commutative properties involve a change of order; that is, $a + b = b + a$. The associative properties involve a change of grouping; that is, $(a + b) + c = a + (b + c)$. The distributive property changes a product to a sum or a sum to a product; that is, $a(b + c) = ab + ac$.

We can identify which of the properties of real numbers has been used to rewrite expressions by closely comparing the expressions and noting any changes.

EXAMPLE 4   Identify Properties of Real Numbers

Identify the property of real numbers illustrated in each of the following:

a. $(2a)b = 2(ab)$

b. $\left(\dfrac{1}{5}\right)11$ is a real number.

c. $4(x + 3) = 4x + 12$

d. $(a + 5b) + 7c = (5b + a) + 7c$

e. $\left(\dfrac{1}{2} \cdot 2\right)a = 1 \cdot a$

f. $1 \cdot a = a$

Solution

a. Associative property of multiplication
b. Closure property of multiplication of real numbers
c. Distributive property
d. Commutative property of addition
e. Inverse for multiplication
f. Identify for multiplication

■ *Try Exercise **26,** page 10.*

Prime numbers and *composite numbers* play an important role in almost every branch of mathematics. A **prime number** is a natural number greater than 1 that has no natural number factors other than itself and 1. The 10 smallest prime numbers are 2, 3, 5, 7, 11, 13, 17, 19, 23, and 29. Each of these numbers has only itself and 1 as factors.

A **composite number** is a natural number greater than 1 that is not a prime number. For example, 10 is a composite number because 10 has both 2 and 5 as factors. The 10 smallest composite numbers are 4, 6, 8, 9, 10, 12, 14, 15, 16, and 18.

Sets are often written using **set-builder notation,** which makes use of a variable and a characteristic property that the elements of the set alone possess. The set of elements x with the property that they are natural numbers less than 4 is written

$$\{x \mid x \text{ is a natural number less than 4}\}$$

and is read as "the set of all elements x such that x is a natural number less than 4." Of course this is the set $\{1, 2, 3\}$.

The **empty set** or **null set** is a set without any elements. The set of numbers that are both prime and also composite is an example of the null set. The null set is denoted by the symbol $\varnothing$.

Just as addition and subtraction are operations performed on real numbers, there are operations performed on sets. Two of these set operations are called *intersection* and *union*. The **intersection** of sets A and B, denoted by $A \cap B$, is the set of all elements belonging both to set A and to set B. The **union** of sets A and B, denoted by $A \cup B$, is the set of all elements belonging to set A or set B or both.

EXAMPLE 5 **Find the Intersection and the Union of Two Sets**

Find each of the following given $A = \{0, 1, 4, 6, 9\}$, $B = \{1, 3, 5, 7, 9\}$ and $P = \{x \mid x \text{ is a single digit prime number}\}$.

a. $A \cap B$ b. $A \cap P$ c. $A \cup B$ d. $A \cup P$

Solution

a. $A \cap B = \{0, 1, 4, 6, 9\} \cap \{1, 3, 5, 7, 9\} = \{1, 9\}$ Only 1 and 9 are common to both sets.

b. First determine that $P = \{2, 3, 5, 7\}$. Therefore:

$A \cap P = \{0, 1, 4, 6, 9\} \cap \{2, 3, 5, 7\} = \varnothing$ There are no common elements.

c. $A \cup B = \{0, 1, 4, 6, 9\} \cup \{1, 3, 5, 7, 9\}$ List the elements of the first
   $= \{0, 1, 3, 4, 5, 6, 7, 9\}$ set. Include elements from the second set that are not already listed.

d. $A \cup P = \{0, 1, 4, 6, 9\} \cup \{2, 3, 5, 7\} = \{0, 1, 2, 3, 4, 5, 6, 7, 9\}$

■ *Try Exercise **42**, page 10.*

The set operations of intersection and union can also be used with sets that have an infinite number of elements. For example, given

$$E = \{x \mid x \text{ is a positive even integer}\} = \{2, 4, 6, 8, 10, \ldots\}$$

$$D = \{x \mid x \text{ is a positive odd integer}\} = \{1, 3, 5, 7, 9, \ldots\}$$

$$N = \{x \mid x \text{ is a natural number}\} = \{1, 2, 3, 4, 5, \ldots\}$$

then $E \cap D = \varnothing$ and $E \cup D = N$.

EXERCISE SET 1.1

In Exercises 1 and 2, determine which of the numbers are **a.** integers, **b.** rational numbers, **c.** irrational numbers, **d.** real numbers.

1. 0   4 $\dfrac{1}{5}$

$\dfrac{11}{3}$ $\sqrt{4}$ $\sqrt{9}$

$3.1\overline{4}$ 3.14 $-0.272272227\ldots$

2. $2.8\overline{10}$ -4.25 $-\dfrac{1}{4}$

$\dfrac{10}{2}$ $\dfrac{2}{10}$ π

$0.131313\ldots$ $0.131131113\ldots$ $\dfrac{0}{4}$

In Exercises 3 to 10, perform the indicated operations.

3. $-7 - (-15)$ **4.** $(6 - 8) - 12$

5. $(-2)(3 - 11)$

6. $(7 - 12)(5 - 21)$

7. $2 + (-5)(-3)$

8. $4 - (-4)3$

9. $(-5)(-2)(-3)$

10. $(-1) + (-3)(4)(-2)$

In Exercises 11 to 20, use the properties of fractions to perform the indicated operations. State each answer in lowest terms. Assume a is a nonzero real number.

11. $\dfrac{2a}{7} - \dfrac{5a}{7}$

12. $\dfrac{2a}{5} + \dfrac{3a}{7}$

13. $\dfrac{-3a}{5} + \dfrac{a}{4}$

14. $\dfrac{7}{8}a - \dfrac{13}{5}a$

15. $\dfrac{-5}{7} \cdot \dfrac{2}{3}$

16. $\dfrac{7}{11} \cdot \dfrac{-22}{21}$

17. $\dfrac{12a}{5} \div \dfrac{-2a}{3}$

18. $\dfrac{2}{5} \div 3\dfrac{2}{3}$

19. $\dfrac{2a}{3} - \dfrac{4a}{5}$

20. $\dfrac{1}{2a} - \dfrac{3}{a}$

21. One pipe can fill a pool in 11 hours. A second pipe can fill the same pool in 15 hours. Assume the first pipe fills $\frac{1}{11}$ of the pool every hour and the second pipe fills $\frac{1}{15}$ of the pool every hour. **a.** Find the amount of the pool the two pipes together fill in 3 hours.   **b.** Find the amount of the pool they fill together in x hours.

22. The relationship between the distance of an object d_0 from a curved mirror, the distance of its image d_i from the mirror, and the focal length f of the mirror is given by the **mirror equation**

$$\frac{1}{f} = \frac{1}{d_0} + \frac{1}{d_i}$$

What is the focal length[1] f of a mirror for which $d_0 = 25$ centimeters and $d_i = -5$ centimeters?

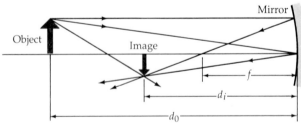

In Exercises 23 to 40, identify the property of equality or the property of real numbers that is illustrated.

23. $1 \cdot a = a$

24. If $a + b = 2$, then $2 = a + b$.

25. $3 + (2 + 5) = (3 + 2) + 5$

[1]For convex mirrors, both the focal length f and the image distance d_i are *negative* quantities.

26. $6 + (2 + 7) = 6 + (7 + 2)$

27. $a(bx) = a(bx)$

28. If $x + 2y = 7$ and $7 = y$, then $x + y/z = 7$.

29. $2 + 3$ is a real number.

30. If $x = 2(y + z)$ and $2(y + z) = 5w$, then $x = 5w$.

31. $(2.3)(5.6)$ is a real number.

32. $p(q + r) = pq + pr$

33. $m + (-m) = 0$

34. $t\left(\dfrac{1}{t}\right) = 1$

35. $7(a + b) = 7(b + a)$

36. $8(gh + 5) = 8(hg + 5)$

37. If $x + 2y = 7$ and $7 = z$, then $x + 2y = z$.

38. $5[x + (y + z)] = 5x + 5(y + z)$

39. $0 + 8z = 8z$

40. $\dfrac{\sqrt{2}}{\sqrt{3} - 1} = \dfrac{\sqrt{2}}{\sqrt{3} - 1} \cdot \dfrac{\sqrt{3} + 1}{\sqrt{3} + 1}$

In Exercises 41 to 52, use $A = \{0, 1, 2, 3, 4\}$, $B = \{1, 3, 5, 11\}$, $C = \{1, 3, 6, 10\}$, and $D = \{0, 2, 4, 6, 8, 10\}$ to find the indicated intersection or union.

41. $A \cap B$

42. $A \cap C$

43. $B \cap C$

44. $B \cap D$

45. $A \cap D$

46. $C \cap D$

47. $A \cup B$

48. $A \cup C$

49. $B \cup C$

50. $B \cup D$

51. $A \cup D$

52. $C \cup D$

53. State the additive inverse of $\sqrt{2} + 7$.

54. State the additive inverse of $1 - \dfrac{\sqrt{2}}{2}$.

55. State the multiplicative inverse of $7\frac{3}{8}$.

56. State the multiplicative inverse of $-4\frac{2}{5}$.

57. State the multiplicative inverse of $1 - \dfrac{\pi}{2}$.

58. State the multiplicative inverse of $\dfrac{3}{\sqrt{3} + 1}$.

59. Show by an example that the operation of subtraction of real numbers is not a commutative operation.

60. Show by an example that the operation of division of nonzero real numbers is not a commutative operation.

61. Show by an example that the operation of subtraction of real numbers is not an associative operation.

62. Show by an example that the operation of division of real numbers is not an associative operation.

63. Is $\{2, 4, 6, 8, 10, \ldots\}$ closed under the operation of
a. addition   **b.** subtraction
c. multiplication   **d.** division

64. Is $\{\ldots, -15, -10, -5, 0, 5, 10, 15, \ldots\}$ closed under the operation of

a. addition **b.** subtraction

c. multiplication **d.** division

In Exercises 65 to 74 classify each statement as true or false.

65. $a/0$ is the multiplicative inverse of $0/a$.

66. $(-1/\pi)$ is the multiplicative inverse of $-\pi$.

67. If $p = q + t/2$ and $q + t/2 = \frac{1}{2}s$, then $\frac{1}{2}s = p$.

68. If $a - b = 7$, then $7 = b - a$.

69. The sum of two composite numbers is a composite number.

70. All integers are natural numbers.

71. Every real number is either a rational or irrational number.

72. Every rational number is either even or odd.

73. 1 is the only positive integer that is not prime and not composite.

74. All repeating decimals are rational numbers.

Calculator Exercises

75. Use a calculator to write each of the following rational numbers as a decimal. If the number is represented by a nonterminating decimal, then use a *bar* over the repeating portion of the decimal.

a. $\dfrac{8}{11}$ **b.** $\dfrac{33}{40}$ **c.** $\dfrac{2}{7}$ **d.** $\dfrac{5}{37}$

76. Use a calculator to determine whether 3.14 or $\frac{22}{7}$ is a closer approximation to π.

77. Use a calculator to complete the following table.

x	0.1	0.01	0.001	0.0000001
$\dfrac{\sqrt{x+9}-3}{x}$				

Now make a guess as to the number the fraction seems to be approaching as x assumes real numbers that are closer and closer to zero.

78. Use a calculator to complete the following table:

x	0.1	0.01	0.001	0.0000001
$\dfrac{\dfrac{1}{2} - \dfrac{1}{x+2}}{x}$				

Now make a guess as to the number the fraction seems to be approaching as x assumes real numbers that are closer and closer to zero.

Supplemental Exercises

In Exercises 79 to 82, determine the elements of the set that is written using set-builder notation.

79. $A = \{x \,|\, x \text{ is a composite number less than 11}\}$

80. $B = \{x \,|\, x \text{ is an even prime number}\}$

81. $C = \{x \,|\, 50 < x < 60 \text{ and } x \text{ is a prime number}\}$

82. $D = \{x \,|\, x \text{ is the smallest odd composite number}\}$

83. Which of the properties of real numbers are satisfied by the set of positive integers?

84. Which of the properties of real numbers are satisfied by the set of integers?

85. Which of the properties of real numbers are satisfied by the set of rational numbers?

86. Which of the properties of real numbers are satisfied by the set of irrational numbers?

87. To prove that the set of prime numbers has an infinite number of elements, Euclid (fl. about 300 B.C.) used a proof by contradiction. His proof was similar to the following. Assume that the set of prime numbers forms a finite set. For convenience, label the largest prime number in this finite set L. Now consider the positive integer T, which is equal to the product of all the prime numbers, plus 1. That is,

$$T = (2)(3)(5)(7) \cdots (L) + 1$$

Explain why

a. T cannot be a prime number. (*Hint:* Which is larger, T or L?)

b. T cannot be a composite number. (*Hint:* Which prime numbers are factors of T?)

The positive integer T must be either a prime or a composite number. The assumption that the prime numbers form a finite set has produced the contradiction that T is not a prime number and T is not a composite number. Thus Euclid reasoned that the set of prime numbers cannot be finite.

88. In 1742, Christian Goldbach conjectured that every even number greater than 2 can be written as the sum of two prime numbers. Many mathematicians have tried to prove or disprove this conjecture without succeeding. Show that Goldbach's conjecture is true for the even numbers **a.** 12 and **b.** 30.

1.2

Intervals, Absolute Value, and Distance

Figure 1.2

Figure 1.3

The real numbers can be represented geometrically by a **coordinate axis** called a **real number line**. Figure 1.2 shows a portion of a real number line. The number associated with a particular point on a real number line is called the **coordinate** of the point. It is customary to label those points whose coordinates are integers. The point corresponding to zero is called the **origin**, denoted 0. Numbers to the right of the origin are **positive real numbers;** numbers to the left of the origin are **negative real numbers.**

A real number line provides a picture of the real numbers. That is, each real number corresponds to one and only one point on the real number line and each point on a real number line corresponds to one and only one real number. This type of correspondence is referred to as a **one-to-one correspondence.**

To **graph** a real number, draw a dot on the point on the real number line that is associated with the real number. The real numbers -3, $-\frac{1}{2}$, 1.75, and $\sqrt{10}$ are graphed in Figure 1.3.

Certain order relationships exist between real numbers. For example, if a and b are real numbers, then:

a **equals** b (denoted by $a = b$) if $a - b = 0$.

a is **greater than** b (denoted by $a > b$) if $a - b$ is positive.

a is **less than** b (denoted by $a < b$) if $b - a$ is positive.

On a horizontal number line, the notation

$a = b$ implies that the point with coordinate a is the same point as the point with coordinate b.

$a > b$ implies that the point with coordinate a is to the right of the point with coordinate b.

$a < b$ implies that the point with coordinate a is to the left of the point with coordinate b

The Trichotomy Property

> Given any two real numbers a and b, exactly one of the following relationships holds:
>
> 1. $a = b$ 2. $a > b$ 3. $a < b$

EXAMPLE 1   Illustrate the Trichotomy Property

Illustrate the Trichotomy Property by replacing ☐ with the appropriate symbol ($<$, $=$, or $>$).

a. $3 \boxed{>} -4$ b. $-2 \boxed{>} -5$ c. $\frac{1}{3} \boxed{=} 0.\overline{3}$ d. $3.14 \boxed{<} \pi$

Solution

a. $3 > -4$ b. $-2 > -5$ c. $\dfrac{1}{3} = 0.\overline{3}$ d. $3.14 < \pi$

■ *Try Exercise* **8,** *page 17.*

The **inequality** symbols $<$ and $>$ are sometimes combined with the equality symbol in the following manner:

$a \geq b$ Read "*a* is greater than or equal to *b*," which means $a > b$ or $a = b$.

$a \leq b$ Read "*a* is less than or equal to *b*," which means $a < b$ or $a = b$.

Inequalities can be used to represent subsets of real numbers. For example, the inequality $x > 2$ represents all real numbers greater than 2; Figure 1.4 shows its graph. The parentheses at 2 means that 2 is not part of the graph.

Figure 1.4

The inequality $x \leq 1$ represents all real numbers less than or equal to 1; Figure 1.5 shows its graph. The bracket at 1 means that 1 is part of the graph.

Figure 1.5

The inequality $-1 \leq x < 3$ represents all real numbers between -1 and 3, including -1 but not including 3. Figure 1.6 shows its graph.

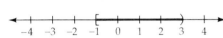

Figure 1.6

Subsets of real numbers can also be represented by a compact form of notation called **interval notation.** For example, $[-1, 3)$ is the interval notation for the subset of real numbers in Figure 1.6.

In general, the interval notation

(a, b) represents all real numbers between *a* and *b*, not including *a* and not including *b*. This is an **open interval.**

$[a, b]$ represents all real numbers between *a* and *b*, including *a* and including *b*. This is a **closed interval.**

$(a, b]$ represents all real numbers between *a* and *b*, not including *a* but including *b*. This is a **half-open interval.**

$[a, b)$ represents all real numbers between *a* and *b*, including *a* but not including *b*. This is a **half-open interval.**

Figure 1.7 shows the four subsets of real numbers that are associated with the four interval notations (a, b), $[a, b]$, $(a, b]$, and $[a, b)$.

Subsets of the real numbers whose graphs extend forever in one or both directions can be represented by interval notation using the **infinity symbol** ∞ or the **negative infinity symbol** $-\infty$.

As Figure 1.8 shows, the interval notation

$(-\infty, a)$ represents all real numbers less than *a*.

(b, ∞) represents all real numbers greater than *b*

$(-\infty, a]$ represents all real numbers less than or equal to *a*.

$[b, \infty)$ represents all real numbers greater than or equal to *b*.

$(-\infty, \infty)$ represents all real numbers.

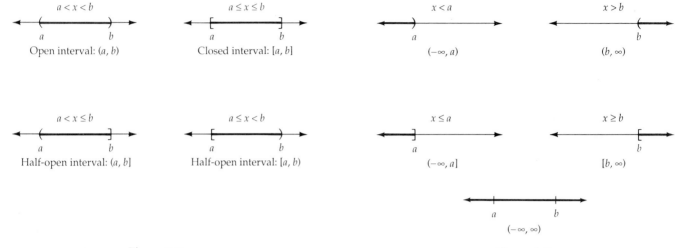

Figure 1.7
Finite intervals

Figure 1.8
Infinite intervals

Caution The infinity symbol ∞ does not represent a real number.

EXAMPLE 2 Express Inequalities as Intervals

Graph the following inequalities and write the inequality using interval notation:

a. $-2 \le x < 3$ b. $x \ge -3$ c. $x < 4$

Solution

	Graph	Interval Notation
a.		$[-2, 3)$
b.		$[-3, \infty)$
c.		$(-\infty, 4)$

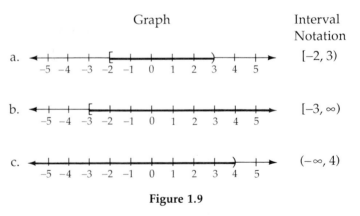

Figure 1.9

■ *Try Exercise* **26**, *page 17.*

EXAMPLE 3 Express Intervals as Inequalities

Graph the following intervals and write each interval as an inequality:

a. $[-4, 2]$ b. $[-3, 1)$ c. $(-\infty, 2)$

Solution

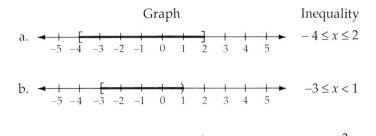

	Graph	Inequality
a.	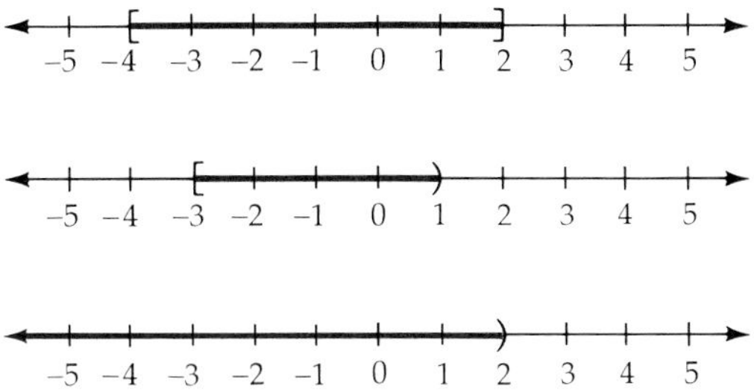	$-4 \le x \le 2$
b.		$-3 \le x < 1$
c.		$x < 2$

Figure 1.10

■ *Try Exercise* **38,** *page 17.*

Figure 1.11

Some graphs consist of more than one interval of the real number line. Figure 1.11 is a graph of the interval $(-\infty, -2)$, along with the interval $[1, \infty)$.

Remark The word "or" is used to denote the union of two sets. The word "and" is used to denote the intersection of two sets. Thus, the graph in Figure 1.11 is denoted by the inequality notation

$$x < -2 \quad \text{or} \quad x \ge 1.$$

To represent this graph using interval notation, use the union symbol $\cup$ and write

$$(-\infty, -2) \cup [1, \infty).$$

Absolute Value and Distance

If a is the coordinate of a point on the real number line, then the **absolute value** of a denoted by $|a|$, is the distance between a and the origin. For example,

$|3| = 3$ because 3 is 3 units from the origin on the real number line.

$|-3| = 3$ because -3 is 3 units from the origin on the real number line.

Remark The absolute value of a real number is never negative. It is always **nonnegative,** which means it is positive or zero.

We can also define absolute value algebraically as follows.

Definition of Absolute Value

The absolute value of the real number a is defined by

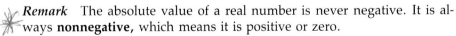

$$|a| = \begin{cases} a \text{ if } a \ge 0 \\ -a \text{ if } a < 0 \end{cases}$$

EXAMPLE 4   **Remove Absolute Value Symbols**

Use the definition of absolute value to write each of the following without absolute value symbols:

a. $|5|$ b. $\left|-\dfrac{3}{2}\right|$ c. $-|0.75|$ d. $|\sqrt{2} - 1|$ e. $|1 - \sqrt{2}|$

Solution

a.   Since $5 > 0$, $|5| = 5$.

b.   Since $-\dfrac{3}{2} < 0$, $\left|-\dfrac{3}{2}\right| = -\left(-\dfrac{3}{2}\right) = \dfrac{3}{2}$.

c.   Since $0.75 > 0$, $-|0.75| = -0.75$.

d.   Since $\sqrt{2} - 1 > 0$, $|\sqrt{2} - 1| = \sqrt{2} - 1$.

e.   Since $1 - \sqrt{2} < 0$, $|1 - \sqrt{2}| = -(1 - \sqrt{2}) = \sqrt{2} - 1$.

■ *Try Exercise 70, page 17.*

The definition of **distance** between any two points on a real number line makes use of the absolute value concept.

Distance Between Points on a Real Number Line

> For any real numbers a and b, the distance between the graph of a and the graph of b is denoted by $d(a, b)$, where
> $$d(a, b) = |a - b| = |b - a|$$

EXAMPLE 5   **Find the Distance Between Points**

Find the distance between the points whose coordinates are given.

a. $5, -2$ b. $-\pi, -2$

Solution

a. $d(5, -2) = |5 - (-2)| = |5 + 2| = |7| = 7$

b. $d(-\pi, -2) = |-\pi - (-2)| = |-\pi + 2| = \pi - 2$

Because $-\pi + 2$ is negative, we can use the definition of absolute value to write $|-\pi + 2|$ as $-(-\pi + 2)$, which simplifies to $\pi - 2$.

■ *Try Exercise 80, page 17.*

EXERCISE SET 1.2 _____

In Exercises 1 to 6, graph each number on a real number line.

1. $-4; -2; \dfrac{7}{4}; 2.5$

2. $-3.5; 0; 3; \dfrac{9}{4}$

3. $\pi; -1.2; 0.25; \dfrac{9}{2}$

4. $-\sqrt{4}; 4; 0.3; \dfrac{8}{3}$

5. $\dfrac{0}{5}; -1; -\sqrt{9}; 4.5$

6. $\dfrac{0}{3}; -3; -\sqrt{16}; 1.\overline{3}$

In Exercises 7 to 24, illustrate the trichotomy property by replacing the ☐ with the appropriate symbol ($<$, $=$, or $>$).

7. $\dfrac{5}{2}$ ☐ 4

8. $-\dfrac{3}{2}$ ☐ -3

9. $\dfrac{2}{3}$ ☐ 0.6666

10. $\dfrac{1}{5}$ ☐ 0.2

11. 1.75 ☐ 2.23

12. 1.25 ☐ 1.3

13. $\sqrt{5}$ ☐ 2

14. $\sqrt{5}/2$ ☐ 2

15. $\sqrt{12}$ ☐ 3

16. $-\sqrt{9}$ ☐ -3

17. 0.4 ☐ $\dfrac{4}{9}$

18. $0.\overline{36}$ ☐ $\dfrac{4}{11}$

19. $\dfrac{0}{2}$ ☐ $-\dfrac{0}{5}$

20. $\dfrac{10}{5}$ ☐ 2

21. $\dfrac{22}{7}$ ☐ π

22. π ☐ 3.14159

23. $\dfrac{22}{7}$ ☐ 3.14159

24. 3.14159 ☐ 3.14

In Exercises 25 to 36, graph each inequality and write the inequality using interval notation.

25. $3 < x < 5$

26. $-2 \le x < 1$

27. $x < 3$

28. $x \ge 4$

29. $x \ge 0$ and $x < 3$

30. $x > -4$ and $x \le 4$

31. $x < -3$ or $x \ge 2$

32. $x \le 2$ or $x > 3$

33. $x > 3$ and $x < 4$

34. $x > -5$ or $x < 1$

35. $x \le 3$ and $x > -1$

36. $x < 5$ and $x \le 2$

In Exercises 37 to 48, graph each interval and write each interval as an inequality.

37. $[-4, 1]$ **38.** $[-2, 3)$ **39.** $(1, 5)$ **40.** $(1, 4]$

41. $[2.5, \infty)$ **42.** $(-\infty, 3]$ **43.** $(-\infty, 2)$ **44.** (π, ∞)

45. $(-\infty, 2] \cup (3, \infty)$ **46.** $(-\infty, 1) \cup (4, \infty)$

47. $(-\infty, 3) \cup (3, \infty)$ **48.** $(-\infty, 1) \cup [2, \infty)$

In Exercises 49 to 56, use the given notation or graph to supply the notation or graph that is marked with a question mark.

	Inequality Notation	Interval Notation	Graph
49.	$x \le 3$	?	?
50.	?	$(-2, \infty)$	?
51.	?	?	$-2\ -1\ \ 0\ \ 1\ \ 2$
52.	$-3 \le x < -1$	?	?
53.	?	$[1, 4]$	?
54.	?	?	$-3\ -2\ -1\ \ 0\ \ 1$
55.	?	$[-2, \pi)$	?
56.	$x < 2$ or $x \ge 4$	?	?

In Exercises 57 to 74, write each real number without absolute value symbols.

57. $|4|$

58. $|-8|$

59. $|-27.4|$

60. $|51.2|$

61. $|4| - |-7|$

62. $-|3| - |-8|$

63. $|5| \cdot |-8|$

64. $|0.5| \cdot |-8|$

65. $|\sqrt{2} + 1|$

66. $|\sqrt{7} + 2|$

67. $-|\sqrt{2} - 1|$

68. $-|2 - \sqrt{7}|$

69. $|\sqrt{3} - 2|$

70. $|2 - \sqrt{3}|$

71. $\left|\dfrac{\pi}{3} - 2\right|$

72. $\left|2 - \dfrac{\pi}{3}\right|$

73. $|3\pi - 10|$

74. $|10 - 3\pi|$

In Exercises 75 to 86, find the distance between the points whose coordinates are given.

75. $8, 1$

76. $-2, -7$

77. $-3, 5$

78. $-5, 8$

79. $16, -34$

80. $-108, 22$

81. $-38, -5$

82. $\pi, 3$

83. $-\pi, 3$

84. $\sqrt{2}, 10$

85. $\sqrt{2}, -10$

86. $\sqrt{5}, -\pi$

In Exercises 87 to 92, use absolute value notation to describe the given expression.

87. Distance between a and 2

88. Distance between b and -7

89. $d(m, n)$

90. $d(p, -q)$

91. The distance between z and 5 is greater than 4.

92. The distance between x and -2 is less than 7.

In Exercises 93 to 96, write interval notation for the given expression.

93. x is a real number and $x \ne 3$ ($\ne$ means "is not equal to").

94. x is a real number whose square is nonnegative.

95. x is a real number whose absolute value is less than 3.

96. x is a real number whose absolute value is greater than 2.

In Exercises 97 to 100, write each expression without absolute value symbols.

97. $|x^2 + 1|$

98. $|y^2 + 10|$

99. $|-w^2 - \pi|$

100. $|-k^2 - \pi|$

In Exercises 101 to 104, classify each statement as true or false.

101. $|x|$ is a positive number.

102. $|-y| = y$

103. If $m < 0$, then $|m| = -m$.

104. For all real numbers a and b,

$$|a - b| = |b - a|.$$

Calculator Exercises

105. Use a calculator to determine the order of the following real numbers. List numbers from smallest to largest: $\dfrac{351}{820}, \dfrac{111}{271}$, and $\dfrac{\sqrt{2}}{\sqrt{11}}$.

106. Use a calculator to determine which two of the following real numbers are equal: $\dfrac{69}{141}, \dfrac{123}{247}, \dfrac{247}{495}$, and $\dfrac{161}{329}$.

Supplemental Exercises

In Exercises 107 to 110, use inequalities to describe the given statement.

107. The interest I is not greater than $120.

108. The rent R will be at least $650 a month.

109. The property has an area A that is at least 2 acres but less than 3 acres.

110. The distance D is greater than 7 miles, and it is not more than 8 miles.

In Exercises 111 to 114, use absolute value notation to describe the given statement.

111. x is closer to 2 than it is to 6.

112. x is farther from 3 than it is to -7.

113. x is within δ units of a.

114. x is not equal to a, but it is within δ units of a.

115. For what values of a and b does $|ab| = |a||b|$?

116. The inequality $|a + b| \le |a| + |b|$ is called the **triangle inequality.** For what values of a and b does

$$|a + b| = |a| + |b|?$$

1.3

Integer Exponents

A compact method of writing $5 \cdot 5 \cdot 5 \cdot 5$ is 5^4. The expression 5^4 is written in **exponential notation.** Similarly, we can write

$$\frac{2x}{3} \cdot \frac{2x}{3} \cdot \frac{2x}{3} \quad \text{as} \quad \left(\frac{2x}{3}\right)^3,$$

and

$$(x + 2y)(x + 2y)(x + 2y)(x + 2y) \quad \text{as} \quad (x + 2y)^4.$$

Exponential notation can be used to express the product of any expression that is used repeatedly as a factor.

Definition of Natural Number Exponents

If b is any real number and n is any natural number, then

$$b^n = \underbrace{b \cdot b \cdot b \cdot \cdots \cdot b}_{n \text{ factors of } b}$$

In the expression b^n, b is the **base,** n is the **exponent,** and b^n is the **nth power of b.**

EXAMPLE 1   **Evaluate Powers**

Evaluate each of the following powers:

a. 3^2 b. $(-5)^4$ c. -5^4 d. $\left(\dfrac{1}{2}\right)^3$

Solution

a. $3^2 = 3 \cdot 3 = 9$ b. $(-5)^4 = (-5)(-5)(-5)(-5) = 625$

c. $-5^4 = -(5 \cdot 5 \cdot 5 \cdot 5) = -625$ d. $\left(\dfrac{1}{2}\right)^3 = \dfrac{1}{2} \cdot \dfrac{1}{2} \cdot \dfrac{1}{2} = \dfrac{1}{8}$

■ *Try Exercise* **4**, *page 25.*

Remark   Notice the difference between $(-5)^4 = 625$ and $-5^4 = -625$. The parentheses in $(-5)^4$ indicates that the base is -5; however, the expression -5^4 means $-(5^4)$. This time the base is 5.

Consider the sequence

$$5^4 = 625$$
$$5^3 = 125$$
$$5^2 = 25$$
$$5^1 = 5$$
$$5^0 = ?$$
$$5^{-1} = ?$$
$$5^{-2} = ?$$

Notice that each power divided by the base 5 yields the power in the row below it. For example, $625 \div 5 = 125$ and $125 \div 5 = 25$. To continue this pattern, the powers 5^0, 5^{-1}, and 5^{-2} must be defined as

$$5^0 = 5 \div 5 = 1$$

$$5^{-1} = 1 \div 5 = \dfrac{1}{5} = \dfrac{1}{5^1}$$

$$5^{-2} = \dfrac{1}{5} \div 5 = \dfrac{1}{5} \cdot \dfrac{1}{5} = \dfrac{1}{5^2}$$

These observations suggest the following definitions.

Definition of b^0

> For any nonzero real number b, $b^0 = 1$.

Any nonzero real number raised to the zero power equals 1. For example,

$$7^0 = 1 \qquad \left(\dfrac{1}{2}\right)^0 = 1 \qquad (-3)^0 = 1 \qquad \pi^0 = 1 \qquad (a^2 + 1)^0 = 1.$$

Definition of b^{-n}

If $b \neq 0$ and n is any natural number, $b^{-n} = 1/b^n$ and $1/b^{-n} = b^n$.

This definition says that if a factor is raised to a negative power, then that factor can be raised to a positive power by moving the factor from the numerator to the denominator or vice versa. For example,

$$3^{-2} = \frac{1}{3^2} = \frac{1}{9}, \qquad \frac{1}{4^{-3}} = 4^3 = 64, \qquad \frac{5^{-2}}{7^{-1}} = \frac{7^1}{5^2} = \frac{7}{25}.$$

Restriction Agreement

The expressions 0^0, 0^n where n is a negative integer, and $x/0$ are all undefined expressions. Therefore, all values of variables in this text have been restricted to avoid any of these undefined expressions. For example, in the expression

$$\frac{x^0 y^{-3}}{z - 4},$$

it should be assumed that $x \neq 0$, $y \neq 0$, and $z \neq 4$, whether or not these restrictions are specifically stated.

The associative property of multiplication can be used to simplify expressions of the type $b^m \cdot b^n$, where m and n are integers. For example,

$$b^3 \cdot b^2 = (b \cdot b \cdot b) \cdot (b \cdot b) \quad \text{By definition.}$$

$$= b \cdot b \cdot b \cdot b \cdot b \qquad \text{Associative property} \\ \text{of multiplication}$$

$$= b^5$$

This process can also be applied if the integers are negative. For example,

$$c^{-2} \cdot c^{-3} = \frac{1}{c \cdot c} \cdot \frac{1}{c \cdot c \cdot c} = \frac{1}{c \cdot c \cdot c \cdot c \cdot c} = \frac{1}{c^5} = c^{-5}.$$

These examples suggest the following property of exponents.

Product Property of Exponential Expressions

If m and n are integers and $b \neq 0$, then $b^m \cdot b^n = b^{m+n}$.

The product property shows that multiplication of powers with like bases can be accomplished by adding the exponents. For example,

$$2^{-2} \cdot 2^5 = 2^{(-2)+5} = 2^3.$$

Remark $2^5 \cdot 3^4$ cannot be simplified using the product property because the bases are not the same.

The definition of negative exponents and the product property can be used to discover a quotient property for exponents.

$$\frac{b^m}{b^n} = b^m \cdot b^{-n} = b^{m+(-n)} = b^{m-n}.$$

Quotient Property of Exponential Expressions

> If m and n are integers and $b \neq 0$, then $b^m/b^n = b^{m-n}$.

The quotient property demonstrates that division of powers with like bases can be accomplished by subtracting the exponents. For example,

$$\frac{5^3}{5^1} = 5^{3-1} = 5^2 = 25 \quad \text{and} \quad \frac{y^3}{y^{-1}} = y^{3-(-1)} = y^4.$$

The definition of exponents and the product property can be used to simplify powers raised to powers. For example, $(4^3)^2 = (4^3)(4^3) = 4^6 = 4^{3 \cdot 2}$. The power property is a generalization of the above result.

Power Property of Exponential Expressions

> If m and n are integers and $b \neq 0$, then $(b^m)^n = b^{mn}$.

The power property says that a power raised to a power can be simplified by multiplying the exponents. For example,

$$[(a + 2b)^{-2}]^{-3} = (a + 2b)^{(-2) \cdot (-3)} = (a + 2b)^6.$$

If the power property is applied to products or quotients, then the following properties result.

Power Property of Products and Quotients

> If m, n, and p are integers and a and b are nonzero real numbers then
>
> $$(ab)^m = a^m b^m \quad \text{and} \quad \left(\frac{a}{b}\right)^m = \frac{a^m}{b^m}.$$
>
> $$(a^m b^n)^p = a^{mp} b^{np} \quad \text{and} \quad \left(\frac{a^m}{b^n}\right)^p = \frac{a^{mp}}{b^{np}}.$$

Using the power property of products, a product raised to a power can be written as the product of the powers of the factors. For example,

$$(3m^3)^4 = 3^4 m^{12} = 81 m^{12}.$$

Using the power property of quotients, a quotient raised to a power can be written as the quotient of the powers of the numerator and the denominator. For example,

$$\left(\frac{3}{z^2}\right)^3 = \frac{3^3}{z^6} = \frac{27}{z^6}.$$

EXAMPLE 2   **Use the Properties of Exponents**

Evaluate each of the following:

a. $(5^5 \cdot 5^{-3})^{-2}$ b. $\left(\dfrac{2^2 \cdot 3^{-2}}{2^{-1} \cdot 5^2}\right)^{-1}$

Solution

a. $(5^5 \cdot 5^{-3})^{-2} = (5^{5+(-3)})^{-2} = (5^2)^{-2} = 5^{-4} = \dfrac{1}{5^4} = \dfrac{1}{625}$

b. $\left(\dfrac{2^{2^{\cdot}} \cdot 3^{-2}}{2^{-1} \cdot 5^2}\right)^{-1} = (2^3 \cdot 3^{-2} \cdot 5^{-2})^{-1} = 2^{-3} \cdot 3^2 \cdot 5^2 = \dfrac{9 \cdot 25}{8} = \dfrac{225}{8}$

■ *Try Exercise* **18,** *page 25.*

To simplify an expression involving exponents, write the expression in a form in which *each base appears at most once* and *no powers of powers or negative exponents appear.*

EXAMPLE 3   **Simplify Exponential Expressions**

Simplify each of the following:

a. $(2x^3y^2)(3xy^5)$ b. $\left(\dfrac{2abc^2}{5a^2b}\right)^3$

c. $(3m^{-2}p^3)^4$ d. $\dfrac{x^n y^{2n}}{x^{n-1} y^n}$

Solution

a. $(2x^3y^2)(3xy^5) = (2 \cdot 3)(x^3 \cdot x)(y^2 \cdot y^5)$ The commutative and

$= 6x^4y^7$ associative properties of multiplication

b. $\left(\dfrac{2abc^2}{5a^2b}\right)^3 = \left(\dfrac{2c^2}{5a}\right)^3$ The quotient property

$= \dfrac{8c^6}{125a^3}$ The power property of products and quotients

c. $(3m^{-2}p^3)^4 = 3^4m^{-8}p^{12}$ The power property of a product

$= \dfrac{81p^{12}}{m^8}$

d. $\dfrac{x^n y^{2n}}{x^{n-1} y^n} = x^{n-(n-1)} y^{2n-n}$ The quotient property

$= xy^n$

■ *Try Exercise* **30,** *page 25.*

Scientific Notation

The properties of exponents provide a compact method of writing and an efficient method of computing with very large or very small numbers. The method is called **scientific notation.** A number written in scientific notation has the form $a \cdot 10^n$, where n is an integer and $1 \leq a < 10$. The following procedure is used to change a number from its decimal form to its scientific notation form.

For numbers greater than 10, move the decimal point to the position to the right of the first digit. The exponent n will equal the number of places the decimal point has been moved. For example,

$$7{,}430{,}000 = 7.43 \times 10^6$$

6 places

For numbers less than 1, move the decimal point to the right of the first nonzero digit. The exponent n will be negative, and its absolute value will equal the number of places the decimal point has been moved. For example,

$$0.00000078 = 7.8 \times 10^{-7}$$

7 places

To change a number from scientific notation to its decimal form, we reverse the above procedure. That is, if the exponent is positive, move the decimal point to the right the same number of places as the exponent. For example,

$$3.5 \times 10^5 = 350{,}000$$

5 places

If the exponent is negative, move the decimal point to the left the same number of places as the absolute value of the exponent. For example,

$$2.51 \times 10^{-8} = 0.0000000251$$

8 places

EXAMPLE 4   Change Form

Write each decimal in scientific notation. Write each number in scientific notation in its decimal form.

a. 3,770,000,000   b. 0.00000000026
c. 2.51×10^5 d. 3.221×10^{-7}

Solution

a. $3{,}770{,}000{,}000 = 3.77 \times 10^9$ b. $0.00000000026 = 2.6 \times 10^{-10}$
c. $2.51 \times 10^5 = 251{,}000$ d. $3.221 \times 10^{-7} = 0.0000003221$

■ *Try Exercise **48**, page 25.*

Most scientific calculators display very large or very small numbers in scientific notation. The number $450,000^2$ is displayed as

$$\boxed{\texttt{2.025\ \ 11}}$$

This means $450,000^2 = 2.025 \times 10^{11}$.

Scientific Notation and Significant Digits

Some numbers are **exact** numbers, and some numbers are **approximate numbers.** Exact numbers are obtained by counting or a definition. For example, 32 students, 47 cents, and 104 pages are exact numbers.

Many numbers used in scientific work are obtained by measuring; as such, they are approximations. For example, if a room is reported to have a length of 80 feet to the nearest 10 feet, this means that the actual length of the room is no less than 75 feet and no more than 85 feet. If the room is measured as 82 feet to the nearest foot, then the actual length is no less than 81.5 feet and no more than 82.5 feet.

A measurement of 80 feet is said to have 1 **significant digit.** A measurement of 82 feet is said to have 2 significant digits. A digit is a significant digit of a number if it meets any of the following conditions:

Exact: Every digit of an exact number is a significant digit.

Approximate: The significant digits of an approximate number are

1. Every nonzero digit
2. The digit 0, provided it is
   a. between two nonzero digits or
   b. to the right of a nonzero digit and the number includes a decimal point

For example,

57,000 has 2 significant digits because all nonzero digits are significant.

57,080 has 4 significant digits because zeros between nonzero digits are significant.

57,100.0 has 6 significant digits. The number includes a decimal point, so all zeros to the right of a nonzero digit are also significant.

0.00230 has 3 significant digits. The last zero is significant because it is to the right of a nonzero digit and the number includes a decimal point.

When a measurement is given as 700 centimeters, there can be confusion. Has this measurement been made to the nearest 1 centimeter, the nearest 10 centimeters, or the nearest 100 centimeters? To avoid confusion, write the number using scientific notation. For example:

Number	Number of Significant Digits	Measured to the Nearest
7.000×10^2	4	one-tenth
7.00×10^2	3	one
7.0×10^2	2	ten
7×10^2	1	hundred

EXERCISE SET 1.3

In Exercises 1 to 26, evaluate each expression.

1. $(-4)^3$ **2.** $(-2)^3$ **3.** -4^3

4. -2^4 **5.** 7^0 **6.** -7^0

7. $(-1)^{18}$ **8.** $(-1)^{19}$ **9.** $3^2 \cdot 3^3$

10. $2^3 \cdot 2^4$ **11.** $\dfrac{3^{-1}}{3^2}$ **12.** $\dfrac{5^6}{5^4}$

13. $2^7 \cdot 2^{-3} \cdot 2$ **14.** $3 \cdot 3^{-12} \cdot 3^8$ **15.** $\dfrac{4^{-8}}{4^{-11}}$

16. $\dfrac{5^{-1}}{5^2}$ **17.** $\left(\dfrac{5^{-3} \cdot 7}{3^{-2}}\right)^{-1}$ **18.** $\left(\dfrac{4 \cdot 5^{-1}}{2^{-3}}\right)^{-2}$

19. $\left(\dfrac{4}{9}\right)^{-2}$ **20.** $\left(\dfrac{4}{6}\right)^{-3}$ **21.** $\left(\dfrac{3^{-2}}{2^{-1} \cdot 5}\right)^2$

22. $\left(\dfrac{2^{-2} \cdot 3^2}{5}\right)^2$ **23.** $\dfrac{(2 \cdot 5)^2}{(2^{-1} \cdot 5)^3}$ **24.** $\dfrac{(2^2 \cdot 3^{-1})^3}{(3 \cdot 5)}$

25. $\left(\dfrac{2^5 \cdot 3^{-5}}{2^{-3}5^4}\right)^0$ **26.** $\left(\dfrac{-3^6 \cdot 2^{-4}}{-4^{-5}}\right)^0$

In Exercises 27 to 46, simplify each exponential expression so that all exponents are positive.

27. $(2x^2y^3)(3x^5y)$ **28.** $(3ab^4)(3ab^3c^2)$

29. $\left(\dfrac{2ab^2c^3}{5ab^2}\right)^3$ **30.** $\left(\dfrac{3pq^2}{-2pq^3r^2}\right)^4$

31. $\dfrac{(3xy^{-3})^2}{(2xy)^{-2}}$ **32.** $\dfrac{(5ab^{-2})^2}{(-3a^2b)^3}$

33. $(2x^{-3}y^0)(3^{-1}xy)^2$ **34.** $(-3abc^2)^2(2ab^{-1})^3$

35. $\left(\dfrac{3x}{y}\right)^{-1}$ **36.** $\left(\dfrac{2x}{5y}\right)^{-2}$

37. $(x^2y^{-3})^{-2}$ **38.** $(x^3y^{-2})^{-3}$

39. $a^{-1} + b^{-2}$ **40.** $a^{-1} - b^{-1}$

41. $\dfrac{4a^2(bc)^{-1}}{(-2)^2a^3b^{-2}c}$ **42.** $\dfrac{6abc^{-2}}{(-3)^{-1}a^{-1}bc^{-3}}$

43. $(2ab^{-3})^2(-2a^{-1}b^3)^2$ **44.** $(3x^{-1}y)^{-1}(3xy)$

45. $\left[\left(\dfrac{b^{-3}}{a^2}\right)^2\left(\dfrac{a^{-2}}{ab}\right)^{-1}\right]^0$ **46.** $\left(\dfrac{x^{-2} + x^3}{x^{-1}}\right)^0$

In Exercises 47 to 58, write each number in scientific notation.

47. 73.4   **48.** 25,600   **49.** 1,900,000

50. 21,000,000   **51.** 163,000,000,000   **52.** 521

53. 0.000032   **54.** 0.00000714   **55.** 0.007

56. 0.00095   **57.** 0.0000000821   **58.** 0.00000000072

In Exercises 59 to 70, change each number from its scientific notation to its decimal form.

59. 6.5×10^3 **60.** 4.2×10^4 **61.** 7.31×10^{-5}

62. 6.85×10^{-9} **63.** 8.0×10^{10} **64.** 9.008×10^{12}

65. 2.17×10^{-4} **66.** 4.007×10^{-3} **67.** 1.0×10^{11}

68. 1.0×10^{-5} **69.** 3.75×10^0 **70.** 8.81×10^0

In Exercises 71 to 82, find the number of significant digits in each of the numbers.

71. 14,300   **72.** 17,010,000   **73.** 20,050.0

74. 40,900.00   **75.** 0.03   **76.** 0.0070

77. 0.00501   **78.** 0.000008   **79.** 8.3×10^3

80. 5.31×10^5 **81.** 1.882×10^{-7} **82.** 4×10^{-5}

Calculator Exercises

The exponential key $\boxed{y^x}$ (or on some calculators $\boxed{x^y}$) on a scientific calculator can be used to evaluate a positive number raised to a natural number power. The following examples illustrate the sequence of key strokes for a calculator that uses algebraic logic.

Power	Key Sequence	Calculator Display
5^2	5 $\boxed{y^x}$ 2 $\boxed{=}$	$\boxed{25.}$
2^{12}	2 $\boxed{y^x}$ 12 $\boxed{=}$	$\boxed{4096.}$

In Exercises 83 to 86, use a calculator to evaluate the exponential expression.

83. 1.08^{10} **84.** 1.12^8

85. $840(1 + 0.09)^{24}$ **86.** $2200\left(1 + \dfrac{0.08}{360}\right)^{900}$

87. Pluto is 5.91×10^{12} meters from the sun. The speed of light is 3.00×10^8 meters per second. Find the time it takes light from the sun to reach Pluto.

88. The earth's mean distance from the sun is 9.3×10^7 miles. This distance is called the astronomical unit (AU). Jupiter is 5.2 AU from the sun. Find the distance in miles from the sun to Jupiter.

89. A principal P invested at a yearly interest rate r compounded n times per year yields a balance B given by the formula

$$B = P\left(1 + \frac{r}{n}\right)^n.$$

Find the balance after one year if $4500 is deposited in an account with a yearly interest rate of 8 percent that is compounded monthly.

90. You plan to save $1 the first day of a month, $2 the second day, $4 the third day, and continue this pattern of saving twice what you saved on the previous day for every day in a month that has 30 days.   **a.** How much money will you need to save on the 30th day?   **b.** How much money will you have after 30 days? (*Hint:* Notice that after 2 days you will have saved $2^2 - 1 = \$3$. After 3 days you have $2^3 - 1 = \$7$.)

Supplemental Exercises

In Exercises 91 to 96, write each expression as an equivalent expression in which the variables x and y occur only once. All the exponents are integers.

91. $\dfrac{x^n y^{n+2}}{x^{n-3} y}$

92. $\dfrac{x^{3n} y^{2n} y^n}{x^{-n+1} y^{-4n}}$

93. $\left(\dfrac{x^n y}{x^{1-n} y^{-1}}\right)^2$

94. $\left(\dfrac{x^n y^{2n}}{y^{3-n}}\right)^{-2}$

95. $\left(\dfrac{x^{3n} y^{2n}}{x^{-2n} y^{3n+1}}\right)^{-1}$

96. $\left(\dfrac{x^{4-n} y^{n+4}}{xy^{n-4}}\right)^2$

97. Which is larger, $3^{(3^3)}$ or $(3^3)^3$? (*Hint:* The parentheses are used to indicate which computation is to be performed first.)

98. Prime numbers of the form $2^p - 1$, where p is a prime number, are called *Mersenne primes*. When $p = 2$, the number $2^p - 1 = 2^2 - 1 = 3$. Thus 3 is a Mersenne prime. Find the next three Mersenne primes.

1.4

Polynomials

A **monomial** is a constant, or a variable, or a product of a constant and one or more variables, with the variables having only nonnegative integer exponents. The constant is called the **numerical coefficient** or simply the **coefficient** of the monomial. The **degree** of the monomial is the sum of the exponents of the variables. For example,

$-5xy^2$ is a monomial with coefficient -5 and degree 3.

$\dfrac{1}{2}u$ is a monomial with coefficient $\dfrac{1}{2}$ and degree 1.

0.25 is a monomial with coefficient 0.25 and degree 0.

The algebraic expressions

$$3x^{-2} \quad \text{and} \quad \frac{5}{x}$$

are not monomials because they cannot be written as a product of a constant and a variable with a nonnegative integer exponent.

A sum of a finite number of monomials is called a **polynomial**. Each monomial is called a **term of the polynomial**. The **degree of a polynomial** is the largest degree of the terms in the polynomial.

Terms of polynomials that have exactly the same variables raised to the same powers are called **like terms**. For example,

$14x^2$ and $-31x^2$ are like terms.

$2x^3y$ and $7xy$ are not like terms because x^3y and xy are not identical.

$4x^2y$ and $-3yx^2$ are like terms because $x^2y = yx^2$ by the commutative property of multiplication.

A polynomial is said to be simplified if all its like terms have been combined. For example, the simplified form of $4x^2 + 3x + 5x$ is $4x^2 + 8x$.

A simplified polynomial that has two terms is called a **binomial**, and a simplified polynomial that has three terms is called a **trinomial**. For example,

$4x + 7$ is a binomial of degree 1.

$2x^3 - 7x^2 + 11$ is a trinomial of degree 3.

$5x^3 - 3x^2y^3 - 7y^4 + 8$ is a polynomial of degree 5.

A nonzero constant, such as 5, is called a **constant polynomial.** It has degree zero since $5 = 5x^0$. The number 0 is defined to be a polynomial with no degree.

General Form of a Polynomial

The **general form of a polynomial** of degree n in the single variable x is

$$a_nx^n + a_{n-1}x^{n-1} + \cdots + a_2x^2 + a_1x + a_0$$

where $a_n \neq 0$ and n is a nonnegative integer. The coefficient a_n is called the **leading coefficient.**

If a polynomial in the single variable x is written with decreasing powers of x, then it is in **standard form.** For example, the polynomial

$$3x^2 - 4x^3 + 7x^4 - 1$$

is written in standard form as

$$7x^4 - 4x^3 + 3x^2 - 1.$$

Remark Because a polynomial is defined as the sum of its terms, the coefficient of each term includes the sign between the terms. For instance, the polynomial $3x^2 - 5x - 2$ has terms of $3x^2$, $-5x$, and -2. The coefficients of the terms are 3, -5, and -2.

The following table shows the leading coefficient, degree, terms, and coefficients of the given polynomials:

Polynomial	Leading Coefficient	Degree	Terms	Coefficients
$9x^2 - x + 5$	9	2	$9x^2$, $-x$, 5	9, -1, 5
$11 - 2x$	-2	1	$-2x$, 11	-2, 11
$x^3 + 5x - 3$	1	3	x^3, $5x$, -3	1, 5, -3

To add polynomials, we add like terms.

EXAMPLE 1   **Add Polynomials**

Find the sum $(3x^2 + 7x - 5) + (4x^2 - 2x + 1)$.

  Solution

$$(3x^2 + 7x - 5) + (4x^2 - 2x + 1) = (3x^2 + 4x^2) + (7x - 2x) + [(-5) + 1]$$
$$= 7x^2 + 5x - 4$$

■ *Try Exercise* **24,** *page 31.*

The **additive inverse of the polynomial** $3x - 7$ is

$$-(3x - 7) = -3x + 7.$$

To subtract a polynomial, we add its additive inverse.

EXAMPLE 2   **Subtract Polynomials**

Find the difference $(2x - 5) - (3x - 7)$.

  Solution

$$(2x - 5) - (3x - 7) = (2x - 5) + (-3x + 7)$$
$$= [2x + (-3x)] + [(-5) + 7]$$
$$= -x + 2$$

■ *Try Exercise* **28,** *page 31.*

The distributive property is also used to find the product of polynomials. For instance, to find the product of $(3x - 4)$ and $(2x^2 + 5x + 1)$, we treat $3x - 4$ as a *single* quantity and *distribute it* over the trinomial $2x^2 + 5x + 1$, as shown in Example 3.

EXAMPLE 3   **Multiply Polynomials**

Find $(3x - 4)(2x^2 + 5x + 1)$.

  Solution

$$(3x - 4)(2x^2 + 5x + 1)$$
$$= (3x - 4)(2x^2) + (3x - 4)(5x) + (3x - 4)(1)$$
$$= (3x)(2x^2) - 4(2x^2) + (3x)(5x) - 4(5x) + (3x)(1) - 4(1)$$
$$= 6x^3 - 8x^2 + 15x^2 - 20x + 3x - 4$$
$$= 6x^3 + 7x^2 - 17x - 4$$

■ *Try Exercise* **32,** *page 31.*

In the following example a vertical format has been used to find the product of $(x^2 + 6x - 7)$ and $(5x - 2)$. Notice that like terms are arranged in the same vertical column.

$$x^2 + 6x - 7$$
$$5x - 2$$
$$\overline{-2x^2 - 12x + 14}$$
$$5x^3 + 30x^2 - 35x$$
$$\overline{5x^3 + 28x^2 - 47x + 14}$$

If the terms of the binomials $(a + b)$ and $(c + d)$ are labeled as in Figure 1.12, then the product of the two binomials can be computed mentally by the **FOIL method.**

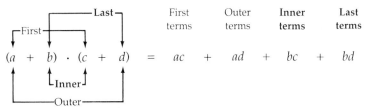

Figure 1.12

EXAMPLE 4 **Multiply Using the FOIL Method**

Find $(7x - 2)(5x + 4)$.

Solution

$$(7x - 2)(5x + 4) = \overset{\text{First}}{(7x)(5x)} + \overset{\text{Outer}}{(7x)(4)} + \overset{\text{Inner}}{(-2)(5x)} + \overset{\text{Last}}{(-2)(4)}$$

$$= 35x^2 + 28x - 10x - 8$$

$$= 35x^2 + 18x - 8$$

■ *Try Exercise **36**, page 32.*

The following products occur so frequently in algebra that they deserve special attention.

Special Product Formulas

Special Forms	Formulas
(Sum)(Difference)	$(x + y)(x - y) = x^2 - y^2$
(Binomial)^2	$(x + y)^2 = x^2 + 2xy + y^2$
	$(x - y)^2 = x^2 - 2xy + y^2$

The variables x and y in these special product formulas can be replaced by other algebraic expressions, as shown in Example 5.

EXAMPLE 5 **Use the Special Product Formulas**

Find each of the following products:

a. $(7x + 10)(7x - 10)$ b. $(2y^2 + 11z)^2$

Solution

a. $(7x + 10)(7x - 10) = (7x)^2 - (10)^2 = 49x^2 - 100$

b. $(2y^2 + 11z)^2 = (2y^2)^2 + 2[(2y^2)(11z)] + (11z)^2$

$= 4y^4 + 44y^2z + 121z^2$

■ *Try Exercise* **60,** *page 32.*

Many applications problems will require you to **evaluate polynomials.** To evaluate a polynomial, you substitute the given value(s) for the variable(s) and then perform the indicated operations using the **Order of Operations Agreement.**

The Order of Operations Agreement

If grouping symbols are present, evaluate by performing the operations within the grouping symbols, innermost grouping symbol first, while observing the order given in steps 1 to 3:

1. First, evaluate each power.
2. Next, do all multiplications and divisions, working from left to right.
3. Last, do all additions and subtractions, working from left to right.

EXAMPLE 6   Evaluate Polynomials

Evaluate the polynomial $2x^3 - 6x^2 + 7$ for $x = -4$.

Solution

$$2x^3 - 6x^2 + 7 = 2(-4)^3 - 6(-4)^2 + 7 \quad \text{Substitute } -4 \text{ for } x.$$

$$= 2(-64) - 6(16) + 7 \quad \text{Evaluate the powers.}$$

$$= -128 - 96 + 7 \quad \text{Perform the multiplications.}$$

$$= -217 \quad \text{Perform the additions and subtractions.}$$

■ *Try Exercise* **72,** *page 32.*

EXAMPLE 7   Solve an Application

The number of singles tennis matches that can be played between n tennis players is given by the polynomial $\frac{1}{2}n^2 - \frac{1}{2}n$. Find the number of singles tennis matches that can be played between 4 tennis players.

Solution

$$\frac{1}{2}n^2 - \frac{1}{2}n = \frac{1}{2}(4)^2 - \frac{1}{2}(4) \qquad \text{Substitute 4 for } n.$$

$$= \frac{1}{2}(16) - \frac{1}{2}(4) = 8 - 2 = 6$$

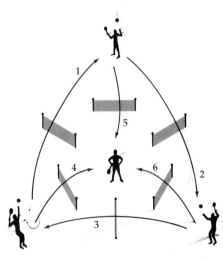

Figure 1.13
4 tennis players can play a total of 6 singles matches.

Therefore 4 tennis players can play a total of 6 singles matches. (See Figure 1.13.)

■ *Try Exercise* **82,** *page 32.*

EXAMPLE 8   Solve an Application

The *stopping distance* for a car is the distance a car travels between the time the driver decides to brake and the time the car comes to a complete stop. The stopping distance in feet, for a car with an antilock brake system, traveling v mph is given by the polynomial $0.04v^2 + 1.1v$. Find the stopping distance for the car when $v = 30$ mph.

Solution

$$0.04v^2 + 1.1v = 0.04(30)^2 + 1.1(30) \qquad \text{Let } v = 30.$$

$$= 0.04(900) + 1.1(30) = 36 + 33 = 69$$

When the car is traveling at 30 mph, it has a stopping distance of 69 feet.

■ *Try Exercise* **84,** *page 32.*

Remark Using the same formula, the car traveling at 60 mph has a stopping distance of 210 feet. Notice that doubling the speed from 30 to 60 mph more than triples the car's stopping distance.

EXERCISE SET 1.4

In Exercises 1 to 10, match the description with the appropriate example. The examples are labeled A, B, C, . . . , J.

1. A monomial of degree 2.
2. A binomial of degree 3.
3. A polynomial of degree 5.
4. A binomial with leading coefficient of -4.
5. A zero-degree polynomial.
6. A fourth-degree polynomial that has a third-degree term.
7. A trinomial with integer coefficients.
8. A trinomial in x and y.
9. A polynomial with no degree.
10. A fourth-degree binomial.

A. $x^3y + xy$ B. $7x^2 + 5x - 11$
C. $\frac{1}{2}x^2 + xy + y^2$ D. $4xy$
E. $8x^3 - 1$ F. $3 - 4x^2$
G. 8 H. $3x^5 - 4x^2 + 7x - 11$
I. $8x^4 - \sqrt{5}x^3 + 7$ J. 0

In Exercises 11 to 16, for each polynomial determine its **a.** standard form, **b.** degree, **c.** coefficients, **d.** leading coefficient, **e.** terms.

11. $2x + x^2 - 7$
12. $-3x^2 - 11 - 12x^4$

13. $x^3 - 1$
14. $4x^2 - 2x + 7$
15. $2x^4 + 3x^3 + 5 + 4x^2$
16. $3x^2 - 5x^3 + 7x - 1$

In Exercises 17 to 22, determine the degree of the given polynomial.

17. $3xy^2 - 2xy + 7x$
18. $x^3 + 3x^2y + 3xy^2 + y^3$
19. $4x^2y^2 - 5x^3y^2 + 17xy^3$
20. $-9x^5y + 10xy^4 - 11x^2y^2$
21. xy
22. $5x^2y - y^4 + 6xy$

In Exercises 23 to 34, perform the indicated operations and simplify if possible by combining like terms. Write the result in standard form.

23. $(3x^2 + 4x + 5) + (2x^2 + 7x - 2)$
24. $(5y^2 - 7y + 3) + (2y^2 + 8y + 1)$
25. $(4w^3 - 2w + 7) + (5w^3 + 8w^2 - 1)$
26. $(5x^4 - 3x^2 + 9) + (3x^3 - 2x^2 - 7x + 3)$
27. $(r^2 - 2r - 5) - (3r^2 - 5r + 7)$
28. $(7s^2 - 4s + 11) - (-2s^2 + 11s - 9)$
29. $(u^3 - 3u^2 - 4u + 8) - (u^3 - 2u + 4)$
30. $(5v^4 - 3v^2 + 9) - (6v^4 + 11v^2 - 10)$
31. $(4x - 5)(2x^2 + 7x - 8)$
32. $(5x - 7)(3x^2 - 8x - 5)$

33. $(3x^2 - 2x + 5)(2x^2 - 5x + 2)$

34. $(2y^3 - 3y + 4)(2y^2 - 5y + 7)$

In Exercises 35 to 52, use the FOIL method to find the indicated product.

35. $(2x + 4)(5x + 1)$ 36. $(5x - 3)(2x + 7)$

37. $(y + 2)(y + 1)$ 38. $(y + 5)(y + 3)$

39. $(4z - 3)(z - 4)$ 40. $(5z - 6)(z - 1)$

41. $(a + 6)(a - 3)$ 42. $(a - 10)(a + 4)$

43. $(b - 4)(b + 6)$ 44. $(b + 5)(b - 2)$

45. $(5x - 11y)(2x - 7y)$ 46. $(3a - 5b)(4a - 7b)$

47. $(9x + 5y)(2x + 5y)$ 48. $(3x - 7z)(5x - 7z)$

49. $(6w - 11x)(2w - 3x)$ 50. $(4m + 5n)(2m - 5n)$

51. $(3p + 5q)(2p - 7q)$ 52. $(2r - 11s)(5r + 8s)$

In Exercises 53 to 58, perform the indicated operations and simplify.

53. $(4d - 1)^2 - (2d - 3)^2$ 54. $(5c - 8)^2 - (2c - 5)^2$

55. $(r + s)(r^2 - rs + s^2)$ 56. $(r - s)(r^2 + rs + s^2)$

57. $(3c - 2)(4c + 1)(5c - 2)$ 58. $(4d - 5)(2d - 1)(3d - 4)$

In Exercises 59 to 68, use the special product formulas to perform the indicated operation.

59. $(3x + 5)(3x - 5)$ 60. $(4x^2 - 3y)(4x^2 + 3y)$

61. $(3x^2 - y)^2$ 62. $(6x + 7y)^2$

63. $(4w + z)^2$ 64. $(3x - 5y^2)^2$

65. $[(x - 2) + y]^2$ 66. $[(x + 3) - y]^2$

67. $[(x + 5) + y][(x + 5) - y]$

68. $[(x - 2y) + 7][(x - 2y) - 7]$

In Exercises 69 to 76, evaluate the given polynomial for the indicated value of the variable.

69. $x^2 + 7x - 1$, for $x = 3$

70. $x^2 - 8x + 2$, for $x = 4$

71. $-x^2 + 5x - 3$, for $x = -2$

72. $-x^2 - 5x + 4$, for $x = -5$

73. $3x^3 - 2x^2 - x + 3$, for $x = -1$

74. $5x^3 - x^2 + 5x - 3$, for $x = -1$

75. $1 - x^5$, for $x = -2$

76. $1 - x^3 - x^5$, for $x = 2$

Calculator Exercises

In Exercises 77 to 80, use a calculator to evaluate the given polynomial for the indicated values of the variable.

77. $4x^2 - 5x - 4$ for
    a. $x = 4.3$ b. $x = 4.4$

78. $8x^3 - 2x^2 + 6.4x - 7.1$ for
    a. $x = 1.2$ b. $x = 1.3$

79. $x^3 - 2x^2 - 5x + 11$ for
    a. $x = 0.001$ b. $x = 0.0001$

80. $-x^2 + 3x - 22.5$ for
    a. $x = 0.0002$ b. $x = 0.0001$

81. Find the number of committees consisting of exactly 2 people that can be formed from a group of 150 people. Use the formula from Example 7.

82. The number of committees consisting of exactly 3 people that can be formed from a group of n people is given by the polynomial

$$\frac{1}{6}n^3 - \frac{1}{2}n^2 + \frac{1}{3}n.$$

Find the number of committees consisting of exactly 3 people that can be formed from a group of 21 people.

83. Find the stopping distance for the car in Example 8 when $v = 100$ mph.

84. On an expressway, the recommended *safe distance* between cars in feet is given by $0.015v^2 + v + 10$ where v is the velocity of the car in miles per hour. Find the safe distance when   a. $v = 30$ mph,   b. $v = 55$ mph.

Supplemental Exercises

The following special product formulas can be used to find the cube of a binomial:

$$(x + y)^3 = x^3 + 3x^2y + 3xy^2 + y^3$$

$$(x - y)^3 = x^3 - 3x^2y + 3xy^2 - y^3$$

In Exercises 85 to 90, make use of the above special product formulas to find the indicated products.

85. $(a + b)^3$ 86. $(a - b)^3$ 87. $(x - 1)^3$

88. $(y + 2)^3$ 89. $(2x - 3y)^3$ 90. $(3x + 5y)^3$

91. If P is a polynomial in x of degree n and Q is a polynomial in x of degree $n - 1$, what is the degree of
    a. $P + Q$ b. $P - Q$ c. $P + P$ d. $P - P$

92. If R and S are each polynomials in x of degree n, what can be said about the degree of
    a. $R + S$ b. RS

93. Many people have tried to find a method of generating prime numbers. One such attempt evaluates the polynomial $n^2 - n + 41$ for $n = 1, 2, 3, \ldots$.
    a. Show that this method does generate a prime number for $n = 1, 2, 3$, and 4.
    b. Find a natural number n for which the polynomial $n^2 - n + 41$ does not generate a prime number.

94. How many prime numbers are generated by the polynomial $2n$, with $n = 1, 2, 3, 4, \ldots$?

1.5
Factoring

Writing a polynomial as a product of polynomials of lower degree is called **factoring**. Factoring is an important procedure, often used to simplify fractional expressions and to solve equations.

In this section we consider only the factorization of polynomials that have integer coefficients. Also, we will be concerned only with **factoring over the integers**. That is, we will only search for polynomial factors that have integer coefficients. In your study of arithmetic, you often factored over the integers. For example, when asked to factor the number 10, you might write $10 = 1 \cdot 10$ or $10 = 2 \cdot 5$. The factorization $10 = \frac{1}{2} \cdot 20$ was not considered because you were only concerned with finding integer factors of 10.

The GCF

The first step in any factorization of a polynomial is to use the distributive property to factor out the **greatest common factor** (GCF) of the terms of the polynomial. The GCF of two or more exponential expressions with the same prime number base or the same variable base is the exponential expression with the smallest exponent. For example:

2^3 is the GCF of 2^3, 2^5, and 2^8,   and a is the GCF of a^4 and a.

The GCF of two or more monomials is the product of the GCF of each *common* base. For example, to find the GCF of $27a^3b^4$ and $18b^3c$, factor the coefficients into prime factors and then write each common base with its smallest exponent:

$$27a^3b^4 = 3^3 \cdot a^3 \cdot b^4$$

$$18b^3c = 2 \cdot 3^2 \cdot b^3$$

The only common bases are 3 and b. The product of these common bases with their smallest exponents is 3^2b^3. The GCF of $27a^3b^4$ and $18b^3c$ is $9b^3$.

Many times you will be able to find the GCF of two or more monomials by inspection. For example, $8x^2$ is the GCF of $24x^3 + 16x^2$ because

8 is the largest number that is a factor of 24 and also a factor of 16.

x^2 is the largest power of x that is a factor of x^3 and also a factor of x^2.

EXAMPLE 1   Factor Out the Greatest Common Factor

Factor out the GCF in each of the following:

a. $10x^3 + 6x$ b. $15x^{2n} + 9x^{n+1} - 3x^n$ (where n is a positive integer)

c. $(m + 5)(x + 3) + (m + 5)(x - 10)$

Solution

a. $10x^3 + 6x = (2x)(5x^2) + (2x)(3)$ Factor each term.

$= (2x)(5x^2 + 3)$ Factor out the GCF.

b. $15x^{2n} + 9x^{n+1} - 3x^n = (3x^n)(5x^n) + (3x^n)(3x) - (3x^n)(1)$ Factor each term.

$$= 3x^n(5x^n + 3x - 1)$$ Factor out the GCF.

c. Use the distributive property to factor out $(m + 5)$.

$$(m + 5)(x + 3) + (m + 5)(x - 10) = (m + 5)[(x + 3) + (x - 10)]$$

$$= (m + 5)[x + 3 + x - 10]$$

$$= (m + 5)(2x - 7)$$

■ *Try Exercise 6, page 41.*

Some polynomials can be **factored by grouping.** Pairs of terms that have a common factor are first grouped together. The process makes repeated use of the distributive property, as shown in the following example.

EXAMPLE 2   Factor by Grouping in Pairs

Factor $6y^3 - 21y^2 - 4y + 14$.

Solution

$$6y^3 - 21y^2 - 4y + 14 = (6y^3 - 21y^2) - (4y - 14)$$ Group the terms in pairs.

$$= 3y^2(2y - 7) - 2(2y - 7)$$ Factor out the common monomial factors.

$$= (2y - 7)(3y^2 - 2)$$ Factor out the common binomial factor.

■ *Try Exercise 14, page 42.*

Some trinomials of the form $x^2 + bx + c$ can be factored by a trial and error procedure. This method makes use of the FOIL method in reverse. For example, consider the following products:

$$(x + 3)(x + 5) = x^2 + 5x + 3x + (3)(5) = x^2 + 8x + 15$$
$$(x - 2)(x - 7) = x^2 - 7x - 2x + (-2)(-7) = x^2 - 9x + 14$$
$$(x + 4)(x - 9) = x^2 - 9x + 4x + (4)(-9) = x^2 - 5x - 36$$

The coefficient of x is the sum of the constant terms of the binomials.

The constant term of the trinomial is the product of the constant terms of the binomials.

Points to Remember to Factor $x^2 + bx + c$.

1. The constant term c of the trinomial is the product of the constant terms of the binomials.
2. The coefficient b in the trinomial is the sum of the constant terms of the binomials.
3. If the constant term c of the trinomial is positive, the constant terms of the binomials have the same sign as the coefficient b of the trinomial.
4. If the constant term c of the trinomial is negative, the constant terms of the binomials have different signs.

To factor $x^2 + 7x - 18$, you must find two binomials whose first terms have a product of x^2, whose last terms have a product of -18; also, the sum of the product of the outer terms and the product of the inner terms must be $7x$.

We begin by listing the possible integer factorizations of -18.

Factors of -18	Sum of the factors
$1 \cdot (-18)$	$1 + (-18) = -17$
$(-1) \cdot 18$	$(-1) + 18 = 17$
$2 \cdot (-9)$	$2 + (-9) = -7$
$(-2) \cdot 9$	$(-2) + 9 = 7$ ←Stop. This is the desired sum.
$3 \cdot (-6)$	
$(-3) \cdot 6$	

Thus, -2 and 9 are the two numbers whose sum is 7 and whose product is -18. Therefore,

$$x^2 + 7x - 18 = (x - 2)(x + 9).$$

The FOIL method can be used to verify that the factorization is correct.

EXAMPLE 3 Factor Trinomials

Factor $x^2 + x - 12$.

Solution Consider factors of -12, which is the constant term of the trinomial.

Factors of -12	Sum of the Factors
$1 \cdot (-12)$	$1 + (-12) = -11$
$(-1) \cdot 12$	$(-1) + 12 = 11$
$2 \cdot (-6)$	$2 + (-6) = -4$
$(-2) \cdot 6$	$(-2) + 6 = 4$
$3 \cdot (-4)$	$3 + (-4) = -1$
$(-3) \cdot 4$	$(-3) + 4 = 1$ ← This is the desired sum.

Thus $x^2 + x - 12$ can be factored as $x^2 + x - 12 = (x - 3)(x + 4)$.

■ *Try Exercise* **18,** *page 42.*

Remark   Sometimes it is impossible to factor a polynomial into the product of two polynomials having integer coefficients. Such polynomials are said to be **nonfactorable** over the integers. For example, $x^2 + 3x + 7$ is nonfactorable over the integers because there are no integers whose product is 7 and whose sum or difference is 3.

The trial and error method can sometimes be used to factor trinomials of the form $ax^2 + bx + c$, which do not have a leading coefficient of 1. We use the factors of a and c to form trial binomial factors.

Factoring trinomials of this type by trial and error may require testing many trial factors. To reduce the number of trial factors, make use of the following points.

Points to Remember to Factor $ax^2 + bx + c$, $a > 0$.

1. If the constant term of the trinomial is positive, the constant terms of the binomials have the same sign as the coefficient b in the trinomial.

2. If the constant term of the trinomial is negative, the constant terms of the binomials have opposite signs.

3. If the terms of the trinomial do not have a common factor, then neither binomial will have a common factor.

EXAMPLE 4   Factor Trinomials

Factor $6x^2 - 11x + 4$.

Solution   Because the constant term of the trinomial is positive and the coefficient of the x term is negative, the constant terms of the binomials will both be negative. This time we find factors of the first term as well as factors of the constant term.

Factors of $6x^2$	Factors of 4 (both negative)
x, $6x$	-1, -4
$2x$, $3x$	-2, -2

Use these factors to write trial factors. Then use the FOIL method to see if any of the trial factors produce the correct middle term. Remember that if the terms of a trinomial do not have a common factor, then a binomial factor cannot have a common factor (point 3). Such trial factors need not be checked.

Trial Factors	Middle Term	
$(x - 1)(6x - 4)$	Common factor	$6x$ and 4 have a common factor.
$(x - 4)(6x - 1)$	$-1x - 24x = -25x$	
$(x - 2)(6x - 2)$	Common factor	
$(2x - 1)(3x - 4)$	$-8x - 3x = -11x$	←This is the correct middle term.

The correct factors have been found. The remaining trial factors need not be checked.

$$6x^2 - 11x + 4 = (2x - 1)(3x - 4)$$

■ *Try Exercise **22**, page 42.*

If a, the leading coefficient of $ax^2 + bx + c$, is negative, then you can factor the trinomial by factoring (-1) from each term of the trinomial and then applying the trial and error procedure.

EXAMPLE 5 **Factor Trinomials of the Form $ax^2 + bxy + cy^2$**

Factor $9x^2 - 14xy - 8y^2$.

Solution Use the factors of $9x^2$ and $-8y^2$ to form the binomial trial factors and then check to see if any produce the correct middle term.

Factors of $9x^2$	Factors of $-8y^2$
$x, 9x$	$y, -8y$
$3x, 3x$	$-y, 8y$
	$2y, -4y$
	$-2y, 4y$

Of all the possible trial factors, only $(x - 2y)$ and $(9x + 4y)$ produce the correct middle term $-14xy$. Thus $9x^2 - 14xy - 8y^2 = (x - 2y)(9x + 4y)$. Check as before.

■ *Try Exercise **26**, page 42.*

EXAMPLE 6 **Factor Trinomials of Degree Greater than 2**

Factor $2x^6 + 9x^3 + 9$.

Solution Because all the signs of the trinomial are positive, the coefficients of the terms in the binomial factors will all be positive.

Factors of $2x^6$	Factors of 9 (both positive)
$x^3, 2x^3$	1, 9
	3, 3

The factors $(x^3 + 3)$ and $(2x^3 + 3)$ are the only trial factors whose product has the correct middle term $9x^3$. Thus $2x^6 + 9x^3 + 9 = (x^3 + 3)(2x^3 + 3)$. Check as before.

■ *Try Exercise* **28,** *page 42.*

Some polynomials can be factored by using the special product formulas in reverse. The following factoring formulas can be verified by multiplying the factors on the right side of each equation.

Factoring Formulas

Difference of two squares	$x^2 - y^2 = (x + y)(x - y)$
Perfect square trinomials	$x^2 + 2xy + y^2 = (x + y)^2$ $x^2 - 2xy + y^2 = (x - y)^2$
Sum of cubes	$x^3 + y^3 = (x + y)(x^2 - xy + y^2)$
Difference of cubes	$x^3 - y^3 = (x - y)(x^2 + xy + y^2)$

The monomial a^2 is a square of a, and a is called a square root of a^2. The factoring formula

$$x^2 - y^2 = (x + y)(x - y)$$

indicates that the **difference of two squares** can be written as the product of the sum and the difference of the square roots of the squares.

EXAMPLE 7   **Factor the Difference of Squares**

Factor   a. $49x^2 - 144$ b. $9s^2 - t^2$

Solution

a. $49x^2 - 144 = (7x)^2 - (12)^2$ Recognize the difference of squares form.

        $= (7x + 12)(7x - 12)$ The binomial factors are the sum and the difference of the square roots of the squares.

b. $9s^2 - t^2 = (3s)^2 - t^2$ Recognize the difference of squares form.

        $= (3s + t)(3s - t)$ Factor.

■ *Try Exercise* **36,** *page 42.*

Caution   The polynomial $x^2 + y^2$ is the *sum* of two squares. You may be tempted to factor it in a manner similar to the method used on the *difference* of two squares; however, it is nonfactorable over the integers. Also notice that $x^2 + y^2 \neq (x + y)^2$.

A **perfect square trinomial** is a trinomial that is the square of a binomial. For example, $x^2 + 6x + 9$ is a perfect square trinomial because

$$(x + 3)^2 = x^2 + 6x + 9.$$

Every perfect square trinomial can be factored by the trial and error method; however, it generally is faster to factor perfect square trinomials by using the factoring formulas.

EXAMPLE 8   Factor Perfect Square Trinomials

Factor   a. $4a^2 + 12a + 9$ b. $16m^2 - 40mn + 25n^2$

Solution

a.   Use the factoring formula

$$x^2 + 2xy + y^2 = (x + y)^2,$$

with $x = 2a$ and $y = 3$.

$$4a^2 + 12a + 9 = (2a)^2 + 2(2a)(3) + 3^2 = (2a + 3)^2$$

b.   Use the formula

$$x^2 - 2xy + y^2 = (x - y)^2,$$

with $x = 4m$ and $y = 5n$.

$$16m^2 - 40mn + 25n^2 = (4m)^2 - 2(4m)(5n) + (5n)^2 = (4m - 5n)^2$$

■ *Try Exercise 46, page 42.*

The product of the same three factors is called a **cube**. For example, $8a^3$ is a cube because $8a^3 = (2a)^3$. The **cube root** of a cube is one of the three equal factors. For example, $3b^5$ is the cube root of $27b^{15}$. To factor the sum or the difference of two cubes, you use the factoring formulas. It helps to use the following patterns that involve the signs of the terms:

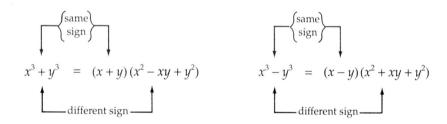

In the factorization of the sum or difference of two cubes, the terms of the binomial factor are the cube roots of the cubes. For example,

$$8a^3 - 27b^3 = (2a)^3 - (3b)^3 = (2a - 3b)(4a^2 + 6ab + 9b^2).$$

The trinomial factor can be obtained from the binomial factor as follows. The first term of the trinomial is the square of the first term of the binomial factor. The second term of the trinomial is the opposite of the product of the two terms of the binomial factor. The third term of the trinomial is the square of the second term of the binomial factor.

EXAMPLE 9   **Factor the Sum or Difference of Cubes**

Factor   a. $8a^3 + b^3$ b. $a^3 - 64$

Solution

a. $8a^3 + b^3 = (2a)^3 + b^3$ Recognize the sum
                                                of cubes form.

  $= (2a + b)(4a^2 - 2ab + b^2)$ Factor.

b. $a^3 - 64 = a^3 - 4^3$ Recognize the difference
                                                of cubes form.

  $= (a - 4)(a^2 + 4a + 16)$ Factor.

■ *Try Exercise* **52**, *page 42.*

General Factoring Strategy

> It is helpful to factor polynomials by using the following steps of the **general factoring strategy:**
>
> 1. Factor out the GCF of all terms.
> 2. Try to factor a binomial as the
>   a.   difference of two squares.
>   b.   sum or difference of two cubes.
> 3. Try to factor a trinomial:
>   a.   as a perfect square trinomial.
>   b.   using the trial and error method.
> 4. Try to factor a polynomial with more than three terms by grouping.
> 5. After each factorization, examine the new factors to see if they can be factored by any of the above methods.

EXAMPLE 10   **Factor Using the General Factoring Strategy**

Completely factor each of the following polynomials:

a. $32x^4 - 162$ b. $x^6 + 7x^3 - 8$

Solution

a. $32x^4 - 162 = 2(16x^4 - 81)$ Factor out the
                                                       common factor.

  $= 2[(4x^2)^2 - 9^2]$ Recognize the difference
                                       of squares form.

  $= 2(4x^2 + 9)(4x^2 - 9)$ Factor the difference
                                       of squares.

  $= 2(4x^2 + 9)(2x + 3)(2x - 3)$ Factor the difference
                                     of squares.

b. Factor $x^6 + 7x^3 - 8$ as the product of two binomials.

$$x^6 + 7x^3 - 8 = (x^3 + 8)(x^3 - 1)$$

Now factor $x^3 + 8$ which is the sum of two cubes, and factor $x^3 - 1$ which is the difference of two cubes.

$$x^6 + 7x^3 - 8 = (x + 2)(x^2 - 2x + 4)(x - 1)(x^2 + x + 1)$$

■ *Try Exercise* **62**, *page 42.*

When factoring by grouping, it may not be clear which terms should be grouped. Some experimentation may be necessary to find a grouping that is of the form of one of the special factoring formulas.

EXAMPLE 11   Factor by Grouping

Use the technique of grouping to factor each of the following:

a. $a^2 + 10ab + 25b^2 - c^2$ b. $p^2 + p - q - q^2$

Solution

a. $a^2 + 10ab + 25b^2 - c^2$

$\quad = (a^2 + 10ab + 25b^2) - c^2$ Group the terms of the perfect square trinomial.

$\quad = (a + 5b)^2 - c^2$ Factor the trinomial.

$\quad = [(a + 5b) + c][(a + 5b) - c]$ Factor the difference of squares.

$\quad = (a + 5b + c)(a + 5b - c)$ Simplify.

b. $p^2 + p - q - q^2 = p^2 - q^2 + p - q$ Rearrange the terms.

$\quad = (p^2 - q^2) + (p - q)$ Regroup.

$\quad = (p + q)(p - q) + (p - q)$ Factor the difference of squares.

$\quad = (p - q)(p + q + 1)$ Factor out the common factor $(p - q)$.

■ *Try Exercise* **72**, *page 42.*

EXERCISE SET 1.5 _____

In Exercises 1 to 8, factor out the GCF from each polynomial.

1. $5x + 20$

2. $8x^2 + 12x - 40$

3. $-15x^2 - 12x$

4. $-6y^2 - 54y$

5. $10x^2y + 6xy - 14xy^2$

6. $6a^3b^2 - 12a^2b + 72ab^3$

7. $(x - 3)(a + b) + (x - 3)(a + 3b)$

8. $(x - 4)(2a - b) + (x + 4)(2a - b)$

In Exercises 9 to 16, factor by grouping in pairs.

9. $3x^3 + x^2 + 6x + 2$

10. $18w^3 + 15w^2 + 12w + 10$

11. $ax^2 - ax + bx - b$

12. $a^2y^2 - ay^3 + ac - cy$

13. $6w^3 + 4w^2 - 15w - 10$

15. $12a^2x^3 - 8abx - 15ax^2 + 10b$

16. $12c^3d - 18c - 10c^2d + 15$

In Exercises 17 to 32, factor each trinomial by the trial and error method.

17. $x^2 + 7x + 12$

18. $x^2 + 9x + 20$

19. $a^2 - 10a - 24$

20. $b^2 + 12b - 28$

21. $6x^2 + 25x + 4$

22. $8a^2 - 26a + 15$

23. $51x^2 - 5x - 4$

24. $57y^2 + y - 6$

25. $6x^2 + xy - 40y^2$

26. $8x^2 + 10xy - 25y^2$

27. $x^4 + 6x^2 + 5$

28. $x^4 + 11x^2 + 18$

29. $6x^4 + 23x^2 + 15$

30. $9x^4 + 10x^2 + 1$

31. $18x^6 - 45x^3 + 28$

32. $4x^6 - 23x^3 - 6$

In Exercises 33 to 42, factor each difference of squares.

33. $x^2 - 9$

34. $x^2 - 64$

35. $4a^2 - 49$

36. $81b^2 - 16c^2$

37. $1 - 100x^2$

38. $1 - 121y^2$

39. $x^4 - 9$

40. $y^4 - 196$

41. $(x + 5)^2 - 4$

42. $(x - 3)^2 - 16$

In Exercises 43 to 50, factor each perfect square trinomial.

43. $x^2 + 10x + 25$

44. $y^2 + 6y + 9$

45. $a^2 - 14a + 49$

46. $b^2 - 24b + 144$

47. $4x^2 + 12x + 9$

48. $25y^2 + 40y + 16$

49. $z^4 + 4z^2w^2 + 4w^4$

50. $9x^4 - 30x^2y^2 + 25y^4$

In Exercises 51 to 58, factor each sum or difference of cubes.

51. $x^3 - 8$

52. $b^3 + 64$

53. $8x^3 - 27y^3$

54. $64u^3 - 27v^3$

55. $8 - x^6$

56. $1 + y^{12}$

57. $(x - 2)^3 - 1$

58. $(y + 3)^3 + 8$

In Exercises 59 to 80, use the general factoring strategy to completely factor each polynomial. If the polynomial does not factor, then state that it is nonfactorable over the integers.

59. $18x^2 - 2$

60. $4bx^3 + 32b$

61. $16x^4 - 1$

62. $81y^4 - 16$

63. $12ax^2 - 23axy + 10ay^2$

64. $6ax^2 - 19axy - 20ay^2$

65. $3bx^3 + 4bx^2 - 3bx - 4b$

66. $2x^6 - 2$

67. $72bx^2 + 24bxy + 2by^2$

68. $64y^3 - 16y^2z + yz^2$

69. $(w - 5)^3 + 8$

70. $5xy + 20y - 15x - 60$

71. $x^2 + 6xy + 9y^2 - 1$

72. $4y^2 - 4yz + z^2 - 9$

73. $8x^2 + 3x - 4$

74. $16x^2 + 81$

75. $a^6 - 1$

76. $3(y + 4)^3 - 81$

77. $5x(2x - 5)^2 - (2x - 5)^3$

78. $6x(3x + 1)^3 - (3x + 1)^4$

79. $4x^2 + 2x - y - y^2$

80. $a^2 + a + b - b^2$

Supplemental Exercises

In Exercises 81 and 82, find all positive values of k such that the trinomial is a perfect square trinomial.

81. $x^2 + kx + 16$

82. $36x^2 + kxy + 100$

In Exercises 83 and 84, find k such that the trinomial is a perfect square trinomial.

83. $x^2 + 16x + k$

84. $x^2 - 14xy + ky^2$

In Exercises 85 and 86, use the general strategy to completely factor each polynomial. In each exercise n represents a positive integer.

85. $x^{4n} - 1$

86. $x^{4n} - 2x^{2n} + 1$

In Exercises 87 to 90, write the area of the shaded portion of each geometric figure in its factored form.

87.

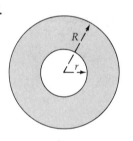

88.

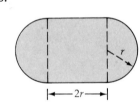

89.

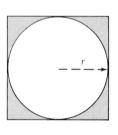

90.

91. The ancient Greeks used geometric figures and the concept of area to illustrate many algebraic concepts. The factoring formula $x^2 - y^2 = (x + y)(x - y)$ can be illustrated by the following figure.

a. Which regions are represented by $(x + y)(x - y)$?

b. Which regions are represented by $x^2 - y^2$?

c. Explain why the area of the regions listed in **a.** must equal the area of the regions listed in **b.**

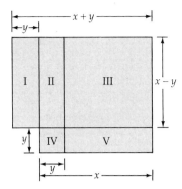

92. What algebraic formula does the following geometric figure illustrate?

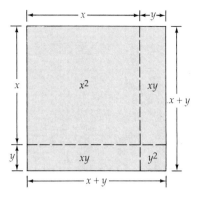

1.6

Rational Expressions

A **rational expression** is a fraction in which the numerator and denominator are polynomials. For example,

$$\frac{3}{x + 1} \quad \text{and} \quad \frac{x^2 - 4x - 21}{x^2 - 9}$$

are rational expressions. However, $\frac{5x + 2}{\sqrt{x} - 3}$ is not a rational expression because the denominator is not a polynomial.

The **domain of a rational expression** is the set of all real numbers that can be used as replacements for the variable. Any value of the variable that causes division by zero is an excluded value from the domain of the rational expression. For example, the domain of

$$\frac{7x}{x^2 - 5x} \qquad x \neq 0, \qquad x \neq 5$$

is the set of all real numbers except 0 and 5. Both 0 and 5 are excluded values since the denominator $x^2 - 5x$ equals zero when $x = 0$ and also when $x = 5$. Sometimes the excluded values are specified to the right of a rational expression, as shown above; however, a rational expression is meaningful only for those real numbers that are not excluded values, regardless of whether or not the excluded values are specifically stated.

Rational expressions have properties similar to the properties of rational numbers.

Properties of Rational Expressions

For all rational expressions P/Q and R/S where $Q \neq 0$ and $S \neq 0$:

Equality $\dfrac{P}{Q} = \dfrac{R}{S}$ if and only if $PS = QR$

Equivalent expressions $\dfrac{P}{Q} = \dfrac{PR}{QR}, \qquad R \neq 0$

Sign properties $-\dfrac{P}{Q} = \dfrac{-P}{Q} = \dfrac{P}{-Q}$

To **simplify a rational expression,** factor the numerator and the denominator. Then use the equivalent expressions property to eliminate factors common to both the numerator and the denominator. A rational expression is **simplified** when 1 is the only common polynomial factor of both the numerator and the denominator.

EXAMPLE 1   **Simplify Rational Expressions**

Simplify each of the following rational expressions:

a. $\dfrac{x^2 - 2x - 15}{3x - 15}$ b. $\dfrac{7 + 20x - 3x^2}{2x^2 - 11x - 21}$

Solution

a. $\dfrac{x^2 - 2x - 15}{3x - 15} = \dfrac{(x + 3)(x - 5)}{3(x - 5)}$

$\qquad\qquad = \dfrac{x + 3}{3} \qquad\qquad\qquad x \neq 5$

b. $\dfrac{7 + 20x - 3x^2}{2x^2 - 11x - 21} = \dfrac{(7 - x)(1 + 3x)}{(x - 7)(2x + 3)}$ 　　　Factor.

$\qquad\qquad = \dfrac{-(x - 7)(1 + 3x)}{(x - 7)(2x + 3)}$ 　　　Use $(7 - x) = -(x - 7)$.

$\qquad\qquad = \dfrac{-(x - 7)(1 + 3x)}{(x - 7)(2x + 3)}$

$\qquad\qquad = \dfrac{-(1 + 3x)}{2x + 3} = -\dfrac{3x + 1}{2x + 3}$

■ *Try Exercise **8**, page 50.*

Caution The rational expression $(x + 3)/3$ does not reduce to $x + 1$ since

$$\frac{x + 3}{3} = \frac{x}{3} + \frac{3}{3} = \frac{x}{3} + 1.$$

Rational expressions can be simplified by eliminating nonzero *factors* common to the numerator and the denominator, but not terms or factors of terms.

Arithmetic operations are defined on rational expressions just as they are on rational numbers.

Arithmetic Operations Defined on Rational Expressions

For all rational expressions P/Q and R/S where $Q \neq 0$ and $S = 0$:

Addition	$\dfrac{P}{Q} + \dfrac{R}{Q} = \dfrac{P + R}{Q}$
Subtraction	$\dfrac{P}{Q} - \dfrac{R}{Q} = \dfrac{P - R}{Q}$
Multiplication	$\dfrac{P}{Q} \cdot \dfrac{R}{S} = \dfrac{PR}{QS}$
Division	$\dfrac{P}{Q} \div \dfrac{R}{S} = \dfrac{P}{Q} \cdot \dfrac{S}{R} = \dfrac{PS}{QR}, \quad R \neq 0$

Factoring often plays a key role in the multiplication and division of rational expressions. Often, the resulting rational expression can be reduced using the equivalent expressions property of rational expressions.

EXAMPLE 2 **Multiply and Divide Rational Expressions**

Perform the indicated operation and then reduce if possible.

a. $\dfrac{x^2 + 7x}{4x^2 - 6x} \cdot \dfrac{x^2 - 16}{x^2 + 3x - 28}$ b. $\dfrac{x^2 + 6x + 9}{x^3 + 27} \div \dfrac{x^2 + 7x + 12}{x^3 - 3x^2 + 9x}$

Solution

a. $\dfrac{x^2 + 7x}{4x^2 - 6x} \cdot \dfrac{x^2 - 16}{x^2 + 3x - 28} = \dfrac{x(x + 7)}{2x(2x - 3)} \cdot \dfrac{(x + 4)(x - 4)}{(x - 4)(x + 7)}$ Factor.

$= \dfrac{x\,(x + 7)\,(x + 4)\,(x - 4)}{2x\,(2x - 3)\,(x - 4)\,(x + 7)}$ Simplify.

$= \dfrac{x + 4}{2(2x - 3)}$

b. $\dfrac{x^2 + 6x + 9}{x^3 + 27} \div \dfrac{x^2 + 7x + 12}{x^3 - 3x^2 + 9x}$

$$= \frac{(x + 3)^2}{(x + 3)(x^2 - 3x + 9)} \div \frac{(x + 4)(x + 3)}{x(x^2 - 3x + 9)} \qquad \text{Factor.}$$

$$= \frac{(x + 3)^2}{(x + 3)(x^2 - 3x + 9)} \cdot \frac{x(x^2 - 3x + 9)}{(x + 4)(x + 3)} \qquad \begin{array}{l}\text{Multiply by the}\\ \text{reciprocal.}\end{array}$$

$$= \frac{\cancel{(x + 3)^2}\, x(\cancel{x^2 - 3x + 9})}{\cancel{(x + 3)}(\cancel{x^2 - 3x + 9})(x + 4)\cancel{(x + 3)}} \qquad \text{Simplify}$$

$$= \frac{x}{x + 4}$$

■ *Try Exercise* **16,** *page 51.*

Addition of rational expressions with a **common denominator** is accomplished by writing the sum of the numerators over the common denominator. For example,

$$\frac{5x}{18} + \frac{x}{18} = \frac{5x + x}{18} = \frac{6x}{18} = \frac{x}{3}.$$

If the rational expressions do not have a common denominator, then they can be written as equivalent rational expressions that have a common denominator by multiplying numerator and denominator of each of the rational expressions by the required polynomials. The following procedure can be used to determine the least common denominator (LCD) of rational expressions. It is similar to the process used to find the LCD of fractions.

Determining the LCD of Two or More Rational Expressions

1. Factor each denominator completely and express repeated factors using exponential notation.

2. Identify the largest power of each factor in any single factorization. The LCD is the product of each factor raised to its largest power.

For example, $\dfrac{1}{x + 3}$ and $\dfrac{5}{2x - 1}$

have a LCD of $(x + 3)(2x - 1)$. The rational expressions

$$\frac{5x}{(x + 5)(x - 7)^3} \quad \text{and} \quad \frac{7}{x(x + 5)^2(x - 7)}$$

have a LCD of $x(x + 5)^2(x - 7)^3$.

Example 3 illustrates the process of adding and subtracting rational expressions by writing each rational expression as an equivalent rational expression having the LCD as its denominator.

EXAMPLE 3   **Add and Subtract Rational Expressions**

Perform the indicated operation and then simplify if possible.

a. $\dfrac{5x}{48} + \dfrac{x}{15}$ b. $\dfrac{x}{x^2 - 4} - \dfrac{2x - 1}{x^2 - 3x - 10}$

Solution

a. Determine the prime factorization of the denominators.

$$48 = 2^4 \cdot 3 \quad \text{and} \quad 15 = 3 \cdot 5$$

The LCD is the product of each of the prime factors raised to its largest power. Thus the LCD is $2^4 \cdot 3 \cdot 5 = 240$. Write each rational expression as an equivalent rational expression with a denominator of 240.

$$\dfrac{5x}{48} + \dfrac{x}{15} = \dfrac{5x \cdot 5}{48 \cdot 5} + \dfrac{x \cdot 16}{15 \cdot 16} \quad \text{Use the equivalent expressions property } \dfrac{P}{Q} = \dfrac{PR}{QR}.$$

$$= \dfrac{25x}{240} + \dfrac{16x}{240} = \dfrac{41x}{240}$$

b. Factor each denominator to determine the LCD of the rational expressions.

$$x^2 - 4 = (x + 2)(x - 2)$$

and

$$x^2 - 3x - 10 = (x + 2)(x - 5)$$

The LCD is $(x + 2)(x - 2)(x - 5)$. Therefore, forming equivalent rational expressions that have the LCD, we have

$$\dfrac{x}{x^2 - 4} - \dfrac{2x - 1}{x^2 - 3x - 10}$$

$$= \dfrac{x(x - 5)}{(x + 2)(x - 2)(x - 5)} - \dfrac{(2x - 1)(x - 2)}{(x + 2)(x - 5)(x - 2)}$$

$$= \dfrac{x^2 - 5x - (2x^2 - 5x + 2)}{(x + 2)(x - 2)(x - 5)} \qquad \text{Subtract.}$$

$$= \dfrac{x^2 - 5x - 2x^2 + 5x - 2}{(x + 2)(x - 2)(x - 5)} \qquad \text{Simplify.}$$

$$= \dfrac{-x^2 - 2}{(x + 2)(x - 2)(x - 5)}$$

$$= -\dfrac{x^2 + 2}{(x + 2)(x - 2)(x - 5)}$$

■ *Try Exercise **30**, page 51.*

Complex Fractions

A **complex fraction** is a fraction whose numerator or denominator contains one or more fractions. Complex fractions can be simplified by using one of the following two methods.

Methods for Simplifying Complex Fractions

Method 1: Multiply by the LCD
1. Determine the LCD of all the fractions in the complex fraction.
2. Multiply both the numerator and the denominator of the complex fraction by the LCD.
3. If possible, simplify the resulting rational expression.

Method 2: Simplify and multiply by the reciprocal of the denominator
1. Simplify the numerator to a single fraction and the denominator to a single fraction.
2. Multiply the numerator by the reciprocal of the denominator.
3. If possible, simplify the resulting rational expression.

EXAMPLE 4   **Simplify Complex Fractions**

Simplify each complex fraction.

a. $\dfrac{3 - \dfrac{2}{a}}{1 + \dfrac{4}{a}}$ b. $\dfrac{\dfrac{2}{x-2} + \dfrac{1}{x}}{\dfrac{3x}{x-5} - \dfrac{2}{x-5}}$

Solution

a. The LCD of all the fractions in the complex fraction is a. Therefore this complex fraction can be simplified by multiplying both the numerator and the denominator of the complex fraction by a.

$$\frac{3 - \dfrac{2}{a}}{1 + \dfrac{4}{a}} = \frac{\left(3 - \dfrac{2}{a}\right)a}{\left(1 + \dfrac{4}{a}\right)a} = \frac{3a - \left(\dfrac{2}{a}\right)a}{a + \left(\dfrac{4}{a}\right)a} \quad \text{Distribute and simplify.}$$

$$= \frac{3a - 2}{a + 4}$$

b. This complex fraction is best simplified by the method of first simplifying the numerator to a single fraction and then simplifying the denominator to a single fraction.

$$\dfrac{\dfrac{2}{x-2}+\dfrac{1}{x}}{\dfrac{3x}{x-5}-\dfrac{2}{x-5}} = \dfrac{\dfrac{2\cdot x}{(x-2)\cdot x}+\dfrac{1\cdot(x-2)}{x\cdot(x-2)}}{\dfrac{3x-2}{x-5}}$$

Simplify numerator and denominator.

$$= \dfrac{\dfrac{2x+(x-2)}{x(x-2)}}{\dfrac{3x-2}{x-5}} = \dfrac{\dfrac{3x-2}{x(x-2)}}{\dfrac{3x-2}{x-5}}$$

$$= \dfrac{\cancel{3x-2}}{x(x-2)}\cdot\dfrac{(x-5)}{\cancel{(3x-2)}}$$

Multiply by the reciprocal of the denominator and simplify.

$$= \dfrac{x-5}{x(x-2)}$$

■ *Try Exercise* **42**, *page 51.*

EXAMPLE 5 **Simplify Complex Fractions with Negative Exponents**

Simplify the complex fraction $\dfrac{c^{-1}}{a^{-1}+b^{-1}}$.

Solution The fraction written without negative exponents becomes

$$\dfrac{\dfrac{1}{c}}{\dfrac{1}{a}+\dfrac{1}{b}}.$$

Using the method of multiplying the numerator and the denominator by abc, which is the LCD of the denominators, yields the following simplification:

$$\dfrac{\dfrac{1}{c}}{\dfrac{1}{a}+\dfrac{1}{b}} = \dfrac{\dfrac{1}{c}\cdot abc}{\left(\dfrac{1}{a}+\dfrac{1}{b}\right)abc} = \dfrac{\dfrac{1}{c}\cdot abc}{\dfrac{1}{a}\cdot abc+\dfrac{1}{b}\cdot abc} = \dfrac{ab}{bc+ac}$$

■ *Try Exercise* **60**, *page 52.*

Remark It is a mistake to write

$$\dfrac{c^{-1}}{a^{-1}+b^{-1}} \quad \text{as} \quad \dfrac{a+b}{c}$$

because a^{-1} and b^{-1} are *terms* and cannot be treated as factors.

EXAMPLE 6   Solve an Application

The *average velocity* for a round trip is given by the complex fraction

$$\frac{2}{\dfrac{1}{v_1} + \dfrac{1}{v_2}}$$

where v_1 is the average velocity going to your destination and v_2 is the average velocity on your return trip. Find the average velocity for a round trip if $v_1 = 50$ mph and $v_2 = 40$ mph.

Solution   Evaluate the complex fraction with $v_1 = 50$ and $v_2 = 40$.

$$\frac{2}{\dfrac{1}{v_1} + \dfrac{1}{v_2}} = \frac{2}{\dfrac{1}{50} + \dfrac{1}{40}} = \frac{2}{\dfrac{1 \cdot 4}{50 \cdot 4} + \dfrac{1 \cdot 5}{40 \cdot 5}}$$ Substitute and simplify the denominator

$$= \frac{2}{\dfrac{4}{200} + \dfrac{5}{200}} = \frac{2}{\dfrac{9}{200}}$$

$$= 2 \cdot \frac{200}{9}$$ Multiply by the reciprocal of the denominator.

$$= \frac{400}{9} = 44\frac{4}{9}$$

The average velocity of the round trip is $44\frac{4}{9}$ mph.

■ *Try Exercise* **64,** *page 52.*

Remark   The average velocity of the round trip is *not* the average of v_1 and v_2. Why?[1]

[1] Because you were traveling slower on the return trip, the return trip took longer than the time spent going to your destination. More time was spent traveling at the slower velocity. Thus the average velocity is less than the average of v_1 and v_2.

EXERCISE SET 1.6

In Exercises 1 to 10, simplify each rational expression.

1. $\dfrac{x^2 - x - 20}{3x - 15}$

2. $\dfrac{2x^2 - 5x - 12}{2x^2 + 5x + 3}$

3. $\dfrac{x^3 - 9x}{x^3 + x^2 - 6x}$

4. $\dfrac{x^3 + 125}{2x^3 - 50x}$

5. $\dfrac{a^3 + 8}{a^2 - 4}$

6. $\dfrac{y^3 - 27}{-y^2 + 11y - 24}$

7. $\dfrac{x^2 + 3x - 40}{-x^2 + 3x + 10}$

8. $\dfrac{2x^3 - 6x^2 + 5x - 15}{9 - x^2}$

9. $\dfrac{4y^3 - 8y^2 + 7y - 14}{-y^2 - 5y + 14}$

10. $\dfrac{x^3 - x^2 + x}{x^3 + 1}$

In Exercises 11 to 40, perform the indicated operation(s). State your results in simplest form.

11. $\left(-\dfrac{4a}{3b^2}\right)\left(\dfrac{6b}{a^4}\right)$

12. $\left(\dfrac{12x^2y}{5z^4}\right)\left(-\dfrac{25x^2z^3}{15y^2}\right)$

13. $\left(\dfrac{6p^2}{5q^2}\right)^{-1}\left(\dfrac{2p}{3q^2}\right)^2$

14. $\left(\dfrac{4r^2s}{3t^3}\right)^{-1}\left(\dfrac{6rs^3}{5t^2}\right)$

15. $\dfrac{x^2 + x}{2x + 3} \cdot \dfrac{3x^2 + 19x + 28}{x^2 + 5x + 4}$

16. $\dfrac{x^2 - 16}{x^2 + 7x + 12} \cdot \dfrac{x^2 - 4x - 21}{x^2 - 4x}$

17. $\dfrac{3x - 15}{2x^2 - 50} \cdot \dfrac{2x^2 + 16x + 30}{6x + 9}$

18. $\dfrac{y^3 - 8}{y^2 + y - 6} \cdot \dfrac{y^2 + 3y}{y^3 + 2y^2 + 4y}$

19. $\dfrac{12y^2 + 28y + 15}{6y^2 + 35y + 25} \div \dfrac{2y^2 - y - 3}{3y^2 + 11y - 20}$

20. $\dfrac{z^2 - 81}{z^2 - 16} \div \dfrac{z^2 - z - 20}{z^2 + 5z - 36}$

21. $\dfrac{a^2 + 9}{a^2 - 64} \div \dfrac{a^3 - 3a^2 + 9a - 27}{a^2 + 5a - 24}$

22. $\dfrac{6x^2 + 13xy + 6y^2}{4x^2 - 9y^2} \div \dfrac{3x^2 - xy - 2y^2}{2x^2 + xy - 3y^2}$

23. $\dfrac{p + 5}{r} + \dfrac{2p - 7}{r}$

24. $\dfrac{2s + 5t}{4t} + \dfrac{-2s + 3t}{4t}$

25. $\dfrac{x}{x - 5} + \dfrac{7x}{x + 3}$

26. $\dfrac{2x}{3x + 1} + \dfrac{5x}{x - 7}$

27. $\dfrac{5y - 7}{y + 4} - \dfrac{2y - 3}{y + 4}$

28. $\dfrac{6x - 5}{x - 3} - \dfrac{3x - 8}{x - 3}$

29. $\dfrac{4z}{2z - 3} + \dfrac{5z}{z - 5}$

30. $\dfrac{3y - 1}{3y + 1} - \dfrac{2y - 5}{y - 3}$

31. $\dfrac{x}{x^2 - 9} - \dfrac{3x - 1}{x^2 + 7x + 12}$

32. $\dfrac{m - n}{m^2 - mn - 6n^2} + \dfrac{3m - 5n}{m^2 + mn - 2n^2}$

33. $\dfrac{1}{x} + \dfrac{2}{3x - 1} \cdot \dfrac{3x^2 + 11x - 4}{x - 5}$

34. $\dfrac{2}{y} - \dfrac{3}{y + 1} \cdot \dfrac{y^2 - 1}{y + 4}$

35. $\dfrac{q + 1}{q - 3} - \dfrac{2q}{q - 3} \div \dfrac{q + 5}{q - 3}$

36. $\dfrac{p}{p + 5} + \dfrac{p}{p - 4} \div \dfrac{p + 2}{p^2 - p - 12}$

37. $\dfrac{1}{x^2 + 7x + 12} + \dfrac{1}{x^2 - 9} + \dfrac{1}{x^2 - 16}$

38. $\dfrac{2}{a^2 - 3a + 2} + \dfrac{3}{a^2 - 1} - \dfrac{5}{a^2 + 3a - 10}$

39. $\left(1 + \dfrac{2}{x}\right)\left(3 - \dfrac{1}{x}\right)$

40. $\left(4 - \dfrac{1}{z}\right)\left(4 + \dfrac{2}{z}\right)$

In Exercises 41 to 58, simplify each complex fraction.

41. $\dfrac{4 + \dfrac{1}{x}}{1 - \dfrac{1}{x}}$

42. $\dfrac{3 - \dfrac{2}{a}}{5 + \dfrac{3}{a}}$

43. $\dfrac{\dfrac{x}{y} - 2}{y - x}$

44. $\dfrac{3 + \dfrac{2}{x - 3}}{4 + \dfrac{1}{2 + \dfrac{1}{x}}}$

45. $\dfrac{5 - \dfrac{1}{x + 2}}{1 + \dfrac{3}{1 + \dfrac{3}{x}}}$

46. $\dfrac{\dfrac{1}{(x + h)^2} - 1}{h}$

47. $\dfrac{1 + \dfrac{1}{b - 2}}{1 - \dfrac{1}{b + 3}}$

48. $r - \dfrac{r}{r + \dfrac{1}{3}}$

49. $\dfrac{1 - \dfrac{1}{x^2}}{1 + \dfrac{1}{x}}$

50. $\dfrac{1}{\dfrac{1}{a} + \dfrac{1}{b}}$

51. $2 - \dfrac{m}{1 - \dfrac{1 - m}{-m}}$

52. $\dfrac{\dfrac{x + h + 1}{x + h} - \dfrac{x}{x + 1}}{h}$

53. $\dfrac{\dfrac{1}{x} - \dfrac{x - 4}{x + 1}}{\dfrac{x}{x + 1}}$

54. $\dfrac{\dfrac{2}{y} - \dfrac{3y - 2}{y - 1}}{\dfrac{y}{y - 1}}$

55. $\dfrac{\dfrac{1}{x + 3} - \dfrac{2}{x - 1}}{\dfrac{x}{x - 1} + \dfrac{3}{x + 3}}$

56. $\dfrac{\dfrac{x + 2}{x^2 - 1} + \dfrac{1}{x + 1}}{\dfrac{x}{2x^2 - x - 1} + \dfrac{1}{x - 1}}$

57. $\dfrac{\dfrac{x^2 + 3x - 10}{x^2 + x - 6}}{\dfrac{x^2 - x - 30}{2x^2 - 15x + 18}}$

58. $\dfrac{\dfrac{2y^2 + 11y + 15}{y^2 - 4y - 21}}{\dfrac{6y^2 + 11y - 10}{3y^2 - 23y + 14}}$

In Exercises 59 to 62, simplify each algebraic fraction. Write all answers with positive exponents.

59. $\dfrac{a^{-1} + b^{-1}}{a - b}$

60. $\dfrac{e^{-2} - f^{-1}}{ef}$

61. $\dfrac{a^{-1}b - ab^{-1}}{a^2 + b^2}$

62. $(a + b^{-2})^{-1}$

63. According to Example 6, the average velocity for a round trip in which the average velocity going to your destination was v_1 and the average velocity on your return was v_2 is given by the complex fraction

$$\frac{2}{\dfrac{1}{v_1} + \dfrac{1}{v_2}}.$$

   a. Find the average velocity for a round trip by helicopter with $v_1 = 180$ mph and $v_2 = 110$ mph.

   b. Simplify the above complex fraction.

64. Using Einstein's theory of relativity, the "sum" of the two velocities v_1 and v_2 is given by the complex fraction

$$\frac{v_1 + v_2}{1 + \dfrac{v_1 v_2}{c^2}},$$

   where c is the speed of light.

   a. Evaluate the complex fraction with $v_1 = 1.2 \times 10^8$ mph, $v_2 = 2.4 \times 10^8$ mph, and $c = 6.7 \times 10^8$ mph.

   b. Simplify the above complex fraction.

65. Find the rational expression in simplest form that represents the sum of the reciprocals of the consecutive integers x and $x + 1$.

66. Find the rational expression in simplest form that represents the positive difference between the reciprocals of the consecutive even integers x and $x + 2$.

67. Find the rational expression in simplest form that represents the sum of the reciprocals of the consecutive even integers $x - 2$, x, and $x + 2$.

68. Find the rational expression in simplest form that represents the sum of the reciprocals of the squares of the consecutive even integers $x - 2$, x, and $x + 2$.

Supplemental Exercises

In Exercises 69 to 72, simplify each algebraic fraction. Write all answers with positive exponents.

69. $\dfrac{(x + 5) - x(x + 5)^{-1}}{x + 5}$

70. $\dfrac{(y + 2) + y^2(y + 2)^{-1}}{y + 2}$

71. $\dfrac{x^{-1} - 4y}{(x^{-1} - 2y)(x^{-1} + 2y)}$

72. $\dfrac{x + y}{x - y} \cdot \dfrac{x^{-1} - y^{-1}}{x^{-1} + y^{-1}}$

The following expression is called a **continued fraction:**

$$\cfrac{1}{1 + \cfrac{1}{1 + \cfrac{1}{1 + \cfrac{1}{1 + \cfrac{1}{1 + \cdots}}}}}$$

In Exercises 73 to 78, the complex fractions are called **convergents** of the above continued fraction. Simplify each convergent, and write it as a decimal.

73. $C_1 = \cfrac{1}{1 + \cfrac{1}{1}}$

74. $C_2 = \cfrac{1}{1 + \cfrac{1}{1 + \cfrac{1}{1}}}$

75. $C_3 = \cfrac{1}{1 + \cfrac{1}{1 + \cfrac{1}{1 + \cfrac{1}{1}}}}$

76. $C_4 = \cfrac{1}{1 + \cfrac{1}{1 + \cfrac{1}{1 + \cfrac{1}{1 + \cfrac{1}{1}}}}}$

77. $C_5 = \cfrac{1}{1 + \cfrac{1}{1 + \cfrac{1}{1 + \cfrac{1}{1 + \cfrac{1}{1 + \cfrac{1}{1}}}}}}$

78. $C_6 = \cfrac{1}{1 + \cfrac{1}{1 + \cfrac{1}{1 + \cfrac{1}{1 + \cfrac{1}{1 + \cfrac{1}{1 + \cfrac{1}{1}}}}}}}$

79. It can be shown that if we continue to evaluate convergents as in exercises 73 to 78, we find that the convergents get closer and closer to the irrational number

$$\frac{-1 + \sqrt{5}}{2}.$$

Use a calculator to approximate this irrational number as a decimal accurate to eight decimal places.

80. Approximate

$$C_6 - \frac{-1 + \sqrt{5}}{2}$$

as a decimal accurate to six decimal places. *Note:* If you worked exercise 78 correctly, you found $C_6 = \frac{13}{21}$, which as a decimal accurate to eight decimal places is 0.61904762.

81. Find

$$\left(1 - \frac{1}{2}\right)\left(1 - \frac{1}{3}\right)\left(1 - \frac{1}{4}\right)\left(1 - \frac{1}{5}\right).$$

82. Find

$$\left(1 - \frac{1}{2}\right)\left(1 - \frac{1}{3}\right)\left(1 - \frac{1}{4}\right)\cdots\left(1 - \frac{1}{100}\right).$$

83. Use $x = \frac{1}{2}$, $y = \frac{1}{3}$, and $z = \frac{1}{4}$ to show that

$$\frac{x^{-1} + y^{-1}}{z^{-1}} \neq \frac{z}{x + y}.$$

84. Rewrite the rational expression

$$\frac{7x + 8y}{2xy}$$

as the sum of two rational expressions, each in simplest form.

85. Evaluate the following complex fraction with $n = 1, 2,$ and 3:

$$\frac{n}{n + \dfrac{n}{n + n}}$$

86. Use the results of Exercise 85 to simplify

$$\frac{2001}{2001 + \dfrac{2001}{2001 + 2001}}.$$

1.7
Rational Exponents and Radicals

To this point, the expression b^n has been defined for real numbers b and integers n, except for the restrictions listed in Section 1.3. Now we wish to extend the definition of exponents to include rational numbers, so that expressions such as $2^{1/2}$ will be meaningful. Not just any definition will do. We want a definition of rational exponents for which the properties of integer exponents are true. The following example shows the direction we can take to accomplish our goal.

If the property for multiplying exponential expressions is to hold for rational exponents, then for rational numbers p and q, $b^p b^q = b^{p+q}$. For example,

$$9^{1/2} \cdot 9^{1/2} \quad \text{must equal} \quad 9^{1/2+1/2} = 9^1 = 9.$$

Thus $9^{1/2}$ must be a square root of 9. That is, $9^{1/2} = 3$.

Also,

$$8^{1/3} \cdot 8^{1/3} \cdot 8^{1/3} \quad \text{must equal} \quad 8^{1/3+1/3+1/3} = 8^1 = 8.$$

Thus $8^{1/3}$ must be a cube root of 8. That is, $8^{1/3} = 2$.

These examples suggest that $b^{1/n}$ can be defined in terms of roots according to the following definition.

Definition of $b^{1/n}$

> If n is an even positive integer and $b \geq 0$, then $b^{1/n}$ is the nonnegative real number such that $(b^{1/n})^n = b$.
>
> If n is an odd positive integer, then $b^{1/n}$ is the real number such that $(b^{1/n})^n = b$.

As examples, $25^{1/2} = 5$ because $5^2 = 25$; $(-64)^{1/3} = -4$ because $(-4)^3 = -64$.

Remark   If n is an even positive integer and $b < 0$, then $b^{1/n}$ is a complex number. We will study complex numbers in Section 1.8.

EXAMPLE 1   Evaluate Exponential Expressions

Evaluate each of the following:

a. $16^{1/2}$ b. $-16^{1/2}$ c. $(-16)^{1/2}$ d. $(-32)^{1/5}$

Solution

a. $16^{1/2} = 4$ because $4^2 = 16$.   b. $-16^{1/2} = -(16^{1/2}) = -4$.

c. $(-16)^{1/2}$ is not a real number.

d. $(-32)^{1/5} = -2$ because $(-2)^5 = 32$.

■ *Try Exercise* **2***, page 62.*

Remark   Notice the difference between $-16^{1/2}$, which equals -4, and $(-16)^{1/2}$, which is not a real number.

To define expressions such as $8^{2/3}$, we will extend our definition of exponents even further. Because we want the power property $(b^p)^q = b^{pq}$ to be true for rational exponents also, we must have $(b^{1/n})^m = b^{m/n}$. With this in mind, we make the following definition.

Definition of $b^{m/n}$

> For all positive integers m and n such that m/n is reduced to lowest terms, and for all real numbers b for which $b^{1/n}$ is a real number,
> $$b^{m/n} = (b^{1/n})^m = (b^m)^{1/n}.$$

Because $b^{m/n}$ is defined as $(b^{1/n})^m$ and also as $(b^m)^{1/n}$, we can evaluate expressions such as $8^{4/3}$ in more than one way. For example, $8^{4/3}$ can be evaluated by using either of the following:

$$8^{4/3} = (8^{1/3})^4 = 2^4 = 16$$

$$8^{4/3} = (8^4)^{1/3} = 4096^{1/3} = 16$$

Of the two methods, the $b^{m/n} = (b^{1/n})^m$ method is usually easier to apply, provided you can evaluate $b^{1/n}$.

EXAMPLE 2 **Evaluate Exponential Expressions**

Evaluate each of the following:

a. $8^{2/3}$ b. $32^{4/5}$ c. $(-9)^{3/2}$ d. $(-64)^{4/3}$

Solution

a. $8^{2/3} = (8^{1/3})^2 = 2^2 = 4$ b. $32^{4/5} = (32^{1/5})^4 = 2^4 = 16$

c. $(-9)^{3/2}$ is not a real number because $(-9)^{1/2}$ is not a real number.

d. $(-64)^{4/3} = [(-64)^{1/3}]^4 = [-4]^4 = 256$

■ *Try Exercise* **6,** *page 62.*

The following properties of exponents were stated in Section 1.3, but they are restated here to remind you that they have now been extended to apply to rational exponents. The bases a and b have been restricted to positive real numbers to avoid the pitfalls illustrated in Exercises 161 and 162.

Properties of Rational Exponents

If p and q represent rational numbers and a and b are positive real numbers, then

Product property $b^p \cdot b^q = b^{p+q}$

Quotient property $\dfrac{b^p}{b^q} = b^{p-q}$

Power properties $(b^p)^q = b^{pq}$ $(ab)^p = a^p b^p$ $\left(\dfrac{a}{b}\right)^p = \dfrac{a^p}{b^p}$

$b^{-p} = \dfrac{1}{b^p}$

Recall from Section 1.3 that an exponential expression is in simplest form if each base appears at most once and no powers of powers or negative exponents appear in the expression.

EXAMPLE 3 **Simplify Exponential Expressions**

Simplify each exponential expression. Assume the variables to be positive real numbers.

a. $x^{1/5} \cdot x^{2/5}$ b. $\left(\dfrac{x^2 y^3}{x^{-3} y^5}\right)^{1/2}$ c. $(x^{1/3} - y^{1/3})(x^{2/3} + x^{1/3} y^{1/3} + y^{2/3})$

Solution

a. $x^{1/5} \cdot x^{2/5} = x^{1/5 + 2/5} = x^{3/5}$

b. $\left(\dfrac{x^2 y^3}{x^{-3} y^5}\right)^{1/2} = (x^5 y^{-2})^{1/2} = x^{5/2} y^{-1} = \dfrac{x^{5/2}}{y}$

c. $(x^{1/3} - y^{1/3})(x^{2/3} + x^{1/3}y^{1/3} + y^{2/3})$
$$= x^1 + x^{2/3}y^{1/3} + x^{1/3}y^{2/3} - x^{2/3}y^{1/3} - x^{1/3}y^{2/3} - y^1$$
$$= x - y$$

■ *Try Exercise* **38,** *page 62.*

Radicals denoted by the notation $\sqrt[n]{b}$ are also used to denote roots. The number b is the **radicand**, and the positive integer n is the **index** of the radical.

Definition of $\sqrt[n]{b}$

> If n is a positive integer and b is a real number such that $b^{1/n}$ is a real number, then $\sqrt[n]{b} = b^{1/n}$.

Remark   If the index n equals 2, then the radical $\sqrt[2]{b}$ is written as simply $\sqrt{b}$, and it is referred to as the **principal square root of** b or simply **the square root of** b.

The symbol $\sqrt{b}$ is reserved to represent the nonnegative square root of b. To represent the negative square root of b, write $-\sqrt{b}$. For example, $\sqrt{25} = 5$, whereas $-\sqrt{25} = -5$.

EXAMPLE 4   **Evaluate Radicals**

Evaluate each of the following radicals:

a. $\sqrt[3]{8}$ b. $\sqrt[4]{81}$ c. $\sqrt{-16}$ d. $\sqrt[5]{-32}$

Solution

a. $\sqrt[3]{8} = 8^{1/3} = 2$ b. $\sqrt[4]{81} = 81^{1/4} = 3$
c. $\sqrt{-16}$ is not a real number.   d. $\sqrt[5]{-32} = (-32)^{1/5} = -2$

■ *Try Exercise* **46,** *page 62.*

The expressions $(\sqrt[n]{b})^m$ and $\sqrt[n]{b^m}$ can be expressed in exponential notation, according to the following definition.

Definition of $(\sqrt[n]{b})^m$

> For all positive integers n, all integers m, and all real numbers b such that $\sqrt[n]{b}$ is a real number, $(\sqrt[n]{b})^m = \sqrt[n]{b^m} = b^{m/n}$.

The equations

$$b^{m/n} = \sqrt[n]{b^m} \quad \text{or} \quad b^{m/n} = (\sqrt[n]{b})^m$$

can be used to write exponential expressions such as $b^{m/n}$ in radical form. Use the denominator n as the index of the radical and the numerator m as the power of the radicand or as the power of the radical. For example,

$(5xy)^{2/3} = (\sqrt[3]{5xy})^2$ Use the denominator 3 as the index of the radical, and the numerator 2 as the power of the radical.

$7a^{1/5}b^{2/5} = 7(ab^2)^{1/5} = 7\sqrt[5]{ab^2}$

$11^{-2/3} = \dfrac{1}{11^{2/3}} = \dfrac{1}{\sqrt[3]{11^2}} = \dfrac{1}{\sqrt[3]{121}}$

$(a^2 + 5)^{3/2} = (\sqrt{a^2 + 5})^3$ or $\sqrt{(a^2 + 5)^3}$

The equations

$$b^{m/n} = \sqrt[n]{b^m} \qquad \text{and} \qquad b^{m/n} = (\sqrt[n]{b})^m$$

can also be used to write radical expressions in exponential form. For example,

$\sqrt{(2ab)^3} = (2ab)^{3/2}$ Use the index 2 as the denominator of the power and the exponent 3 as the numerator of the power.

$5\sqrt[3]{x^2} = 5x^{2/3}$

$\sqrt[6]{m^4n^2} = (m^4n^2)^{1/6} = m^{4/6}n^{2/6} = m^{2/3}n^{1/3}$

The definition of $\sqrt[n]{b^m}$ can often be used to evaluate radical expressions.

EXAMPLE 5   Evaluate Radical Expressions

Evaluate

a. $(\sqrt[3]{8})^4$ b. $(\sqrt[4]{9})^2$ c. $(\sqrt{7})^2$

Solution

a. $(\sqrt[3]{8})^4 = 8^{4/3} = (8^{1/3})^4 = (2)^4 = 16$ b. $(\sqrt[4]{9})^2 = 9^{2/4} = 9^{1/2} = 3$

c. $(\sqrt{7})^2 = 7^{2/2} = 7$

■ *Try Exercise 52, page 62.*

Remark You might think $\sqrt{x^2} = x$, but the following example shows that this is *not* true for all values of x.

Case 1 If x is a positive number such as 5, then
$$\sqrt{5^2} = \sqrt{25} = 5 = x.$$

Case 2 If x is a negative number such as -5, then
$$\sqrt{(-5)^2} = \sqrt{25} = 5 = -x.$$

In summary, if $x \geq 0$, then $\sqrt{x^2} = x$. If $x < 0$, then $\sqrt{x^2} = -x$. This property of radicals can be expressed compactly using absolute value notation. For any real number x, $\sqrt{x^2} = |x|$.

Agreement   All variables used in radical expressions in the remainder of this text represent only nonnegative real numbers.

   With this agreement in force, we will seldom need to concern ourselves with the use of absolute value symbols. That is, we shall write $\sqrt{x^2} = x$.

   Since radicals are defined in terms of rational powers, it is not surprising that radicals have properties similar to those of exponential expressions.

Properties of Radicals

If m and n are natural numbers greater than or equal to 2 and a and b are nonnegative real numbers, then

Product property $\sqrt[n]{a} \cdot \sqrt[n]{b} = \sqrt[n]{ab}$

Quotient property $\dfrac{\sqrt[n]{a}}{\sqrt[n]{b}} = \sqrt[n]{\dfrac{a}{b}}$ $(b \neq 0)$

Index properties $\sqrt[m]{\sqrt[n]{b}} = \sqrt[mn]{b}$ $(\sqrt[n]{b})^n = b$ $\sqrt[n]{b^n} = b$

A radical is in **simplest radical form** if it meets all the following:

1.   The radicand contains only powers less than the index. ($\sqrt{x^5}$ does not satisfy this requirement.)
2.   The index of the radical is reduced as small as possible. ($\sqrt[6]{x^3}$ does not satisfy this requirement since $\sqrt[6]{x^3} = x^{3/6} = x^{1/2} = \sqrt{x}$.)
3.   The denominator has been rationalized. That is, no radicals appear in a denominator. ($1/\sqrt{2}$ does not satisfy this requirement.)
4.   No fractions appear in the radicand. ($\sqrt{2/5}$ does not satisfy this requirement.)

EXAMPLE 6   **Simplify Radicals by Removing Factors**

Simplify each of the following radicals. Assume that all the variables represent nonnegative real numbers.

a. $\sqrt[3]{32}$ b. $\sqrt{12y^7}$ c. $\sqrt{162x^2y^5z^7}$ d. $\sqrt[3]{\sqrt{x^8y}}$

   Solution

a.   Factor the radicand into prime factors and simplify using the index property $\sqrt[n]{b^n} = b$.

$$\sqrt[3]{32} = \sqrt[3]{2^5} = \sqrt[3]{2^3 \cdot 2^2} = \sqrt[3]{2^3} \cdot \sqrt[3]{2^2} = 2\sqrt[3]{4}$$

b. $\sqrt{12y^7} = \sqrt{2^2 \cdot 3 \cdot y^6 \cdot y^1} = \sqrt{(2y^3)^2(3y)} = \sqrt{(2y^3)^2}\sqrt{3y} = 2y^3\sqrt{3y}$

c. $\sqrt{162x^2y^5z^7} = \sqrt{2 \cdot 3^4 \cdot x^2 \cdot y^4 \cdot y \cdot z^6 \cdot z}$

$$= \sqrt{(3^2xy^2z^3)^2(2yz)} = 9xy^2z^3\sqrt{2yz}$$

d. $\sqrt[3]{\sqrt{x^8 y}} = \sqrt[6]{x^8 y} = \sqrt[6]{(x^6)(x^2 y)} = x\sqrt[6]{x^2 y}$

■ *Try Exercise 70, page 62.*

We can often reduce the index of a radical by writing the radical in its equivalent exponential form.

EXAMPLE 7 Reduce the Index of a Radical

Simplify each of the following radicals. When possible, reduce the index of the radicals.

a. $\sqrt[4]{b^2}$ b. $\sqrt[6]{81a^8}$ c. $\sqrt{\sqrt[3]{27x^3}}$

Solution

a. $\sqrt[4]{b^2} = b^{2/4} = b^{1/2} = \sqrt{b}$ Change to exponential form, reduce the exponent, and change to radical form.

b. $\sqrt[6]{81a^8} = (3^4 a^8)^{1/6} = 3^{4/6} a^{8/6} = 3^{2/3} a^{4/3} = a(3^{2/3} a^{1/3}) = a\sqrt[3]{9a}$

c. $\sqrt{\sqrt[3]{27x^3}} = \sqrt[6]{3^3 x^3} = [(3x)^3]^{1/6} = (3x)^{3/6} = (3x)^{1/2} = \sqrt{3x}$

■ *Try Exercise 86, page 62.*

Arithmetic Operations on Radicals

Like radicals have the same radicand and the same index. For example,

$$3\sqrt[3]{x^2 y} \qquad \text{and} \qquad -2\sqrt[3]{x^2 y}$$

are like radicals. Addition and subtraction of like radicals is accomplished by using the distributive property. For example,

$$4\sqrt{3x} + 7\sqrt{3x} = (4 + 7)\sqrt{3x} = 11\sqrt{3x}$$
$$2x\sqrt[3]{y^2} - 7x\sqrt[3]{y^2} + x\sqrt[3]{y^2} = (2x - 7x + x)\sqrt[3]{y^2} = -4x\sqrt[3]{y^2}$$

Remark   The sum $2\sqrt{3} + 5\sqrt{2}$ cannot be simplified any further. The radicals are not like radicals.

It is possible to combine radicals that do not appear to be like radicals if they can be simplified to be like radicals. For example,

$$3\sqrt{2} + \sqrt{8} = 3\sqrt{2} + 2\sqrt{2} = 5\sqrt{2}.$$

EXAMPLE 8 Combine Radicals

Simplify and perform the indicated operations in each of the following:

a. $5\sqrt{32} + 2\sqrt{128}$ b. $4\sqrt[3]{4x^4} - 5x\sqrt[3]{32x}$

Solution

a. $5\sqrt{32} + 2\sqrt{128} = 5\sqrt{2^5} + 2\sqrt{2^7} = 5 \cdot 4\sqrt{2} + 2 \cdot 8\sqrt{2}$
$$= 20\sqrt{2} + 16\sqrt{2} = 36\sqrt{2}$$

b. $4\sqrt[3]{4x^4} - 5x\sqrt[3]{32x} = 4\sqrt[3]{2^2x^4} - 5x\sqrt[3]{2^5x} = 4 \cdot x\sqrt[3]{2^2x} - 5x \cdot 2\sqrt[3]{2^2x}$

$$= 4x\sqrt[3]{4x} - 10x\sqrt[3]{4x} = -6x\sqrt[3]{4x}$$

■ *Try Exercise* **98,** *page 62.*

Multiplication of radical expressions is very similar to the multiplication procedures used to multiply polynomials. Sometimes we can use the FOIL method.

EXAMPLE 9   **Multiply Radical Expressions**

Find the product of each of the following. Simplify where possible.

a. $(\sqrt{3} + 5)(\sqrt{3} - 2)$ b. $(\sqrt{5x} - \sqrt{2y})(\sqrt{5x} + \sqrt{2y})$

Solution

a. $(\sqrt{3} + 5)(\sqrt{3} - 2) = (\sqrt{3})^2 - 2\sqrt{3} + 5\sqrt{3} - 10$ The FOIL method

$$= 3 - 2\sqrt{3} + 5\sqrt{3} - 10 = -7 + 3\sqrt{3}$$

b. $(\sqrt{5x} - \sqrt{2y})(\sqrt{5x} + \sqrt{2y}) = (\sqrt{5x})^2 - (\sqrt{2y})^2 = 5x - 2y$

■ *Try Exercise* **102,** *page 62.*

To **rationalize the denominator** of a fraction means to write it in an equivalent form that does not involve any radicals in its denominator.

EXAMPLE 10   **Rationalize the Denominator**

Rationalize the denominator of each of the following:

a. $\dfrac{3}{\sqrt{2}}$ b. $\dfrac{5}{\sqrt[3]{a}}$ c. $\sqrt{\dfrac{3x}{32y}}$

Solution

a. $\dfrac{3}{\sqrt{2}} = \dfrac{3}{\sqrt{2}} \cdot \dfrac{\sqrt{2}}{\sqrt{2}} = \dfrac{3\sqrt{2}}{2}$ Multiply numerator and denominator by $\sqrt{2}$.

b. $\dfrac{5}{\sqrt[3]{a}} = \dfrac{5}{\sqrt[3]{a}} \cdot \dfrac{\sqrt[3]{a^2}}{\sqrt[3]{a^2}} = \dfrac{5\sqrt[3]{a^2}}{\sqrt[3]{a^3}} = \dfrac{5\sqrt[3]{a^2}}{a}$ Use $\sqrt[3]{a} \cdot \sqrt[3]{a^2} = \sqrt[3]{a^3} = a$.

c. $\sqrt{\dfrac{3x}{32y}} = \dfrac{\sqrt{3x}}{\sqrt{32y}} = \dfrac{\sqrt{3x}}{4\sqrt{2y}} = \dfrac{\sqrt{3x}}{4\sqrt{2y}} \cdot \dfrac{\sqrt{2y}}{\sqrt{2y}} = \dfrac{\sqrt{6xy}}{8y}$

■ *Try Exercise* **116,** *page 62.*

To rationalize the denominator of a fractional expression such as

$$\frac{1}{\sqrt{m} + \sqrt{n}},$$

we make use of the conjugate of $\sqrt{m} + \sqrt{n}$, which is $\sqrt{m} - \sqrt{n}$. The product of these conjugate pairs does not involve a radical.

$$(\sqrt{m} + \sqrt{n})(\sqrt{m} - \sqrt{n}) = m - n.$$

In Example 11 we make use of the conjugate of the denominator to rationalize the denominator.

EXAMPLE 11 Rationalize the Denominator

Rationalize the denominator of each of the following:

a. $\dfrac{2}{\sqrt{3} + \sqrt{a}}$ b. $\dfrac{a + \sqrt{5}}{a - \sqrt{5}}$

Solution

a. $\dfrac{2}{\sqrt{3} + \sqrt{a}} = \dfrac{2}{\sqrt{3} + \sqrt{a}} \cdot \dfrac{\sqrt{3} - \sqrt{a}}{\sqrt{3} - \sqrt{a}} = \dfrac{2\sqrt{3} - 2\sqrt{a}}{3 - a}$

b. $\dfrac{a + \sqrt{5}}{a - \sqrt{5}} = \dfrac{a + \sqrt{5}}{a - \sqrt{5}} \cdot \dfrac{a + \sqrt{5}}{a + \sqrt{5}} = \dfrac{a^2 + 2a\sqrt{5} + 5}{a^2 - 5}$

■ *Try Exercise **124**, page 63.*

Recall that a radical is in simplest radical form if

1. The radicand contains only powers less than the index.
2. The index of the radical is reduced as small as possible.
3. The denominator has been rationalized.
4. No fractions appear in the radicand.

In the following example, we use several of the properties of real numbers and radicals to write a radical in its simplest radical form.

$$\sqrt{\dfrac{36x^3}{x^4}} = \sqrt{\dfrac{36}{x}} \qquad \text{Reduce the radicand.}$$

$$= \dfrac{\sqrt{36}}{\sqrt{x}} \qquad \text{The quotient property } \sqrt{\dfrac{a}{b}} = \dfrac{\sqrt{a}}{\sqrt{b}}.$$

$$= \dfrac{6}{\sqrt{x}}$$

$$= \dfrac{6\sqrt{x}}{\sqrt{x}\sqrt{x}} \qquad \text{Rationalize the denominator.}$$

$$= \dfrac{6\sqrt{x}}{x}$$

Therefore $\dfrac{6\sqrt{x}}{x}$ is the simplest radical form of $\sqrt{\dfrac{36x^3}{x^4}}$.

EXERCISE SET 1.7

In Exercises 1 to 24, evaluate each expression.

1. $9^{1/2}$ **2.** $49^{1/2}$ **3.** $-9^{1/2}$ **4.** $-25^{1/2}$

5. $4^{3/2}$ **6.** $16^{3/2}$ **7.** $-64^{2/3}$ **8.** $-125^{2/3}$

9. $(-64)^{2/3}$ **10.** $(-125)^{2/3}$ **11.** $16^{-1/2}$ **12.** $9^{-1/2}$

13. $27^{-2/3}$ **14.** $4^{-3/2}$ **15.** $\left(\dfrac{9}{16}\right)^{1/2}$

16. $\left(\dfrac{4}{25}\right)^{1/2}$ **17.** $\left(\dfrac{4}{25}\right)^{3/2}$ **18.** $\left(-\dfrac{1}{27}\right)^{-2/3}$

19. $10^{3/2} \cdot 10^{1/2}$ **20.** $3^{1/2} \cdot 3^{1/2}$ **21.** $7^{-1/4} \cdot 7^{5/4}$

22. $6^{2/3} \cdot 6^{-1/3}$ **23.** $\dfrac{5^{4/3}}{5^{1/3}}$ **24.** $\dfrac{11^{5/4}}{11^{-3/4}}$

In Exercises 25 to 42, simplify each expression. Write your final result using only positive exponents. Assume the variables to be positive real numbers and any variables that appear as exponents to be rational numbers.

25. $(x^{1/2})(x^{3/5})$ **26.** $(y^{4/3})(y^{1/4})$

27. $(8a^3)^{2/3}$ **28.** $(27b^6)^{2/3}$

29. $(81x^4y^{12})^{1/4}$ **30.** $(625a^8b^4)^{1/4}$

31. $\dfrac{a^{3/4} \cdot b^{1/2}}{a^{1/4} \cdot b^{1/5}}$ **32.** $\dfrac{x^{1/2} \cdot y^{5/6}}{x^{3/2} \cdot y^{1/6}}$

33. $a^{1/3}(a^{5/3} + 7a^{2/3})$ **34.** $m^{3/4}(m^{1/4} - 8m^{5/4})$

35. $(p^{1/2} + q^{1/2})(p^{1/2} - q^{1/2})$ **36.** $(c + d^{1/3})(c - d^{1/3})$

37. $\left(\dfrac{m^2n^4}{m^{-2}n}\right)^{1/2}$ **38.** $\left(\dfrac{r^3s^{-2}}{rs^4}\right)^{1/2}$

39. $\dfrac{(x^{n+1/2}) \cdot x^{-n}}{x^{1/2}}$ **40.** $\dfrac{r^{n/2} \cdot r^{2n}}{r^{-n}}$

41. $\dfrac{r^{1/n}}{r^{1/m}}$ **42.** $\dfrac{s^{2/n}}{s^{-n/2}}$

In Exercises 43 to 56, make use of the properties of radicals to evaluate each radical without the aid of a calculator.

43. $\sqrt{4}$ **44.** $\sqrt{36}$ **45.** $\sqrt[3]{-216}$ **46.** $\sqrt[3]{-64}$

47. $\sqrt{\dfrac{9}{16}}$ **48.** $\sqrt{\dfrac{25}{49}}$ **49.** $\sqrt[5]{32}$ **50.** $\sqrt[6]{729}$

51. $(\sqrt[4]{4})^2$ **52.** $(\sqrt[4]{25})^2$ **53.** $(\sqrt[4]{6})^4$

54. $(\sqrt[5]{14})^5$ **55.** $(\sqrt{7})^4$ **56.** $(\sqrt{11})^4$

In Exercises 57 to 62, write each exponential expression in radical form.

57. $(3x)^{1/2}$ **58.** $(6y)^{1/3}$ **59.** $5(xy)^{1/4}$

60. $2a(bc)^{1/5}$ **61.** $(5w)^{2/3}$ **62.** $(a + b)^{3/4}$

In Exercises 63 to 68, write each radical in exponential form.

63. $\sqrt[3]{17k}$ **64.** $4\sqrt{3m}$ **65.** $\sqrt[5]{a^2}$

66. $3\sqrt[4]{5n}$ **67.** $\sqrt{\dfrac{7a}{3}}$ **68.** $\sqrt[3]{\dfrac{5b^2}{7}}$

In Exercises 69 to 84, simplify each radical by removing from the radical all possible factors of the radicand.

69. $\sqrt{45}$ **70.** $\sqrt{75}$ **71.** $\sqrt[3]{24}$

72. $\sqrt[3]{135}$ **73.** $\sqrt[3]{-81}$ **74.** $\sqrt[3]{-250}$

75. $-\sqrt[3]{32}$ **76.** $-\sqrt[3]{243}$ **77.** $\sqrt{3^2 \cdot 5^3}$

78. $\sqrt{2^2 \cdot 3^5}$ **79.** $\sqrt[3]{2^3 \cdot 5^4 \cdot 7}$ **80.** $\sqrt[3]{3^4 \cdot 5^5}$

81. $\sqrt{24x^3y^2}$ **82.** $\sqrt{18x^2y^5}$ **83.** $-\sqrt[3]{16a^3y^7}$

84. $-\sqrt[3]{54c^2d^5}$

In Exercises 85 to 90, simplify each radical by reducing the index of the radical.

85. $\sqrt[4]{9x^2}$ **86.** $\sqrt[4]{25a^2b^2}$ **87.** $\sqrt[6]{16m^4n^2}$

88. $\sqrt[6]{27r^3}$ **89.** $\sqrt[8]{81x^6y^2}$ **90.** $\sqrt[8]{64a^2b^4}$

In Exercises 91 to 100 simplify each expression by simplifying the radicals and combining like radicals.

91. $4\sqrt{2} + 3\sqrt{2}$ **92.** $6\sqrt{5} + 2\sqrt{5}$

93. $7\sqrt{3} - \sqrt{3}$ **94.** $8\sqrt{11} - \sqrt{11}$

95. $\sqrt{8} - 5\sqrt{2}$ **96.** $\sqrt{27} + 4\sqrt{3}$

97. $2\sqrt[3]{2} - \sqrt[3]{16}$ **98.** $5\sqrt[3]{3} + 2\sqrt[3]{81}$

99. $\sqrt{8x^3y} + x\sqrt{2xy}$ **100.** $4\sqrt{a^5b} - a^2\sqrt{ab}$

In Exercises 101 to 110, find the indicated product of the radical expressions. Express each term in simplest radical form.

101. $(\sqrt{5} + 8)(\sqrt{5} + 3)$ **102.** $(\sqrt{7} + 4)(\sqrt{7} - 1)$

103. $(\sqrt{2x} + 3)(\sqrt{2x} - 3)$ **104.** $(7 - \sqrt{3a})(7 + \sqrt{3a})$

105. $(5\sqrt{2y} + \sqrt{3z})^2$ **106.** $(3\sqrt{5y} - 4)^2$

107. $(\sqrt{x - 3} + 5)^2$ **108.** $(\sqrt{x + 7} - 3)^2$

109. $(\sqrt{2x + 5} + 7)^2$ **110.** $(\sqrt{9x - 2} + 11)^2$

In Exercises 111 to 126, simplify each expression by rationalizing the denominator. When possible, reduce the resulting fraction to lowest terms. Assume that each radicand is a positive real number and that each denominator is a nonzero real number.

111. $\dfrac{2}{\sqrt{2}}$ **112.** $\dfrac{3x}{\sqrt{3}}$ **113.** $\sqrt{\dfrac{5}{18}}$ **114.** $\sqrt{\dfrac{7}{40}}$

115. $\dfrac{3}{\sqrt[3]{2}}$ **116.** $\dfrac{2}{\sqrt[3]{4}}$ **117.** $\dfrac{4}{\sqrt[3]{8x^2}}$ **118.** $\dfrac{2}{\sqrt[4]{4y}}$

119. $\sqrt{\dfrac{10}{18}}$ **120.** $\sqrt{\dfrac{14}{40}}$ **121.** $\sqrt{\dfrac{2x}{27y}}$ **122.** $\sqrt{\dfrac{4c}{50d}}$

123. $\dfrac{3}{\sqrt{5}+\sqrt{x}}$ **124.** $\dfrac{5}{\sqrt{y}-\sqrt{3}}$

125. $\dfrac{\sqrt{7}}{2-\sqrt{7}}$ **126.** $\dfrac{6\sqrt{6}}{5+\sqrt{6}}$

In Exercises 127 to 130, rationalize the numerator of each radical expression. When possible, reduce the resulting fraction to lowest terms.

127. $\dfrac{\sqrt{5}}{3}$ **128.** $\dfrac{\sqrt{7}}{4}$

129. $\dfrac{3+\sqrt{5}}{7}$ **130.** $\dfrac{2-\sqrt{6}}{10}$

In Exercises 131 to 138, write each radical in simplest radical form.

131. $\sqrt{\dfrac{20a^5}{6a}}$ **132.** $\sqrt{\dfrac{12b^4}{10b}}$ **133.** $\sqrt{\dfrac{45xy^2}{10y^3}}$

134. $\sqrt{\dfrac{14xz^5}{6xz}}$ **135.** $\sqrt{\dfrac{30xy}{4xy^2}}$ **136.** $\sqrt{\dfrac{48a}{20a^4}}$

137. $\dfrac{\sqrt[3]{24xy^2}}{\sqrt[3]{2x^2y}}$ **138.** $\dfrac{\sqrt[3]{60}}{\sqrt{14a}}$

Calculator Exercises

In Exercises 139 to 148, use a calculator to evaluate each exponential expression or radical. Express your answers using scientific notation with three significant digits. If necessary, use the $\boxed{y^x}$ key.

139. $8^{1/5}$ **140.** $10^{1/10}$ **141.** $(-12)^{3/5}$
142. $(-15)^{2/3}$ **143.** $\sqrt{437}$ **144.** $\sqrt{511}$
145. $\sqrt{7.81 \times 10^4}$ **146.** $\sqrt{6.23 \times 10^{-6}}$
147. $\sqrt[3]{-251}$ **148.** $\sqrt[3]{-344}$

Supplemental Exercises

The percent P of light that will pass through an opaque material is given by the equation $P = 10^{-kd}$, where d is the thickness of the material in centimeter, and k is a constant that depends upon the material.

149. Find the percent (to the nearest 1 percent) of light that will pass through opaque glass for which
   a. $k = 0.15$ and $d = 0.6$ centimeter.
   b. $k = 0.15$ and $d = 1.2$ centimeter.

150. The number of hours h needed to cook a pot roast which weighs p pounds can be approximated by using the formula $h = 0.9(p)^{0.6}$.
   a. Find the time (to the nearest hundredth of an hour) required to cook a 12-pound pot roast.
   b. If pot roast A weighs twice as much as pot roast B, then pot roast A should be cooked for a period of time that is how many times longer than the time at which pot roast B is cooked?

In Exercises 151 to 154, find the value of p for which the statement is true.

151. $a^{2/5}a^p = a^2$ **152.** $b^{-3/4}b^{2p} = b^3$

153. $\dfrac{x^{-3/4}}{x^{3p}} = x^4$ **154.** $(x^4 x^{2p})^{1/2} = x$

In Exercises 155 to 158, factor each expression over the set of real numbers. *Example:*

$$x^2 - 5 = x^2 - (\sqrt{5})^2 = (x + \sqrt{5})(x - \sqrt{5})$$

155. $x^2 - 7$ **156.** $y^2 - 11$
157. $x^2 + 6\sqrt{2}x + 18$ **158.** $y^2 - 8\sqrt{5}y + 80$
159. *Prove:* $\sqrt{a^2 + b^2} \neq a + b$. (*Hint:* Find a counterexample.)
160. When does $\sqrt[3]{a^3 + b^3} = a + b$? (*Hint:* Cube each side of the equation.)
161. Which step in the following demonstration is the incorrect step?

$$3 = 3$$
$$= 3^{2/2} \quad \text{because } 1 = 2/2$$
$$= (3^2)^{1/2} \quad \text{because } b^{m/n} = (b^m)^{1/n}$$
$$= 9^{1/2} \quad \text{because } 3^2 = 9$$
$$= [(-3)^2]^{1/2} \quad \text{substitution of } (-3)^2 \text{ for } 9$$
$$= (-3)^{2(1/2)} \quad \text{the power property } (b^p)^q = b^{pq}$$
$$= (-3)^1$$
$$= -3$$

162. Which step in the following demonstration is the incorrect step?

$$4 = 16^{1/2}$$
$$= [(-4)(-4)]^{1/2}$$
$$= (-4)^{1/2}(-4)^{1/2}$$

Which is impossible because $(-4)^{1/2}$ is not a real number.

1.8

Complex Numbers

There is no real number whose square is a negative number. For example, there is no real number x such that $x^2 = -1$. In the seventeenth century, a new number, called an **imaginary number,** was defined. The square of an imaginary number is a negative real number. The letter i was chosen to represent an imaginary number whose square is -1.

Definition of i

> The number i, called the **imaginary unit,** is a number such that $i^2 = -1$.

Many of the solutions to equations in the remainder of this text will involve radicals such as $\sqrt{-a}$, where a is a positive real number. The expression $\sqrt{-a}$, with $a > 0$, is defined as follows.

Definition of $\sqrt{-a}$

> For any positive real number a, $\sqrt{-a} = i\sqrt{a}$.

Remark   This definition with $a = 1$ implies

$$\sqrt{-1} = i.$$

The above definition is often used to write the square root of a negative real number as the product of the imaginary unit i and a positive real number. For example,

$$\sqrt{-4} = i\sqrt{4} = 2i \quad \text{and} \quad \sqrt{-7} = i\sqrt{7}.$$

Definition of a Complex Number

> If a and b are real numbers and i is the imaginary unit, then $a + bi$ is called a **complex number.** The real number a is called the **real part** and the real number b is called the **imaginary part** of the complex number.

Remark   Even though b is a real number, it is called the imaginary part of the complex number $a + bi$. For example, the complex number $3 + 8i$ has the real number 8 as its imaginary part.

The real numbers are a subset of the complex numbers. This can be observed by letting $b = 0$. Then $a + bi = a + 0i = a$, which is a real number. Any number that can be written in the form $0 + bi = bi$, where b is a nonzero real number, is an **imaginary number** (or a pure imaginary number). For example, i, $3i$, and $-0.5i$ are all imaginary numbers.

A complex number is in **standard form** when it is written in the form $a + bi$.

EXAMPLE 1   Write Complex Numbers in Standard Form

Write each complex number in standard form.

a. $3 + \sqrt{-4}$ b. $\sqrt{-37} - 3$ c. $-\sqrt{-121}$

Solution   Use the definition $\sqrt{-a} = i\sqrt{a}$ and the definition of standard form.

a. $3 + \sqrt{-4} = 3 + i\sqrt{4} = 3 + 2i$ $a + bi$ form with $a = 3$ and $b = 2$.

b. $\sqrt{-37} - 3 = i\sqrt{37} - 3 = -3 + i\sqrt{37}$

c. $-\sqrt{-121} = -(i\sqrt{121}) = -11i = 0 - 11i$

■ *Try Exercise **2**, page 69.*

Remark   The expression $\sqrt{a}i$ is often written as $i\sqrt{a}$ so that it is not mistaken for $\sqrt{ai}$.

Arithmetic Operations on Complex Numbers

To compute with complex numbers, we must first define the arithmetic operations of addition, subtraction, multiplication, and division of two complex numbers.

Definition of Addition and Subtraction of Complex Numbers

If $a + bi$ and $c + di$ are complex numbers, then

Addition $(a + bi) + (c + di) = (a + c) + (b + d)i$

Subtraction $(a + bi) - (c + di) = (a - c) + (b - d)i$

Thus to add two numbers, you add their real parts to produce the real part of the sum and you add their imaginary parts to produce the imaginary part of the sum.

EXAMPLE 2   Add Complex Numbers

Write each sum as a complex number.

a. $(4 + 2i) + (3 + 7i)$ b. $2 + (6 - 5i)$

Solution

a. $(4 + 2i) + (3 + 7i) = (4 + 3) + (2 + 7)i = 7 + 9i$

b. $2 + (6 - 5i) = (2 + 0i) + (6 - 5i) = (2 + 6) + (0 + (-5))i = 8 - 5i$

■ *Try Exercise **12**, page 69.*

To subtract two complex numbers, subtract their real parts to produce the real part of the difference and subtract their imaginary parts to produce the imaginary part of difference.

EXAMPLE 3   Subtract Complex Numbers

Write each difference as a complex number.

a. $(5 + 7i) - (-3 + 9i)$ b. $i - (3 - 4i)$

Solution

a. $(5 + 7i) - (-3 + 9i) = [5 - (-3)] + (7 - 9)i = 8 - 2i$

b. $i - (3 - 4i) = (0 + 1i) - (3 - 4i)$

$$= (0 - 3) + [1 - (-4)]i = -3 + 5i$$

■ *Try Exercise* **16,** *page 69.*

Definition of Multiplication of Complex Numbers

> If $a + bi$ and $c + di$ are complex numbers, then
>
> $$(a + bi)(c + di) = (ac - bd) + (ad + bc)i$$

Since every complex number can be written as a sum of two terms, it is natural to perform multiplication on complex numbers in a manner consistent with the operation of multiplication defined on binomials and the definition $i^2 = -1$. Thus to multiply complex numbers it is not necessary to memorize the definition of multiplication.

EXAMPLE 4   Multiply Complex Numbers

Write each product as a complex number.

a. $(3 + 5i)(2 - 4i)$ b. $4i(3 + 7i)$

Solution

a. $(3 + 5i)(2 - 4i) = 6 - 12i + 10i - 20i^2$ The FOIL method

$$= 6 - 12i + 10i - 20(-1)$$ Substitute (-1) for i^2.

$$= 6 + 20 - 12i + 10i$$ Simplify.

$$= 26 - 2i$$

b. $4i(3 + 7i) = 12i + 28i^2$ Distribute the $4i$

$$= 12i + 28(-1)$$

$$= -28 + 12i$$

■ *Try Exercise* **24,** *page 69.*

The complex numbers $a + bi$ and $a - bi$ are called **complex conjugates** or **conjugates** of each other. The conjugate of the complex number z is denoted by $\bar{z}$. For example,

$$\overline{3 + 2i} = 3 - 2i \quad \text{and} \quad \overline{7 - 11i} = 7 + 11i.$$

The Product of z and $\bar{z}$

The product of a complex number and its conjugate is always a real number.

Proof: We can verify this theorem by using the complex number $a + bi$ and its conjugate $a - bi$.

$$(a + bi)(a - bi) = a^2 - abi + abi - b^2 i^2$$
$$= a^2 - b^2(-1)$$
$$= a^2 + b^2 \qquad \leftarrow \text{a real number}$$

The fact that the product of a complex number and its conjugate is always a real number can be used to find the quotient of two complex numbers. For example, to find the quotient $(a + bi)/(c + di)$, multiply numerator and denominator by the conjugate of the denominator.

EXAMPLE 5 **Divide Complex Numbers**

Write each quotient as a complex number.

a. $\dfrac{3 + 2i}{5 - i}$ b. $\dfrac{2 - 7i}{-i}$

Solution

a. $\dfrac{3 + 2i}{5 - i} = \dfrac{(3 + 2i)(5 + i)}{(5 - i)(5 + i)}$ Multiply numerator and denominator by $5 + i$, which is the conjugate of the denominator.

$$= \dfrac{15 + 3i + 10i + 2i^2}{25 + 1}$$

$$= \dfrac{13 + 13i}{26} = \dfrac{1}{2} + \dfrac{1}{2}i \quad \text{Write in standard form.}$$

b. $\dfrac{2 - 7i}{-i} = \dfrac{(2 - 7i)(i)}{-i(i)}$ Multiply numerator and denominator by i, which is the conjugate of the denominator.

$$= \dfrac{2i - 7i^2}{-i^2}$$

$$= \dfrac{7 + 2i}{1} = 7 + 2i$$

■ *Try Exercise **34**, page 69.*

There are several theorems about complex numbers and their conjugates. For example, the following theorem concerns the sum of a complex number $z = a + bi$ and its conjugate $\bar{z} = a - bi$.

The Sum of z and $\bar{z}$

> The sum of a complex number z and its conjugate $\bar{z}$ is a real number.

Proof: To verify this, we let $z = a + bi$ and $\bar{z} = a - bi$.

$$+ \bar{z} = (a + bi) + (a - bi) = (a + a) + [b + (-b)]i$$

$$= 2a + 0i = 2a \qquad\qquad\qquad \leftarrow \text{a real number}$$

The following powers of i illustrate a pattern:

$$i^1 = i \qquad\qquad\qquad i^5 = i^4 \cdot i = (1)i = i$$

$$i^2 = -1 \qquad\qquad\qquad i^6 = i^4 \cdot i^2 = (1)(-1) = -1$$

$$i^3 = i^2 \cdot i = (-1)i = -i \qquad i^7 = i^4 \cdot i^3 = (1)(-i) = -i$$

$$i^4 = i^2 \cdot i^2 = (-1)((-1) = 1 \qquad i^8 = (i^4)^2 = 1^2 = 1$$

Since $(i^4)^n = 1$ for any integer n, it is possible to evaluate powers of i by factoring out powers of i^4, as shown in the following example:

$$i^{25} = (i^4)^6(i) = 1^6(i) = i.$$

The following theorem is often used to evaluate powers of i. Essentially it makes use of division to eliminate powers of i^4.

Powers of i

> If n is a positive integer, then $i^n = i^r$, where r is the remainder of the division of n by 4.

This theorem is particularly useful when evaluating large powers of i.

EXAMPLE 6   **Evaluate Powers of i**

Evaluate each of the following: a. i^{14} b. i^{543}

Solution   Use the theorem on powers of i.

a. $i^{14} = i^2 = -1$ Remainder of $14 \div 4$ is 2.

b. $i^{543} = i^3 = -i$ Remainder of $543 \div 4$ is 3.

■ *Try Exercise* **54,** *page 69.*

Caution   To compute $\sqrt{a}\,\sqrt{b}$ when both a and b are negative numbers, write each radical in terms of i before multiplying. For example,

Correct method $\sqrt{-1}\,\sqrt{-1} = i \cdot i = i^2 = -1$

Incorrect method $\sqrt{-1}\,\sqrt{-1} = \sqrt{(-1)(-1)} = \sqrt{1} = 1.$

EXAMPLE 7 **Simplify Products Involving Radicals with Negative Radicands**

Simplify each of the following:

a. $\sqrt{-16}\sqrt{-25}$ b. $\sqrt{-9}\sqrt{-7}$ c. $(2 + \sqrt{-5})(2 - \sqrt{-5})$

Solution

a. $\sqrt{-16}\sqrt{-25} = (4i)(5i) = 20i^2 = -20$

b. $\sqrt{-9}\sqrt{-7} = (3i)(i\sqrt{7}) = 3i^2\sqrt{7} = -3\sqrt{7}$

c. $(2 + \sqrt{-5})(2 - \sqrt{-5}) = (2 + i\sqrt{5})(2 - i\sqrt{5}) = 4 + 5 = 9$

■ *Try Exercise* **68**, *page 69.*

EXERCISE SET 1.8

In Exercises 1 to 10, write the complex number in standard form.

1. $2 + \sqrt{-9}$ **2.** $3 + \sqrt{-25}$

3. $4 - \sqrt{-121}$ **4.** $5 - \sqrt{-144}$

5. $8 + \sqrt{-3}$ **6.** $9 - \sqrt{-75}$

7. $\sqrt{-16} + 7$ **8.** $\sqrt{-49} + 3$

9. $\sqrt{-81}$ **10.** $-\sqrt{-100}$

In Exercises 11 to 28, simplify and write the complex number in standard form.

11. $(2 + 5i) + (3 + 7i)$ **12.** $(1 - 3i) + (6 + 2i)$

13. $(-5 - i) + (9 - 2i)$ **14.** $5 + (3 - 2i)$

15. $(8 - 6i) - (10 - i)$ **16.** $(-3 + i) - (-8 + 2i)$

17. $(7 - 3i) - (-5 - i)$ **18.** $7 - (3 - 2i)$

19. $8i - (2 - 3i)$ **20.** $(4i - 5) - 2$

21. $3(2 + 7i) + 5(2 - i)$ **22.** $8(4 - i) - (4 - 3i)$

23. $(2 + 3i)(4 - 5i)$ **24.** $(5 - 3i)(-2 - 4i)$

25. $(5 + 7i)(5 - 7i)$ **26.** $(-3 - 5i)(-3 + 5i)$

27. $(8i + 11)(-7 + 5i)$ **28.** $(9 - 12i)(15i + 7)$

In Exercises 29 to 48, write each expression as a complex number in standard form.

29. $\dfrac{4 + i}{3 + 5i}$ **30.** $\dfrac{5 - i}{4 + 5i}$ **31.** $\dfrac{1}{7 - 3i}$ **32.** $\dfrac{1}{-8 + i}$

33. $\dfrac{3 + 2i}{3 - 2i}$ **34.** $\dfrac{5 - 7i}{5 + 7i}$ **35.** $\dfrac{2i}{11 + i}$ **36.** $\dfrac{3i}{5 - 2i}$

37. $\dfrac{6 + i}{i}$ **38.** $\dfrac{5 - i}{-i}$

39. $(3 - 5i)^2$ **40.** $(-5 + 7i)^2$

41. $(1 - i) - 2(4 + i)^2$ **42.** $(4 - i) - 5(2 + 3i)^2$

43. $(1 - i)^3$ **44.** $(2 + i)^3$

45. $(2i)(8i)$ **46.** $(-5)(7i)$

47. $(5i)^2(-3i)$ **48.** $(-6i)(-5i)^2$

In Exercises 49 to 64, simplify and write the complex number as either i, $-i$, 1, or -1.

49. i^3 **50.** $-i^3$ **51.** i^5 **52.** $-i^5$

53. i^{10} **54.** i^{28} **55.** $-i^{40}$ **56.** i^{40}

57. i^{223} **58.** i^{553} **59.** i^{2001} **60.** i^{5000}

61. i^{5042} **62.** i^0 **63.** i^{-1} **64.** $i^{10,000}$

In Exercises 65 to 72, simplify each product.

65. $\sqrt{-1}\sqrt{-4}$ **66.** $\sqrt{-16}\sqrt{-49}$

67. $\sqrt{-64}\sqrt{-5}$ **68.** $\sqrt{-3}\sqrt{-121}$

69. $(3 + \sqrt{-2})(3 - \sqrt{-2})$

70. $(4 + \sqrt{-81})(4 - \sqrt{-81})$

71. $(5 + \sqrt{-16})^2$ **72.** $(3 - \sqrt{-144})^2$

In Exercises 73 to 80, evaluate

$$\frac{-b \pm \sqrt{b^2 - 4ac}}{2a}$$

for the given values of a, b, and c. Write your final answer as a complex number in standard form.

73. $a = 3$, $b = -3$, $c = 3$ **74.** $a = 1$, $b = -3$, $c = 10$

75. $a = 2$, $b = 4$, $c = 4$ **76.** $a = 4$, $b = -4$, $c = 2$

77. $a = 2$, $b = 6$, $c = 6$ **78.** $a = 6$, $b = -5$, $c = 5$

79. $a = 2$, $b = 1$, $c = 3$ **80.** $a = 3$, $b = 2$, $c = 4$

The **absolute value of the complex number** $a + bi$ is denoted by $|a + bi|$ and defined as the real number $\sqrt{a^2 + b^2}$.

In Exercises 81 to 88, find the indicated absolute value of each complex number.

81. $|3 + 4i|$ **82.** $|5 + 12i|$ **83.** $|2 - 5i|$ **84.** $|4 - 4i|$

85. $|7 - 4i|$ **86.** $|11 - 2i|$ **87.** $|-3i|$ **88.** $|18i|$

Supplemental Exercises

In Exercises 89 to 92, use the complex number $z = a + bi$ and its conjugate $\bar{z} = a - bi$ to establish each result.

89. Prove that the absolute value of a complex number and the absolute value of its conjugate are equal.

90. Prove that the difference of a complex number and its conjugate is a pure imaginary number.

91. Prove that the conjugate of the sum of two complex numbers equals the sum of the conjugates of the two numbers.

92. Prove that the conjugate of the product of two complex numbers equals the product of the conjugates of the two numbers.

93. Show that if $x = 1 + i\sqrt{3}$, then $x^2 - 2x + 4 = 0$.

94. Show that if $x = 1 - i\sqrt{3}$, then $x^2 - 2x + 4 = 0$.

95. Is $T = \{1, -1, i, -i\}$ closed under the operation of addition?

96. Is $T = \{1, -1, i, -i\}$ closed under the operation of multiplication?

97. Simplify
$$[(3 + \sqrt{5}) + (7 - \sqrt{3})i][(3 + \sqrt{5}) - (7 - \sqrt{3})i].$$

98. Simplify $[2 - (3 - \sqrt{5})i][2 + (3 - \sqrt{5})i]$.

99. Simplify $\left(\dfrac{-1}{2} + \dfrac{\sqrt{3}}{2}i\right)^3$.

100. Simplify $(a + bi)^3$, where a and b are real numbers.

101. Simplify $i + i^2 + i^3 + i^4 + \cdots + i^{28}$.

102. Simplify $i + i^2 + i^3 + i^4 + \cdots + i^{100}$.

The product $(a + bi)(a - bi) = a^2 + b^2$ can be used to factor the sum of two squares over the set of complex numbers. *Example:*

$$x^2 + 25 = x^2 + 5^2 = (x + 5i)(x - 5i)$$

In Exercises 103 to 106, factor each polynomial over the set of complex numbers.

103. $x^2 + 9$ **104.** $y^2 + 121$

105. $4x^2 + 81$ **106.** $144y^2 + 625$

In Exercises 107 to 110, evaluate the polynomial for the given value of x.

107. $x^2 + 36; \quad x = 6i$ **108.** $x^2 + 100; \quad x = -10i$

109. $x^2 - 6x + 10; \quad x = 3 + i$

110. $x^2 + 10x + 29; \quad x = -5 + 2i$

Chapter 1 Review

The reviews at the end of each chapter list important definitions, properties, and procedures from each section in the chapter. They are designed to help you review your knowledge of major concepts.

1.1   **The Real Number System**

The following sets of numbers are used extensively in the study of algebra:

Integers	$\{\ldots, -3, -2, -1, 0, 1, 2, 3, \ldots\}$
Rational numbers	{all terminating or repeating decimals}
Irrational numbers	{all nonterminating, nonrepeating decimals}
Real numbers	{all rational or irrational numbers}

Properties of Fractions   See page 5.

Properties of Equality   See page 6.

Properties of the Set of Real Numbers   See page 7.

1.2   Intervals, Absolute Value, and Distance

Absolute Value

The absolute value of the real number a is defined by

$$|a| = \begin{cases} a & \text{if } a \geqq 0 \\ -a & \text{if } a < 0. \end{cases}$$

Distance Between Points on the Real Number Line

For any real numbers a and b, the distance between the graph of a and the graph of b is denoted by $d(a, b)$, where

$$d(a, b) = |a - b| = |b - a|.$$

1.3   Integer Exponents

Definition of Natural Number Exponents

If b is any real number and n is any natural number, then

$$b^n = \underbrace{b \cdot b \cdot b \cdots \cdot b}_{n \text{ factors of } b}$$

Definition of b^0

For any nonzero base b, $b^0 = 1$.

Definition of b^{-n}

If $b \neq 0$ and n is any natural number, $b^{-n} = 1/b^n$ and $1/b^{-n} = b^n$.

Scientific Notation

A number is written in **scientific notation** if it is of the form $a \cdot 10^n$, where n is an integer and a is a decimal such that $1 \leq a < 10$.

1.4   Polynomials

Special Product Formulas

Special Forms	Formulas
(Sum)(difference)	$(x + y)(x - y) = x^2 - y^2$
(Binomial)^2	$(x + y)^2 = x^2 + 2xy + y^2$
	$(x - y)^2 = x^2 - 2xy + y^2$

The Order of Operations Agreement
If grouping symbols are present, evaluate by performing the operations within the grouping symbols, innermost grouping symbol first, while observing the order given in steps 1 to 3.

1.   First, evaluate each power.
2.   Next, do all multiplications and divisions, working from left to right.
3.   Last, do all additions and subtractions, working from left to right.

1.5   Factoring Formulas

Factoring Formulas

Difference of two squares	$x^2 - y^2 = (x + y)(x - y)$
Perfect square trinomials	$x^2 + 2xy + y^2 = (x + y)^2$
	$x^2 - 2xy + y^2 = (x - y)^2$
Sum of cubes	$x^3 + y^3 = (x + y)(x^2 - xy + y^2)$
Difference of cubes	$x^3 - y^3 = (x - y)(x^2 + xy + y^2)$

General Factoring Strategy   See page 40.

1.6   Rational Expressions

A **rational expression** is a fraction in which the numerator and denominator are polynomials.

Properties of Rational Expressions   See page 44.

Methods for Simplifying Complex Fractions   See page 48.

1.7   Rational Exponents and Radicals

Definition of $b^{1/n}$
If n is an even positive integer and $b \geq 0$, then $b^{1/n}$ is the nonnegative real number such that $(b^{1/n})^n = b$.

If n is an odd positive integer, then $b^{1/n}$ is the real number such that $(b^{1/n})^n = b$.

Definition of $b^{m/n}$
For all integers m and n such that m/n is reduced to lowest terms, and for all real numbers b for which $b^{1/n}$ is a real number, $b^{m/n} = (b^{1/n})^m = (b^m)^{1/n}$

Properties of Rational Exponents
If p and q represent rational numbers and a and b are positive real numbers, then

Product property $b^p \cdot b^q = b^{p+q}$

Quotient property $\dfrac{b^p}{b^q} = b^{p-q}$

Power properties $(b^p)^q = b^{pq}$ $(ab)^p = a^p b^p$ $\left(\dfrac{a}{b}\right)^p = \dfrac{a^p}{b^p}$ $b^{-p} = \dfrac{1}{b^p}$

Definition of $\sqrt[n]{b}$
If n is a positive integer and b is a real number such that $b^{1/n}$ is a real number, then $\sqrt[n]{b} = b^{1/n}$.

Properties of Radicals
If m and n are natural numbers greater than or equal to 2 and a and b are nonnegative real numbers, then

Product property $\sqrt[n]{a} \cdot \sqrt[n]{b} = \sqrt[n]{ab}$

Quotient property $\dfrac{\sqrt[n]{a}}{\sqrt[n]{b}} = \sqrt[n]{\dfrac{a}{b}}$ $(b \neq 0)$

Index properties $\sqrt[m]{\sqrt[n]{b}} = \sqrt[mn]{b}$ $(\sqrt[n]{b})^n = b$ $\sqrt[n]{b^n} = b$

1.8 Complex Numbers

Definition of i
The number i, called the **imaginary unit,** is the number such that $i^2 = -1$.

Definition of a Complex Number
If a and b are real numbers and i is the imaginary unit, then $a + bi$ is called a complex number. The complex numbers $a + bi$ and $a - bi$ are called complex conjugates or conjugates of each other.

Powers of i
If n is a positive integer, then $i^n = i^r$, where r is the remainder of the division of n by 4.

CHAPTER 1 CHALLENGE EXERCISES

In Exercises 1 to 10, answer true or false. If the statement is false, give an example.

1. If a and b are real numbers, then $|a - b| = |b - a|$.

2. If a is a real number, then $a^2 \geqq a$.

3. The set of rational numbers is closed under the operation of addition.

4. The set of irrational numbers is closed under the operation of addition.

5. Let $x \oplus y$ denote the average of the two real numbers x and y. That is,
$$x \oplus y = \frac{x + y}{2}.$$

The operation $\oplus$ is an associative operation since $(x \oplus y) \oplus z = x \oplus (y \oplus z)$ for all real numbers x, y, and z.

6. Using interval notation, the inequality $x > a$ is written as $[a, \infty)$.

7. If n is a real number, then $\sqrt{n^2} = n$.

8. Every real number is a complex number.

9. The sum of a complex number z and its conjugate $\bar{z}$ is a real number.

10. The product of a complex number z and its conjugate $\bar{z}$ is a real number.

CHAPTER 1 REVIEW EXERCISES

In Exercises 1 to 4, classify each number as one or more of the following: integer, rational number, irrational number, real number.

1. 3   **2.** $\sqrt{7}$ **3.** $-\dfrac{1}{2}$ **4.** $0.\overline{5}$

In exercises 5 and 6, illustrate the trichotomy property by replacing the □ with the appropriate symbol (<, =, *or* >).

5. $\dfrac{7}{2}\ \square\ 3$ **6.** $\sqrt{2}\ \square\ 1.5$

In Exercises 7 to 10, identify the property of equality or the real number property that is illustrated.

7. $5(x + 3) = 5x + 15$

8. $a(3 + b) = a(b + 3)$

9. $(6c)d = 6(cd)$

10. $\sqrt{2} + 3$ is a real number.

11. Is the set of integers closed under the operation of
  a. addition   **b.** subtraction
  c. multiplication   **d.** division

12. Is $\{0, 3, 6, 9, 12, \ldots\}$ closed under the operation of
  a. addition   **b.** subtraction
  c. multiplication   **d.** division

13. Which properties of real numbers are satisfied by $\{-4, -2, 0, 2, 4\}$?

14. Which properties of real numbers are satisfied by $\{\ldots, -5, -3, -1, 1, 3, 5, \ldots\}$?

In Exercises 15 and 16, graph each inequality and write the inequality using interval notation.

15. $-4 < x \leqq 2$ **16.** $x \leqq -1$ or $x > 3$

In Exercises 17 and 18, graph each interval and write each interval as an inequality.

17. $[-3, 2)$ **18.** $(-1, \infty)$

In Exercises 19 to 22, write each real number without absolute value symbols.

19. $|7|$ **20.** $|2 - \pi|$

21. $|4 - \pi|$ **22.** $|-11|$

In Exercises 23 and 24, find the distance on the real number line between the points whose coordinates are given.

23. $-3,\ \ 14$ **24.** $\sqrt{5},\ \ -\sqrt{2}$

In Exercises 25 and 26, evaluate each expression.

25. $-5^2 + (-11)$ **26.** $\dfrac{(2^2 \cdot 3^{-2})^2}{3^{-1} \cdot 2^3}$

In Exercises 27 and 28, simplify each expression.

27. $(3x^2y)(2x^3y)^2$ **28.** $\left(\dfrac{2a^2b^3c^{-2}}{3ab^{-1}}\right)^2$

In Exercises 29 and 30, write each number in scientific notation.

29. 620,000   **30.** 0.0000017

In Exercises 31 and 32, change each number from its scientific notation to its decimal form.

31. 3.5×10^4 **32.** 4.31×10^{-7}

In Exercises 33 to 36, perform the indicated operation and express each result as a polynomial in standard form.

33. $(2a^2 + 3a - 7) + (-3a^2 - 5a + 6)$

34. $(5b^2 - 11) - (3b^2 - 8b - 3)$

35. $(2x^2 + 3x - 5)(3x^2 - 2x + 4)$

36. $(3y - 5)^3$

In Exercises 37 to 40, completely factor each polynomial over the integers.

37. $3x^2 + 30x + 75$ **38.** $25x^2 - 30xy + 9y^2$

39. $20a^2 - 4b^2$ **40.** $16a^3 + 250$

In Exercises 41 to 50, simplify each radical expression. Assume the variables are positive real numbers.

41. $\sqrt{48a^2b^7}$ **42.** $\sqrt{12a^3b}$ **43.** $\sqrt{72x^2y}$

44. $\sqrt{18x^3y^5}$ **45.** $\sqrt{\dfrac{54xy^3}{10x}}$ **46.** $-\sqrt{\dfrac{24xyz^3}{15z^6}}$

47. $\dfrac{7x}{\sqrt[3]{2x^2}}$ **48.** $\dfrac{5y}{\sqrt[3]{9y}}$

49. $\sqrt[3]{-135x^2y^7}$ **50.** $\sqrt[3]{-250xy^6}$

In Exercises 51 and 52, write the complex number in standard form and give its conjugate.

51. $3 - \sqrt{-64}$ **52.** $\sqrt{-4} + 6$

In Exercises 53 to 56, simplify and write the complex number in standard form.

53. $(3 + 7i) + (2 - 5i)$ **54.** $(6 - 8i) - (9 - 11i)$

55. $(5 + 3i)(2 - 5i)$ **56.** $\dfrac{4 + i}{7 - 2i}$

In Exercises 57 to 60, simplify and write each complex number as either i, $-i$, 1, or -1.

57. i^{20} **58.** i^{57} **59.** $\dfrac{1}{i^{28}}$ **60.** i^{-200}

2

Equations and Inequalities

How can it be that mathematics, being after all a product of human thought independent of experience, is so admirably adopted to the objects of reality?

ALBERT EINSTEIN (1879–1955)

DNA, KNOTS, LINKS, AND POLYNOMIALS

Many practical applications are a direct result of esoteric mathematics. For example, biologists are now starting to apply the mathematics known as *knot theory* to the study of DNA.

Due to the nature of DNA, the problems involved in the study of DNA are very complex. The average person has over 9 million miles of DNA in their cells. For DNA to fit in a human cell, it must be coiled, linked, and knotted. Because this linking and knotting can be accomplished in billions of ways, it is no wonder that biologists use knot theory to help classify different DNA knots and links as a first step toward understanding how DNA knots and links during replication and recombination.

The following figure illustrates a knot and a link. The knot on the left consists of a single string in which the ends have been made to vanish by being fastened together. The link on the right consists of two strings or components linked together.

One of the main concerns of knot theory is to determine whether two links are equivalent or distinct. Two links are equivalent if one can be moved into the same configuration as the second. If two links are not equivalent, then they are said to be distinct. It can be difficult in even simple cases to determine if two links are equivalent by observation. Consider the links shown in the following figure. Since the link on the left can be moved into the same configuration as the link on the right, the links are equivalent. With more complicated links, it is generally very difficult to determine if the links are equivalent or distinct by observation.

Two versions of the same link; the link is known as the Whitehead link, named after the topologist J. H. C. Whitehead.

In 1984, Vaughan Jones, a professor at the University of California at Berkeley, discovered polynomials that can be associated with links. A particular link has associated with it one of Jones' polynomials. Even if the link is moved, it still has the same polynomial. Thus biologists now have a mathematical method for the study of links. If the Jones' polynomials for two links are different, then the two links are distinct and no amount of movement can make the two links have the same configuration.

2.1

Linear Equations

Recall from Section 1.1 that an equation is a statement about the equality of two expressions. If neither expression contains a variable, it is easy to determine whether the equation is a true or a false statement. For example, the equation $2 + 3 = 5$ is a true statement. If the expressions contain a variable, the equation may be a true statement for some values of the variable and a false statement for other values of the variable. For example, $3x + 2 = 14$ is a true statement when $x = 4$, but it is false for any number except 4.

A number is said to **satisfy** an equation if substituting the number for the variable produces an equation that is a true statement. To **solve** an equation means to find all values of the variable that satisfy the equation. The values that make the equation true are called **solutions** or **roots** of the equation. For example, 5 is a solution or root of $2x - 10 = 0$ because $2(5) - 10 = 0$ is a true statement.

The **solution set** of an equation is the set of all solutions of the equation. The solution set of $x^2 - 5x + 6 = 0$ is $\{2, 3\}$ because 2 and 3 are the only numbers that satisfy the equation.

Equivalent equations have the same solution set. The process of solving an equation is generally accomplished by producing *simpler* but equivalent equations until the solutions are easy to observe. To produce these simpler equivalent equations, we often apply the following properties.

Addition and Multiplication Properties of Equality

For real numbers a, b, and c, $a = b$ and $a + c = b + c$ are equivalent equations.
If $c \neq 0$, then $a = b$ and $ac = bc$ are equivalent equations.

Essentially, these properties state that an equivalent equation is produced by either adding the same expression to each side of an equation or multiplying each side of an equation by the same nonzero expression. For example,

$$x - 2 = 7 \quad \text{and} \quad x - 2 + 2 = 7 + 2 \text{ are equivalent equations.}$$

$$3x = 12 \quad \text{and} \quad \frac{1}{3}(3x) = \frac{1}{3}(12) \text{ are equivalent equations.}$$

Definition of a Linear Equation

A **linear equation** in one variable is an equation that can be written in the form

$$ax + b = 0$$

where a and b are real numbers, with $a \neq 0$.

The addition and multiplication properties of equality can be used to solve a linear equation.

EXAMPLE 1   **Solve a Linear Equation**

Solve the linear equation $\frac{3}{4}x - 6 = 0$.

Solution   To isolate the x term on the left side of the equation, add 6 to each side of the equation.

$$\frac{3}{4}x - 6 = 0$$

$$\frac{3}{4}x - 6 + 6 = 0 + 6$$

$$\frac{3}{4}x = 6$$

To get the variable x alone on the left side of the equation, multiply each side by 4/3 (the reciprocal of 3/4).

$$\left(\frac{4}{3}\right)\left(\frac{3}{4}x\right) = \left(\frac{4}{3}\right)(6)$$

$$x = 8$$

Check by substituting 8 for x in the original equation.

$$\frac{3}{4}x - 6 = 0$$

$$\frac{3}{4}(8) - 6 \stackrel{?}{=} 0$$

$$0 = 0 \quad \text{True}$$

The proposed solution, 8, satisfies the original equation. The solution set of $\frac{3}{4}x - 6 = 0$ is $\{8\}$.

■ *Try Exercise* **2**, *page 83.*

If an equation involves fractions, it is convenient to multiply each side of the equation by the LCD of all the denominators to produce an equivalent equation that does not contain fractions.

EXAMPLE 2   **Solve by Clearing Fractions**

Solve the equation $\dfrac{2}{3}x + 10 - \dfrac{x}{5} = \dfrac{36}{5}$.

Solution   Multiply each side of the equation by 15, which is the LCD of all the denominators.

$$15\left(\dfrac{2}{3}x + 10 - \dfrac{x}{5}\right) = 15\left(\dfrac{36}{5}\right)$$

$$10x + 150 - 3x = 108$$

Combine like terms on each side of the equation.

$$7x + 150 = 108$$

To isolate the term that involves the variable x, use the addition property of equality to add -150 to each side of the equation. Subtracting 150 is equivalent to adding -150.

$$7x + 150 - 150 = 108 - 150$$

$$7x = -42$$

To get the variable x *alone* on the left side of the equation, use the multiplication property of equality to multiply each side by 1/7. Dividing both sides by 7 is equivalent to multiplying by 1/7.

$$\dfrac{7x}{7} = \dfrac{-42}{7}$$

$$x = -6$$

Check as before.   ■ *Try Exercise* **12,** *page 83.*

EXAMPLE 3   **Solve an Equation by Applying Properties**

Solve the equation $(x + 2)(5x + 1) = 5x(x + 1)$.

Solution

$(x + 2)(5x + 1) = 5x(x + 1)$

$5x^2 + 11x + 2 = 5x^2 + 5x$ The FOIL method and the distributive property.

$11x + 2 = 5x$ Subtract $5x^2$ from each side.

$6x + 2 = 0$ Subtract $5x$ from each side.

$6x = -2$ Subtract 2 from each side.

$x = -\dfrac{1}{3}$ Divide each side of the equation by 6.

Check as before.   ■ *Try Exercise* **20,** *page 83.*

An equation that has no solutions is called a **contradiction.** The equation $x = x + 1$ is a contradiction. No number is equal to itself increased by 1.

An **identity** is an equation that is true for *every* real number for which all terms of the equation are defined. Examples of identities include the equations $x + x = 2x$, and $(x + 3)^2 = x^2 + 6x + 9$.

An equation that is true for some values of the variable but not true for other values of the variable is called a **conditional equation.** For example, $x + 2 = 8$ is a conditional equation because it is true for $x = 6$ and false for any number not equal to 6.

EXAMPLE 4   **Classify an Equation as an Identity, a Conditional Equation, or a Contradiction**

Determine whether each of the following equations is an identity, a conditional equation, or a contradiction.

a. $3(x + 2) = 3x + 6$ b. $2x + 5 = 17$

c. $(x + 3)(x - 3) = x^2 + 1$

Solution

a.   The equation $3(x + 2) = 3x + 6$ is an identity because the left side of the equation simplifies to $3x + 6$, which is identical to the right side. The original equation is true for all values of x.

b.   The equation $2x + 5 = 17$ is a conditional equation because it is true for $x = 6$ and false for $x = 0$. Actually, the equation is false for all numbers other than 6; however, to prove that an equation is a conditional equation, we only need to show that there is at least one number that makes it true and at least one number that makes it false.

c.   The following sequence of equivalent equations shows that the equation $(x + 3)(x - 3) = x^2 + 1$ is a contradiction.

$$(x + 3)(x - 3) = x^2 + 1$$

$$x^2 - 9 = x^2 + 1 \quad \text{Use } (a + b)(a - b) = a^2 - b^2.$$

$$-9 = 1 \qquad \text{Subtract } x^2 \text{ from each side of the equation.}$$

Notice that the last equation is equivalent to $0x - 9 = 1$, which is not true for any value of x. Since it is equivalent to the original equation, the original equation is a contradiction.

■ *Try Exercise* **30,** *page 83.*

The multiplication property of equality states that you can multiply each side of an equation by the same *nonzero* number. If you multiply each side of an equation by an algebraic expression that involves a variable, you must restrict the variable so that the expression is not equal to zero.

EXAMPLE 5   **Solve Equations That Have Restrictions**

Solve the following equations:

a. $\dfrac{x}{x - 3} = \dfrac{9}{x - 3} - 5$

b. $1 + \dfrac{x}{x - 5} = \dfrac{5}{x - 5}$

Solution

a.   First, note that the denominator $x - 3$ would equal zero if $x = 3$. To produce a simpler equivalent equation, multiply each side by $x - 3$, with the restriction that $x \neq 3$.

$$(x - 3)\left(\frac{x}{x - 3}\right) = (x - 3)\left(\frac{9}{x - 3} - 5\right)$$

$$x = (x - 3)\left(\frac{9}{x - 3}\right) - (x - 3)5$$

$$x = 9 - 5x + 15$$

Adding $5x$ to each side of the equation produces

$$6x = 24$$

$$x = 4$$

Since $4 \neq 3$, we check by substituting 4 for x in the original equation to show that 4 is indeed a solution. The solution set of our original equation is {4}.

b.   To produce a simpler equivalent equation, multiply each side of the equation by $x - 5$, with the restriction that $x \neq 5$.

$$(x - 5)\left(1 + \frac{x}{x - 5}\right) = (x - 5)\left(\frac{5}{x - 5}\right)$$

$$(x - 5)1 + (x - 5)\left(\frac{x}{x - 5}\right) = 5$$

$$x - 5 + x = 5$$

$$2x - 5 = 5$$

$$2x = 10$$

$$x = 5$$

Although we have obtained 5 as a proposed solution, 5 is *not* a solution of the original equation since it contradicts our restriction $x \neq 5$. Substitution of 5 for x in the original equation results in denominators of 0. In this case, the solution set of the original equation is the empty set.

■ *Try Exercise **38,** page 83.*

A **literal equation** is an equation that involves more than one variable. To solve a literal equation for one of its variables, you treat all the other variables as constants.

EXAMPLE 6   Solve a Literal Equation

Solve the literal equation $2x - 5y = 10$ for y.

Solution

$$-5y = -2x + 10 \quad \text{Add } -2x \text{ to each side of the equation.}$$

$$y = \frac{2}{5}x - 2 \quad \text{Divide each side of the equation by } -5.$$

■ *Try Exercise* **58,** *page 83.*

Many literal equations can be solved by using the methods for solving linear equations, provided you list necessary restrictions.

EXAMPLE 7   Solve a Literal Equation and State Restrictions

Solve $xy + z = yz$ for

a. x b. y.

Solution

a.   To solve for x, subtract z from each side of the equation to isolate the term that contains x.

$$xy + z = yz$$

$$xy = yz - z$$

$$x = \frac{yz - z}{y} \quad \text{Divide each side of the equation by } y.$$

Restriction: $y \neq 0$.

b.   To solve for y, first isolate the terms that involve the variable y on the left side of the equation.

$$xy + z = yz$$

$$xy - yz = -z \quad \text{Subtract } z \text{ and } yz \text{ from each side of the equation to isolate the terms that contain } y.$$

$$y(x - z) = -z \quad \text{Factor } y \text{ from each term on the left side of the equation.}$$

$$y = \frac{-z}{x - z} \quad \text{Divide each side of the equation by } x - z.$$

Restriction: $(x - z) \neq 0$.

■ *Try Exercise* **66,** *page 83.*

EXERCISE SET 2.1

In Exercises 1 to 28, solve and check each equation.

1. $2x + 10 = 40$

2. $-3y + 20 = 2$

3. $5x + 2 = 2x - 10$

4. $4x - 11 = 7x + 20$

5. $2(x - 3) - 5 = 4(x - 5)$

6. $5(x - 4) - 7 = -2(x - 3)$

7. $4(2r - 17) + 5(3r - 8) = 0$

8. $6(5s - 11) - 12(2s + 5) = 0$

9. $\frac{3}{4}x + \frac{1}{2} = \frac{2}{3}$

10. $\frac{x}{4} - 5 = \frac{1}{2}$

11. $\frac{2}{3}x - 5 = \frac{1}{2}x - 3$

12. $\frac{1}{2}x + 7 - \frac{1}{4}x = \frac{19}{2}$

13. $0.2x + 0.4 = 3.6$

14. $0.04x - 0.2 = 0.07$

15. $x + 0.08(60) = 0.20(60 + x)$

16. $6(t + 1.5) = 12t$

17. $\frac{3}{5}(n + 5) - \frac{3}{4}(n - 11) = 0$

18. $-\frac{5}{7}(p + 11) + \frac{2}{5}(2p - 5) = 0$

19. $3(x + 5)(x - 1) = (3x + 4)(x - 2)$

20. $5(x + 4)(x - 4) = (x - 3)(5x + 4)$

21. $5[x - (4x - 5)] = 3 - 2x$

22. $6[3y - 2(y - 1)] - 2 + 7y = 0$

23. $\frac{40 - 3x}{5} = \frac{6x + 7}{8}$

24. $\frac{12 + x}{-4} = \frac{5x - 7}{3} + 2$

25. $0.08x + 0.12(4000 - x) = 432$

26. $0.075y + 0.06(10,000 - y) = 727.50$

27. $0.115x + 0.0975(8000 - x) = 823.75$

28. $0.145x + 0.109(4000) = 0.12(4000 + x)$

In Exercises 29 to 36, determine if the equation is an identity, a conditional equation, or a contradiction.

29. $-3(x - 5) = -3x + 15$

30. $2x + \frac{1}{3} = \frac{6x + 1}{3}$

31. $2y + 7 = 3(y - 1)$

32. $x^2 + 10x = x(x + 10)$

33. $\frac{4y + 7}{4} = y + 7$

34. $(x + 3)^2 = x^2 + 9$

35. $2x + 5 = x + 9 + x$

36. $(x - 3)(x + 4) = x^2 + 4x - 11$

In Exercises 37 to 56, solve and check each equation.

37. $\frac{3}{x + 2} = \frac{5}{2x - 7}$

38. $\frac{4}{y + 2} = \frac{7}{y - 4}$

39. $\frac{30}{10 + x} = \frac{20}{10 - x}$

40. $\frac{6}{8 + x} = \frac{4}{8 - x}$

41. $\frac{3x}{x + 4} = 2 - \frac{12}{x + 4}$

42. $\frac{8}{2m + 1} - \frac{1}{m - 2} = \frac{5}{2m + 1}$

43. $2 + \frac{9}{r - 3} = \frac{3r}{r - 3}$

44. $\frac{t}{t - 4} + 3 = \frac{4}{t - 4}$

45. $\frac{5}{x - 3} - \frac{3}{x - 2} = \frac{4}{x - 3}$

46. $\frac{4}{x - 1} + \frac{7}{x + 7} = \frac{5}{x - 1}$

47. $\frac{2x + 5}{3x - 1} = 1$

48. $\frac{4x - 1}{3x + 2} = \frac{5}{6}$

49. $(y + 3)^2 = (y + 4)^2 + 1$

50. $(z - 7)^2 = (z - 2)^2 + 9$

51. $\frac{x}{x - 3} = \frac{x + 4}{x + 2}$

52. $\frac{x}{x - 5} = \frac{x + 7}{x + 1}$

53. $\frac{x + 3}{x + 5} = \frac{x - 3}{x - 4}$

54. $\frac{x - 6}{x + 4} = \frac{x - 1}{x + 2}$

55. $\frac{4x - 3}{2x} = \frac{2x - 4}{x - 2}$

56. $\frac{x + 3}{x + 1} = \frac{x + 6}{x + 4}$

In Exercises 57 to 68, solve each equation for x in terms of the other letters. State any necessary restrictions.

57. $2x + 3y = 6$

58. $4x - 7y = -15$

59. $2cx - d = 5(x - c)$

60. $2rx + 7 = 8(r - x)$

61. $\frac{x}{a} + \frac{y}{b} = 1$

62. $y = mx + b$

63. $y - y_1 = m(x - x_1)$

64. $Ax + By + C = 0$

65. $\frac{3}{x + 4} = l + w$

66. $\frac{2}{x - 2} = \frac{y}{x - 3}$

67. $mx - m^2 - nx = -n^2$

68. $px - p^2 + 6q = 3qx + 9q^2$

Calculator Exercises

In Exercises 69 to 74, use a calculator to solve each equation. State your final answer as a decimal with three significant digits.

69. $2.77x - 5.47 = 9.68$

70. $3.21x + 7.14 = 7.82x$

71. $\dfrac{1.62}{3.14x} = \dfrac{2.66}{9.21x}$

72. $\dfrac{3.84}{2.45x} = \dfrac{-1.92}{4.46 - 5.78x}$

73. $\dfrac{30.45x + 12.45}{6.71 - 2.34x} = 1.86$

74. $8.53x + 7.34(125 - 2.00x) = 108$

Supplemental Exercises

In Exercises 75 to 78, determine whether the given pair of equations is equivalent.

75. $3x - 11 = -5$, $\dfrac{3x - 11}{x - 2} = \dfrac{-5}{x - 2}$

76. $3x - 9 = x - 3$, $\dfrac{3x - 9}{x - 3} = \dfrac{x - 3}{x - 3}$

77. $\dfrac{1}{t} = \dfrac{1}{a} + \dfrac{1}{b}$, $t = \dfrac{ab}{a + b}$,   where t is a variable and a and b are positive constants.

78. $\dfrac{2}{x} = \dfrac{1}{x - 1}$, $2(x - 1) = x$

79. Let a, b, and c be real constants. Show that an equation of the form $ax + b = c$ has $x = \dfrac{c - b}{a}$ $(a \neq 0)$ as its solution.

80. Let a, b, c, and d be real constants. Show that an equation of the form $ax + b = cx + d$ has $x = \dfrac{d - b}{a - c}$ $(a - c \neq 0)$ as its solution.

In Exercises 81 to 86, solve each equation for x.

81. $\sqrt{7x} - 3 = 7$

82. $\sqrt{8x} + 2 = 14$

83. $\sqrt{3x} - 5 = \sqrt{27x} + 2$

84. $\sqrt{20x} + 14 = \sqrt{5x} - 8$

85. $a^2x - b = b^2x + a$ (assume $a \neq \pm b$)

86. $a^3x - a^2 + ab = b^2 - b^3x$ (assume $a \neq -b$)

2.2

Formulas and Applications

A **formula** is an equation or inequality that expresses known relationships between two or more variables. Following is a table of formulas from geometry that will be used frequently throughout this text. In each formula, the variable P represents perimeter, A represents area, C represents circumference of a circle, and V represents volume.

Formulas from Geometry

Rectangle	Square	Triangle	Circle
$P = 2l + 2w$	$P = 4s$	$P = a + b + c$	$C = \pi d = 2\pi r$
$A = lw$	$A = s^2$	$A = \frac{1}{2}bh$	$A = \pi r^2$

Rectangular Solid	Cube	Right Circular Cone	Cylinder
$V = lwh$	$V = s^3$	$V = \frac{1}{3}\pi r^2 h$	$V = \pi r^2 h$

It is often necessary to solve a formula for a specified variable. The process consists of first isolating all terms that contain the specified variable on one side of the equation and all terms that do not contain the specified variable on the other side.

EXAMPLE 1   Solve a Formula for a Specified Variable

Solve $2l + 2w = P$ for l.

Solution $2l + 2w = P$

$2l = P - 2w$ Subtract $2w$ from each side to isolate the $2l$ term.

$l = \dfrac{P - 2w}{2}$ Divide each side by 2.

■ *Try Exercise* **2,** *page 91.*

Remark   The previous solution can also be expressed in the equivalent form $l = P/2 - w$.

EXAMPLE 2   Solve a Formula By Factoring

Solve $S = 2lw + 2hl + 2hw$ for h.

Solution   First, we isolate all terms with the variable h on one side and all terms without h on the other side.

$S = 2lw + 2hl + 2hw$

$S - 2lw = 2hl + 2hw$ Subtract $2lw$ from each side.

$S - 2lw = h(2l + 2w)$ Factor h from the right side.

$\dfrac{S - 2l\text{w}}{2l + 2w} = h$ Divide each side by $2l + 2w$.

■ *Try Exercise* **4,** *page 91.*

People with good problem-solving skills generally work application problems by applying specific techniques in a series of small steps.

Guidelines for Solving Application Problems

1. Read the problem carefully. If necessary, reread the problem several times.
2. When appropriate, draw a sketch and label parts of the drawing with the specific information given in the problem.
3. Determine the unknown quantities, and label them with variables. Write down any equation that relates the variables.
4. Use the information from step 3, along with a known formula or some additional information given in the problem, to write an equation.
5. Solve the equation obtained in step 4, and check to see if these results satisfy all the conditions of the original problem.

EXAMPLE 3   **Solve a Geometric Application Problem**

The length of a rectangle is 2 feet longer than three times its width. If the perimeter of the rectangle is 92 feet, find the width and the length of the rectangle.

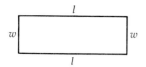

Figure 2.1

Solution

1. Read the problem carefully.
2. Draw a rectangle as shown in Figure 2.1.
3. Label the length of the rectangle l and the width of the rectangle w. The problem states that the length l is 2 feet longer than three times its width w. Thus l and w are related by the equation

$$l = 3w + 2.$$

4. Since the problem involves the length, width, and perimeter of a rectangle, we use the geometric formula $2l + 2w = P$. To write an equation that involves only constants and a single variable, say, w, substitute 92 for P and $3w + 2$ for l.

$$2l + 2w = P$$
$$2(3w + 2) + 2w = 92$$

5. Now solve for the unknown w.

$$6w + 4 + 2w = 92$$
$$8w + 4 = 92$$
$$8w = 88$$
$$w = 11$$

Since the length l is two more than three times the width, we find

$$l = 3(11) + 2 = 35.$$

A check verifies that 35 is two more than three times 11. Also, twice the length (70) plus twice the width (22) gives the perimeter (92). The width of the rectangle is 11 feet, and its length is 35 feet.

■ *Try Exercise* **22,** *page 91.*

Many *uniform motion* problems can be solved using the formula $d = rt$, where d is the distance traveled, r is the rate of speed, and t is the time.

EXAMPLE 4   **Solve a Uniform Motion Problem**

A runner runs a course at an average speed of 6 mph. One hour later, a cyclist starts on the same course at an average speed of 15 mph. How long after the runner started did the cyclist overtake the runner?

Solution   If we represent the the time the runner has spent on the course by t, then the time the cyclist takes to overtake the runner is $t - 1$.

The following table organizes our information and helps us determine how to write the distances each person travels.

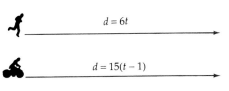

$d = 6t$

$d = 15(t - 1)$

Figure 2.2

	rate r	·	time t	=	distance d
Runner	6	·	t	=	$6t$
Cyclist	15	·	$t - 1$	=	$15(t - 1)$

Figure 2.2 indicates that the runner and the cyclist cover the same distance. Thus our equation is

$$6t = 15(t - 1)$$

Solving this equation yields

$$6t = 15t - 15$$

$$-9t = -15$$

$$t = 1\frac{2}{3}$$

A check will verify that the cyclist does overtake the runner $1\frac{2}{3}$ hours after the runner starts.

■ *Try Exercise* **30**, *page 92.*

Many business problems can be solved by using linear equations, as shown in Examples 5, 6 and 7.

EXAMPLE 5 **Solve a Business Application**

It costs a tennis shoe manufacturer $26.55 to produce a pair of tennis shoes that sell for $49.95. How many pairs of tennis shoes must be produced and sold for a profit of $14,274.00 to be made?

Solution The *profit* is equal to the *revenue* minus the *cost*. If x equals the number of pairs of tennis shoes to be sold, then the revenue will be $49.95x$ and the cost will be $26.55x$. Using

$$(\text{Profit}) = (\text{revenue}) - (\text{cost})$$

gives us

$$14{,}274.00 = 49.95x - 26.55x$$

$$14{,}274.00 = 23.40x$$

$$\frac{14{,}274.00}{23.40} = x$$

$$610 = x$$

The manufacturer must produce and sell 610 pairs of tennis shoes to produce the desired profit.

■ *Try Exercise* **38**, *page 92.*

EXAMPLE 6   **Solve an Application Involving Percent**

The price of a car rose 2 percent this month. If the car now costs $18,425, how much to the nearest dollar did the car cost last month?

Solution   Using x as the price last month, we have

$$\left(\begin{matrix} \text{Price last} \\ \text{month} \end{matrix}\right) + 2\% \text{ increase} = \left(\begin{matrix} \text{present} \\ \text{cost} \end{matrix}\right)$$

$$x \quad + \quad 0.02x \quad = 18{,}425$$

$$1.02x = 18{,}425$$

$$x = \frac{18{,}425}{1.02}$$

$$x \approx 18{,}064 \quad \text{Rounded to the nearest dollar.}$$

Last month the price of the car was $18,064.

■ *Try Exercise* **40**, *page 92.*

Simple interest problems can be solved by using the formula $I = Prt$, where I is the interest, P is the principal, r is the simple interest rate per period, and t is the number of periods.

EXAMPLE 7   **Solve a Simple Interest Problem**

An accountant invests part of a $6000 bonus in a 5 percent simple interest account and the remainder of the money at 8.5 percent simple interest. If together the investments earn $370 per year, find the amount invested at each rate.

Solution   Let x be the amount invested at 5 percent. The remainder of the money is $6000 − x$, which will be the amount invested at 8.5 percent. Using the interest formula $I = Prt$, with $t = 1$ year, yields

$$\text{Interest at } 5\% = x \cdot 0.05 = 0.05x$$

$$\text{Interest at } 8.5\% = (6000 - x) \cdot (0.085) = 510 - 0.085x$$

Now the interest earned on the two accounts must equal the total interest $370, so

$$0.05x + (510 - 0.085x) = 370.$$

Solving this equation, we get

$$-0.035x = -140$$

$$x = \frac{-140}{-0.035}$$

$$x = 4000$$

Therefore, the accountant invested $4000 at 5 percent and the remaining $2000 at 8.5 percent. Check as before.

■ *Try Exercise* **42**, *page 92.*

Percent mixture problems involve combining solutions or alloys that have different concentrations of a common substance. Percent mixture problems can be solved using the formula $pA = Q$, where p is the percent of concentration, A is the amount of the solution or alloy, and Q is the quantity of a substance in the solution or alloy. For example, in 4 liters of a 25% acid solution, p is the percent of acid (25%), A is the amount of solution (4 liters), and Q is the amount of acid in the solution which equals $((0.25) \cdot (4)$ liters $= 1$ liter).

EXAMPLE 8   Solve a Percent Mixture Problem

A chemist mixes an 11% hydrochloric acid with a 6% hydrochloric acid solution. How many milliliters (ml) of each solution should the chemist use to make a 600-milliliter solution that is 8% hydrochloric acid?

Solution   Let x be the number of milliliters of the 11% solution. Since the final solution will have a total of 600 milliliters of fluid, $600 - x$ is the number of milliliters of the 6% solution. Use a table to organize the information given in the problem:

% solution	% of concentration p		Amount (ml) of solution A	=	Quantity (ml) of acid Q
11	0.11	·	x	=	$0.11x$
6	0.06	·	$600 - x$	=	$0.06(600 - x)$
8	0.08	·	600	=	$0.08(600)$

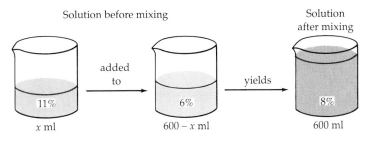

Figure 2.3

Because all the hydrochloric acid in the final solution comes from either the 11% solution or the 6% solution, the number of milliliters of hydrochloric acid in the 11% solution added to the number of milliliters of hydrochloric acid in the 6% solution must equal the number of milliliters of hydrochloric acid in the 8% solution. That is,

$$\begin{pmatrix} \text{ml of acid in} \\ \text{11\% solution} \end{pmatrix} + \begin{pmatrix} \text{ml of acid in} \\ \text{6\% solution} \end{pmatrix} = \begin{pmatrix} \text{ml of acid in} \\ \text{8\% solution} \end{pmatrix}$$

$$0.11x \quad + \quad 0.06(600 - x) \quad = \quad 0.08(600)$$

Solving for x yields

$$0.11x + 36 - 0.06x = 48$$
$$0.05x + 36 = 48$$
$$0.05x = 12$$
$$x = 240$$

Therefore the chemist should use 240 milliliters of the 11% solution and 360 milliliters of the 6% solution to make a 600 milliliter solution, which is 8% hydrochloric acid.

■ *Try Exercise 46, page 92.*

To solve a *work problem*, we use the equation

rate of work × time worked = part of task completed

For example, if a painter can paint a wall in 15 minutes, then the painter can paint $\frac{1}{15}$ of the wall in 1 minute. The painter's *rate of work* is $\frac{1}{15}$ of the wall each minute. In general, if a task can be completed in x minutes, then the rate of work is $1/x$ of the task each minute.

EXAMPLE 9   **Solve an Application Problem Involving Work**

Pump A can fill a pool in 6 hours and pump B can fill the same pool in 3 hours. How long will it take to fill the pool if both pumps are used?

Solution

Since pump A fills the pool in 6 hours, $\frac{1}{6}$ represents the part of the pool filled by pump A in 1 hour.

Since pump B fills the pool in 3 hours, $\frac{1}{3}$ represents the part of the pool filled by pump B in 1 hour.

Let $t =$ the number of hours to fill the pool together. Then

$$t \cdot \frac{1}{6} = \frac{t}{6} \quad \text{Part of the pool filled by pump } A.$$

$$t \cdot \frac{1}{3} = \frac{t}{3} \quad \text{Part of the pool filled by pump } B.$$

Because $\begin{pmatrix} \text{Part filled} \\ \text{by pump } A \end{pmatrix} + \begin{pmatrix} \text{Part filled} \\ \text{by pump } B \end{pmatrix} = \begin{pmatrix} \text{1 filled} \\ \text{pool} \end{pmatrix}$

$$\frac{t}{6} \quad + \quad \frac{t}{3} \quad = \quad 1$$

Multiplying each side of the equation by 6 produces

$$t + 2t = 6$$
$$3t = 6$$
$$t = 2$$

Together the two pumps can fill the pool in 2 hours.

Check   Since pump A fills $\frac{2}{6}$ or $\frac{1}{3}$ of the pool in 2 hours and pump B fills $\frac{2}{3}$ of the pool in 2 hours, the result is correct.

■ *Try Exercise* **56,** *page 93.*

EXERCISE SET 2.2

In Exercises 1 to 18, solve the formula for the specified variable.

1. $V = \dfrac{1}{3}\pi r^2 h;$ h (geometry)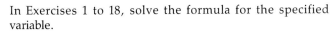

2. $P = S - Sdt;$ t (business)

3. $I = Prt;$ t (business)

4. $A = P + Prt;$ P (business)

5. $F = \dfrac{Gm_1 m_2}{d^2};$ m_1 (physics)

6. $A = \dfrac{1}{2}h(b_1 + b_2);$ b_1 (geometry)

7. $s = v_0 t - 16t^2;$ v_0 (physics)

8. $\dfrac{1}{f} = \dfrac{1}{d_o} + \dfrac{1}{d_i};$ f (astronomy)

9. $Q_w = m_w c_w (T_f - T_w);$ T_w (physics)

10. $T \Delta t = Iw_f - Iw_i;$ I (physics)

11. $a_n = a_1 + (n - 1)d;$ d (mathematics)

12. $y - y_1 = m(x - x_1);$ x (mathematics)

13. $S = \dfrac{a_1}{1 - r};$ r (mathematics)

14. $\dfrac{P_1 V_1}{T_1} = \dfrac{P_2 V_2}{T_2};$ V_2 (chemistry)

15. $\dfrac{w_1}{w_2} = \dfrac{f_2 - f}{f - f_1};$ f_1 (hydrostatics)

16. $v = \dfrac{v_1 + v_2}{1 + \dfrac{v_1 v_2}{c^2}};$ v_1 (physics)

17. $f_{LC} = f_v \dfrac{v + v_{LC}}{v};$ v_{LC} (physics)

18. $F_1 d_1 + F_2 d_2 = F_3 d_3 + F_4 d_4;$ F_3 (physics)

In Exercises 19 to 60, solve by using the Guidelines for Solving Application Problems.   Pg 85

19. One-fifth of a number plus one-fourth of the number is five less than one-half the number. What is the number?

20. The numerator of a fraction is 4 less than the denominator. If the numerator is increased by 14 and the denominator is decreased by 10, the resulting number is 5. What is the original fraction?

21. The length of a rectangle is 3 feet less than twice the width of the rectangle. If the perimeter of the rectangle is 174 feet, find the width and the length.

22. The width of a rectangle is 1 meter more than half the length of the rectangle. If the perimeter of the rectangle is 110 meters, find the width and the length.

23. A triangle has a perimeter of 84 centimeters. The two longer sides of the triangle are each three times as long as the shortest side. Find the length of each side of the triangle.

24. A triangle has a perimeter of 185 miles. The two smaller sides of the triangle are each one-third the length of the longest side. Find the length of each side of the triangle.

25. Find two consecutive natural numbers whose sum is 1745.

26. Find three consecutive odd integers whose sum is 2001.

27. The difference of the squares of two consecutive positive even integers is 76. Find the integers.

28. The product of two consecutive integers is 90 less than the product of the next two integers. Find the four integers.

29. Running at an average rate of 6 meters per second, a sprinter ran to the end of a track and then jogged back to the starting point at an average rate of 2 meters per second. The total time for the sprint and the jog back was 2 minutes 40 seconds. Find the length of the track.

30. A motorboat leaves a harbor and travels to an island at an average rate of 15 knots. The average speed on the return trip was 10 knots. If the total trip took 7.5 hours, how far is the harbor from the island?

31. A plane leaves an airport traveling at an average speed of 240 kilometers per hour. How long will it take a second plane traveling the same route at an average speed of 600 kilometers per hour to catch up with the first plane if it leaves 3 hours later?

32. A plane leaves Chicago headed for Los Angeles at 540 mph. One hour later, a second plane leaves Los Angeles headed for Chicago at 660 mph. If the air route from Chicago to Los Angeles is 1800 miles, how long will it take for the planes to pass by each other? How far from Chicago will they be at that time?

33. A motor boat takes twice as long to go 18 miles upstream as it takes to return. If the boat cruises at 6 mph in still water, find the rate of the current.

34. An airplane has an airspeed (the rate in still air) of 420 mph. The plane can travel 96 miles with the wind in the same time that it can travel 86 miles against the wind. What is the velocity of the wind?

35. A student has test scores of 80, 82, 94, and 71. What score does the student need on the next test to produce an average score of 85?

36. A student has test scores of 90, 74, 82, and 90. The next examination is the final examination, which will count as two tests. What score does the student need on the final examination to produce an average score of 85?

37. It costs a manufacturer of sunglasses $8.95 to produce sunglasses that sell for $29.99. How many sunglasses must be produced and sold for a profit of $17,884.00 to be made?

38. It costs a restaurant owner 18 cents per glass for orange juice, which is sold for 75 cents per glass. How many glasses of orange juice must be served and sold for a profit of $2337 to be made?

39. The price of a computer fell 20 percent this year. If the computer now costs $750, how much did it cost last year?

40. The price of a magazine subscription rose 4 percent this year. If the subscription now costs $26, how much did the subscription cost last year?

41. An investment adviser invested $14,000 in two accounts. One investment earned 8 percent annual simple interest, and the other investment earned 6.5 percent annual simple interest. The amount of interest earned for 1 year was $1024. How much was invested in each account?

42. A total of $7500 is deposited into two simple interest accounts. On one account the annual simple interest rate is 5 percent, and on the second account the annual simple interest rate is 7 percent. The amount of interest earned for 1 year was $405. How much was invested in each account?

43. An investment of $2500 is made at an annual simple interest rate of 5.5 percent. How much additional money must be invested at an annual simple interest rate of 8 percent so that the total interest earned is 7 percent of the total investment?

44. An investment of $4600 is made at an annual simple interest rate of 6.8 percent. How much additional money must be invested at an annual simple interest rate of 9 percent so that the total interest earned is 8 percent of the total investment?

45. How many grams of pure silver must a silversmith mix with a 45% silver alloy to produce 200 grams of a 50% alloy?

46. How many liters of a 40% sulfuric acid solution should be mixed with 4 liters of a 24% sulfuric acid solution to produce a 30% solution?

47. How many liters of water should be evaporated from 160 liters of a 12% saline solution so that the solution that remains is a 20% saline solution?

48. A radiator contains 6 liters of a 25% antifreeze solution. How much should be drained and replaced with pure antifreeze to produce a 33% antifreeze solution?

49. A ballet performance brought in $61,800 on the sale of 3000 tickets. If the tickets sold for $14 and $25, how many of each were sold?

50. A vending machine contains $41.25. The machine contains 255 coins, which consist of only nickels, dimes, and quarters. If the machine contains twice as many dimes as nickels, how many of each type of coin does the machine contain?

51. A coffee shop decides to blend a $12 per pound coffee with a $9 per pound coffee to produce a blend that will sell at $10 per pound. How much of each should be used to yield 20 pounds of the new blend?

52. A bag contains 42 coins, with a total weight of 246 grams. If the bag contains only gold coins that weigh 8 grams each and silver coins that weigh 5 grams each, how many gold and how many silver coins are in the bag?

53. How much pure gold should be melted with 15 grams of 14 karat gold to produce 18 karat gold? *Hint:* A karat is a measure of the purity of gold in an alloy. Pure gold measures 24 karats. An alloy that measures x karats is $x/24$ gold. For example, 18 karat gold is $18/24 = 3/4$ gold.

54. How much 14 karat gold should be melted with 4 ounces of pure gold to produce 18 karat gold? (*Hint:* See Exercise 53.)

55. An electrician can install the electric wires in a house in 14 hours. A second electrician requires 18 hours. How long would it take both electricians working together to install the wires?

56. Printer A can print a report in 3 hours. Printer B can print the same report in 4 hours. How long would it take both printers working together to print the report?

57. A worker can build a fence in 8 hours. With the help of an assistant, the fence can be built in 5 hours. How long would it take the assistant to build the fence alone?

58. A roofer and an assistant can repair a roof together in 6 hours. The assistant can complete the repair alone in 14 hours. If both the roofer and the assistant work together for 2 hours and then the assistant is left alone to finish the job, how much longer will the assistant need to finish the repairs?

59. A book and a bookmark together sell for $10.10. If the price of the book is $10.00 more than the price of the bookmark, find the price of the book and the price of the bookmark.

60. Three people decide to share the cost of a yacht. By bringing in an additional partner, the cost for each would be reduced by $4000. What is the total cost of the yacht?

Supplemental Exercises

The *Archimedean law of the lever* states that for a lever to be in a state of balance with respect to a point called the fulcrum, the sum of the downward forces times their respective distances from the fulcrum on one side of the fulcrum must equal the sum of the downward forces times their respective distances from the fulcrum on the other side of the fulcrum, as the accompanying figure shows:

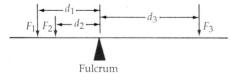

$$F_1 d_1 + F_2 d_2 = F_3 d_3$$

61. A 100-pound person 8 feet from the fulcrum and a 40-pound person 5 feet from the fulcrum balance with a 160-pound person on a teeter-totter. How far from the fulcrum is the 160-pound person?

62. A lever 21 feet long has a force of 117 pounds applied to one end of the lever and a force of 156 pounds to the other end. Where should the fulcrum be located to produce a state of balance?

63. How much force applied 5 feet from the fulcrum is needed to lift 400 pounds that are 0.5 feet on the other side of the fulcrum?

64. Two workers need to lift a 1440-pound rock. They use a 6-foot steel bar with the fulcrum 1 foot from the rock, as the figure below shows. One worker applies 180 pounds to the other end of the lever. How much force will the second worker need to apply 1 foot from that end to lift the rock?

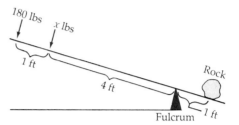

65. One pump can fill a pool in 10 hours. A second pump can fill the same pool in 8 hours. The pool has a drain that can drain the pool in 6 hours. Both pumps work together for 2 hours with the drain closed. The drain is then opened. How much longer will it take the two pumps to fill the pool?

66. If a pump can fill a pool in A hours and another pump can fill the pool in B hours, then the formula

$$T = \frac{AB}{A + B}$$

can be used to determine the total time T in hours that will be required to fill the pool if the pumps are allowed to work together. Now consider the case where the pool is filled by three pumps. One can fill it in A hours, a second in B hours, and a third in C hours. Derive a formula for the total time T needed to fill the pool in terms of A, B, and C.

67. Two seconds after firing a rifle at a target, the shooter hears the impact of the bullet. If sound travels at 1100 feet per second and the bullet at 1865 feet per second, determine the distance to the target.

68. Sound travels through sea water 4.62 times faster than through air. The sound of an exploding mine on the surface of the water and partially submerged reaches a ship through the water 4 seconds before it reaches the ship through the air. How far is the ship from the explosion? Use 1100 feet per second as the speed of sound through the air.

69. If a parade 2 miles long is preceding at 3 mph, how long will it take a runner, jogging at 6 mph, to travel from the front of the parade to the end of the parade?

70. How long would the runner in Exercise 69 take to jog from the end of the parade to the start of the parade?

71. The work of the ancient Greek mathematician Diophantus had great influence on later European number theorists. Nothing is known about his personal life, except for the information given in the following epigram. "Diophantus passed $\frac{1}{6}$ of his life in childhood, $\frac{1}{12}$ in youth, and $\frac{1}{7}$ more as a bachelor. Five years after his marriage was born a son who died four years before his father, at $\frac{1}{2}$ his father's (final) age." How old was Diophantus when he died?

72. The relationship between the Fahrenheit temperature (F) and the Celsius temperature (C) is given by the formula

$$F = \frac{9}{5}C + 32.$$

At what temperature will a Fahrenheit thermometer and a Celsius thermometer read the same?

2.3

Quadratic Equations and Applications

A **quadratic equation** in the variable x is an equation that can be written in the **standard quadratic form** $ax^2 + bx + c = 0$, $a \neq 0$.

Several methods can be used to solve quadratic equations. If the quadratic polynomial $ax^2 + bx + c$ in a quadratic equation can be factored over the integers, then the equation can be solved by factoring and using the **zero product property.**

Zero Product Property

> If A and B are algebraic expressions, then
>
> $$AB = 0 \quad \text{if and only if} \quad A = 0 \text{ or } B = 0.$$

Therefore, if a quadratic equation can be written as the product of two factors set equal to zero, then at least one of the factors is zero.

EXAMPLE 1 **Solve by Using the Zero Product Property**

Solve each of the following quadratic equations:

a. $3x^2 + 10x - 8 = 0$ b. $x^2 + 10x + 25 = 0$

Solution

a. $3x^2 + 10x - 8 = 0$

$(3x - 2)(x + 4) = 0$ Factor.

$3x - 2 = 0 \quad$ or $\quad x + 4 = 0 \quad$ Apply the zero product property.

$3x = 2 \quad$ or $\quad x = -4$

$x = \dfrac{2}{3} \quad$ or $\quad x = -4 \quad$ Check as before.

The solution set of $3x^2 + 10x - 8 = 0$ is $\left\{-4, \dfrac{2}{3}\right\}$.

b.　$x^2 + 10x + 25 = 0$

$$(x + 5)^2 = 0 \qquad \text{Factor.}$$

$x + 5 = 0 \qquad \text{or} \qquad x + 5 = 0 \qquad$ Apply the zero product property.

$x = -5 \qquad \text{or} \qquad x = -5 \qquad$ Check as before.

The solution set of $x^2 + 10x + 25 = 0$ is $\{-5\}$.

■ *Try Exercise 4, page 104.*

Remark In Example 1b, the solution or root -5 is called a **double solution** or a **double root** of the equation because the application of the zero product property produced the two identical equations $x + 5 = 0$, both of which have a root of -5.

The Square Root Theorem

The quadratic equation $x^2 = c$ can be solved by factoring and applying the zero product property to yield the roots $\sqrt{c}$ and $\sqrt{-c}$ as shown:

$$x^2 = c$$

$$x^2 - c = 0$$

$$(x + \sqrt{c})(x - \sqrt{c}) = 0$$

$x + \sqrt{c} = 0 \qquad \text{or} \qquad x - \sqrt{c} = 0$

$x = -\sqrt{c} \qquad \text{or} \qquad x = \sqrt{c}$

This result is known as the square root theorem, which we will use to solve quadratic equations that can be written in the form $A^2 = B$.

The Square Root Theorem

> If A and B are algebraic expressions such that
> $$A^2 = B, \qquad \text{then} \quad A = \pm\sqrt{B}.$$

EXAMPLE 2　Solve by Using the Square Root Theorem

Use the square root theorem to solve each quadratic equation.

a.　$(x + 1)^2 = 49$　　b.　$(x - 3)^2 = -28$

Solution

a.　$(x + 1)^2 = 49$

$x + 1 = \pm\sqrt{49} \qquad$ The square root theorem

$x + 1 = \pm 7$

$x = -1 \pm 7$

Thus $x = -1 - 7 = -8$ or $x = -1 + 7 = -6$.

The solution set of $(x + 1)^2 = 49$ is $\{-8, 6\}$.

b. $(x - 3)^2 = -28$

$\qquad x - 3 = \pm\sqrt{-28}$ $\qquad\qquad$ The square root theorem

$\qquad x - 3 = \pm 2i\sqrt{7}$

$\qquad\qquad x = 3 \pm 2i\sqrt{7}$

The solution set of $(x - 3)^2 = -28$ is $\{3 - 2i\sqrt{7}, 3 + 2i\sqrt{7}\}$.

■ *Try Exercise **18**, page 104.*

Completing the Square

Consider the following binomial squares and their perfect square trinomial products:

Square of a Binomial		Perfect Square Trinomial
$(x + 6)^2$	$=$	$x^2 + 12x + 36$
$(x - 3)^2$	$=$	$x^2 - 6x + 9$

In each perfect square trinomial, the coefficient of x^2 is 1, and the constant term of the perfect square trinomial is the square of one-half of the coefficient of its x term.

$$x^2 + 12x + 36, \qquad \left(\frac{1}{2} \cdot 12\right)^2 = 36$$

$$x^2 - 6x + 9, \qquad \left(\frac{1}{2}(-6)\right)^2 = 9$$

Adding to a binomial of the form $x^2 + bx$, the constant that makes it a perfect square trinomial is called **completing the square.** For example, to complete the square of $x^2 + 14x$, add

$$\left(\frac{1}{2} \cdot 14\right)^2 = 49$$

to produce the perfect square trinomial $x^2 + 14x + 49$. The method of completing the square can be used to solve quadratic equations.

EXAMPLE 3   Solve by Completing the Square

Solve the quadratic equation $x^2 + 6x - 7 = 0$.

$\qquad$ *Solution*   First isolate the variable terms on one side of the equation:

$$x^2 + 6x - 7 = 0$$

$$x^2 + 6x = 7 \quad \text{Isolate the terms in } x.$$

Complete the square by adding the square of half the coefficient of the x term to each side of the equation.

$$x^2 + 6x + 9 = 7 + 9 \qquad \text{Think:} \quad \left(\frac{1}{2} \cdot 6\right)^2 = 3^2 = 9.$$

$$(x + 3)^2 = 16 \qquad \text{Factor the perfect square trinomial.}$$
$$x + 3 = \pm\sqrt{16} \qquad \text{Apply the square root theorem.}$$
$$x + 3 = \pm 4$$
$$x = -3 \pm 4$$
$$x = 1 \text{ or } -7 \qquad \text{Check as before.}$$

The solution set of $x^2 + 6x - 7 = 0$ is $\{-7, 1\}$.

■ *Try Exercise* **30,** *page 104.*

Remark   Completing the square by adding the square of half the coefficient of the x term requires that the coefficient of the x^2 term be 1. If the coefficient of the x^2 term is not 1, multiply each term on each side of the equation by the reciprocal of the coefficient of x^2.

EXAMPLE 4   **Solve by Completing the Square**

Solve the quadratic equation $2x^2 + 8x - 15 = 0$.

Solution $2x^2 + 8x - 15 = 0$

$$2x^2 + 8x = 15$$

$$\frac{1}{2}(2x^2 + 8x) = \frac{1}{2}(15) \qquad \begin{array}{l}\text{Multiply both sides of the} \\ \text{equation by the reciprocal of} \\ \text{the leading coefficient.}\end{array}$$

$$x^2 + 4x = \frac{15}{2}$$

$$x^2 + 4x + 4 = \frac{15}{2} + 4 \qquad \begin{array}{l}\text{Add one half the square of} \\ \text{the } x \text{ coefficient to both sides.}\end{array}$$

$$(x + 2)^2 = \frac{23}{2} \qquad \text{Factor and simplify.}$$

$$x + 2 = \pm\sqrt{\frac{23}{2}} \qquad \begin{array}{l}\text{Apply the square} \\ \text{root theorem.}\end{array}$$

$$x = -2 \pm \frac{\sqrt{46}}{2} \qquad \begin{array}{l}\text{Add } -2 \text{ to each side of the} \\ \text{equation, and rationalize} \\ \text{the denominator.}\end{array}$$

$$x = \frac{-4 \pm \sqrt{46}}{2}$$

The solution set of $2x^2 + 8x - 15 = 0$ is $\left\{\dfrac{-4 - \sqrt{46}}{2}, \dfrac{-4 + \sqrt{46}}{2}\right\}$.

■ *Try Exercise* **38,** *page 104.*

Completing the square is a powerful method because it can be used to solve any quadratic equation. In Example 5, the solution set consists of two complex numbers.

EXAMPLE 5   Solve by Completing the Square

Solve the quadratic equation $x^2 = 2x - 6$.

Solution
$$x^2 - 2x = -6$$
$$x^2 - 2x + 1 = -6 + 1 \quad \text{Complete the square.}$$
$$(x - 1)^2 = -5$$
$$x - 1 = \pm\sqrt{-5} \quad \text{Apply the square root theorem.}$$
$$x = 1 \pm i\sqrt{5}$$

The solution set of $x^2 = 2x - 6$ is $\{1 - i\sqrt{5}, 1 + i\sqrt{5}\}$.

■ *Try Exercise* **40,** *page 104.*

The Quadratic Formula

Completing the square on the standard quadratic form $ax^2 + bx + c = 0$, $(a \neq 0)$, produces a formula for x in terms of the coefficients a, b, and c. The formula is known as the **quadratic formula,** and it is another method for solving quadratic equations.

The Quadratic Formula

If $ax^2 + bx + c = 0$, $\quad a \neq 0$, $\quad$ then
$$x = \frac{-b \pm \sqrt{b^2 - 4ac}}{2a}.$$

Proof: We assume a is a positive real number. If a were a negative real number, then we could multiply each side of the equation by -1 to make it positive.

$$ax^2 + bx + c = 0 \quad (a \neq 0) \qquad \text{Given}$$
$$ax^2 + bx = -c \qquad \text{Isolate the variable terms.}$$
$$x^2 + \frac{b}{a}x = -\frac{c}{a} \qquad \text{Multiply each term on each side of the equation by } \frac{1}{a}.$$
$$x^2 + \frac{b}{a}x + \left(\frac{b}{2a}\right)^2 = \left(\frac{b}{2a}\right)^2 - \frac{c}{a} \qquad \text{Complete the square.}$$
$$\left(x + \frac{b}{2a}\right)^2 = \left(\frac{b}{2a}\right)^2 - \frac{c}{a} \qquad \text{Factor the left side.}$$

$$\left(x + \frac{b}{2a}\right)^2 = \frac{b^2}{4a^2} - \frac{4a}{4a} \cdot \frac{c}{a} \qquad \text{Simplify the right side.}$$

$$x + \frac{b}{2a} = \pm\sqrt{\frac{b^2 - 4ac}{4a^2}} \qquad \text{Apply the square root theorem.}$$

$$x + \frac{b}{2a} = \pm\frac{\sqrt{b^2 - 4ac}}{2a} \qquad \text{Since } a > 0, \sqrt{4a^2} = 2a.$$

$$x = -\frac{b}{2a} \pm \frac{\sqrt{b^2 - 4ac}}{2a} \qquad \text{Add } -\frac{b}{2a} \text{ to each side.}$$

$$x = \frac{-b \pm \sqrt{b^2 - 4ac}}{2a}$$

To solve a quadratic equation using the quadratic formula, first write the quadratic equation in the standard quadratic form $ax^2 + bx + c = 0$.

EXAMPLE 6   Solve by Using the Quadratic Formula

Solve the quadratic equation $3x^2 = -2x + 5$.

Solution   First write the equation in standard quadratic form and then identify the values of a, b, and c. The equation $3x^2 + 2x - 5 = 0$ has coefficients $a = 3$, $b = 2$, and $c = -5$.

$$x = \frac{-(2) \pm \sqrt{2^2 - 4(3)(-5)}}{2(3)} \qquad \text{Substitute in the quadratic formula}$$

$$x = \frac{-2 \pm \sqrt{64}}{6} = \frac{-2 \pm 8}{6}$$

$$= 1 \text{ or } -\frac{5}{3}$$

The solution set of $3x^2 = -2x + 5$ is $\left\{-\frac{5}{3}, 1\right\}$.

■ *Try Exercise **46**, page 104.*

The equation $3x^2 = -2x + 5$ could have been solved by factoring and using the zero product property. As a general rule, you should first try to solve quadratic equations by factoring. If the factoring process proves difficult, then solve by using the quadratic formula.

EXAMPLE 7   Solve Using the Quadratic Formula

Solve the quadratic equation $4x^2 - 8x + 1 = 0$.

Solution   In this example the coefficients are $a = 4$, $b = -8$, and $c = 1$.

$$x = \frac{-(-8) \pm \sqrt{(-8)^2 - 4(4)(1)}}{2(4)} = \frac{8 \pm \sqrt{48}}{8} = \frac{8 \pm 4\sqrt{3}}{8} = \frac{2 \pm \sqrt{3}}{2}$$

The solution set of $4x^2 - 8x + 1 = 0$ is $\left\{ \dfrac{2 - \sqrt{3}}{2}, \dfrac{2 + \sqrt{3}}{2} \right\}$.

■ *Try Exercise 50, page 104.*

The Discriminant

In the quadratic formula

$$x = \frac{-b \pm \sqrt{b^2 - 4ac}}{2a},$$

the expression $b^2 - 4ac$ is called the **discriminant** of the quadratic formula. The quadratic formula involves the square root of the discriminant. If $b^2 - 4ac \geq 0$, then $\sqrt{b^2 - 4ac}$ is a real number; if $b^2 - 4ac < 0$, then $\sqrt{b^2 - 4ac}$ is a complex number.

Thus the sign of the discriminant determines whether the roots of a quadratic equation are real numbers or complex numbers.

The Discriminant and Roots of a Quadratic Equation

The quadratic equation $ax^2 + bx + c = 0$, with real coefficients and $a \neq 0$, has discriminant $b^2 - 4ac$.

If $b^2 - 4ac > 0$, then the quadratic equation has *two distinct real roots.*

If $b^2 - 4ac = 0$, then the quadratic equation has *a real root* that is a double root.

If $b^2 - 4ac < 0$, then the quadratic equation has *two distinct nonreal complex roots.*

By examining the discriminant, it is possible to determine whether the roots of a quadratic equation are real numbers or complex numbers without actually finding the roots.

EXAMPLE 8   Use the Discriminant to Classify Roots

Classify the roots of each quadratic equation as real numbers or complex numbers.

a. $2x^2 - 5x + 1 = 0$ b. $3x^2 + 6x + 7 = 0$ c. $x^2 + 6x + 9 = 0$

Solution

a. $2x^2 - 5x + 1 = 0$ has coefficients $a = 2$, $b = -5$, and $c = 1$.

$$b^2 - 4ac = (-5)^2 - 4(2)(1) = 25 - 8 = 17$$

Because the discriminant 17 is *positive*, the equation $2x^2 - 5x + 1 = 0$ has *two distinct real roots.*

b. $3x^2 + 6x + 7 = 0$ has coefficients $a = 3$, $b = 6$, and $c = 7$.

$$b^2 - 4ac = 6^2 - 4(3)(7) = 36 - 84 = -48$$

Because the discriminant -48 is *negative*, $3x^2 + 6x + 7 = 0$ has *two distinct complex roots*.

c. $x^2 + 6x + 9 = 0$ has coefficients $a = 1$, $b = 6$, and $c = 9$.

$$b^2 - 4ac = 6^2 - 4(1)(9) = 36 - 36 = 0$$

Because the discriminant is 0, the equation $x^2 + 6x + 9 = 0$ has *a real root*. The root is a double root.

■ *Try Exercise 62, page 104.*

The Sum and Product of the Roots of a Quadratic Equation

The roots of $x^2 + 2x - 15 = 0$ are -5 and 3. Notice that the sum of the roots equals the opposite of the coefficient of the x term and the product of the roots equals the constant term. This example is a special case of the following theorem.

The Sum and Product of the Roots Theorem

If $a \neq 0$ and r_1 and r_2 are roots of $ax^2 + bx + c = 0$, which has the equivalent form

$$x^2 + \frac{b}{a}x + \frac{c}{a} = 0,$$

then

$$\text{the sum of the roots } r_1 + r_2 = -\frac{b}{a},$$

$$\text{the product of the roots } r_1 r_2 = \frac{c}{a}.$$

One method of checking the solutions or roots of an equation is to substitute the roots into the original equation. In Example 9, the sum and product of the roots theorem is used to check the proposed roots of quadratic equations.

EXAMPLE 9 Check by Using the Sum and Product of the Roots Theorem

Check the proposed roots of each quadratic equation.

a. $x^2 - 2x - 15 = 0$ b. $x^2 - 4x - 4 = 0$
Proposed roots: $-3, 5$ Proposed roots: $2 \pm 2\sqrt{2}$

Solution

a. The sum of the proposed roots is $(-3) + 5 = 2$.
Their product is $(-3)(5) = -15$.

The proposed roots check because

$$\text{their sum, 2, equals } -\frac{b}{a} = -\frac{-2}{1} = 2, \quad \text{and}$$

$$\text{their product, } -15, \text{ equals } \frac{c}{a} = \frac{-15}{1} = -15.$$

b. The sum of the proposed roots is $(2 - 2\sqrt{2}) + (2 + 2\sqrt{2}) = 4.$
    Their product is $(2 - 2\sqrt{2})(2 + 2\sqrt{2}) = 4 - 8 = -4.$
    The proposed roots check because

$$\text{their sum, 4, equals } -\frac{b}{a} = -\frac{-4}{1} = 4, \quad \text{and}$$

$$\text{their product, } -4, \text{ equals } \frac{c}{a} = \frac{-4}{1} = -4.$$

■ *Try Exercise **72**, page 104.*

As a matter of comparison, try checking Example 9b by the substitution method.

Quadratic Equations and Applications

Many application problems lead to quadratic equations. Be sure to use the guidelines for solving application problems from Section 2.2 when you work the application problems throughout the remainder of this text.

EXAMPLE 10   Solve an Application

A cylindrical container (Figure 2.4) is to be constructed so that it has the capacity of 62 cubic inches and a height of 5.5 inches. Find the inner radius of the container.

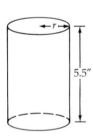

Figure 2.4

    Solution   Recall that the volume V of a cylinder is given by the formula $V = \pi r^2 h$, where r is the radius and h is the height. Substituting 5.5 for h and 62 for V produces the quadratic equation $62 = \pi r^2(5.5)$. Simplifying and applying the square root theorem produces

$$\frac{62}{\pi(5.5)} = r^2$$

$$3.6 \approx r^2 \quad \text{Simplify using } \pi \approx 3.14.$$

$$\pm 1.9 \approx r \quad \text{Apply the square root theorem.}$$

Thus the radius r is about 1.9 inches. The solution $r \approx -1.9$ inches should be discarded. Why?[1]

■ *Try Exercise **78**, page 104.*

---

[1]Although -1.9 is an approximate solution to the quadratic equation $62 = \pi r^2(5.5)$, it is not a solution to the given problem because the radius of a circle must be a positive real number.

EXAMPLE 11   **Solve an Application**

A veterinarian wishes to use 132 feet of chain-link fencing to enclose a rectangular region and subdivide the region into two smaller rectangles, as shown in Figure 2.5. If the total enclosed area is 576 square feet, find the dimensions of the enclosed region.

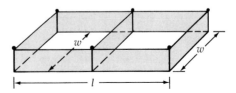

Figure 2.5

Solution   Let x be the width of the enclosed region. Then $3x$ represents the amount of fencing used to construct the three widths. The amount of fencing left for the two lengths is $132 - 3x$. Thus each length must be half of the remaining fencing, or $\dfrac{132 - 3x}{2}$.

Now we have variable expressions in x for both the width and the length. Substituting these into the area formula $lw = A$ produces a quadratic equation in x.

$$lw = A$$

$$\left(\frac{132 - 3x}{2}\right)(x) = 576 \quad \text{Substitute.}$$

$$132x - 3x^2 = 1152 \quad \text{Simplify.}$$

$$-3x^2 + 132x - 1152 = 0$$

$$x^2 - 44x + 384 = 0 \quad \text{Divide each term by } -3.$$

Although this quadratic formula can be solved by factoring, the following solution makes use of the quadratic formula. What reason can you give for using the quadratic formula rather than the factoring method?[2]

$$x = \frac{-(-44) \pm \sqrt{(-44)^2 - 4(1)(384)}}{2(1)} \quad \text{Apply the quadratic formula.}$$

$$x = \frac{44 \pm \sqrt{400}}{2} = \frac{44 \pm 20}{2} = 12 \text{ or } 32$$

Thus there are two valid solutions to the problem:

1. If the width $x = 12$ feet, then the length is $\dfrac{132 - 3(12)}{2} = 48$ feet.

2. If the width $x = 32$ feet, then the length is $\dfrac{132 - 3(32)}{2} = 18$ feet.

■ *Try Exercise* **80,** *page 104.*

[2]Factoring $x^2 - 44x + 384$ by the trial-and-error method may be time consuming because 384 has several integer factors.

EXERCISE SET 2.3

In Exercises 1 to 16, solve each quadratic equation by factoring and applying the zero product property.

1. $x^2 - 2x - 15 = 0$

2. $y^2 + 3y - 10 = 0$

3. $8y^2 + 189y - 72 = 0$

4. $12w^2 - 41w + 24 = 0$

5. $3x^2 - 7x = 0$

6. $5x^2 = -8x$

7. $8 + 14t - 15t^2 = 0$

8. $12 - 26w + 10w^2 = 0$

9. $12 - 21s - 6s^2 = 0$

10. $-144 + 320y + 9y^2 = 0$

11. $(x - 5)^2 - 9 = 0$

12. $(3x + 4)^2 - 16 = 0$

13. $(2x - 5)^2 - (4x - 11)^2 = 0$

14. $(5x + 3)^2 - (x + 7)^2 = 0$

15. $14x = x^2 + 49$

16. $41x = 12x^2 + 35$

In Exercises 17 to 28, use the square root theorem to solve each quadratic equation.

17. $x^2 = 81$

18. $y^2 = 225$

19. $2x^2 = 48$

20. $3x^2 = 144$

21. $3x^2 + 12 = 0$

22. $4y^2 + 20 = 0$

23. $(x - 5)^2 = 36$

24. $(x + 4)^2 = 121$

25. $(x - 8)^2 = (x + 1)^2$

26. $(x + 5)^2 = (2x + 1)^2$

27. $x^2 = (x + 1)^2$

28. $4x^2 = (2x + 3)^2$

In Exercises 29 to 44, solve by completing the square.

29. $x^2 + 6x + 1 = 0$

30. $x^2 + 8x - 10 = 0$

31. $x^2 - 2x - 15 = 0$

32. $x^2 + 2x - 8 = 0$

33. $x^2 + 10x = 0$

34. $x^2 - 6x = 0$

35. $x^2 + 3x - 1 = 0$

36. $x^2 + 7x - 2 = 0$

37. $2x^2 + 4x - 1 = 0$

38. $2x^2 + 10x - 3 = 0$

39. $3x^2 - 8x + 1 = 0$

40. $4x^2 - 4x + 15 = 0$

41. $5x^2 + 3x - 1 = 0$

42. $4x^2 + 7x - 2 = 0$

43. $5 - 6x - 3x^2 = 0$

44. $2 + 10x - 5x^2 = 0$

In Exercises 45 to 60, solve by using the quadratic formula.

45. $x^2 - 2x - 15 = 0$

46. $x^2 - 5x - 24 = 0$

47. $x^2 + x - 1 = 0$

48. $x^2 + x + 1 = 0$

49. $2x^2 + 4x + 1 = 0$

50. $2x^2 + 4x - 1 = 0$

51. $3x^2 - 5x + 3 = 0$

52. $3x^2 - 5x + 4 = 0$

53. $5x^2 + x - 1 = 0$

54. $4x^2 + 6x - 3 = 0$

55. $\frac{1}{2}x^2 + \frac{3}{4}x - 1 = 0$

56. $\frac{2}{3}x^2 - 5x + \frac{1}{2} = 0$

57. $\sqrt{2}\,x^2 + 3x + \sqrt{2} = 0$

58. $2x^2 + \sqrt{5}\,x - 3 = 0$

59. $x^2 = 3x - 5$

60. $-x^2 = 7x - 1$

In Exercises 61 to 66, determine the discriminant of the quadratic equation, and then classify the roots of the equation as **a.** two distinct real numbers, **b.** one real number (which is a double root), or **c.** two distinct complex numbers. Do not solve the equations.

61. $2x^2 - 5x - 7 = 0$

62. $x^2 + 3x - 11 = 0$

63. $3x^2 - 2x + 10 = 0$

64. $x^2 + 3x + 3 = 0$

65. $x^2 - 20x + 100 = 0$

66. $4x^2 + 12x + 9 = 0$

In Exercises 67 to 70, find all values of k so that each quadratic equation has exactly one real root. *Hint:* The quadratic equation $ax^2 + bx + c = 0$ will have exactly one real root if and only if $b^2 - 4ac = 0$.

67. $16x^2 + kx + 9 = 0$

68. $x^2 + kx + 81 = 0$

69. $y^2 - 7y + k = 0$

70. $x^2 + 15x + k = 0$

In Exercises 71 to 76, use the sum and product of the roots theorem to determine if the given numbers are roots of the quadratic equation.

71. $x^2 - 5x - 24 = 0$, $-3, 8$

72. $x^2 + 4x - 21 = 0$, $-7, 3$

73. $2x^2 - 7x - 30 = 0$, $-5/2, 6$

74. $9x^2 - 12x - 1 = 0$, $(2 + \sqrt{5})/3$, $(2 - \sqrt{5})/3$

75. $x^2 - 2x + 2 = 0$, $1 + i$, $1 - i$

76. $x^2 - 4x + 12 = 0$, $2 + 3i, 2 - 3i$

77. The perimeter of a rectangle is 27 centimeters and its area is 35 square centimeters. Find the length and the width of the rectangle.

78. The perimeter of a rectangle is 34 feet and its area is 60 square feet. Find the length and the width of the rectangle.

79. A gardener wishes to use 600 feet of fencing to enclose a rectangular region and subdivide the region into two smaller rectangles. If the total enclosed area is 15,000 square feet, find the dimensions of the enclosed region.

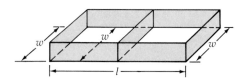

80. A farmer wishes to use 400 yards of fencing to enclose a rectangular region and subdivide the region into three

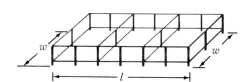

smaller rectangles. If the total enclosed area is 4800 square yards, find the dimensions of the enclosed region.

81. The sum of the squares of two consecutive positive even integers is 244. Find the numbers.

82. The sum of the squares of three consecutive integers is 302. Find the numbers.

83. Find a positive real number that is 5 larger than its reciprocal.

84. Find a positive real number that is 2 smaller than its reciprocal.

85. A salesperson drove the first 105 miles of a trip in 1 hour more than it took to drive the last 90 miles. If the average rate during the last 90 miles was 10 mph faster than the average rate during the first 105 miles, find the average rate for each portion of the trip.

86. A car and a bus both completed a 240-mile trip. The car averaged 10 mph faster than the bus and completed the trip in 48 minutes less time than the bus. Find the average rate in miles per hour of the bus.

87. A mason can build a wall in 6 hours less than an apprentice. Together they can build the wall in 4 hours. How long would it take the apprentice working alone to build the wall?

88. Pump A can fill a pool in 2 hours less time than pump B. Together the pumps can fill the pool in 2 hours 24 minutes. Find how long it takes pump A to fill the pool.

Calculator Exercises

In Exercises 89 to 94, use a calculator to solve each quadratic equation. Round your answers to the nearest hundredth.

89. $2.37x^2 + 4.15 = 0$

90. $5.21x^2 - 6.28x = 0$

91. $x^2 + 2.78x - 4.61 = 0$

92. $2.34x^2 - 3.21x + 7.31 = 0$

93. $15x^2 - 41x - 31 = 0$

94. $16x^2 - 3x + 5 = 0$

Supplemental Exercises

In Exercises 95 to 102, use the quadratic formula to solve each equation for the indicated variable, in terms of the other variables. Assume that none of the denominators are zero.

95. $s = -gt^2 + v_0 t + s_0,$ for t

96. $S = 2\pi rh + 2\pi r^2,$ for r

97. $-xy^2 + 4y + 3 = 0,$ for y

98. $D = \dfrac{n}{2}(n - 3),$ for n

99. $3x^2 + xy + 4y^2 = 0,$ for x

100. $3x^2 + xy + 4y^2 = 0,$ for y

101. $x = y^2 + y - 8,$ for y

102. $P = \dfrac{E^2 R}{(r + R)^2},$ for R

103. Prove that the equation $ax^2 + bx + c = 0$ with real coefficients such that $ac < 0$ has two distinct real roots.

104. Prove that the equation $ax^2 + c = 0$ with real coefficients such that $ac > 0$ has two distinct complex roots.

105. The ancient Greeks defined a rectangle as a "golden rectangle" provided its length l and its width w satisfy the equation

$$\frac{l}{w} = \frac{w}{l - w}.$$

a. Solve the above formula for w.
b. The Parthenon is an ancient Greek temple. The rectangular portion of the front of the Parthenon is a golden rectangle. If the length l of the front of the Parthenon measures 101 feet, how tall is the Parthenon?

106. Use the quadratic formula to prove that if r_1 and r_2 are roots of the quadratic equation $ax^2 + bx + c = 0$, then

$$r_1 + r_2 = -\frac{b}{a} \quad \text{and} \quad r_1 r_2 = \frac{c}{a}.$$

107. The sum S of the first n natural numbers $1, 2, 3, \ldots, n$, is given by the formula

$$S = \frac{n}{2}(n + 1).$$

How many consecutive natural numbers starting with 1 produces a sum of 253?

108. The number of diagonals D of a polygon with n sides is given by the formula

$$D = \frac{n}{2}(n - 3).$$

Determine the number of sides of a polygon with 464 diagonals.

2.4

Other Types of Equations

Some equations that are neither linear nor quadratic can be solved by the various techniques presented in this section. For instance, the **third-degree equation** or **cubic equation** in Example 1 can be solved by factoring the polynomial on the left side of the equation and using the zero product property to set each factor equal to zero.

EXAMPLE 1   Solve an Equation by Factoring

Solve the equation $x^3 - 16x = 0$.

> **Solution** $x^3 - 16x = 0$
>
> $x(x^2 - 16) = 0$ Factor out the GCF, x.
>
> $x(x + 4)(x - 4) = 0$ Factor the difference of squares.

Set each factor equal to zero.

$$x = 0 \quad \text{or} \quad x + 4 = 0 \quad \text{or} \quad x - 4 = 0$$

$$x = 0 \quad \text{or} \quad x = -4 \quad \text{or} \quad x = 4$$

A check will show that -4, 0, and 4 are roots of the original equation. The solution set of $x^3 - 16x = 0$ is $\{-4, 0, 4\}$.

■ *Try Exercise* **6,** *page 111.*

Caution   If you had attempted to solve Example 1 by dividing each side by x, you would have produced the equation $x^2 - 16 = 0$, which only has roots of -4 and 4. In this case the division of each side of the equation by the variable x has not produced an equivalent equation. Why?[3] To avoid this common mistake, factor out any variable factors that are common to each term instead of dividing each side of the equation by the factor.

Some equations that involve radical expressions can be solved by using the following result.

The Power Principle

> If P and Q are algebraic expressions and n is a positive integer, then every solution of $P = Q$ is a solution of $P^n = Q^n$.

EXAMPLE 2   Solve a Radical Equation

Use the power principle to solve $\sqrt{x + 4} = 3$.

[3]To divide each side of an equation by a variable, the variable must be restricted so that it is not equal to 0. However, $x = 0$ is a solution of the original equation.

Solution   Square each side of the equation.

$$\sqrt{x + 4} = 3$$

$$(\sqrt{x + 4})^2 = 3^2 \quad \text{The power principle: If } P = Q,$$
$$\text{then } P^n = Q^n, \text{ with } n = 2.$$

$$x + 4 = 9$$

$$x = 5$$

Check:
$$\sqrt{x + 4} = 3$$

$$\sqrt{5 + 4} \stackrel{?}{=} 3 \quad \text{Substitute 5 for } x.$$

$$\sqrt{9} \stackrel{?}{=} 3$$

$$3 = 3 \quad \text{5 checks.}$$

The solution set is {5}.

■ *Try Exercise* **14**, *page 111.*

Caution   Some care must be taken when using the power principle. The principle states that every solution of the original equation $P = Q$ is a solution of $P^n = Q^n$. However, the equation $P^n = Q^n$ may have more solutions than the original equation $P = Q$. As an example, consider the equation $x = 3$. The only solution of the equation $x = 3$ is the real number 3. Square each side of the equation to produce the equation $x^2 = 9$. This quadratic equation, $x^2 = 9$, has both 3 and -3 as solutions. The -3 is called an **extraneous solution** because it is not a solution of the original equation $x = 3$. In general, any solution of the equation $P^n = Q^n$, which is not a solution of the original equation $P = Q$, is called an extraneous solution. Extraneous solutions may be introduced whenever we raise each side of an equation to an *even* power.

EXAMPLE 3   Solve a Radical Equation

Solve the radical equation $x = 2 + \sqrt{2 - x}$. Check all proposed solutions.

Solution   Isolate the radical before squaring each side.

$$x = 2 + \sqrt{2 - x}$$

$$x - 2 = \sqrt{2 - x}$$

$$(x - 2)^2 = (\sqrt{2 - x})^2 \qquad \text{Square each side of the equation.}$$

$$x^2 - 4x + 4 = 2 - x$$

$$x^2 - 3x + 2 = 0 \qquad \text{Collect and combine like terms.}$$

$$(x - 2)(x - 1) = 0 \qquad \text{Factor.}$$

$$x - 2 = 0 \quad \text{or} \quad x - 1 = 0$$

$$x = 2 \quad \text{or} \qquad x = 1 \quad \text{Proposed solutions}$$

Check for $x = 2$:
$$x = 2 + \sqrt{2 - x}$$
$$2 \stackrel{?}{=} 2 + \sqrt{2 - (2)} \quad \text{Substitute 2 for } x.$$
$$2 \stackrel{?}{=} 2 + \sqrt{0}$$
$$2 = 2 \qquad\qquad\qquad 2 \text{ is a solution.}$$

Check for $x = 1$:
$$x = 2 + \sqrt{2 - x}$$
$$1 \stackrel{?}{=} 2 + \sqrt{2 - (1)} \quad \text{Substitute 1 for } x.$$
$$1 \stackrel{?}{=} 2 + \sqrt{1}$$
$$1 \neq 3 \qquad\qquad\qquad 1 \text{ is not a solution.}$$

The check shows that 1 is not a solution. It is an extraneous solution that was created by squaring each side of the equation. The solution set is {2}.

■ *Try Exercise **16**, page 111.*

In Example 4 it will be necessary to square $(1 + \sqrt{2x - 5})$. Recall the special product formula
$$(x + y)^2 = x^2 + 2xy + y^2.$$
Using this special product formula to square $(1 + \sqrt{2x - 5})$ produces
$$(1 + \sqrt{2x - 5})^2 = 1 + 2\sqrt{2x - 5} + (2x - 5).$$

EXAMPLE 4   Solve a Radical Equation

Solve the radical equation $\sqrt{x + 1} - \sqrt{2x - 5} = 1$. Check all proposed solutions.

Solution   First write an equivalent equation in which one radical is isolated on one side of the equation.
$$\sqrt{x + 1} - \sqrt{2x - 5} = 1$$
$$\sqrt{x + 1} = 1 + \sqrt{2x - 5}$$

The next step is to square each side. Using the result from the discussion preceding this example, we get
$$x + 1 = 1 + 2\sqrt{2x - 5} + (2x - 5)$$
$$-x + 5 = 2\sqrt{2x - 5} \qquad \text{Isolate the remaining radical.}$$

The right side still contains a radical, so square each side again.
$$x^2 - 10x + 25 = 4(2x - 5)$$
$$x^2 - 10x + 25 = 8x - 20$$
$$x^2 - 18x + 45 = 0$$
$$(x - 3)(x - 15) = 0$$
$$x = 3 \quad \text{or} \quad x = 15 \quad \text{Proposed solutions}$$

Check for x = 3: $\sqrt{x+1} - \sqrt{2x-5} = 1$

$\sqrt{3+1} - \sqrt{2(3)-5} \stackrel{?}{=} 1$

$\sqrt{4} - \sqrt{1} \stackrel{?}{=} 1$

$2 - 1 \stackrel{?}{=} 1$

$1 = 1$ 3 is a solution.

Check for x = 15: $\sqrt{x+1} - \sqrt{2x-5} = 1$

$\sqrt{15+1} - \sqrt{2(15)-5} \stackrel{?}{=} 1$

$\sqrt{16} - \sqrt{25} \stackrel{?}{=} 1$

$4 - 5 \stackrel{?}{=} 1$

$-1 \neq 1$ 15 is not a solution.

Therefore, the solution set is {3}.

■ *Try Exercise* **20,** *page 112.*

Some equations that involve fractional exponents can be solved by raising each side to a reciprocal power. For example, to solve $x^{1/3} = 4$, raise each side to the third power to find that $x = 64$. Be sure to check all proposed solutions to determine if they are actual solutions or extraneous solutions.

EXAMPLE 5   Solve Equations That Involve Fractional Exponents

Solve the equation $(x^2 + 4x + 52)^{3/2} = 512$.

Solution   Because the equation involves a three-halves power, start by raising each side of the equation to the two-thirds power.

$$[(x^2 + 4x + 52)^{3/2}]^{2/3} = 512^{2/3} \quad \text{The reciprocal of } \frac{2}{3} \text{ is } \frac{3}{2}.$$

$$x^2 + 4x + 52 = 64 \quad \text{Think: } 512^{2/3} = (\sqrt[3]{512})^2 = 8^2 = 64.$$

$$x^2 + 4x - 12 = 0 \quad \text{Subtract 64 from each side.}$$

$$(x - 2)(x + 6) = 0$$

$$x - 2 = 0 \quad \text{or} \quad x + 6 = 0$$

$$x = 2 \quad \text{or} \quad x = -6$$

A check will verify that 2 and -6 are both solutions of the original equation. The solution set is $\{-6, 2\}$.

■ *Try Exercise* **32,** *page 112.*

The equation $4x^4 - 25x^2 + 36 = 0$ is said to be **quadratic in form,** which means it can be written in the form

$$au^2 + bu + c = 0 \qquad a \neq 0$$

where u is an algebraic expression involving x. For example, if we make the substitution $u = x^2$ (which implies $u^2 = x^4$), then our original equation can be written as

$$4u^2 - 25u + 36 = 0.$$

This quadratic equation can be solved for u, and then using the relationship $u = x^2$, we can determine the solutions of the original equation.

EXAMPLE 6   Solve an Equation That Is Quadratic in Form

Solve the equation $4x^4 - 25x^2 + 36 = 0$.

Solution   Make the substitutions $u = x^2$ and $u^2 = x^4$ to produce the quadratic equation $4u^2 - 25u + 36 = 0$. Factor the quadratic polynomial on the left side of the equation.

$$(4u - 9)(u - 4) = 0$$

$$4u - 9 = 0 \quad \text{or} \quad u - 4 = 0$$

$$u = \frac{9}{4} \quad \text{or} \quad u = 4$$

Now use the relationship $u = x^2$ to produce

$$x^2 = \frac{9}{4} \quad \text{or} \quad x^2 = 4$$

$$x = \pm\sqrt{\frac{9}{4}} \quad \text{or} \quad x = \pm\sqrt{4}$$

$$x = \pm\frac{3}{2} \quad \text{or} \quad x = \pm2 \qquad \text{Check as before.}$$

The solution set is $\{-2, -\frac{3}{2}, \frac{3}{2}, 2\}$.

■ *Try Exercise* **42**, *page 112.*

Following is a table of equations that are quadratic in form, along with an appropriate substitution that will allow them to be written in the form $au^2 + bu + c = 0$.

Equations That Are Quadratic in Form

Original Equation	Substitution	$au^2 + bu + c = 0$ form
$x^4 - 8x^2 + 15 = 0$	$u = x^2$	$u^2 - 8u + 15 = 0$
$x^6 + x^3 - 12 = 0$	$u = x^3$	$u^2 + u - 12 = 0$
$x^{1/2} - 9x^{1/4} + 20 = 0$	$u = x^{1/4}$	$u^2 - 9u + 20 = 0$
$2x^{2/3} + 7x^{1/3} - 4 = 0$	$u = x^{1/3}$	$2u^2 + 7u - 4 = 0$
$15x^{-2} + 7x^{-1} - 2 = 0$	$u = x^{-1}$	$15u^2 + 7u - 2 = 0$

EXAMPLE 7 **Solve an Equation That Is Quadratic in Form**

Solve the equation $3x^{2/3} - 5x^{1/3} - 2 = 0$.

Solution Substituting u for $x^{1/3}$ gives us

$$3u^2 - 5u - 2 = 0$$

$$(3u + 1)(u - 2) = 0$$

$$3u + 1 = 0 \quad\text{or}\quad u - 2 = 0$$

$$u = -\frac{1}{3} \quad\text{or}\quad u = 2$$

Replacing u with $x^{1/3}$ yields

$$x^{1/3} = -\frac{1}{3} \quad\text{or}\quad x^{1/3} = 2$$

Cubing each side of each equation produces the following proposed solutions:

$$x = -\frac{1}{27} \quad\text{or}\quad x = 8$$

A check will verify that both proposed solutions are actual solutions. The solution set is $\{-\frac{1}{27}, 8\}$.

■ *Try Exercise* **52,** *page 112.*

It is possible to solve equations that are quadratic in form without making a formal substitution. For example, to solve $x^4 + 5x^5 - 36 = 0$, factor the equation and apply the zero product property as shown:

$$x^4 + 5x^2 - 36 = 0$$

$$(x^2 + 9)(x^2 - 4) = 0$$

$$x^2 + 9 = 0 \quad\text{or}\quad x^2 - 4 = 0$$

$$x^2 = -9 \quad\text{or}\quad x^2 = 4$$

$$x = \pm\sqrt{-9} \quad\text{or}\quad x = \pm\sqrt{4}$$

$$x = \pm 3i \quad\text{or}\quad x = \pm 2$$

A check will show that $-3i$, $3i$, -2, and 2 are all solutions of the original equation.

EXERCISE SET 2.4

In Exercises 1 to 12, factor to solve each equation.

1. $x^3 - 25x = 0$

2. $x^3 - x = 0$

3. $x^3 - 2x^2 - x + 2 = 0$

4. $x^3 - 4x^2 - 2x + 8 = 0$

5. $2x^5 - 18x^3 = 0$

6. $x^4 - 36x^2 = 0$

7. $x^4 - 3x^3 - 40x^2 = 0$

8. $x^4 + 3x^3 - 8x - 24 = 0$

9. $x^4 - 16x^2 = 0$

10. $x^4 - 16 = 0$

11. $x^3 - 8 = 0$

12. $x^3 + 8 = 0$

In Exercises 13 to 30, use the power principle to solve each radical equation. Check all proposed solutions.

13. $\sqrt{x - 4} - 6 = 0$

14. $\sqrt{10 - x} = 4$

15. $x = 3 + \sqrt{3 - x}$

16. $x = \sqrt{5 - x} + 5$

17. $\sqrt{3x - 5} - \sqrt{x + 2} = 1$

18. $\sqrt{6 - x} + \sqrt{5x + 6} = 6$

19. $\sqrt{2x + 11} - \sqrt{2x - 5} = 2$

20. $\sqrt{x + 7} - 2 = \sqrt{x - 9}$

21. $\sqrt{x + 7} + \sqrt{x - 5} = 6$

22. $x = \sqrt{12x - 35}$ **23.** $2x = \sqrt{4x + 15}$

24. $\sqrt[3]{7x - 3} = \sqrt[3]{2x + 7}$

25. $\sqrt[3]{2x^2 + 5x - 3} = \sqrt[3]{x^2 + 3}$

26. $\sqrt[4]{x^2 + 20} = \sqrt[4]{9x}$

27. $\sqrt{3\sqrt{5x + 16}} = \sqrt{5x - 2}$

28. $\sqrt{4\sqrt{2x - 5}} = \sqrt{x + 5}$

29. $\sqrt{3x + 1} + \sqrt{2x - 1} = \sqrt{10x - 1}$

30. $\sqrt{x - 3} + \sqrt{x + 3} = \sqrt{9 - x}$

In Exercises 31 to 40, solve each equation that involves fractional exponents. Check all proposed solutions.

31. $(3x + 5)^{1/3} = (-2x + 15)^{1/3}$

32. $(4z + 7)^{1/3} = 2$

33. $(x + 4)^{2/3} = 9$ **34.** $(x - 5)^{3/2} = 125$

35. $(4x)^{2/3} = (30x + 4)^{1/3}$ **36.** $z^{2/3} = (3z - 2)^{1/3}$

37. $4x^{3/4} = x^{1/2}$ **38.** $x^{3/5} = 2x^{1/5}$

39. $(3x - 5)^{2/3} + 6(3x - 5)^{1/3} = -8$

40. $2(x + 1)^{1/2} - 11(x + 1)^{1/4} + 12 = 0$

In Exercises 41 to 60, find all the real solutions of each equation by first rewriting each equation as a quadratic equation.

41. $x^4 - 9x^2 + 14 = 0$ **42.** $x^4 - 10x^2 + 9 = 0$

43. $2x^4 - 11x^2 + 12 = 0$ **44.** $6x^4 - 7x^2 + 2 = 0$

45. $x^6 + x^3 - 6 = 0$ **46.** $6x^6 + x^3 - 15 = 0$

47. $21x^6 + 22x^3 = 8$ **48.** $-3x^6 + 377x^3 - 250 = 0$

49. $x^{1/2} - 3x^{1/4} + 2 = 0$ **50.** $2x^{1/2} - 5x^{1/4} - 3 = 0$

51. $3x^{2/3} - 11x^{1/3} - 4 = 0$ **52.** $6x^{2/3} - 7x^{1/3} - 20 = 0$

53. $9x^4 = 30x^2 - 25$ **54.** $4x^4 - 28x^2 = -49$

55. $x^{2/5} - 1 = 0$ **56.** $2x^{2/5} - x^{1/5} = 6$

57. $\dfrac{1}{x^2} + \dfrac{3}{x} - 10 = 0$

58. $10\left(\dfrac{x - 2}{x}\right)^2 + 9\left(\dfrac{x - 2}{x}\right) - 9 = 0$

59. $9x - 52\sqrt{x} + 64 = 0$ **60.** $8x - 38\sqrt{x} + 9 = 0$

Calculator Exercises

In Exercises 61 to 64, use a calculator to solve each equation. Round each solution to the nearest hundredth.

61. $x^4 - 3x^2 + 1 = 0$ **62.** $x - 4\sqrt{x} + 1 = 0$

63. $x^2 - \sqrt{9x^2 - 1} = 0$ **64.** $2x^2 = \sqrt{10x^2 - 3}$

Supplemental Exercises

In Exercises 65 to 70, solve for x in terms of the other variables.

65. $x^2 + y^2 = 9$ **66.** $\dfrac{x^2}{a^2} + \dfrac{y^2}{b^2} = 1$

67. $\sqrt{x} - \sqrt{y} = \sqrt{z}$ **68.** $x - y = \sqrt{x^2 + y^2 + 5}$

69. $x + y = \sqrt{x^2 - y^2 + 7}$ **70.** $x + \sqrt{x} = -y$

71. Solve $(\sqrt{x} - 2)^2 - 5\sqrt{x} + 14 = 0$ for x. (*Hint:* use the substitution $u = \sqrt{x} - 2$, and then rewrite so that the equation is quadratic in terms of the variable u.)

72. Solve $(\sqrt[3]{x} + 3)^2 - 8\sqrt[3]{x} = 12$ for x. (*Hint:* use the substitution $u = \sqrt[3]{x} + 3$, and then rewrite so that the equation is quadratic in terms of the variable u.)

73. A conical funnel has a height h of 4 inches and a lateral surface area L of 15π square inches. Find the radius r of the cone. (*Hint:* Use the formula $L = \pi r \sqrt{r^2 + h^2}$.)

74. As flour is poured onto a table, it forms a right circular cone whose height is one-third the diameter of the base. What is the diameter of the base when the cone has a volume of 192 cubic inches?

75. A silver sphere has a diameter of 8 millimeters and a second silver sphere has a diameter of 12 millimeters. The spheres are melted down and recast to form a single cube. How long is the length s of each edge of the cube? Round your answer to the nearest tenth of a millimeter.

76. The period of a pendulum T is the time it takes a pendulum to complete one swing from left to right and back. For a pendulum near the surface of the earth,

$$T = 2\pi \sqrt{\dfrac{L}{32}},$$

where T is measured in seconds and L is the length of the pendulum in feet. Find the length of a pendulum that has a period of 4 seconds. Round to the nearest tenth of a foot.

77. On a ship, the distance d that you can see to the horizon is given by $d = 1.5\sqrt{h}$, where h is the height of your eye measured in feet above sea level and d is measured in miles. How high is the eye level of a navigator who can see 14 miles to the horizon? Round to the nearest foot.

78. The radius r of a circle inscribed in a triangle with sides of length a, b, and c is given by

$$r = \sqrt{\dfrac{(s - a)(s - b)(s - c)}{s}}$$

where $s = \frac{1}{2}(a + b + c)$. **a.** Find the length of the radius of a circle inscribed in a triangle with sides 5 inches, 6 inches, and 7 inches. **b.** The radius of a circle inscribed in an equilateral triangle measures 2 inches. How long is the length of each side of the equilateral triangle?

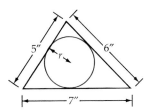

79. The radius r of a circle that is circumscribed about a triangle with sides a, b, and c is given by

$$r = \frac{abc}{4\sqrt{s(s-a)(s-b)(s-c)}}$$

where $s = \frac{1}{2}(a + b + c)$. **a.** Find the radius of a circle that is circumscribed about a triangle with sides of length 7 inches, 10 inches, and 15 inches. **b.** A circle with radius 5 inches is circumscribed about an equilateral triangle. What is the length of each side of the equilateral triangle?

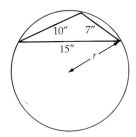

The depth s from the opening of a well to the water can be determined by measuring the total time between the instant you drop a stone and the time you hear it hit the water. The time (in seconds) it takes the stone to hit the water is given by $\sqrt{s}/4$, where s is measured in feet. The time (also in seconds) required for the sound of the impact to travel up to your ears is given by $s/1100$. Thus the total time T (in seconds) between the instant you drop a stone and the moment you hear its impact is

$$T = \frac{\sqrt{s}}{4} + \frac{s}{1100}.$$

80. One of the world's deepest water wells is 7320 feet deep. Find the time between the instant a stone is dropped and the time you hear it hit the water if the surface of the water is 7100 feet below the opening of the well. Round your answer to the nearest tenth of a second.

81. Solve $T = \dfrac{\sqrt{s}}{4} + \dfrac{s}{1100}$ for s.

82. Use the result of Exercise 81 to determine how deep a well is if the time between the instant you drop a stone and the moment you hear its impact is 3 seconds. Round your answer to the nearest foot.

2.5

Inequalities

In Section 1.2 we used the concept of an inequality to describe the order of real numbers on the real number line, and we also used inequalities to represent subsets of real numbers. In this section we consider inequalities that involve a variable. In particular, we consider how to determine which real values of the variable make the inequality a true statement.

The set of all solutions of an inequality is called the **solution set of the inequality.** For example, the solution set of $x + 1 > 4$ is the set of all real numbers greater than 3. **Equivalent inequalities** have the same solution set. The process of solving an inequality can be accomplished by producing *simpler* but equivalent inequalities until the solutions are found. To produce these simpler but equivalent inequalities, we apply the following properties.

Properties of Inequalities

For real numbers a, b, and c,

1. $a < b$ and $a + c < b + c$ are equivalent inequalities. (*Adding the same number to each side of an inequality preserves the order of the inequality.*)

2. If $c > 0$, then $a < b$ and $ac < bc$ are equivalent inequalities. (*Multiplying each side of an inequality by the same positive number preserves the order of the inequality.*)

3. If $c < 0$, then $a < b$ and $ac > bc$ are equivalent inequalities. (*Multiplying each side of an inequality by the same negative number reverses the order of the inequality.*)

Similar properties are also valid if the $<$ symbol in the three properties of inequalities is replaced with the $\leq$, $>$, or $\geq$ symbol.

Notice the difference between Properties 2 and 3. Property 2 states that an equivalent inequality will be produced by multiplying each side of an inequality by the same *positive* number provided the inequality symbol is not changed. For example, if each side of the inequality $3 < 4$ is multiplied by 2, then the following inequalities are all equivalent:

$$3 < 4$$

$$2 \cdot 3 < 2 \cdot 4$$

$$6 < 8$$

However, Property 3 states that an equivalent inequality will be produced by multiplying each side of an inequality by the same *negative* number provided the inequality symbol is reversed. For example, if each side of $3 < 4$ is multiplied by -2, then the following inequalities are all equivalent:

$$3 < 4$$

$$(-2) \cdot 3 > (-2) \cdot 4$$

$$-6 > -8$$

Example 1 illustrates the process of solving an inequality by producing a sequence of equivalent inequalities.

EXAMPLE 1   **Solve an Inequality**

Solve the inequality $2(x + 3) < 4x + 10$.

Solution $2(x + 3) < 4x + 10$

$2x + 6 < 4x + 10$ The distributive property

$-2x < 4$ Add $-4x$ and -6 to each side of the inequality.

$x > -2$ Multiply each side by the reciprocal of -2 and reverse the inequality symbol.

Thus the original inequality is true for all real numbers greater than -2. The solution set is $\{x \mid x > -2\}$. Using interval notation, the solution set is written as $(-2, \infty)$. (See Section 1.2 to review interval notation.)

■ *Try Exercise* **8**, *page 121.*

EXAMPLE 2   Solve an Application That Involves an Inequality

You can rent a car from Company A for $26 per day plus $0.09 a mile. Company B charges $12 per day plus $0.14 a mile. Find the number of miles for which it is cheaper to rent from Company A if you rent a car for 1 day.

Solution   Let m equal the number of miles the car is to be driven. Then the cost of renting the car will be

$$\$26 + \$0.09m \quad \text{from Company A}$$

$$\$12 + \$0.14m \quad \text{from Company B}$$

If renting from Company A is to be cheaper than renting from Company B, then we must have

$$26 + 0.09m < 12 + 0.14m$$

Solving for m produces

$$14 < 0.05m$$

$$\frac{14}{0.05} < m$$

$$280 < m$$

Thus renting from Company A is cheaper if you drive more than 280 miles.

■ *Try Exercise* **12**, *page 121.*

To solve $-3 < 2x + 1 < 11$ for x, you apply the properties of inequalities to all three parts of the inequality.

EXAMPLE 3   Solve an Inequality

Solve the inequality $-3 < 2x + 1 < 11$.

Solution $-3 < 2x + 1 < 11$

$-4 < \quad 2x \quad < 10$ Add -1 to each of the three parts of the inequality.

$-2 < \quad x \quad < 5$ Multiply each of the three parts of the inequality by $1/2$.

The last inequality implies that $x > -2$ and $x < 5$. Using interval notation, the solution set is written as $(-2, 5)$.

■ *Try Exercise* **16**, *page 121.*

EXAMPLE 4   **Solve an Application That Involves Inequalities**

A photographic developer needs to be kept at a temperature between 15°C and 25°C. What would be the Fahrenheit temperature range?

Solution   The formula that relates the Celsius temperature (C) to the Fahrenheit temperature (F) is

$$C = \frac{5}{9}(F - 32).$$

We are given that

$$15 < C < 25.$$

Substituting $\frac{5}{9}(F - 32)$ for C yields

$$15 < \frac{5}{9}(F - 32) < 25$$

$27 <$ $F - 32$ < 45 Multiply each of the three parts of the inequality by 9/5.

$59 <$ F < 77 Add 32 to each of the three parts of the inequality.

Thus, the developer needs to be kept between 59°F and 77°F.

■ *Try Exercise 22, page 121.*

The Critical Value Method for Solving Inequalities

Any value of x that causes a polynomial in x to equal zero is called a **zero** of the polynomial. For example, -4 and 1 are both zeros of the polynomial $x^2 + 3x - 4$.

A Sign Property of Polynomials

> Nonzero polynomials in x have the property that for any x between two consecutive real zeros, either all values of the polynomial are positive or all values of the polynomial are negative.

In our work with inequalities that involve polynomials, the real zeros of the polynomial are also referred to as **critical values of the inequality** because on a number line they separate the real numbers that make an inequality involving a polynomial true from those that make it false. In Example 5 we use critical values and the Sign Property of Polynomials to solve an inequality.

EXAMPLE 5   **Solve a Polynomial Inequality**

Solve the inequality $x^2 + 3x - 4 < 0$.

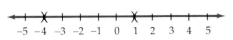

Figure 2.6

Solution   Factoring the polynomial $x^2 + 3x - 4$ produces the equivalent inequality

$$(x + 4)(x - 1) < 0.$$

Thus the zeros of the polynomial $x^2 + 3x - 4$ are -4 and 1. They are the critical values of the inequality $x^2 + 3x - 4 < 0$. They separate the real number line into the three intervals shown in Figure 2.6.

To determine the intervals on which $x^2 + 3x - 4 < 0$, you can pick a number called a **test value** from each of the three intervals and then determine whether $x^2 + 3x - 4 < 0$ for each of these test values. For example, in the interval $(-\infty, -4)$, pick a test value of, say, -5. Then

$$x^2 + 3x - 4 = (-5)^2 + 3(-5) - 4 = 6.$$

Since 6 is not less than 0, by the Sign Property of Polynomials, no number in the interval $(-\infty, -4)$ makes $x^2 + 3x - 4 < 0$.

Now pick a test value from the interval $(-4, 1)$, say, 0. When $x = 0$,

$$x^2 + 3x - 4 = 0^2 + 3(0) - 4 = -4.$$

Since -4 is less than 0, by the Sign Property of Polynomials, all numbers in the interval $(-4, 1)$ make $x^2 + 3x - 4 < 0$.

If we pick a test value of 2 from the interval $(1, \infty)$, then

$$x^2 + 3x - 4 = (2)^2 + 3(2) - 4 = 6.$$

Since 6 is not less than 0, by the Sign Property of Polynomials, no number in the interval $(1, \infty)$ makes $x^2 + 3x - 4 < 0$.

The following table is a summary of our work:

Interval	$(-\infty, -4)$	$(-4, 1)$	$(1, \infty)$
Test value x	-5	0	2
$x^2 + 3x - 4 \overset{?}{<} 0$	$(-5)^2 + 3(-5) - 4 < 0$ $6 < 0$ False	$(0)^2 + 3(0) - 4 < 0$ $-4 < 0$ True	$(2)^2 + 3(2) - 4 < 0$ $6 < 0$ False

Figure 2.7

The solution set of $x^2 + 3x - 4 < 0$ is graphed in Figure 2.7. Note that in this case the critical values -4 and 1 are not included in the solution set because they do not make $x^2 + 3x - 4$ less than 0.

■ *Try Exercise* **30**, *page 121.*

To avoid the arithmetic in Example 5, we often use a *sign diagram.* For example, note that the factor $(x + 4)$ is negative for all $x < -4$ and positive for all $x > -4$. The factor $(x - 1)$ is negative for all $x < 1$ and positive for all $x > 1$. These results are shown in Figure 2.8.

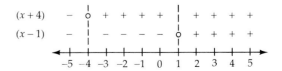

Figure 2.8
Sign diagram for $(x + 4)(x - 1)$.

To determine on which intervals the product $(x + 4)(x - 1)$ is negative, we examine the sign diagram to see where the factors have opposite signs. Since this occurs only on the interval $(-4, 1)$, where $(x + 4)$ is positive and $(x - 1)$ is negative, the original inequality is only true on the interval $(-4, 1)$.

Following is a summary of the steps used to solve polynomial inequalities by the critical value method.

Solving a Polynomial Inequality by the Critical Value Method

1. Write the inequality so that one side of the inequality is a non-zero polynomial and the other side is 0.

2. Find the real zeros of the polynomial.[4] They are the critical values of the original inequality.

3. Use a sign diagram or test values to determine which of the intervals formed by the critical values are to be included in the solution set.

EXAMPLE 6   Use the Critical Value Method to Solve an Application

A manufacturer of tennis racquets finds that the yearly revenue R from a particular type of racquet is given by $R = 160x - x^2$, where x is the price in dollars of each racquet. Find the interval in terms of x for which the yearly revenue is greater than $6000. That is, solve

$$160x - x^2 > 6000 .$$

Solution   Write the inequality so that 0 appears on the right side of the inequality.

$$160x - x^2 - 6000 > 0$$

Arrange the terms in descending powers and multiply each side of the inequality by -1 to produce

$$x^2 - 160x + 6000 < 0 .$$

Factor the left side.

$$(x - 60)(x - 100) < 0$$

Use the zero product property to find the zeros.

$(x - 60)(x - 100) = 0$ Replace the inequality with an equal sign.

$x = 60$ or $x = 100$ Set each factor equal to 0 and solve for x.

The zeros are 60 and 100. They separate the real number line into the intervals $(-\infty, 60)$, $(60, 100)$, and $(100, \infty)$. The sign diagram in Figure 2.9 shows that the inequality $(x - 60)(x - 100) < 0$ is true on the interval $(60, 100)$, and it is false on the other intervals.

[4]In Chapter 4, additional methods are developed to find the zeros of a polynomial; however, for the present we will find the zeros by factoring or by using the quadratic formula.

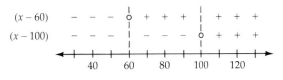

Figure 2.9

Thus the revenue is greater than $6000 per year when the price of each racquet is between $60 and $100.

■ *Try Exercise* **54,** *page 121.*

Remark In Example 6 it is easy to see that *x* cannot equal zero because no revenue will be produced if the racquets are given away. Also, if the price of the racquets increases above $100, the revenue becomes less than $6000. Why does this seem reasonable?[5]

Rational Inequalities

A rational expression is the quotient of two polynomials. **Rational inequalities** involve rational expressions, and they can be solved by an extension of the critical value method. The **critical values of a rational expression** are the numbers that cause the numerator of the rational expression to equal zero or the denominator of the rational expression to equal zero.

Rational expressions also have the property that they remain either positive for all values of the variable between consecutive critical values or negative for all values of the variable between consecutive critical values.

EXAMPLE 7   Solve by Using the Critical Values of a Rational Inequality

Solve the inequality $\dfrac{(x - 2)(x + 3)}{x - 4} \geq 0$.

Solution The critical values include the zeros of the numerator, namely, 2 and −3, and the zero of the denominator, which is 4. These three critical numbers separate the real number line into four intervals. The sign diagram in Figure 2.10 shows the sign of each of the factors $(x - 2)$, $(x + 3)$, and $(x - 4)$ on each of the four intervals.

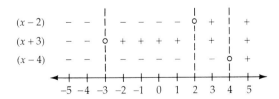

Figure 2.10

[5]As the price increases above $100, the number of people that decide to purchase the racquet decreases. The manufacturer makes more money on each sale, but since there are fewer sales, the revenue decreases.

Figure 2.11

The sign diagram shows that the rational expression is positive on the two intervals $(-3, 2)$ and $(4, \infty)$. The critical values -3 and 2 are solutions because they satisfy the original inequality. However, the critical value 4 is not a solution because the denominator $x - 4$ is zero when $x = 4$. Therefore the solution set is $[-3, 2] \cup (4, \infty)$. The graph of the solution set is shown in Figure 2.11.

■ *Try Exercise* **38**, *page 121.*

EXAMPLE 8   Solve a Rational Inequality

Solve the rational inequality $\dfrac{3x + 4}{x + 1} \leq 2$.

Solution   Write the inequality so that 0 appears on the right side of the inequality.

$$\frac{3x + 4}{x + 1} \leq 2$$

$$\frac{3x + 4}{x + 1} - 2 \leq 0$$

Write the left side as a rational expression.

$$\frac{3x + 4}{x + 1} - \frac{2(x + 1)}{x + 1} \leq 0 \quad \text{The LCD is } x + 1.$$

$$\frac{3x + 4 - 2x - 2}{x + 1} \leq 0 \quad \text{Simplify.}$$

$$\frac{x + 2}{x + 1} \leq 0$$

The critical values of the above inequality are -2 and -1 because the numerator $x + 2$ is equal to zero when $x = -2$, and the denominator $x + 1$ is equal to zero when $x = -1$. The critical values -2 and -1 separate the real number line into the three intervals $(-\infty, -2)$, $(-2, -1)$, and $(-1, \infty)$.

All values of x on the interval $(-2, -1)$, make $(x + 2)/(x + 1)$ negative, as desired. On the other intervals the quotient $(x + 2)/(x + 1)$ is positive. See the sign diagram in Figure 2.12.

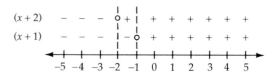

Figure 2.12

Figure 2.13

The graph of the solution set is shown in Figure 2.13. Note that -2 is included in the solution set because $(x + 2)/(x + 1) = 0$ when $x = -2$.

However, -1 is not included in the solution set because the denominator $(x + 1)$ is zero when $x = -1$.

■ *Try Exercise **46**, page 121.*

EXERCISE SET 2.5

In Exercises 1 to 10, use the properties of inequalities to solve each inequality.

1. $2x + 3 < 11$ **2.** $3x - 5 > 16$

3. $x + 4 > 3x + 16$ **4.** $5x + 6 < 2x + 1$

5. $-6x + 1 \geq 19$ **6.** $-5x + 2 \leq 37$

7. $-3(x + 2) \leq 5x + 7$ **8.** $-4(x - 5) \geq 2x + 15$

9. $-4(3x - 5) > 2(x - 4)$ **10.** $3(x + 7) \leq 5(2x - 8)$

11. A bank offers two checking account plans. Under plan A, you pay $5.00 a month plus $0.01 a check. Under plan B, you pay $1.00 a month plus $0.08 a check. Under what conditions is it less expensive to use plan A?

12. You can rent a car for the day from Company A for $19.00 plus $0.12 a mile. Company B charges $12.00 plus $0.21 a mile. Find the number of miles m (to the nearest mile) for which it is cheaper to rent from Company A.

13. A salesclerk has a choice between two payment plans. Plan A pays $100.00 a week plus $8.00 a sale. Plan B pays $250.00 a week plus $3.50 a sale. How many sales per week must be made for plan A to yield the greater paycheck?

14. A video store offers two rental plans. Plan A requires a $15.00 yearly membership fee and charges $1.49 per video per day. Plan B does not have a membership fee but charges $1.99 per video per day. How many videos can be rented per year if Plan B is to be the least expensive of the plans?

In Exercises 15 to 20, use the properties of inequalities to solve each inequality.

15. $-2 < 4x + 1 \leq 17$ **16.** $-16 < 2x + 5 < 9$

17. $10 \geq 3x - 1 \geq 0$ **18.** $0 \leq 2x + 6 \leq 54$

19. $20 > 8x - 2 \geq -5$ **20.** $4 \leq 10x + 1 \leq 51$

21. The average daily minimum-maximum temperatures for the city of Palm Springs during the month of September is 68 to 104 degrees Fahrenheit. What is the corresponding temperature range measured on the Celsius temperature scale? (*Hint:* Let F be the average daily temperature. Then $68 \leq F \leq 104$. Now substitute $\frac{9}{5}C + 32$ for F and solve the resulting inequality for C.)

22. The average daily minimum-maximum temperatures for the city of Palm Springs during the month of January is 41 to 68 degrees Fahrenheit. What is the corresponding temperature range measured on the Celsius temperature scale?

23. The sum of three consecutive even integers is between 36 and 54. Find all possible sets of integers that satisfy these conditions.

24. The sum of three consecutive odd integers is between 63 and 81. Find all possible sets of integers that satisfy these conditions.

In Exercises 25 to 52, use the critical value method to solve each inequality. Use interval notation to write each solution set.

25. $x^2 + 7x > 0$ **26.** $x^2 - 5x \leq 0$

27. $x^2 - 16 \leq 0$ **28.** $x^2 - 49 > 0$

29. $x^2 + 7x + 10 < 0$ **30.** $x^2 + 5x + 6 < 0$

31. $x^2 - 3x \geq 28$ **32.** $x^2 < -x + 30$

33. $6x^2 - 4 \leq 5x$ **34.** $12x^2 + 8x \geq 15$

35. $8x^2 \geq 2x + 15$ **36.** $12x^2 - 16x < -5$

37. $\dfrac{x + 4}{x - 1} < 0$ **38.** $\dfrac{x - 2}{x + 3} > 0$

39. $\dfrac{x - 5}{x + 8} \geq 0$ **40.** $\dfrac{x - 4}{x + 6} \leq 0$

41. $\dfrac{x}{2x + 7} \geq 0$ **42.** $\dfrac{x}{3x - 5} \leq 0$

43. $\dfrac{(x + 1)(x - 4)}{x - 2} < 0$ **44.** $\dfrac{x(x - 4)}{x + 5} > 0$

45. $\dfrac{x + 2}{x - 5} \leq 2$ **46.** $\dfrac{3x + 1}{x - 2} \geq 4$

47. $\dfrac{6x^2 - 11x - 10}{x} > 0$ **48.** $\dfrac{3x^2 - 2x - 8}{x - 1} \geq 0$

49. $\dfrac{x^2 - 6x + 9}{x - 5} \leq 0$ **50.** $\dfrac{x^2 + 10x + 25}{x + 1} \geq 0$

51. $\dfrac{x^2 - 3x - 4}{x + 1} \geq 0$ **52.** $\dfrac{x^2 + 6x + 9}{x + 3} \leq 0$

53. The monthly revenue R for a product is given by $R = 420x - 2x^2$, where x is the price in dollars of each unit produced. Find the interval in terms of x for which the monthly revenue is greater than zero.

54. A shoe manufacturer finds that the monthly revenue R from a particular syle of running shoe is given by $R = 312x - 3x^2$, where x is the price in dollars of each pair of shoes sold. Find the interval in terms of x for which the monthly revenue is greater than or equal to $5925.

In Exercises 55 to 64, determine the set of all real numbers x such that y will be a real number. *Hint:* $\sqrt{a}$ is a real number if and only if $a \geq 0$.

55. $y = \sqrt{x + 9}$

56. $y = \sqrt{x - 3}$

57. $y = \sqrt{9 - x^2}$

58. $y = \sqrt{25 - x^2}$

59. $y = \sqrt{x^2 - 16}$

60. $y = \sqrt{x^2 - 81}$

61. $y = \sqrt{x^2 - 2x - 15}$

62. $y = \sqrt{x^2 + 4x - 12}$

63. $y = \sqrt{x^2 + 1}$

64. $y = \sqrt{x^2 - 1}$

Supplemental Exercises

In Exercises 65 to 68, use the critical value method to solve each inequality. Use interval notation to write each solution set.

65. $\dfrac{(x - 3)^2}{(x - 6)^2} > 0$

66. $\dfrac{(x - 1)^2}{(x - 4)^4} \geq 0$

67. $\dfrac{(x - 4)^2}{(x + 3)^3} \geq 0$

68. $\dfrac{(2x - 7)}{(x - 1)^2 (x + 2)^2} \geq 0$

In Exercises 69 to 74, determine the set of all real numbers x such that y will be a real number.

69. $y = \sqrt[4]{x^3 - 3x}$

70. $y = \sqrt[4]{x^4 - 4x^3 + 4x^2}$

71. $y = \sqrt[6]{5 + x^2}$

72. $y = \sqrt[6]{(x + 3)^6}$

73. $y = \sqrt{x(x + 2)(x - 5)}$

74. $y = \sqrt{\dfrac{x - 3}{(x + 2)(x - 4)}}$

In Exercises 75 to 78, find the values of k such that the given equation will have at least one real solution.

75. $x^2 + kx + 6 = 0$

76. $x^2 + kx + 11 = 0$

77. $2x^2 + kx + 7 = 0$

78. $-3x^2 + kx - 4 = 0$

79. The equation $s = -16t^2 + v_0 t + s_0$ gives the height s in feet above ground level, at the time t seconds of an object thrown directly upward from a height s_0 feet above the ground and with an initial velocity of v_0 feet per second. If a ball is thrown directly upward from ground level with an initial velocity of 64 feet per second, find the time interval for which the ball has a height of more than 48 feet.

80. A ball is thrown directly upward from a height of 32 feet above the ground with an initial velocity of 80 feet per second. Find the time interval for which the ball will be more than 96 feet above the ground. (*Hint:* See Exercise 79.)

81. In any triangle, the sum of the lengths of the two shorter sides must be greater than the length of the longest side. Find all possible values of x if a triangle has sides of length
   a. x, $x + 5$, and $x + 9$ **b.** x, $x^2 + x$, and $2x^2 + x$
   c. $\dfrac{1}{x + 2}$, $\dfrac{1}{x + 1}$, and $\dfrac{1}{x}$

82. A hiker decides to walk through a train tunnel. Two-thirds of the way through the tunnel, the hiker notices an approaching train. If the train is two tunnel lengths away from the hiker, which way should the hiker run to maximize the chance of escaping the tunnel before being overrun by the train?

83. Find the solution set of $a^2 > a$, where $a > 0$.

84. If $a > b > 0$, show that $\dfrac{1}{a} < \dfrac{1}{b}$.

2.6

Absolute Value Equations and Inequalities

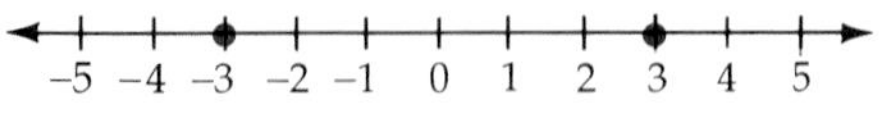

Recall that the absolute value of a real number x is the distance between the number x and 0 on the real number line. For example, the solution set of the absolute value equation $|x| = 3$ is the set of all real numbers that are 3 units from 0. Figure 2.14 illustrates that there are only two numbers that are 3 units from 0, namely, 3 and -3. Therefore the solution set of $|x| = 3$ is $\{-3, 3\}$.

Recall from Section 1.2 that $|a - b|$ is the distance on a real number line between the graph of a and the graph of b. Thus $|x - 2|$ is the dis-

Figure 2.14
$|x| = 3.$

tance between the real number x and 2. The equation $|x - 2| = 5$ is satisfied by real numbers x that are 5 units from 2. Therefore $|x - 2| = 5$ has $\{-3, 7\}$ as its solution set.

The following property is used to solve absolute value equations.

A Property of Absolute Value Equations

> For any variable expression E and any nonnegative real number k,
> $$|E| = k \quad \text{if and only if} \quad E = k \ \text{ or } \ E = -k.$$

EXAMPLE 1   **Solve an Absolute Value Equation**

Solve the equation $|2x - 5| = 21$.

Solution $|2x - 5| = 21$ implies $2x - 5 = 21$ or $2x - 5 = -21$. Solving each of these equations produces

$$2x - 5 = 21 \qquad \text{or} \qquad 2x - 5 = -21$$
$$2x = 26 \qquad\qquad\qquad 2x = -16$$
$$x = 13 \qquad\qquad\qquad x = -8$$

Therefore the solution set of $|2x - 5| = 21$ is $\{-8, 13\}$.

■ *Try Exercise* **10,** *page 126.*

Remark   Some absolute value equations have an empty solution set. For example, $|x + 2| = -5$ is false for all values of x. Notice that the left side of the equation is an absolute value. Because the absolute value of any real number is nonnegative, the equation is never true.

Absolute Value Inequalities

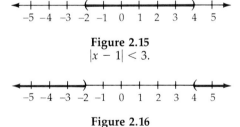

Figure 2.15
$|x - 1| < 3$.

Figure 2.16
$|x - 1| > 3$.

The solution set of the absolute value inequality $|x - 1| < 3$ is the set of all real numbers whose distance from 1 is *less* than 3. Therefore the solution set consists of all numbers greater than -2 and less than 4. See Figure 2.15. Using interval notation, the solution set is $(-2, 4)$.

The solution set of the absolute value inequality $|x - 1| > 3$ is the set of all real numbers whose distance from 1 is *greater* than 3. Therefore the solution set consists of all real numbers less than -2 *or* greater than 4. See Figure 2.16. Using interval notation, the solution set is $(-\infty, -2) \cup (4, \infty)$.

The following properties are used to solve absolute value inequalities.

Properties of Absolute Value Inequalities

> For any variable expression E and any nonnegative number k,
> $$|E| \le k \quad \text{if and only if} \quad -k \le E \le k.$$
> $$|E| \ge k \quad \text{if and only if} \quad E \le -k \ \text{ or } \ E \ge k.$$

EXAMPLE 2   **Solve an Absolute Value Inequality**

Solve the inequality $|2 - 3x| < 7$.

Solution $|2 - 3x| < 7$ implies $-7 < 2 - 3x < 7$. Solving this inequality yields the following:

$$-7 < 2 - 3x < 7$$

$$-9 < \quad -3x \quad < 5 \qquad \text{Add } -2 \text{ to each of the three parts of the inequality.}$$

$$3 > \quad x \quad > -\frac{5}{3} \qquad \text{Multiply each part of the inequality by } -\frac{1}{3} \text{ and reverse the inequality symbols.}$$

Using interval notation, the solution set of $|2 - 3x| < 7$ is given by $(-5/3, 3)$.

■ *Try Exercise* **30**, *page 126.*

EXAMPLE 3   **Solve an Absolute Value Inequality**

Solve the inequality $|4x - 3| \geq 5$.

Solution $|4x - 3| \geq 5$ implies $4x - 3 \leq -5$ or $4x - 3 \geq 5$. Solving each of these inequalities produces

$$4x - 3 \leq -5 \qquad \text{or} \qquad 4x - 3 \geq 5$$

$$4x \leq -2 \qquad\qquad\qquad 4x \geq 8$$

$$x \leq -\frac{1}{2} \qquad\qquad\qquad x \geq 2$$

Therefore the solution set of $|4x - 3| \geq 5$ is $(-\infty, -1/2] \cup [2, \infty)$.

■ *Try Exercise* **34**, *page 126.*

Remark   Some absolute value inequalities have a solution set that consists of all real numbers. For example, $|x + 9| \geq 0$ is true for all values of x. Notice that the left side of the equation is an absolute value. Because the absolute value of any real number is nonnegative, the equation is always true. Recall from Section 1.2 that the interval notation for the set of all real numbers is $(-\infty, \infty)$.

The graph of the solutions set of $|x - a| < \delta$ is called the **delta neighborhood** of a. The delta symbol, δ, is used to represent a positive real number, and a represents a constant. Example 4 shows why the graph of the solution set of $|x - a| < \delta$ is called a delta neighborhood of a. The graph consists of all points on a number line that are within a distance delta of a.

EXAMPLE 4   **Solve an Absolute Value Inequality**

Solve the inequality $|x - a| < \delta$ for x. Assume $\delta > 0$.

Figure 2.17
$|x - a| < \delta$.

Solution $|x - a| < \delta$ means $-\delta < x - a < \delta$. Adding a to each of the three parts of the inequality produces $a - \delta < x < a + \delta$. Therefore the solution set of $|x - a| < \delta$ is the open interval $(a - \delta, a + \delta)$. The graph of $|x - a| < \delta$ is shown in Figure 2.17.

■ *Try Exercise* **46,** *page 126.*

We can also solve absolute value inequalities using the critical value method. To solve an absolute value inequality of the form $|P| > k$, where P is a polynomial such as $x^2 - 5$, we first find all values of x such that the left side of the inequality is *equal* to the right side.

EXAMPLE 5 **Solve an Absolute Value Inequality**

Solve the inequality $|x^2 - 5| > 4$.

Solution $|x^2 - 5| = 4$ implies $x^2 - 5 = 4$ or $x^2 - 5 = -4$. Solving each of these equations produces

$$x^2 - 5 = 4 \qquad \text{or} \qquad x^2 - 5 = -4$$
$$x^2 = 9 \qquad\qquad\qquad x^2 = 1$$
$$x = \pm 3 \qquad\qquad\qquad x = \pm 1$$

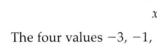

Figure 2.18

The four values -3, -1, 1, and 3 separate a real number line into the 5 intervals as shown in Figure 2.18. They are the critical values of the inequality because they separate the real numbers that make the inequality true from those that make it false.

We can use a test value from each of the intervals to determine on which intervals the original inequality is true. For example, if we choose -4 from the interval $(-\infty, -3)$, then the inequality $|x^2 - 5| > 4$ is true because

$$|(-4)^2 - 5| = |16 - 5| = |11| = 11$$

and 11 is greater than 4. Therefore the interval $(-\infty, -3)$ is part of the solution set. Continuing in a similar manner will produce the results shown in the following table:

Interval	$(-\infty, -3)$	$(-3, -1)$	$(-1, 1)$	$(1, 3)$	$(3, \infty)$
Test value x	-4	-2	0	2	4
$|x^2 - 5| \overset{?}{>} 4$	$|(-4)^2 - 5| > 4$	$|(-2)^2 - 5| > 4$	$|(0)^2 - 5| > 4$	$|(2)^2 - 5| > 4$	$|(4)^2 - 5| > 4$
	$11 > 4$	$1 > 4$	$5 > 4$	$1 > 4$	$11 > 4$
	True	False	True	False	True

Thus the solution set of $|x^2 - 5| > 4$ is $(-\infty, -3) \cup (-1, 1) \cup (3, \infty)$.

■ *Try Exercise* **48,** *page 126.*

EXERCISE SET 2.6

In Exercises 1 to 22, solve each absolute value equation for x.

1. $|x| = 4$

2. $|x| = 7$

3. $|x - 5| = 2$

4. $|x - 8| = 3$

5. $|x + 6| = 1$

6. $|x + 9| = 5$

7. $|x + 14| = 20$

8. $|x - 3| = 14$

9. $|2x - 5| = 11$

10. $|2x - 3| = 21$

11. $|2x + 6| = 10$

12. $|2x + 14| = 60$

13. $\left|\dfrac{x - 4}{2}\right| = 8$

14. $\left|\dfrac{x + 3}{4}\right| = 6$

15. $|2x + 5| = -8$

16. $|4x - 1| = -17$

17. $2|x + 3| + 4 = 34$

18. $3|x - 5| - 16 = 2$

19. $|2x - a| = b$ (assume $b > 0$)

20. $3|x - d| = c$ (assume $c > 0$)

21. $|x - a| = \delta$ (assume $\delta > 0$)

22. $|x + m| = m$ (assume $m > 0$)

In Exercises 23 to 46, use interval notation to express the solution set of each inequality.

23. $|x| < 4$

24. $|x| > 2$

25. $|x - 1| < 9$

26. $|x - 3| < 10$

27. $|x + 3| > 30$

28. $|x + 4| < 2$

29. $|2x - 1| > 4$

30. $|2x - 9| < 7$

31. $|x + 3| \geq 5$

32. $|x - 10| \geq 2$

33. $|3x - 10| \leq 14$

34. $|2x - 5| \geq 1$

35. $|4 - 5x| \geq 24$

36. $|3 - 2x| \leq 5$

37. $|x - 5| \geq 0$

38. $|x - 7| \geq 0$

39. $|x - 4| \leq 0$

40. $|2x + 7| \leq 0$

41. $|5x - 1| < -4$

42. $|2x - 1| < -9$

43. $|2x + 7| \geq -5$

44. $|3x + 11| \geq -20$

45. $|x - 3| < b$ (assume $b > 0$)

46. $|x - c| < d$ (assume $d > 0$)

In Exercises 47 to 54, use the critical value method to solve each inequality. Use interval notation to write the solution sets.

47. $|x^2 - 1| < 1$

48. $|x^2 - 2| > 1$

49. $|x^2 - 10| < 6$

50. $|x^2 + 4| \geq 10$

51. $|x^2 + 7x + 11| \geq 1$

52. $|x^2 - 5x + 6| \leq 1$

53. $|x^2 - 21.5| \geq 4.5$

54. $|x^2 - 6.5| \leq 2.5$

In Exercises 55 to 60, determine whether the statement is true or false. If it is false, explain why.

55. $|x + 2| = |x| + |2|$

56. $|x - 5| = |x| - |5|$

57. $|x - 7| \geq 0$

58. $|x| |5| = |5x|$

59. If $t < 0$, then $|t| = -t$.

60. The absolute value of any real number is a positive number.

Supplemental Exercises

In Exercises 61 to 68, find the values of x that make the equation true.

61. $|x + 4| = x + 4$

62. $|x - 1| = x - 1$

63. $|x + 7| = -(x + 7)$

64. $|x - 3| = -(x - 3)$

65. $|2x + 7| = 2x + 7$

66. $|3x - 11| = -3x + 11$

67. $|x - 2| + |x + 4| = 8$

68. $|x + 1| - |x + 3| = 4$

In Exercises 69 to 80, use interval notation to express the solution set of each inequality.

69. $1 < |x| < 5$

70. $2 < |x| < 3$

71. $3 \leq |x| < 7$

72. $0 < |x| \leq 3$

73. $0 < |x - a| < \delta$ (assume $\delta > 0$)

74. $0 < |x - 5| < 2$

75. $2 < |x - 6| < 4$

76. $1 \leq |x - 3| < 5$

77. $\left|1 - \dfrac{3x}{4}\right| \geq 6$

78. $\left|2 + \dfrac{3x}{5}\right| < 10$

79. $|x| > |x - 1|$

80. $|x - 2| \leq |x + 4|$

81. Write an absolute value inequality to represent all real numbers within **a.** 8 units of 3;   **b.** k units of j (assume $k > 0$).

82. Write an absolute value inequality to represent all real numbers that are more than **a.** 5 units away from 1; **b.** k units away from j (assume $k > 0$).

83. The length of the sides of a square have been measured accurately to within 0.01 feet. If this measured length is 4.25 feet,
    a. Write an absolute value inequality that describes the relationship between the actual length of each side of the square s and its measured length.
    b. Solve the absolute value inequality from **a** for s.

Chapter 2 Review

2.1 Linear Equations

A number is said to satisfy an equation if a substitution of the number for the variable results in an equation that is a true statement. To solve an equation means to find all values of the variable that satisfy the equation. These values that make the equation true are called solutions or roots of the equation. The set of all solutions of an equation is called the solution set of the equation. Equivalent equations have the same solution set.

A linear equation in the variable x is an equation that can be written in the form $ax + b = 0$, where a and b are real numbers, with $a \neq 0$. A literal equation is an equation that involves more than one variable.

2.2 Formulas and Applications

A formula is an equation or inequality that expresses known relationships between two or more variables. Application problems are solved by using the guidelines on page 85.

2.3 Quadratic Equations and Applications

A quadratic equation in the variable x is an equation that can be written in the form $ax^2 + bx + c = 0$, where $a \neq 0$.

If the quadratic polynomial in a quadratic equation is factorable over the set of integers, then the equation can be solved by factoring and using the zero product property (see page 94). Every quadratic equation can be solved by completing the square or the quadratic formula.

The Quadratic Formula
If $ax^2 + bx + c = 0$, $a \neq 0$, then

$$x = \frac{-b \pm \sqrt{b^2 - 4ac}}{2a}.$$

2.4 Other Types of Equations

The Power Principle
If P and Q are algebraic expressions and n is a positive integer, then every solution of $P = Q$ is a solution of $P^n = Q^n$.

An equation is said to be quadratic in form if it can be written in the form $au^2 + bu + c = 0$, $a \neq 0$, where u is an algebraic expression involving x.

2.5   Inequalities

The set of all solutions of an inequality is the **solution set of the inequality. Equivalent inequalities** have the same solution set. To solve an inequality, use the **Properties of Inequalities,** page 114, or the critical value method, page 116.

2.6   Absolute Value Equations and Inequalities

Absolute value equations and inequalities can be solved by applying the following properties:

For any variable expression E and any nonnegative real number k,

$$|E| = k \quad \text{if and only if} \quad E = k \quad \text{or} \quad E = -k$$

$$|E| \le k \quad \text{if and only if} \quad -k \le E \le k$$

$$|E| \ge k \quad \text{if and only if} \quad E \le -k \quad \text{or} \quad E \ge k$$

CHAPTER 2 CHALLENGE EXERCISES

In Exercises 1 to 10, answer true or false. If the statement is false, give an example.

1. If $x^2 = 9$, then $x = 3$.

2. The equations

$$x = \sqrt{12 - x} \quad \text{and} \quad x^2 = 12 - x$$

are equivalent equations.

3. Adding the same constant to each side of a given equation produces an equation that is equivalent to the given equation.

4. If $a > b$, then $-a < -b$.

5. If $a \ne 0$, $b \ne 0$, and $a > b$, then $\dfrac{1}{a} > \dfrac{1}{b}$.

6. The discriminant of $ax^2 + bx + c = 0$ is $\sqrt{b^2 - 4ac}$.

7. If $\sqrt{a} + \sqrt{b} = c$, then $a + b = c^2$.

8. The solution set of $|x - a| < b$ with $b > 0$ is given by the interval $(a - b, a + b)$.

9. The only quadratic equation that has roots of 4 and -4 is $x^2 - 16 = 0$.

10. Every quadratic equation $ax^2 + bx + c = 0$, with real coefficients such that $ac < 0$ has two distinct real roots.

CHAPTER 2 REVIEW EXERCISES

In Exercises 1 to 24, solve each equation.

1. $x - 2(5x - 3) = -3(-x + 4)$

2. $3x - 5(2x - 7) = -4(5 - 2x)$

3. $\dfrac{4x}{3} - \dfrac{4x - 1}{6} = \dfrac{1}{2}$

4. $\dfrac{3x}{4} - \dfrac{2x - 1}{8} = \dfrac{3}{2}$

5. $\dfrac{x}{x + 2} + \dfrac{1}{4} = 5$

6. $\dfrac{y - 1}{y + 1} - 1 = \dfrac{2}{y}$

7. $3x^3 - 5x^2 = 0$

8. $2x^3 - 8x = 0$

9. $6x^4 - 23x^2 + 20 = 0$

10. $3x + 16\sqrt{x} - 12 = 0$

11. $\sqrt{x^2 - 15} = \sqrt{-2x}$

12. $\sqrt{x^2 - 24} = \sqrt{2x}$

13. $\sqrt{3x + 4} + \sqrt{x - 3} = 5$

14. $\sqrt{2x + 2} - \sqrt{x + 2} = \sqrt{x - 6}$

15. $\sqrt{4 - 3x} - \sqrt{5 - x} = \sqrt{5 + x}$

16. $\sqrt{3x + 9} - \sqrt{2x + 4} = \sqrt{x + 1}$

17. $\dfrac{1}{(y + 3)^2} = 1$

18. $\dfrac{1}{(2s - 5)^2} = 4$

19. $|x - 3| = 2$

20. $|x + 5| = 4$

21. $|2x + 1| = 5$

22. $|3x - 7| = 8$

23. $(x + 2)^{1/2} + x(x + 2)^{3/2} = 0$

24. $x^2(3x - 4)^{1/4} + (3x - 4)^{5/4} = 0$

In Exercises 25 to 42, solve each inequality. Express your solutions sets using interval notation.

25. $-3x + 4 \geq -2$

26. $-2x + 7 \leq 5x + 1$

27. $x^2 + 3x - 10 \leq 0$

28. $x^2 - 2x - 3 > 0$

29. $61 \leq \dfrac{9}{5}(C + 32) \leq 95$

30. $30 < \dfrac{5}{9}(F - 32) < 65$

31. $x^3 - 7x^2 + 12x \leq 0$

32. $x^3 + 4x^2 - 21x > 0$

33. $\dfrac{x + 3}{x - 4} > 0$

34. $\dfrac{x(x - 5)}{x + 7} \leq 0$

35. $\dfrac{2x}{3 - x} \leq 10$

36. $\dfrac{x}{5 - x} \geq 1$

37. $|3x - 4| < 2$

38. $|2x - 3| \geq 1$

39. $0 < |x| < 2$

40. $0 < |x| \leq 1$

41. $0 < |x - 2| < 1$

42. $0 < |x - a| < b$ (assume $b > 0$)

In Exercises 43 to 48, solve each equation for the indicated unknown.

43. $V = \pi r^2 h$,   for h

44. $P = \dfrac{A}{1 + rt}$,   for t

45. $A = \dfrac{h}{2}(b_1 + b_2)$,   for b_1

46. $P = 2(l + w)$,   for w

47. $e = mc^2$,   for m

48. $F = G\dfrac{m_1 m_2}{s^2}$,   for m_1

49. One-half of a number minus one-fourth of the number is four more than one-fifth of the number. What is the number?

50. The length of a rectangle is 9 feet less than twice the width of the rectangle. If the perimeter of the rectangle is 54 feet, find the width and the length.

51. A motorboat leaves a harbor and travels to an island at an average rate of 8 knots. The average speed on the return trip was 6 knots. If the total trip took 7 hours, how far is it from the harbor to the island?

52. The price of a magazine subscription rose 5 percent this year. If the subscription now costs $21, how much did the subscription cost last year?

53. A total of $5500 is deposited into two simple interest accounts. On one account the annual simple interest rate is 4 percent, and on the second account the annual simple interest rate is 6 percent. The amount of interest earned for 1 year was $295. How much is invested in each account?

54. A calculator and a battery together sell for $21. If the price of the calculator is $20 more than the price of the battery, find the price of the calculator and the price of the battery.

55. Eighteen owners share the maintenance cost of a condominium complex. If six more units are sold, the maintenance cost will be reduced by $12 per month for each of the present owners. What is the total monthly maintenance cost for the condominium complex?

56. The perimeter of a rectangle is 40 inches and its area is 96 square inches. Find the length and the width of the rectangle.

57. A mason can build a wall in 9 hours less than an apprentice. Together they can build the wall in 6 hours. How long would it take the apprentice working alone to build the wall?

58. An art show brought in $33,196 on the sale of 4526 tickets. If the adult tickets sold for $8 and student tickets sold for $2, how many of each type of ticket were sold?

59. As sand is poured from a chute, it forms a right circular cone whose height is one-fourth the diameter of the base. What is the diameter of the base when the cone has a volume of 144 cubic feet?

60. A manufacturer of calculators finds that the monthly revenue R from a particular style of calculator is given by $R = 72x - 2x^2$, where x is the price in dollars of each calculator. Find the interval in terms of x for which the monthly revenue is greater than $576.

3

Functions and Graphs

So that we may say the door is now opened, for the first time, to a new method fraught with numerous and wonderful results which in future years will command the attention of other minds.

GALILEO GALILEI (1564–1642)

TRAP-DOOR FUNCTIONS, CRYPTOGRAPHY, AND SMART CARDS

Some activities are naturally easier to *reverse* than others. Activities that are not easy to reverse include squeezing the toothpaste out of the tube and cooking fresh eggs to make an omelet. These activities can be thought of as *one-way activities.*

The activity of using your car to pull a trailer is a *trap-door activity.* It is said to be a trap-door activity, because although you can drive either forward or backward, the activity of backing up is very difficult until you have mastered the necessary skills.

There are mathematical functions called *trap-door functions.* For example, the function of cubing a number is a trap-door function. It is easy to cube a number, but more difficult to determine what number has been cubed, by examining the cube.

Ideally, a mathematical trap-door function would be such that it and its inverse can be evaluated in a matter of seconds. However, just to discover how to evaluate the inverse function requires several years of intense effort.

Mathematicians have recently found interesting and useful applications of trap-door functions. One of these applications is the art of writing and deciphering secret codes. This art, called cryptography, is vital to the military and commerce of every large nation.

Adi Shamir, Amos Fiat, and Uriel Feige, of the Weizmann Institute have recently used trap-door functions to develop the concept of zero-knowledge proofs. This has lead to the development of credit cards called "smart" cards. These credit cards contain electronic chips that use zero-knowledge proofs. The cards can identify the user of the card as the owner of the card, without giving a sales clerk the card number or any other information that could be used to make unauthorized purchases.

3.1

A Two-Dimensional Coordinate System and Graphs

Each point on a coordinate axis is associated with a number called its co-ordinate. Each point on a flat **two-dimensional** surface, called a **plane**, is associated with an **ordered pair** of numbers called **coordinates** of the point. Ordered pairs are denoted by (a, b), where the real number a is called the **x-coordinate** or **abscissa** and the real number b is called the **y-coordinate** or **ordinate**.

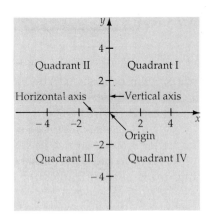

Figure 3.1

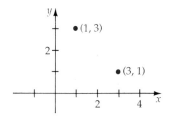

Figure 3.2

The coordinates of a point are determined by the point's position relative to a horizontal coordinate axis called the **x-axis** and a vertical coordinate axis called the **y-axis.** The axes intersect at the point $(0, 0)$, called the **origin.** In Figure 3.1, the axes are labeled so that positive numbers appear to the right of the origin on the x-axis and above the origin on the y-axis. The four regions formed by the axes are called **quadrants** and are numbered counterclockwise. This two-dimensional coordinate system is referred to as a **Cartesian coordinate system** in honor of René Descartes (1596–1650), who was the first to use this concept to study mathematical relationships.

Caution   In Section 1.2, the notation (a, b) was used to denote an interval on a one-dimensional number line. In this section, (a, b) denotes an ordered pair in a two-dimensional plane. This should not cause confusion because as each mathematical topic is introduced, it will be clear whether a one-dimensional or a two-dimensional coordinate system is involved.

The order in which the coordinates of an ordered pair are listed is important. Figure 3.2 shows that $(1, 3)$ and $(3, 1)$ do not denote the same point.

Equality of Ordered Pairs

> The ordered pairs (a, b) and (c, d) are equal if and only if $a = c$ and $b = d$.

EXAMPLE 1   Plot Points on a Cartesian Coordinate System

Plot the points $(4, 3)$, $(-3, 1)$, $(-2, -3)$, $(3, -2)$, and $(0, 1)$ on a Cartesian coordinate system.

Solution

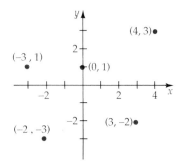

Figure 3.3

■ *Try Exercise* **2,** *page 136.*

The distance between two points on a horizontal line is the absolute value of the difference between the x-coordinates of the two points. The distance between two points on a vertical line is the absolute value of the difference between the y-coordinates of the two points. For example,

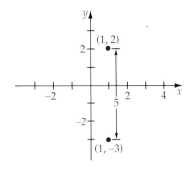

Figure 3.4

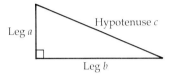

Figure 3.5

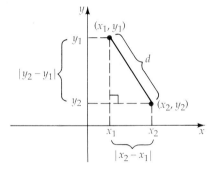

Figure 3.6

the distance d between the points with coordinates $(1, 2)$ and $(1, -3)$ is

$$d = |2 - (-3)| = 5.$$

If two points are not on a horizontal or vertical line, then a *distance formula* for the distance between the two points can be developed by using the *Pythagorean Theorem.*

The Pythagorean Theorem

> If a and b denote the lengths of the legs of a right triangle and c the length of the hypotenuse, then
> $$c^2 = a^2 + b^2$$

The Pythagorean Theorem states that the square of the length of the hypotenuse of a right triangle is equal to the sum of the squares of the lengths of the two legs. Recall that a **right triangle** contains one 90° angle. The side opposite the 90° angle is called the **hypotenuse**. The two other sides are called **legs**. See Figure 3.5.

The Pythagorean Theorem can be used to find the distance between the points (x_1, y_1) and (x_2, y_2) in a Cartesian coordinate system. In Figure 3.6, the distance between the two points is the length of the hypotenuse of a right triangle whose sides are horizontal and vertical line segments that measure $|x_2 - x_1|$ and $|y_2 - y_1|$, respectively. Applying the Pythagorean Theorem to this triangle produces

$$d^2 = |x_2 - x_1|^2 + |y_2 - y_1|^2$$

$$d = \sqrt{|x_2 - x_1|^2 + |y_2 - y_1|^2}$$ The square root theorem. Since d is nonnegative, the negative root is not listed.

$$= \sqrt{(x_2 - x_1)^2 + (y_2 - y_1)^2}$$ Since $|x_2 - x_1|^2 = (x_2 - x_1)^2$ and $|y_2 - y_1|^2 = (y_2 - y_1)^2$

Thus, we have established the following theorem.

The Distance Formula

> The distance d between the points $P_1(x_1, y_1)$ and $P_2(x_2, y_2)$ is
> $$d = \sqrt{(x_2 - x_1)^2 + (y_2 - y_1)^2}.$$

The distance d between the points $P_1(x_1, y_1)$ and $P_2(x_2, y_2)$ is denoted by $d(P_1, P_2)$.

EXAMPLE 2   Use the Distance Formula

Find the distance $d(P_1, P_2)$ between the points $P_1(-3, 4)$ and $P_2(7, 2)$.

Solution   Apply the distance formula to the points $P_1(-3, 4)$ and

$P_2(7,2)$. Thus $x_1 = -3$, $y_1 = 4$, $x_2 = 7$, and $y_2 = 2$.

$$d(P_1, P_2) = \sqrt{(x_2 - x_1)^2 + (y_2 - y_1)^2}$$
$$= \sqrt{[7 - (-3)]^2 + (2 - 4)^2}$$
$$= \sqrt{104} = 2\sqrt{26} \approx 10.2$$

■ *Try Exercise 12, page 136.*

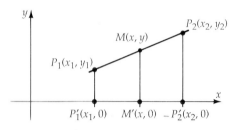

Figure 3.7

The **midpoint** M of a line segment is the point on the line segment that is equidistant from the endpoints $P_1(x_1, y_1)$ and $P_2(x_2, y_2)$ of the segment. See Figure 3.7. To determine a formula for the midpoint $M(x, y)$ of the line segment from $P_1(x_1, y_1)$ to $P_2(x_2, y_2)$, draw vertical lines through P_1, M, and P_2. Notice that the vertical lines intersect the x-axis at $P_1'(x_1, 0)$, $M'(x, 0)$, and $P_2'(x_2, 0)$. A theorem from geometry (if three parallel lines intercept equal parts on one transversal, they intercept equal parts on every transversal) gives us

$$d(P_1', M') = d(M', P_2').$$

Now, $d(P_1', P_2') = x_2 - x_1$ (assuming $x_1 < x_2$). Since M' is the midpoint of line segment from P_1' to P_2', the x-coordinate of M' is

$$x = x_1 + \frac{1}{2}(x_2 - x_1) = x_1 + \frac{1}{2}x_2 - \frac{1}{2}x_1$$
$$= \frac{1}{2}x_1 + \frac{1}{2}x_2 = \frac{x_1 + x_2}{2}$$

Thus the x-coordinate of M is $(x_1 + x_2)/2$. A similar argument can be used to show that the y-coordinate of M is $(y_1 + y_2)/2$. In the above proof, we assumed that $x_1 < x_2$. If $x_1 > x_2$, we would have obtained the same result. Thus we have established the following theorem.

The Midpoint Formula

> The midpoint M of the line segment from $P_1(x_1, y_1)$ to $P_2(x_2, y_2)$ is given by
>
> $$\left(\frac{x_1 + x_2}{2}, \frac{y_1 + y_2}{2} \right).$$

The midpoint formula is an easy formula to remember because it says that the x-coordinate of the midpoint of a line segment will be the average of the x-coordinates of the endpoints of the line segment, and the y-coordinate of the midpoint of a line segment will be the average of the y-coordinates of the endpoints of the line segment.

EXAMPLE 3   **Use the Midpoint Formula**

Find the midpoint M of the line segment connecting the points $P_1(-2, 6)$ and $P_2(3, 4)$.

Solution   Use the midpoint formula.

$$M = \left(\frac{x_1 + x_2}{2}, \frac{y_1 + y_2}{2}\right)$$

$$= \left(\frac{(-2) + 3}{2}, \frac{6 + 4}{2}\right) = \left(\frac{1}{2}, 5\right) \quad \text{Let } x_1 = -2, y_1 = 6, x_2 = 3,$$
$$\text{and } y_2 = 4.$$

■ *Try Exercise* **24,** *page 137.*

The Cartesian coordinate system makes it possible to combine the concepts and methods of algebra and geometry in a way that is useful to both branches of mathematics. One mathematical tool that makes this possible is called a *graph* and is defined as follows.

The Graph of an Equation

> The **graph of an equation** in the two variables x and y is the set of all points whose coordinates satisfy the equation.

To graph an equation, we first determine several ordered pairs whose coordinates satisfy the equation. It is often convenient to solve the equation for y and then substitute values for x. The resulting ordered pairs are listed in a table.

EXAMPLE 4   **Graph by Plotting Points**

Graph $-2x + y = 1$ by plotting points.

Solution   First solve the equation for y.

$$-2x + y = 1$$

$$y = 2x + 1$$

Now choose an x value and use the equation to determine the corresponding y value. For example, if $x = -3$, then the corresponding y value is

$$y = 2(-3) + 1 = -6 + 1 = -5.$$

Continuing in this manner produces the following table:

When x is:	−3	−2	−1	0	1	2	3
y is:	−5	−3	−1	1	3	5	7

The table represents the ordered pairs

$$(-3, -5) \quad (-2, -3) \quad (-1, -1) \quad (0, 1) \quad (1, 3) \quad (2, 5) \quad \text{and} \quad (3, 7)$$

Now plot the ordered pairs as points on a Cartesian coordinate system. The points appear to lie on a straight line, so connect them as shown in Figure 3.8.

■ *Try Exercise* **38,** *page 137.*

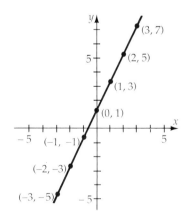

Figure 3.8
$-2x + y = 1$

EXAMPLE 5   **Graph by Plotting Points**

Graph $y = |x - 2|$ by plotting points.

Solution   This equation is already solved for y, so start by choosing an x value and using the equation to determine the corresponding y value. For example, $x = -3$, then

$$y = |(-3) - 2| = |-5| = 5 .$$

Continuing in this manner produces the following table:

When x is:	−3	−2	−1	0	1	2	3	4	5
y is:	5	4	3	2	1	0	1	2	3

Now plot the points listed in the table. The points form a V shape, as shown in Figure 3.9.

■ *Try Exercise 42, page 137.*

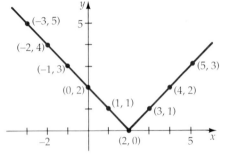

Figure 3.9
$y = |x - 2|$

EXAMPLE 6   **Graph by Plotting Points**

Graph $y = x^2$ by plotting points.

Solution   This equation is already solved for y, so begin by finding ordered pairs that satisfy the equation.

When x is:	−3	−2	−1	0	1	2	3
y is:	9	4	1	0	1	4	9

Plot the points and draw a *smooth curve* through them, as shown in Figure 3.10. The graph is called a *parabola;* it is studied in more detail in Section 6.

■ *Try Exercise 46, page 137.*

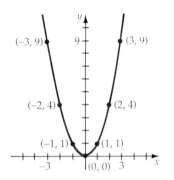

Figure 3.10
$y = x^2$

EXERCISE SET 3.1

In Exercises 1 to 4, plot the points whose coordinates are given on a Cartesian coordinate system.

1. $(2, 4)$, $(0, -3)$, $(-2, 1)$, $(-5, -3)$

2. $(-3, -5)$, $(-4, 3)$, $(0, 2)$, $(-2, 0)$

3. $(20, 5)$, $(-10, 5)$, $(14, 8)$, $(15, -15)$

4. $(100, 200)$, $(-50, 300)$, $(-100, -250)$

In Exercises 5 to 10, use the Pythagorean Theorem to find the length of the unknown side of each right triangle.

5. $a = 5$ and $b = 12$. Find c.

6. $a = 6$ and $b = 8$. Find c.

7. $a = 10$ and $c = 14$. Find b.

8. $a = 15$ and $c = 25$. Find b.

9. $b = 2$ and $c = 12$. Find a.

10. $b = 7$ and $c = 11$. Find a.

In Exercises 11 to 22, find the distance between the points whose coordinates are given.

11. $(6, 4)$, $(-8, 11)$ **12.** $(-5, 8)$, $(-10, 14)$

13. $(-4, -20)$, $(-10, 15)$ **14.** $(40, 32)$, $(36, 20)$

15. $(5, -8)$, $(0, 0)$ **16.** $(0, 0)$, $(5, 13)$

17. $(\sqrt{3}, \sqrt{8})$, $(\sqrt{12}, \sqrt{27})$ **18.** $(\sqrt{125}, \sqrt{20})$, $(6, 2\sqrt{5})$

19. (a, b), $(-a, -b)$

20. $(a - b, b)$, $(a, a + b)$

21. $(x, 4x)$, $(-2x, 3x)$ given that $x < 0$

22. $(x, 4x)$, $(-2x, 3x)$ given that $x > 0$

In Exercises 23 to 30, find the midpoint of the line segment with endpoints as given.

23. $(1, -1), (5, 5)$

24. $(-5, -2), (6, 10)$

25. $(6, -3), (6, 11)$

26. $(4, 7), (-10, 7)$

27. $(5, 12), (0, 0)$

28. $(0, 0), (-3, 4)$

29. $(1.75, 2.25), (-3.5, 5.57)$

30. $(-8.2, 10.1), (-2.4, -5.7)$

In Exercises 31 to 36, use the Pythagorean Theorem to determine which of the triangles whose vertices are given are right triangles.

31. $(1, -3), (4, 1), (-8, 10)$

32. $(4, 3), (10, -2), (-5, -20)$

33. $(5, 1), (9, -5), (6, -7)$

34. $(-5, 0), (6, 3), (2, 15)$

35. $(0, -3), (-4, 2), (1, 7)$

36. $(-7, 8), (2, 3), (-8, -15)$

In Exercises 37 to 52, graph each equation by plotting points that satisfy the equation.

37. $x - y = 4$

38. $2x + y = -1$

39. $y = 0.25x^2$

40. $3x^2 + 2y = -4$

41. $y = -2|x - 3|$

42. $y = |x + 3| - 2$

43. $y = |x^2 - 4|$

44. $y = 7 - 2|x|$

45. $y = x^2 - 3$

46. $y = x^2 + 1$

47. $y = \dfrac{1}{2}(x - 1)^2$

48. $y = 2(x + 2)^2$

49. $y = x^2 + 2x - 8$

50. $y = x^2 - 2x - 8$

51. $y = -x^2 + 2$

52. $y = -x^2 - 1$

53. Find all points on the x-axis that are 10 units from $(4, 6)$. (*Hint:* First write the distance formula with $(4, 6)$ as one of the points and $(x, 0)$ as the other point.)

54. Find all points on the y-axis that are 12 units from $(5, -3)$.

Calculator Exercises

55. Three positive integers x, y, and z are **Pythagorean triples** if they satisfy the equation $x^2 + y^2 = z^2$. Use a calculator to determine which of the following are Pythagorean triples.

   a. 24, 7, 25   **b.** 80, 39, 89

   c. 420, 29, 421   **d.** 52, 165, 173

56. Pythagorean triples x, y, and z (see Exercise 55) are produced by the equations

$$x = m^2 - n^2, \quad y = 2mn, \quad z = m^2 + n^2$$

where m and n are positive integers, with $m > n$. Use the above equations and the following values of m and n to produce the three positive integers x, y, and z. Then use your calculator to verify that they are Pythagorean triples.

   a. $m = 2, n = 1$ **b.** $m = 4, n = 3$

   c. $m = 8, n = 3$ **d.** $m = 15, n = 8$

Supplemental Exercises

In Exercises 57 to 68, graph the set of all points whose x- and y-coordinates satisfy the given conditions.

57. $x = 3$

58. $y = 2$

59. $x = 1, y \geq 1$

60. $y = -3, x \geq -2$

61. $y \leq 3$

62. $x \geq 2$

63. $xy \geq 0$

64. $|y| \geq 1, \dfrac{x}{y} \leq 0$

65. $|x| = 2, |y| = 3$

66. $|x| = 4, |y| = 1$

67. $|x| \leq 2, y \geq 2$

68. $x \geq 1, |y| \leq 3$

In Exercises 69 to 72, find the other endpoint of the line segment that has the given endpoint and midpoint.

69. Endpoint $(5, 1)$, midpoint $(9, 3)$

70. Endpoint $(4, -6)$, midpoint $(-2, 11)$

71. Endpoint $(-3, -8)$, midpoint $(2, -7)$

72. Endpoint $(5, -4)$, midpoint $(0, 0)$

73. Use the distance formula to determine if the points given by $(1, -4), (3, 2), (-3, 4)$, and $(-5, -2)$ are the vertices of a square.

74. Use the distance formula to determine if the points given by $(2, -1), (5, 0), (6, 3)$, and $(3, 2)$ are the vertices of a **a.** parallelogram, **b.** rhombus, **c.** square.

75. Find a formula for the set of all points (x, y) for which the distance from (x, y) to $(3, 4)$ is 5.

76. Find a formula for the set of all points (x, y) for which the distance from (x, y) to $(-5, 12)$ is 13.

77. Find a formula for the set of all points (x, y) for which the sum of the distances from (x, y) to $(4, 0)$ and from (x, y) to $(-4, 0)$ is 10.

78. Find a formula for the set of all points for which the absolute value of the differences of the distance from (x, y) to $(0, 4)$ and from (x, y) to $(0, -4)$ is 6.

3.2

Graphing Techniques

To fully understand many of the concepts in this text, you must be able to quickly sketch the graph of an equation. In this section we introduce several concepts that will help you sketch the graph of certain equations in an efficient manner.

Symmetry with Respect to a Line

You can quickly sketch the graph of some equations by using the concept of **symmetry**. A two-dimensional graph can have *symmetry with respect to a line* or *symmetry with respect to a point*.

Definition of Symmetry with Respect to a Line

> A graph is **symmetric with respect to a line** L if for each point P on the graph there is a point P' on the graph such that the line L is the perpendicular bisector of the line segment PP'.

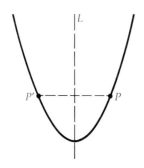

Figure 3.11

In Figure 3.11, the graph in color is symmetric with respect to the line L. The line L is called a **line of symmetry.** The points P and P' are reflections or images of each other with respect to the line L.

The graph in Figure 3.12 is symmetric with respect to the line l. Notice that the graph has the property that if the paper is folded along the dotted line l, the point A' will coincide with the point A, the point B' will coincide with the point B, and the point C' will coincide with the point C. One part of the graph is a *mirror image* of the rest of the graph across the line l.

A graph is **symmetric with respect to the y-axis** if, whenever the point given by (x, y) is on the graph, then $(-x, y)$ is also on the graph. The graph in Figure 3.13 is symmetric with respect to the y-axis. A graph is **symmetric with respect to the x-axis** if, whenever the point given by (x, y) is on the graph, then $(x, -y)$ is also on the graph. The graph in Figure 3.14 is symmetric with respect to the x-axis.

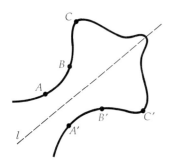

Figure 3.12

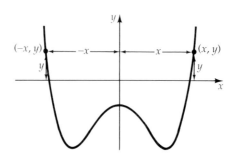

Figure 3.13
Symmetry with respect to the y-axis

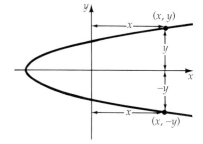

Figure 3.14
Symmetry with respect to the x-axis

The following *tests* allow us to determine whether a graph will be symmetric to a coordinate axis by examining its equation.

Tests for Symmetry with Respect to a Coordinate Axis

> The graph of an equation is symmetric with respect to the
>
> 1. y-axis if the replacement of x with $-x$ leaves the equation unaltered.
> 2. x-axis if the replacement of y with $-y$ leaves the equation unaltered.

EXAMPLE 1 **Determine Symmetry with Respect to a Coordinate Axis**

Determine if the graph of the given equations has symmetry with respect to either the x- or the y-axis. a. $y = x^2 + 2$ b. $x = |y| - 2$

Solution

a. The equation $y = x^2 + 2$ is unaltered by the replacement of x with $-x$. That is, the simplification of

$y = (-x)^2 + 2$ yields the original equation $y = x^2 + 2$.

Thus the graph of $y = x^2 + 2$ is symmetric with respect to the y-axis. The equation $y = x^2 + 2$ *is altered* by the replacement of y with $-y$. That is, the simplification of

$-y = x^2 + 2$ *does not* yield the original equation $y = x^2 + 2$.

Thus the graph of $y = x^2 + 2$ is not symmetric with respect to the x-axis. See Figure 3.15.

b. The equation $x = |y| - 2$ *is altered* by the replacement of x with $-x$. That is, the simplification of

$-x = |y| - 2$ *does not* yield the original equation $x = |y| - 2$.

This implies that the graph of $x = |y| - 2$ is not symmetric with respect to the y-axis.

The equation $x = |y| - 2$ is unaltered by the replacement of y with $-y$. That is, the simplification of

$x = |-y| - 2$ yields the original equation $x = |y| - 2$.

Thus the graph of $x = |y| - 2$ is symmetric with respect to the x-axis. See Figure 3.16.

■ *Try Exercise* **14**, *page 146.*

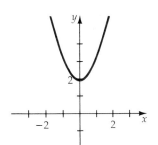

Figure 3.15
$y = x^2 + 2$

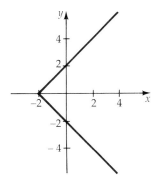

Figure 3.16
$x = |y| - 2$

Symmetry with Respect to a Point

> A graph is **symmetric with respect to a point** Q if for each point P on the graph there is a point P' on the graph such that Q is the midpoint of the line segment PP'.

The graph in Figure 3.17 is symmetric with respect to the point Q. For any point P on the graph, there exists a point P' on the graph such that the distance $d(P, Q)$ is the same as the distance $d(Q, P')$.

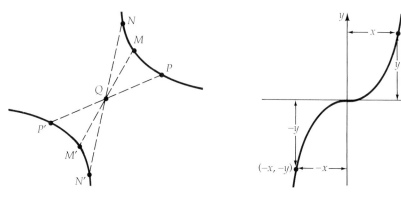

Figure 3.17
Symmetry with respect to the point Q

Figure 3.18

Symmetry with Respect to the Origin

A graph is symmetric with respect to the origin if, whenever the point given by (x, y) is on the graph, then $(-x, -y)$ is also on the graph. The graph in Figure 3.18 is symmetric with respect to the origin.

To determine whether the graph of an equation is symmetric with respect to the origin, we use the following test.

Test for Symmetry with Respect to the Origin

> The graph of an equation is symmetric with respect to the origin if the replacement of x with $-x$ and y with $-y$ leaves the equation unaltered.

EXAMPLE 2   **Determine Symmetry with Respect to the Origin**

Determine if the graph of each of the following equations has symmetry with respect to the origin: a. $xy = 4$ b. $y = x^3 + 1$

Solution

a.   The equation $xy = 4$ is unaltered by the replacement of x with $-x$ and y with $-y$. That is, the simplification of

$$(-x)(-y) = 4 \text{ yields the original equation } xy = 4.$$

Thus the graph of $xy = 4$ is symmetric with respect to the origin. See Figure 3.19.

b.   The equation $y = x^3 + 1$ *is altered* by the replacement of x with $-x$ and y with $-y$. That is, the simplification of

$$-y = (-x)^3 + 1 \text{ does not yield the original equation } y = x^3 + 1.$$

Thus the graph of $y = x^3 + 1$ is not symmetric with respect to the origin. See Figure 3.20.

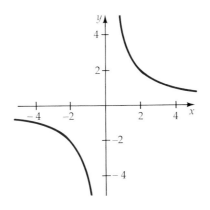

Figure 3.19
$xy = 4$

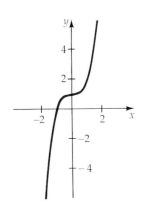

Figure 3.20
$y = x^3 + 1$

■ *Try Exercise 26, page 146.*

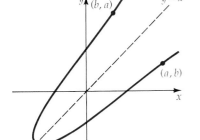

Figure 3.21
Symmetry with respect to the line $y = x$

Symmetry with Respect to the Line $y = x$

Another important type of symmetry is the symmetry with respect to the diagonal line that is the graph of $y = x$. The graph in Figure 3.21 is symmetric with respect to the line $y = x$. The image of any point (a, b) on the graph with respect to the line $y = x$ is the point (b, a). That is, if a point is on a graph symmetric to the line $y = x$, then the point obtained by interchanging the x-coordinate with the y-coordinate will also be on the graph. Thus to determine whether the graph of an equation is symmetric with respect to the line, given by the graph of $y = x$, we use the following test.

Test for Symmetry with Respect to the Line $y = x$

> The graph of an equation is symmetric with respect to the line $y = x$ if the replacement of x with y and y with x leaves the equation unaltered.

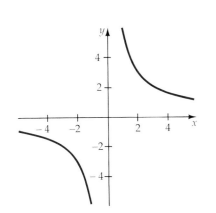

Figure 3.22
$xy = 6$

EXAMPLE 3   Determine Symmetry with Respect to the Line $y = x$

Determine if the graph of each of the folowing equations has symmetry with respect to the line $y = x$:   a. $xy = 6$ b. $x + y^2 = 4$

Solution

a.   The equation $xy = 6$ is unaltered by the replacement of x with y and y with x. That is, the rewriting of

$$yx = 6 \quad \text{yields the original equation} \quad xy = 6.$$

Thus the graph of $xy = 6$ is symmetric with respect to the graph of $y = x$. See Figure 3.22.

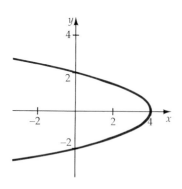

Figure 3.23
$x + y^2 = 4$

b. The equation $x + y^2 = 4$ *is altered* by the replacement of x with y and y with x. That is, the rewriting of

$$y + x^2 = 4 \quad does\ not\ \text{yield the original equation} \quad x + y^2 = 4.$$

Thus the graph of $x + y^2 = 4$ *is not* symmetric with respect to the graph of $y = x$. See Figure 3.23.

■ *Try Exercise* **40,** *page 146.*

Intercepts

Any point that has an x- or a y-coordinate of zero is called an **intercept** of the graph of an equation because it is at these points that the graph intersects the x- or the y-axis.

Definition of x-Intercepts and y-Intercepts

> If $(x_1, 0)$ satisfies an equation, then the point $(x_1, 0)$ is called an **x-intercept** of the graph of the equation.
> If $(0, y_1)$ satisfies an equation, then the point $(0, y_1)$ is called a **y-intercept** of the graph of the equation.

To find the x-intercepts of the graph of an equation, let $y = 0$ and solve the equation for x. To find the y-intercepts of the graph of an equation, let $x = 0$ and solve the equation for y.

For example, to find the x-intercepts of the graph of $|x + 2y| = 4$, let $y = 0$ and solve for x.

$$|x + 0| = 4$$
$$|x| = 4$$
$$x = \pm 4$$

Thus, the x-intercepts are $(-4, 0)$, and $(4, 0)$. To find the y-intercepts of the graph of $|x + 2y| = 4$, let $x = 0$ and solve for y.

$$|0 + 2y| = 4$$
$$|2y| = 4$$
$$2y = -4 \quad \text{or} \quad 2y = 4$$
$$y = -2 \quad \text{or} \quad y = 2$$

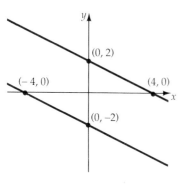

Figure 3.24
$|x + 2y| = 4$

Thus, the y-intercepts are $(0, -2)$ and $(0, 2)$. Figure 3.24 shows the graph of $|x + 2y| = 4$ and its intercepts.

Since every x-intercept $(x_1, 0)$ has 0 for its y-coordinate, it is convenient to refer to the x-intercept as the real number x_1. Also, since every y-intercept $(0, y_1)$ has 0 for its x-coordinate, it is convenient to refer to the y-intercept as the real number y_1. For example, to say that $3x + 5y = 15$ has an x-intercept of 5 and a y-intercept of 3 means that the graph of $3x + 5y = 15$ intercepts the x-axis at $(5, 0)$ and the y-axis at $(0, 3)$.

EXAMPLE 4   Use Symmetry and Intercepts to Graph

Graph each of the following: a. $y = x^3 - 4x$ b. $y = 2|x| - 4$

Solution

a. First determine if the graph has symmetry with respect to a line or a point. Since the equation $y = x^3 - 4x$ is unaltered by replacing x with $-x$ and y with $-y$, the graph is symmetric with respect to the origin. Further testing shows that the graph is not symmetric with respect to the x-axis, the y-axis, or the graph of $y = x$. The following table lists ordered pairs that satisfy the equation. We have used only ordered pairs whose x-coordinate is nonnegative since we will use symmetry with respect to the origin to graph the portion of the graph to the left of the y-axis.

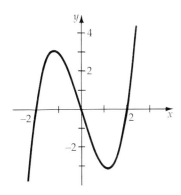

Figure 3.25
$y = x^3 - 4x$

x	0	$\frac{1}{2}$	1	$\frac{3}{2}$	2	$\frac{5}{2}$	3
$y = x^3 - 4x$	0	$-\frac{15}{8}$	-3	$-\frac{21}{8}$	0	$\frac{45}{8}$	15

Now plot the points. Since the graph is symmetric with respect to the origin, we know that if (x, y) is on the graph, then its image through the origin, $(-x, -y)$ is also on the graph. For example, since the point $(1, -3)$ is on the graph, its image through the origin $(-1, 3)$ is also on the graph. Continuing in this manner produces the graph in Figure 3.25. The intercepts are $(-2, 0)$, $(0, 0)$, and $(2, 0)$.

b. Since the graph is symmetric with respect to the y-axis, we have chosen only values of $x \geq 0$, as shown in the following table:

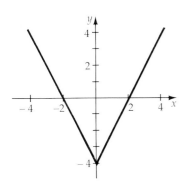

Figure 3.26
$y = 2|x| - 4$

x	0	1	2	3	4
$y = 2|x| - 4$	-4	-2	0	2	4

Plotting the above points and their images with respect to the y-axis produces the graph in Figure 3.26. The graph has a y-intercept of $(0, -4)$ and x-intercepts of $(-2, 0)$, and $(2, 0)$.

■ *Try Exercise* **70,** *page 146.*

Some graphs will have more than one line of symmetry. For example, the graph of $|x| + |y| = 2$ has symmetry with respect to the x-axis, the y-axis, the origin, and the line given by $y = x$. Figure 3.27 is the graph of $|x| + |y| = 2$.

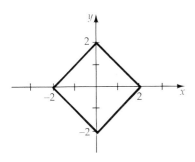

Figure 3.27
$|x| + |y| = 2$

Circles and Their Graphs

Frequently you will sketch graphs by plotting points and using the concepts of symmetry and intercepts. However, some graphs can be sketched by merely recognizing the form of the equation. A *circle* is an example of a curve whose graph can be sketched easily after you have inspected its equation.

Definition of a Circle

A **circle** is the set of points in a plane that are a fixed distance from a specified point. The distance is the **radius** of the circle, and the specified point is the **center** of the circle.

Standard Form of the Equation of a Circle

The **standard form of the equation of a circle** with center at (h, k) and radius r is

$$(x - h)^2 + (y - k)^2 = r^2.$$

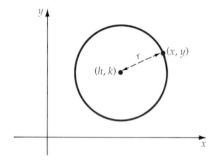

Figure 3.28

To derive the standard form, we use the distance formula. Figure 3.28 is a circle with center (h, k) and radius r. The point (x, y) is on the circle if and only if it is a distance of r units from the center (h, k). Thus (x, y) is on the circle if and only if

$$\sqrt{(x - h)^2 + (y - k)^2} = r$$

$$(x - h)^2 + (y - k)^2 = r^2 \quad \text{Square both sides}.$$

For example, the equation $(x - 3)^2 + (y + 1)^2 = 4$ is the equation of a circle. The standard form of the equation is

$$(x - 3)^2 + (y - (-1))^2 = 2^2$$

from which it can be determined that $h = 3$, $k = -1$, and $r = 2$. Thus, the graph is a circle centered at $(3, -1)$ with a radius of 2.

If a circle is centered at the origin $(0, 0)$ (that is, if $h = 0$ and $k = 0$), then the standard form of the equation of the circle simplifies to

$$x^2 + y^2 = r^2.$$

For example, the graph of $x^2 + y^2 = 9$ is a circle with center at the origin and radius $r = 3$.

EXAMPLE 5   **Find the Standard Form**

Find the standard form of the equation of a circle that has center $C(-4, -2)$ and contains the point $P(-1, 2)$.

Solution   See the sketch of the circle in Figure 3.29. Since the point P is on the circle, the radius r of the circle must equal the distance from C to P. Thus

$$r = \sqrt{(-1 - (-4))^2 + (2 - (-2))^2}$$

$$= \sqrt{9 + 16} = \sqrt{25} = 5$$

Using the standard form with $h = -4$, $k = -2$, and $r = 5$, we obtain

$$(x + 4)^2 + (y + 2)^2 = 5^2.$$

■ *Try Exercise **90**, page 146.*

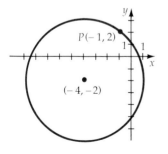

Figure 3.29
$(x + 4)^2 + (y + 2)^2 = 5^2$

If we rewrite $(x + 4)^2 + (y + 2)^2 = 5^2$ by squaring and combining like terms, we produce

$$x^2 + 8x + 16 + y^2 + 4y + 4 = 25$$
$$x^2 + y^2 + 8x + 4y - 5 = 0.$$

The above form of the equation is known as the **general form of the equation of a circle.** By completing the square it is always possible to write the equation $x^2 + y^2 + Ax + By + C = 0$ in the form

$$(x - h)^2 + (y - k)^2 = s$$

for some number s. If $s > 0$, the graph is a circle with radius $r = \sqrt{s}$. If $s = 0$, the graph is the point (h, k), and if $s < 0$, then the equation has no real solutions and there is no graph.

EXAMPLE 6   Find the Center and Radius of a Circle

Find the center and the radius of the circle that has the general form

$$x^2 + y^2 - 6x + 4y - 3 = 0.$$

Solution   First rearrange and group the terms as shown.

$$(x^2 - 6x) + (y^2 + 4y) = 3$$

Now complete the square of $(x^2 - 6x)$ and $(y^2 + 4y)$.

$$(x^2 - 6x + 9) + (y^2 + 4y + 4) = 3 + 9 + 4 \qquad \text{Add 9 and 4 to}$$
$$\text{the right side also.}$$
$$(x - 3)^2 + (y + 2)^2 = 16$$
$$(x - 3)^2 + (y - (-2))^2 = 4^2$$

From the above standard form, the graph of the equation is a circle with center $(3, -2)$ and radius 4.

■ *Try Exercise* **96**, *page 147.*

EXERCISE SET 3.2

In Exercises 1 to 6, plot the image of the given point with respect to the

a. *y*-axis. Label this point A.

b. *x*-axis. Label this point B.

c. origin. Label this point C.

1. $P(5, -3)$ **2.** $Q(-4, 1)$ **3.** $R(-2, 3)$

4. $S(-5, 3)$ **5.** $T(-4, -5)$ **6.** $U(5, 1)$

In Exercises 7 and 8, sketch a graph that is symmetric to the given graph with respect to the *x*-axis.

7.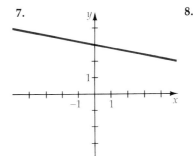

8.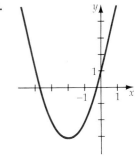

In Exercises 9 and 10, sketch a graph that is symmetric to the given graph with respect to the y-axis.

9.

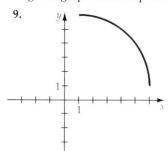

In Exercises 11 and 12, sketch a graph that is symmetric to the given graph with respect to the origin.

11.

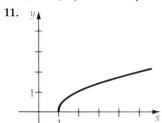

12.

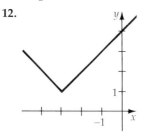

In Exercises 13 to 24, determine if the graph of each equation has symmetry with respect to the **a.** x-axis, **b.** y-axis.

13. $y = 2x^2 - 5$
14. $x = 3y^2 - 7$
15. $y = x^3 + 2$
16. $y = x^5 - 3x$
17. $x^2 + y^2 = 9$
18. $x^2 - y^2 = 10$
19. $x^2 = y^4$
20. $xy = 8$
21. $|x| - |y| = 6$
22. $|x - y| = 6$
23. $4x + 5y = 2$
24. $4x + 5y^2 = 2$

In Exercises 25 to 38, determine if the graph of each equation is symmetric with respect to the origin.

25. $y = x + 1$
26. $y = 3x - 2$
27. $y = x^3 - x$
28. $y = -x^3$
29. $y = \dfrac{9}{x}$
30. $y = -\dfrac{25}{x}$
31. $x^2 + y^2 = 10$
32. $x^2 - y^2 = 4$
33. $y = 4x^2 + x$
34. $y = x^2 - x$
35. $y = 6 - x$
36. $x + y = 0$
37. $y = \dfrac{x}{|x|}$
38. $|y| = |x|$

In Exercises 39 to 48, determine if the graph of each equation is symmetric with respect to the graph of the line $y = x$.

39. $y = -x + 1$
40. $2y = 5 - 2x$
41. $y = x^2$
42. $y = -x^2$
43. $y = \dfrac{9}{x}$
44. $y = -\dfrac{25}{x}$
45. $x^2 + y^2 = 10$
46. $x^2 - y^2 = 4$
47. $|y| = |x|$
48. $x^3 + y^3 = 8$

In Exercises 49 to 58, determine the x- and the y-intercepts of the graph of each equation.

49. $2x + 5y = 12$
51. $x = -y^2 + 5$
52. $x = y^2 - 6$
53. $x = |y| - 4$
54. $x = y^3 - 2$
55. $x^2 + y^2 = 4$
56. $x^2 = y^2$
57. $|x| + |y| = 4$
58. $|x - 4y| = 8$

In Exercises 59 to 74, use symmetry to graph the given equations. Label each intercept.

59. $y = x^2 - 1$
60. $x = y^2 - 1$
61. $y = 6 - x$
62. $x + y = 0$
63. $y = -x + 1$
64. $y = -x - 1$
65. $y = x^3 - x$
66. $y = -x^3$
67. $xy = 4$
68. $xy = -8$
69. $y = 2|x - 4|$
70. $y = |x - 2| - 1$
71. $y = (x - 2)^2 - 4$
72. $y = (x - 1)^2 - 4$
73. $y = x - |x|$
74. $|y| = |x|$

In Exercises 75 to 84, determine the center and radius of the circle with the given equation.

75. $x^2 + y^2 = 36$
76. $x^2 + y^2 = 49$
77. $x^2 + y^2 = 10^2$
78. $x^2 + y^2 = 4^2$
79. $(x - 1)^2 + (y - 3)^2 = 7^2$
80. $(x - 2)^2 + (y - 4)^2 = 5^2$
81. $(x + 2)^2 + (y + 5)^2 = 25$
82. $(x + 3)^2 + (y + 5)^2 = 121$
83. $(x - 8)^2 + y^2 = \dfrac{1}{4}$
84. $x^2 + (y - 12)^2 = 1$

In Exercises 85 to 94, find an equation of a circle satisfying the given conditions. Write your answer in standard form.

85. Center $(4, 1)$, radius $r = 2$
86. Center $(5, -3)$, radius $r = 4$
87. Center $\left(\dfrac{1}{2}, \dfrac{1}{4}\right)$, radius $r = \sqrt{5}$
88. Center $\left(0, \dfrac{2}{3}\right)$, radius $r = \sqrt{11}$
89. Center $(0, 0)$, passing through $(-3, 4)$
90. Center $(0, 0)$, passing through $(5, 12)$
91. Center $(1, 3)$, passing through $(4, -1)$
92. Center $(-2, 5)$, passing through $(1, 7)$
93. Center $(4, 1)$, radius $r = 2$
94. Center $(-5, 3)$, radius $r = \sqrt{2}$

In Exercises 95 to 104, find the center and the radius of the circle whose equation is written in the general form.

95. $x^2 + y^2 - 6x + 5 = 0$

96. $x^2 + y^2 - 6x - 4y + 12 = 0$

97. $x^2 + y^2 - 4x - 10y + 20 = 0$

98. $x^2 + y^2 + 4x - 2y - 11 = 0$

99. $x^2 + y^2 - 14x + 8y + 56 = 0$

100. $x^2 + y^2 - 10x + 2y + 25 = 0$

101. $4x^2 + 4y^2 + 4x - 63 = 0$

102. $9x^2 + 9y^2 - 6y - 17 = 0$

103. $x^2 + y^2 - x + \dfrac{2}{3}y + \dfrac{1}{3} = 0$

104. $x^2 + y^2 - 2x + 2y + \dfrac{7}{4} = 0$

Supplemental Exercises

In Exercises 105 to 110, sketch the graph of each equation. Use symmetry when possible.

105. $|x| + |y| = 4$

106. $|2x| + |y| = 4$

107. $y = |x| + |x - 2| - 4$

108. $y = -|x| + |x - 2| - 4$

109. $y = 0.5|x + 3| - |x - 2| - 1$

110. $y = |x + 4| - 0.5|x| - 2$

111. Find an equation of a circle that has a diameter with endpoints $(2, 3)$ and $(-4, 11)$. Write your answer in standard form.

112. Find an equation of a circle that has a diameter with endpoints $(7, -2)$ and $(-3, 5)$. Write your answer in standard form.

113. Find an equation of a circle that has center at $(7, 11)$ and is tangent to the x-axis. Write your answer in standard form.

114. Find an equation of a circle that has center at $(-2, 3)$ and is tangent to the y-axis. Write your answer in standard form.

115. Find an equation of a circle that is tangent to both axes, has its center in the second quadrant, and a radius of 3.

116. Find an equation of a circle that is tangent to both axes, has its center in the third quadrant, and a diameter of $\sqrt{5}$.

3.3

Functions

Table 3.1

Score	Grade
90–100	A
80–89	B
70–79	C
60–69	D
0–59	F

In many situations in science, business, and mathematics, a correspondence exists between two sets of objects. The correspondence is often described by a table, an equation, or a graph. For example, Table 3.1 describes a grading scale that defines a correspondence between the set of percent scores and the set of letter grades. For any percent score, the table assigns *only one* letter grade. For example, a score of 84 percent receives a letter grade of B. It is convenient to record this example as the ordered pair (84, B).

The equation $d = 16t^2$ indicates the distance (d) a rock will fall (neglecting air resistance) and the time (t) of its fall. According to this equation, in 3 seconds a rock will fall 144 feet, which we denote by the ordered pair (3, 144). For each nonnegative number t, the equation assigns *only one* nonnegative value for the distance. Therefore the equation defines a correspondence between the set of nonnegative real numbers and the set of nonnegative real numbers. Several of the other ordered pairs determined by $d = 16t^2$ are $(0, 0)$, $(1, 16)$, $(2, 64)$, and $(\frac{5}{2}, 100)$.

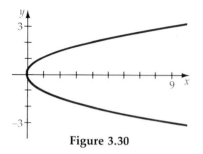

Figure 3.30

Correspondences are represented by graphs like in Figure 3.30 which describes a correspondence between a nonnegative number x and its positive and negative square root. The graph defines a correspondence between the set of nonnegative real numbers and the set of real numbers. The graph assigns to each positive number x two real numbers y_1 and y_2. For example, the points $(9, 3)$ and $(9, -3)$ are both on the graph.

Table 3.1, the equation $d = 16t^2$, and the graph in Figure 3.30 are each an example of a correspondence between two sets. The first set is the *domain* D of the correspondence, and the second set is the *range* R of the correspondence. For example, the domain of the correspondence defined by Table 3.1 is the set of percent scores $\{0, 1, 2, \ldots, 100\}$, and its range is the set of grades $\{A, B, C, D, F\}$. In this chapter we are primarily interested in correspondences in which each element in the domain of the correspondence is assigned to *only one* element in the range.

Definition of a Function

> A **function** f from a set D to a set R is a correspondence, or rule, that pairs each element of D with exactly one element of R. The set D is called the **domain** of f, and the set R is called the **range** of f.

The correspondence described by Table 3.1 and the equation $d = 16t^2$ are both functions. However, the correspondence given by the graph in Figure 3.30 is *not* a function because some elements of its domain are paired with more than one element of its range.

Functions are designated by letters or a combination of letters, such as f, g, A, log, or tan. If x is an element of the domain of a function f, then $f(x)$, read "f of x," is the element in the range of f that is associated with x. For example, if

$$f(x) = x + 3 \quad \text{and} \quad 2 \text{ is in the domain of } f,$$

then $f(2) = 5$. The **value** of the function f at $x = 2$ is 5, and thus 5 is an element of the range of f. The process of determining the value of $f(x)$ is referred to as *evaluating the function* f at x.

EXAMPLE 1   Evaluate Functions

For the function f defined by $f(x) = x^2 - 11$, evaluate each of the following:

a. $f(7)$ b. $f(-5)$ c. $f(3h)$ d. $f(w + 3)$

Solution

a. $f(7) = 7^2 - 11 = 49 - 11 = 38$

b. $f(-5) = (-5)^2 - 11 = 25 - 11 = 14$

c. $f(3h) = (3h)^2 - 11 = 9h^2 - 11$

d. $f(w + 3) = (w + 3)^2 - 11 = w^2 + 6w + 9 - 11 = w^2 + 6w - 2$

■ *Try Exercise 2, page 152.*

If the equation $3x + y = 5$ is solved for y, we obtain

$$y = -3x + 5.$$

Because the value of the variable y depends on the value of the variable x, we call y the **dependent variable** and x the **independent variable.** Using $f(x)$ as a symbol for the dependent variable produces the function

$$f(x) = -3x + 5.$$

Some equations do not define functions. Consider, for example, the equation

$$y^2 = 25 - x^2.$$

This equation does not define a function because given $x = 3$, y can be 4 or -4. Thus 3 is an element of the domain, that is not paired with *exactly* one element of the range R.

To determine if an equation defines a function,

1. Solve (if possible) for the dependent variable.
2. Determine if each value of the independent variable produces exactly one value of the dependent variable.

EXAMPLE 2 Identify Functions

Identify which of the following equations define y as a function of x:

a. $3x + y = 1$ b. $-4x^2 + y^2 = 9$

Solution

a. $3x + y = 1$ Solve for y.

$$y = -3x + 1$$

Because $-3x + 1$ is a unique real number for each value of x, this equation defines y as a function of x.

b. $-4x^2 + y^2 = 9$

$$y^2 = 4x^2 + 9$$

$$y = \pm\sqrt{4x^2 + 9}$$

The right side of this equation, $\pm\sqrt{4x^2 + 9}$, produces two values for each value of x. For example, when $x = 0$, $y = 3$ or $y = -3$. Therefore this equation does not define y as a function of x.

■ *Try Exercise* **14,** *page 153.*

If the domain of a function is not specifically stated, then the domain is the largest set of real numbers for which the defining expression in x leads to a real number range value. Thus, the domain of the function is restricted to avoid division by zero and even roots of negative numbers.

EXAMPLE 3   **Find the Domain of a Function**

Find the domain of the functions represented by the following equations:

a. $f(x) = \sqrt{x + 1}$ b. $S(t) = \dfrac{1}{t - 4}$

Solution

a. To avoid taking the square root of a negative number, we must require that the radicand $x + 1$ be greater than or equal to zero. However, $x + 1 \geq 0$ implies that $x \geq -1$. Thus the domain of the function $f(x) = \sqrt{x + 1}$ is $\{x \mid x \geq -1\}$.

b. The real number 4 must be excluded from the domain since it causes division by zero. Therefore the domain of S is all real numbers except 4. Using set notation, the domain is $\{t \mid t \neq 4\}$.

■ *Try Exercise* **22,** *page 153.*

It is possible to determine the range of a function from its equation. For example, consider the function $f(x) = \sqrt{x + 1}$. As x assumes all values greater than or equal to -1, the radicand $x + 1$ assumes all values greater than or equal to 0. Thus the function $f(x) = \sqrt{x + 1}$ will assume all values greater than or equal to zero. The range is $\{y \mid y \geq 0\}$.

Finding the range of a function can be difficult using the definition of range and the method described above. Fortunately, the graphing methods introduced in Section 4 will help to find the range of many functions.

Some functions are classified as either *even* or *odd*. Knowing a function is even or odd is an aid to drawing its graph. The following definition provides a procedure that can be used to determine if a function is an even or odd function.

Definition of Even and Odd Functions

> The function f is an **even function** if
> $$f(-x) = f(x) \quad \text{for all } x \text{ in the domain of } f.$$
> The function f is an **odd function** if
> $$f(-x) = -f(x) \quad \text{for all } x \text{ in the domain of } f.$$

EXAMPLE 4   **Identify Even or Odd Functions**

Determine whether the following functions are even, odd, or neither:

a. $f(x) = x^3$ b. $F(x) = |x|$ c. $h(x) = x^4 + 2x$

Solution   Replace x with $-x$ and simplify.

a. $f(-x) = (-x)^3 = -x^3 = -(x^3) = -f(x)$

This function is an odd function because $f(-x) = -f(x)$.

b. $F(-x) = |-x| = |x| = F(x)$

This function is an even function because $F(-x) = F(x)$.

c. $h(-x) = (-x)^4 + 2(-x) = x^4 - 2x$

This function is neither an even nor an odd function because

$$h(-x) = x^4 - 2x,$$

which is not equal to either $h(x)$ or $-h(x)$.

■ *Try Exercise* **34,** *page 153.*

EXAMPLE 5   Solve an Application

A car was purchased for $16,500. Assuming the car depreciates at a constant rate of $2200 per year (*straight-line depreciation*) for the first 7 years, write the value v of the car as a function of time, and calculate the value of the car 3 years after the purchase.

Solution   Let t represent the number of years that have elapsed since the car was purchased. Then $2200t$ is the amount that the car has depreciated after t years. The value of the car at time t is given by

$$v(t) = 16,500 - 2200t \quad 0 \le t \le 7.$$

When $t = 3$, the value of the car is

$$v(3) = 16,500 - 2200(3) = 16,500 - 6600 = \$9900.$$

■ *Try Exercise* **52,** *page 153.*

EXAMPLE 6   Solve an Application

Studies show that a particular type of peach tree will yield 275 peaches per tree if only 25 peach trees are planted per acre. For each additional tree planted per acre, the yield of a tree decreases by 5 peaches. Find the function that describes the correspondence between the yield per acre and the number of trees planted if 25 or more trees are planted per acre.

Solution   Let x represent the number of additional trees planted over 25. (Thus x is a natural number.) Since the yield per tree decreases by 5 for each additional tree planted, we know that

$275 - 5x$ is the yield per tree and

$25 + x$ is the number of trees per acre.

The total yield $Y(x)$ is the product of the yield per tree and the number of trees per acre.

$$\begin{pmatrix} \text{Total} \\ \text{Yield} \end{pmatrix} = \begin{pmatrix} \text{yield} \\ \text{per tree} \end{pmatrix} \cdot \begin{pmatrix} \text{number of} \\ \text{trees per acre} \end{pmatrix}$$

$$Y(x) = (275 - 5x) \cdot (25 + x) = -5x^2 + 150x + 6875$$

■ *Try Exercise* **54,** *page 153.*

Note that if 36 trees are planted per acre ($x = 11$), the yield per acre will be

$$Y(11) = -5(11)^2 + 150(11) + 6875 = 7920.$$

If 45 trees are planted per acre ($x = 20$), the yield per acre will be

$$Y(20) = -5(20)^2 + 150(20) + 6875 = 7875.$$

Thus 36 trees per acre yield more fruit than 45 trees per acre. In Section 6 we introduce a method that will allow us to determine the optimal number of trees to plant.

Functions play an important role in applied mathematics. Often in applied mathematics, formulas are used to determine the functional relationship that exists between two variables.

EXAMPLE 7   Solve an Application

A lighthouse is 2 miles south of a port. A ship leaves port sailing east at a rate of 7 mph. If t is the number of hours the ship has been sailing, express the distance d between the ship and the lighthouse as a function of time.

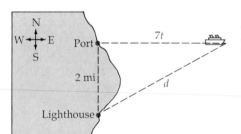

Figure 3.31

Solution   Draw a diagram and label it as shown in Figure 3.31. Note that because distance = (rate)(time) and the rate is 7, in t hours the ship has sailed a distance of $7t$.

$$d^2 = (7t)^2 + 2^2 \quad \text{The Pythagorean Theorem}$$

$$d^2 = 49t^2 + 4$$

$$d = \sqrt{49t^2 + 4} \quad \text{The } \pm \text{ sign is not used since } d \text{ must be nonnegative.}$$

$$f(t) = \sqrt{49t^2 + 4} \quad \text{Since } d = f(t)$$

■ *Try Exercise* **60,** *page 154.*

EXERCISE SET 3.3

In Exercises 1 to 10, find the indicated functional values.

1. Given $f(x) = 3x - 1$, find
   a. $f(2)$ **b.** $f(-1)$ **c.** $f(0)$
   d. $f\left(\dfrac{2}{3}\right)$ **e.** $f(k)$ **f.** $f(k + 2)$

2. Given $g(x) = 2x^2 + 3$, find
   a. $g(3)$ **b.** $g(-1)$ **c.** $g(0)$
   d. $g\left(\dfrac{1}{2}\right)$ **e.** $g(c)$ **f.** $g(c + 5)$

3. Given $A(w) = \sqrt{w^2 + 5}$, find
   a. $A(0)$ **b.** $A(2)$ **c.** $A(-2)$
   d. $A(4)$ **e.** $A(r + 1)$ **f.** $A(-c)$

4. Given $J(t) = 3t^2 - t$, find
   a. $J(-4)$ **b.** $J(0)$ **c.** $J\left(\dfrac{1}{3}\right)$
   d. $J(-c)$ **e.** $J(x + 1)$ **f.** $J(x + h)$

5. Given $f(x) = \dfrac{1}{|x|}$, find
   a. $f(2)$ **b.** $f(-2)$ **c.** $f\left(\dfrac{-3}{5}\right)$
   d. $f(-\pi)$ **e.** $f(c^2 + 4)$ **f.** $f(2 + h)$

6. Given $T(x) = 5$, find

a. $T(-3)$ **b.** $T(0)$ **c.** $T\left(\dfrac{2}{7}\right)$

d. $T(\pi)$ **e.** $T(x + h)$ **f.** $T(3k + 5)$

7. Given $L(t) = -\sqrt{16 - t^2}$, find
a. $L(0)$ **b.** $L(2)$ **c.** $L(\sqrt{2})$

d. $L\left(\dfrac{5}{2}\right)$ **e.** $L(a)$ **f.** $L(-a)$

8. Given $a(t) = -16t^2 + 20t - 2$, find
a. $a(0)$ **b.** $a(2)$ **c.** $a(-2)$
d. $a(3)$ **e.** $a(k - 1)$ **f.** $a(x + h)$

9. Given $s(x) = \dfrac{x}{|x|}$, find

a. $s(4)$ **b.** $s(5)$ **c.** $s(-2)$
d. $s(-3)$ **e.** $s(t), \quad t > 0$ **f.** $s(t), \quad t < 0$

10. Given $r(x) = \dfrac{x}{x + 4}$, find

a. $r(0)$ **b.** $r(-1)$ **c.** $r(-3)$
d. $r\left(\dfrac{1}{2}\right)$ **e.** $r(0.1)$ **f.** $r(10,000)$

In Exercises 11 to 20, identify the equations that define y as a function of x.

11. $2x + 3y = 7$ **12.** $5x + y = 8$
13. $-x + y^2 = 2$ **14.** $x^2 - 2y = 2$
15. $y = 4 \pm \sqrt{x}$ **16.** $x^2 + y^2 = 9$.
17. $y = \sqrt[3]{x}$ **18.** $y = |x| + 5$
19. $y^2 = x^2$ **20.** $y^3 = x^3$

In Exercises 21 to 32, determine the domain of the function represented by the given equation.

21. $f(x) = 3x - 4$ **22.** $f(x) = -2x + 1$
23. $f(x) = x^2 + 2$ **24.** $f(x) = 3x^2 + 1$

25. $f(x) = \dfrac{4}{x + 2}$ **26.** $f(x) = \dfrac{6}{x - 5}$

27. $f(x) = \sqrt{7 + x}$ **28.** $f(x) = \sqrt{4 - x}$
29. $f(x) = \sqrt{4 - x^2}$ **30.** $f(x) = \sqrt{12 - x^2}$

31. $f(x) = \dfrac{1}{\sqrt{x + 4}}$ **32.** $f(x) = \dfrac{1}{\sqrt{5 - x}}$

In Exercises 33 to 48, identify whether the given function is an even function, an odd function, or neither.

33. $g(x) = x^2 - 7$ **34.** $h(x) = x^2 + 1$
35. $F(x) = x^5 + x^3$ **36.** $G(x) = 2x^5 - 10$
37. $A(x) = \pi x^2$ **38.** $P(x) = 2x + 2(8 - x)$

39. $H(x) = 3|x|$ **40.** $T(x) = |x| + 2$
41. $f(x) = 1$ **42.** $k(x) = 2 + x + x^2$
43. $r(x) = \sqrt{x^2 + 4}$ **44.** $u(x) = \sqrt{3 - x^2}$
45. $s(x) = 16x^2$ **46.** $v(x) = 16x^2 + x$

47. $w(x) = 4 + \sqrt[3]{x}$ **48.** $z(x) = \dfrac{x^3}{x^2 + 1}$

49. A rectangle has length of l feet and a perimeter of 50 feet.

a. Write the width w of the rectangle as a function of its length.

b. Write the area A of the rectangle as a function of its length.

50. The sum of two numbers is 20. Let x represent one of the numbers.

a. Write the second number y as a function of x.

b. Write the product P of the two numbers as a function of x.

51. A bus was purchased for $80,000. Assuming the bus depreciates at a rate of $6,500 per year (*straight-line depreciation*) for the first 10 years, write the value v of the bus as a function of the time t (measured in years) for $0 \le t \le 10$.

52. A boat was purchased for $44,000. Assuming the boat depreciates at a rate of $4,200 per year (*straight-line depreciation*) for the first 8 years, write the value v of the boat as a function of the time t (measured in years) for $0 \le t \le 8$.

53. A watch manufacturer charges $19.95 per watch if less than 50 watches are ordered. If more than 50 but less than 200 watches are ordered, the manufacturer reduces the charge per watch by $0.05 for each watch over 50. Thus, for example, if 60 watches are ordered, the charge per watch is $19.45. Find a function that describes the correspondence between the number of watches x ordered and the cost C of the order if $50 < x < 200$.

54. A particular type of avocado tree yields 240 avocados per tree if only 30 trees are planted per acre. For each additional tree planted per acre, the yield of a tree decreases by 10 avocados. Find the function that describes the correspondence between the yield per acre and the number of trees planted.

55. A manufacturer produces a product at a cost of $22.80 per unit. The manufacturer has a fixed cost of $400.00 per day. Each unit retails for $37.00. Let x represent the number of units produced in a 5-day period.

a. Write the total cost C as a function of x.

b. Write the revenue R as a function of x.

c. Write the profit P as a function of x. (*Hint:* Recall that $P(x) = R(x) - C(x)$.)

56. An open box is to be made from a square piece of cardboard having dimensions 30 inches by 30 inches by cutting out squares of area x^2 from each corner, as shown in the figure. Express the volume V of the box as a function of x.

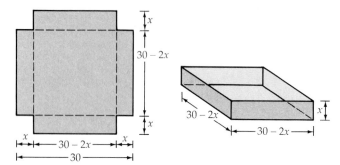

57. A cone has an altitude of 15 centimeters and a radius of 3 centimeters. A right circular cylinder of radius r and height h is inscribed in the cone as shown in the figure. Use similar triangles to write h as a function of r.

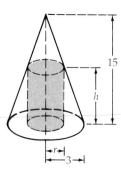

58. Water is running out of a conical funnel that has an altitude of 20 inches and a radius of 10 inches, as shown in the figure.
  a. Write the radius r of the water as a function of its depth h.
  b. Write the volume V of the water as a function of its depth h.

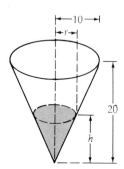

59. For the first minute of flight, a hot air balloon rises vertically at a rate of 3 meters per second. If t is the time in

seconds that the balloon has been airborne, write the distance d between the balloon and a point on the ground 50 meters from the point of lift-off as a function of t.

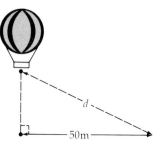

60. An athlete swims from point A to point B at the rate of 2 mph and runs from point B to point C at a rate of 8 mph. Use the dimensions in the figure to write the time t required to reach point C as a function of x.

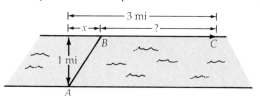

Calculator Exercises

61. An airplane consumes fuel at the rate of

$$f(s) = \frac{1}{425}\left(\frac{8000}{s} + s\right), \quad 200 \le s \le 400$$

gallons per mile when flying at a speed of s miles per hour. Complete the following table by evaluating f (nearest hundredth of a gallon) for the indicated speeds:

s	200	250	300	350	400
$f(s)$					

62. A business finds that the number of feet f of pipe it can sell per week is a function of the price p in cents per foot as given by

$$f(p) = \frac{320{,}000}{p + 25} \quad 40 \le p \le 90$$

Complete the following table by evaluating f (nearest 100 feet) for the indicated values of p:

p	40	50	60	75	90
$f(p)$					

63. The yield Y of apples per tree is related to the amount x of a particular type of fertilizer applied (in pounds per year) by the function

$$Y(x) = 400[1 - 5(x - 1)^{-2}] \quad 5 \le x \le 20$$

Complete the following table by evaluating Y (nearest apple) for the indicated applications:

x	5	10	12.5	15	20
$Y(x)$					

64. A manufacturer finds that the cost C in dollars of producing x items of a product is given by

$$C(x) = (225 + 1.4\sqrt{x})^2 \quad 100 \leq x \leq 1000$$

Complete the following table by evaluating C (nearest dollar) for the indicated number of items

x	100	200	500	750	1000
$C(x)$					

Supplemental Exercises

The notation $f(x)\big|_a^b$ is used to denote the difference $f(b) - f(a)$. That is,

$$f(x)\big|_a^b = f(b) - f(a)$$

In Exercises 65 to 70, evaluate $f(x)\big|_a^b$ for the given function f and the indicated values of a and b.

65. $f(x) = x^2 - x; f(x)\big|_2^3$

66. $f(x) = -3x + 2; f(x)\big|_4^7$

67. $f(x) = 2x^3 - 3x^2 - x; f(x)\big|_0^2$

68. $f(x) = \sqrt{8 - x}; f(x)\big|_0^8$

69. $f(x) = x^2 - 4; f(x)\big|_{-3}^3$

70. $f(x) = 2|6 - x| + 3; f(x)\big|_1^{10}$

In Exercises 71 to 74, each function has two or more independent variables.

71. Given $f(x, y) = 3x + 5y - 2$, find
   a. $f(1, 7)$ **b.** $f(0, 3)$ **c.** $f(-2, 4)$
   d. $f(4, 4)$ **e.** $f(k, 2k)$ **f.** $f(k + 2, k - 3)$

72. Given $g(x, y) = 2x^2 - |y| + 3$, find
   a. $g(3, -4)$ **b.** $g(-1, 2)$
   c. $g(0, -5)$ **d.** $g\left(\dfrac{1}{2}, -\dfrac{1}{4}\right)$
   e. $g(c, 3c), c > 0$ **f.** $g(c + 5, c - 2), c < 0$

73. The area of a triangle with sides a, b, and c is given by the function

$$A(a, b, c) = \sqrt{s(s - a)(s - b)(s - c)}$$

where s is the semiperimeter

$$s = \frac{a + b + c}{2}$$

Find $A(5, 8, 11)$.

74. The cost in dollars to hire a house painter is given by the function

$$C(h, g) = 15h + 14g$$

where h is the number of hours it takes to paint the house and g is the number of gallons of paint required to paint the house. Find $C(18, 11)$.

75. Let $g(x)$ be the xth digit in the decimal representation of π. For example, $g(1) = 3$. Find $g(6)$.

76. Let $d(x)$ be the xth digit in the decimal representation of $\sqrt{2}$. Find $d(7)$.

77. Let $P(n)$ be the nth prime number. For example, $P(1) = 2$. Find $P(7)$.

78. Let $T(n)$ be the total number of digits it takes to number from page 1 to page n of a book. Find $T(250)$.

3.4
Graphs of Functions

The following alternate definition of a function is concise and lends itself to a geometric interpretation of a function.

Alternate Definition of a Function

> A function is a set of ordered pairs in which no two ordered pairs that have the same first component have different second components.

Remark   Because of the notation $y = f(x)$, the ordered pairs of the function f can be written as (x, y) or $(x, f(x))$.

You can determine if a finite set of ordered pairs is a function by examination of the x and y values. For example,

$$\{(1, 2), (2, 4), (3, 6), (4, 8)\}$$

is a function because each first component is paired with exactly one second component. However, the set

$$\{(4, 3), (4, -2), (5, 7)\}$$

is not a function because the ordered pairs $(4, 3)$ and $(4, -2)$ have the same first component and different second components.

The **graph of a function** is the graph of all the ordered pairs of the function. The definition that a function is a set of ordered pairs in which no two ordered pairs that have the same first component have different second components implies that a graph is the graph of a function if any vertical line intersects the graph at no more than one point. This is formally known as the **vertical line test.**

The Vertical Line Test for Functions

> A graph is the graph of a function if and only if no vertical line intersects the graph at more than one point.

EXAMPLE 1   **Use the Vertical Line Test**

Which of the following graphs are graphs of functions?

a.
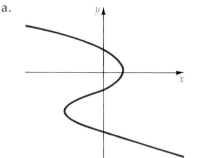

b.
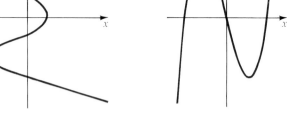

Figure 3.32   Figure 3.33

Solution

a.   This graph *is not* the graph of a function since some vertical lines intersect the graph in more than one point.

b.   This graph is the graph of a function since every vertical line intersects the graph in at most one point.

■ *Try Exercise **8**, page 162.*

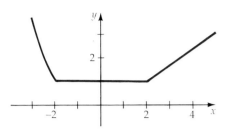

Figure 3.34

Consider the graph in Figure 3.34. As a point on the graph moves from left to right, this graph falls for values of $x \leq -2$, it remains at the same height from $x = -2$ to $x = 2$ and it rises for $x \geq 2$. The function represented by the graph is said to be **decreasing** on the interval $(-\infty, -2]$, **constant** on the interval $[-2, 2]$, and **increasing** on the interval $[2, \infty)$.

The concepts of a function increasing, decreasing, or remaining constant are made more precise by the following definition.

Definition of Increasing, Decreasing, and Constant Functions

If a and b are elements of an interval I that is a subset of the domain of a function f, then

f is **increasing** on I if $f(a) < f(b)$ whenever $a < b$.

f is **decreasing** on I if $f(a) > f(b)$ whenever $a < b$.

f is **constant** on I if $f(a) = f(b)$ for all a and b.

Recall that a function is a set of ordered pairs in which no two ordered pairs that have the same first component have different second components. This means that given any x, there is only one y that can be paired with that x. A **one-to-one function** satisfies the additional condition that given any y, there is only one x that can be paired with the given y. In a manner similar to the vertical line test, we can state a horizontal line test for one-to-one functions.

Horizontal Line Test for a One-to-one Function

If any horizontal line intersects the graph of a function at most once, then the graph is the graph of a one-to-one function.

For example, some horizontal lines intersect the graph in Figure 3.35 at more than one point. It is *not* the graph of a one-to-one function. Every horizontal line intersects the graph in Figure 3.36 at most once. This is the graph of a one-to-one function.

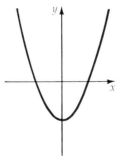

Figure 3.35
Some horizontal lines intersect this graph at more than one point. It is *not* the graph of a one-to-one function.

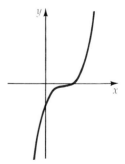

Figure 3.36
Every horizontal line intersects this graph at most once. This is the graph of a one-to-one function.

Recall that a function f is an even function if $f(-x) = f(x)$ for all x in the domain of f, and a function f is an odd function if $f(-x) = -f(x)$ for all x in the domain of f. The following properties are a result of the tests for symmetry in Section 2:

1. The graph of an even function is symmetric with respect to the y-axis.
2. The graph of an odd function is symmetric with respect to the origin.

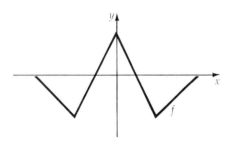

Figure 3.37

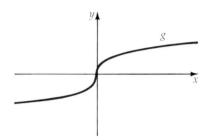

Figure 3.38

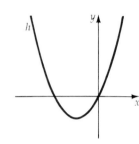

Figure 3.39

The graph of f in Figure 3.37 is symmetric with respect to the y-axis. It is the graph of an even function. The graph of g in Figure 3.38 is symmetric with respect to the origin. It is the graph of an odd function. The graph of h in Figure 3.39 is not symmetric to the y-axis and is not symmetric to the origin. It is neither an even nor an odd function.

Translations

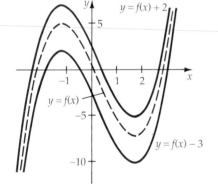

Figure 3.40

The shape of a graph may be exactly the same as the shape of another graph; only their position relative to the origin may differ. For example, the graph of $y = f(x) + 2$ is the graph of $y = f(x)$, with each point moved up vertically two units. The graph of $y = f(x) - 3$ is the graph of $y = f(x)$, with each point moved down vertically three units. See Figure 3.40.

The graphs of $y = f(x) + 2$ and $y = f(x) - 3$ in Figure 3.40 are called **vertical translations** of the graph of $y = f(x)$.

Vertical Translations

> If f is a function and c is a positive constant, then
>
> $y = f(x) + c$ is the graph of $y = f(x)$ shifted up *vertically* c units.
>
> $y = f(x) - c$ is the graph of $y = f(x)$ shifted down *vertically* c units.

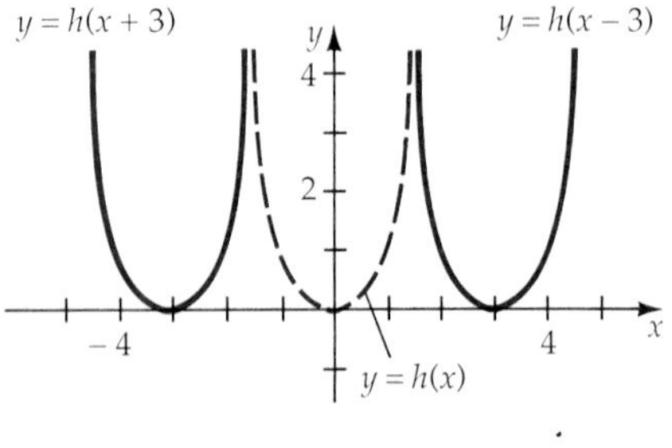

In Figure 3.41, the graph of $y = h(x + 3)$ is the graph of $y = h(x)$, with each point shifted to the left horizontally three units. Similarly, the graph of $y = h(x - 3)$ is the graph of $y = h(x)$, with each point shifted to the right horizontally three units.

The graphs of $y = h(x + 3)$ and $y = h(x - 3)$ in Figure 3.41 are called **horizontal translations** of the graph of $y = h(x)$.

Figure 3.41

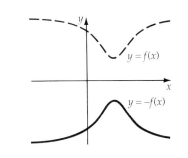

Figure 3.42

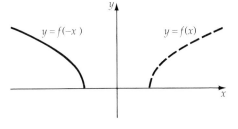

Figure 3.43

Horizontal Translations

> If f is a function and c is a positive constant, then
>
> $y = f(x + c)$ is the graph of $y = f(x)$ shifted left *horizontally* c units.
>
> $y = f(x - c)$ is the graph of $y = f(x)$ shifted right *horizontally* c units.

The graph of $y = -f(x)$ cannot be obtained from the graph of $y = f(x)$ by a combination of vertical and/or horizontal shifts. Figure 3.42 illustrates that the graph of $y = -f(x)$ is the reflection of the graph of $y = f(x)$ across the x-axis.

Figure 3.43 illustrates that the graph of $y = f(-x)$ is the reflection of the graph of $y = f(x)$ across the y-axis.

Reflections

> The graph of
>
> $y = -f(x)$ is the graph of $y = f(x)$ reflected across the x-axis.
>
> $y = f(-x)$ is the graph of $y = f(x)$ reflected across the y-axis.

Some graphs of functions can be sketched by using a combination of translations and reflections.

EXAMPLE 2   Graph by Using Translations and Reflections

Use the graph of $y = f(x)$ shown in Figure 3.44 to sketch the graph of each of the following:

a. $y = f(x - 1) + 2$ b. $y = -f(x) + 3$

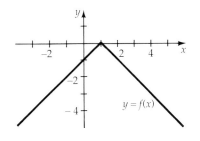

Figure 3.44

Solution

a. The equation $y = f(x - 1) + 2$ indicates that we can sketch the desired graph by shifting the graph of $y = f(x)$ one unit to the right and two units up. See Figure 3.45.

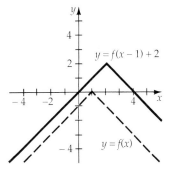

Figure 3.45

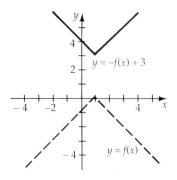

Figure 3.46

b. The equation $y = -f(x) + 3$ indicates that we can sketch the desired graph by reflecting the graph of $y = f(x)$ with respect to the x-axis and then shifting that graph up three units. See Figure 3.46.

■ *Try Exercise* **48,** *page 164.*

Vertical Shrinking and Stretching

The graph of the equation $y = c \cdot f(x)$ for $c \neq 1$ shrinks or stretches the graph of $y = f(x)$. To determine the points on the graph of $y = c \cdot f(x)$, multiply each y-coordinate of the points on the graph of $y = f(x)$ by c. For example, Figure 3.47 shows that the graph of $y = \frac{1}{2}|x|$ can be obtained by plotting points that have a y-coordinate that is one-half the y-coordinate of those found on the graph of $y = |x|$.

If $0 < c < 1$, then the graph of $y = c \cdot f(x)$ is obtained by **shrinking** the graph of $y = f(x)$. Figure 3.47 illustrates that the graph of $y = |x|$ is shrunk toward the x-axis, to form the graph of $y = \frac{1}{2}|x|$.

If $c > 1$, then the graph of $y = c \cdot f(x)$ is obtained by **stretching** the graph of $y = f(x)$. For example, if $f(x) = |x|$, then we obtain graph of

$$y = 2f(x) = 2|x|$$

by stretching the graph of f away from the x-axis. See Figure 3.48.

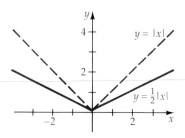

Figure 3.47

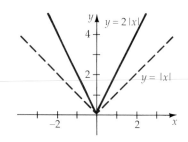

Figure 3.48

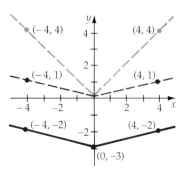

Figure 3.49
$H(x) = \frac{1}{4}|x| - 3$

EXAMPLE 3   Graph By Using Vertical Shrinking and Shifting

Graph the function $H(x) = \frac{1}{4}|x| - 3$.

Solution   The graph of $y = |x|$ has a V shape with its lowest point at $(0, 0)$ and passing through $(4, 4)$ and $(-4, 4)$. The graph of $y = \frac{1}{4}|x|$ is a shrinking of the graph of $y = |x|$. The y-coordinates of $(0, 0)$, $(4, 1)$, and $(-4, 1)$ are obtained by multiplying the y-coordinates of the ordered pairs $(0, 0)$, $(4, 4)$, and $(4, -4)$ by $1/4$. To find the points on the graph of H, we still need to subtract 3 from each y-coordinate. Thus the graph of H is a V shape with its lowest point at $(0, -3)$ and passing through $(4, -2)$ and $(-4, -2)$.

■ *Try Exercise* **50d,** *page 164.*

The domain of a function can be determined by observing the x-coordinates of the points on its graph. The range of a function can be determined by observing the y-coordinates of the points on its graph.

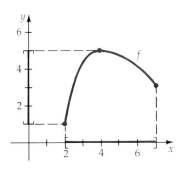

Figure 3.50

EXAMPLE 4   Find Domain and Range From a Graph

Find the domain and the range of the function f defined by the graph in Figure 3.50.

Solution   Because the graph extends from $x = 2$ on the left to $x = 7$ on the right, the domain of f is the interval $[2, 7]$. Using set notation, the domain is written as $\{x \mid 2 \leq x \leq 7\}$. Because the graph extends from a height of $y = 1$ to a height of $y = 5$, the range is the interval $[1, 5]$. Using set notation, the range is written as $\{y \mid 1 \leq y \leq 5\}$.

■ *Try Exercise* **52,** *page 164.*

Piecewise-defined functions are functions represented by more than one equation. To graph a piecewise-defined function, graph each equation over the indicated domain.

EXAMPLE 5   Graph Piecewise-Defined Functions

Graph the function $m(x) = \begin{cases} 2 & \text{if } x \leq 1 \\ x & \text{if } x > 1 \end{cases}$

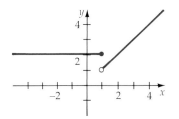

Figure 3.51

$$m(x) = \begin{cases} 2 & \text{if } x \leq 1 \\ x & \text{if } x > 1 \end{cases}$$

Solution   First draw the graph of $m(x) = 2$ for all domain values $x \leq 1$. This gives you the horizontal ray shown in the graph. The solid dot indicates that the point $(1, 2)$ *is* part of the graph. Then draw the graph of $m(x) = x$ for the domain values $x > 1$. The open dot at $(1, 1)$ indicates that it *is not* part of the graph. [See Figure 3.51].

■ *Try Exercise* **70,** *page 164.*

The **greatest integer function** is the piecewise function defined by

$$f(x) = \begin{cases} \vdots & \vdots \\ -1 & \text{if } -1 \leq x < 0 \\ 0 & \text{if } 0 \leq x < 1 \\ 1 & \text{if } 1 \leq x < 2 \\ \vdots & \vdots \end{cases}$$

It can be written more compactly as $f(x) = [\![x]\!]$ by using the following definition.

Definition of the Greatest Integer Function $f(x) = [\![x]\!]$

For any real number x, the greatest integer function $f(x) = [\![x]\!]$ is equal to x if x is an integer, or, the greatest integer less than x if x is not an integer.

For example,

$$[\![3]\!] = 3 \qquad [\![2.72]\!] = 2 \qquad [\![0.954]\!] = 0 \qquad [\![-3.7]\!] = -4.$$

The greatest integer function is sometimes referred to as a **step function.** Figure 3.52 shows that the graph of $f(x) = [\![x]\!]$ resembles a set of steps. The domain of the greatest integer function is the set of all real numbers, and its range is the set of integers.

You can quickly sketch the graph of $f(x) = -[\![x - 2]\!]$ by using the graph of $f(x) = [\![x]\!]$. For example, Figure 3.53 shows that the graph of $f(x) = -[\![x - 2]\!]$ is the graph of $f(x) = [\![x]\!]$, shifted two units to the right and reflected across the x-axis.

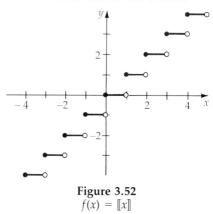

Figure 3.52
$f(x) = [\![x]\!]$

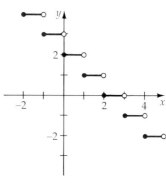

Figure 3.53
$f(x) = -[\![x - 2]\!]$

EXERCISE SET 3.4

In Exercises 1 to 6, identify the sets of the ordered pairs (x, y) that define y as a function of x.

1. $\{(2, 3), (5, 1), (-4, 3), (7, 11)\}$

2. $\{(5, 10), (3, -2), (4, 7), (5, 8)\}$

3. $\{(4, 4), (6, 1), (5, -3)\}$

4. $\{(2, 2), (3, 3), (7, 7)\}$

5. $\{(1, 0), (2, 0), (3, 0)\}$

6. $\left\{\left(-\dfrac{1}{3}, \dfrac{1}{4}\right), \left(-\dfrac{1}{4}, \dfrac{1}{3}\right), \left(-\dfrac{1}{4}, \dfrac{2}{3}\right)\right\}$

In Exercises 7 to 16, use the vertical line test for functions to determine which of the following graphs are graphs of functions.

7.
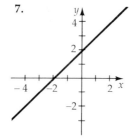

8.
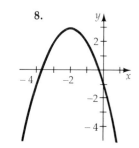

9.
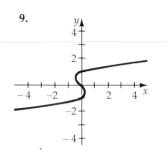

10.
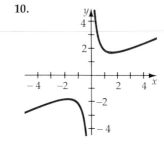

11.
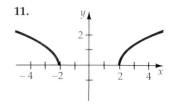

12.
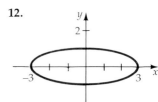

13.
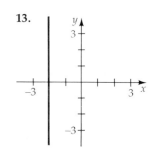

14.
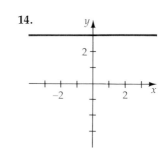

15.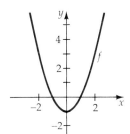

16.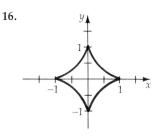

25.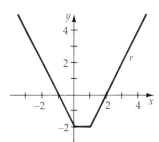

26.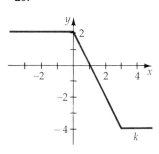

In Exercises 17 to 26, use the graph to identify the intervals over which the function is increasing, constant, or decreasing.

17.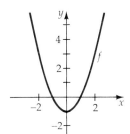

18.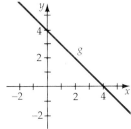

19.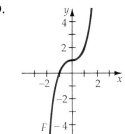

20.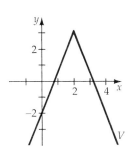

21.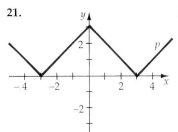

22.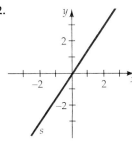

23.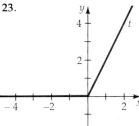

24.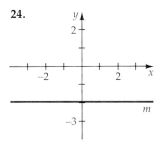

In Exercises 27 to 36, use the indicated graph to determine if the function is a one-to-one function.

27. f as shown in Exercise 17

28. g as shown in Exercise 18

29. F as shown in Exercise 19

30. V as shown in Exercise 20

31. p as shown in Exercise 21

32. s as shown in Exercise 22

33. t as shown in Exercise 23

34. m as shown in Exercise 24

35. r as shown in Exercise 25

36. k as shown in Exercise 26

In Exercises 37 to 46, use the indicated graph to determine if the function is even, odd, or neither.

37. f as shown in Exercise 17

38. g as shown in Exercise 18

39. F as shown in Exercise 19

40. V as shown in Exercise 20

41. p as shown in Exercise 21

42. s as shown in Exercise 22

43. t as shown in Exercise 23

44. m as shown in Exercise 24

45. r as shown in Exercise 25

46. k as shown in Exercise 26

47. Use the graph of $f(x) = \sqrt{4 - x^2}$ to sketch the graph of each of the following:

    a. $y = f(x) + 3$ **b.** $y = f(x - 3)$

    c. $y = f(x + 3)$ **d.** $y = -f(x)$

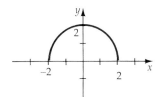

48. Use the graph of $g(x) = x^{2/3}$ to sketch the graph of each of the following:

a. $y = g(x) + 2$

b. $y = g(x - 3) - 1$

c. $y = -g(x) - 1$

d. $y = -g(x + 2)$

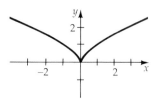

49. Use the graph of $h(x) = x^2 - 2x - 3$ to sketch the graph of each of the following:

a. $y = h(x) + 2$

b. $y = -h(x) + 3$

c. $y = h(x - 3) - 1$

d. $y = -h(x + 2) + 2$

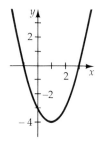

50. Use the graph of $k(x) = -x^2 - 2x + 8$ to sketch the graph of each of the following:

a. $y = k(x) - 2$

b. $y = k(x - 2)$

c. $y = \dfrac{1}{2}k(x)$

d. $y = -k(x) + 2$

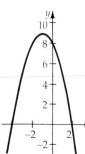

In Exercises 51 to 60, find the domain and range of the indicated graph.

51. f in Exercise 17

52. g in Exercise 18

53. F in Exercise 19

54. V in Exercise 20

55. p in Exercise 21

56. s in Exercise 22

57. t in Exercise 23

58. m in Exercise 24

59. r in Exercise 25

60. k in Exercise 26

In Exercises 61 to 72, sketch the graph of the function.

61. $f(x) = |x| - 3$

62. $f(x) = |x| + 1$

63. $f(x) = 2x$

64. $f(x) = 2x + 3$

65. $f(x) = -|x| + 1$

66. $f(x) = |x - 4|$

67. $f(x) = (x - 2)^2$

68. $f(x) = (x + 3)^2 - 1$

69. $f(x) = \begin{cases} |x| & \text{if } x \le 1 \\ 2 & \text{if } x > 1 \end{cases}$

70. $g(x) = \begin{cases} -4 & \text{if } x \le 0 \\ x^2 - 4 & \text{if } 0 < x \le 1 \\ -x & \text{if } x > 1 \end{cases}$

71. $J(x) = \begin{cases} 4 & \text{if } x \le -2 \\ x^2 & \text{if } -2 < x < 2 \\ -x + 6 & \text{if } x \ge 2 \end{cases}$

72. $K(x) = \begin{cases} 1 & \text{if } x \le -2 \\ x^2 - 3 & \text{if } -2 < x < 2 \\ \dfrac{1}{2}x & \text{if } x \ge 2 \end{cases}$

Supplementary Exercises

In Exercises 73 and 74, sketch the graph of the piecewise function.

73. $s(x) = \begin{cases} 1 & \text{if } x \text{ is an integer} \\ 2 & \text{if } x \text{ is not an integer} \end{cases}$

74. $v(x) = \begin{cases} 2x - 2 & \text{if } x \ne 3 \\ 1 & \text{if } x = 3 \end{cases}$

75. Use the graph of $f(x) = 2/(x^2 + 1)$ to determine an equation for the graphs shown in **a** and **b**.

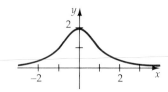

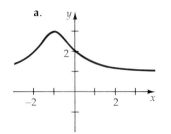

a.

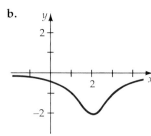

b.

76. Use the graph of $f(x) = x\sqrt{2 + x}$ to determine an equation for the graphs shown in **a** and **b**.

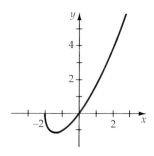

a.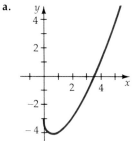

b.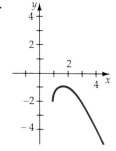

In Exercises 77 to 82, express the height of the given rectangle as a function of x. The midpoint of the upper base of the rectangle is on the graph of f, and the midpoint of the lower base is on the graph of g.

77.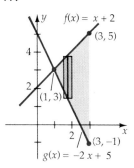

78.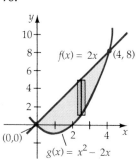

79.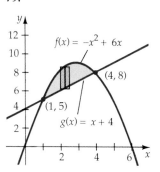

80.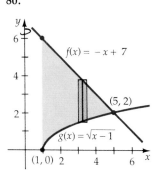

81.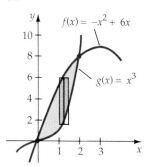

82.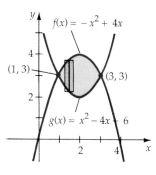

3.5

Linear Functions and Lines

Much of the remainder of this text is concerned with the study of various functions and their graphs. The first function we discuss is important because it is both simple and useful.

Definition of a Linear Function

> A linear function is a function that can be represented by an equation of the form
>
> $$f(x) = mx + b$$
>
> where m and b are real constants.

The graph of a linear function is a nonvertical straight line. If $m = 0$, then $f(x) = mx + b$ simplifies to $f(x) = b$, which is called a **constant function**.

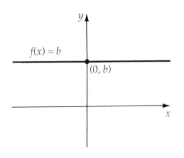

Figure 3.54
$f(x) = b$

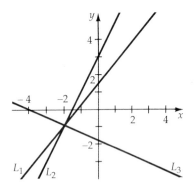

Figure 3.55

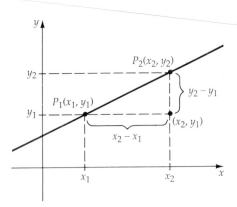

Figure 3.56

The graph of a constant function is a horizontal line, as shown in Figure 3.54.

The graphs shown in Figure 3.55 are the graphs of linear functions for various values of m. The graphs intersect at the point $(-2, -1)$ but the *steepness* of each graph is different. The steepness of a line is called the **slope** of the line and is denoted by the symbol m. The slope of a line is the ratio of the change in y values of any two points on the line to the change in the x values of the same two points. For example, the graph of the line L_1 in Figure 3.55 passes through the points $(-2, -1)$, and $(3, 5)$. The change in the y values of these two points is determined by subtracting the two y-coordinates.

$$\text{Change in } y = 5 - (-1) = 6$$

The change in the x values is determined by subtracting the two x-coordinates

$$\text{Change in } x = 3 - (-2) = 5.$$

The slope m of L_1 is the ratio of the change in the y values of the two points to the change in the x values of the two points. That is,

$$m = \frac{\text{change in } y}{\text{change in } x} = \frac{6}{5}.$$

Since the slope of a nonvertical line can be calculated by using any two arbitrary points on the line, we have the following formula.

Slope of a Nonvertical Line

The slope m of the line passing through the points $P_1(x_1, y_1)$ and $P_2(x_2, y_2)$ with $x_1 \neq x_2$ is given by

$$m = \frac{y_2 - y_1}{x_2 - x_1}.$$

Since the numerator $y_2 - y_1$ is the vertical **rise** and the denominator $x_2 - x_1$ is the horizontal **run** from P_1 to P_2, slope is often referred to as the *rise over the run* or the *change in y divided by the change in x*. See Figure 3.56. Lines that have a positive slope slant upward from left to right. Lines that have a negative slope slant downward from left to right.

EXAMPLE 1   **Find the Slope of a Line**

Find the slope of the line passing through the points whose coordinates are given.   a. $(1, 2)$ and $(3, 6)$ b. $(-3, 4)$ and $(1, -2)$

Solution

a.   The slope of the line passing through $(1, 2)$ and $(3, 6)$ is

$$m = \frac{y_2 - y_1}{x_2 - x_1} = \frac{6 - 2}{3 - 1} = \frac{4}{2} = 2.$$

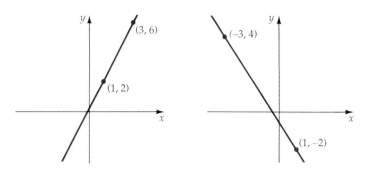

Figure 3.57   **Figure 3.58**

Since $m > 0$, the line slants upward from left to right. See the graph of the line in Figure 3.57.

b.   The slope of the line passing through $(-3, 4)$ and $(1, -2)$ is

$$m = \frac{y_2 - y_1}{x_2 - x_1} = \frac{-2 - 4}{1 - (-3)} = \frac{-6}{4} = -\frac{3}{2}.$$

Since $m < 0$, the line slants downward from left to right. See the graph of the line in Figure 3.58.

■ Try Exercise 2, page 174.

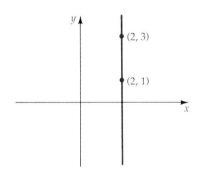

Figure 3.59

The definition of slope does not apply to vertical lines. Consider, for example, the points $(2, 1)$ and $(2, 3)$ on the vertical line in Figure 3.59. Applying the definition of slope to this line produces

$$m = \frac{3 - 1}{2 - 2}$$

which is undefined because it requires division by zero. Since division by zero is undefined, we say that the slope of any vertical line is also undefined.

When computing the slope of a line, it does not matter which point is labeled P_1 and which is labeled P_2 because

$$\frac{y_2 - y_1}{x_2 - x_1} = \frac{y_1 - y_2}{x_1 - x_2}.$$

For example, the slope of a line that passes through $(1, 2)$ and $(5, 10)$ is given by

$$\frac{10 - 2}{5 - 1} = \frac{8}{4} = 2, \quad \text{or} \quad \frac{2 - 10}{1 - 5} = \frac{-8}{-4} = 2.$$

Using functional notation, the points P_1 and P_2 can be represented by

$$P_1(x_1, f(x_1)) \quad \text{and} \quad P_2(x_2, f(x_2)).$$

Using this notation, the slope formula

$$m = \frac{y_2 - y_1}{x_2 - x_1} \text{ is expressed as } m = \frac{f(x_2) - f(x_1)}{x_2 - x_1}.$$

EXAMPLE 2   Use Functional Notation to Find Slope

Find the slope of the line passing through:

a. $(2, f(2))$ and $(-3, f(-3))$ b. $(x, f(x))$ and $(x + h, f(x + h))$

Solution

a.
$$m = \frac{f(x_2) - f(x_1)}{x_2 - x_1} = \frac{f(-3) - f(2)}{(-3) - 2} = \frac{f(-3) - f(2)}{-5}$$

b.
$$m = \frac{f(x_2) - f(x_1)}{x_2 - x_1} = \frac{f(x + h) - f(x)}{(x + h) - x} = \frac{f(x + h) - f(x)}{h}$$

■ *Try Exercise 14, page 174.*

Remark   The expression $\dfrac{f(x + h) - f(x)}{h}$

from Example 2b is called the **difference quotient.** It represents the slope of a nonvertical line.

The linear function $f(x) = mx + b$ is often written as $y = mx + b$. The equation $y = mx + b$ is called the **slope-intercept form** because of the following theorem.

Slope-Intercept Form

> The graph of the function $f(x) = mx + b$ has slope m and y intercept $(0, b)$.

Proof: The slope of the function $f(x) = mx + b$ is given by

$$\frac{f(x_2) - f(x_1)}{x_2 - x_1} = \frac{(mx_2 + b) - (mx_1 + b)}{x_2 - x_1} = \frac{m(x_2 - x_1)}{x_2 - x_1} = m, \quad x_1 \neq x_2.$$

The y-intercept of the graph of $f(x) = mx + b$ is found by letting $x = 0$ and solving for y:

$$y = m(0) + b = b.$$

Thus $(0, b)$ is the y-intercept, and m is the slope of the graph of $y = mx + b$.

If an equation is written in the form $y = mx + b$, then its graph can be drawn by first plotting the y-intercept $(0, b)$ and then using its slope m to determine another point on the line.

EXAMPLE 3   Graph Linear Equations

Graph the following linear equation $y = 2x - 1$.

Solution

The equation $y = 2x - 1$ is in slope-intercept form, with $b = -1$ and

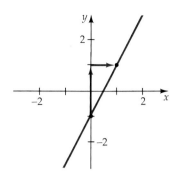

Figure 3.60
$y = 2x - 1$

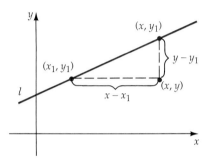

Figure 3.61
The slope of line l is $m = \dfrac{y - y_1}{x - x_1}$.

$m = 2$ or $\dfrac{2}{1}$. To graph the equation, first plot the y-intercept $(0, -1)$ and then use the slope to plot a second point, which is two units up and one unit over from the y-intercept. (See Figure 3.60.)

■ *Try Exercise* **16**, *page 174.*

The previous example was concerned with sketching the graph of a linear equation. Let us now consider the problem of finding an equation of a line provided we know its slope and at least one point on the line. Figure 3.61 suggests that if (x_1, y_1) is a point on a line l of slope m and (x, y) is *any other* point on the line, then

$$\frac{y - y_1}{x - x_1} = m.$$

Multiplying each side of this equation by $x - x_1$ produces

$$y - y_1 = m(x - x_1).$$

This equation is called the **point-slope form** of the line l.

Point-Slope Form

> The graph of the equation
>
> $$y - y_1 = m(x - x_1)$$
>
> is a line that has slope m and passes through (x_1, y_1).

EXAMPLE 4   Use the Point-Slope Form

Find an equation of a line with slope -3 that passes through $(-1, 4)$. Write your answer in slope-intercept form.

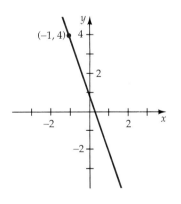

Figure 3.62
$y = -3x + 1$

> ***Solution***   Use the point-slope form with $m = -3$, $x_1 = -1$, and $y_1 = 4$.
>
> $$y - y_1 = m(x - x_1)$$
> $$y - 4 = -3(x - (-1))$$
> $$y - 4 = -3x - 3$$
> $$y = -3x + 1$$

Figure 3.62 shows the graph of $y = -3x + 1$ which has slope of -3 and passes through $(-1, 4)$.

■ *Try Exercise* **28**, *page 174.*

To determine an equation of a nonvertical line that passes through two points, first determine the slope of the line and then use the coordinates of either one of the points in the point-slope form.

EXAMPLE 5   **Equation of a Line Through Two Points**

Find the equation in slope-intercept form of a line that passes through $(-3, 2)$ and $(2, -4)$.

Solution   The slope is

$$m = \frac{y_2 - y_1}{x_2 - x_1} = \frac{-4 - 2}{2 - (-3)} = \frac{-6}{5}.$$

Now use the point-slope form. Let $(x_1, y_1) = (-3, 2)$ or $(2, -4)$.

Case 1: $m = \dfrac{-6}{5}$, $x_1 = -3$, $y_1 = 2$ *Case 2:* $m = -\dfrac{6}{5}$, $x_1 = 2$, $y_1 = -4$.

$$y - y_1 = m(x - x_1) \qquad \text{or} \qquad y - y_1 = m(x - x_1)$$

$$y - 2 = \frac{-6}{5}(x - (-3)) \qquad\qquad y - (-4) = \frac{-6}{5}(x - 2)$$

$$y - 2 = \frac{-6x}{5} - \frac{18}{5} \qquad\qquad y + 4 = \frac{-6x}{5} + \frac{12}{5}$$

$$y = -\frac{6}{5}x - \frac{18}{5} + 2 \qquad\qquad y = -\frac{6}{5}x + \frac{12}{5} - 4$$

$$y = -\frac{6}{5}x - \frac{8}{5} \qquad\qquad y = -\frac{6}{5}x - \frac{8}{5}$$

Notice that we obtained the same result in both cases.

■ *Try Exercise* **40**, *page 174.*

An equation of the form $Ax + By + C = 0$, where A, B, and C are real numbers and both A and B are not zero, is called the **general form** of the equation of a line. For example, the equation $y = -\frac{6}{5}x - \frac{8}{5}$ in Example 5 is written in general form as

$$6x + 5y + 8 = 0.$$

Parallel Lines and Perpendicular Lines

Two lines in a plane that have no points in common are said to be **parallel**. In Figure 3.63, two right triangles have been constructed by using the lines l_1 and l_2 and drawing line segments parallel to the coordinate axes. The lines l_1 and l_2 are parallel if and only if the right triangles are similar. Also, the triangles are similar if and only if the sides are proportional. This suggests the following theorem.

Parallel Lines and Slope

> Two nonvertical lines are parallel if and only if their slopes are equal.

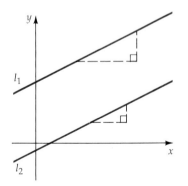

Figure 3.63

Recall that the phrase "if and only if" in this theorem means that both the following are true:

1. If the two nonvertical lines l_1 and l_2 are parallel, then their slopes are equal.

2. If the two nonvertical lines l_1 and l_2 have equal slopes, then they are parallel.

EXAMPLE 6   Find an Equation of a Parallel Line

Find the general form of the equation of the line l that passes through the point $(-2, 3)$ and is parallel to the graph of

$$3x + 4y - 12 = 0.$$

Solution   First express the equation in its slope-intercept form.

$$3x + 4y - 12 = 0$$

$$4y = -3x + 12$$

$$y = -\frac{3}{4}x + 3$$

Since parallel lines have equal slopes, the slope of the line l is $m = -\dfrac{3}{4}$.

Use the point-slope form with $m = -\dfrac{3}{4}$, $x_1 = -2$, and $y_1 = 3$.

$$y - y_1 = m(x - x_1)$$

$$y - 3 = -\frac{3}{4}(x + 2)$$

$$4y - 12 = -3(x + 2)$$

$$4y - 12 = -3x - 6$$

$$3x + 4y - 6 = 0$$

■ *Try Exercise* **48,** *page 174.*

Two lines that intersect at a right angle (90°) are **perpendicular** to each other. For example, every vertical line is perpendicular to every horizontal line. If the lines l_1 and l_2 are neither vertical nor horizontal, then you can determine if they are perpendicular by the following theorem.

Perpendicular Lines

Two lines with slopes m_1 and m_2 are perpendicular if and only if

$$m_1 m_2 = -1$$

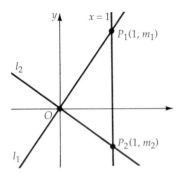

Figure 3.64

Proof: To simplify computations, consider the case of the two lines l_1 and l_2 that intersect at the origin O, as shown in Figure 3.64. The vertical line $x = 1$ intersects l_1 and l_2 at the points $(1, m_1)$ and $(1, m_2)$, respectively. Line l_1 is perpendicular to line l_2 if and only if angle $P_1 O P_2$ is a right angle. By the Pythagorean Theorem, angle $P_1 O P_2$ is a right angle if and only if

$$\left(\begin{array}{c} \text{Distance} \\ \text{from } P_1 \text{ to } P_2 \end{array}\right)^2 = \left(\begin{array}{c} \text{Distance} \\ \text{from } P_1 \text{ to } O \end{array}\right)^2 + \left(\begin{array}{c} \text{Distance} \\ \text{from } O \text{ to } P_2 \end{array}\right)^2$$

or by the distance formula,

$$(m_1 - m_2)^2 = (1^2 + m_1^2) + (1^2 + m_2^2)$$

$$m_1^2 - 2m_1 m_2 + m_2^2 = 2 + m_1^2 + m_2^2$$

$$-2m_1 m_2 = 2$$

$$m_1 m_2 = -1$$

A similar proof can be applied to lines that do not intersect at the origin.

Remark $m_1 m_2 = -1$ means that m_1 and m_2 are *negative reciprocals* of each other, that is $m_1 = -1/m_2$.

EXAMPLE 7   Find an Equation of a Perpendicular Line

Find the equation of the line l through $(2, 1)$ that is perpendicular to the graph of $3x + y - 6 = 0$. Write the answer in the general form of the equation of a line.

Solution   Find the slope m_1 of the given line. The slope m_2 of the line l is the negative reciprocal of m_1.

$$3x + y - 6 = 0$$

$$y = -3x + 6 \quad \text{Thus } m_1 = -3.$$

$$m_2 = -\frac{1}{m_1} = -\frac{1}{-3} = \frac{1}{3}$$

Now use the point-slope form to find an equation of the line l, and write the equation in its general form.

$$y - y_1 = m(x - x_1)$$

$$y - 1 = \frac{1}{3}(x - 2)$$

$$3y - 3 = x - 2$$

$$-x + 3y - 1 = 0$$

The general form $-x + 3y - 1 = 0$ is not unique. For instance, it could also be expressed as $x - 3y + 1 = 0$.

■ *Try Exercise 52, page 175.*

The equation

$$\frac{x}{a} + \frac{y}{b} = 1$$

is referred to as the intercept form of a line because its graph is a line through the x-intercept $(a, 0)$ and the y-intercept $(0, b)$. In algebra courses, all the following forms of the equation of a line are introduced and studied. The derivation of the **two-point form** and the **intercept form** are left for Exercise Set 3.5. They are listed here to provide a complete summary.

Forms of Linear Equations

General form:	$Ax + By + C = 0$
Vertical line:	$x = a$
Horizontal line:	$y = b$
Slope-intercept form:	$y = mx + b$
Point-slope form:	$y - y_1 = m(x - x_1)$
Two-point form:	$y - y_1 = \left(\dfrac{y_2 - y_1}{x_2 - x_1}\right)(x - x_1)$
Intercept form:	$\dfrac{x}{a} + \dfrac{y}{b} = 1$

EXAMPLE 8   Solve An Application

A business purchases a duplicating machine for $4200. It is estimated that after 8 years the duplicating machine will have a value v of $400. If straight-line depreciation is used, find

a. a linear function that expresses the value of the duplicating machine v as a function of the machines age x, where $0 \leq x \leq 8$.

b. the value of the machine after $2\frac{1}{2}$ years.

Solution

a. Since the value of the machine is determined by straight-line depreciation, we need to find a linear function $v(x)$ such that $v(0) = 4200$ and $v(8) = 400$. The slope m of the line given by the linear function v is found by computing the ratio of the change in v divided by the change in x. That is,

$$m = \frac{v(0) - v(8)}{0 - 8} = \frac{4200 - 400}{0 - 8} = \frac{3800}{-8} = -475$$

Now, using the point-slope formula

$$v - v_1 = m(x - x_1)$$

with $m = -475$, $v_1 = 4200$, and $x_1 = 0$, we get

$$v - 4200 = -475(x - 0)$$

$$v - 4200 = -475x$$

$$v = -475x + 4200$$

Using functional notation, we have $v(x) = -475x + 4200$.

b. To find the value of the duplicating machine after $2\frac{1}{2}$ years, evaluate $v(x)$ with $x = 2.5$.

$$v(2.5) = -475(2.5) + 4200 = -1187.50 + 4200 = 3012.50$$

The value of the duplicating machine after $2\dfrac{1}{2}$ years will be $3012.50.

■ *Try Exercise* **60**, *page 175.*

EXERCISE SET 3.5

In Exercises 1 to 10, find the slope of the line that passes through the given points.

1. $(3, 4)$ and $(1, 7)$ **2.** $(-2, 4)$ and $(5, 1)$

3. $(4, 0)$ and $(0, 2)$ **4.** $(-3, 4)$ and $(2, 4)$

5. $(0, 0)$ and $(0, 4)$ **6.** $(0, 0)$ and $(3, 0)$

7. $(-3, 4)$ and $(-4, -2)$ **8.** $(-5, -1)$ and $(-3, 4)$

9. $\left(-4, \dfrac{1}{2}\right)$ and $\left(\dfrac{7}{3}, \dfrac{7}{2}\right)$ **10.** $\left(\dfrac{1}{2}, 4\right)$ and $\left(\dfrac{7}{4}, 2\right)$

In Exercises 11 to 14, find the slope of a line that passes through the given points.

11. $(3, f(3))$ and $(3 + h, f(3 + h))$

12. $(-2, f(-2 + h))$ and $(-2 + h, f(-2 + h))$

13. $(0, f(0))$ and $(h, f(h))$

14. $(a, f(a))$ and $(a + h, f(a + h))$

In Exercises 15 to 26, graph the lines whose equations are given by finding the slope and y-intercept of each line.

15. $y = 2x - 4$ **16.** $y = -x + 1$

17. $y = -\dfrac{1}{3}x + 4$ **18.** $y = \dfrac{2}{3}x - 2$

19. $y = 3$ **20.** $y = x$

21. $y = 2x$ **22.** $y = -3x$

23. $2x + y = 5$ **24.** $x - y = 4$

25. $4x + 3y - 12 = 0$ **26.** $2x + 3y + 6 = 0$

In Exercises 27 to 54, find the equation of the indicated line. Write your answer in slope-intercept form.

27. y-intercept $(0, 3)$, slope 1

28. y-intercept $(0, 5)$, slope -2

29. y-intercept $(0, -1)$, slope 3

30. y-intercept $(0, -2)$, slope -4

31. y-intercept $\left(0, \dfrac{1}{2}\right)$, slope $\dfrac{3}{4}$

32. y-intercept $\left(0, \dfrac{3}{4}\right)$, slope $-\dfrac{2}{3}$

33. y-intercept $(0, 4)$, slope 0

34. y-intercept $(0, -1)$, slope $\dfrac{1}{2}$

35. Through $(1, 3)$, slope 2

36. Through $(2, -1)$, slope 3

37. Through $(-3, 2)$, slope -4

38. Through $(-5, -1)$, slope -3

39. Through $(3, 1)$ and $(-1, 4)$

40. Through $(5, -6)$ and $(2, -8)$

41. Through $(7, 11)$ and $(2, -1)$

42. Through $(-5, 6)$ and $(-3, -4)$

43. Through $\left(\dfrac{1}{2}, 0\right)$ and $\left(\dfrac{3}{4}, \dfrac{1}{5}\right)$

44. Through $\left(-\dfrac{3}{4}, \dfrac{1}{5}\right)$ and $\left(\dfrac{2}{3}, \dfrac{1}{4}\right)$

45. Through $(1.5, 2.4)$ and $(3.6, -5.1)$

46. Through $(-4, -3.5)$ and $(2.5, 6.5)$

47. Through $(1, 3)$ parallel to $3x + 4y = -24$

48. Through $(2, -1)$ parallel to $x + y = 10$

49. Through $(-3, 4)$ parallel to $2x - y = 7$

50. Through $(-1, -5)$ parallel to $x + 3y = 9$

51. Through $(1, 2)$ perpendicular to $x + y = 4$

52. Through $(-3, 4)$ perpendicular to $2x - y = 7$

53. Through $(-2, 1)$ perpendicular to $3x - 5y = 11$

54. Through $(-3, 5)$ perpendicular to $4x - 7y = 8$

55. Use the graph to find the slope of the line that passes through P_1 and P_2.

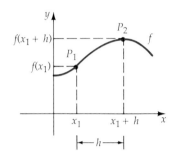

56. Use the graph to find the slope of the line that passes through P_3 and P_4.

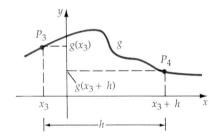

57. Find the linear equation that expresses the relationship between the temperature in degrees Fahrenheit, F, and degrees Celsius, C. Use the fact that water freezes at 32°F which is 0°C, and boils at 212°F which is 100°C.

58. Use the result of Exercise 57 to find the

a. Celsius temperature that corresponds to 81°F.

b. Fahrenheit temperature that corresponds to 60°C.

59. A business purchases a computer for $8280. After 12 years the computer will be obsolete and have no value.

a. Find a linear function that expresses the value V of the computer in terms of the number of years n that it has been used, where $0 \le n \le 12$.

b. Evaluate $V(3)$ to determine the value of the computer after 3 years.

60. A car has a purchase price of $40,090. It is estimated that after 15 years the car will have a value V of $8500. If *straight-line depreciation* is used,

a. Find a linear function that expresses the value V of the car in terms of the car's age n, where $0 \le n \le 15$.

b. Evaluate $V(4)$ to determine the value of the car after 4 years.

61. In business, *marginal cost* is a phrase used to represent the rate of change or slope of a cost function that relates the cost C with respect to the number of units x produced. If a cost function is given by $C(x) = 8x + 275$, find

a. $C(0)$,   **b.** $C(1)$,   **c.** $C(10)$,   **d.** marginal cost

62. In business, *marginal revenue* is a phrase used to represent the rate of change or slope of a revenue function that relates the revenue R with respect to the number of units x sold. If a revenue function is given by the function $R(x) = 210x$, find

a. $R(0)$,   **b.** $R(1)$,   **c.** $R(10)$,   **d.** marginal revenue

Supplemental Exercises

63. A rental company purchases a truck for $19,500. The truck requires an average of $6.75 per day in maintenance.

a. Find the linear function that expresses the total cost C of owning the truck after t days.

b. The truck rents for $55.00 a day. Find the linear function that expresses the revenue R when the truck has been rented for t days.

c. The profit after t days, $P(t)$, is given by the function $P(t) = R(t) - C(t)$. Find the linear function $P(t)$.

d. The **break-even point** is the value of t for which $P(t) = 0$. Use the function $P(t)$ obtained in **c** to determine the break-even point.

64. A magazine company had a profit of $98,000 per year when it had 32,000 subscribers. When it obtained 35,000 subscribers, it had a profit of $117,500. Assuming that the profit P is a linear function of the number of subscribers s,

a. Find the function P.

b. What would the profit be if the company obtains 50,000 subscribers?

c. What is the number of subscribers needed to break even?

65. Use the point-slope form to derive the following equation, called the two-point form:

$$y - y_1 = \left(\frac{y_2 - y_1}{x_2 - x_1} \right)(x - x_1).$$

66. Use the two-point form from Exercise 65 to show that the line with intercepts $(a, 0)$ and $(0, b)$, $a \ne 0$ and $b \ne 0$, has the equation

$$\frac{x}{a} + \frac{y}{b} = 1.$$

In Exercises 67 to 70, use the two-point form to find an equation of the line that passes through the indicated points. Write your answers in slope-intercept form.

67. $(5, 1)$, $(4, 3)$ **68.** $(2, 7)$, $(-1, 6)$

69. $(-11, 8)$, $(7, -5)$ **70.** $(-3, 4)$, $(-7, -11)$

In Exercises 71 to 76, use the equation from Exercise 66 (called the intercept form) to write an equation of a line with the indicated intercepts.

71. x-intercept $(3, 0)$, y-intercept $(0, 5)$

72. x-intercept $(-2, 0)$, y-intercept $(0, 7)$

73. x-intercept $\left(\frac{1}{2}, 0\right)$, y-intercept $\left(0, \frac{1}{4}\right)$

74. x-intercept $\left(\frac{2}{3}, 0\right)$, y-intercept $\left(0, \frac{7}{8}\right)$

75. x-intercept $(a, 0)$, y-intercept $(0, 3a)$,
    point on the line $(5, 2)$, $a \neq 0$

76. x-intercept $(-b, 0)$, y-intercept $(0, 2b)$,
    point on the line $(-3, 10)$, $b \neq 0$

3.6

Quadratic Functions

In the previous section, we studied linear functions and their applications. Many applications cannot be accurately modeled by a linear function. For example, to determine the distance that a rock will fall in t seconds requires a *quadratic function*.

Definition of a Quadratic Function

> A **quadratic function** is a function that can be represented by an equation of the form
>
> $$f(x) = ax^2 + bx + c$$
>
> where a, b, and c are real numbers and $a \neq 0$.

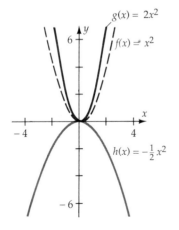

g(x) = 2x²

f(x) = x²

h(x) = -½x²

Figure 3.65

The graph of the quadratic function $f(x) = ax^2 + bx + c$ is a **parabola**. If b and c are both zero, then $f(x) = ax^2 + bx + c$ simplifies to $f(x) = ax^2$. The graph of $f(x) = ax^2$ is a parabola that is symmetric with respect to the y-axis and

1. Opens up if $a > 0$.

2. Opens down if $a < 0$.

If a parabola opens up, then the lowest point on the parabola is called the **vertex of the parabola.** If a parabola opens down, then the vertex is the highest point on the parabola. The graph of $f(x) = ax^2$ has its vertex at the origin $(0, 0)$.

Quadratic functions can be graphed by plotting points and drawing a smooth curve through these points. The graphs of $f(x) = x^2$, $g(x) = 2x^2$, and $h(x) = -\frac{1}{2}x^2$ are shown in Figure 3.65.

The coefficient a in $f(x) = ax^2$ stretches or shrinks the parabola. For small values of $|a|$, the graph of $f(x) = ax^2$ is wider than it is for larger values of $|a|$. This concept was introduced in Section 4 when we discussed the stretching and shrinking of a graph.

EXAMPLE 1   Graph Quadratic Functions Using Translations

Sketch the graph of   a. $f(x) = (x - 3)^2 - 2$ b. $g(x) = 2(x + 1)^2 - 4$

Solution

a. The graph of $f(x) = (x - 3)^2 - 2$ is the graph of $f(x) = x^2$ shifted horizontally three units to the right and vertically downward two units. See Figure 3.66.

b. The graph of $g(x) = 2(x + 1)^2 - 4$ is the graph of $f(x) = 2x^2$, shifted horizontally one unit to the left and vertically downward four units. See Figure 3.67.

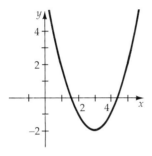

Figure 3.66
$f(x) = (x - 3)^2 - 2$

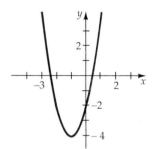

Figure 3.67
$g(x) = 2(x + 1)^2 - 4$

■ *Try Exercise **2**, page 183.*

Standard Form of Quadratic Functions

Every quadratic function $f(x) = ax^2 + bx + c$ can be written in the **standard form**

$$f(x) = a(x - h)^2 + k \quad a \neq 0.$$

The graph of f is a parabola with vertex (h, k). The parabola is symmetric with respect to the vertical line $x = h$, which is called the **axis of the parabola**. The parabola opens up if $a > 0$, and it opens down if $a < 0$.

The standard form is useful because it readily gives information about the vertex of the graph of the function. For example, the graph of $f(x) = 2(x - 4)^2 - 3$ is a parabola with vertex $(4, -3)$. Since a is the positive number 2, the parabola opens upward. It is symmetric to the vertical line $x = 4$, which is its axis.

If a quadratic function is not written in standard form, you can find its standard form by completing the square.

EXAMPLE 2   Find the Standard Form of a Parabola

Use the technique of completing the square to find the standard form of the quadratic function $g(x) = 2x^2 - 12x + 19$. Sketch the graph.

Solution $g(x) = 2x^2 - 12x + 19$

$$= 2(x^2 - 6x) + 19 \qquad \text{Factor 2 from variable terms.}$$

$$= 2(x^2 - 6x + 9 - 9) + 19 \qquad \text{Complete the square.}$$

$$= 2(x^2 - 6x + 9) - 2(9) + 19 \qquad \text{Regroup.}$$

$$= 2(x - 3)^2 - 18 + 19 \qquad \text{Factor and simplify.}$$

$$= 2(x - 3)^2 + 1 \qquad \text{Standard form}$$

The graph of g is the graph of $f(x) = 2x^2$ translated three units to the right and one unit up. It has vertex $(3, 1)$, and it opens upward with the vertical line $x = 3$ as its axis. See Figure 3.68.

■ *Try Exercise* **10**, *page 183.*

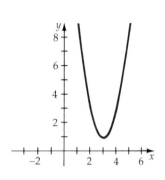

Figure 3.68
$g(x) = 2x^2 - 12x + 19$

In many applications it is important to be able to determine the maximum or minimum value of a function.

Maximum or Minimum Value of a Quadratic Function

> If $a > 0$, then the vertex (h, k) is the lowest point on the graph of $f(x) = a(x - h)^2 + k$ and the y-coordinate k of the vertex is the **minimum value** of the function f.
>
> If $a < 0$, then the vertex (h, k) is the highest point on the graph of $f(x) = a(x - h)^2 + k$ and the y-coordinate k is the **maximum value** of the function f.
>
> In either case, the maximum or minimum is achieved when $x = h$.

EXAMPLE 3   Determine Maximum or Minimum Values

A peach tree yields 275 peaches per tree when 25 trees are planted per acre. For each additional tree planted per acre, the yield of a tree decreases by 5 peaches. Find the number of peach trees that should be planted per acre to produce a maximum yield.

Solution   As in Example 6, Section 3, let x represent the number of additional trees planted over 25. Then

$$25 + x = \text{number of trees per acre}$$

$$275 - 5x = \text{yield per tree}.$$

The total yield $Y(x)$ is the product of the yield per tree and the number of trees per acre.

$$Y(x) = (275 - 5x)(25 + x) = -5x^2 + 150x + 6875$$

To determine the maximum yield, complete the square to get the equation in standard form.

$$\begin{aligned} Y(x) &= -5x^2 + 150x + 6875 \\ &= -5(x^2 - 30x) + 6875 & &\text{Factor } -5 \text{ from the} \\ & & &\text{variable terms.} \\ &= -5(x^2 - 30x + 225 - 225) + 6875 & &\text{Complete the square.} \\ &= -5(x^2 - 30x + 225) + 1125 + 6875 & &\text{Regroup.} \\ &= -5(x - 15)^2 + 8000 & &\text{Factor and simplify.} \end{aligned}$$

From the standard form, we see that the vertex point of the graph of Y is $(15, 8000)$. Thus the maximum yield of 8000 peaches per acre will be achieved when $x = 15$. Since x is the number of trees planted per acre above 25, the maximum yield occurs when 40 trees are planted per acre.

■ *Try Exercise* **20,** *page 183.*

EXAMPLE 4 Determine Maximum or Minimum Values

A long sheet of tin 20 inches wide is to be made into a trough by bending up two sides so they are perpendicular to the bottom. How many inches should be turned up so that the trough will achieve its maximum carrying capacity?

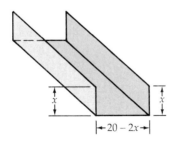

Figure 3.69

Solution The trough is shown in Figure 3.69. If x is the number of inches to be turned up on each side, then the width of the base is $20 - 2x$ inches. The maximum carrying capacity of the trough will occur when the cross-sectional area is a maximum. The cross-sectional area $A(x)$ is given by

$$\begin{aligned} A(x) &= x(20 - 2x) & &\text{Area} = (\text{length})(\text{width}) \\ &= -2x^2 + 20x \end{aligned}$$

To find when A obtains its maximum value, find the x-coordinate of the vertex of the graph of A.

The graph of $A(x) = -2x^2 + 20x$ is a parabola that opens down. Complete the square to produce the equation in its standard form.

$$\begin{aligned} A(x) &= -2x^2 + 20x \\ &= -2(x^2 - 10x) & &\text{Factor.} \\ &= -2(x^2 - 10x + 25 - 25) & &\text{Complete the square.} \\ &= -2(x^2 - 10x + 25) + 50 & &\text{Regroup.} \\ &= -2(x - 5)^2 + 50 & &\text{Factor to produce} \\ & & &\text{the standard form.} \end{aligned}$$

Therefore, when $x = 5$ inches are turned up, the maximum carrying capacity will be achieved.

■ *Try Exercise* **30,** *page 183.*

The Vertex of the Graph of $f(x) = ax^2 + bx + c$

The vertex of the graph of $f(x) = ax^2 + bx + c$ can be determined by the method of completing the square.

$$f(x) = ax^2 + bx + c$$

$$= (ax^2 + bx) + c \qquad \text{Regroup.}$$

$$= a\left(x^2 + \frac{b}{a}x\right) + c \qquad \text{Since } bx = a\frac{b}{a}x,$$
$$\qquad\qquad\qquad\qquad\qquad \text{factor an } a \text{ out.}$$

$$= a\left(x^2 + \frac{b}{a}x + \frac{b^2}{4a^2} - \frac{b^2}{4a^2}\right) + c \quad \text{Complete the square.}$$

$$= a\left(x^2 + \frac{b}{a}x + \frac{b^2}{4a^2}\right) - \frac{b^2}{4a} + c \quad \text{Multiply } a \text{ times } -\frac{b^2}{4a^2} \text{ to}$$
$$\qquad\qquad\qquad\qquad\qquad\qquad \text{produce the } -\frac{b^2}{4a} \text{ term.}$$

$$= a\left(x + \frac{b}{2a}\right)^2 + c - \frac{b^2}{4a} \qquad \text{Factor.}$$

$$= a\left(x - \left(-\frac{b}{2a}\right)\right)^2 + \frac{4ac - b^2}{4a}$$

From the standard form, we see that the vertex of the graph of the function $f(x) = ax^2 + bx + c$ is

$$\left(-\frac{b}{2a}, \frac{4ac - b^2}{4a}\right).$$

Although the y-coordinate of the vertex is $\dfrac{4ac - b^2}{4a}$, it is often easier to determine the y value of the vertex point by evaluating f at $x = -b/2a$. Thus we have the following formula.

Vertex Formula

The vertex of the graph of $f(x) = ax^2 + bx + c$ is

$$\left(-\frac{b}{2a}, f\left(-\frac{b}{2a}\right)\right).$$

EXAMPLE 5   Use the Vertex Formula

Find the vertex of the graph of $f(x) = 2x^2 - 8x + 3$.

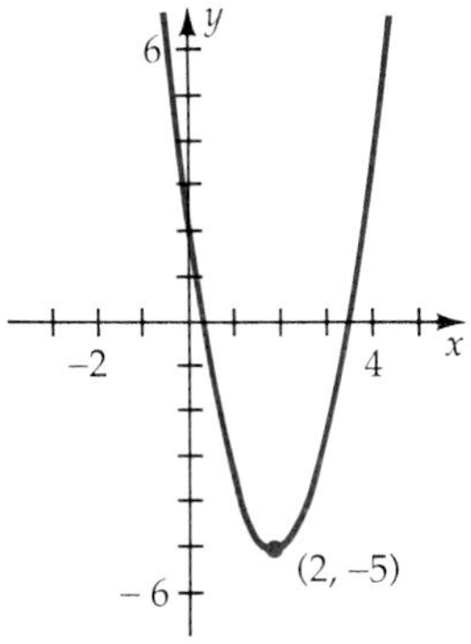

Figure 3.70
$f(x) = 2x^2 - 8x + 3.$

Solution   The *x*-coordinate of the vertex is given by

$$x = -\frac{b}{2a} = -\frac{-8}{2(2)} = \frac{8}{4} = 2.$$

The *y*-coordinate of the vertex is found by evaluating $f(2)$.

$$f(2) = 2(2)^2 - 8(2) + 3 = 8 - 16 + 3 = -5$$

Thus the vertex of the graph of $f(x) = 2x^2 - 8x + 3$ is $(2, -5)$. See Figure 3.70.

■ *Try Exercise* **36,** *page 184.*

Recall that the *y*-intercept of the graph of an equation can be determined by setting $x = 0$ and solving the equation for *y*. Therefore the *y*-intercept of the graph of $f(x) = ax^2 + bx + c$ is $(0, c)$ since

$$f(0) = a(0)^2 + b(0) + c = c.$$

To find the *x*-intercepts (if any) of a quadratic function, set $f(x) = 0$ and solve the equation $ax^2 + bx + c = 0$ by factoring or the quadratic formula.

EXAMPLE 6   Find Intercepts

Determine the intercepts and sketch the graph of each quadratic function.

a. $f(x) = x^2 + 2x - 1$ b. $f(x) = x^2 + 3x + 3$

Solution

a.   The *y*-intercept of $f(x) = x^2 + 2x - 1$ is $(0, -1)$. Solve $x^2 + 2x - 1 = 0$ to determine the *x*-intercepts. Using the quadratic formula,

$$x = \frac{-(2) \pm \sqrt{2^2 - 4(1)(-1)}}{2(1)} = \frac{-2 \pm \sqrt{8}}{2} = \frac{-2 \pm 2\sqrt{2}}{2} = -1 \pm \sqrt{2}.$$

The *x*-intercepts are $(-1 + \sqrt{2}, 0)$ and $(-1 - \sqrt{2}, 0)$. See Figure 3.71.

b.   The *y*-intercept of $f(x) = x^2 + 3x + 3$ is $(0, 3)$. There are no *x*-intercepts of $f(x) = x^2 + 3x + 3$ (see Figure 3.72) because the quadratic

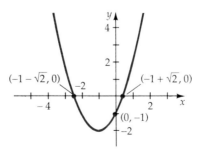

Figure 3.71
$f(x) = x^2 + 2x - 1$

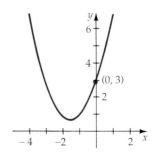

Figure 3.72
$f(x) = x^2 + 3x + 3$

formula applied to the equation $x^2 + 3x + 3 = 0$ yields

$$x = \frac{-3 \pm \sqrt{9 - 4(1)(3)}}{2(1)} = \frac{-3 \pm \sqrt{-3}}{2}$$

which is not a real number.

■ *Try Exercise* **46,** *page 184.*

EXAMPLE 7   Solve an Application

A ball is thrown vertically upward with an initial velocity of 48 feet per second. If the ball started its flight at a height of 8 feet, then its height s at time t can be determined by the function $s(t) = -16t^2 + 48t + 8$ where $s(t)$ is measured in feet above ground level and t is the number of seconds of flight.

a. Determine the time it takes the ball to attain its maximum height.

b. Determine the maximum height the ball attains.

c. Determine the time it takes the ball to hit the ground.

Solution

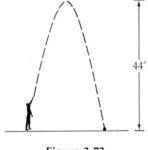

Figure 3.73

a. The graph of the function $s(t) = -16t^2 + 48t + 8$ is a parabola that opens downward. Therefore s will attain its maximum value at the vertex of its graph. Using the vertex formula with $a = -16$ and $b = 48$, we get

$$t = -\frac{b}{2a} = -\frac{48}{2(-16)} = \frac{3}{2}.$$

Therefore the ball attains its maximum height one and one-half seconds into its flight.

b. When $t = \frac{3}{2}$, the height of the ball is

$$s\left(\frac{3}{2}\right) = -16\left(\frac{3}{2}\right)^2 + 48\left(\frac{3}{2}\right) + 8 = 44 \text{ feet}.$$

c. The ball will hit the ground when its height $s(t) = 0$. Therefore, solve $-16t^2 + 48t + 8 = 0$ for t.

$$-16t^2 + 48t + 8 = 0$$

$$-2t^2 + 6t + 1 = 0 \qquad \text{Divide each side by 8.}$$

$$t = \frac{-(6) \pm \sqrt{6^2 - 4(-2)(1)}}{2(-2)} \qquad \text{The quadratic formula}$$

$$= \frac{-6 \pm \sqrt{44}}{-4} = \frac{-3 \pm \sqrt{11}}{-2}$$

Using a calculator to approximate the positive root, we find that the ball will hit the ground in $t \approx 3.16$ seconds.

■ *Try Exercise* **56,** *page 184.*

EXERCISE SET 3.6

In Exercises 1 to 8 match the quadratic function with its graph.

1. $f(x) = x^2 - 3$

2. $f(x) = x^2 + 2$

3. $f(x) = (x - 4)^2$

4. $f(x) = (x + 3)^2$

5. $f(x) = -2x^2 + 2$

6. $f(x) = -\frac{1}{2}x^2 + 3$

7. $f(x) = (x + 1)^2 + 3$

8. $f(x) = -2(x - 2)^2 + 2$

a.

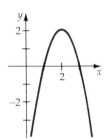

b.

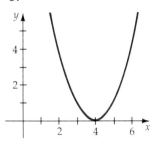

c.

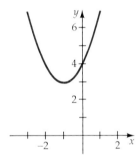

d.

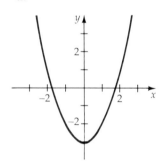

e.

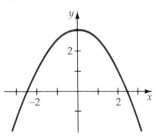

f.

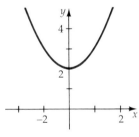

g.

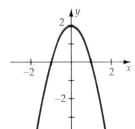

h.

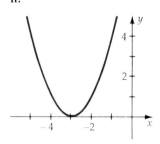

In Exercises 9 to 18, use the method of completing the square to find the standard form of the quadratic function and then sketch its graph.

9. $f(x) = x^2 + 4x + 1$

10. $f(x) = x^2 + 6x - 1$

11. $f(x) = x^2 - 8x + 5$

12. $f(x) = x^2 - 10x + 3$

13. $f(x) = x^2 + 3x + 1$

14. $f(x) = x^2 + 7x + 2$

15. $f(x) = -x^2 + 4x + 2$

16. $f(x) = -x^2 - 2x + 5$

17. $f(x) = -3x^2 + 3x + 7$

18. $f(x) = -2x^2 - 4x + 5$

In Exercises 19 to 28, find the maximum or minimum value of the quadratic function. State whether this value is a maximum or a minimum.

19. $f(x) = x^2 + 8x$

20. $f(x) = -x^2 - 6x$

21. $f(x) = -x^2 + 6x + 2$

22. $f(x) = -x^2 + 10x - 3$

23. $f(x) = 2x^2 + 3x + 1$

24. $f(x) = 3x^2 + x - 1$

25. $f(x) = 5x^2 - 11$

26. $f(x) = 3x^2 - 41$

27. $f(x) = -\frac{1}{2}x^2 + 6x + 17$

28. $f(x) = -\frac{3}{4}x^2 - \frac{2}{5}x + 7$

29. The height of an arch is given by the equation

$$h(x) = -\frac{3}{64}x^2 + 27 \quad -24 \le x \le 24$$

where $|x|$ is the horizontal distance in feet from the center of the arch.

a. What is the maximum height of the arch?

b. What is the height of the arch 10 feet to the right of center?

c. How far from the center is the arch 8 feet tall?

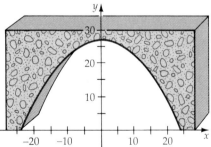

30. The profit function P for a company selling x items is given by $P(x) = -3x^2 + 96x - 368$.

a. What value of x will maximize the profit?

b. What is the maximum profit?

31. A company sells tennis rackets at a price of $60 per racket if 40 or less are ordered. If a buyer orders between 41 and 400 rackets, then the price is reduced $0.10 per racket ordered. Find the size of the order that will give the company the most money.

32. The sum of the length l and width w of a rectangular area is 240 meters.

 a. Write w as a function of l.

 b. Write the area A as a function of l.

 c. Find the dimensions that produce the greatest area.

33. A veterinarian uses 600 feet of chain link fencing to enclose a rectangular region and subdivide the region into two smaller rectangular regions by placing a fence parallel to one of the sides, as shown in the figure.

 a. Write the width w as a function of the length x.

 b. Write the total area A as a function of x.

 c. Find the dimensions that produce the greatest enclosed area.

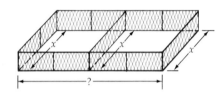

34. A farmer uses 1200 feet of fence to enclose a rectangular region and also subdivide the region into three smaller rectangular regions by placing two fences parallel to one of the sides. Find the dimensions that produce the greatest enclosed area.

In Exercises 35 to 44, use the vertex formula to determine the vertex of the graph of the quadratic function.

35. $f(x) = x^2 - 10x$

36. $f(x) = x^2 - 6x$

37. $f(x) = x^2 - 10$

38. $f(x) = x^2 - 4$

39. $f(x) = -x^2 + 6x + 1$

40. $f(x) = -x^2 + 4x + 1$

41. $f(x) = 2x^2 - 3x + 7$

42. $f(x) = 3x^2 - 10x + 2$

43. $f(x) = -4x^2 + x + 1$

44. $f(x) = -5x^2 - 6x + 3$

In Exercises 45 to 54, determine the y- and the x-intercepts (if any) of the quadratic function.

45. $f(x) = x^2 + 6x$

46. $f(x) = -x^2 + 4x$

47. $f(x) = x^2 + 6x + 9$

48. $f(x) = x^2 + 4x + 4$

49. $f(x) = -x^2 - 3x + 4$

50. $f(x) = -x^2 + x + 6$

51. $f(x) = x^2 + x + 1$

52. $f(x) = x^2 - 2x + 7$

53. $f(x) = 2x^2 + 3x + 4$

54. $f(x) = 3x^2 - 5x + 6$

55. If the initial velocity of a projectile is 128 feet per second, then its height h is a function of time given by the equation $h(t) = -16t^2 + 128t$.

 a. Find the maximum height of the projectile.

 b. Find the time t when the projectile achieves its maximum height.

 c. Find the time t when the projectile hits the ground.

56. The height of a projectile with an initial velocity of 64 feet per second and an initial height of 80 feet is a function of time given by $h(t) = -16t^2 + 64t + 80$.

 a. Find the maximum height of the projectile.

 b. Find the time t when the projectile achieves its maximum height.

 c. Find the time t when the projectile has a height of 0 feet.

Supplemental Exercises

57. Find the quadratic function whose graph has a minimum at $(2, 1)$ and passes through $(0, 4)$.

58. Find the quadratic function whose graph has a maximum at $(-3, 2)$ and passes through $(0, -5)$.

59. A wire 32 inches long is bent so that it has the shape of a rectangle. The length of the rectangle is x and the width is w.

 a. Write w as a function of x.

 b. Write the area A of the rectangle as a function of x.

60. Use the function A from Exercise 59b to prove that the area A is greatest if the rectangle is a square.

61. Show that the function $f(x) = x^2 + bx - 1$ has a real zero for any value b.

62. Show that the function $g(x) = -x^2 + bx + 1$ has a real zero for any value b.

63. What effect does increasing the constant c have on the graph of $f(x) = ax^2 + bx + c$?

64. If $a > 0$, what effect does decreasing the coefficient a have on the graph of $f(x) = ax^2 + bx + c$?

65. Find two numbers whose sum is 8 and whose product is a maximum.

66. Find two numbers whose difference is 12 and whose product is a minimum.

3.7

The Algebra of Functions

Functions can be defined in terms of other functions. For example, the function $h(x) = x^2 + 8x$ could be considered the sum of the functions

$$f(x) = x^2 \quad \text{and} \quad g(x) = 8x.$$

Thus if we are given any two functions, f and g, we can define the four new functions $f + g$, $f - g$, fg, and f/g as follows.

Operations on Functions

For all values of x for which both $f(x)$ and $g(x)$ are defined, we define the following functions:

Sum: $(f + g)(x) = f(x) + g(x)$

Difference: $(f - g)(x) = f(x) - g(x)$

Product: $(fg)(x) = f(x) \cdot g(x)$

Quotient: $\left(\dfrac{f}{g}\right)(x) = \dfrac{f(x)}{g(x)} \quad g(x) \neq 0$

Domain of $f + g$, $f - g$, fg, f/g

For the given functions f and g, the domains of $f + g$, $f - g$, and $f \cdot g$ consist of all real numbers formed by the intersection of the domains of f and g. The domain of f/g is the set of all real numbers formed by the intersection of the domains of f and g, except for those real numbers x such that $g(x) = 0$.

EXAMPLE 1   Determine Domains and Operate with Functions

If $f(x) = \sqrt{x - 1}$ and $g(x) = x^2 - 4$, find the domain of each of the following: $f + g$, $f - g$, fg, and $\dfrac{f}{g}$.

Solution   Notice that f has domain $\{x \mid x \geq 1\}$ and g has the domain of all real numbers. Therefore the domain of $f + g$, $f - g$, and fg will be $\{x \mid x \geq 1\}$. Since $g(x) = 0$ when $x = -2$ or $x = 2$, neither -2 nor 2 is in the domain of f/g. The domain of f/g will be $\{x \mid x \geq 1 \text{ and } x \neq 2\}$.

■ *Try Exercise 10, page 189.*

EXAMPLE 2   **Evaluate Functions**

Let $f(x) = x^2 - 9$ and $g(x) = 2x + 6$. Find each of the following.

a. $(f + g)(5)$ b. $(fg)(-1)$,   c. $\left(\dfrac{f}{g}\right)(4)$

Solution

a. $(f + g)(x) = f(x) + g(x)$

$= (x^2 - 9) + (2x + 6)$

$= x^2 + 2x - 3$

Therefore

$(f + g)(5) = (5)^2 + 2(5) - 3 = 25 + 10 - 3 = 32$

b. $(fg)(x) = f(x) \cdot g(x)$

$= (x^2 - 9)(2x + 6)$

$\cdot \quad = 2x^3 + 6x^2 - 18x - 54$

Therefore

$(fg)(-1) = 2(-1)^3 + 6(-1)^2 - 18(-1) - 54$

$= -2 + 6 + 18 - 54 = -32$

c. $\left(\dfrac{f}{g}\right)(x) = \dfrac{f(x)}{g(x)} = \dfrac{x^2 - 9}{2x + 6}$

$= \dfrac{(x + 3)(x - 3)}{2(x + 3)}$

$= \dfrac{x - 3}{2}, \qquad x \neq -3$

Therefore,

$\left(\dfrac{f}{g}\right)(4) = \dfrac{4 - 3}{2} = \dfrac{1}{2}.$

■ *Try Exercise* **14**, *page 190.*

The Difference Quotient

The expression

$$\dfrac{f(x + h) - f(x)}{h}, \quad h \neq 0$$

is called the **difference quotient** of f. It is an important function because it enables us to study the manner in which a function changes in value as the independent variable changes.

EXAMPLE 3   **Find a Difference Quotient**

Find the difference quotient of the function f defined by $f(x) = x^2 + 7$.

Solution

$$\frac{f(x + h) - f(x)}{h} = \frac{[(x + h)^2 + 7] - [x^2 + 7]}{h}$$ Apply the difference quotient.

$$= \frac{[x^2 + 2xh + h^2 + 7] - [x^2 + 7]}{h}$$

$$= \frac{x^2 + 2xh + h^2 + 7 - x^2 - 7}{h}$$

$$= \frac{2xh + h^2}{h} = \frac{\cancel{h}(2x + h)}{\cancel{h}} = 2x + h$$

■ *Try Exercise **30**, page 190.*

Composition of Functions

Composition of functions is yet another method of constructing a function from two given functions. The process consists of using the range element of one function as the domain element of another function.

Composite functions occur in many business transactions. For example, suppose the manufacturing cost (in dollars) per compact disc player is given by the function

$$m(x) = \frac{180x + 2600}{x}$$

where x is the number of compact disc players to be manufactured. An electronics outlet agrees to sell the compact discs by marking up the manufacturing cost per player $m(x)$ by 30 percent. Notice that the selling price s will be a function of $m(x)$. More specifically,

$$s[m(x)] = 1.30(m(x)).$$

Simplifying $s[m(x)]$ produces the following:

$$s[m(x)] = 1.30\left(\frac{180x + 2600}{x}\right)$$

$$= 1.30(180) + 1.30\frac{2600}{x} = 234 + \frac{3380}{x}$$

The function produced in this manner is referred to as the composition of m by s. The notation $s \circ m$ is used to denote this composition function. That is,

$$(s \circ m)(x) = 234 + \frac{3380}{x}.$$

Composition of Functions

> For the functions f and g, the **composite function** or **composition** of f by g, is given by
>
> $$(g \circ f)(x) = g[f(x)]$$
>
> for all x in the domain of f such that $f(x)$ is in the domain of g.

If f and g are specified by equations, you can use substitution to find equations that specify $(g \circ f)$ and $(f \circ g)$.

EXAMPLE 4   Form Composite Functions

If $f(x) = x^2 - 3x$ and $g(x) = 2x + 1$, find   a. $(g \circ f)$ b. $(f \circ g)$.

Solution

a. $(g \circ f) = g[f(x)] = 2(f(x)) + 1$ Substitute $f(x)$ for x in g.

$= 2(x^2 - 3x) + 1$

$= 2x^2 - 6x + 1$

b. $(f \circ g) = f[g(x)] = (g(x))^2 - 3(g(x))$ Substitute $g(x)$ for x in f.

$= (2x + 1)^2 - 3(2x + 1)$

$= 4x^2 - 2x - 2$

Note that in this example, $(g \circ f) \neq (f \circ g)$.

■ *Try Exercise* **38**, *page 190.*

To evaluate $(f \circ g)(c)$ for some constant c, you can use either of the following methods.

Method 1   First evaluate $g(c)$, substitute this result for x in the f function, and evaluate.

Method 2   First determine $f[g(x)]$ and then substitute c for x.

EXAMPLE 5   Evaluate Composite Functions

Evaluate $(f \circ g)(3)$, where $f(x) = 2x - 5$ and $g(x) = 4x^2 + 1$.

Solution

Method 1 $(f \circ g)(3) = f[g(3)]$ Evaluate $g(3)$.

$= f[4(3)^2 + 1]$

$= f(37)$ Substitute in f.

$= 2(37) - 5 = 69$

Method 2 $(f \circ g)(x) = 2(g(x)) - 5$ Form $f[(g(x)]$.

$$= 2(4x^2 + 1) - 5$$

$$= 8x^2 + 2 - 5 = 8x^2 - 3$$

$(f \circ g)(3) = 8(3)^2 - 3 = 69$ Substitute 3 for x.

■ *Try Exercise* **54,** *page 190.*

Remark   Both methods 1 and 2 produced the same result. Although method 2 required the most work, it is the best method to use if you have to evaluate $(f \circ g)(x)$ for several values of x since you just evaluate $8x^2 - 3$.

Some care must be used when forming the composition of functions. For instance, if $f(x) = x + 1$ and $g(x) = \sqrt{x - 4}$, then $(g \circ f)(2)$ is not defined since $f(2) = 3$ and 3 is not in the domain of g. We can avoid this problem by making restrictions. For example, if the domain of the function $f(x) = x + 1$ is restricted to $[3, \infty)$, then the range of f will consist of $[4, \infty)$, and this set of numbers is a subset of the domain of g. Many of the functions in this text have restrictions listed so that the forming of composition functions will be meaningful. See Figure 3.74.

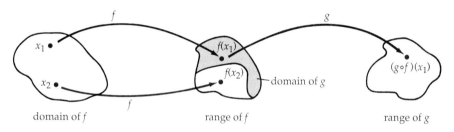

Figure 3.74

The composition function $g \circ f$ is only defined for those x in the domain of f for which $f(x)$ is in the domain of g. Therefore, $(g \circ f)(x_1)$ is a real number, but $(g \circ f)(x_2)$ is not defined.

In general, $f \circ g \neq g \circ f$. This was verified in Example 4. The special class of functions such that $f \circ g = g \circ f$ is discussed in the next section.

EXERCISE SET 3.7

In Exercises 1 to 12, use the given functions f and g to find $f + g$, $f - g$, fg, and f/g. State the domain of each.

1. $f(x) = x^2 - 2x - 15$, $g(x) = x + 3$

2. $f(x) = x^2 - 25$, $g(x) = x - 5$

3. $f(x) = 2x + 8$, $g(x) = x + 4$

4. $f(x) = 5x - 15$, $g(x) = x - 3$

5. $f(x) = x^3 + 2x^2 + 7x$, $g(x) = x$

6. $f(x) = x^2 - 5x - 8$, $g(x) = -x$

7. $f(x) = 2x^2 + 4x - 7$, $g(x) = 2x^2 + 3x - 5$

8. $f(x) = 6x^2 + 10$, $g(x) = 3x^2 + x - 10$

9. $f(x) = \sqrt{x - 3}$, $g(x) = x$

10. $f(x) = \sqrt{x - 4}$, $g(x) = -x$

11. $f(x) = \sqrt{4 - x^2}$, $g(x) = 2 + x$

12. $f(x) = \sqrt{x^2 - 9}$, $g(x) = x - 3$

In Exercises 13 to 28, evaluate the indicated function, where $f(x) = x^2 - 3x + 2$ and $g(x) = 2x - 4$.

13. $(f + g)(5)$

14. $(f + g)(-7)$

15. $(f + g)\left(\dfrac{1}{2}\right)$

16. $(f + g)\left(\dfrac{2}{3}\right)$

17. $(f - g)(-3)$

18. $(f - g)(24)$

19. $(f - g)(-1)$

20. $(f - g)(0)$

21. $(fg)(7)$

22. $(fg)(-3)$

23. $(fg)\left(\dfrac{2}{5}\right)$

24. $(fg)(-100)$

25. $\left(\dfrac{f}{g}\right)(-4)$

26. $\left(\dfrac{f}{g}\right)(11)$

27. $\left(\dfrac{f}{g}\right)\left(\dfrac{1}{2}\right)$

28. $\left(\dfrac{f}{g}\right)\left(\dfrac{1}{4}\right)$

In Exercises 29 to 36, find the difference quotient of the given function.

29. $f(x) = 2x + 4$

30. $f(x) = 4x - 5$

31. $f(x) = x^2 - 6$

32. $f(x) = x^2 + 11$

33. $f(x) = 2x^2 + 4x - 3$

34. $f(x) = 2x^2 - 5x + 7$

35. $f(x) = -4x^2 + 6$

36. $f(x) = -5x^2 - 4x$

In Exercises 37 to 48, find $g \circ f$ and $f \circ g$ for the given functions f and g.

37. $f(x) = 3x + 5$, $g(x) = 2x - 7$

38. $f(x) = 2x - 7$, $g(x) = 3x + 2$

39. $f(x) = x^2 + 4x - 1$, $g(x) = x + 2$

40. $f(x) = x^2 - 11x$, $g(x) = 2x + 3$

41. $f(x) = x^3 + 2x$, $g(x) = -5x$

42. $f(x) = -x^3 - 7$, $g(x) = x + 1$

43. $f(x) = \dfrac{2}{x + 1}$, $g(x) = 3x - 5$

44. $f(x) = \sqrt{x + 4}$, $g(x) = \dfrac{1}{x}$

45. $f(x) = \dfrac{1}{x^2}$, $g(x) = \sqrt{x - 1}$

46. $f(x) = \dfrac{6}{x - 2}$, $g(x) = \dfrac{3}{5x}$

47. $f(x) = \dfrac{3}{|5 - x|}$, $g(x) = -\dfrac{2}{x}$

48. $f(x) = |2x + 1|$, $g(x) = 3x^2 - 1$

In Exercises 49 to 64, evaluate each composite function, where $f(x) = 2x + 3$, $g(x) = x^2 - 5x$, and $h(x) = 4 - 3x^2$.

49. $(g \circ f)(4)$

50. $(f \circ g)(4)$

51. $(f \circ g)(-3)$

52. $(g \circ f)(-1)$

53. $(g \circ h)(0)$

54. $(h \circ g)(0)$

55. $(f \circ f)(8)$

56. $(f \circ f)(-8)$

57. $(h \circ g)\left(\dfrac{2}{5}\right)$

58. $(g \circ h)\left(-\dfrac{1}{3}\right)$

59. $(g \circ f)(\sqrt{3})$

60. $(f \circ g)(\sqrt{2})$

61. $(g \circ f)(2c)$

62. $(f \circ g)(3k)$

63. $(g \circ h)(k + 1)$

64. $(h \circ g)(k - 1)$

Calculator Exercises

In Exercises 65 to 68, use a calculator to determine each functional value. Give your answers accurate to four significant digits. Use $f(x) = x^2$, $g(x) = \sqrt{x}$, and $h(x) = 1/x$.

65. $g(644.5)$

66. $h(0.2354)$

67. $(f \circ h)(427.4)$

68. $(h \circ f)(9101)$

Supplemental Exercises

In Exercises 69 to 74, show that $(g \circ f)(x) = x$ and $(f \circ g)(x) = x$.

69. $f(x) = 2x + 3$, $g(x) = \dfrac{x - 3}{2}$

70. $f(x) = 4x - 5$, $g(x) = \dfrac{x + 5}{4}$

71. $f(x) = \dfrac{4}{x + 1}$, $g(x) = \dfrac{4 - x}{x}$

72. $f(x) = \dfrac{2}{1 - x}$, $g(x) = \dfrac{x - 2}{x}$

73. $f(x) = x^3 - 1$, $g(x) = \sqrt[3]{x + 1}$

74. $f(x) = -x^3 + 2$, $g(x) = \sqrt[3]{2 - x}$

75. Let x be the number of computer monitors to be manufactured. The manufacturing cost (in dollars) per computer monitor is given by the function

$$m(x) = \dfrac{60x + 34,000}{x}.$$

A computer store will sell the monitors by marking up the manufacturing cost per player $m(x)$ by 45 percent. Thus the selling price s is a function of $m(x)$ given by the equation $s[m(x)] = 1.45(m(x))$.

a. Express the selling price as a function of the number of monitors to be manufactured. That is, find $s \circ m$.

b. Find $(s \circ m)(24,650)$.

76. The number of bookcases b that a factory can produce per day is a function of the number of hours t it operates.

$$b(t) = 40t \quad \text{for} \quad 0 \le t \le 12$$

The daily cost c to manufacture b bookcases is given by the function $c(b) = 0.1b^2 + 90b + 800$. Evaluate and interpret your answer to each of the following:

a. $b(5)$ **b.** $c(5)$

c. $(c \circ b)(t)$ **d.** $(c \circ b)(10)$

77. Let $f(x) = \dfrac{x + 1}{x}$. Evaluate each of the following:

a. $f(1)$ **b.** $(f \circ f)(1)$

c. $(f \circ f \circ f)(1)$ **d.** $(f \circ f \circ f \circ f)(1)$

78. Let $f(x) = \sqrt{x}$. Evaluate each of the following:

a. $f(65536)$ **b.** $(f \circ f)(65536)$

c. $(f \circ f \circ f)(65536)$ **d.** $(f \circ f \circ f \circ f)(65536)$

3.8
Inverse Functions

The operations of addition and subtraction are said to be **inverse operations,** because one *undoes* the other. For example, if you start with a number, say, 10, and then add 4 and subtract 4, you will have the number 10 that you started with. Some functions are inverses of each other in the sense that one will undo the other. The function $g(x) = \frac{1}{2}x - 4$ undoes the function $f(x) = 2x + 8$. To illustrate, let $x = 10$. Then

$$f(x) = f(10) = 2(10) + 8 = 28.$$

Now evaluate $g[f(10)] = g(28)$.

$$g(f(10)) = g(28) = \frac{1}{2}(28) - 4 = 14 - 4 = 10.$$

Thus we started with $x = 10$, we evaluated $f(10)$, we evaluated $g[f(10)]$, and the end result was the number that we started with, namely, 10. This was not a coincidence. In Example 1 we show $f[g(x)] = g[f(x)] = x$ for all values of x where f and g are inverse functions.

Definition of an Inverse Function

> If f is a one-to-one function with domain X and range Y and g is a function with domain Y and range X, then g is the **inverse function** of f if and only if
>
> $$(f \circ g)(x) = x \quad \text{for all } x \text{ in the domain of } g$$
>
> and
>
> $$(g \circ f)(x) = x \quad \text{for all } x \text{ in the domain of } f.$$

EXAMPLE 1   **Verify That Functions Are Inverse Functions**

Verify that $g(x) = \dfrac{1}{2}x - 4$ is the inverse of $f(x) = 2x + 8$.

Solution   We need to show that $(f \circ g)(x) = x$ and that $(g \circ f)(x) = x$.

$$(f \circ g)(x) = f[g(x)] = f\left[\frac{1}{2}x - 4\right] = 2\left(\frac{1}{2}x - 4\right) + 8 = x - 8 + 8 = x$$

Also,

$$(g \circ f)(x) = g[f(x)] = g[2x + 8] = \frac{1}{2}[2x + 8] - 4 = x + 4 - 4 = x$$

Therefore the function g is the inverse function of f. This work also shows that f is the inverse function of g.

■ *Try Exercise* **2,** *page 196.*

Notice that the above definition of an inverse function requires f to be a one-to-one function. The reason for this restriction is now explained.

   If a one-to-one function is given as a set of ordered pairs, then its inverse will be the set of ordered pairs with their components interchanged. For example, the inverse of

$$\{(4, 7), (5, 2), (6, 11)\} \quad \text{is} \quad \{(7, 4), (2, 5), (11, 6)\}.$$

Now consider the function j defined by $j(x) = x^2 - 1$. Some of the ordered pairs of j are

$$(-2, 3), \quad (-1, 0), \quad (0, -1), \quad (1, 0), \quad \text{and} \quad (2, 3).$$

The inverse of j would contain the ordered pairs

$$(3, -2), \quad (0, -1), \quad (-1, 0), \quad (0, 1), \quad \text{and} \quad (3, 2).$$

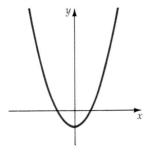

Figure 3.75
$j(x) = x^2 - 1$

This set of ordered pairs does *not* satisfy the definition of a function because there are ordered pairs with the same first component and *different* second components. For example, the ordered pairs $(3, -2)$ and $(3, 2)$ both have 3 as their first component, but they have different second components. This example illustrates that not all functions have inverses that are functions.

   Figure 3.75 is the graph of the function j. The horizontal line test from Section 7 indicates that j is *not* the graph of a *one-to-one* function. The horizontal line test can be used to show that the function $h(x) = \frac{1}{2}x^3$ is a one-to-one function. See Figure 3.76. Some of the ordered pairs of h are

$$(-2, -4), \quad \left(-1, -\frac{1}{2}\right), \quad (0, 0), \quad \left(1, \frac{1}{2}\right), \quad \text{and} \quad (2, 4).$$

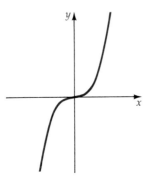

Figure 3.76
$h(x) = \frac{1}{2}x^3$

Because h is a one-to-one function, given any y in the range of h, there corresponds exactly one x in the domain of h. Thus, interchanging the coordinates of each ordered pair defined by h will yield a set of ordered pairs that is a function. This function with the coordinates interchanged is called the inverse function of h.

The previous examples illustrate that the one-to-one property is exactly what is required for a function to have an inverse function.

Condition for a Function to Have an Inverse Function

A function f will have an inverse function if and only if it is a one-to-one function.

The inverse of the function f is often denoted by f^{-1}. In Example 1, we verified that g was the inverse of f, so in this case the function g could be written as f^{-1}.

Caution   The notation f^{-1} for an inverse function does not mean $1/f$. The function denoted by $1/f$ is called the **reciprocal function** and is an entirely different function from f^{-1}. For example, from Example 1, we showed that $f^{-1}(x) = g(x) = \frac{1}{2}x - 4$, whereas $1/f(x) = 1/(2x + 8)$.

If a one-to-one function f is defined by an equation, then we use the following method to find the equation of the inverse f^{-1}.

Find the Equation for f^{-1}

To find the inverse f^{-1} of the one-to-one function f,

1. Substitute y for $f(x)$.
2. Interchange x and y.
3. Solve, if possible, for y in terms of x.
4. Substitute $f^{-1}(x)$ for y.
5. Verify that the domain of f is the range of f^{-1} and that the range of f is the domain of f^{-1}.

EXAMPLE 2   Find the Inverse of a One-to-One Function

Find the inverse of the one-to-one function $f(x) = 2x - 6$.

Solution   Begin by substituting y for $f(x)$:

$$y = 2x - 6$$

Interchanging x and y, we get

$$x = 2y - 6.$$

Now solve for y.

$$x + 6 = 2y$$

$$\frac{x + 6}{2} = y$$

This equation can be written as

$$y = \frac{1}{2}x + 3.$$

Using inverse notation, we write

$$f^{-1}(x) = \frac{1}{2}x + 3.$$

In this example, the function f has a domain of all real numbers and a range of all real numbers, so the inverse f^{-1} will also have a domain of all real numbers and a range of all real numbers.

■ *Try Exercise* **10,** *page 197.*

Sometimes it is necessary to collect the terms that contain a y factor together on the same side of the equation and then factor those terms to solve for y.

EXAMPLE 3   **Find the Inverse of a One-to-One Function**

Find the inverse of the function defined by $g(x) = \dfrac{2x}{x + 3}$.

Solution

$$g(x) = \frac{2x}{x + 3}$$

$$y = \frac{2x}{x + 3}$$

$$x = \frac{2y}{y + 3} \qquad \text{Interchange } x \text{ and } y.$$

$$x(y + 3) = 2y \qquad \text{Multiply by } (y + 3).$$

$$xy + 3x = 2y$$

$$xy - 2y = -3x \qquad \text{Collect the terms that contain}$$
$$\text{a factor of } y \text{ on one side.}$$

$$y(x - 2) = -3x \qquad \text{Factor out the } y.$$

$$y = \frac{-3x}{x - 2} \qquad \text{Solve for } y.$$

$$g^{-1}(x) = \frac{-3x}{x - 2} \quad \text{or} \quad \frac{3x}{2 - x}$$

■ *Try Exercise* **18,** *page 197.*

The function defined by $f(x) = x^2 - 4x$ is not a one-to-one function and therefore does not have an inverse function. However, the function

$G(x) = x^2 - 4x$ with domain restricted to $\{x | x \geq 2\}$ is a one-to-one function. It has an inverse function denoted by G^{-1}.

EXAMPLE 4   Find the Inverse Function and State Its Domain and Range

Find the inverse G^{-1} of the function $G(x) = x^2 - 4x$, for $x \geq 2$. State the domain and range of both G and G^{-1}.

Solution   First note that the domain of G is given as $\{x | x \geq 2\}$. The graph of G in Figure 3.77 shows that G has range: $\{y | y \geq -4\}$. Because the domain of G^{-1} is the range of G and the range of G^{-1} is the domain of G^{-1}, G^{-1} has domain: $\{x | x \geq -4\}$ and range: $\{y | y \geq 2\}$.

Now we proceed to find G^{-1}. The method shown uses the technique of completing the square.

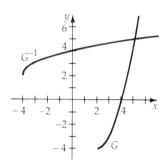

Figure 3.77
$G(x) = x^2 - 4x, \quad x \geq 2$
$G^{-1}(x) = 2 + \sqrt{x + 4}$

$$G(x) = x^2 - 4x \qquad \text{for } x \geq 2$$

$$y = x^2 - 4x$$

$$x = y^2 - 4y \qquad \text{Interchange } x \text{ and } y.$$

$$x + 4 = y^2 - 4y + 4 \qquad \text{To complete the square of } y^2 - 4y \text{ we need to add 4 to each side.}$$

$$x + 4 = (y - 2)^2 \qquad \text{Factor.}$$

$$\pm\sqrt{x + 4} = y - 2 \qquad \text{Apply the Square Root Theorem.}$$

$$2 \pm \sqrt{x + 4} = y$$

The range of G^{-1} is $\{y | y \geq 2\}$. Recall that the radical $\sqrt{x + 4}$ is a nonnegative number. Therefore, to make $G^{-1}(x) = 2 \pm \sqrt{x + 4}$ a real number greater than or equal to 2 requires that we consider only the nonnegative square root. Thus G^{-1} is given by

$$G^{-1}(x) = 2 + \sqrt{x + 4}$$

■ *Try Exercise* **32**, *page 197.*

The graphs of G and G^{-1} are in Figure 3.77. Note that the graphs appear to be symmetric with respect to the line $y = x$. This is in fact the case for a function and its inverse.

Symmetry Property of f and f^{-1}

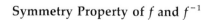

The graph of a function f and the graph of the inverse function f^{-1} are symmetric with respect to the line given by $y = x$.

The symmetry property of f and f^{-1} can be used to graph the inverse of a one-to-one function.

EXAMPLE 5   **Graph the Inverse of a Function**

Graph f^{-1} if f is the function defined by the following graph.

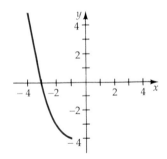

Figure 3.78

Solution   Sketch the graph of f^{-1} by drawing the reflection of f with respect to the line given by $y = x$.

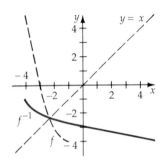

Figure 3.79

■ *Try Exercise* **36,** *page 197.*

Remark   In the graph the diagonal line given by $y = x$ is not a part of the graph of f or its inverse f^{-1}. It is included to illustrate that the graph of f and f^{-1} are symmetric with respect to this diagonal line.

The concept of an inverse function will play an important role as new functions are introduced in the following chapters.

EXERCISE SET 3.8

In Exercises 1 to 8, verify that f and g are inverse functions by showing that $(f \circ g)(x) = x$ and $(g \circ f)(x) = x$.

1. $f(x) = 2x + 1$, $g(x) = \dfrac{x - 1}{2}$

2. $f(x) = \dfrac{1}{2}x - 3$, $g(x) = 2x + 6$

3. $f(x) = 3x - 5$, $g(x) = \dfrac{x + 5}{3}$

4. $f(x) = -2x + 1$, $g(x) = -\dfrac{1}{2}x + \dfrac{1}{2}$

5. $f(x) = \dfrac{1}{x + 1}$, $g(x) = \dfrac{1 - x}{x}$

6. $f(x) = \dfrac{1}{x} + 1$, $g(x) = \dfrac{1}{x-1}$

7. $f(x) = \sqrt[3]{x-1}$, $g(x) = x^3 + 1$

8. $f(x) = x^3 - 2$, $g(x) = \sqrt[3]{x+2}$

In Exercises 9 to 24, find the inverse of the given function.

9. $f(x) = 4x + 1$

10. $g(x) = \dfrac{2}{3}x + 4$

11. $F(x) = -6x + 1$

12. $h(x) = -3x - 2$

13. $j(t) = 2t + 1$

14. $m(s) = -3s + 8$

15. $f(v) = 1 - v^3$

16. $u(t) = 2t^3 + 5$

17. $f(x) = \dfrac{-3x}{x+4}$

18. $G(x) = \dfrac{3x}{x-5}$

19. $M(t) = \dfrac{t-5}{t}$

20. $P(v) = \dfrac{2v}{v+1}$

21. $r(t) = \dfrac{1}{t^2}$, $t < 0$

22. $F(x) = \dfrac{1}{x}$, $x > 0$

23. $J(x) = x^2 + 4$, $x \geq 0$

24. $N(x) = 2x^2 + 1$, $x \leq 0$

In Exercises 25 to 34, find the inverse of f. State the domain and range of both f and f^{-1}.

25. $f(x) = x^2 + 3$, $x \geq 0$

26. $f(x) = x^2 - 4$, $x \geq 0$

27. $f(x) = \sqrt{x}$, $x \geq 0$

28. $f(x) = \sqrt{16 - x}$, $x \leq 16$

29. $f(x) = \sqrt{9 - x^2}$, $0 \leq x \leq 3$

30. $f(x) = \sqrt{16 - x^2}$, $-4 \leq x \leq 0$

31. $f(x) = x^2 - 4x + 1$, $x \geq 2$

32. $f(x) = x^2 + 6x - 6$, $x \geq -3$

33. $f(x) = x^2 + 8x - 9$, $x \leq -4$

34. $f(x) = x^2 - 2x - 2$, $x \leq 1$

In Exercises 35 to 40, graph f^{-1} if f is the function defined by the graph.

35.

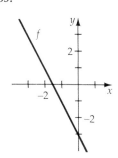

36.

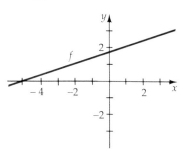

37.

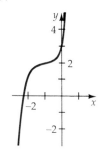

38.

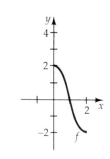

39.

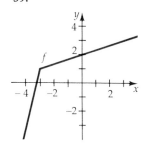

40.

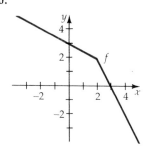

In Exercises 41 to 48, graph each function f and its inverse f^{-1} on the same coordinate plane. Note that the graphs are symmetric with respect to the line $y = x$.

41. $f(x) = 3x + 3$, $f^{-1}(x) = \dfrac{1}{3}x - 1$

42. $f(x) = x - 4$, $f^{-1}(x) = x + 4$

43. $f(x) = \dfrac{1}{2}x$, $f^{-1}(x) = 2x$

44. $f(x) = 2x - 4$, $f^{-1}(x) = \dfrac{1}{2}x + 2$

45. $f(x) = x^2 + 2$, $x \geq 0$, $f^{-1}(x) = \sqrt{x - 2}$, $x \geq 2$

46. $f(x) = x^2 - 3$, $x \geq 0$, $f^{-1}(x) = \sqrt{x + 3}$, $x \geq -3$

47. $f(x) = (x - 2)^2$, $x \leq 2$, $f^{-1}(x) = -\sqrt{x} + 2$, $x \geq 0$

48. $f(x) = (x + 3)^2$, $x \geq -3$, $f^{-1}(x) = \sqrt{x} - 3$, $x \geq 0$

Supplemental Exercises

In Exercises 49 to 52, find the inverse of the given function.

49. $f(x) = ax + b$, $a \neq 0$

50. $f(x) = ax^2 + bx + c$; $a \neq 0$, $x > -\dfrac{b}{2a}$

51. $f(x) = \dfrac{x-1}{x+1}$, $x \neq -1$

52. $f(x) = \dfrac{2-x}{x+2}$, $x \neq -2$

Only one-to-one functions have inverses that are functions. In Exercises 53 to 60, determine whether or not the given function is a one-to-one function.

53. $f(x) = x^2 + 8$

54. $v(s) = s^2 - 4$

55. $p(t) = \sqrt{9 - t}$

56. $v(t) = \sqrt{16 + t}$

57. $G(x) = -\sqrt{x}$

58. $K(x) = 1 - \sqrt{x - 5}$

59. $F(x) = |x| + x$

60. $T(x) = |x| - x$

In Exercises 61 to 64, assume that the given function has an inverse function.

61. If $f(5) = 2$, find $f^{-1}(2)$.   **62.** If $v(3) = 11$, find $v^{-1}(11)$.

63. If $s(4) = 60$, find $s^{-1}(60)$.   **64.** If $F(-8) = 5$, find $F^{-1}(5)$.

65. If the ordered pair (a, b) belongs to the graph of the function f and f has an inverse, then (b, a) belongs to the graph of f^{-1}. Prove that the points $P(a, b)$ and $Q(b, a)$ are symmetric with respect to the graph of the line $y = x$ by using the definition of symmetry with respect to a line found on page 138.

66. Graph $f(x) = -x + 3$. Use the graph to explain why f is its own inverse.

3.9

Variation and Applications

Many real-life situations involve variables that are related by a type of function called a **variation.** For example, a fish jumping or a stone thrown into a pond ge•erates circular ripples whose circumference and diameter are increasing. The equation $C = \pi d$ expresses the relationship between the circumference C of a circle and its diameter d. Notice that as d increases, C increases. In fact, if d doubles in size, then C also doubles in size. The circumference C is said to **vary directly** as the diameter d.

Definition of Direct Variation

> The variable y **varies directly** as the variable x, or y **is directly proportional** to x, if and only if
>
> $$y = kx,$$
>
> where k is a constant called the **constant of proportionality** or the **variation constant.**

Direct variations occur in many daily applications. For example, the cost of a newspaper is 25 cents. The cost C to purchase n newspapers is directly proportional to the number n. That is, $C = 25n$. In this example the variation constant is 25.

To solve a problem that involves a variation, we typically write a general equation that relates the variables and then use given information to solve for the variation constant.

EXAMPLE 1   Solve a Direct Variation

The distance sound travels varies directly as the time it travels. If sound travels 1340 meters in 4 seconds, find the distance sound will travel in 5 seconds.

Solution The first step is to write an equation that relates the distance d to the time t. Since d varies directly as t, our equation is

$$d = kt.$$

Because $d = 1340$ when $t = 4$, we obtain

$$1340 = k(4) \quad \text{which implies} \quad k = \frac{1340}{4} = 335.$$

Therefore, the specific equation that relates the distance d sound travels in t seconds is

$$d = 335t.$$

To find the distance sound travels in 5 seconds, replace t with 5 to produce

$$d = 335(5) = 1675.$$

Therefore, under the same conditions sound will travel 1675 meters in 5 seconds

■ *Try Exercise* **14,** *page 203.*

Remark If x and y are related by the equation

$$y = kx^n,$$

then y is said to **vary directly as the nth power of x.**

EXAMPLE 2   Solve a Variation of the Form $y = kx^2$

The distance s that an object falls from rest (neglecting air resistance) varies directly as the square of the time t that it has been falling. If an object falls 64 feet in 2 seconds, how far will it fall in 10 seconds?

Solution Since s varies directly as the square of t,

$$s = kt^2.$$

The variable s is 64 when t equals 2, so

$$64 = k(2^2) \quad \text{which implies} \quad k = \frac{64}{4} = 16.$$

The specific equation that relates the distance s an object falls in t seconds is

$$s = 16t^2.$$

Letting $t = 10$ yields

$$s = 16(10^2) = 16(100) = 1600.$$

Therefore, under the same conditions an object will fall 1600 feet in 10 seconds.

■ *Try Exercise* **24,** *page 203.*

Definition of Inverse Variation

> The variable y **varies inversely** as the variable x, or y **is inversely proportional to** x, if and only if
> $$y = \frac{k}{x},$$
> where k is the variation constant.

Many applications involve inverse variations.

EXAMPLE 3 Solve an Inverse Variation

Boyle's Law states that the pressure P of a sample of gas at a constant temperature varies inversely as the volume V. The pressure of a gas in a balloon with volume 8 cubic inches is found to be 12 pounds per square inch. If the pressure is reduced, the volume of the balloon increases to 20 cubic inches. Find the new pressure of the gas.

Solution   Since P varies inversely as V,

$$P = \frac{k}{V}.$$

The pressure P equals 12 when the volume V is 8, so

$$12 = \frac{k}{8} \quad \text{which implies} \quad k = 96.$$

Consequently, the specific formula for P is

$$P = \frac{96}{V}.$$

When the volume is 20 cubic inches, we have

$$P = \frac{96}{20} = 4.8 \text{ pounds per square inch}.$$

■ *Try Exercise 28, page 203.*

Remark   If x and y are related by the equation $y = k/x^n$, then y is said to **vary inversely as the nth power of x.**

Some variations involve more than two variables.

Definition of Joint Variation

> The variable z **varies jointly** as the variables x and y if and only if
> $$z = kxy,$$
> where k is a constant.

EXAMPLE 4   **Solve a Joint Variation**

The cost of insulating the ceiling of a house varies jointly with the thickness of the insulation and the area of the ceiling. It costs $175 to insulate a 2100-square foot ceiling with 4-inch thick insulation. Find the cost of insulating a 2400-square foot ceiling with insulation that is 6 inches thick.

Solution   Since the cost C varies jointly as the area A of the ceiling and the thickness T of the insulation, we can write the variation

$$C = kAT.$$

Using the fact that $C = 175$ when $A = 2100$ and $T = 4$ gives us

$$175 = k(2100)(4) \quad \text{which implies} \quad k = \frac{175}{(2100)(4)} = \frac{1}{48}.$$

Consequently, the specific formula for C is

$$C = \frac{1}{48}AT.$$

Now when $A = 2400$ and $T = 6$, we have

$$C = \frac{1}{48}(2400)(6) = 300.$$

Thus the cost of insulating the 2400-square foot ceiling with 6-inch insulation is $300.

■ *Try Exercise* **30**, *page 203.*

The definition of a joint variation can be extended to include powers.

Remark   If $z = kx^n y^m$, then z is said to **vary jointly as the nth power of x and the mth power of y.**

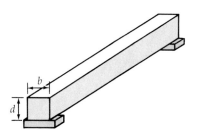

Figure 3.80

EXAMPLE 5   **Solve a Joint Variation Involving Powers**

The load L that can be safely supported by a horizontal beam of given length varies jointly as the width w and the square of the depth d. See Figure 3.80. If a beam with width 2 inches and depth 4 inches can safely support a load of 400 pounds, determine the load that a beam (of the same length and material) of width 4 inches and depth 6 inches can safely support.

Solution   The general formula for the safe load is

$$L = kwd^2.$$

Substitution of known values produces

$$400 = k(2)(4^2) = 32k.$$

Solving for k yields

$$k = \frac{400}{32} = 12.5.$$

Therefore $L = 12.5(wd^2)$.

Substituting 4 for w and 6 for d produces

$$L = 12.5(4)(6^2) = 1800 \text{ pounds}.$$

■ *Try Exercise* **32,** *page 204.*

Combined variations involve more than one type of variation.

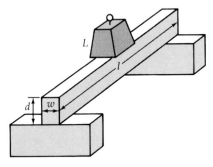

Figure 3.81

EXAMPLE 6   Solve a Combined Variation

The weight that a horizontal beam with rectangular cross section can safely support varies jointly as the width and square of the depth of the cross section and inversely as the length of the beam. See Figure 3.81. If a 4- by 4-inch beam 10 feet long safely supports a load of 256 pounds, what load L can a beam made of the same material and with a width w of 4 inches, a depth d of 6 inches, and a length l of 16 feet safely support?

Solution   The general variation equation is

$$L = k\frac{wd^2}{l}.$$

Using the given data yields

$$256 = k\frac{4(4^2)}{10}.$$

Solving for k produces $k = 40$, and thus the specific formula for L is

$$L = 40\frac{wd^2}{l}.$$

Substituting 4 for w, 6 for d, and 16 for l gives

$$L = 40\frac{4(6^2)}{16} = 360 \text{ pounds}.$$

■ *Try Exercise* **34,** *page 204.*

EXERCISE SET 3.9

In Exercises 1 to 12, write an equation that represents the relationship between the given variables. Use k as the variation constant.

1. d varies directly as t.

2. r varies directly as the square of s.

3. y varies inversely as x.

4. p is inversely proportional to q.

5. m varies jointly as n and p.

6. t varies jointly as r and the cube of s.

7. V varies jointly as l, w, and h.

8. u varies directly as v and inversely as the square of w.

9. A is directly proportional to the square of s.

10. A varies jointly as h and the square of r.

11. F varies jointly as m_1 and m_2 and inversely as the square of d.

12. T varies jointly as t and r and the square of a.

In Exercises 13 to 20, write the equation that relates the variables and then use the given data to solve for the variation constant.

13. y varies directly as x and $y = 64$ when $x = 48$.

14. m is directly proportional to n and $m = 92$ when $n = 23$.

15. r is directly proportional to the square of t and $r = 144$ when $t = 108$.

16. C varies directly as r and $C = 94.2$ when $r = 15$.

17. T varies jointly as r and the square of s and $T = 210$ when $r = 30$ and $s = 5$.

18. u varies directly as v and inversely as the square root of w and $u = 0.04$ when $v = 8$ and $w = 0.04$.

19. V varies jointly as l, w, and h and $V = 240$ when $l = 8$ and $w = 6$ and $h = 5$.

20. t varies directly as the cube of r and inversely as the square root of s and $t = 10$ when $r = 5$ and $s = 0.09$.

21. Charles's law states that the volume V occupied by a gas (at a constant pressure) is directly proportional to its absolute temperature T. An experiment with a balloon shows that the volume of the balloon is 0.85 liters at 270 K (absolute temperature).[1] What will the volume of the balloon be when its temperature is 324 K?

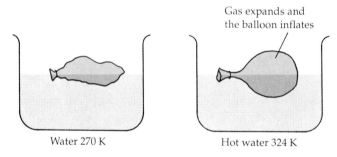

Gas expands and the balloon inflates

Water 270 K   Hot water 324 K

22. Hooke's law states that the distance a spring stretches varies directly as the weight on the spring. A weight of 80 pounds stretches a spring 6 inches. How far will a weight of 100 pounds stretch the spring?

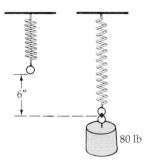

6"

80 lb

[1] Absolute temperature is a measure using the Kelvin scale. A degree on the Kelvin scale is the same size as a Celsius degree; however, 0 on the Kelvin scale corresponds to $-273°$ on the Celsius scale.

23. The pressure a liquid exerts at a given point on a submarine is directly proportional to the depth of the point below the surface of the liquid. If the pressure at a depth of 3 feet is 187.5 pounds per square foot, find the pressure at a depth of 7 feet.

24. The range of a projectile is directly proportional to the square of its velocity. If a motorcyclist can make a jump of 140 feet by coming off a ramp at 60 mph, find the distance the motorcyclist could expect to jump if the speed coming off the ramp is increased to 65 mph.

25. The period T (the time it takes a pendulum to make one complete oscillation) varies directly as the square root of its length L. A pendulum 3 feet long has a period of 1.8 seconds.

   a. Find the period of a pendulum 10 feet long.

   b. What is the length of a pendulum that *beats seconds* (that is, it has a 2-second period)?

26. The area of a projected picture on a movie screen varies directly as the square of the distance from the projector to the screen. If a distance of 20 feet produces a picture with an area of 64 square feet, what distance will produce an area of 100 square feet?

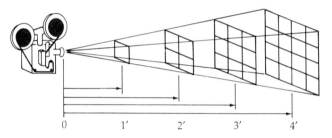

0   1'   2'   3'   4'

27. The loudness measured in decibels of a stereo speaker is inversely proportional to the square of the distance of the listener from the speaker. The loudness is 28 decibels at a distance of 8 feet. What is the loudness when you are 4 feet from the speaker.

28. The illumination a source of light provides is inversely proportional to the square of the distance from the source. If the illumination at a distance of 10 feet from the source is 50 footcandles, what is the illumination at a distance of 15 feet from the source?

29. The cost of insulating the exterior walls of a house varies jointly with the thickness of the insulation and the area of the walls. It cost $180 to insulate 1600 square feet of walls with insulation 3 inches thick. Find the cost of insulating 2200 square feet of exterior walls with insulation 4 inches thick.

30. The simple interest earned in a given time period varies jointly with the principal and the interest rate. A principal of $2600 at an interest rate of 4 percent yields $78. During the same time period, determine the simple interest earned on $4400 at an interest rate of 6 percent.

31. The volume V of a right circular cone varies jointly as the square of the radius r and the height h. What happens to V if:

   a. r is tripled?

   b. h is tripled?

   c.   Both r and h are tripled?

32. The load L a horizontal beam can safely support varies jointly as the breadth b and the square of the depth d. If a beam with breadth 2 inches and depth 6 inches will safely support up to 200 pounds, how many pounds will a beam of the same length that has breadth 4 inches and depth 4 inches be expected to support?

33. The **Ideal Gas Law** states that the volume V of a gas varies jointly as the number of moles of gas n and the absolute temperature T and inversely as the pressure P. What happens to V if n is tripled and P is reduced by a factor of one-half?

34. The maximum load a cylindrical column of circular cross section can support varies directly as the fourth power of the diameter and inversely as the square of the height. If a column 2 feet in diameter and 10 feet high will support up to 6 tons, how much of a load will a column 3 feet in diameter and 14 feet high support?

Calculator Problems

35. A meteorite approaching the earth has a velocity that varies inversely as the square root of the distance from the center of the earth. If the meteorite has a velocity of 3 miles per second at 4900 miles from the center of the earth, find the velocity of the meteorite when it is 4225 miles from the center of the earth.

36. A commuter airline has found that the average number of passengers per month between any two cities on its service routes is directly proportional to each city's population and inversely proportional to the square of the distance between them. Alameda has a population of 50,000 and it is 80 miles from Baltic, which has a population of 64,000. Airline records indicate an average of 3840 passengers per month travel between Alameda and Baltic. Estimate the average number of passengers per day the airline could expect between Alameda and Crystal Lake, which has a population of 144,000 and is 160 miles from Alameda.

37. The frequency f of vibration of a piano string varies directly as the square root of the tension T on the string and inversely as the length L of the string. If the middle a string has a frequency of 440 vibrations per second, find the frequency of a string that has 1.25 times as much tension and is six-fifths as long.

39. The load L a horizontal beam can safely support varies jointly as the breadth b and the square of the depth d and inversely as the length l. If a 12-foot beam with breadth 4 inches and depth 8 inches will safely support 800 pounds, how many pounds will a 16-foot beam with a breadth of 3.5 inches and depth of 6 inches be expected to support?

40. The force needed to keep a car from skidding on a curve varies jointly as the weight of the car and the square of the speed and inversely as the radius of the curve. It takes 2800 pounds of force to keep a 1800-pound car from skidding on a curve with radius 425 feet at 45 mph. What force is needed to keep the same car from skidding when it takes a similar curve with radius 450 feet at 55 mph?

Supplemental Exercises

41. **Kepler's Third Law** states that the time T needed for a planet to make one complete revolution about the sun is directly proportional to the $\frac{3}{2}$ power of the average distance d between the planet and the sun. The earth, which averages 93 million miles from the sun, completes one revolution in 365 days. Find the average distance from the sun to Mars if Mars completes one revolution about the sun in 686 days.

42. The weight W of an object varies inversely as the square of the distance d from the object to the center of the earth. At what altitude will a rocket weigh half of what it weighs at sea level? Assume sea level to be 4000 miles from the center of the earth.

43. If $f(x)$ varies directly as x, prove that $f(x_2) = f(x_1)\dfrac{x_2}{x_1}$.

44. Use the formula in Exercise 43 to solve the following direct variation *without* solving for the variation constant.

The distance a spring stretches varies directly as the force applied. An experiment shows that a force of 17 kilograms stretches the spring 8.5 centimeters. How far will a 22-kilogram force stretch the spring?

45. If $f(x)$ varies inversely as x, prove that $f(x_2) = f(x_1)\dfrac{x_1}{x_2}$.

46. Use the formula in Exercise 45 to solve the following inverse variation *without* solving for the variation constant. The volume of a gas varies inversely as pressure (assuming the temperature remains constant). An experiment shows that a particular gas has a volume of 2.4 liters under a pressure of 280 grams per square centimeter. What volume will the gas have when a pressure of 330 grams per square centimeter is applied?

Chapter 3 Review

3.1 A Two-Dimensional Coordinate System and Graphs

The Pythagorean Theorem   If a and b denote the lengths of the legs of a right triangle and c the length of the hypotenuse, then $c^2 = a^2 + b^2$.

The Distance Formula   The distance d between the points represented by (x_1, y_1) and (x_2, y_2), is $d = \sqrt{(x_2 - x_1)^2 + (y_2 - y_1)^2}$.

3.2 Graphing Techniques

The graph of an equation is symmetric with respect to the

- y-axis if the replacement of x with $-x$ leaves the equation unaltered.
- x-axis if the replacement of y with $-y$ leaves the equation unaltered.
- origin if the replacement of x with $-x$ and y with $-y$ leaves the equation unaltered.
- line $y = x$ if the replacement of x with y and y with x leaves the equation unaltered.

The standard form of the equation of a circle with center at (h, k) and radius r is $(x - h)^2 + (y - k)^2 = r^2$.

3.3 Functions

Definition of a Function
A function f from a set D to a set R is a correspondence, or rule, that pairs each element of D with exactly one element of R. The set D is called the domain of f, and the set R is called the range of f.

The function f is an even function if $f(-x) = f(x)$ for all x in the domain of f.

The function f is an odd function if $f(-x) = -f(x)$ for all x in the domain of f.

3.4 Graphs of Functions

Alternate Definition of a Function
A function is a set of ordered pairs in which no two ordered pairs that have the same first component have different second components.

A graph is the graph of a function if and only if no vertical line intersects the graph at more than one point. If any horizontal line intersects the graph of a function at most once, then the graph is the graph of a one-to-one function.

If f is a function and c is a positive constant, then

- $y = f(x) + c$ is the graph of $y = f(x)$ shifted *vertically* c units up.
- $y = f(x) - c$ is the graph of $y = f(x)$ shifted *vertically* c units down.
- $y = f(x + c)$ is the graph of $y = f(x)$ shifted *horizontally* c units to the left.
- $y = f(x - c)$ is the graph of $y = f(x)$ shifted *horizontally* c units to the right.

The graph of

- $y = -f(x)$ is the graph of $y = f(x)$ reflected across the x-axis.
- $y = f(-x)$ is the graph of $y = f(x)$ reflected across the y-axis.

3.5 Linear Functions and Lines

A function is a linear function if it can be written in the form $f(x) = mx + b$ where m and b are real constants.

The slope m of the line passing through the points $P_1(x_1, y_1)$ and $P_2(x_2, y_2)$ with $x_1 \neq x_2$ is given by

$$m = \frac{y_2 - y_1}{x_2 - x_1}.$$

The graph of the equation $y = mx + b$ has slope m and y intercept $(0, b)$.

Two nonvertical lines are parallel if and only if their slopes are equal. Two lines with slopes m_1 and m_2 are perpendicular if and only if $m_1 m_2 = -1$.

3.6 Quadratic Functions

A quadratic function is a function that can be represented by an equation of the form $f(x) = ax^2 + bx + c$ where a, b, and c are real numbers and $a \neq 0$.

The vertex of the graph of $f(x) = ax^2 + bx + c$ is

$$\left(-\frac{b}{2a}, f\left(-\frac{b}{2a} \right) \right).$$

Every quadratic function $f(x) = ax^2 + bx + c$ can be written in the standard form $f(x) = a(x - h)^2 + k$, $a \neq 0$. The graph of f is a parabola with vertex (h, k). The parabola is symmetric with respect to the vertical line $x = h$, which is called the axis of the parabola. The parabola opens up if $a > 0$, and it opens down if $a < 0$.

3.7 The Algebra of Functions

For all values of x for which both $f(x)$ and $g(x)$ are defined, we define the following functions:

Sum: $(f + g)(x) = f(x) + g(x)$

Difference: $(f - g)(x) = f(x) - g(x)$

Product: $(fg)(x) = f(x) \cdot g(x)$

Quotient: $\left(\dfrac{f}{g}\right)(x) = \dfrac{f(x)}{g(x)}, \quad g(x) \neq 0$

For the functions f and g, the composite function or composition of f by g, is given by $(g \circ f)(x) = g[f(x)]$ for all x in the domain of f such that $f(x)$ is in the domain of g.

3.8 Inverse Functions

If f is a one-to-one function with domain X and range Y, and g is a function with domain Y and range X, then g is the inverse function of f if and only if $(f \circ g)(x) = x$ for all x in the domain of g and $(g \circ f)(x) = x$ for all y in the domain of f.

A function f will have an inverse function if and only if it is a one-to-one function. The graph of a function f and the graph of the inverse function f^{-1} are symmetric with respect to the line given by $y = x$.

3.9 Variation and Applications

The variable y varies directly as the variable x, if and only if $y = kx$, where k is a constant called the variation constant.

The variable y varies inversely as the variable x, if and only if $y = k/x$, where k, is the variation constant.

The variable z varies jointly as the variables x and y if and only if $z = kxy$, where k, is the variation constant.

CHALLENGE EXERCISES

In Exercises 1 to 10, answer true or false. If the statement is false, give an example.

1. Let f be any function. Then $f(a) = f(b)$ implies that $a = b$.

2. Every function has an inverse function.

3. If $(f \circ g)(a) = a$ and $(g \circ f)(a) = a$ for some constant a, then f and g are inverse functions.

4. Let f be a function such that $f(x) = f(x + 4)$ for all real numbers x. If $f(2) = 3$, then $f(18) = 3$.

5. For all functions f, $[f(x)]^2 = f[f(x)]$.

6. Let f be any function. Then for all a and b such that $f(b) \neq 0$ and $b \neq 0$,

$$\frac{f(a)}{f(b)} = \frac{a}{b}.$$

7. The **identity function** $f(x) = x$ is its own inverse.

8. If f is the function given by $f(x) = |x|$, then $f(a + b) = f(a) + f(b)$ for all real numbers a and b.

9. If f is the function given by $f(x) = |x|$, then $f(ab) = f(a)f(b)$ for all real numbers a and b.

10. If f is a one-to-one function and a and b are real numbers in the domain of f with $a < b$, then $f(a) \neq f(b)$.

REVIEW EXERCISES

In Exercises 1 to 4, use the Pythagorean Theorem to find the unknown. The letters a and b represent the lengths of the legs of a right triangle, and c represents the length of the hypotenuse.

1. $a = 6$, $b = 8$. Find c.   **2.** $a = 11$, $c = 20$. Find b.

3. $b = 12$, $c = 13$. Find a.   **4.** $a = 7$, $b = 14$. Find c.

In Exercises 5 and 6, find the distance between the points whose coordinates are given.

5. $(-3, 2)$, $(7, 11)$ **6.** $(5, -4)$, $(-3, -8)$

In Exercises 7 and 8, find the midpoint of the line segment with endpoints as given.

7. $(2, 8)$, $(-3, 12)$ **8.** $(-4, 7)$, $(8, -11)$

In Exercises 9 and 10, sketch a graph that is symmetric to the given graph with respect to the **a.** x-axis, **b.** y-axis, **c.** origin.

9.   **10.**

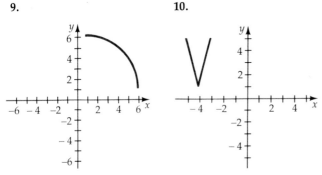

In Exercises 11 to 18, determine if the graph of each equation is symmetric with respect to the **a.** x-axis, **b.** y-axis, **c.** origin.

11. $y = x^2 - 7$

12. $x = y^2 + 3$

13. $y = x^3 - 4x$

14. $y^2 = x^2 + 4$

15. $\dfrac{x^2}{3^2} + \dfrac{y^2}{4^2} = 1$

16. $xy = 8$

17. $|y| = |x|$

18. $|x + y| = 4$

In Exercises 19 and 20, determine the center and radius of the circle with the given equation.

19. $(x - 3)^2 + (y + 4)^2 = 81$

20. $x^2 + y^2 + 10x + 4y + 20 = 0$

In Exercises 21 and 22, find the equation in standard form of a circle that satisfies the given conditions.

21. Center $C = (2, -3)$,   radius $r = 5$

22. Center $C = (-5, 1)$,   passing through $(3, 1)$

23. If $f(x) = 3x^2 + 4x - 5$, find

    a. $f(1)$ **b.** $f(-3)$ **c.** $f(t)$

    d. $f(x + h)$ **e.** $3f(t)$ **f.** $f(3t)$

24. If $g(x) = \sqrt{64 - x^2}$, find

    a. $g(3)$ **b.** $g(-5)$ **c.** $g(8)$

    d. $g(-x)$ **e.** $2g(t)$ **f.** $g(2t)$

25. If $f(x) = x^2 + 4x$ and $g(x) = x - 8$, find

    a. $(f \circ g)(3)$ **b.** $(g \circ f)(-3)$

    c. $(f \circ g)(x)$ **d.** $(g \circ f)(x)$

26. If $f(x) = 2x^2 + 7$ and $g(x) = |x - 1|$, find

    a. $(f \circ g)(-5)$ **b.** $(g \circ f)(-5)$

    c. $(f \circ g)(x)$ **d.** $(g \circ f)(x)$

27. If $f(x) = 4x^2 - 3x - 1$, find the difference quotient

$$\frac{f(x + h) - f(x)}{h}.$$

28. If $g(x) = x^3 - x$, find the difference quotient

$$\frac{g(x + h) - g(x)}{h}.$$

In Exercises 29 to 32, determine the domain of the function represented by the given equation.

29. $f(x) = -2x^2 + 3$ **30.** $f(x) = \sqrt{6 - x}$

31. $f(x) = \sqrt{25 - x^2}$ **32.** $f(x) = \dfrac{3}{x^2 - 2x - 15}$

In Exercises 33 to 38, sketch the graph of f. Find the interval(s) in which f is **a.** increasing, **b.** constant, **c.** decreasing.

33. $f(x) = |x - 3| - 2$ **34.** $f(x) = x^2 - 5$

35. $f(x) = |x + 2| - |x - 2|$ **36.** $f(x) = [\![x + 3]\!]$

37. $f(x) = \dfrac{1}{2}x - 3$ **38.** $f(x) = \sqrt[3]{x}$

In Exercises 39 to 44, sketch the graph of g. **a.** Find the domain and the range of g. **b.** State whether g is even, odd, or neither even nor odd.

39. $g(x) = -x^2 + 4$ **40.** $g(x) = -2x - 4$

41. $g(x) = |x - 2| + |x + 2|$ **42.** $g(x) = \sqrt{16 - x^2}$

43. $g(x) = x^3 - x$ **44.** $g(x) = 2[\![x]\!]$

In Exercises 45 and 46, find the slope-intercept form of the equation of the line through the two points.

45. $(-1, 3)$ $(4, -7)$ **46.** $(0, 0)$ $(7, 11)$

47. Find the slope-intercept form of the equation of the line parallel to the graph of $3x - 4y = 8$ and passing through $(2, 11)$.

48. Find the slope-intercept form of the equation of the line perpendicular to the graph of $2x = -5y + 10$ and passing through $(-3, -7)$.

In Exercises 49 to 54, find the vertex of the graph of the quadratic function.

49. $f(x) = 3x^2 - 6x + 11$ **50.** $h(x) = 4x^2 - 10$

51. $k(x) = -6x^2 + 60x + 11$ **52.** $m(x) = 14 - 8x - x^2$

53. $s(t) = -16t^2 + 1050$ **54.** $d(t) = 2t^2 - 10t$

In Exercises 55 to 60, use the method of completing the square to write each quadratic equation in its standard form.

55. $f(x) = x^2 + 6x + 10$ **56.** $f(x) = 2x^2 + 4x + 5$

57. $f(x) = -x^2 - 8x + 3$ **58.** $f(x) = 4x^2 - 6x + 1$

59. $f(x) = -3x^2 + 4x - 5$ **60.** $f(x) = x^2 - 6x + 9$

In Exercises 61 to 66, first write the function in standard form, and then make use of transformations to graph the function.

61. $F(x) = x^2 + 4x - 7$ **62.** $A(x) = x^2 - 6x - 5$

63. $P(x) = 3x^2 - 4$ **64.** $G(x) = 2x^2 - 8x + 3$

65. $W(x) = -4x^2 - 6x + 6$ **66.** $T(x) = -2x^2 - 10x$

In Exercises 67 and 68, use the given functions f and g to find $f + g$, $f - g$, fg, and f/g. State the domain of each.

67. $f(x) = x^2 - 9$, $g(x) = x + 3$

68. $f(x) = x^3 + 8$, $g(x) = x^2 - 2x + 4$

In Exercises 69 to 72, determine if the given functions are inverses.

69. $F(x) = 2x - 5$, $G(x) = \dfrac{x + 5}{2}$

70. $h(x) = \sqrt{x}$, $k(x) = x^2$, $x \geq 0$

71. $l(x) = \dfrac{x + 3}{x}$, $m(x) = \dfrac{3}{x - 1}$

72. $p(x) = \dfrac{x - 5}{2x}$, $q(x) = \dfrac{2x}{x - 5}$

In Exercises 73 to 76, find the inverse of the function. Sketch the graph of the function and its inverse on the same set of coordinates axes.

73. $f(x) = 3x - 4$ **74.** $g(x) = -2x + 3$

75. $h(x) = -\frac{1}{2}x - 2$ **76.** $k(x) = \frac{1}{x}$

77. Find two numbers whose sum is 50 and whose product is a maximum.

78. Find two numbers whose difference is 10 and whose sum of their squares is a minimum.

79. The roadway of the Golden Gate Bridge is 220 feet above the water. The height h of a rock dropped from the bridge is a function of the time t it has fallen. If h is measured in feet and t is measured in seconds, then the function is given by $h(t) = -16t^2 + 220$. Use the function to find the time it will take the rock to hit the water.

80. The suspension cables of the Golden Gate Bridge approximate the shape of a parabola. If h is the height of the cables above the roadway in feet, and $|x|$ is the distance in feet from the center of the bridge, then the parabolic shape of the cables is represented by

$$h(x) = \frac{1}{8820}x^2 + 25, \quad -2100 \leq x \leq 2100$$

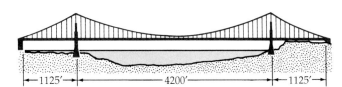

├—1125'—┤├————4200'————┤├—1125'—┤

a. Find the height of the cables 1050 feet from the center of the bridge.

b. The towers that support the cables are 2100 feet from the center of the bridge. Find the height (above the roadway) of the towers that support the cables.

4

Polynomial and Rational Functions

Euler calculated without apparent effort, just as men breathe, as eagles sustain themselves in the wind.

DOMINIQUE FRANÇOIS ARAGO (1786–1853)

PERFECT NUMBERS

A natural number is a **perfect number** if it is the sum of its factors other than itself. The factors of 6, other than 6, are 1, 2, and 3. Thus 6 is a perfect number because

$$1 + 2 + 3 = 6.$$

Ancient Greek philosophers knew of the first four perfect numbers:

$$6, \quad 28, \quad 496, \quad \text{and} \quad 8128.$$

They conjectured that there are infinitely many perfect numbers and that every perfect number is an even number. After more than 2000 years, we still do not know whether these conjectures are true or false.

The fifth, sixth, and seventh perfect numbers, which were not discovered until 1603, are

$$33{,}550{,}336, \quad 8{,}589{,}896{,}056, \quad \text{and} \quad 137{,}438{,}691{,}328.$$

The eighth perfect number, discovered by Leonhard Euler in 1772, is

$$2{,}305{,}483{,}008{,}139{,}952{,}128.$$

The search for perfect numbers still continues. To date, only thirty-five perfect numbers have been discovered. The largest of these perfect numbers has so many digits that it would fill a book of several hundred pages.

4.1

Polynomial Division and Synthetic Division

Division of a polynomial by another polynomial plays a major role in this chapter. Dividing a polynomial by another polynomial is similar to the long division process used for dividing positive integers. For example, to divide $(x^2 + 9x - 16)$ by $(x - 3)$, we use the following procedure.

Step 1

$$
\begin{array}{r}
x \\
x - 3 \overline{) x^2 + 9x - 16} \\
\underline{x^2 - 3x} \\
12x - 16
\end{array}
$$

Divide: $x \overline{)x^2} = \dfrac{x^2}{x} = x$

Multiply: $x(x - 3) = x^2 - 3x$

Subtract: $(x^2 + 9x) - (x^2 - 3x) = 12x$

$$
\begin{array}{r}
x +\ 12 \\
x - 3\overline{)\, x^2 +\ 9x\ -\ 16} \\
\underline{x^2 -\ 3x}
\end{array}
$$

Step 2

$12x - 16$ Divide: $x\overline{)\,12x} = \dfrac{12x}{x} = 12$

$\underline{12x - 36}$ Multiply: $12(x - 3) = 12x - 36$

20 Subtract: $(12x - 16) - (12x - 36) = 20$

Thus $(x^2 + 9x - 16) \div (x - 3) = x + 12$, with a remainder of 20.

In this example, $x^2 + 9x - 16$ is called the **dividend**, $x - 3$ is the **divisor**, $x + 12$ is the **quotient**, and 20 is the **remainder**. The dividend is equal to the product of the divisor and the quotient, plus the remainder. That is,

$$
\underbrace{x^2 + 9x - 16}_{\text{Dividend}} = \underbrace{(x - 3)}_{\text{Divisor}} \cdot \underbrace{(x + 12)}_{\text{Quotient}} + \underbrace{20}_{\text{Remainder}}
$$

The above result is a special case of a theorem known as the **Division Algorithm for Polynomials.**

The Division Algorithm for Polynomials

> If $P(x)$ and $D(x)$ are polynomials such that $D(x) \neq 0$, then there exists unique polynomials $Q(x)$ and $R(x)$ such that $P(x) = D(x)Q(x) + R(x)$ where either $R(x) = 0$, or the degree of $R(x)$ is less than the degree of $D(x)$.

The polynomial $P(x)$ is the dividend, $D(x)$ is the divisor, $Q(x)$ is the quotient and the polynomial $R(x)$ is the remainder.

$$
\underbrace{P(x)}_{\text{Dividend}} = \underbrace{D(x)}_{\text{Divisor}} \cdot \underbrace{Q(x)}_{\text{Quotient}} + \underbrace{R(x)}_{\text{Remainder}}
$$

Multiplying both sides of $P(x) = D(x)Q(x) + R(x)$ by $1/D(x)$ produces the fractional form

$$
\frac{P(x)}{D(x)} = Q(x) + \frac{R(x)}{D(x)}.
$$

Remark 1   If $R(x) = 0$, then $D(x)$ is a factor of $P(x)$.

Remark 2   If the degree of $D(x)$ is greater than the degree of $P(x)$, then $Q(x) = 0$, and $R(x) = P(x)$.

EXAMPLE 1   **Divide Polynomials**

Perform the indicated division.

$$
\frac{x^4 + 3x^2 - 6x - 10}{x^2 + 3x - 5}
$$

Solution

$$
\begin{array}{r}
x^2 - 3x + 17 \\
x^2 + 3x - 5\overline{\smash{)}\,x^4 + 0x^3 + 3x^2 - 6x - 10} \\
\underline{x^4 + 3x^3 - 5x^2} \\
-3x^3 + 8x^2 - 6x \\
\underline{-3x^3 - 9x^2 + 15x} \\
17x^2 - 21x - 10 \\
\underline{17x^2 + 51x - 85} \\
-72x + 75
\end{array}
$$

Writing $0x^3$ for the missing term helps align like terms in the same column.

Thus $\dfrac{x^4 + 3x^2 - 6x - 10}{x^2 + 3x - 5} = x^2 - 3x + 17 + \dfrac{-72x + 75}{x^2 + 3x - 5}.$

■ *Try Exercise* **6,** *page 217.*

Synthetic Division

The procedure for dividing a polynomial by a binomial of the form $x - c$ can be condensed by a method called **synthetic division.** To understand the synthetic division method, consider the following division:

$$
\begin{array}{r}
3x^2 - 2x + 3 \\
x - 2\overline{\smash{)}\,3x^3 - 8x^2 + 7x + 2} \\
\underline{3x^3 - 6x^2} \\
-2x^2 + 7x \\
\underline{-2x^2 + 4x} \\
3x + 2 \\
\underline{3x - 6} \\
8
\end{array}
$$

No essential data are lost by omitting the variables since the position of a term indicates the power of the term.

$$
\begin{array}{r}
3 \quad -2 \quad 3 \\
-2\overline{\smash{)}\,3 \quad -8 \quad 7 \quad 2} \\
\underline{3 \quad -6} \\
-2 \quad 7 \\
\underline{-2 \quad 4} \\
3 \quad 2 \\
\underline{3 \quad -6} \\
8
\end{array}
$$

The coefficients shown in color are duplicates of those directly above them. Omitting these repeated coefficients (in color) allows the vertical spacing to be condensed as:

$$
\begin{array}{r}
\phantom{-2\overline{)}}3 \quad -2 \quad 3 \\
-2\overline{)\,3 \quad -8 \quad 7 \quad 2} \\
-6 \quad 4 \quad -6 \\
\hline
-2 \quad 3 \quad 8
\end{array}
$$

The coefficients in color in the top row can be omitted since they are duplicates of those in the bottom row. The leading coefficient of the quotient (top row) could be written in the bottom row with the coefficients of the other terms so that the vertical spacing can be condensed even more.

$$
\begin{array}{r|rrrr}
-2 & 3 & -8 & 7 & 2 \\
 & & -6 & 4 & -6 \\
\hline
 & 3 & -2 & 3 & 8
\end{array}
$$

So that we may add the numbers in each column instead of subtracting them, we change the sign of the divisor. This changes the sign of each number in the second row.

$$
\begin{array}{r|rrrr}
2 & 3 & -8 & 7 & 2 \\
 & & 6 & -4 & 6 \\
\hline
 & 3 & -2 & 3 & 8
\end{array}
\quad \longleftarrow \text{Remainder}
$$

Coefficients
of the quotient

The following example illustrates step by step the synthetic division procedure.

$$
\frac{2x^3 - 9x^2 + 5}{x - 3} = ?
$$

Coefficients
of the dividend

$$
\begin{array}{r|rrrr}
3 & 2 & -9 & 0 & 5 \\
 & & & & \\
\hline
\end{array}
$$
Synthetic division form with 0 inserted for the missing x term.

$$
\begin{array}{r|rrrr}
3 & 2 & -9 & 0 & 5 \\
 & & & & \\
\hline
 & 2 & & &
\end{array}
$$
Bring down the leading coefficient 2.

$$
\begin{array}{r|rrrr}
3 & 2 & -9 & 0 & 5 \\
 & & 6 & & \\
\hline
 & 2 & & &
\end{array}
$$
Multiply $3 \cdot 2$ and place the product (6) in the middle row and in the next column to the right.

$$
\begin{array}{r|rrrr}
3 & 2 & -9 & 0 & 5 \\
 & & 6 & & \\
\hline
 & 2 & -3 & &
\end{array}
$$
Add -9 and 6 and place the sum in the bottom row.

$$\begin{array}{r|rrrr}
3 & 2 & -9 & 0 & 5 \\
  &   & 6 & -9 & -27 \\
\hline
  & 2 & -3 & -9 & -22
\end{array}$$

Repeat the previous steps for columns 3 and 4.

$\longleftarrow$ Remainder

Coefficients of the quotient

$$\frac{2x^3 - 9x^2 + 5}{x - 3} = 2x^2 - 3x - 9 + \frac{-22}{x - 3}$$

Remark The synthetic division method shown in the previous example is only used to divide by a polynomial of the form $x - c$, where the coefficient of x is 1. To divide a polynomial by a polynomial that is not a binomial, use the long division method.

EXAMPLE 2   Use Synthetic Division to Divide Polynomials

Use synthetic division to perform the indicated division.

$$\frac{x^4 - 4x^2 + 7x + 15}{x + 4}$$

Solution

Since the divisor is $x + 4$, we perform the synthetic division with $c = -4$.

$$\begin{array}{r|rrrrr}
-4 & 1 & 0 & -4 & 7 & 15 \\
   &   & -4 & 16 & -48 & 164 \\
\hline
   & 1 & -4 & 12 & -41 & 179
\end{array}$$

The quotient is $x^3 - 4x^2 + 12x - 41$ and the remainder is 179.

$$\frac{x^4 - 4x^2 + 7x + 15}{x + 4} = x^3 - 4x^2 + 12x - 41 + \frac{179}{x + 4}$$

■ *Try Exercise* **12**, *page 218.*

The following theorem shows that synthetic division can be used to find the value $P(c)$ for any polynomial function P and constant c.

The Remainder Theorem

If a polynomial $P(x)$ is divided by $x - c$, then the remainder is $P(c)$.

Proof: The Division Algorithm states that

$$P(x) = (x - c)Q(x) + R(x)$$

where $R(x)$ is zero or the degree of $R(x)$ is less than the degree of $x - c$. Since the degree of $x - c$ is 1, the remainder $R(x)$ must be some constant, say, r. Therefore

$$P(x) = (x - c)Q(x) + r.$$

The above equality evaluated at $x = c$ produces

$$P(c) = (c - c)Q(c) + r = (0)Q(c) + r = r.$$

EXAMPLE 3   Use the Remainder Theorem to Evaluate a Polynomial

Use the Remainder Theorem to evaluate $P(x) = 2x^3 - 3x^2 + 4x - 1$ for $x = -1$ and $x = 3$.

Solution   Perform synthetic divisions and examine the remainders.

$$
\begin{array}{r|rrrr}
-1 & 2 & -3 & 4 & -1 \\
   &   & -2 & 5 & -9 \\
\hline
   & 2 & -5 & 9 & -10
\end{array}
$$

The remainder is -10. By the Remainder Theorem, $P(-1) = -10$.

$$
\begin{array}{r|rrrr}
3 & 2 & -3 & 4 & -1 \\
  &   & 6 & 9 & 39 \\
\hline
  & 2 & 3 & 13 & 38
\end{array}
$$

The remainder is 38. By the Remainder Theorem, $P(3) = 38$.

■ *Try Exercise* **32**, *page 218.*

Remark   We can check our work by substituting -1 for x and 3 for x.

$$P(-1) = 2(-1)^3 - 3(-1)^2 + 4(-1) - 1 = -10$$

$$P(3) = 2(3)^3 - 3(3)^2 + 4(3) - 1 = 38$$

The following theorem is a result of the Remainder Theorem.

The Factor Theorem

A polynomial $P(x)$ has a factor $(x - c)$ if and only if $P(c) = 0$.

Proof: Part 1: Given $P(x)$ has a factor of $(x - c)$, show that $P(c) = 0$. If $(x - c)$ is a factor of $P(x)$, then $P(x) = (x - c) \cdot Q(x)$ for some $Q(x)$. Thus the division of $P(x)$ by $(x - c)$ has a remainder of zero, and the Remainder Theorem implies that $P(c) = 0$.

Part 2: Given $P(c) = 0$, show that $(x - c)$ is a factor of $P(x)$. The division algorithm applied to the polynomial $P(x)$ with divisor $(x - c)$ produces

$$P(x) = (x - c)Q(x) + R(x).$$

Since $P(c) = 0$, the Remainder Theorem implies that $R(x) = 0$. Thus

$$P(x) = (x - c)Q(x)$$

which shows $x - c$ is a factor of $P(x)$.

EXAMPLE 4   **Find Factors of a Polynomial**

Determine whether $(x - 3)$ or $(x + 5)$ are factors of

$$P(x) = x^4 + x^3 - 21x^2 - x + 20.$$

Solution

$$
\begin{array}{r|rrrrr}
3 & 1 & 1 & -21 & -1 & 20 \\
  &   & 3 & 12 & -27 & -84 \\
\hline
  & 1 & 4 & -9 & -28 & -64
\end{array}
$$

A remainder of -64 implies that $(x - 3)$ is *not* a factor of $P(x)$.

$$
\begin{array}{r|rrrrr}
-5 & 1 & 1 & -21 & -1 & 20 \\
   &   & -5 & 20 & 5 & -20 \\
\hline
   & 1 & -4 & -1 & 4 & 0
\end{array}
$$

A remainder of 0 implies that $(x + 5)$ is a factor of $P(x)$.

■ *Try Exercise* **42,** *page 218.*

Remark   Because $(x + 5)$ is a factor of $x^4 + x^3 - 21x^2 - x + 20$, it is also true that -5 is a zero of $x^4 + x^3 - 21x^2 - x + 20$. Why?[1]

The synthetic division

$$
\begin{array}{r|rrrrr}
-5 & 1 & 1 & -21 & -1 & 20 \\
   &   & -5 & 20 & 5 & -20 \\
\hline
   & 1 & -4 & -1 & 4 & \mathbf{0}
\end{array}
$$

also implies that the quotient $Q(x) = x^3 - 4x^2 - x + 4$ is also a factor of $P(x) = x^4 + x^3 - 21x^2 - x + 20$. That is,

$$P(x) = (x + 5)(x^3 - 4x^2 - x + 4).$$

The quotient $Q(x) = x^3 - 4x^2 - x + 4$ is called a **reduced polynomial** because it is 1 degree less than the degree of $P(x)$. Reduced polynomials play an important role in Section 3.

---

[1] Because the Factor Theorem states that a polynomial $P(x)$ has a factor $(x - c)$ if and only if $P(c) = 0$.

EXERCISE SET 4.1

In Exercises 1 to 10, use long division to divide the first polynomial by the second.

1. $5x^3 + 6x^2 - 17x + 20,\ \ x + 3$

2. $6x^3 + 15x^2 - 8x + 2,\ \ x + 4$

3. $2x^4 + 15x^3 + 7x^2 - 135x - 225,\ \ 2x + 5$

4. $6x^4 + 3x^3 - 11x^2 - 3x + 9,\ \ 2x - 3$

5. $3x^4 + x^3 - 99x^2 - 30,\ \ 3x^2 + x + 1$

6. $2x^4 - x^3 - 23x^2 + 9x + 45,\ \ 2x^2 - x - 5$

7. $20x^4 - 3x^2 + 9,\ \ 5x^2 - 2$

8. $24x^5 + 20x^3 - 16x^2 - 15,\ \ 6x^2 + 5$

9. $x^3 + 5x^2 + 6x - 19,\ \ x^2 + x - 4$

10. $2x^4 + 3x^3 - 7x - 10,\ \ x^2 - 2x - 5$

In Exercises 11 to 30, use synthetic division to divide the first polynomial by the second.

11. $4x^3 - 5x^2 + 6x - 7, \quad x - 2$

12. $5x^3 + 6x^2 - 8x + 1, \quad x - 5$

13. $4x^3 - 2x + 3, \quad x + 1$

14. $6x^3 - 4x^2 + 17, \quad x + 3$

15. $x^5 - 10x^3 + 5x - 1, \quad x - 4$

16. $6x^4 - 2x^3 - 3x^2 - x, \quad x - 5$

17. $x^5 - 1, \quad x - 1$

18. $x^4 + 1, \quad x + 1$

19. $8x^3 - 4x^2 + 6x - 3, \quad x - \dfrac{1}{2}$

20. $12x^3 + 5x^2 + 5x + 6, \quad x + \dfrac{3}{4}$

21. $x^8 + x^6 + x^4 + x^2 + 4, \quad x - 2$

22. $-x^7 - x^5 - x^3 - x - 5, \quad x + 1$

23. $x^6 + x - 10, \quad x + 3$

24. $2x^5 - 3x^4 - 5x^2 - 10, \quad x - 4$

25. $3x^2 - 4x + 5, \ x - 0.3$ **26.** $2x^2 - 12x + 1, \ x + 0.4$

27. $2x^3 - 11x^2 - 17x + 3, \ x$ **28.** $5x^4 - 2x^2 + 6x - 1, \ x$

29. $x + 8, \ x + 2$ **30.** $3x - 17, \ x - 3$

In Exercises 31 to 40, use the Remainder Theorem to find $P(c)$.

31. $P(x) = 3x^3 + x^2 + x - 5, \ c = 2$

32. $P(x) = 2x^3 - x^2 + 3x - 1, \ c = 3$

33. $P(x) = 4x^4 - 6x^2 + 5, \ c = -2$

34. $P(x) = 6x^3 - x^2 + 4x, \ c = -3$

35. $P(x) = -2x^3 - 2x^2 - x - 20, \ c = 10$

36. $P(x) = -x^3 + 3x^2 + 5x + 30, \ c = 8$

37. $P(x) = -x^4 + 1, \ c = 3$

38. $P(x) = x^5 - 1, \ c = 1$

39. $P(x) = x^4 - 10x^3 + 2, \ c = 3$

40. $P(x) = x^5 + 20x^2 - 1, \ c = -5$

In Exercises 41 to 52, use synthetic division and the Factor Theorem to determine if the given binomial is a factor of $P(x)$.

41. $P(x) = x^3 + 2x^2 - 5x - 6, \ x - 2$

42. $P(x) = x^3 + 4x^2 - 27x - 90, \ x + 6$

43. $P(x) = 2x^3 + x^2 - 2x - 1, \ x + 1$

44. $P(x) = 3x^3 + 4x^2 - 27x - 36, \ x - 4$

45. $P(x) = x^4 - 25x^2 + 144, \ x + 3$

46. $P(x) = x^4 - 25x^2 + 144, \ x - 3$

47. $P(x) = x^5 + 2x^4 - 22x^3 - 50x^2 - 75x, \ x - 5$

48. $P(x) = 9x^4 - 6x^3 - 23x^2 - 4x + 4, \ x + 1$

49. $P(x) = 16x^4 - 8x^3 + 9x^2 + 14x - 4, \ x - \dfrac{1}{4}$

50. $P(x) = 10x^4 + 9x^3 - 4x^2 + 9x + 6, \ x + \dfrac{1}{2}$

51. $P(x) = ^2 - 4x - 1, \ x - (2 + \sqrt{5})$

52. $P(x) = x^2 - 4x - 1, \ x - (2 - \sqrt{5})$

In Exercises 53 to 62, use synthetic division to show that c is a zero of $P(x)$.

53. $P(x) = 3x^3 - 8x^2 - 10x + 28, \ c = 2$

54. $P(x) = 4x^3 - 10x^2 - 8x + 6, \ c = 3$

55. $P(x) = x^4 - 1, \ c = 1$

56. $P(x) = x^3 + 8, \ c = -2$

57. $P(x) = 3x^4 + 8x^3 + 10x^2 + 2x - 20, \ c = -2$

58. $P(x) = x^4 - 2x^2 - 100x - 75, \ c = 5$

59. $P(x) = 2x^3 - 18x^2 - 50x + 66, \ c = 11$

60. $P(x) = 2x^4 - 34x^3 + 70x^2 - 153x + 45, \ c = 15$

61. $P(x) = 3x^2 - 8x + 4, \ c = \dfrac{2}{3}$

62. $P(x) = 5x^2 + 12x + 4, \ c = -\dfrac{2}{5}$

Supplemental Exercises

63. Use the Factor Theorem to prove that for any positive odd integer n, $x^n + 1$ has $x + 1$ as a factor.

64. Use the Factor Theorem to prove that for any positive integer n, $x^n - 1$ has $x - 1$ as a factor.

65. Find the remainder of $5x^{48} + 6x^{10} - 5x + 7$ divided by $x - 1$.

66. Find the remainder of $18x^{80} - 6x^{50} + 4x^{20} - 2$ divided by $x + 1$.

67. Prove that $P(x) = 4x^4 + 7x^2 + 12$ has no factor of the form $x - c$, where c is a real number.

68. Prove that $P(x) = -5x^6 - 4x^2 - 10$ has no factor of the form $x - c$, where c is a real number.

69. Use synthetic division to show that $(x - i)$ is a factor of $x^3 - 3x^2 + x - 3$.

70. Use synthetic division to show that $(x + 2i)$ is a factor of $x^4 - 2x^3 + x^2 - 8x - 12$.

4.2

Graphs of Polynomial Functions

Table 4.1 summarizes information developed in Chapter 3 concerning graphs of polynomial functions of degree 0, 1, or 2. Polynomial functions of degree 3 or higher can be graphed by the technique of point-plotting. However, some additional knowledge about polynomial functions will make graphing easier.

All polynomial functions have graphs that are **smooth continuous curves.** The terms smooth and continuous are defined rigorously in calculus, but for the present a smooth curve is a curve that does not have sharp corners, as shown in Figure 4.1(a). A continuous curve does not have a break or hole, as shown in Figure 4.1(b).

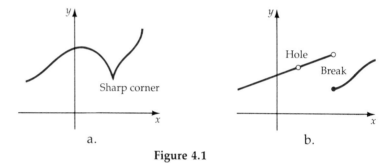

Figure 4.1

The Leading Term Test

The graph of a polynomial function may have several up and down fluctuations; however, the graph of every polynomial function will eventually increase or decrease without bound as the graph moves far to the left or far to the right. The **leading term** $a_n x^n$ is said to **dominate** the polynomial function $P(x) = a_n x^n + a_{n-1} x^{n-1} + \cdots + a_1 x + a_0$ as $|x|$ becomes large, because the absolute value of $a_n x^n$ will be much larger than the absolute value of any of the other terms. Because of this condition, you can determine the far left and far right behavior of the polynomial by examining the leading coefficient a_n and the degree n of the polynomial.

Table 4.2 indicates the far left and the far right behavior of a polynomial function P with leading term $a_n x^n$.

TABLE 4.1

Polynomial function $P(x)$	Graph
$P(x) = a$ (degree 0)	Horizontal line through $(0, a)$
$P(x) = ax + b$ (degree 1)	Line with y-intercept $(0, b)$ and slope a
$P(x) = ax^2 + bx + c$ (degree 2)	Parabola with vertex $\left(-\dfrac{b}{2a}, P\left(-\dfrac{b}{2a} \right) \right)$

TABLE 4.2 The graph of polynomial $P(x)$ with leading term $a_n x^n$ has the following far right and far left behavior.

	n is even	*n* is odd
$a_n > 0$	Up to left and up to right	Down to left and up to right
$a_n < 0$	Down to left and down to right	Up to left and down to right

EXAMPLE 1 Determine the Left and Right Behavior of a Polynomial Function

Examine the leading term to determine the far left and the far right behavior of the graphs of each of the following polynomial functions:

a. $P(x) = x^3 - x$

b. $S(x) = \dfrac{1}{2}x^4 - \dfrac{5}{2}x^2 + 2$

c. $T(x) = -2x^3 + x^2 + 7x - 6$

d. $U(x) = -x^4 + 8x^2 + 9$

Solution

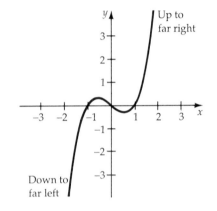

a. Since $a_n = 1$ is *positive* and $n = 3$ is *odd*, the graph of P goes down to its far left and up to its far right. See Figure 4.2.

b. Since $a_n = \dfrac{1}{2}$ is *positive* and $n = 4$ is *even*, the graph of S goes up to its far left and up to its far right. See Figure 4.3.

c. Since $a_n = -2$ is *negative* and $n = 3$ is *odd*, the graph of T goes up to its far left and down to its far right. See Figure 4.4.

d. Since $a_n = -1$ is *negative* and $n = 4$ is *even*, the graph of U goes down to its far left and down to its far right. See Figure 4.5.

Figure 4.2

$P(x) = x^3 - x$

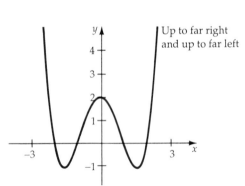

Figure 4.3

$S(x) = \dfrac{1}{2}x^4 - \dfrac{5}{2}x^2 + 2$

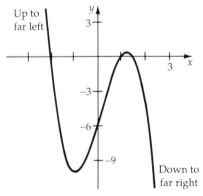

Figure 4.4

$T(x) = -2x^3 + x^2 + 7x - 6$

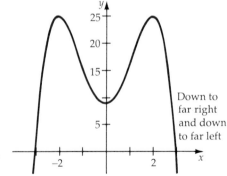

Figure 4.5

$U(x) = -x^4 + 8x^2 + 9$

■ *Try Exercise* **2**, *page 224.*

Recall that the Remainder Theorem states that if the polynomial $P(x)$ is divided by $x - c$, then the remainder is $P(c)$. Because the arithmetic in synthetic division is often easier than the arithmetic involved in evaluating a polynomial at $x = c$, by substituting c for x, we will often use the synthetic division process to evaluate a polynomial.

EXAMPLE 2   Use Synthetic Division to Evaluate a Polynomial Function

Use synthetic division to find points on the graph of the polynomial function $P(x) = 2x^3 + 5x^2 - x - 5$, and then sketch the graph of $P(x)$.

Solution   Choose convenient values of x, say, -3, -2, -1, 1, and 2.

For $x = -3$,

$$
\begin{array}{r|rrrr}
-3 & 2 & 5 & -1 & -5 \\
   &   & -6 & 3 & -6 \\
\hline
   & 2 & -1 & 2 & -11
\end{array}
$$

The remainder is -11. Thus, by the Remainder Theorem, $P(-3) = -11$.

For $x = -2$,

$$
\begin{array}{r|rrrr}
-2 & 2 & 5 & -1 & -5 \\
   &   & -4 & -2 & 6 \\
\hline
   & 2 & 1 & -3 & 1 \quad P(-2) = 1.
\end{array}
$$

For $x = -1$,

$$
\begin{array}{r|rrrr}
-1 & 2 & 5 & -1 & -5 \\
   &   & -2 & -3 & 4 \\
\hline
   & 2 & 3 & -4 & -1 \quad P(-1) = -1.
\end{array}
$$

For $x = 1$,

$$
\begin{array}{r|rrrr}
1 & 2 & 5 & -1 & -5 \\
  &   & 2 & 7 & 6 \\
\hline
  & 2 & 7 & 6 & 1 \quad P(1) = 1.
\end{array}
$$

For $x = 2$,

$$
\begin{array}{r|rrrr}
2 & 2 & 5 & -1 & -5 \\
  &   & 4 & 18 & 34 \\
\hline
  & 2 & 9 & 17 & 29 \quad P(2) = 29.
\end{array}
$$

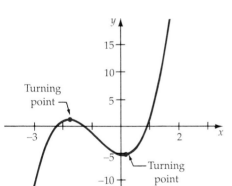

Figure 4.6
$P(x) = 2x^3 + 5x^2 - x - 5$

To find the y-intercept of the graph of $P(x) = 2x^3 + 5x^2 - x - 5$, substitute 0 for x to produce $P(0) = -5$. Thus the y-intercept is $(0, -5)$. Drawing a smooth continuous curve through the points $(-3, -11)$, $(-2, 1)$, $(-1, -1)$, $(0, -5)$, $(1, 1)$ and $(2, 29)$ produces the graph in Figure 4.6.

■ *Try Exercise 12, page 224.*

Figure 4.6 illustrates a polynomial function of degree 3 with two **turning points,** where the function changes from an increasing function to a decreasing function or vice versa. In general, a polynomial function of degree n has at most $n - 1$ turning points. To determine the exact location of turning points requires concepts and techniques from calculus.

Zeros and x-Intercepts

If a polynomial function can be factored into linear factors, then it can be graphed by plotting its x-intercepts and a few intermediate points. Before this procedure is illustrated, it is important to review some equivalent terminology.

If P is a polynomial function and c is a real number, then each of the following statements are equivalent in the sense that if any one statement is true, then they are all true, and if any one statement is false, then they are all false.

- $(x - c)$ is a *factor* of P.

- $x = c$ is a *solution* or *root* of the equation $P(x) = 0$.

- $x = c$ is a *zero* of P.

- $(c, 0)$ is an *x-intercept* of the graph of $y = P(x)$.

EXAMPLE 3   Find Intercepts and Graph a Polynomial Function

Sketch the graph $P(x) = (x + 1)(x - 1)(2x - 3)$.

Solution   Because $(x + 1)$, $(x - 1)$ and $(2x - 3)$ are factors of P,

$$(-1, 0), \ (1, 0), \ \text{and} \ \left(\frac{3}{2}, 0\right)$$

are x-intercepts of the graph of P. The x-intercepts separate the x-axis into the four intervals shown in Figure 4.7.

Synthetic division can now be used to determine additional points. We will use convenient x-values from each of the intervals. First, we rewrite $P(x)$ in its standard form.

$$P(x) = (x - 1)(x + 1)(2x - 3)$$
$$= 2x^3 - 3x^2 - 2x + 3$$

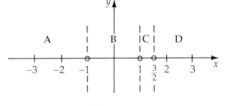

Figure 4.7

$x = -2$:

$$\begin{array}{r|rrrr} -2 & 2 & -3 & -2 & 3 \\ & & -4 & 14 & -24 \\ \hline & 2 & -7 & 12 & -21 \end{array}$$

$(-2, -21)$ is a point on the graph of P.

$x = 1.2$:

$$\begin{array}{r|rrrr} 1.2 & 2 & -3 & -2 & 3 \\ & & 2.4 & -0.72 & -3.264 \\ \hline & 2 & -0.6 & -2.72 & -0.264 \end{array}$$

$(1.2, -0.264)$ is a point on the graph of P.

$x = 2$:

$$\begin{array}{r|rrrr} 2 & 2 & -3 & -2 & 3 \\ & & 4 & 2 & 0 \\ \hline & 2 & 1 & 0 & 3 \end{array}$$

$(2, 3)$ is a point on the graph of P.

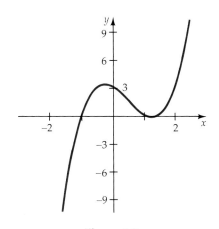

Figure 4.8
$$P(x) = (x - 1)(x + 1)(2x - 3)$$
$$= 2x^3 - 3x^2 - 2x + 3$$

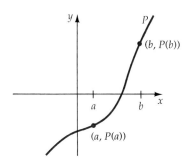

Figure 4.9
$P(a) < 0, P(b) > 0$

$x = 3:$

$$\begin{array}{r|rrrr} 3 & 2 & -3 & -2 & 3 \\ & & 6 & 9 & 21 \\ \hline & 2 & 3 & 7 & 24 \end{array}$$ $(3, 24)$ is a point on the graph of P.

Because $P(0) = 3$, the y-intercept is $(0, 3)$.

Sketch the graph by drawing a smooth continuous curve through each of the points determined above and the x-intercepts as shown in Figure 4.8.

■ *Try Exercise* **32**, *page 225.*

The graph of every polynomial function P is a smooth continuous curve, and if the graph of P changes sign on an interval, then the function P must equal zero at least once in the interval. This result is known as the Zero Location Theorem. Although we will not prove the Zero Location Theorem, we often will use it in our search for the zeros of polynomial functions.

The Zero Location Theorem

> Let $P(x)$ be a polynomial. If $a < b$, and if $P(a)$ and $P(b)$ have different signs, then there is at least one value c between a and b such that $P(c) = 0$.

If, for instance, a polynomial function is negative at $x = a$ and positive at $x = b$, then there must be at least one point between a and b where the polynomial function is zero. See Figure 4.9.

EXAMPLE 4   **Apply the Zero Location Theorem**

Use the Zero Location Theorem to verify that

a. $P(x) = x^3 - x - 25$ has a real zero between $a = 3$ and $b = 4$.

b. $S(x) = -x^3 + 5x + 1$ has a real zero between $a = -2.2$ and $b = -2.1$.

Solution

a.

$$\begin{array}{r|rrrr} 3 & 1 & 0 & -1 & -25 \\ & & 3 & 9 & 24 \\ \hline & 1 & 3 & 8 & -1 \end{array}$$ $\leftarrow P(3)$ is negative.

$$\begin{array}{r|rrrr} 4 & 1 & 0 & -1 & -25 \\ & & 4 & 16 & 60 \\ \hline & 1 & 4 & 15 & 35 \end{array}$$ $\leftarrow P(4)$ is positive.

Because $P(3)$ and $P(4)$ have different signs, P must have a real zero between 3 and 4.

b.

$$-2.2 \begin{array}{|rrrr} -1 & 0 & 5 & 1 \\ & 2.2 & -4.84 & -0.352 \\ \hline -1 & 2.2 & 0.16 & 0.648 \end{array}$$ ← $S(-2.2)$ is positive.

$$-2.1 \begin{array}{|rrrr} -1 & 0 & 5 & 1 \\ & 2.1 & -4.41 & -1.239 \\ \hline -1 & 2.1 & 0.59 & -0.239 \end{array}$$ ← $S(-2.1)$ is negative.

Because $S(-2.2)$ and $S(-2.1)$ have different signs, S, must have a real zero between -2.2 and -2.1.

■ *Try Exercise* **48**, *page 225.*

EXERCISE SET 4.2

In Exercises 1 to 10, examine the leading term and determine the far left and the far right behavior of the graph of the polynomial function.

1. $P(x) = 3x^4 - 2x^2 - 7x + 1$

2. $P(x) = -2x^3 - 6x^2 + 5x - 1$

3. $P(x) = 5x^5 - 4x^3 - 17x^2 + 2$

4. $P(x) = -6x^4 - 3x^3 + 5x^2 - 2x + 5$

5. $P(x) = 2 - 3x - 4x^2$

6. $P(x) = -16 + x^4$

7. $P(x) = \dfrac{1}{2}(x^3 + 5x^2 - 2)$

8. $P(x) = -\dfrac{1}{4}(x^4 + 3x^3 - 2x + 6)$

9. $P(x) = -\dfrac{2}{3}(x + 1)^3$

10. $P(x) = \dfrac{1}{5}(x - 1)^4$

In Exercises 11 to 20, use synthetic division to find the indicated functional values for the given polynomial function.

11. $f(x) = 3x^2 - 10x + 1$
    a. $f(2)$ b. $f(-2)$ c. $f(3)$
    d. $f(-3)$ e. $f(4)$ f. $f(-6)$

12. $g(x) = 2x^3 - 5x^2 + 4x + 10$
    a. $g(1)$ b. $g(-1)$ c. $g(2)$
    d. $g(3)$ e. $g(0)$ f. $g(-5)$

13. $h(x) = -3x^3 - 2x + 10$
    a. $h(3)$ b. $h(-3)$ c. $h(-1)$
    d. $h(1)$ e. $h(2)$ f. $h(-2)$

14. $j(x) = x^4 - 2x^3 - 3x^2 + x - 8$
    a. $j(2)$ b. $j(-2)$ c. $j(4)$
    d. $j(3)$ e. $j(0)$ f. $j(-1)$

15. $k(x) = -x^4 - 8x^3 + 2x^2 - 5x + 8$
    a. $k(-1)$ b. $k(-2)$ c. $k(3)$
    d. $k(-3)$ e. $k(4)$ f. $k(5)$

16. $f(t) = t^3 - t - 21$
    a. $f(2)$ b. $f(-2)$ c. $f(3)$
    d. $f(-3)$ e. $f(-4)$ f. $f(6)$

17. $s(t) = -16t^2 + 40t + 110$
    a. $s(2)$ b. $s(3)$ c. $s(4)$
    d. $s\left(\dfrac{1}{2}\right)$ e. $s\left(\dfrac{3}{2}\right)$ f. $s\left(\dfrac{5}{2}\right)$

18. $F(x) = 2x^5 - 3x^3 - 2x^2 + x + 7$
    a. $F(1)$ b. $F(-2)$ c. $F(3)$
    d. $F(-3)$ e. $F(0)$ f. $F(4)$

19. $P(x) = -3x^5 + x^3 + 80$
    a. $P(2)$ b. $P(-1)$ c. $P(-2)$
    d. $P(-3)$ e. $P\left(\dfrac{1}{3}\right)$ f. $P(-6)$

20. $v(t) = -t^3 + 2t^2 - 8t - 6$
    a. $v(-2)$ b. $v(-1)$ c. $v\left(\dfrac{1}{2}\right)$
    d. $v(1)$ e. $v(2)$ f. $v(7)$

In Exercises 21 to 30, find the real zeros of the polynomial function.

21. $P(x) = (x + 2)(x - 3)(2x + 7)$

22. $P(x) = (x - 5)(x - 1)(4x + 1)$

23. $P(x) = x(x - 1)(5x - 2)$

24. $P(x) = x(x - 4)(x - 1)(x + 7)$

25. $P(x) = (3x + 7)(2x - 11)(x + 5)^2$

26. $P(x) = (x - 3)^2(2x - 1)$

27. $P(x) = x^3 + x^2 - 6x$

28. $P(x) = 2x^3 - 7x^2 - 15x$

29. $P(x) = x^3 - 1$

30. $P(x) = x^3 + 8$

In Exercises 31 to 46, use the sketching techniques of this section to sketch the graph of each polynomial function.

31. $P(x) = (x - 1)(x + 1)(x - 3)$

32. $P(x) = (x - 2)(x + 3)(x + 1)$

33. $P(x) = -(x + 4)(x - 1)(x + 2)$

34. $P(x) = -x(x + 3)(x - 3)$

35. $P(x) = (x - 3)(x - 1)(2x + 7)$

36. $P(x) = (3x - 1)(x + 3)(x + 1)$

37. $P(x) = x^2(x^2 - 4)$ **38.** $P(x) = -x^2(x^2 - 1)$

39. $P(x) = (x - 2)^2(x + 1)$ **40.** $P(x) = (x - 3)(x + 1)^2$

41. $P(x) = x^3 + 2x^2 - 3x$ **42.** $P(x) = x^3 - 6x^2 + 9x$

43. $P(x) = x^3 - x^2 + x - 1$

44. $P(x) = 2x^3 - x^2 - 20x + 28$

45. $P(x) = x^4 + 3x^3 + 4x^2$

46. $P(x) = 2x^4 + 6x^3 - 25x^2 - 43x + 30$

In Exercises 47 to 52, use the Zero Location Theorem to verify that P has a zero between a and b.

47. $P(x) = 2x^3 + 3x^2 - 23x - 42$; $a = 3, b = 4$

48. $P(x) = 4x^3 - x^2 - 6x + 1$; $a = 0, b = 1$

49. $P(x) = 3x^3 + 7x^2 + 3x + 7$; $a = -3, b = -2$

50. $P(x) = 2x^3 - 21x^2 - 2x + 21$; $a = 10, b = 11$

51. $P(x) = 4x^4 + 7x^3 - 11x^2 + 7x - 15$; $a = 1, b = 1\frac{1}{2}$

52. $P(x) = 5x^3 - 16x^2 - 20x + 64$; $a = 3, b = 3\frac{1}{2}$

Supplemental Exercises

53. Let $f(x) = x^3 + c$. On the same coordinate axes, sketch the graph of f for each value of c.

   a. $c = 0$ **b.** $c = 2$ **c.** $c = -3$

54. Let $f(x) = ax^3$. On the same coordinate axes, sketch the graph of f for each value of a.

   a. $a = 2$ **b.** $a = \frac{1}{2}$ **c.** $a = -1$

55. Let $f(x) = (x - h)^3$. On the same coordinate axes, sketch the graph of f for each value of h.

   a. $h = 2$ **b.** $h = -1$ **c.** $h = -5$

56. Explain how the graph of $f(x) = a(x - h)^3 + c$ compares with the graph of $g(x) = x^3$.

57. On the same coordinate axes, sketch the graph of the function $f(x) = x^n$ over the interval $-1 \le x \le 1$, for each value of n.

   a. $n = 2$ **b.** $n = 4$ **c.** $n = 6$

58. Use the result of Exercise 57 to make a conjecture about the graph of $y = x^n$, where n is a large, positive, even integer.

59. Graph $P(x) = x^5 - x^4 - 2x^3$.

60. Graph $P(x) = -x^5 + 4x^3$.

4.3

Zeros of Polynomial Functions

Recall that if $P(x)$ is a polynomial function, then the values of x for which $P(x)$ is to equal 0 are called the *zeros* of $P(x)$ or the *roots* of the equation $P(x) = 0$. A zero of a polynomial may be a **multiple zero.** For example, the polynomial $x^2 + 6x + 9$ can be expressed in factored form as $(x + 3)(x + 3)$. Setting each factor equal to zero yields $x = -3$ in both cases. Thus $x^2 + 6x + 9$ has a zero of -3 that occurs twice. The following definition will be most useful when discussing multiple zeros.

Definition of Multiple Zeros of a Polynomial

> If a polynomial function P has $(x - r)$ as a factor exactly k times, then r is said to be a **zero of multiplicity k** of the polynomial P.

The polynomial

$$P(x) = (x - 5)(x - 5)(x + 2)(x + 2)(x + 2)(x + 4)$$

has

- 5 as a zero of multiplicity 2
- -2 as a zero of multiplicity 3
- -4 as a zero of multiplicity 1

A zero of multiplicity 1 is generally referred to as a simple zero.

When searching for the zeros of a polynomial function, it is important to know how many zeros to expect. This question is answered completely in Section 4. For the work in this section, the following result is valuable.

Number of Zeros of a Polynomial Function

> A polynomial function P of degree n has at most n zeros, where each zero of multiplicity k is counted k times.

The rational zeros of polynomials with integer coefficients can be found with the aid of the following theorem.

The Rational Zero Theorem

> If $P(x) = a_n x^n + a_{n-1} x^{n-1} + \cdots + a_1 x + a_0$ has integer coefficients and p/q (where p and q have no common prime factors) is a rational zero of $P(x)$, then p is a factor of a_0 and q is a factor of a_n.

Proof: Since p/q is a zero of $P(x)$,

$$a_n \left(\frac{p}{q}\right)^n + a_{n-1} \left(\frac{p}{q}\right)^{n-1} + \cdots + a_1 \left(\frac{p}{q}\right) + a_0 = 0.$$

Multiplying both sides by q^n produces

$$a_n p^n + a_{n-1} p^{n-1} q + \cdots + a_1 p q^{n-1} + a_0 q^n = 0,$$

which can be written as

$$p(a_n p^{n-1} + a_{n-1} p^{n-2} q + \cdots + a_1 q^{n-1}) = -a_0 q^n.$$

This implies that p is a factor of the integer $a_0 q^n$. Also, since p and q have no common prime factors, p is a factor of a_0. A similar procedure may be used to establish that q is a factor of a_n.

The Rational Zero Theorem often is used to make a list of all possible rational zeros of a polynomial. The list consists of all rational numbers of the form p/q, where p is an integer factor of the constant term a_0, and q is an integer factor of the leading coefficient a_n.

EXAMPLE 1   Apply the Rational Zero Theorem

Use the Rational Zero Theorem to list all possible rational zeros of each of the following polynomials:

a. $2x^3 + x^2 - 18x - 9$ b. $4x^4 + 5x^3 + 7x^2 - 34x + 8$

Solution

a. List all integers p that are factors of 9 and all integers q that are factors of 2.

$$\text{Factors of 9} \quad \pm1, \pm3, \pm9$$

$$\text{Factors of 2} \quad \pm1, \pm2$$

Form all possible rational numbers using ±1, ±3, or ±9 as the numerator and ±1 or ±2 as the denominator.

$$\frac{p}{q}: \quad \pm1, \pm3, \pm9, \pm\frac{1}{2}, \pm\frac{3}{2}, \pm\frac{9}{2}$$

By the Rational Zero Theorem, if $P(x) = 2x^3 + x^2 - 18x - 9$ has a rational zero, then it must be one of the twelve numbers in this list.

b. List all integers p that are factors of 8 and all integers q that are factors of 4.

$$\text{Factors of 8} \quad \pm1, \pm2, \pm4, \pm8$$

$$\text{Factors of 4} \quad \pm1, \pm2, \pm4$$

By the Rational Zero Theorem, the possible rational zeros are

$$\pm1, \pm2, \pm4, \pm8, \pm\frac{1}{2}, \pm\frac{1}{4}$$

■ *Try Exercise* **12,** *page 232.*

Remark   It is not necessary to list a factor that is already listed in reduced form. For example, $\pm\frac{4}{2}$ is not listed since it simplifies to ±2.

Caution   The Rational Zero Theorem gives the possible rational zeros. That is, if $P(x)$ has a rational zero p/q, where p and q have no common prime factors, then p is a factor of a_0 and q is a factor of a_n. However, it is possible that $P(x)$ has no rational zeros.

Upper and Lower Bounds for Real Zeros

A real number b is called an **upper bound** of the zeros of the polynomial function P if no zero is greater than b. A real number a is called a **lower bound** of the zeros of P if no zero is less than a. The following theorem is often used to find positive upper bounds and negative lower bounds for the real zeros of a polynomial function.

Upper and Lower Bound Theorem

> **Upper bound** If $b > 0$ and all the numbers in the bottom row of the synthetic division of P by $x - b$ are either positive or zero, then b is an upper bound for the real zeros of P.
>
> **Lower bound** If $a < 0$ and the numbers in the bottom row of the synthetic division of P by $x - a$ alternate in sign (the number zero can be considered positive or negative), then a is a lower bound for the real zeros of P.

Upper and lower bounds are not unique. For example, if b is an upper bound for the real zeros of P, then any number greater than b is also an upper bound. Also, if a is a lower bound for the real zeros of P, then any number less than a is also a lower bound.

EXAMPLE 2   Find Upper and Lower Bounds

According to the upper and lower bound theorem, what is the smallest positive integer that is an upper bound and the largest negative integer that is a lower bound of $P(x) = 2x^3 + 7x^2 - 4x - 14$?

Solution To find the smallest positive integer upper bound, use synthetic division with $1, 2, \ldots,$ as test values.

$$
\begin{array}{r|rrrr}
1 & 2 & 7 & -4 & -14 \\
  &   & 2 & 9 & 5 \\
\hline
  & 2 & 9 & 5 & -9
\end{array}
\qquad
\begin{array}{r|rrrr}
2 & 2 & 7 & -4 & -14 \\
  &   & 4 & 22 & 36 \\
\hline
  & 2 & 11 & 18 & 22
\end{array}
\quad \leftarrow \text{All positive signs}
$$

Thus, 2 is the smallest positive integer upper bound.

Now find the largest negative integer lower bound.

$$
\begin{array}{r|rrrr}
-1 & 2 & 7 & -4 & -14 \\
   &   & -2 & -5 & 9 \\
\hline
   & 2 & 5 & -9 & -5
\end{array}
\qquad
\begin{array}{r|rrrr}
-2 & 2 & 7 & -4 & -14 \\
   &   & -4 & -6 & 20 \\
\hline
   & 2 & 3 & -10 & 6
\end{array}
$$

$$
\begin{array}{r|rrrr}
-3 & 2 & 7 & -4 & -14 \\
   &   & -6 & -3 & 21 \\
\hline
   & 2 & 1 & -7 & 7
\end{array}
\qquad
\begin{array}{r|rrrr}
-4 & 2 & 7 & -4 & -14 \\
   &   & -8 & 4 & 0 \\
\hline
   & 2 & -1 & 0 & -14
\end{array}
\quad \leftarrow \text{Alternating signs}
$$

Thus, -4 is the largest negative integer lower bound.

■ *Try Exercise* **24**, *page 232.*

Remark Since -4 is a lower bound and 2 is an upper bound, the real zeros of $2x^3 + 7x^2 - 4x - 14$ must be in the interval $(-4, 2)$. The graph of P is shown in Figure 4.10. Notice that its x-intercepts are between -4 and 2.

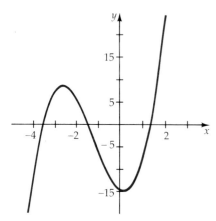

Figure 4.10
$P(x) = 2x^3 + 7x^2 - 4x - 14$

A proof of the upper and lower bound theorem uses the Division Algorithm and the argument shown in the following example.

Consider the polynomial $2x^3 + 7x^2 - 4x - 14$. By the Division Algorithm,

$$2x^3 + 7x^2 - 4x - 14 = (x - 2)(2x^2 + 11x + 18) + 22$$

If $x > 2$, then $(x - 2) > 0$ and $(2x^2 + 11x + 18) > 0$.

Thus $(x - 2)(2x^2 + 11x + 18) + 22$ is greater than zero and consequently so is $2x^3 + 7x^2 - 4x - 14$. Hence if $x > 2$, then $2x^3 + 7x^2 - 4x - 14 \neq 0$, which implies that 2 is an upper bound of the real zeros of the polynomial $2x^3 + 7x^2 - 4x - 14$.

Descartes' Rule of Signs

Descartes' Rule of Signs is another theorem often used to obtain information about the zeros of a polynomial. In Descartes' Rule of Signs, the number of *variations in sign* of the coefficients of a polynomial $P(x)$ or $P(-x)$ refers to sign changes of the coefficients from positive to negative or negative to positive as you examine successive terms of the polynomial. The terms of the polynomial are assumed to be in descending powers of x. For example, the polynomial

$$P(x) = +3x^4 - 5x^3 - 7x^2 + x - 7$$

$$\qquad\qquad 1 \qquad\quad 2 \quad 3$$

has three variations of sign. The polynomial

$$P(-x) = +3(-x)^4 - 5(-x)^3 - 7(-x)^2 + (-x) - 7$$

$$= +\quad 3x^4 \ + \ 5x^3 \ - \ 7x^2 \ - \ x \ - 7$$

$$1$$

has one variation in sign.

Terms that have a coefficient of 0 are not counted as a variation of sign and may be ignored. For example,

$$P(x) = -x^5 + 4x^2 + 1$$

$$1$$

has one variation in sign.

Descartes' Rule of Signs

Let $P(x)$ be a polynomial with terms arranged in decreasing powers of x.

1. The number of positive real zeros is equal to the number of variations in sign of $P(x)$ or is equal to that number decreased by an even integer.

2. The number of negative real zeros is equal to the number of variations in sign of $P(-x)$ or is equal to that number decreased by an even integer.

The proof of Descartes' Rule of Signs is beyond the scope of this course and not given in this text.

EXAMPLE 3   Apply Descartes' Rule of Signs

Determine both the number of possible positive and negative real zeros of each of the following polynomials:

a. $x^4 - 5x^3 + 5x^2 + 5x - 6$ b. $2x^5 + 3x^3 + 5x^2 + 8x + 7$

Solution

a.
$$P(x) = x^4 - 5x^3 + 5x^2 + 5x - 6$$
$$1 \qquad 2 \qquad\qquad 3$$

There are three variations of sign. By Descartes' Rule of Signs, there are either three or one positive real zeros. Now examine the variations of sign of $P(-x)$.

$$P(-x) = x^4 + 5x^3 + 5x^2 - 5x - 6$$
$$1$$

There is one variation of sign of $P(-x)$. By Descartes' Rule of Signs, there is one negative real zero.

b. $P(x) = 2x^5 + 3x^3 + 5x^2 + 8x + 7$ has no variation of sign, so there are no positive real zeros.

$$P(-x) = -2x^5 - 3x^3 + 5x^2 - 8x + 7$$
$$1 \qquad 2 \qquad 3$$

$P(-x)$ has three variations of sign, so there are either three or one negative real zeros.

■ *Try Exercise 36, page 233.*

In applying Descartes' Rule of Signs, each zero of multiplicity k is counted as k zeros. For instance, the polynomial

$$P(x) = x^2 - 10x + 25$$

has two variations in sign. Thus by Descartes' Rule of Signs it must have either two or zero positive real zeros. Factoring the polynomial produces $(x - 5)^2$, from which it can be observed that 5 is a positive zero of multiplicity two.

Example 4 uses the theorems of this section to determine the zeros of polynomial functions.

EXAMPLE 4   Find the Zeros of a Polynomial

Find the zeros of each of the following polynomials:

a. $P(x) = 3x^4 + 23x^3 + 56x^2 + 52x + 16$ b. $P(x) = 3x^3 + 13x^2 - 16$

Solution

a. By Descartes' Rule of Signs there are no positive real zeros, and there
    are either four, two, or zero negative real zeros. By the Rational Zero
    Theorem, the possible negative rational zeros of the polynomial
    $3x^4 + 23x^3 + 56x^2 + 52x + 16$ are

$$\frac{p}{q}: \quad -1, -2, -4, -8, -16, -\frac{1}{3}, -\frac{2}{3}, -\frac{4}{3}, -\frac{8}{3}, -\frac{16}{3}.$$

Synthetic division is used to test the above possible zeros.

$$x = -8: \qquad -8 \begin{array}{|rrrrr} 3 & 23 & 56 & 52 & 16 \\ & -24 & 8 & -512 & 3680 \\ \hline 3 & -1 & 64 & -460 & 3696 \end{array} \longleftarrow -8 \text{ is a lower bound}$$

$$x = -4: \qquad -4 \begin{array}{|rrrrr} 3 & 23 & 56 & 52 & 16 \\ & -12 & -44 & -48 & -16 \\ \hline 3 & 11 & 12 & 4 & 0 \end{array} \longleftarrow -4 \text{ is a zero}$$

Since -4 is a zero, the factors of $P(x)$ are $(x + 4)$ and the reduced
polynomial $(3x^3 + 11x^2 + 12x + 4)$. Thus

$$P(x) = (x + 4)(3x^3 + 11x^2 + 12x + 4).$$

All remaining zeros must be zeros of $3x^3 + 11x^2 + 12x + 4$. The Rational Zero Theorem indicates that the only possible negative rational
zeros are

$$\frac{p}{q}: \quad -1, -2, -4, -\frac{1}{3}, -\frac{2}{3}, -\frac{4}{3}.$$

Synthetic division is again used to test possible zeros.

$$-2 \begin{array}{|rrrr} 3 & 11 & 12 & 4 \\ & -6 & -10 & -4 \\ \hline 3 & 5 & 2 & 0 \end{array} \longleftarrow -2 \text{ is a zero}$$

Since -2 is a zero, $(x + 2)$ is also a factor of P. Thus we have

$$P(x) = (x + 4)(x + 2)(3x^2 + 5x + 2).$$

All the remaining zeros must be zeros of the reduced polynomial
$3x^2 + 5x + 2$.

$$3x^2 + 5x + 2 = 0$$

$$(3x + 2)(x + 1) = 0$$

$$x = -\frac{2}{3} \quad \text{or} \quad -1$$

The zeros of $3x^4 + 23x^3 + 56x^2 + 52x + 16$ are $-4, -2, -1, -\frac{2}{3}$.

b. By Descartes' Rule of Signs $P(x) = 3x^3 + 13x^2 - 16$ has one positive
    real zero and either two or zero negative real zeros. By the Rational

Zero Theorem, the possible rational zeros are

$$\frac{p}{q}: \quad \pm 1, \ \pm 2, \ \pm 4, \ \pm 8, \ \pm 16, \ \pm\frac{1}{3}, \ \pm\frac{2}{3}, \ \pm\frac{4}{3}, \ \pm\frac{8}{3}, \ \pm\frac{16}{3}.$$

Synthetic division is used to test the above possible zeros.

$x = 2$:

$$\begin{array}{r|rrrr}
2 & 3 & 13 & 0 & -16 \\
  &   & 6 & 38 & 76 \\
\hline
  & 3 & 19 & 38 & 60
\end{array}$$

Because all numbers in the bottom row are positive, 2 is an upper bound.

$x = 1$:

$$\begin{array}{r|rrrr}
1 & 3 & 13 & 0 & -16 \\
  &   & 3 & 16 & 16 \\
\hline
  & 3 & 16 & 16 & 0
\end{array}$$

$\longleftarrow$ 1 is the positive real zero

Find the zeros of the reduced polynomial $3x^2 + 16x + 16$ to determine the remaining zeros.

$$3x^2 + 16x + 16 = 0$$

$$(3x + 4)(x + 4) = 0$$

$$3x + 4 = 0 \qquad \text{or} \qquad x + 4 = 0$$

$$x = -\frac{4}{3} \qquad \text{or} \qquad x = -4$$

The zeros of $3x^3 + 13x^2 - 16$ are $1, -\dfrac{4}{3}, -4$.

■ *Try Exercise* **48,** *page 233.*

EXERCISE SET 4.3

In Exercises 1 to 10, find the zeros of the polynomial and state the multiplicity of each zero.

1. $P(x) = (x - 3)^2(x + 5)$

2. $P(x) = (x + 4)^3(x - 1)^2$

3. $P(x) = x^2(3x + 5)^2$

4. $P(x) = x^3(2x + 1)(3x - 12)^2$

5. $P(x) = (x^2 - 4)(x + 3)^2$

6. $P(x) = (x + 4)^3(x^2 - 9)^2$

7. $P(x) = (x^2 - 3x - 10)^2$

8. $P(x) = (x^3 - 4x)(2x - 7)^2$

9. $P(x) = x^4 - 10x^2 + 9$

10. $P(x) = x^4 - 12x^2 + 32$

In Exercises 11 to 22, use the Rational Zero Theorem to list possible rational zeros for each polynomial.

11. $x^3 + 3x^2 - 6x - 8$

12. $x^3 - 19x - 30$

13. $2x^3 + x^2 - 25x + 12$

14. $3x^3 + 11x^2 - 6x - 8$

15. $6x^4 + 23x^3 + 19x^2 - 8x - 4$

16. $6x^4 + 23x^3 + 15x^2 - 23x - 21$

17. $2x^3 + 9x^2 - 2x - 9$

18. $2x^4 + 11x^3 + 21x^2 + 17x + 5$

19. $4x^4 - 12x^3 - 3x^2 + 12x - 7$

20. $x^5 - x^4 - 7x^3 + 7x^2 - 12x - 12$

21. $x^5 - 32$ \qquad\qquad **22.** $x^4 - 1$

In Exercises 23 to 34, find the smallest positive integer and the largest negative integer that by the upper and lower bound theorem are upper and lower bounds for the real zeros of the following polynomials.

23. $x^3 + 3x^2 - 6x - 6$ \qquad **24.** $x^3 - 19x - 28$

25. $2x^3 + x^2 - 25x + 10$ \qquad **26.** $3x^3 + 11x^2 - 6x - 9$

27. $6x^4 + 23x^3 + 19x^2 - 8x - 4$

28. $6x^4 + 23x^3 + 15x^2 - 23x - 21$

29. $2x^3 + 9x^2 - 2x - 9$

30. $2x^4 + 11x^3 + 21x^2 + 17x + 5$

31. $4x^4 - 12x^3 - 3x^2 + 12x - 7$

32. $x^5 - x^4 - 7x^3 + 7x^2 - 12x - 12$

33. $x^5 - 32$ 34. $x^4 - 1$

In Exercises 35 to 46, use Descartes' Rule of Signs to state the number of possible positive and negative real zeros of each polynomial.

35. $x^3 + 3x^2 - 6x - 8$ 36. $x^3 - 19x - 30$

37. $2x^3 + x^2 - 25x + 12$ 38. $3x^3 + 11x^2 - 6x - 8$

39. $6x^4 + 23x^3 + 19x^2 - 8x - 4$

40. $6x^4 + 23x^3 + 15x^2 - 23x - 21$

41. $2x^3 + 9x^2 - 2x - 9$

42. $2x^4 + 11x^3 + 21x^2 + 17x + 5$

43. $4x^4 - 12x^3 - 3x^2 + 12x - 7$

44. $x^5 - x^4 - 7x^3 + 7x^2 - 12x - 12$

45. $x^5 - 32$ 46. $x^4 - 1$

In Exercises 47 to 64, find the zeros of each polynomial.

47. $x^3 + 3x^2 - 6x - 8$ 48. $x^3 - 19x - 30$

49. $2x^3 + x^2 - 25x + 12$ 50. $3x^3 + 11x^2 - 6x - 8$

51. $6x^4 + 23x^3 + 19x^2 - 8x - 4$

52. $6x^4 + 23x^3 + 15x^2 - 23x - 21$

53. $2x^3 + 9x^2 - 2x - 9$

54. $2x^4 + 11x^3 + 21x^2 + 17x + 5$

55. $2x^4 - 9x^3 - 2x^2 + 27x - 12$

56. $3x^3 - x^2 - 6x + 2$

57. $x^3 - 3x - 2$

58. $3x^4 - 4x^3 - 11x^2 + 16x - 4$

59. $x^4 - 5x^2 - 2x$

60. $x^3 - 2x + 1$

61. $x^4 + x^3 - 3x^2 - 5x - 2$

62. $6x^4 - 17x^3 - 11x^2 + 42x$

63. $2x^4 - 17x^3 + 4x^2 + 35x - 24$

64. $x^5 + 5x^4 + 10x^3 + 10x^2 + 5x + 1$

Supplemental Exercises

In Exercises 65 to 70, verify that each polynomial has no rational zeros.

65. $x^4 - 2x^3 + 11x^2 - 2x + 10$

66. $x^4 - 2x^3 + 21x^2 - 2x + 20$

67. $2x^4 + x^2 + 5$

68. $4x^4 + 14x^2 + 5$

69. $x^4 - 4x^3 + 14x^2 - 4x + 13$

70. $x^6 + 3x^4 + 3x^2 + 1$

In Exercises 71 to 74, determine if the given polynomial satisfies the following theorem.

Theorem Let $P(x) = a_n x^n + a_{n-1}x^{n-1} + \cdots + a_1 x + a_0$ be a polynomial with integer coefficients and $n \geq 2$. If a_n, a_0, and $f(1)$ are all odd, then $P(x)$ has no rational zeros.

71. $x^5 + 2x^4 + x^3 - x^2 + x + 945$

72. $5x^3 - 2x^2 - x + 1815$

73. $3x^4 - 5x^3 + 6x^2 - 2x + 9009$

74. $15x^7 - 4x^3 + x^2 - 6075$

75. Prove that $\sqrt{2}$ is an irrational number. (*Hint:* First show that $P(x) = x^2 - 2$ has a positive zero. Then show that P has no rational zeros.)

4.4

The Fundamental Theorem of Algebra

The German mathematician Carl Friedrich Gauss (1777–1855) was the first to prove that every polynomial has at least one complex zero. This concept is so basic to the study of algebra that it is called the **Fundamental Theorem of Algebra.** The proof of the Fundamental Theorem is beyond the scope of this text; however, it is important to understand the theorem and its consequences. In each of the following theorems, keep in mind that the terms *complex coefficients* and *complex zeros* include real coefficients and real zeros since the set of real numbers is a subset of the set of complex numbers.

The Fundamental Theorem of Algebra

> If P is a polynomial of degree of $n \geq 1$ with complex coefficients, then P has at least one complex zero.

Let P be a polynomial of degree $n \geq 1$, with complex coefficients. The Fundamental Theorem implies that P has a complex zero, say, c_1. The Factor Theorem implies that

$$P(x) = (x - c_1)Q(x)$$

where $Q(x)$ is a polynomial of degree 1 less than the degree of $P(x)$. Recall that the polynomial $Q(x)$ is called a reduced polynomial. Assuming the degree of $Q(x)$ is 1 or more, the Fundamental Theorem, implies that it must also have a zero. A continuation of this reasoning process leads to the following theorem, which is a corollary of the Fundamental Theorem.

The Number of Zeros of a Polynomial

> If P is a polynomial of degree $n \geq 1$ with complex coefficients, then P has exactly n complex zeros provided each zero is counted according to its multiplicity.

Even though every polynomial of nth degree has exactly n zeros, the zeros may not be distinct. For example, the third-degree polynomial

$$x^3 - 5x^2 + 3x + 9$$

factors into

$$(x + 1)(x - 3)(x - 3)$$

which has zeros -1, 3, and 3. The zero 3 is a zero of multiplicity two.

Although the Fundamental Theorem and its corollary give information about the existence and the number of zeros of a polynomial, they do not provide a method of actually finding the zeros. If a polynomial has real coefficients, then the following theorem can help determine the zeros of the polynomial.

The Conjugate Pair Theorem

> If $a + bi$ $(b \neq 0)$ is a complex zero of the polynomial P, *with real coefficients*, then the conjugate $a - bi$ is also a complex zero of the polynomial.

Proof: If the complex number z is a zero of the polynomial P, then

$$P(z) = a_n z^n + a_{n-1} z^{n-1} + \cdots + a_1 z + a_0 = 0,$$

where each coefficient a_i is a real number. Since the complex number on the left side of the equation is equal to the complex number on the right side, their conjugates are also equal. That is,

$$\overline{a_n z^n + a_{n-1} z^{n-1} + \cdots + a_1 z + a_0} = \overline{0} = 0.$$

Because the conjugate of a sum of complex numbers is the sum of the conjugates (see Exercise 91, Exercise Set 1.8),

$$\overline{a_n z^n} + \overline{a_{n-1} z^{n-1}} + \cdots + \overline{a_1 z} + \overline{a_0} = 0.$$

Because the conjugate of a product of complex numbers is the product of the conjugates (see Exercise 92, Exercise Set 1.8),

$$\overline{a_n}\,\overline{z^n} + \overline{a_{n-1}}\,\overline{z^{n-1}} + \cdots + \overline{a_1}\,\overline{z} + \overline{a_0} = 0.$$

It can also be shown that if z is a complex number, $\overline{z^n} = \overline{z}^n$ for every positive integer n. This combined with the fact that for each real number a, $\overline{a} = a$ produces the following:

$$a_n \overline{z}^n + a_{n-1} \overline{z}^{n-1} + \cdots + a_1 \overline{z} + a_0 = 0.$$

This equation can be written as $P(\overline{z}) = 0$, and the theorem is established.

EXAMPLE 1   Use the Conjugate Pair Theorem to Find Zeros

Find all the zeros of $x^4 - 4x^3 + 14x^2 - 36x + 45$ given that $2 + i$ is a zero.

Solution   Because the coefficients are real numbers and $2 + i$ is a zero, the Conjugate Pair Theorem implies that $2 - i$ must also be a zero. Synthetic division can now be used to find the other two zeros.

$$
\begin{array}{r|rrrrr}
2+i & 1 & -4 & 14 & -36 & 45 \\
& & 2+i & -5 & 18+9i & -45 \\
\hline
& 1 & -2+i & 9 & -18+9i & 0
\end{array}
$$

The coefficients of the reduced polynomial.

$$
\begin{array}{r|rrrr}
2-i & 1 & -2+i & 9 & -18+9i \\
& & 2-i & 0 & 18-9i \\
\hline
& 1 & 0 & 9 & 0
\end{array}
$$

← The coefficients of the reduced polynomial.

The coefficients of the next reduced polynomial.

The resulting reduced polynomial is $x^2 + 9$, which has $3i$ and $-3i$ as zeros. Therefore the four zeros of

$$x^4 - 4x^3 + 14x^2 - 36x + 45$$

are $2 + i$, $2 - i$, $3i$, and $-3i$.

■ *Try Exercise 2, page 238.*

Factors of a Polynomial

The following theorem is a result of the Conjugate Pair Theorem.

Linear and Quadratic Factors of a Polynomial

> Every polynomial with real coefficients and positive degree n can be written as the product of linear and quadratic factors with real coefficients, where the quadratic factors have no real zeros.

Proof: If P is a polynomial of degree n, then it has precisely n complex zeros $c_1 c_2, \ldots, c_n$. It may be written in the factored form

$$P(x) = a(x - c_1)(x - c_2) \cdots (x - c_n),$$

where a is the leading coefficient of P. If any zero c_k is a real number, then $(x - c_k)$ is a linear factor as referred to in the statement of the theorem. If any zero c_k is a complex number, say,

$$c_k = a + bi, \quad b \neq 0.$$

then by the Conjugate Pair Theorem,

$$c_j = a - bi$$

is also a zero of P. The product of $(x - c_k)$ and $(x - c_j)$ is a quadratic factor with real coefficients of 1, $-2a$, and $a^2 + b^2$ as demonstrated below.

$$(x - c_k)(x - c_j) = [x - (a + bi)][x - (a - bi)]$$
$$= x^2 - 2ax + (a^2 + b^2)$$

Thus $x^2 - 2ax + (a^2 + b^2)$ is a quadratic factor with real coefficients.

A quadratic factor with no real zeros is said to be **irreducible over the reals.**

EXAMPLE 2   Factor a Polynomial into Linear and Quadratic Factors

Write each polynomial as a product of linear factors and quadratic factors that are irreducible over the reals:

a. $P(x) = x^3 - 3x^2 + x - 3$ b. $P(x) = x^3 - 6x^2 + 13x - 10$

Solution

a.   Factoring by grouping produces:

$$P(x) = x^3 - 3x^2 + x - 3 = (x^3 - 3x^2) + (x - 3)$$
$$= x^2(x - 3) + 1(x - 3) = (x - 3)(x^2 + 1)$$

Since each binomial factor is irreducible over the reals, the above factorization is complete.

b.   Since $x^3 - 6x^2 + 13x - 10$ cannot be factored by grouping, synthetic division is used to determine zeros that also determine factors. By the Rational Zero Theorem, we know that ± 1, ± 2, ± 5, and ± 10 are possible rational zeros. Testing each of these, we find

$$\begin{array}{r|rrrr} 2 & 1 & -6 & 13 & -10 \\ & & 2 & -8 & 10 \\ \hline & 1 & -4 & 5 & 0 \end{array} \leftarrow 2 \text{ is a zero}$$

Using the quadratic formula, we find that the reduced polynomial $x^2 - 4x + 5$ has zeros of $2 \pm i$, so it cannot be factored over the field of real numbers. Thus $x^3 - 6x^2 + 13x - 10$ factors into

$$(x - 2)(x^2 - 4x + 5),$$

which is a product of a linear and a quadratic factor that is irreducible over the reals.

■ *Try Exercise* **20,** *page 238.*

Many of the problems in this section and Section 3 dealt with the process of finding the zeros of a given polynomial. Example 3 considers the reverse process of finding a polynomial when the zeros are given.

EXAMPLE 3   **Determine a Polynomial Given Its Zeros**

Find each of the following:

a.   A polynomial of degree 3 that has 1, 2, and -3 as zeros

b.   A polynomial of degree 4 that has real coefficients and zeros $2i$ and $3 - 7i$

Solution

a.   Since 1, 2, and -3 are zeros, $(x - 1)$, $(x - 2)$, and $(x + 3)$ are factors. Multiplying these factors produces a polynomial that has the indicated zeros.

$$(x - 1)(x - 2)(x + 3) = (x^2 - 3x + 2)(x + 3) = x^3 - 7x + 6$$

b.   By the Conjugate Pair Theorem, the polynomial also must have $-2i$ and $3 + 7i$ as zeros. The product of the factors $x - 2i$, $x - (-2i)$, $x - (3 - 7i)$, and $x - (3 + 7i)$ will produce the desired polynomial.

$$(x - 2i)(x + 2i)[x - (3 - 7i)][x - (3 + 7i)]$$
$$= (x^2 + 4)(x^2 - 6x + 58)$$
$$= x^4 - 6x^3 + 62x^2 - 24x + 232$$

■ *Try Exercise* **40,** *page 238.*

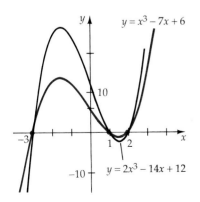

Figure 4.11

A polynomial that has a given set of zeros is not unique. For example, $x^3 - 7x + 6$ has zeros 1, 2, and -3, but so does any nonzero multiple of the polynomial, such as $2x^3 - 14x + 12$. This concept is illustrated in Figure 4.11. The graphs of the two polynomials are different; however, they have the same x-intercepts.

EXERCISE SET 4.4 _____

In Exercises 1 to 12, use the given zero to find the remaining zeros of each polynomial.

1. $2x^3 - 5x^2 + 6x - 2$; $1 + i$
2. $3x^3 - 29x^2 + 92x + 34$; $5 + 3i$
3. $x^3 + 3x^2 + x + 3$; $-i$
4. $x^4 - 6x^3 + 71x^2 - 146x + 530$; $2 + 7i$
5. $x^5 - x^4 - 3x^3 + 3x^2 - 10x + 10$; $i\sqrt{2}$
6. $x^4 - 4x^3 + 14x^2 - 4x + 13$; $2 - 3i$
7. $12x^3 - 28x^2 + 23x - 5$; $\frac{1}{3}$
8. $8x^4 - 2x^3 + 199x^2 - 50x - 25$; $-5i$
9. $x^4 - 4x^3 + 19x^2 - 30x + 50$; $1 + 3i$
10. $12x^4 - 52x^3 + 19x^2 - 13x + 4$; $\frac{1}{2}i$
11. $x^5 - x^4 - 4x^3 - 4x^2 - 5x - 3$; i
12. $x^5 - 3x^4 + 7x^3 - 13x^2 + 12x - 4$; $-2i$

In Exercises 13 to 18, find all the zeros of the polynomial. *Hint:* First determine the rational zeros.

13. $x^4 + x^3 - 2x^2 + 4x - 24$
14. $x^4 - 3x^3 + 5x^2 - 27x - 36$
15. $2x^4 + x^3 + 39x^2 + 136x - 78$
16. $x^3 - 13x^2 + 65x - 125$
17. $x^5 - 9x^4 + 34x^3 - 58x^2 + 45x - 13$
18. $x^4 - 4x^3 + 53x^2 - 196x + 196$

In Exercises 19 to 28, factor each polynomial into linear factors and/or quadratic factors that are irreducible over the reals.

19. $x^3 - x^2 - 2x$
20. $6x^3 - 23x^2 - 4x$
21. $x^3 + 9x$
22. $x^3 + 10x$
23. $x^4 + 2x^2 - 24$
24. $x^4 - 8x^2 - 20$
25. $x^4 + 3x^2 + 2$
26. $x^5 + 11x^3 + 18x$
27. $x^4 - 2x^3 + x^2 - 8x - 12$
28. $x^4 + 2x^3 + 6x^2 + 32x + 40$

In Exercises 29 to 38, find a polynomial of lowest degree that has the given zeros.

29. $4, -3, 2$
30. $-1, 1, -5$
31. $3, 2i, -2i$
32. $0, i, -i$
33. $3 + i, 3 - i, 2 + 5i, 2 - 5i$
34. $2 + 3i, 2 - 3i, -5, 2$
35. $6 + 5i, 6 - 5i, 2, 3, 5$
36. $\frac{1}{2}, 4 - i, 4 + i$
37. $\frac{3}{4}, 2 + 7i, 2 - 7i$
38. $\frac{1}{4}, -\frac{1}{5}, i, -i$

In Exercises 39 to 46, find a polynomial $P(x)$ with real coefficients that has the indicated zeros and satisfies the given conditions.

39. Zeros: $2 - 5i, -4$, degree 3
40. Zeros: $3 + 2i, 7$, degree 3
41. Zeros: $4 + 3i, 5 - i$, degree 4
42. Zeros: $i, 3 - 5i$, degree 4
43. Zeros: $-1, 2, 3$, degree 3, $P(1) = 12$
44. Zeros: $3i, 2$, degree 3, $P(3) = 27$
45. Zeros: $3, -5, 2 + i$, degree 4, $P(1) = 48$
46. Zeros: $\frac{1}{2}, 1 - i$, degree 3, $P(4) = 140$

Supplemental Exercises

47. Verify that $x^3 - x^2 - ix^2 - 9x + 9 + 9i$ has $1 + i$ as a zero and that its conjugate $1 - i$ is not a zero. Explain why this does not contradict the Conjugate Pair Theorem.

48. Verify that $x^3 - x^2 - ix^2 - 20x + ix + 20i$ has a zero of i but not of its conjugate $-i$. Explain why this does not contradict the Conjugate Pair Theorem.

49. Show that 2 is a zero of multiplicity 3 of the polynomial

$$P(x) = x^5 - 6x^4 + 21x^3 - 62x^2 + 108x - 72,$$

and express $P(x)$ as a product of linear factors and/or quadratic factors that are irreducible over the reals.

50. Show that -1 is a zero of multiplicity 4 of the polynomial

$$P(x) = x^6 + 5x^5 + 11x^4 + 14x^3 + 11x^2 + 5x + 1,$$

and express $P(x)$ as a product of linear factors and/or quadratic factors that are irreducible over the reals.

51. Find a polynomial $P(x)$ of degree 5 such that 1 is a zero of multiplicity 2, 2 is a zero of multiplicity 3, and $P(-1) = -54$.

52. Find a polynomial $P(x)$ of degree 5 such that -4 is a zero of multiplicity 4, $\frac{1}{2}$ is a zero of multiplicity 1, and $P(1) = 125$.

4.5

Rational Functions and Their Graphs

If $P(x)$ and $Q(x)$ are polynomials, then the function F given by

$$F(x) = \frac{P(x)}{Q(x)}$$

is called a **rational function.** The domain of F is the set of all real numbers except for those for which $Q(x) = 0$. For example, the domain of

$$F(x) = \frac{x^2 - x - 5}{x(2x - 5)(x + 3)}$$

is the set of all real numbers except 0, 5/2, and −3.

The graph of the rational function $G(x) = \dfrac{x + 1}{x - 2}$ is shown in Figure 4.12. The graph shows that G has the following behavior:

1. The graph does not exist at $x = 2$. That is, 2 is not in the domain of G.
2. The graph has an x-intercept at $(-1, 0)$ and a y-intercept at $(0, -\frac{1}{2})$.
3. The functional values $G(x)$ approach 1 as x increases or decreases without bound.
4. The functional values $G(x)$ increase without bound as x approaches 2 from the right.
5. The functional values $G(x)$ decrease without bound as x approaches 2 from the left.

When discussing graphs that increase or decrease without bound, it is convenient to use mathematical notation. The notation

$$f(x) \to \infty \text{ as } x \to a^+$$

means that the functional values $f(x)$ increase without bound as x approaches a from the right. Recall that the symbol ∞ does not represent a real number but is used merely to describe the concept of a variable taking on larger and larger values without bound. See Figure 4.13 (a).

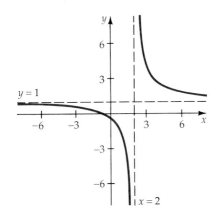

Figure 4.12

$$G(x) = \frac{x + 1}{x - 2}$$

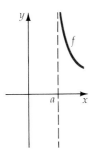

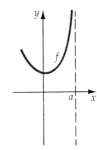
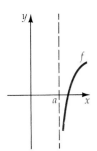
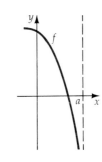

a. $f(x) \to \infty$
  as $x \to a^+$

b. $f(x) \to \infty$
  as $x \to a^-$

c. $f(x) \to -\infty$
  as $x \to a^+$

d. $f(x) \to -\infty$
  as $x \to a^-$

Figure 4.13

The notation

$$f(x) \rightarrow \infty \text{ as } x \rightarrow a^-$$

means that the functional values $f(x)$ increase without bound as x approaches a from the left. See Figure 4.13 (b).

The notation

$$f(x) \rightarrow -\infty \text{ as } x \rightarrow a^+$$

means that the functional values $f(x)$ decrease without bound as x approaches a from the right. See Figure 4.13 (c).

The notation

$$f(x) \rightarrow -\infty \text{ as } x \rightarrow a^-$$

means that the functional values $f(x)$ decrease without bound as x approaches a from the left. See Figure 4.13 (d).

Asymptotes

Each graph in Figure 4.13 approaches a vertical line through $(a, 0)$ as $x \rightarrow a^+$ or a^-. The line is said to be a *vertical asymptote* to the graph.

Definition of a Vertical Asymptote

> The line $x = a$ is a **vertical asymptote** of the graph of a function F provided
>
> $$F(x) \rightarrow \infty \quad \text{or} \quad F(x) \rightarrow -\infty$$
>
> as x approaches a from either the left or right.

In Figure 4.12, the line $x = 2$ is a vertical asymptote of the graph of G. Notice that the graph of G in Figure 4.12 also approaches the horizontal line $y = 1$ as $x \rightarrow \infty$ and as $x \rightarrow -\infty$. The line $y = 1$ is a *horizontal asymptote* of the graph of G.

Definition of a Horizontal Asymptote

> The line $y = b$ is a **horizontal asymptote** of the graph of a function F provided
>
> $$F(x) \rightarrow b \text{ as } x \rightarrow \infty \text{ or } x \rightarrow -\infty .$$

Figure 4.14 illustrates some of the ways the graph of a rational function may approach its horizontal asymptote. It is common practice to display the asymptotes of the graph of a rational function using dashed lines. Although a rational function may have several vertical asymptotes, it can have at most one horizontal asymptote. The graph of a rational function will never intersect any of its vertical asymptotes. Why?[2] However, the graph may intersect its horizontal asymptote.

---

[2] If $x = a$ is a vertical asymptote of a rational function R, then $R(a)$ is undefined.

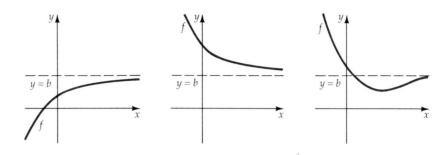

Figure 4.14
$f(x) \to b$ as $x \to \infty$.

Geometrically, a line is an asymptote to a curve if the distance between the line and a point $P(x, y)$ on the curve approaches zero as the distance between the origin and the point P increase without bound.

Vertical asymptotes of the graph of a rational function can be found using the following theorem.

Theorem on Vertical Asymptotes

> If the real number a is a zero of the denominator $Q(x)$, then the graph of $F(x) = P(x)/Q(x)$, where $P(x)$ and $Q(x)$ have no common factors, has the vertical asymptote $x = a$.

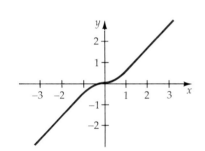

Figure 4.15

$f(x) = \dfrac{x^3}{x^2 + 1}$

EXAMPLE 1   **Find the Vertical Asymptotes of a Rational Function**

Find the vertical asymptotes of each rational function.

a. $f(x) = \dfrac{x^3}{x^2 + 1}$ b. $g(x) = \dfrac{x}{x^2 - x - 6}$

Solution

a. To find the vertical asymptotes, set the denominator equal to zero. The denominator $x^2 + 1$ has no real zeros, so the graph of f has no vertical asymptotes. See Figure 4.15.

b. The denominator $x^2 - x - 6 = (x - 3)(x + 2)$ has zeros of 3 and -2. Since the numerator x has no common factors with the denominator $x^2 - x - 6$, $x = 3$ and $x = -2$ are both vertical asymptotes of the graph of g, as shown in Figure 4.16.

■ *Try Exercise **2**, page 249.*

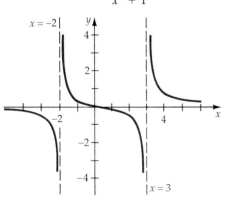

Figure 4.16

$g(x) = \dfrac{x}{x^2 - x - 6}$

The following theorem implies that a horizontal asymptote can be determined by examining the leading terms of the numerator and the denominator of a rational function.

Theorem on Horizontal Asymptotes

Let $$F(x) = \frac{a_n x^n + a_{n-1} x^{n-1} + \cdots + a_1 x + a_0}{b_m x^m + b_{m-1} x^{m-1} + \cdots + b_1 x + b_0}$$

be a rational function with numerator of degree n and denominator of degree m.

1. If $n < m$, then the x-axis is a horizontal asymptote of the graph of F.
2. If $n = m$, then the line $y = a_n/b_m$ is a horizontal asymptote of the graph of F.
3. If $n > m$, the graph of F has no horizontal asymptote.

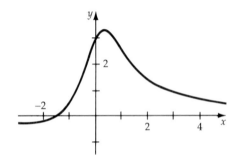

Figure 4.17

$$f(x) = \frac{2x + 3}{x^2 + 1}$$

EXAMPLE 2 **Find the Horizontal Asymptote of a Rational Function**

Find the horizontal asymptote of each rational function:

a. $f(x) = \dfrac{2x + 3}{x^2 + 1}$ b. $g(x) = \dfrac{4x^2 + 1}{3x^2}$ c. $h(x) = \dfrac{x^3 + 1}{x - 2}$

Solution

a. The degree of the numerator $2x + 3$ is less than the degree of the denominator $x^2 + 1$. By the Theorem on Horizontal Asymptotes, the x-axis is the horizontal asymptote of f. See the graph of f in Figure 4.17.

b. The numerator $4x^2 + 1$ and denominator $3x^2$ of g are both of degree 2. By the Theorem on Horizontal Asymptotes, the line $y = \frac{4}{3}$ is the horizontal asymptote of g. See the graph of g in Figure 4.18.

c. The degree of the numerator $x^3 + 1$ is larger than the degree of the denominator $x - 2$, so by the Theorem on Horizontal Asymptotes, the graph of h has no horizontal asymptote.

■ *Try Exercise **6**, page 249.*

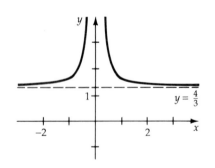

Figure 4.18

$$g(x) = \frac{4x^2 + 1}{3x^2}$$

The proof of the Theorem on Horizontal Asymptotes makes use of the technique used in the following verification. To verify that

$$g(x) = \frac{4x^2 + 1}{3x^2 + 8x + 7}$$

has a horizontal asymptote of $y = \frac{4}{3}$, divide the numerator and the denominator by the largest power of the variable x (x^2 in this case):

$$g(x) = \frac{\dfrac{4x^2 + 1}{x^2}}{\dfrac{3x^2 + 8x + 7}{x^2}} = \frac{4 + \dfrac{1}{x^2}}{3 + \dfrac{8}{x} + \dfrac{7}{x^2}} \qquad x \neq 0.$$

Now, as x increases without bound or decreases without bound, the fractions $1/x^2$, $8/x$, and $7/x^2$ approach zero. Thus

$$g(x) \to \frac{4 + 0}{3 + 0 + 0} = \frac{4}{3} \text{ as } x \to \pm\infty$$

and hence the line $y = \frac{4}{3}$ is the horizontal asymptote of the graph of the rational function g.

The zeros and vertical asymptotes of a rational function F divide the x-axis into intervals. In each interval,

1. $F(x)$ is positive for all x in the interval, or
2. $F(x)$ is negative for all x in the interval.

For example, consider the rational function

$$g(x) = \frac{x + 1}{x^2 + 2x - 3}$$

which has vertical asymptotes of $x = -3$ and $x = 1$ and a zero of -1. These three numbers divide the x-axis into the four intervals

$$(-\infty, -3), (-3, -1), (-1, 1), \text{ and } (1, \infty).$$

Notice in Figure 4.19, that $g(x)$ is

- Negative for all x such that $x < -3$.
- Positive for all x such that $-3 < x < -1$.
- Negative for all x such that $-1 < x < 1$.
- Positive for all x such that $x > 1$.

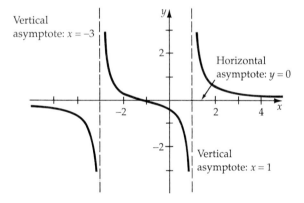

Figure 4.19

$$g(x) = \frac{x + 1}{x^2 + 2x - 3}$$

General Graphing Procedure

If $F(x) = P(x)/Q(x)$, where $P(x)$ and $Q(x)$ are polynomials that have no common factor, then the following general procedure offers useful guidelines for graphing $F(x)$.

**General Procedure for Graphing Rational Functions
That Have No Common Factors**

1. *Asymptotes:* Find the real zeros of the denominator $Q(x)$. For each zero a, draw the dashed line $x = a$. Each line is a vertical asymptote of the graph of F. Graph any horizontal asymptote. These can be found by using the Theorem on Horizontal Asymptotes. If the degree of the numerator $P(x)$ is larger than the degree of the denominator $Q(x)$, then the graph of F does not have a horizontal asymptote.

2. *Intercepts:* Find the real zeros of the numerator $P(x)$. For each zero a, plot the point $(a, 0)$. Each point is an x-intercept of the graph of F. Evaluate $F(0)$. Plot $(0, F(0))$, the y-intercept of the graph of F.

3. *Additional points:* Plot at least two points that lie in the intervals between and beyond the vertical asymptotes and the x-intercepts.

4. *Behavior near asymptotes:* If $x = a$ is a vertical asymptote, determine whether $F(x) \to \infty$ or $F(x) \to -\infty$ as $x \to a^-$ and also as $x \to a^+$.

5. *Complete the sketch:* Use all the information obtained above to sketch the graph of F. Plot additional points if necessary to gain additional knowledge about the function.

EXAMPLE 3   **Graph a Rational Function**

Sketch the graph of $f(x) = \dfrac{4x^2}{x^2 + 3}$.

Solution   *Asymptotes:* The denominator $x^2 + 3$ has no real zeros, so the graph of f has no vertical asymptotes.

The numerator and denominator both have degree 2. The leading coefficients of the numerator and denominator are 4 and 1, respectively. By the Theorem on Horizontal Asymptotes, the graph of f has horizontal asymptote $y = \frac{4}{1} = 4$.

Intercepts: The numerator $4x^2$ has 0 as its only zero. Therefore the graph of f has an x-intercept at the origin. Since $f(0) = 0$, f has y-intercept $(0, 0)$.

Additional points: The intervals determined by the x-intercept are $x < 0$ and $x > 0$. Generally, it is necessary to determine points in all intervals. Since f is an even function, its graph is symmetrical with respect to the y-axis. The following table lists a few points for $x > 0$. Symmetry can be used to locate corresponding points for $x < 0$.

x	1	2	6
$f(x)$	1	$\dfrac{16}{7} \approx 2.29$	$\dfrac{48}{13} \approx 3.69$

Behavior near asymptotes: As x increases or decreases without bound, $f(x)$ approaches the horizontal asymptote $y = 4$. To determine whether the graph of f intersects the horizontal asymptote at any point, solve the equation $f(x) = 4$.

There are no solutions of $f(x) = 4$ because

$$\frac{4x^2}{x^2 + 3} = 4 \quad \text{implies} \quad 4x^2 = 4x^2 + 12.$$

This is not possible. Thus the graph of f does not intersect the horizontal asymptote but approaches it from below as x increases or decreases without bound.

Vertical Asymptote	Horizontal Asymptote	x-intercept	y-intercept	Additional Points
None	$y = 4$	$(0, 0)$	$(0, 0)$	$(1, 1)$, $(2, 2.29)$, $(6, 3.69)$

Complete the sketch: The previous information can now be used to finish the sketch. The completed graph is shown in Figure 4.20.

■ *Try Exercise 10, page 249.*

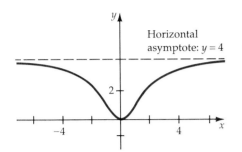

Figure 4.20

$$f(x) = \frac{4x^2}{x^2 + 3}$$

EXAMPLE 4 **Graph a Rational Function**

Sketch the graph of $h(x) = \dfrac{x^2 + 1}{x^2 + x - 2}$.

Solution *Asymptotes:* The denominator $x^2 + x - 2 = (x + 2)(x - 1)$ has zeros -2 and 1; thus the lines $x = -2$ and $x = 1$ are vertical asymptotes.

The numerator and denominator both have degree 2. The leading coefficients of the numerator and denominator are both 1. Thus, h has horizontal asymptote $y = \frac{1}{1} = 1$.

Intercept(s): The numerator $x^2 + 1$ has no real zeros, so the graph of h has no x-intercepts. Since $h(0) = -0.5$, h has y-intercept $(0, -0.5)$.

Additional points: The intervals determined by the vertical asymptotes are $(-\infty, -2)$, $(-2, 1)$, and $(1, \infty)$. Plot a few points from each interval:

x	-5	-3	-1	0.5	2	3	4
$h(x)$	$\dfrac{13}{9}$	2.5	-1	-1	1.25	1	$\dfrac{17}{18}$

The graph of h will intersect the horizontal asymptote $y = 1$ exactly once. This can be determined by solving the equation $h(x) = 1$, as shown below.

$$\frac{x^2 + 1}{x^2 + x - 2} = 1$$

$$x^2 + 1 = x^2 + x - 2 \quad \text{Multiply both sides by } x^2 + x - 2.$$

$$1 = x - 2$$

$$3 = x$$

The only solution is $x = 3$. Therefore the graph of h intersects the horizontal asymptote at $(3, 1)$.

Behavior near asymptotes: As x approaches -2 from the left, the denominator $(x + 2)(x - 1)$ approaches 0 but remains positive. The numerator $x^2 + 1$ approaches 5, which is positive, so the quotient $h(x)$ increases without bound. Stated in mathematical notation,

$$h(x) \to \infty \text{ as } x \to -2^-.$$

Similarly, it can be determined that

$$h(x) \to -\infty \text{ as } x \to -2^+,$$
$$h(x) \to -\infty \text{ as } x \to 1^-,$$
$$h(x) \to \infty \quad \text{as } x \to 1^+.$$

Vertical Asymptotes	Horizontal Asymptote	x-intercept	y-intercept	Additional Points
$x = -2$, $x = 1$	$y = 1$	None	$(0, -0.5)$	$(-5, 1.\overline{4})$, $(-3, 2.5)$, $(-1, -1)$, $(0.5, -1)$ $(2, 1.25)$, $(3, 1)$, $(4, 0.9\overline{4})$

Complete the sketch: Use the previous information to obtain the graph sketched in Figure 4.21.

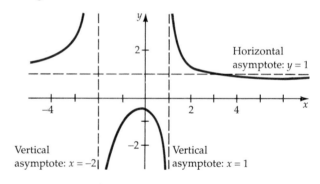

Figure 4.21

$$h(x) = \frac{x^2 + 1}{x^2 + x - 2}$$

■ *Try Exercise* **26,** *page 249.*

Slant Asymptotes

Some rational functions have an asymptote that is neither vertical nor horizontal but slanted.

Theorem on Slant Asymptotes

> The rational function given by $F(x) = P(x)/Q(x)$, where $P(x)$ and $Q(x)$ have no common factors, will have a **slant asymptote** if the degree of the polynomial $P(x)$ in the numerator is one greater than the degree of the polynomial $Q(x)$ in the denominator.

To find the slant asymptote, use division to express $F(x)$ in the form

$$F(x) = \frac{P(x)}{Q(x)} = (mx + b) + \frac{r(x)}{Q(x)},$$

where the degree of $r(x)$ is less than the degree of $Q(x)$. Since

$$\frac{r(x)}{Q(x)} \to 0 \quad \text{as} \quad x \to \pm\infty,$$

$F(x) \to mx + b$ as $x \to \pm\infty$.

The line represented by $y = mx + b$ is called the slant asymptote of the graph of F.

EXAMPLE 5   Find the Slant Asymptote of a Rational Function

Find the slant asymptote of $f(x) = \dfrac{2x^3 + 5x^2 + 1}{x^2 + x + 3}$.

Solution   Because the degree of the numerator $2x^3 + 5x^2 + 1$ is exactly one larger than the degree of the denominator $x^2 + x + 3$, f has a slant asymptote. To find the asymptote, divide $2x^3 + 5x^2 + 1$ by $x^2 + x + 3$.

$$
\begin{array}{r}
2x + 3 \\
x^2 + x + 3 \overline{)\, 2x^3 + 5x^2 + 0x + 1} \\
\underline{2x^3 + 2x^2 + 6x } \\
3x^2 - 6x + 1 \\
\underline{3x^2 + 3x + 9} \\
- 9x - 8
\end{array}
$$

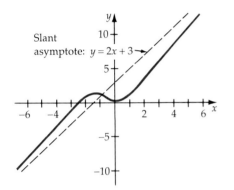

Figure 4.22

$f(x) = \dfrac{2x^3 + 5x^2 + 1}{x^2 + x + 3}$

Therefore

$$f(x) = \frac{2x^3 + 5x^2 + 1}{x^2 + x + 3} = (2x + 3) + \frac{-9x - 8}{x^2 + x + 3},$$

and the line $y = 2x + 3$ is the slant asymptote for the graph of f. Figure 4.22 shows the graph of f and its slant asymptote.

■ *Try Exercise 34, page 249.*

Remark   The function f in Example 5 does not have a vertical asymptote because the denominator $x^2 + x + 3$ does not have any real zeros. However, the function

$$g(x) = \frac{2x^2 - 4x + 5}{3 - x}$$

has both a slant asymptote and a vertical asymptote. Why?[3] Figure 4.23 shows the graph of g and its asymptotes.

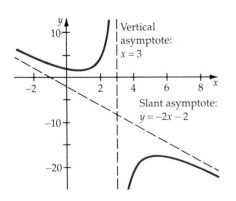

Figure 4.23

$g(x) = \dfrac{2x^2 - 4x + 5}{3 - x}$

[3] It has a slant asymptote because the degree of the numerator is one greater than the degree of the denominator and they have no common factors. It has a vertical asymptote because the denominator equals zero when $x = 3$.

EXAMPLE 6 **Graph a Rational Function That Has a Slant Asymptote**

Sketch the graph of $j(x) = \dfrac{x^3 - 1}{x^2}$.

Solution *Asymptotes:* The denominator x^2 has 0 as its only zero. Thus the y-axis is the vertical asymptote of the graph of j.

The degree of the numerator $x^3 - 1$ is exactly one more than the degree of the denominator x^2, so j has a slant asymptote. Dividing $x^3 - 1$ by x^2 shows that j can be expressed as

$$j(x) = \frac{x^3}{x^2} - \frac{1}{x^2} = x - \frac{1}{x^2}.$$

From this we see that $j(x) \to x$ as $x \to \pm\infty$. Therefore j has a slant asymptote of $y = x$.

Intercepts: The numerator $x^3 - 1$ has a real zero of 1. Therefore $(1, 0)$ is the only x-intercept of the graph of j. Since $j(0)$ is undefined, the graph of j does not have a y-intercept.

Additional points: The intervals determined by the vertical asymptote and the x-intercept are $x < 0$, $0 < x < 1$, and $x > 1$. The following table lists a few points from each interval:

x	-5	-2	-1	-0.5	0.5	0.8	2	5
$j(x)$	-5.04	-2.25	-2	-4.5	-3.5	-0.7625	1.75	4.96

Vertical Asymptote	Slant Asymptote	x-intercept	y-intercept	Additional Points
y-axis	$y = x$	$(1, 0)$	None	$(-5, -5.04)$, $(-2, -2.25)$, $(-1, -2)$ $(-0.5, -4.5)$, $(0.5, -3.5)$, $(0.8, -0.7625)$ $(2, 1.75)$, $(5, 4.96)$

Complete the sketch: Use all the previous information to complete the sketch of j as shown in Figure 4.24.

■ *Try Exercise* **40,** *page 250.*

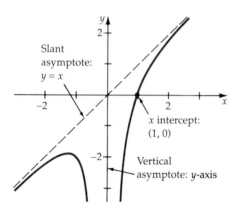

Figure 4.24

$j(x) = \dfrac{x^3 - 1}{x^2}$

If a rational function has a numerator and denominator that have a common factor, then the rational function should be reduced to lowest terms before you apply the general procedure for sketching the graph of a rational function.

EXAMPLE 7 **Graph a Rational Function That Has a Common Factor**

Sketch the graph of $f(x) = \dfrac{x^2 - 3x - 4}{x^2 - 6x + 8}$.

Solution   Factor the numerator and denominator to obtain

$$f(x) = \frac{x^2 - 3x - 4}{x^2 - 6x + 8}$$

$$= \frac{(x + 1)(x - 4)}{(x - 2)(x - 4)} \quad x \neq 2, \, x \neq 4.$$

Thus for all x values other than $x = 4$, the graph of f is the same as the graph of

$$G(x) = \frac{x + 1}{x - 2}$$

Figure 4.12 shows a graph of G. The graph of f will be the same as this graph, except it will have an open circle at $(4, 2.5)$ to indicate that it is undefined for $x = 4$. See the graph of f in Figure 4.25.

■ *Try Exercise 50, page 250.*

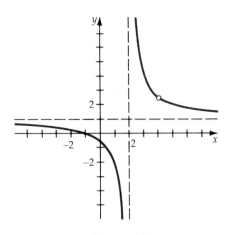

Figure 4.25

$$f(x) = \frac{x^2 - 3x - 4}{x^2 - 6x + 8}$$

EXERCISE SET 4.5

In Exercises 1 to 4, find all vertical asymptotes of each rational function.

1. $F(x) = \dfrac{2x - 1}{x^2 + 3x}$

2. $F(x) = \dfrac{3x^2 + 5}{x^2 - 4}$

3. $F(x) = \dfrac{x^2 + 11}{6x^2 - 5x - 4}$

4. $F(x) = \dfrac{3x - 5}{x^3 - 8}$

In Exercises 5 to 8, find the horizontal asymptote of each rational function.

5. $F(x) = \dfrac{4x^2 + 1}{x^2 + x + 1}$

6. $F(x) = \dfrac{3x^3 - 27x^2 + 5x - 11}{x^5 - 2x^3 + 7}$

7. $F(x) = \dfrac{15{,}000x^3 + 500x - 2000}{700 + 500x^3}$

8. $F(x) = 6000\left(1 - \dfrac{25}{(t + 5)^2}\right)$

In Exercises 9 to 32, determine the vertical and horizontal asymptotes and sketch the graph of the rational function F. Label all intercepts and asymptotes.

9. $F(x) = \dfrac{1}{x + 4}$

10. $F(x) = \dfrac{1}{x - 2}$

11. $F(x) = \dfrac{-4}{x - 3}$

12. $F(x) = \dfrac{-3}{x + 2}$

13. $F(x) = \dfrac{4}{x}$

14. $F(x) = \dfrac{-4}{x}$

15. $F(x) = \dfrac{x}{x + 4}$

16. $F(x) = \dfrac{x}{x - 2}$

17. $F(x) = \dfrac{x + 4}{2 - x}$

18. $F(x) = \dfrac{x + 3}{1 - x}$

19. $F(x) = \dfrac{1}{x^2 - 9}$

20. $F(x) = \dfrac{-2}{x^2 - 4}$

21. $F(x) = \dfrac{1}{x^2 + 2x - 3}$

22. $F(x) = \dfrac{1}{x^2 - 2x - 8}$

23. $F(x) = \dfrac{x}{9 - x^2}$

24. $F(x) = \dfrac{x}{x^2 - 16}$

25. $F(x) = \dfrac{x^2}{x^2 + 4x + 4}$

26. $F(x) = \dfrac{x^2}{x^2 - 6x + 9}$

27. $F(x) = \dfrac{10}{x^2 + 2}$

28. $F(x) = \dfrac{-20}{x^2 + 4}$

29. $F(x) = \dfrac{2x^2 - 2}{x^2 - 9}$

30. $F(x) = \dfrac{6x^2 - 5}{2x^2 + 6}$

31. $F(x) = \dfrac{x^2 + x + 4}{x^2 + 2x - 1}$

32. $F(x) = \dfrac{2x^2 - 14}{x^2 - 6x + 5}$

In Exercises 33 to 36, find the slant asymptote of each rational function.

33. $F(x) = \dfrac{3x^2 + 5x - 1}{x + 4}$

34. $F(x) = \dfrac{x^3 - 2x^2 + 3x + 4}{x^2 - 3x + 5}$

35. $F(x) = \dfrac{x^3 - 1}{x^2}$

36. $F(x) = \dfrac{4000 + 20x + 0.0001x^2}{x}$

In Exercises 37 to 46, determine the vertical and slant asymptotes and sketch the graph of the rational function *F*.

37. $F(x) = \dfrac{x^2 - 4}{x}$

38. $F(x) = \dfrac{x^2 + 10}{2x}$

39. $F(x) = \dfrac{x^2 - 3x - 4}{x + 3}$

40. $F(x) = \dfrac{x^2 - 4x - 5}{2x + 5}$

41. $F(x) = \dfrac{2x^2 + 5x + 3}{x - 4}$

42. $F(x) = \dfrac{4x^2 - 9}{x + 3}$

43. $F(x) = \dfrac{x^2 - x}{x + 2}$

44. $F(x) = \dfrac{x^2 + x}{x - 1}$

45. $F(x) = \dfrac{x^3 + 1}{x^2 - 4}$

46. $F(x) = \dfrac{x^3 - 1}{3x^2}$

In Exercises 47 to 56, sketch the graph of the rational function *F*. (*Hint:* First examine the numerator and the denominator to determine if there are any common factors.)

47. $F(x) = \dfrac{x^2 + x}{x + 1}$

48. $F(x) = \dfrac{x^2 - 3x}{x - 3}$

49. $F(x) = \dfrac{2x^3 + 4x^2}{2x + 4}$

50. $F(x) = \dfrac{x^2 - x - 12}{x^2 - 2x - 8}$

51. $F(x) = \dfrac{-2x^3 + 6x}{2x^2 - 6x}$

52. $F(x) = \dfrac{x^3 + 3x^2}{x(x + 3)(x - 1)}$

53. $F(x) = \dfrac{x^2 - 3x - 10}{x^2 + 4x + 4}$

54. $F(x) = \dfrac{2x^2 + x - 3}{x^2 - 2x + 1}$

55. $F(x) = \dfrac{x^3 + x^2 - 14x - 24}{x + 2}$

56. $F(x) = \dfrac{2x^3 + 5x^2 - 4x - 3}{x - 1}$

Supplemental Exercises

57. The cost *C* in dollars to remove *p* percent of the salt in a tank of sea water is given by

$$C(p) = \dfrac{2000p}{100 - p}, \quad 0 \le p < 100$$

  a. Find the cost of removing 40 percent of the salt.

  b. Find the cost of removing 80 percent of the salt.

  c. Sketch the graph of *C*.

58. The temperature F (measured in degrees Fahrenheit) of a dessert placed in a freezer for *t* hours is given by the rational function

$$F(t) = \dfrac{60}{t^2 + 2t + 1}, \quad t \ge 0$$

  a. Find the temperature of the dessert after it has been in the freezer for 1 hour.

  b. Find the temperature of the dessert after 4 hours.

  c. Sketch the graph of F.

59. A large electronic firm finds that the number of computers it can produce per week after *t* weeks of production is approximated by

$$C(t) = \dfrac{2000t^2 + 20,000t}{t^2 + 10t + 25} \quad 0 \le t \le 50$$

  a. Find the number of computers it produced during the first week.

  b. Find the number of computers it produced during the tenth week.

  c. What is the equation of the horizontal asymptote of the graph of *C*?

  d. Sketch the graph of *C* and then use the graph to estimate how many weeks pass until it can produce 1900 computers in a single week.

60. The cost of publishing *x* books is given by

$$C(x) = 40,000 + 20x + 0.0001x^2.$$

The average cost per book is given by

$$A(x) = \dfrac{C(x)}{x} = \dfrac{40,000 + 20x + 0.0001x^2}{x}$$

where $1000 \le x \le 100,000$.

  a. What is the average cost per book if 5000 books are published?

  b. What is the average cost per book if 10,000 books are published?

  c. What is the equation of the slant asymptote of the graph of the average cost function?

  d. Sketch the graph of *A* and estimate the number of books that should be published to minimize the average cost per book.

61. One of Poiseuille's Laws states that the resistance *R* encountered by blood flowing through a blood vessel is given by the rational function

$$R(r) = C\dfrac{L}{r^4}$$

where *C* is a positive constant determined by the viscosity of the blood, *L* is the length of the blood vessel, and *r* is the radius.

  a. Explain the meaning of $R(r) \to \infty$ as $r \to 0$.

b. Explain the meaning of $R(r) \to 0$ as $r \to \infty$.

c. Sketch the graph of R for $0 < r \le 4$ millimeters given that $C = 1$ and $L = 100$ millimeters.

62. A cylindrical soft drink can is to be made so that it will have a volume of 354 millimeters. If r is the radius of the can in centimeters, then the total surface area A of the can is given by the rational function

$$A(r) = \frac{2\pi r^3 + 708}{r}$$

a. Use the graph of A to estimate the value of r that produces the minimum value of A.

b. Does the graph of A have a slant asymptote?

c. Explain the meaning of the following statement as it applies to the graph of A:

$$\text{As } r \to \infty, A \to 2\pi r^2.$$

63. Determine the point where the graph of

$$F(x) = \frac{2x^2 + 3x + 4}{x^2 + 4x + 7}$$

intersects its horizontal asymptote.

64. Determine the point where the graph of

$$F(x) = \frac{3x^3 + 2x^2 - 8x - 12}{x^2 + 4}$$

intersects its slant asymptote.

65. Determine the two points where the graph of

$$F(x) = \frac{x^3 + x^2 + 4x + 1}{x^3 + 1}$$

intersects its horizontal asymptote.

66. Give an example of a rational function that intersects its slant asymptote at two points.

4.6
Approximation of Zeros

Finding the zeros of polynomials can be difficult. For example, finding the zeros of the polynomial $x^3 - 9x - 12$ is challenging because it does not have rational zeros. Thus the Rational Zeros Theorem is not helpful. The Zero Location Theorem from Section 4.2 can be used to establish that there is an irrational zero of $x^3 - 9x - 12$ in the interval between $x = 3$ and $x = 4$. For convenience, the Zero Location Theorem is now restated.

The Zero Location Theorem

> Let P be a polynomial with real coefficients. If $a < b$ and if $P(a)$ and $P(b)$ have different signs, then there is at least one value c between a and b such that $P(c) = 0$.

EXAMPLE 1 **Apply the Zero Location Theorem**

Show that $P(x) = x^3 - 9x - 12$ has a zero between 3 and 4.

Solution Since there is one variation of sign of $P(x)$, Descartes' Rule of Signs implies that there is one positive zero. Compare the sign of $P(3)$ with the sign of $P(4)$.

$$P(3) = -12 \quad \text{and} \quad P(4) = 16$$

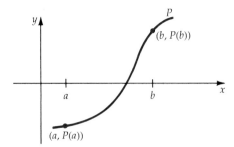

Figure 4.26

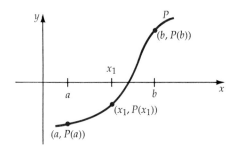

Figure 4.27

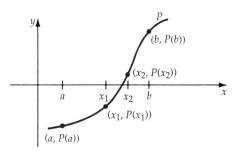

Figure 4.28

Because $P(3)$ and $P(4)$ have different signs, we know by the Zero Location Theorem that P has a zero between 3 and 4.

■ *Try Exercise 2, page 254.*

The Bisection Method

The **bisection method** is often used to approximate the real zeros of a polynomial. The strategy of the bisection method is to establish that exactly one real zero of a polynomial P is in an interval (a, b) and then reduce the size of the interval.

Consider for example a polynomial P such that $P(a) < 0$ and $P(b) > 0$. In Figure 4.26, the graph of $y = P(x)$ is below the x-axis at $x = a$ and above the x-axis at $x = b$. Because the graph of $y = P(x)$ is continuous, there must be a zero of P between a and b. Now bisect the interval (a, b) and denote the midpoint of the interval by x_1, where

$$x_1 = \frac{a + b}{2}.$$

If $P(x_1) = 0$, then x_1 is a zero of P. However, if $P(x_1) < 0$, as in Figure 4.27, then by the Zero Location Theorem, P has a zero between x_1 and b. The zero is now located in the interval (x_1, b), whose length is half the length of the original interval.

Evaluate $P(x_2)$, where x_2 is the midpoint of this new interval. Suppose that this time $P(x_2) > 0$, as in Figure 4.28. By the Zero Location Theorem, P has a zero between x_1 and x_2. Once again we have located the zero in an interval half the length of the previous interval. Each application of the bisection method that locates the zero in a smaller interval is referred to as an **iteration.**

By repeating this bisection procedure, you can approximate a zero as close as desired because each iteration will halve the length of the interval in which the zero lies. Three iterations of the bisection method will reduce the length of the initial interval that contains the zero by a factor of $2^3 = 8$. Ten iterations will reduce the length of the initial interval by a factor of 2^{10}, which is about 1000.

EXAMPLE 2 Use the Bisection Method to Approximate a Zero

Use three iterations of the bisection method to approximate the real zero of $P(x) = x^3 - 9x - 12$, which is located between 3 and 4.

Solution First iteration: The desired zero is in the interval $I_1 = (3, 4)$. The midpoint of I_1 is 3.5. Because $P(3.5) = -0.625$ is negative and $P(4)$ is positive, we know by the Zero Location Theorem that the zero is located between 3.5 and 4.

Second iteration: The desired zero is in the interval $I_2 = (3.5, 4)$. The midpoint of I_2 is 3.75. Because the value $P(3.5) = -0.625$ is negative and the value $P(3.75) = 6.984375$ is positive, we know by the Zero Location Theorem that the zero is located between 3.5 and 3.75.

Third iteration: The desired zero is in the interval $I_3 = (3.5, 3.75)$. The midpoint of I_3 is 3.625. Because the value $P(3.5) = -0.625$ is negative and the value $P(3.625) = 3.009765625$ is positive, we know by the Zero Location Theorem that the zero is located between 3.5 and 3.625.

■ *Try Exercise* **8**, *page 254.*

Remark We demonstrated that a zero of P in Example 2 is in the interval $(3.5, 3.625)$. If the midpoint of $(3.5, 3.625)$, which is 3.5625, is used as the approximation of the zero, then the maximum error of the approximation can only be one-half of the length of the interval. That is, 3.5625 approximates the actual zero with a maximum error of

$$\frac{3.625 - 3.5}{2} = 0.0625.$$

Caution Sometimes the bisection method is not applicable. For example, the polynomial P in Figure 4.29 has two zeros in the interval (a, b). The bisection method is not applicable for finding the root closest to b. To apply the Zero Location Theorem, we must isolate a single zero between a and b such that $P(a)$ and $P(b)$ have different signs.

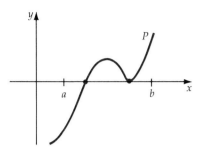

Figure 4.29

Variation of the Bisection Method

The following iteration procedure is a variation of the bisection method. It can be used to approximate a real zero of a polynomial to within 10^{-k} units by applying exactly k iterations.

Subdivide into Ten Equal Subintervals

> The following method will approximate a real zero of a polynomial to within 10^{-k} units. Use the Zero Location Theorem to determine consecutive integers a and b such that the polynomial P has a single zero between a and b.
>
> 1. Subdivide the interval (a, b) into ten equal subintervals. Use the Zero Location Theorem to determine which of these subintervals contains the zero.
> 2. Replace a and b with the endpoints of the subinterval from step 1. Repeat step 1 k times, and stop. The midpoint of the resulting interval will be within 10^{-k} units of the zero.

EXAMPLE 3 Use the Variation of the Bisection Method to Approximate a Zero

Approximate the real zero of $P(x) = x^3 - x - 1$ to within 10^{-2}.

Solution Since there is one variation in sign, Descartes' Rule of Signs implies that there is one positive zero. Because $P(1) = -1$ and

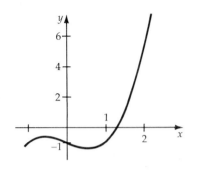

Figure 4.30
$P(x) = x^3 - x - 1$

$P(2) = 5$, we know by the Zero Location Theorem that P has a zero in the interval $(1, 2)$. See Figure 4.30.

Step 1: Subdivide the interval $(1, 2)$ into ten equal subintervals, as shown in Figure 4.31. Evaluate $P(x)$ for $x = 1.1, 1.2, \ldots$ until there is a change of sign of P. Because $P(1.3) < 0$ and $P(1.4) > 0$, there is zero in the interval $(1.3, 1.4)$.

Step 2: Let $a = 1.3$ and $b = 1.4$; repeat the iteration process.

Step 1': Subdivide the interval $(1.3, 1.4)$ into ten equal subintervals. Evaluate $P(x)$ for $x = 1.31, 1.32, \ldots$ until there is a change of sign of P. Because $P(1.32) < 0$ and $P(1.33) > 0$, there is a zero in the interval $(1.32, 1.33)$. See Figure 4.32.

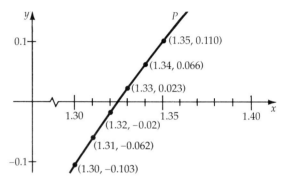

Figure 4.32

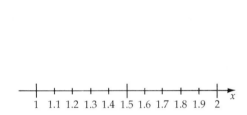

Figure 4.31

Step 2': Stop. After two iterations, we know that P has a zero in the interval $(1.32, 1.33)$. Every real number in the interval $(1.32, 1.33)$ is within 10^{-2} of the zero. The midpoint of this interval is 1.325, which is the desired approximation.

■ *Try Exercise **14**, page 254.*

EXERCISE SET 4.6

In Exercises 1 to 6, use the Zero Location Theorem to verify the given statement.

1. $P(x) = x^3 - 2x - 5$ has a zero between 2 and 3.

2. $P(x) = x^3 + 18x - 30$ has a zero between 1 and 2.

3. $P(x) = x^3 - 36x - 96$ has a zero between 7 and 8.

4. $P(x) = x^3 - 27x - 90$ has a zero between 6 and 7.

5. $P(x) = 3x^4 - 6x^2 + 8x - 3$ has a zero between -2 and -1 and a zero between 0 and 1.

6. $P(x) = x^4 - 8x^3 + 25x^2 - 36x + 8$ has a zero between 0 and 1 and a zero between 3 and 4.

In Exercises 7 to 12, the given polynomials have one zero that satisfies the given condition. Locate the zero between two consecutive integers and then use three iterations of the bisection method to approximate the zero.

7. $P(x) = x^3 - 2x - 5$, $x > 0$

8. $P(x) = x^3 + 18x - 10$, $x > 0$

9. $P(x) = 2x^3 + 3x^2 - 8$, $x > 0$

10. $P(x) = x^3 - 4x^2 + 10$, $x < 0$

11. $P(x) = x^4 + x - 5$, $x < 0$

12. $P(x) = x^4 + x - 20$, $x > 0$

In Exercises 13 to 18, use the variation of the bisection method to approximate the indicated zero of the given polynomial to within one-hundredth of a unit.

13. $P(x) = x^3 + 10x^2 - 8$, $0 < x < 1$

14. $P(x) = x^3 - 2x - 20$, $2 < x < 3$

15. $P(x) = x^4 - x^2 - 1$, $1 < x < 2$

16. $P(x) = x^4 + x - 10$, $1 < x < 2$

17. $P(x) = x^5 - 4x^3 - 5$, $x > 0$

18. $P(x) = 3x^4 + x^2 - 7$, $x < 0$

Supplemental Exercises

19. Use the bisection method to approximate $\sqrt{2}$ to the nearest tenth. (*Hint:* $\sqrt{2}$ is the positive zero of the polynomial $x^2 - 2$.)

20. Use the bisection method to approximate $\sqrt[3]{5}$ to the nearest tenth. (*Hint:* $\sqrt[3]{5}$ is the real zero of the polynomial $x^3 - 5$.)

In Exercises 21 and 22, use the bisection method to approximate to the nearest tenth the x-coordinate of the point(s) of intersection of the graphs of the given polynomials. (*Hint:* The graphs intersect where $P(x) = Q(x)$; therefore find the roots of $P(x) - Q(x) = 0$ or the zeros of $P(x) - Q(x)$.)

21. $P(x) = x^3 - 5$ and $Q(x) = -2x^3 + x^2 - 1$

22. $P(x) = x^4 + x^2 + 2$ and $Q(x) = x^4 + x^3 - 5x^2 + 1$

Chapter 4 Review

4.1 Polynomial Division and Synthetic Division

The Remainder Theorem   If a polynomial $P(x)$ is divided by $x - c$, then the remainder is $P(c)$.

The Factor Theorem   A polynomial $P(x)$ has a factor $(x - c)$ if and only if $P(c) = 0$.

4.2 Graphs of Polynomial Functions

The following characterstcs and properties are used in graphing polynomial functions:

1. Continuity — Polynomial functions are smooth continuous curves.
2. Leading term test — Determines the behavior of the graph of a polynomial function at the far right or the far left.
3. Zeros of the function determine the x-intercepts.

4.3 Zeros of Polynomial Functions

Values of x that satisfy $P(x) = 0$ are called the zeros of P.

Definition of Multiple Zeros of a Polynomial
If a polynomial function has $(x - r)$ as a factor exactly k times, then r is said to be a zero of multiplicity k of the polynomial P.

Number of Zeros of a Polynomial Function
A polynomial function of degree n has at most n zeros, where each zero of multiplicity k is counted k times.

The Rational Zero Theorem If $P(x) = a_n x^n + a_{n-1} x^{n-1} + \cdots + a_1 x + a_0$ has integer coefficients and p/q (where p and q have no common prime factors) is a rational zero of $P(x)$, then p is a factor of a_0 and q is a factor of a_n.

The upper and lower bound theorem determines the upper and lower bounds of the zeros of a polynomial.

Descartes' Rule of Signs

1. The number of positive real zeros is equal to the number of variations in sign of $P(x)$ or is equal to that number decreased by an even integer.

2. The number of negative real zeros is equal to the number of variations in sign of $P(-x)$ or is equal to that number decreased by an even integer.

4.4 The Fundamental Theorem of Algebra

The Fundamental Theorem of Algebra If $P(x)$ is a polynomial of degree $n \geq 1$ with complex coefficients, then $P(x)$ has at least one complex zero.

The Number of Zeros of a Polynomial If $P(x)$ is a polynomial of degree $n \geq 1$ with complex coefficients, then $P(x)$ has exactly n complex zeros provided each zero is counted according to its multiplicity.

The Conjugate Pair Theorem If $a + bi$ ($b \neq 0$) is a complex zero of the polynomial $P(x)$, with real coefficients, then the conjugate $a - bi$ is also a complex zero of the polynomial.

Linear and Quadratic Factors of a Polynomial Every polynomial with real coefficients and positive degree n can be written as the product of linear and quadratic factors with real coefficients, where the quadratic factors have no real zeros.

4.5 Rational Functions and Their Graphs

If $P(x)$ and $Q(x)$ are polynomials, then the function F given by

$$F(x) = \frac{P(x)}{Q(x)}$$

is called a rational function.

General Procedure for Graphing Rational Functions That Have No Common Factors

1. Find the zeros in the denominator. The vertical asymptotes will occur at these points. Graph any horizontal asymptote.

2. Find the zeros in the numerator. These points will give the x-intercepts.

3. Find additional points that lie in the intervals between the x-intercepts and the vertical asymptotes.

4. Determine the behavior near the asymptotes.

5. Complete the sketch by using the information above.

4.6 Approximation of Zeros

Iteration methods are used to approximate irrational zeros.

CHALLENGE EXERCISES

In Exercises 1 to 14, answer true or false. If the statement is false, given an example.

1. The complex zeros of a polynomial with complex coefficients always occur in conjugate pairs.

2. Descartes' Rule of Signs indicates that $x^3 - x^2 + x - 1$ must have three positive zeros.

3. The polynomial $2x^5 + x^3 - 7x^4 - 5x^2 + 4x + 10$ has two variations in sign.

4. If 4 is an upper bound of the zeros of the polynomial P, then 5 is also an upper bound of the zeros of P.

5. The graph of every rational function has a vertical asymptote.

6. The graph of the rational function

$$F(x) = \frac{x^2 - 4x + 4}{x^2 - 5x + 6}$$

has a vertical asymptote of $x = 2$.

7. If 7 is a zero of the polynomial P, then $x - 7$ is a factor of P.

8. According to the Zero Location Theorem, the polynomial function $P(x) = x^3 + 6x - 2$ has a real zero between 0 and 1.

9. Synthetic division can be used to show that $3i$ is a zero of $x^3 - 2x^2 + 9x - 18$.

10. Every fourth-degree polynomial with complex coefficients has exactly four complex zeros provided each zero is counted according to its multiplicity.

11. The graph of a rational function never intersects any of its vertical asymptotes.

12. The graph of a rational function can have at most one horizontal asymptote.

13. Descartes' Rule of Signs indicates that the polynomial function $P(x) = x^3 + 2x^2 + 4x - 7$ does have a positive zero.

14. Every polynomial has at least one real zero.

REVIEW EXERCISES

In Exercises 1 to 6, use long division to divide the first polynomial by the second.

1. $x^3 + 5x^2 + 2x - 17, x^2 + x + 3$

2. $2x^3 - 5x + 1, x^2 + 4$

3. $-x^4 + 2x^2 - 12x - 3, x^3 + x$

4. $x^3 - 5x^2 - 6x - 11, x^2 - 6x - 1$

5. $6x^4 + 8x^3 - 47x^2 + 19x + 5, 2x^2 + 6x - 5$

6. $x^4 + 3x^3 - 6x^2 - 13x + 15, x^2 + 2x - 3$

In Exercises 7 to 12, use synthetic division to divide the first polynomial by the second.

7. $4x^3 - 11x^2 + 5x - 2, x - 3$

8. $5x^3 - 18x + 2, x - 1$

9. $3x^3 - 5x + 1, x + 2$

10. $2x^3 + 7x^2 + 16x - 10, x - \frac{1}{2}$

11. $3x^3 - 10x^2 - 36x + 55, x - 5$

12. $x^4 + 9x^3 + 6x^2 - 65x - 63, x + 7$

In Exercises 13 to 16, use the Remainder Theorem to find $P(c)$.

13. $P(x) = x^3 + 2x^2 - 5x + 1, c = 4$

14. $P(x) = -4x^3 - 10x + 8, c = -1$

15. $P(x) = 6x^4 - 12x^2 + 8x + 1, c = -2$

16. $P(x) = 5x^5 - 8x^4 + 2x^3 - 6x^2 - 9, c = 3$

In Exercises 17 to 20, use synthetic division to show that c is a zero of the given polynomial.

17. $x^3 + 2x^2 - 26x + 33, c = 3$

18. $2x^4 + 8x^3 - 8x^2 - 31x + 4$, $c = -4$

19. $x^5 - x^4 - 2x^2 + x + 1$, $c = 1$

20. $2x^3 + 3x^2 - 8x + 3$, $c = \dfrac{1}{2}$

In Exercises 21 to 26, graph the polynomial function.

21. $P(x) = x^3 - x$

22. $P(x) = -x^3 - x^2 + 8x + 12$

23. $P(x) = x^4 - 6$ **24.** $P(x) = x^5 - x$

25. $P(x) = x^4 - 10x^2 + 9$ **26.** $P(x) = x^5 - 5x^3$

In Exercises 27 to 32, use the Rational Zero Theorem to list all possible rational zeros for each polynomial.

27. $x^3 - 7x - 6$ **28.** $2x^3 + 3x^2 - 29x - 30$

29. $15x^3 - 91x^2 + 4x + 12$

30. $x^4 - 12x^3 + 52x^2 - 96x + 64$

31. $x^3 + x^2 - x - 1$ **32.** $6x^5 + 3x - 2$

In Exercises 33 to 36, use Descartes' Rule of Signs to state the number of possible positive and negative real zeros of each polynomial.

33. $x^3 + 3x^2 + x + 3$

34. $x^4 - 6x^3 - 5x^2 + 74x - 120$

35. $x^4 - x - 1$

36. $x^5 - 4x^4 + 2x^3 - x^2 + x - 8$

In Exercises 37 to 42, find the zeros of the polynomial.

37. $x^3 + 6x^2 + 3x - 10$ **38.** $x^3 - 10x^2 + 31x - 30$

39. $6x^4 + 35x^3 + 72x^2 + 60x + 16$

40. $2x^4 + 7x^3 + 5x^2 + 7x + 3$

41. $x^4 - 4x^3 + 6x^2 - 4x + 1$ **42.** $2x^3 - 7x^2 + 22x + 13$

43. Find a third-degree polynomial with zeros of 4, −3, and 1/2.

44. Find a fourth-degree polynomial with zeros of 2, −3, i, and $-i$.

45. Find a fourth-degree polynomial with real coefficients that has zeros of 1, 2, and $5i$.

46. Find a fourth-degree polynomial with real coefficients that has −2 as a zero of multiplicity two and also has $1 + 3i$ as a zero.

In Exercises 47 to 50, find the vertical, horizontal, and slant asymptotes for each rational function.

47. $f(x) = \dfrac{3x + 5}{x + 2}$ **48.** $f(x) = \dfrac{2x^2 + 12x + 2}{x^2 + 2x - 3}$

49. $f(x) = \dfrac{2x^2 + 5x + 11}{x + 1}$ **50.** $f(x) = \dfrac{6x^2 - 1}{2x^2 + x + 7}$

In Exercises 51 to 58, graph each rational function.

51. $f(x) = \dfrac{3x - 2}{x}$ **52.** $f(x) = \dfrac{x + 4}{x - 2}$

53. $f(x) = \dfrac{6}{x^2 + 2}$ **54.** $f(x) = \dfrac{4x^2}{x^2 + 1}$

55. $f(x) = \dfrac{2x^3 - 4x + 6}{x^2 - 4}$ **56.** $f(x) = \dfrac{x}{x^3 - 1}$

57. $f(x) = \dfrac{3x^2 - 6}{x^2 - 9}$ **58.** $f(x) = \dfrac{-x^3 + 6}{x^2}$

5

Exponential and Logarithmic Functions

The science of Pure Mathematics, in its modern developments, may claim to be the most original creation of the human spirit.

A. N. WHITEHEAD (1861–1947)

1, 2, 4, 8, 16, ?

Inductive reasoning is the process of reasoning from particular facts to a general conclusion. Use inductive reasoning to see if you can discover a function between the number of points on a circle and the number of regions in the interior of the circle that are partitioned off by the line segments connecting the points. For example, consider the following five circles. For each circle, count the number of points on the circle and the number of regions that the line segments between the points partition each circle.

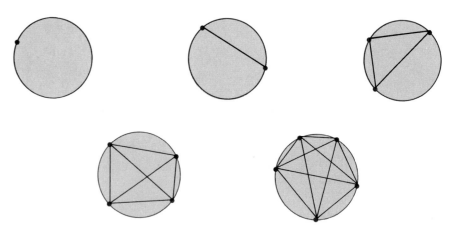

The line segments between points on each circle partition the circle into a number of regions. Can you guess how many regions a circle with six points has?

Your results should agree with the following results:

Number of points on the circle	1	2	3	4	5	6
Maximum number of partitioned regions	1	2	4	8	16	?

Now use inductive reasoning to guess the maximum number of partitioned regions you expect a circle with six points to have. To verify your guess, place six points on a large circle. Connect the six points in all possible ways. Count the regions formed. How does this result compare with your guess?

Completion of this experiment should convince you that thirty-one is the maximum number of partitioned regions that can be formed by the line segments that connect six points on a circle. The purpose of this experiment is to show that inductive reasoning may not lead to valid conclusions.

You will be encouraged to use inductive reasoning to develop mathematical ideas. However, you need to be aware that conclusions based on inductive reasoning may prove to be incorrect.

5.1

Exponential Functions and Their Graphs

In Chapter 1 we defined the real number b^x for every positive base b and every rational number x. For example,

$$2^3 = 8, \qquad 2^{-4} = \frac{1}{2^4} = \frac{1}{16}, \qquad \text{and} \qquad 2^{2/3} = \sqrt[3]{2^2} = \sqrt[3]{4}$$

The exponential key $\boxed{y^x}$ (or on some calculators $\boxed{x^y}$) on a scientific calculator can be used to evaluate a positive number that is raised to a rational power. The following examples illustrate the sequence of key strokes for a calculator that uses algebraic logic.

Power	Key Sequence	Calculator Display
3^2	3 $\boxed{y^x}$ 2 $\boxed{=}$	$9.$
$2^{1.4}$	2 $\boxed{y^x}$ 1.4 $\boxed{=}$	2.6390158
$1.4^{-3.21}$	1.4 $\boxed{y^x}$ 3.21 $\boxed{+/-}$ $\boxed{=}$	0.3395697
$5^{7/3}$	5 $\boxed{y^x}$ $\boxed{(}$ 7 $\boxed{\div}$ 3 $\boxed{)}$ $\boxed{=}$	42.749398

To define powers of the form b^x, where b is a positive real number and x is a real number, we will require a definition that includes powers with irrational exponents, such as $2^{\sqrt{2}}$, 3^π, and $10^{-\sqrt{5}}$.

For our purposes it is convenient to define $2^{\sqrt{2}}$ as the unique real number that can be approximated as closely as desired using an exponent that takes on closer and closer rational approximations of $\sqrt{2}$.

Using a scientific calculator, the key stroke sequence

$$2 \ \boxed{y^x} \ 2 \ \boxed{\sqrt{x}} \ \boxed{=}$$

yields $2^{\sqrt{2}} = 2.6651441$ (to the nearest ten millionth).

Definition of Exponential Functions

> The **exponential function f with base b** is defined by
> $$f(x) = b^x$$
> where b is a positive constant other than 1 and x is any real number.

The following graphs are representative of the graphs of exponential functions. Each one was graphed by plotting points and then connecting the points with a smooth curve.

EXAMPLE 1   **Graph an Exponential Function**

Sketch the graph of the following exponential functions:

a. $f(x) = 2^x$ and   b. $f(x) = \left(\dfrac{1}{2}\right)^x$

Solution

a.

x	$f(x) = 2^x$
-2	$2^{-2} = \dfrac{1}{4}$
-1	$2^{-1} = \dfrac{1}{2}$
0	$2^0 = 1$
1	$2^1 = 2$
2	$2^2 = 4$
3	$2^3 = 8$

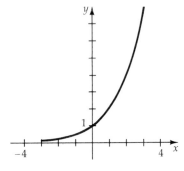

Figure 5.1
$f(x) = 2^x$

b.

x	$f(x) = \left(\dfrac{1}{2}\right)^x = 2^{-x}$
-3	$2^3 = 8$
-2	$2^2 = 4$
-1	$2^1 = 2$
0	$2^0 = 1$
1	$2^{-1} = \dfrac{1}{2}$
2	$2^{-2} = \dfrac{1}{4}$

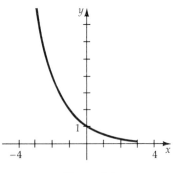

Figure 5.2
$f(x) = \left(\dfrac{1}{2}\right)^x$

■ *Try Exercise* **26,** *page 266.*

To graph an exponential function over a certain portion of its domain, you may need to use different scales on the *x*- and *y*-axes. For example, the graph of $f(x) = 10^x$, for $-2 \le x \le 1$, is shown in Figure 5.3. Observe that each unit on the *y*-axis represents a distance of five units and that each unit on the *x*-axis represents one unit.

For positive real numbers b, $b \ne 1$, the exponential function $f(x) = b^x$ has the following properties:

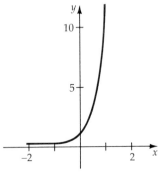

Figure 5.3
$f(x) = 10^x$

- ■ *f* has the set of real numbers as its domain.
- ■ *f* has the set of positive real numbers as its range.
- ■ *f* has a graph with a *y*-intercept of $(0, 1)$.
- ■ *f* has a graph asymptotic to the *x*-axis.

- *f* is a one-to-one function.

  The exponential function $f(x) = b^x$ is

- An increasing function if $b > 1$. See Figure 5.4(a).
- A decreasing function if $0 < b < 1$. See Figure 5.4(b).

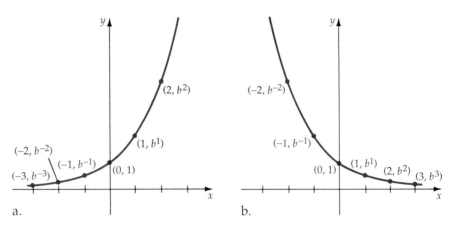

Figure 5.4

Many applications involve functions such as

$$f(x) = ab^p,$$

where *a* is a constant and *p* is some expression involving *x*. Functions of this type can be graphed by plotting points. Be sure to make use of symmetry when possible.

EXAMPLE 2   **Graph a Function of the Form $f(x) = ab^p$**

Sketch the graph of $f(x) = 3 \cdot 2^{-x^2}$.

Solution   Since $f(x) = f(-x)$, this function is an even function. We can sketch its graph by plotting points to the right of the *y*-axis and then using symmetry to determine points to the left of the *y*-axis.

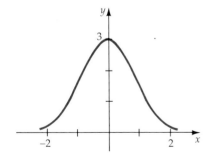

Figure 5.5
$f(x) = 3 \cdot 2^{-x^2}$

x	$f(x) = 3 \cdot 2^{-x^2}$
0	$3 \cdot 2^{-0^2} = 3$
1	$3 \cdot 2^{-1^2} = \dfrac{3}{2}$
2	$3 \cdot 2^{-2^2} = \dfrac{3}{16}$

■ *Try Exercise* **36,** *page 266.*

Remark   A calculator could be used to compute additional points on the graph of f. However, writing f in the fractional form

$$f(x) = 3 \cdot \frac{1}{2^{x^2}}$$

allows us to determine that as $|x|$ increases without bound, the denominator increases without bound while the numerator remains constant. Therefore $f(x)$ approaches 0 as $|x|$ increases without bound. This implies that the graph of f has the x-axis as a horizontal asymptote. Also, the maximum value of f is reached when the denominator 2^{x^2} is its smallest. This occurs when $x = 0$ and $y = 3$. The graph of f is a *bell-shaped* curve similar to the graph of the normal distribution curve, which plays a major role in statistics.

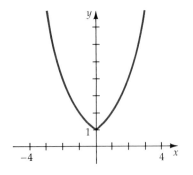

Figure 5.6
$f(x) = 2^{|x|}$

EXAMPLE 3   **Graph a Function of the Form $f(x) = ab^p$**

Sketch the graph of $f(x) = 2^{|x|}$.

Solution

x	$f(x) = 2^{|x|}$
0	$2^{|0|} = 1$
1	$2^{|1|} = 2$
2	$2^{|2|} = 4$
3	$2^{|3|} = 8$

Because $f(x) = 2^{|x|}$ is an even function, the points to the left of the y-axis can be determined using symmetry.

■ *Try Exercise* **38,** *page 266.*

Remark   Note that Figure 5.6 is not the graph of a parabola.

The irrational number π is often used in applications that involve circles. Using techniques developed in a calculus course, we can verify that as n increases without bound,

$$\left(1 + \frac{1}{n}\right)^n$$

approaches an irrational number that is denoted by e. The number e often occurs in applications involving growth or decay. It is denoted by e in honor of the mathematician Leonhard Euler (1707–1783). Although e has a nonterminating and nonrepeating decimal representation, Euler was able to compute e to several decimal places by using large values of n to evaluate $(1 + 1/n)^n$. The entries in Table 5.1 illustrate the process. The value of e accurate to eight decimal places is 2.71828183.

TABLE 5.1

Value of n	Value of $\left(1 + \frac{1}{n}\right)^n$
1	2
10	2.59374246
100	2.704813829
1000	2.716923932
10,000	2.718145927
100,000	2.718268237
1,000,000	2.718280469
10,000,000	2.718281693

The Natural Exponential Function

For all real numbers x, the function defined by

$$f(x) = e^x$$

is called the **natural exponential function.**

To evaluate e^x for specific values of x, you use a calculator with an $\boxed{e^x}$ key, or you can use the $\boxed{y^x}$ key with y as 2.7182818. For example,

Power	Key Sequence	Calculator Display
e^2	2 $\boxed{e^x}$	7.389056099
$e^{4.21}$	4.21 $\boxed{e^x}$	67.35653981
$e^{-1.8}$	1.8 $\boxed{+/-}$ $\boxed{e^x}$	0.165298888

To graph the natural exponential function, use a calculator to approximate e^x for the desired domain values. The resulting points can then be plotted and connected with a smooth curve.

EXAMPLE 4   Graph the Natural Exponential Function

Graph $f(x) = e^x$.

Solution   Use a calculator to find the values of e^x. The values in the table have been rounded to the nearest hundredth. Plot the points and then connect the points with a smooth curve.

x	$f(x) = e^x$
-3	0.05
-2	0.14
-1	0.37
0	1.00
1	2.72
2	7.39

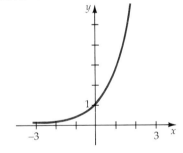

Figure 5.7
$f(x) = e^x$

■ *Try Exercise **44**, page 266.*

Notice in Figure 5.8 how the graph of $f(x) = e^x$ compares with the graphs of $g(x) = 2^x$ and $h(x) = 3^x$. You may have anticipated that the graph of f would be between the graph of g and h because e is between 2 and 3.

Functions of the type

$$f(x) = \frac{b^x + b^{-x}}{2}$$

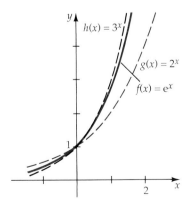

Figure 5.8

occur in calculus. These functions are difficult to graph by the method of point-plotting. Example 5 uses the technique of *averaging the y values* of known graphs to sketch the desired graph.

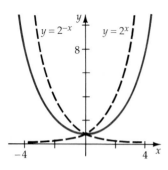

Figure 5.9
$$f(x) = \frac{2^x + 2^{-x}}{2}$$

EXAMPLE 5   **Graph by Averaging y Values**

Graph the function $f(x) = \dfrac{2^x + 2^{-x}}{2}$.

Solution   Notice that f is the average of 2^x and 2^{-x}. Therefore sketch the graph of f by drawing a curve *halfway between* the graph of $y = 2^x$ and $y = 2^{-x}$. See Figure 5.9.

■ *Try Exercise 46, page 266.*

From the sketch it appears that the graph of f is symmetric with respect to the y-axis, which would imply that f is an even function. This is indeed the case because

$$f(-x) = \frac{2^{-x} + 2^{-(-x)}}{2} = \frac{2^{-x} + 2^x}{2} = f(x).$$

EXERCISE SET 5.1

In Exercises 1 to 12, use a calculator to determine each power accurate to six significant digits.

1. $3^{\sqrt{2}}$ **2.** $5^{\sqrt{3}}$ **3.** $10^{\sqrt{7}}$ **4.** $10^{\sqrt{11}}$

5. $\sqrt{3}^{\sqrt{2}}$ **6.** $\sqrt{5}^{\sqrt{7}}$ **7.** $e^{5.1}$ **8.** $e^{-3.2}$

9. $e^{\sqrt{3}}$ **10.** $e^{\sqrt{5}}$ **11.** $e^{-0.031}$ **12.** $e^{-0.42}$

In Exercises 13 to 24, use a calculator to evaluate each functional value, accurate to six significant digits, given that $f(x) = 3^x$ and $g(x) = e^x$.

13. $f(\sqrt{15})$ **14.** $f(\pi)$ **15.** $f(e)$ **16.** $f(-\sqrt{15})$

17. $g(\sqrt{7})$ **18.** $g(\pi)$ **19.** $g(e)$ **20.** $g(-3.4)$

21. $f[g(2)]$ **22.** $f[g(-1)]$ **23.** $g[f(2)]$ **24.** $g[f(-1)]$

In Exercises 25 to 34, sketch the graph of each exponential function.

25. $f(x) = 3^x$ **26.** $f(x) = 4^x$

27. $f(x) = \left(\dfrac{3}{2}\right)^x$ **28.** $f(x) = \left(\dfrac{4}{3}\right)^x$

29. $f(x) = \left(\dfrac{1}{3}\right)^x$ **30.** $f(x) = \left(\dfrac{2}{3}\right)^x$

31. $f(x) = \left(\dfrac{1}{2}\right)^{-x}$ **32.** $f(x) = \left(\dfrac{1}{3}\right)^{-x}$

33. $f(x) = \dfrac{5^x}{2}$ **34.** $f(x) = \dfrac{10^x}{10}$

In Exercises 35 to 48, sketch the graph of each function.

35. $f(x) = \left(\dfrac{1}{3}\right)^{|x|}$ **36.** $f(x) = 3^{-(x^2)}$

37. $f(x) = 3^{x^2}$ **38.** $f(x) = 2^{-|x|}$

39. $f(x) = 2^{x-3}$ **40.** $f(x) = 2^{x+3}$

41. $f(x) = 3^x - 1$ **42.** $f(x) = 3^x + 1$

43. $f(x) = -e^x$ **44.** $f(x) = \dfrac{1}{2}e^x$

45. $f(x) = \dfrac{3^x + 3^{-x}}{2}$ **46.** $f(x) = \dfrac{e^x + e^{-x}}{2}$

47. $f(x) = -(3^x)$ **48.** $f(x) = -(0.5^x)$

In Exercises 49 to 52, graph each pair of functions on the same set of coordinate axes.

49. $f(x) = 2^x$, $g(x) = 2^{-x}$

50. $f(x) = \left(\dfrac{2}{3}\right)^x$, $g(x) = \left(\dfrac{2}{3}\right)^{-x}$

51. $f(x) = 2^{x+1}$, $g(x) = 2^x + 1$

52. $f(x) = 3^x$, $g(x) = \left(\dfrac{1}{3}\right)^{-x}$

Calculator Exercises

53. Use a calculator to complete the following table:

Value of n	Value of $\left(1 + \dfrac{1}{n}\right)^n$
5	
50	
500	
5000	
50,000	
500,000	
5,000,000	

54. Use a calculator to complete the following table.

  a. How do the entries in this table compare with those in Exercise 53?

  b. As n approaches 0, what value does $(1 + n)^{1/n}$ appear to approach?

Value of n	Value of $\left(1 + n\right)^{1/n}$
0.2	
0.02	
0.002	
0.0002	
0.00002	
0.000002	
0.0000002	

Supplemental Exercises

55. Evaluate $h(x) = (-2)^x$, for **a.** $x = 1$, **b.** $x = 2$, and **c.** $x = 1.5$. Explain why h is not an exponential function.

56. Graph $j(x) = 1^x$. Explain why j is not an exponential function.

57. Graph $f(x) = e^x$, and then sketch the graph of f reflected about the graph of the line given by $y = x$.

58. Graph $g(x) = 10^x$, and then sketch the graph of g reflected about the graph of the line given by $y = x$.

59. Prove that

$$F(x) = \frac{e^x - e^{-x}}{2}$$

is an odd function.

60. Prove that $G(x) = e^x$ is neither an odd function nor an even function.

61. The number of bacteria present in a culture is given by $N(t) = 10{,}000(2^t)$, where $N(t)$ is the number of bacteria present after t hours. Find the number of bacteria present when   **a.** $t = 1$ hour,   **b.** $t = 2$ hours,   **c.** $t = 5$ hours.

62. The production function for an oil well is given by the function $B(t) = 100{,}000(e^{-0.2t})$, where $B(t)$ is the number of barrels of oil the well can produce per month after t years. Find the number of barrels of oil the well can produce per month when   **a.** $t = 1$ year,   **b.** $t = 2$ years,   **c.** $t = 5$ years.

63. Which of the following powers, e^π or π^e, is larger?

64. Let $f(x) = x^{(x^x)}$ and $g(x) = (x^x)^x$. Which is larger, $f(3)$ or $g(3)$?

65. Graph $f(x) = \dfrac{2^x - 2^{-x}}{2}$.

66. Graph $f(x) = \dfrac{e^x - e^{-x}}{2}$.

5.2

Logarithms and Logarithmic Properties

Every exponential function is a one-to-one function, and therefore, has an inverse function. Recall from Section 3.8 that we determined the inverse of a function represented by an equation by interchanging the variables and then solving for the dependent variable. If we attempt to use this procedure for the exponential function $f(x) = b^x$, we get

$$f(x) = b^x$$

$$y = b^x$$

$$x = b^y \quad \text{Interchange the variables.}$$

How do we solve $x = b^y$ for the exponent y? None of our previous methods can be used to solve the equation $x = b^y$ for the exponent y.

Thus we must develop a new procedure. One method would be merely to write

$$y = \text{exponent of } b \text{ that produces } x.$$

This procedure would work, but it is not concise. We need compact notation to represent y as the exponent of b that produces x. For historical reasons, we use the notation in the following definition.

Definition of a Logarithm

> If $x > 0$ and b is a positive constant ($b \neq 1$), then
>
> $$y = \log_b x \quad \text{if and only if} \quad b^y = x.$$
>
> In the equation $y = \log_b x$, y is referred to as the **logarithm**, b is the **base**, and x is the **argument**.

The notation $\log_b x$ is read as "the logarithm (or log) base b of x." The definition of a logarithm indicates that *a logarithm is an exponent*.

The equations

$$y = \log_b x \qquad \text{and} \qquad b^y = x$$

are different ways of expressing the same thing.

$$y = \log_b x \text{ is the logarithmic form of } b^y = x$$

$$b^y = x \text{ is the exponential form of } y = \log_b x.$$

EXAMPLE 1   **Change from Logarithmic to Exponential Form**

Write each of the following equations in their exponential form:

a. $2 = \log_7 x$ b. $3 = \log_{10}(x + 8)$ c. $\log_5 125 = x$

Solution   Use the definition $y = \log_b x$ if and only if $b^y = x$.

a.

$$\overbrace{2 = \log_7 x \qquad \text{implies} \qquad 7^2 = x}^{\text{Logarithms are exponents}}$$

$$\underbrace{\phantom{2 = \log_7 x \qquad \text{implies} \qquad 7^2 = x}}_{\text{Base}}$$

b. $3 = \log_{10}(x + 8)$ implies $10^3 = (x + 8)$.

c. $\log_5 125 = x$ implies $5^x = 125$.

■ *Try Exercise **2**, page 275.*

EXAMPLE 2   **Change from Exponential to Logarithmic Form**

Write each of the following equations in their logarithmic form:

a. $x = 25^{1/2}$ b. $\dfrac{1}{16} = x^{-4}$ c. $27^x = 3$

Solution   Use $x = b^y$ if and only if $y = \log_b x$.

a.

$$\overset{\displaystyle \ulcorner\; \text{Exponent} \;\urcorner}{x = 25^{1/2} \quad \text{implies} \quad \tfrac{1}{2} = \log_{25} x}$$

$$\underset{\text{Base}}{\rule{0pt}{0pt}}$$

b. $\dfrac{1}{16} = x^{-4}$ implies $-4 = \log_x \dfrac{1}{16}$.

c. $27^x = 3$ implies $\log_{27} 3 = x$.

■ *Try Exercise* **12**, *page 275.*

Some logarithms can be evaluated by using the definition of a logarithm and the following theorem.

Equality of Exponents Theorem

> If b is a positive real number ($b \neq 1$) such that $b^x = b^y$, then $x = y$.

EXAMPLE 3   **Evaluate Logarithms**

Evaluate each logarithm.

a. $\log_2 32 = x$ b. $\log_5 125 = x$ c. $\log_{10} \dfrac{1}{100} = x$

Solution

a. $\log_2 32 = x$ implies $2^x = 32$ Change to exponential form.

$\qquad\qquad\qquad\qquad\quad\; 2^x = 2^5$ Factor.

$\qquad\qquad\qquad\qquad\qquad x = 5$ Equality of Exponents Theorem

b. $\log_5 125 = x$ implies $5^x = 125$

$\qquad\qquad\qquad\qquad\quad\; 5^x = 5^3$

$\qquad\qquad\qquad\qquad\qquad x = 3$

c. $\log_{10} \dfrac{1}{100} = x$ implies $10^x = \dfrac{1}{100}$

$\qquad\qquad\qquad\qquad\qquad 10^x = 10^{-2}$

$\qquad\qquad\qquad\qquad\qquad\; x = -2$

■ *Try Exercise* **22**, *page 275.*

Since logarithms are exponents, they have many properties that can be established by using the properties of exponents.

Properties of Logarithms

In the following properties, b, M, and N are positive real numbers ($b \neq 1$), and p is any real number.

$\log_b b = 1$

$\log_b 1 = 0$

$\log_b(b^p) = p$

$\log_b(MN) = \log_b M + \log_b N$ Product property

$\log_b\left(\dfrac{M}{N}\right) = \log_b M - \log_b N$ Quotient property

$\log_b(M^p) = p \log_b M$ Power property

$\log_b M = \log_b N$ implies $M = N$ One-to-one property

$M = N$ implies $\log_b M = \log_b N$ Logarithm of each side property

$b^{\log_b p} = p$ (for $p > 0$)   Inverse property

The first three properties of logarithms can be proven by using the definition of a logarithm. That is,

$$\log_b b = 1 \quad \text{because} \quad b^1 = b.$$

$$\log_b 1 = 0 \quad \text{because} \quad b^0 = 1.$$

$$\log_b(b^p) = p \quad \text{because} \quad b^p = b^p.$$

To prove the product property $\log_b(MN) = \log_b M + \log_b N$, let

$$\log_b M = x \quad \text{and} \quad \log_b N = y.$$

Writing each equation in its equivalent exponential form produces

$$M = b^x \quad \text{and} \quad N = b^y.$$

Forming the product of the respective right and left sides produces

$$MN = b^x b^y \quad \text{or} \quad MN = b^{x+y}.$$

Applying the definition of a logarithm yields the equivalent form

$$\log_b(MN) = x + y.$$

Since $x = \log_b M$ and $y = \log_b N$, this becomes

$$\log_b(MN) = \log_b M + \log_b N.$$

The proofs of the quotient property and the power property are similar to the proof of the product property. They are left as exercises.

The logarithm of each side property will be used several times in this chapter. It states that if two positive real numbers are equal, their logarithms base b are also equal.

Proof: Let $M > 0$, $N > 0$ and $M = N$.

$$\log_b M = \log_b M \quad \text{The reflexive property}$$

$$\log_b M = \log_b N \quad \text{Substitute } N \text{ for } M.$$

The proofs of the one-to-one property and the inverse property are left as exercises.

The properties of logarithms are often used to rewrite logarithms and expressions that involve logarithms.

EXAMPLE 4   Use Logarithmic Properties

Use the properties of logarithms to express the following logarithms in terms of logarithms of x, y, and z:

a. $\log_b xy^2$ b. $\log_b \dfrac{x^2\sqrt{y}}{z^5}$

Solution

a. $\log_b xy^2 = \log_b x + \log_b y^2 \quad \text{Product property}$

$\qquad\qquad = \log_b x + 2\log_b y \quad \text{Power property}$

b. $\log_b \dfrac{x^2\sqrt{y}}{z^5} = \log_b x^2\sqrt{y} - \log_b z^5 \qquad\qquad \text{Quotient property}$

$\qquad\qquad\quad = \log_b x^2 + \log_b \sqrt{y} - \log_b z^5 \quad \text{Product property}$

$\qquad\qquad\quad = 2\log_b x + \dfrac{1}{2}\log_b y - 5\log_b z \quad \text{Power property}$

■ *Try Exercise* **32,** *page 275.*

Sometimes it is possible to use known logarithmic values and the properties of logarithms to evaluate logarithms.

EXAMPLE 5   Use Logarithmic Properties

Given $\log_8 2 \approx 0.3333$, $\log_8 3 \approx 0.5283$, and $\log_8 5 \approx 0.7740$, evaluate the following:

a. $\log_8 15$ b. $\log_8 \dfrac{5}{2}$ c. $\log_8 \sqrt[3]{9}$

Solution

a. $\log_8 15 = \log_8(3 \cdot 5) = \log_8 3 + \log_8 5 \quad \text{Product property}$

$\qquad\qquad \approx 0.5283 + 0.7740 = 1.3023$

b. $\log_8 \dfrac{5}{2} = \log_8 5 - \log_8 2 \qquad\qquad\qquad \text{Quotient property}$

$\qquad\qquad \approx 0.7740 - 0.3333 = 0.4407$

c. $\log_8 \sqrt[3]{9} = \log_8 3^{2/3} = \dfrac{2}{3} \log_8 3$ Power property

$\approx \dfrac{2}{3}(0.5283) = 0.3522$

■ *Try Exercise* **42,** *page 275.*

The properties of logarithms are also used to rewrite expressions that involve logarithms as a single logarithm.

EXAMPLE 6   **Use Logarithmic Properties**

Use the properties of logarithms to rewrite the following expressions as a single logarithm:

a. $2 \log_b x + \dfrac{1}{2} \log_b(x + 4)$ b. $4 \log_b(x + 2) - 3 \log_b(x - 5)$

Solution

a. $2 \log_b x + \dfrac{1}{2} \log_b(x + 4) = \log_b x^2 + \log_b(x + 4)^{1/2}$ Power property

$= \log_b x^2(x + 4)^{1/2}$ Product property

b. $4 \log_b(x + 2) - 3 \log_b(x - 5) = \log_b(x + 2)^4 - \log_b(x - 5)^3$ Power property

$= \log_b \dfrac{(x + 2)^4}{(x - 5)^3}$ Quotient property

■ *Try Exercise* **52,** *page 275.*

Definition of Common Logarithm and Natural Logarithm

> Logarithms with a base of 10 are called **common logarithms.** It is customary to write $\log_{10} x$ as $\log x$.
>
> Logarithms with a base of e are called **natural logarithms.** They are often used in calculus. It is customary to write $\log_e x$ as $\ln x$.

Most scientific calculators have a key marked $\boxed{\log}$ for evaluating common logarithms and a key marked $\boxed{\ln}$ for evaluating natural logarithms. For example,

Logarithm	Key Sequence	Calculator Display
log 24	24 $\boxed{\log}$	1.380211242
ln 81	81 $\boxed{\ln}$	4.394449155
log 0.58	.58 $\boxed{\log}$	-0.236572006

If you use a calculator to try to evaluate the logarithm of a negative number, it will give you an error indication. Recall that the definition of $y = \log_b x$ required x to be a positive real number.

Appendix II contains a table of common logarithms and a table of natural logarithms. Use the tables to evaluate common and natural logarithms if a scientific calculator is not available. Appendix I explains the use of the common logarithmic table. All the properties of logarithms apply to both common and natural logarithms.

Logarithms that are not common logarithms or natural logarithms can be evaluated by using the following theorem.

Change-of-Base Formula

If x, a, and b are positive real numbers with $a \neq 1$ and $b \neq 1$, then

$$\log_b x = \frac{\log_a x}{\log_a b}.$$

Proof: Let $\log_b x = y$

Then $b^y = x$ By definition of $\log_b x$

Now take the logarithm with base a of each side.

$\log_a(b^y) = \log_a x$ Logarithm of each side property

$y \log_a b = \log_a x$ Power property

$y = \dfrac{\log_a x}{\log_a b}$ Solve for y by dividing by $\log_a b$.

$\log_b x = \dfrac{\log_a x}{\log_a b}$ Substitute $\log_b x$ for y.

EXAMPLE 7   **Use the Change-of-Base Formula**

Evaluate each of the following logarithms:

a. $\log_3 18$ b. $\log_{12} 400$ c. $\log_9 4$

Solution   In each case we use the change-of-base formula with $a = 10$. That is, we will evaluate these logarithms by using the $\boxed{\log}$ key on a scientific calculator.

a.

$$\log_3 18 = \frac{\log 18}{\log 3} \approx \frac{1.25527}{0.47712} \approx 2.63093$$

b.

$$\log_{12} 400 = \frac{\log 400}{\log 12} \approx \frac{2.60206}{1.07918} \approx 2.41115$$

c.

$$\log_9 4 = \frac{\log 4}{\log 9} \approx \frac{0.60206}{0.95424} \approx 0.63093$$

■ *Try Exercise* **62,** *page 275.*

Remark   The logarithms could also have been evaluated by using the
$\boxed{\ln}$ key. For example, for part a,

$$\log_3 18 = \frac{\ln 18}{\ln 3} \approx \frac{2.89037}{1.09861} \approx 2.63093$$

The change-of-base formula can be used to evaluate common logarithms using natural logarithms and to evaluate natural logarithms using
common logarithms. For example, if we substitute e for a and 10 for b in
the change-of-base formula, we get

$$\log x = \frac{\ln x}{\ln 10} \approx \frac{\ln x}{2.3026} \approx 0.4343 \ln x.$$

Substituting e for b and 10 for a in the change-of-base formula yields

$$\ln x = \frac{\log x}{\log e} \approx \frac{\log x}{0.4343}.$$

Antilogarithms

Given $M = \log N$, it is often necessary to determine the value of N. In
this case the number N is called the **antilogarithm of** M.

Definition of Antilogarithms

> If M and N are real numbers with $N > 0$, such that
>
> $$\log_b N = M$$
>
> then N is the **antilogarithm of** M for the base b.

Using the definition of a logarithm, we see that N can be evaluated by the
formula $N = b^M$. It is convenient to evaluate antilogarithms using a scientific calculator. The following examples demonstrate the method used by
some calculators.

Logarithm	Key Sequence	Calculator Display
$\log N = 2.4031$	2.4031 $\boxed{\text{INV}}$ $\boxed{\log}$	252.9880456
$\ln N = 3.8067$	3.8067 $\boxed{\text{INV}}$ $\boxed{\ln}$	45.00168799

Remark   Some calculators have keys marked $\boxed{10^x}$ and $\boxed{e^x}$. These keys
can be used to find the antilogarithms of a number with base 10 or base e.
If $\log N = 2.4031$, you can evaluate N by entering the number 2.4031 and
pressing the $\boxed{10^x}$ key. Also, if $\ln N = 3.8067$, then N can be evaluated by
entering the number 3.8067 and pressing the $\boxed{e^x}$ key.

Antilogarithms can also be found using the tables in Appendix II. Appendix I explains the process of using the tables to find logarithms and
antilogarithms.

EXERCISE SET 5.2

In Exercises 1 to 10, change each equation to its exponential form.

1. $\log_{10} 100 = 2$

2. $\log_{10} 1000 = 3$

3. $\log_5 125 = 3$

4. $\log_5 \dfrac{1}{25} = -2$

5. $\log_3 81 = 4$

6. $\log_3 1 = 0$

7. $\log_b r = t$

8. $\log_b(s + t) = r$

9. $-3 = \log_3 \dfrac{1}{27}$

10. $-1 = \log_7 \dfrac{1}{7}$

In Exercises 11 to 20, change each equation to its logarithmic form.

11. $2^4 = 16$

12. $3^5 = 243$

13. $7^3 = 343$

14. $7^{-4} = \dfrac{1}{2401}$

15. $10{,}000 = 10^4$

16. $\dfrac{1}{1000} = 10^{-3}$

17. $b^k = j$

18. $p = m^n$

19. $b^1 = b$

20. $b^0 = 1$

In Exercises 21 to 30, evaluate each logarithm. Do not use a calculator.

21. $\log_{10} 1{,}000{,}000$

22. $\log_{10} \dfrac{1}{1000}$

23. $\log_2 32$

24. $\log_3 243$

25. $\log_{3/2} \dfrac{27}{8}$

26. $\log_{0.5} 16$

27. $\log_5 \dfrac{1}{25}$

28. $\log_{0.3} \dfrac{100}{9}$

29. $\log_b 1$

30. $\log_b b$

In Exercises 31 to 40, write the given logarithm in terms of logarithms of x, y, and z.

31. $\log_b xyz$

32. $\log_b x^2 y^3$

33. $\log_3 \dfrac{x}{z^4}$

34. $\log_5 \dfrac{x^2}{yz^3}$

35. $\log_b \dfrac{\sqrt{x}}{y^3}$

36. $\log_b \dfrac{\sqrt{x}}{\sqrt[3]{z}}$

37. $\log_b x \sqrt[3]{\dfrac{y^2}{z}}$

38. $\log_b \sqrt[3]{x^2 z \sqrt{y}}$

39. $\log_7 \dfrac{\sqrt{x + z^2}}{x^2 - y}$

40. $\log_5 \left(\dfrac{x^2 - y}{z^2}\right)$

In Exercises 41 to 50, evaluate the logarithm using the values $\log_b 2 \approx 0.3562$, $\log_b 3 \approx 0.5646$, and $\log_b 5 \approx 0.8271$, and the properties of logarithms. Do not use a calculator.

41. $\log_b 6$

42. $\log_b 20$

43. $\log_b 9$

44. $\log_b 4$

45. $\log_b \dfrac{2}{5}$

46. $\log_b \dfrac{3}{2}$

47. $\log_b 30$

48. $\log_b 45$

49. $\log_b 2b$

50. $\log_b \dfrac{b^2}{3}$

In Exercises 51 to 60, write each logarithmic expression as a single logarithm.

51. $\log_{10}(x + 5) + 2 \log_{10} x$

52. $5 \log_3 x - 4 \log_3 y + 2 \log_3 z$

53. $\dfrac{1}{2}[3 \log_b(x - y) + \log_b(x + y) - \log_b z]$

54. $\log_b(y^3 z^2) - 3 \log_b(x\sqrt{y}) + 2 \log_b\left(\dfrac{x}{z}\right)$

55. $\log_8(x^2 - y^2) - \log_8(x - y)$

56. $\log_4(x^3 - y^3) - \log_4(x - y)$

57. $4 \ln(x - 3) + 2 \ln x$

58. $3 \ln z - 2 \ln(z + 1)$

59. $\ln x - \ln y + \ln z$

60. $\dfrac{1}{2} \log x + 2 \log y$

Calculator Exercises

In Exercises 61 to 70, use the change-of-base formula and a calculator to approximate the following logarithms accurate to five significant digits.

61. $\log_7 20$

62. $\log_5 37$

63. $\log_{11} 8$

64. $\log_{50} 22$

65. $\log_6 0.045$

66. $\log_4 \sqrt{7}$

67. $\log_{0.5} 5$

68. $\log_{0.2} 17$

69. $\log_\pi e$

70. $\log_\pi \sqrt{15}$

In Exercises 71 to 82, use a calculator (or the tables in Appendix II) to approximate the antilogarithm N to three significant digits.

71. $\log N = 0.4857$

72. $\log N = 0.9557$

73. $\log N = 3.5038$

74. $\log N = 7.8476$

75. $\log N = -2.4760$

76. $\log N = -4.3536$

77. $\ln N = 2.001$

78. $\ln N = 2.262$

79. $\ln N = 0.693$

80. $\ln N = 0.531$

81. $\ln N = -1.204$

82. $\ln N = -0.511$

Supplemental Exercises

In Exercises 83 to 88, find all the real numbers that are solutions of the given inequality. Use interval notation to write your answers.

83. $0 \le \log x \le 1000$

84. $-3 \le \log x \le -2$

85. $e \le \ln x \le e^3$

86. $-2 \le \ln x \le 3$

87. $-\log x > 0$

88. $100 - 10 \log(x + 1) > 0$

89. Verify the quotient property of logarithms. *Hint:* Use a method similar to the proof of the product property.

90. Verify the power property of logarithms.

91. Give the reason for each step in the proof of the inverse property of logarithms.

$$\log_b x = \log_b x \qquad \underline{\quad ? \quad}$$
$$b^{\log_b x} = x \qquad \underline{\quad ? \quad}$$

92. Give the reason for each step in the proof of the one-to-one property of logarithms.

$$\log_b M = \log_b N \qquad \text{Given}$$
$$b^{\log_b N} = M \qquad \underline{\quad ? \quad}$$
$$N = M \qquad \underline{\quad ? \quad}$$

5.3

Logarithmic Functions and Their Graphs

Section 2 developed the concept of a logarithm and the properties of logarithms. With this background we can now introduce the concept of a *logarithmic function*.

Definition of a Logarithmic Function

> The **logarithmic function** f **with base** b is defined by
> $$f(x) = \log_b x$$
> where b is a positive constant $b \ne 1$, and x is any *positive* real number.

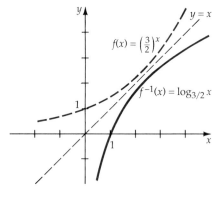

Figure 5.10

Recall that logarithms were defined so that we could write the inverse of $g(x) = b^x$ in a convenient manner. Thus the logarithmic function given by $f(x) = \log_b x$ is the inverse of the exponential function $g(x) = b^x$. The graph of $y = \log_b x$ can be obtained by reflecting the graph of $y = b^x$ across the graph of the line $y = x$. This is illustrated in Figure 5.10 for the exponential function $f(x) = \left(\frac{3}{2}\right)^x$ and its inverse $f^{-1}(x) = \log_{3/2} x$.

If you rewrite a logarithmic function in its equivalent exponential form, then the logarithmic function can be graphed by plotting points.

EXAMPLE 1   Graph a Logarithmic Function

Graph the logarithmic function $f(x) = \log_2 x$.

Solution   Changing $y = \log_2 x$ to its exponential form $2^y = x$ allows us to evaluate x for convenient integer values of y.

y	-2	-1	0	1	2	3
$x = 2^y$	$2^{-2} = \frac{1}{4}$	$2^{-1} = \frac{1}{2}$	$2^0 = 1$	$2^1 = 2$	$2^2 = 4$	$2^3 = 8$

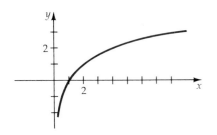

Figure 5.11
$f(x) = \log_2 x$

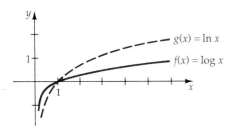

Figure 5.12

The above table was constructed by first picking a y value and then computing the corresponding x value. Drawing a smooth curve through the resulting points produces the graph of $f(x) = \log_2 x$ shown in Figure 5.11.

■ *Try Exercise **2**, page 280.*

The logarithmic functions $f(x) = \log x$ and $g(x) = \ln x$ are often used in applications and advanced mathematics. Their graphs can be drawn by using the techniques developed in Example 1. However, most scientific calculators have the $\boxed{\log}$ and the $\boxed{\ln}$ keys that can be used to determine points on the graphs. For example, the following table was made by entering convenient values of x and then evaluating $\log x$ and $\ln x$ by pressing the $\boxed{\log}$ and the $\boxed{\ln}$ keys. Figure 5.12 shows the graph of $f(x) = \log x$ and the graph of $g(x) = \ln x$.

x	0.5	1	5	10
$f(x) = \log x$	−0.3010	0	0.6990	1
$g(x) = \ln x$	−0.6931	0	1.6094	2.3026

For all positive real numbers $b \neq 1$, the function $f(x) = \log_b x$ has the following properties:

■ f has the set of positive real numbers as its domain.

■ f has the set of real numbers as its range.

■ f has a graph with an x-intercept of $(1, 0)$.

■ f has a graph asymptotic to the y-axis.

■ f is a one-to-one function.

The logarithmic function $f(x) = \log_b x$ is:

■ an increasing function if $b > 1$. See Figure 5.13a.

■ a decreasing function if $0 < b < 1$. See Figure 5.13b.

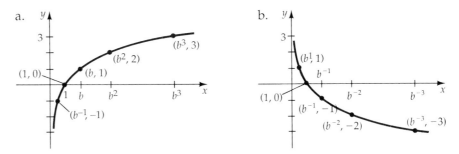

Figure 5.13

Many applied problems involve functions that are defined in terms of

$$f(x) = \log_b c,$$

where c is an algebraic expression that involves x. Example 2 illustrates that the graph of such a function may differ considerably from the graph of the logarithmic function $f(x) = \log_b x$.

EXAMPLE 2   Graph a Function That Is Logarithmic in Form

Graph the function $f(x) = \log_2|x|$.

Solution   This graph can be sketched by using the method of plotting points, but the procedure developed in the following discussion allows us to sketch the graph with little or no computation.

The domain of $f(x) = \log_2 x$ is the set of positive real numbers. Since $|x| > 0$ for all $x \neq 0$, the domain of $f(x) = \log_2|x|$ is the set of all nonzero real numbers. If $x > 0$, then $|x| = x$, and the graph of $f(x) = \log_2|x|$ will be the same as the graph of $f(x) = \log_2 x$ shown in Figure 5.11.

For $x < 0$, we use the fact that $f(x) = \log_2|x|$ is an even function. Its graph is symmetrical to the y-axis; thus we reflect the right-hand part of the graph across the y-axis to produce the graph in Figure 5.14.

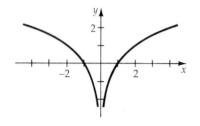

Figure 5.14
$f(x) = \log_2 |x|$

■ *Try Exercise **10**, page 280.*

EXAMPLE 3   Graph a Function That Is Logarithmic in Form

Graph the function $f(x) = \log_4(-2x)$.

Solution   Writing $y = \log_4(-2x)$ in its exponential form produces $4^y = -2x$ or $(-\frac{1}{2})4^y = x$. Choosing convenient values for y yields the following table:

y	-2	-1	0	1	2
$x = \left(-\dfrac{1}{2}\right)4^y$	$\left(-\dfrac{1}{2}\right)4^{-2} = -\dfrac{1}{32}$	$\left(-\dfrac{1}{2}\right)4^{-1} = -\dfrac{1}{8}$	$\left(-\dfrac{1}{2}\right)4^0 = -\dfrac{1}{2}$	$\left(-\dfrac{1}{2}\right)4^1 = -2$	$\left(-\dfrac{1}{2}\right)4^2 = -8$

The domain of $f(x) = \log_4(-2x)$ is the set of all negative real numbers because $-2x > 0$ if and only if $x < 0$.

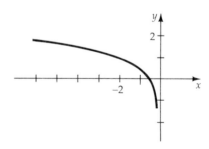

Figure 5.15
$f(x) = \log_4(-2x)$

■ *Try Exercise **12**, page 280.*

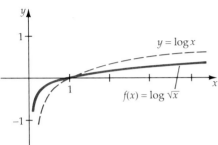

Figure 5.16

EXAMPLE 4   Graph a Function That Is Logarithmic in Form

Graph the function $f(x) = \log\sqrt{x}$.

Solution   Since $f(x) = \log\sqrt{x}$ is equivalent to $f(x) = \frac{1}{2}\log x$, the graph of $f(x) = \log\sqrt{x}$ can be obtained by *shrinking* $f(x) = \log x$. That is, we determine points (a, b) on the graph of $f(x) = \log x$. Then plot the points $(a, \frac{1}{2}b)$ to sketch the graph of $f(x) = \log\sqrt{x}$. (For a review of stretching and shrinking, see Section 3.4.)

Figure 5.16 shows the graph of $f(x) = \log\sqrt{x}$. Notice that the point $(a, \frac{1}{2}b)$ is on the graph of $f(x) = \log\sqrt{x}$ if and only if the point (a, b) is on the graph of $f(x) = \log x$.

■ *Try Exercise* **14,** *page 280.*

Horizontal and/or vertical translations of the graph of the logarithmic function $f(x) = \log_b x$ sometimes can be used to obtain the graph of functions that involve logarithms.

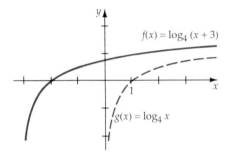

Figure 5.17

EXAMPLE 5   Use Translations to Graph

Graph   a. $f(x) = \log_4(x + 3)$ b. $f(x) = \log_4 x + 3$.

Solution

a. The graph of $f(x) = \log_4(x + 3)$ can be obtained by shifting the graph of $g(x) = \log_4 x$ three units to the left. (For a review of *horizontal translations,* see Section 3.4.) Figure 5.17 shows the graph of $g(x) = \log_4 x$ and the graph of $f(x) = \log_4(x + 3)$.

b. The graph of $f(x) = \log_4 x + 3$ can be obtained by shifting the graph of $g(x) = \log_4 x$ three units upward. (For a review of *vertical translations,* see Section 3.4.) Figure 5.18 shows the graph of $g(x) = \log_4 x$ and the graph of $f(x) = \log_4 x + 3$.

■ *Try Exercise* **20,** *page 280.*

Seismologists measure the magnitude of earthquakes using the Richter scale. On this scale, the magnitude R of an earthquake is determined by the function.

$$R = \log \frac{I}{I_0}$$

where I_0 is the measure of a *zero-level earthquake,* and I is the intensity of the earthquake being measured.

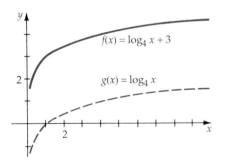

Figure 5.18

EXAMPLE 6   Find the Richter Scale Measure of an Earthquake

Find the Richter scale measure of

a. an earthquake that had an intensity of $I = 10{,}000 \, I_0$;

b. the San Francisco earthquake of 1906, which had an intensity of $I = 199{,}526{,}000 I_0$.

Solution

a. $R = \log \dfrac{10{,}000 I_0}{I_0} = \log 10{,}000 = \log 10^4 = 4$ (on the Richter scale).

b. $R = \log \dfrac{199{,}526{,}000 I_0}{I_0} = \log 199{,}526{,}000 \approx 8.3$ (on the Richter scale).

■ *Try Exercise* **42**, *page 281.*

EXAMPLE 7   Compare the Richter Scale Measures of Earthquakes

If an earthquake has an intensity 100 times the intensity of a second earthquake, then how much larger is the Richter scale measure of the larger earthquake than that of the smaller?

Solution   Let I represent the intensity of the smaller earthquake and $100I$ represent the intensity of the larger earthquake. The Richter scale measures of the small earthquake R_1 and the larger earthquake R_2 are given by

$$R_1 = \log \frac{I}{I_0} \quad \text{and} \quad R_2 = \log \frac{100I}{I_0}.$$

Using the properties of logarithms, we can write R_2 as

$$R_2 = \log 100\frac{I}{I_0} = \log 100 + \log \frac{I}{I_0} = \log 10^2 + R_1 = 2 + R_1$$

Thus an earthquake that is 100 times as intense as a smaller earthquake will have a Richter scale measure that is 2 more than that of the smaller earthquake.

■ *Try Exercise* **44**, *page 281.*

EXERCISE SET 5.3 _____

In Exercises 1 to 28, graph the given logarithmic functions. If possible, make use of translations.

1. $f(x) = \log_3 x$

2. $f(x) = \log_5 x$

3. $f(x) = \log_{1/2} x$

4. $f(x) = \log_{1/4} x$

5. $f(x) = -2 \ln x$

6. $f(x) = -\log x$

7. $f(x) = \log_4 x^2$

8. $f(x) = 2 \log_5 x$

9. $f(x) = |\ln x|$

10. $f(x) = \ln|x|$

11. $f(x) = -|\ln x|$

12. $f(x) = -|\log_2 x^3|$

13. $f(x) = \log \sqrt[3]{x}$

14. $f(x) = \ln \sqrt{x}$

15. $f(x) = 3 + \log_2 x$

16. $f(x) = -2 + \log_4 x$

17. $f(x) = \log(x + 10)$

18. $f(x) = \ln(x + 3)$

19. $f(x) = \ln(x - 5)$

20. $f(x) = \log(x - 2)$

21. $f(x) = \log_5(x + 2)^2$

22. $f(x) = \log_5(x - 1)^2$

23. $f(x) = -\ln(x - 4)$

24. $f(x) = \ln(x + 3)^{-1}$

25. $f(x) = 4 - 2 \ln x$

26. $f(x) = x + \ln x$

27. $f(x) = \log(4 - x)$

28. $f(x) = -\ln(e - x)$

In Exercises 29 to 34, determine the domain of the given logarithmic function.

29. $J(x) = \dfrac{1}{\ln x}$

30. $K(x) = \log(x^2 + 4)$

31. $L(x) = \log(x^2 - 9)$

32. $f(x) = \dfrac{1}{\ln(4x + 10)}$

33. $h(x) = \log(x^2) + 4$

34. $Q(x) = \log(|x| + 1)$

In Exercises 35 to 40, find the range of the given logarithmic function.

35. $J(x) = \dfrac{1}{\ln x}$

36. $K(x) = \log(x^2 + 4)$

37. $L(x) = \log(x^2 - 9)$

38. $V(x) = |\ln x|$

39. $R(x) = \ln|x|$

40. $B(x) = 4 + \log(x^2)$

41. A furniture outlet finds that the number of sofas it can sell in a month is given by

$$S(d) = 15 + 10 \ln(0.01d + 1),$$

where $S(d)$ is the number of sofas sold and d is the dollar amount spent for advertising.

  a. How many sofas will be sold with no advertising expenditure?

  b. How many sofas (to the nearest unit) will be sold with an advertising expenditure of $5000?

  c. How much advertising expenditure is necessary to sell 195 sofas?

42. What will an earthquake measure on the Richter scale if it has an intensity of $I = 100,000I_0$?

43. The Colombia earthquake of 1906 had an intensity of $I = 398,107,000I_0$. What did it measure on the Richter scale?

44. If an earthquake has an intensity 1000 times the intensity of a second earthquake, then how much larger is the Richter scale measure of the larger earthquake than that of the smaller?

Supplemental Exercises

45. Given $f(x) = \ln x$, evaluate

  a. $f(e^3)$,   **b.** $f(e^{\ln 4})$,   **c.** $f(e^{3\ln 3})$.

46. Given $f(x) = \log_5 x$, evaluate

  a. $f(5^2)$,   **b.** $f(5^{\log_5 4})$,   **c.** $f(5^{3\log_5 3})$.

47. Explain why the graph of $F(x) = \log_b x^2$ and the graph of $G(x) = 2\log_b x$ are not identical.

48. Explain why the graph of $F(x) = |\log_b x|$ and the graph of $G(x) = \log_b|x|$ are not identical.

49. Graph $f(x) = e^{-x}(\ln x)$ for $1 \le x \le e^2$.

50. Graph $g(x) = \log[\![x]\!]$ for $1 \le x \le 10$. Recall that $[\![x]\!]$ represents the greatest integer function.

In Exercises 51 to 56, determine the domain and the range of the given function.

51. $f(x) = \sqrt{\log x}$

52. $f(x) = \sqrt{\ln x^3}$

53. $f(x) = 100 - \ln\sqrt{1 - x^2}$

54. $f(x) = 10 + |\ln(x - e)|$

55. $f(x) = \log(\log x)$

56. $f(x) = |\ln(-\ln x)|$

57. The Coalinga, California, earthquake of May 2, 1983, had a Richter scale measure of 6.5. Find the Richter scale measure of an earthquake that has an intensity 200 times the intensity of the Coalinga quake.

58. The earthquake just south of Concepción, Chile, on May 22, 1960, had a Richter scale measure of 9.5. Find the Richter scale measure of an earthquake that has an intensity one-half the intensity of this quake.

5.4
Exponential and Logarithmic Equations

If a variable appears as an exponent in a term of an equation, then the equation is called an **exponential equation**. Example 1 uses the Equality of Exponents Theorem to solve exponential equations.

EXAMPLE 1   Solve Exponential Equations

Solve the following exponential equations: **a.** $5^{x-1} = 125$ **b.** $49^{2x} = \dfrac{1}{7}$

  Solution   Write each side of the equation as a power of the same base and then equate the exponents.

a. $5^{x-1} = 125$

$\quad$ $5^{x-1} = 5^3$ Write each side as a power of 5.

$\quad$ $x - 1 = 3$ Equate the exponents.

$\quad\quad$ $x = 4$

b. $49^{2x} = \dfrac{1}{7}$

$\quad$ $(7^2)^{2x} = 7^{-1}$ Write each side as a power of 7.

$\quad$ $7^{4x} = 7^{-1}$

$\quad$ $4x = -1$ Equate the exponents.

$\quad$ $x = -\dfrac{1}{4}$

■ *Try Exercise **2**, page 286.*

Some exponential equations can be solved by taking the logarithm of both sides of the equation.

EXAMPLE 2   Use the Logarithm of Each Side Theorem to Solve an Equation

Solve the following exponential equations: a. $5^x = 40$ b. $3^{2x-1} = 5^{x+2}$

Solution   Start by taking the logarithm of each side of the equation.

a. $\quad\quad$ $5^x = 40$

$\quad$ $\log(5^x) = \log 40$

$\quad$ $x \log 5 = \log 40$

$\quad$ $x = \dfrac{\log 40}{\log 5}$ $\quad\quad$ Exact solution

$\quad$ $x \approx \dfrac{1.60206}{0.69897} \approx 2.29203$ $\quad$ Decimal approximation

b. $\quad\quad$ $3^{2x-1} = 5^{x+2}$

$\quad$ $\log 3^{2x-1} = \log 5^{x+2}$

$\quad$ $(2x - 1) \log 3 = (x + 2) \log 5$ $\quad\quad$ Power property

$\quad$ $2x \log 3 - \log 3 = x \log 5 + 2 \log 5$ $\quad\quad$ Distributive property

Collecting terms involving the variable x on the left side yields

$$2x \log 3 - x \log 5 = 2 \log 5 + \log 3$$

$$x(2 \log 3 - \log 5) = 2 \log 5 + \log 3$$

$$x = \dfrac{2 \log 5 + \log 3}{2 \log 3 - \log 5}$$

Using a calculator, a decimal approximation to the solution can be obtained:

$$x \approx \frac{1.87506}{0.25527} \approx 7.34540$$

■ *Try Exercise* **12,** *page 286.*

Logarithmic Equations

Equations that involve logarithms are called **logarithmic equations.** The properties of logarithms, along with the definition of a logarithm, are valuable aids to solving a logarithmic equation.

EXAMPLE 3 **Solve a Logarithmic Equation**

Solve the logarithmic equation $\log 2x - \log(x - 3) = 1$.

Solution $\log 2x - \log(x - 3) = 1$

$$\log \frac{2x}{x - 3} = 1 \qquad \text{Quotient property}$$

$$\frac{2x}{x - 3} = 10^1 \qquad \text{Definition of logarithm}$$

$$2x = 10x - 30$$

$$-8x = -30$$

$$x = \frac{15}{4}$$

Check the solution by substituting 15/4 into the original equation.

■ *Try Exercise* **24,** *page 286.*

Example 4 uses the one-to-one property of logarithms to solve a logarithmic equation.

EXAMPLE 4 **Solve a Logarithmic Equation**

Solve the logarithmic equation $\ln(3x + 8) = \ln(2x + 2) + \ln(x - 2)$.

Solution $\ln(3x + 8) = \ln(2x + 2) + \ln(x - 2)$

$\ln(3x + 8) = \ln(2x + 2)(x - 2) \qquad \text{Product property}$

$\ln(3x + 8) = \ln(2x^2 - 2x - 4)$

$\quad\; 3x + 8 = 2x^2 - 2x - 4 \qquad\quad \text{One-to-one property}$
$\qquad\qquad\qquad\qquad\qquad\qquad\quad\; \text{of logarithms}$

$\qquad\; 0 = 2x^2 - 5x - 12$

$\qquad\; 0 = (2x + 3)(x - 4)$

Thus $-3/2$ and 4 are possible solutions. The number $-3/2$ does not check in the original equation. Why?[1] It can be shown that 4 checks and thus the only solution is $x = 4$.

■ *Try Exercise* **32,** *page 286.*

Equations That Involve $b^x \pm b^{-x}$

Recall from Section 1 that the graph of $f(x) = (2^x + 2^{-x})/2$ can be sketched by drawing a curve *halfway* between the graph of $y = 2^x$ and $y = 2^{-x}$ since $(2^x + 2^{-x})/2$ is the average of 2^x and 2^{-x}. The solutions of the equation $(2^x + 2^{-x})/2 = 3$ are represented by the x-coordinates of the points of intersection of the graph of $y = 3$ and the graph of $y = (2^x + 2^{-x})/2$ as shown in Figure 5.19.

Example 5 uses an algebraic method to solve $(2^x + 2^{-x})/2 = 3$.

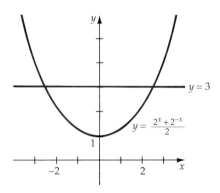

Figure 5.19
$$y = \frac{2^x + 2^{-x}}{2}$$

EXAMPLE 5   **Solve an Equation Involving $b^x + b^{-x}$**

Solve the exponential equation $\dfrac{2^x + 2^{-x}}{2} = 3$.

Solution   Multiplying each side by 2 produces

$$2^x + 2^{-x} = 6$$

$$2^{2x} + 2^0 = 6 \cdot (2^x) \quad \text{Multiply by } 2^x \text{ to clear negative exponents.}$$

$$(2^x)^2 - 6(2^x) + 1 = 0 \quad \text{Write in quadratic form.}$$

Substituting u for 2^x produces the quadratic equation

$$(u)^2 - 6(u) + 1 = 0.$$

By the quadratic formula,

$$u = \frac{6 \pm \sqrt{36 - 4}}{2} = \frac{6 \pm 4\sqrt{2}}{2} = 3 \pm 2\sqrt{2}.$$

Replacing u with 2^x produces

$$2^x = 3 \pm 2\sqrt{2}.$$

Now take the common logarithm of each side.

$$\log 2^x = \log(3 \pm 2\sqrt{2})$$

$$x \log 2 = \log(3 \pm 2\sqrt{2}) \quad \text{Power property of logarithms}$$

$$x = \frac{\log(3 \pm 2\sqrt{2})}{\log 2} \approx \pm 2.54$$

■ *Try Exercise* **42,** *page 286.*

---

[1] If $x = -3/2$, the original equation becomes $\ln(7/2) = \ln(-1) + \ln(-7/2)$. This cannot be true because the function $f(x) = \ln x$ is not defined for negative values of x.

Remark If natural logarithms had been used in Example 5, then the exact solutions would be

$$x = \frac{\ln(3 \pm 2\sqrt{2})}{\ln 2}.$$

The pH of a Solution

Whether an aqueous solution is acidic or basic depends on its hydronium-ion concentration. Thus acidity is a function of hydronium-ion concentration. Since these hydronium-ion concentrations may be very small, it is convenient to measure acidity in terms of **pH**, which is defined as the negative of the common logarithm of the molar hydronium-ion concentration. As a mathematical formula, this is stated as

$$pH = -\log[H_3O^+].$$

EXAMPLE 6 **Find the pH of a Solution**

Find the pH of the following: a. Orange juice with $[H_3O^+] = 2.80 \times 10^{-4}$ M
b. milk with $[H_3O^+] = 3.97 \times 10^{-7}$ M

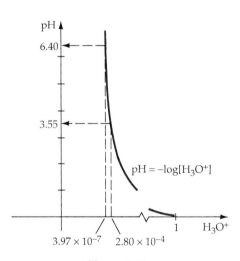

Figure 5.20

Solution

a. $pH = -\log[H_3O^+] = -\log(2.80 \times 10^{-4}) \approx -(-3.55) = 3.55$

The orange juice has a pH of 3.55 (to the nearest hundredth).

b. $pH = -\log[H_3O^+] = -\log(3.97 \times 10^{-7}) \approx -(-6.40) = 6.40$

The milk has a pH of 6.40 (to the nearest hundredth).

■ *Try Exercise* **50,** *page 286.*

Figure 5.20 shows how the pH function *maps* small positive numbers that are relatively close together on the hydronium-ion concentration axis into numbers that are farther apart on the pH axis. The pH of pure water is 7.0.

EXAMPLE 7 **Solve an Application**

Determine the hydronium-ion concentration of a sample of blood with pH = 7.41.

Solution Substitute 7.41 for the pH and solve for H_3O^+.

$$pH = -\log[H_3O^+]$$
$$7.41 = -\log[H_3O^+] \quad \text{Substitute 7.41 for pH.}$$
$$-7.41 = \log[H_3O^+] \quad \text{Multiply both sides by } -1.$$
$$10^{-7.41} = [H_3O^+] \quad \text{Definition of } y = \log_b x.$$
$$3.9 \times 10^{-8} \approx [H_3O^+]$$

The hydronium-ion concentration of the blood sample is 3.9×10^{-8} M.

■ *Try Exercise* **52,** *page 286.*

EXERCISE SET 5.4

In Exercises 1 to 40, solve for x.

1. $2^x = 64$

2. $3^x = 243$

3. $8^x = 512$

4. $25^x = 3125$

5. $49^x = \dfrac{1}{343}$

6. $9^x = \dfrac{1}{243}$

7. $2^{5x+3} = \dfrac{1}{8}$

8. $3^{4x-7} = \dfrac{1}{9}$

9. $\left(\dfrac{2}{5}\right)^x = \dfrac{8}{125}$

10. $\left(\dfrac{2}{5}\right)^x = \dfrac{25}{4}$

11. $5^x = 70$

12. $6^x = 50$

13. $3^{-x} = 120$

14. $7^{-x} = 63$

15. $\left(\dfrac{3}{5}\right)^x = 0.92$

16. $\left(\dfrac{7}{3}\right)^x = 22$

17. $10^{2x+3} = 315$

18. $10^{6-x} = 550$

19. $e^x = 10$

20. $e^{x+1} = 20$

21. $\left(1 + \dfrac{0.08}{12}\right)^{12x} = 1.5$

22. $\left(1 + \dfrac{0.05}{365}\right)^{365x} = 2$

23. $\log_2 x + \log_2(x - 4) = 2$

24. $\log_3 x + \log_3(x + 6) = 3$

25. $\log(5x - 1) = 2 + \log(x - 2)$

26. $1 + \log(3x - 1) = \log(2x + 1)$

27. $\log \sqrt{x^3 - 17} = \dfrac{1}{2}$

28. $\log(x^3) = (\log x)^2$

29. $\log(\log x) = 1$

30. $2 \ln \dfrac{e}{\sqrt{3}} = 3 - \ln x$

31. $\dfrac{1}{3} \ln 125 + \dfrac{1}{2} \ln x = \ln x$

32. $\ln x = \dfrac{1}{2} \ln\left(2x + \dfrac{5}{2}\right) + \dfrac{1}{2} \ln 2$

33. $\ln(e^{3x}) = 6$

34. $\log_b(b^{5x+2}) = 4$

35. $\ln x^2 = \ln 9$

36. $\log_x 9 = 3$

37. $\log_x 8 = 2$

38. $4 \log_x 2 - \dfrac{1}{2} \log_x 4 = 2 - \dfrac{1}{3} \log_x 8$

39. $e^{\ln(x-1)} = 4$

40. $10^{\log(2x+7)} = 8$

In Exercises 41 to 44, use common logarithms to solve for x.

41. $\dfrac{10^x - 10^{-x}}{2} = 20$

42. $\dfrac{10^x + 10^{-x}}{2} = 8$

43. $\dfrac{10^x + 10^{-x}}{10^x - 10^{-x}} = 5$

44. $\dfrac{10^x - 10^{-x}}{10^x + 10^{-x}} = \dfrac{1}{2}$

In Exercises 45 to 48, use natural logarithms to solve for x.

45. $\dfrac{e^x + e^{-x}}{2} = 15$

46. $\dfrac{e^x - e^{-x}}{2} = 15$

47. $\dfrac{1}{e^x - e^{-x}} = 4$

48. $\dfrac{e^x + e^{-x}}{e^x - e^{-x}} = 3$

49. Find the pH of a sample of lemon juice that has a hydronium-ion concentration of 6.3×10^{-3}.

50. An *acidic solution* has a pH of less than 7, whereas a *basic solution* has a pH of greater than 7. Household ammonia has an hydronium-ion concentration of 1.26×10^{-12}. Determine the pH of the ammonia, and state whether it is an acid or a base.

51. Find the hydronium-ion concentration of beer, which has a pH of 4.5.

52. Normal rain has a pH of 5.6. A recent acid rain had a pH of 3.1. Find the hydronium concentration of this rain.

53. The population P of a city grows exponentially according to the function

$$P(t) = 8500(1.1)^t, \qquad 0 \le t \le 8$$

where t is measured in years.

a. Find the population at time $t = 0$ and also at time $t = 2$.

b. When, to the nearest year, will the population reach 15,000?

54. After a race, a runner's pulse rate R in beats per minute decreases according to the function $R(t) = 145e^{-0.092t}$, $(0 \le t \le 15)$ where t is measured in minutes.

a. Find the runner's pulse rate at the end of the race and also 1 minute after the end of the race.

b. How long, to the nearest minute, after the end of the race will the runner's pulse rate be 80 beats per minute?

55. A can of soda at 70°F is placed into a refrigerator that maintains a constant temperature of 36°F. The temperature T of the soda t minutes after it is placed into the refrigerator is

$$T(t) = 36 + 43e^{-0.058t}$$

a. Find the temperature of the soda 10 minutes after it is placed into the refrigerator.

b. When, to the nearest minute, will the temperature of the soda be 45°F?

56. During surgery, a patient's circulatory system requires at least 50 milligrams of an anesthetic. The amount of anesthetic present t hours after 80 milligrams of anesthetic are administered is

$$A(t) = 80(0.727)^t.$$

a. How much of the anesthetic is present in the patient's circulatory system 30 minutes after the anesthetic is administered?

b. How long, to the nearest minute can the operation last if the patient does not receive an additional dosage of anesthetic?

Supplemental Exercises

57. The following argument shows that $0.125 > 0.25$. Find the incorrect step.

$$3 > 2$$
$$3(\log 0.5) > 2(\log 0.5)$$
$$\log 0.5^3 > \log 0.5^2$$
$$0.5^3 > 0.5^2$$
$$0.125 > 0.25$$

58. The following argument shows that $4 = 6$. Find the incorrect step.

$$4 = \log_2 16$$
$$= \log_2(8 + 8)$$
$$= \log_2 8 + \log_2 8$$
$$= 3 + 3$$
$$= 6$$

In Exercises 59 to 66, determine the *number* of solutions of the given equation. (*Hint:* Sketch the graph of the function on the left side of the equation and the graph of the function on the right side of the equation on the same coordinate axis. Then determine the *number* of intersections of their graphs to determine the *number* of solutions to the equation.)

59. $2^x = \log x$

60. $10^{-x} = \log x$

61. $e^x - 4 = \ln x$

62. $x = -\log x$

63. $\ln x = x^2 - 5$

64. $\log x = x^3$

65. $\dfrac{2^x + 2^{-x}}{2} - 2 = \log|x|$

66. $\dfrac{2^x + 2^{-x}}{2} = \dfrac{3^x + 3^{-x}}{2}$

67. A common mistake that students make is to write $\log(x + y)$ as $\log x + \log y$. For what values of x and y does $\log(x + y) = \log x + \log y$? (*Hint:* Solve for x in terms of y.)

68. Which is larger, 500^{501} or 506^{500}? (*Hint:* Let $x = 500^{501}$ and $y = 506^{500}$ and then compare $\ln x$ with $\ln y$.)

69. Explain why the functions $F(x) = (1.4)^x$ and $G(x) = e^{0.336x}$ essentially represent the same functions.

70. Find the constant k that will make $f(t) = (2.2)^t$ and $g(t) = e^{-kt}$ essentially represent the same function.

71. Solve $e^{1/x} > 2$. Write your answer using interval notation.

72. Solve $\log(x^2) > (\log x)^2$. Write your answer using interval notation.

5.5

Applications of Exponential and Logarithmic Functions

In many applications, a quantity N grows or decays according to the function $N(t) = N_0 e^{kt}$. In this function, N is a function of time t and N_0 is the value of N at time $t = 0$. If k is a *positive* constant, then $N(t) = N_0 e^{kt}$ is called an exponential **growth function.** If k is a *negative* constant, then $N(t) = N_0 e^{kt}$ is called an exponential **decay function.** The following examples will give you an understanding of how growth and decay functions arise naturally in the investigation of certain phenomena.

Interest is money paid for the use of money. The interest I is called **simple interest** if it is a fixed percent r per time period t of the amount of money invested. The amount of money invested is called the **principal** P. Simple interest is computed using the formula $I = Prt$. For example, if $1000 is invested at 12% for 3 years, the simple interest is

$$I = Prt = \$1000(0.12)(3) = \$360.$$

The balance after t years is $B = P + I = P + Prt$. In the previous example, the $1000 invested for 3 years produced $360 interest. Thus the balance after 3 years is $1360.

Compound Interest

In many financial transactions, interest is added to the principal at regular intervals so that interest is paid on interest as well as the principal. Interest earned in this manner is called **compound interest.** For example, if $1000 is invested at 12% annual interest compounded annually for 3 years, then the total interest after 3 years is

First-year interest $1000(0.12) = $120.00

Second-year interest $1120(0.12) = $134.40

Third-year interest $\underline{$1254.40(0.12) \approx $150.53}$

            $404.93   ←Total interest

This method of computing the balance can be tedious and time-consuming. A *compound interest formula* that can be used to determine the balance due after n periods of compounding can be developed as follows.

Note that if P dollars is invested at an interest rate of r per period, then the balance after one period is $B_1 = P + Pr = P(1 + r)$, where Pr represents the interest earned for the period. Observe that B_1 is the product of the original principal P and $(1 + r)$. If the amount B_1 is reinvested for another period, then the balance after the second period is

$$B_2 = (B_1)(1 + r) = P(1 + r)(1 + r) = P(1 + r)^2.$$

Successive reinvestments lead to the following results:

Number of periods	Balance
3	$B_3 = P(1 + r)^3$
4	$B_4 = P(1 + r)^4$
.	.
.	.
.	.
n	$B_n = P(1 + r)^n$

The equation $B_n = P(1 + r)^n$ is valid if r is the interest rate paid during each of the n compounding periods. If r is an annual interest rate and n is the number of compounding periods per year, then the interest rate each period is r/n and the number of compounding periods after t years is nt. Thus the compound interest formula is expressed as follows.

The Compound Interest Formula

A principal P invested at an annual interest rate r compounded n times per year for t years produces the balance

$$B = P\left(1 + \frac{r}{n}\right)^{nt}.$$

EXAMPLE 1   **Solve an Application**

Find the balance if $1000 is invested at an annual interest rate of 10%, for 2 years compounded   a.   annually   b.   daily   c.   hourly

Solution

a.   Use the compound interest formula, with $P = 1000$, $r = 0.1$, $t = 2$, and $n = 1$.

$$B = \$1000\left(1 + \frac{0.1}{1}\right)^{1 \cdot 2} = \$1000(1.1)^2 = \$1210.00$$

b.   Since there are 365 days in a year, use $n = 365$.

$$B = \$1000\left(1 + \frac{0.1}{365}\right)^{365 \cdot 2} \approx \$1000(1.000273973)^{730} \approx \$1221.37$$

c.   Since there are 8760 hours in a year, use $n = 8760$.

$$B = \$1000\left(1 + \frac{0.1}{8760}\right)^{8760 \cdot 2} \approx \$1000(1.000011416)^{17520} \approx \$1221.41$$

■ *Try Exercise **4**, page 295.*

Remark   As the number of compounding periods increases, the balance seems to approach some upper limit. Even if the interest is compounded each *second*, the balance to the nearest cent remains $1221.40.

To **compound continuously** means to increase the number of compounding periods without bound. We can better understand the concept of compounding continuously if we evaluate the balance B in the compound interest formula $B = P(1 + r/n)^{nt}$ with a principal of $1, a yearly interst rate of 100% = 1, for 1 year, as n increases without bound. That is, evaluate

$$\left(1 + \frac{1}{n}\right)^n \text{ as } n \to \infty.$$

Recall that in Section 1 e was defined as the number that $(1 + 1/n)^n$ approaches as n increases without bound. Thus we see that the number e plays an important role in continuous compounding.

To derive a continuous interest formula, substitute $1/m$ for r/n in the compound interest formula

$$B = P\left(1 + \frac{r}{n}\right)^{nt} \tag{1}$$

to produce

$$B = P\left(1 + \frac{1}{m}\right)^{nt} \tag{2}$$

This substitution is motivated by the desire to express $(1 + r/n)^n$ as $[(1 + 1/m)^m]^r$, which will approach e^r as m gets large without bound.

Solving the equation $1/m = r/n$ for n yields $n = mr$, and thus the exponent nt can be written as mrt. Therefore Equation (2) can be expressed as

$$B = P\left(1 + \frac{1}{m}\right)^{mrt} = P\left[\left(1 + \frac{1}{m}\right)^m\right]^{rt} \qquad (3)$$

By the definition of e, we know that as m gets larger without bound,

$$\left(1 + \frac{1}{m}\right)^m \text{ approaches } e.$$

Thus, using continuous compounding, Equation (3) simplifies to $B = Pe^{rt}$.

Continuous Compounding Interest Formula

> If an account with principal P and annual interest rate r is compounded continuously for t years, then the balance is $B = Pe^{rt}$.

EXAMPLE 2   **Compound Continuously**

Find the balance after 4 years if $800 is invested at an annual rate of 6% compounded continuously.

Solution   Use the continuous compounding formula.

$$B = Pe^{rt} = 800e^{0.06(4)} = 800e^{0.24}$$

$$\approx \$800(1.27124915) = \$1017.00 \quad \text{To nearest cent}$$

■ *Try Exercise 6, page 295.*

Exponential Growth

Given any two points on the graph of a growth function $N(t) = N_0 e^{kt}$, you can use the given data to solve for the constants N_0 and k.

EXAMPLE 3   **Find the Exponential Growth Function That Models Given Data**

Find the exponential growth function for a town whose population was 16,400 in 1970 and 20,200 in 1980.

Solution   We need to determine N_0 and k in $N(t) = N_0 e^{kt}$. If we represent the year 1970 by $t = 0$, then our given data are $N(0) = 16,400$ and $N(10) = 20,200$. Because N_0 is defined to be $N(0)$, we know $N_0 = 16,400$. To determine k, substitute $t = 10$, and $N_0 = 16,400$ in $N(t) = N_0 e^{kt}$ to produce

$$N(10) = 16,400e^{k \cdot 10}$$

$$20,200 = 16,400e^{10k} \quad \text{Substitute 20,200 for } N(10).$$

$$\frac{20,200}{16,400} = e^{10k}$$

To solve this equation for k, take the natural logarithm of each side.

$$\ln\left(\frac{20{,}200}{16{,}400}\right) = \ln e^{10k}$$

$$\ln\left(\frac{20{,}200}{16{,}400}\right) = 10k \quad \text{Use } \log_b(b^p) = p.$$

$$\tfrac{1}{10}\ln\left(\frac{20{,}200}{16{,}400}\right) = k$$

$$0.0208 \approx k$$

The exponential growth function is $N(t) = 16{,}400e^{0.0208t}$

■ *Try Exercise **10**, page 295.*

EXAMPLE 4   Solve an Application

Use the exponential growth function from Example 3 to

a.   estimate the population of the town in the year 1995 and

b.   estimate when the population will be double its 1970 population.

   Solution

a.   Let $t = 25$.

$$N(t) = 16{,}400e^{0.0208t}$$

$$N(25) = 16{,}400e^{0.0208 \cdot 25}$$

$$\approx 27{,}600 \qquad \text{To nearest 100}$$

b.   We need to solve for t when $N(t) = 2 \cdot 16{,}400 = 32{,}800$.

$$32{,}800 = 16{,}400e^{0.0208t}$$

$$2 = e^{0.0208t} \qquad \text{Divide each side by 16,400.}$$

$$\ln 2 = 0.0208t \qquad \text{Take the natural logarithm of each side.}$$

$$\frac{\ln 2}{0.0208} = t$$

$$33 \approx t \qquad \text{To nearest year.}$$

■ *Try Exercise **12**, page 295.*

Exponential Decay

Many radioactive materials *decrease* exponentially. This decrease, called radioactive decay, is measured in terms of **half-life,** which is defined as the time required for the disintegration of half the atoms in a sample of a radioactive substance. Following are the half-lives of selected radioactive isotopes:

Isotope	Half-Life
Carbon (C^{14})	5730 years
Radium (Ra^{226})	1660 years
Polonium (Po^{210})	138 days
Phosphorus (P^{32})	14 days
Polonium (Po^{214})	1/10,000th of a second

EXAMPLE 5   Find the Exponential Decay Function That Models Given Data

Find the exponential decay function for the amount of phosphorus (P^{32}) that remains in a sample after t days.

Solution   When $t = 0$, $N(0) = N_0 e^{k(0)} = N_0$. Thus, we know that $N(0) = N_0$. Also, because the phosphorus has a half-life of 14 days, $N(14) = 0.5N_0$. To find k, substitute $t = 14$ in $N(t) = N_0 e^{kt}$ and solve for k.

$$N(14) = N_0 \cdot e^{k \cdot 14}$$

$$0.5N_0 = N_0 e^{14k} \qquad \text{Substitute } 0.5N_0 \text{ for } N(14).$$

$$0.5 = e^{14k} \qquad \text{Divide each side by } N_0.$$

Taking the natural logarithm of each side produces

$$\ln 0.5 = \ln e^{14k}$$

$$\ln 0.5 = 14k$$

$$\frac{1}{14} \ln 0.5 = k$$

$$-0.0495 \approx k$$

The exponential decay function is $N(t) = N_0 e^{-0.0495t}$.

■ *Try Exercise **14**, page 295.*

Remark   Since $e^{-0.0495} \approx (0.5)^{1/14}$, the decay function $N(t) = N_0 e^{-0.0495t}$ can also be written as $N(t) = N_0(0.5)^{t/14}$. In this form it is easy to see that if t is increased by 14, N will decrease by a factor of 0.5.

EXAMPLE 6   Solve an Application

Use $N(t) = N_0(0.5)^{t/14}$ to estimate the amount of phosphorus (P^{32}) that remains in a sample after 50 days.

Solution $N(t) = N_0(0.5)^{t/14}$

$$N(50) = N_0(0.5)^{50/14} \approx 0.0841N_0$$

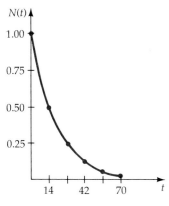

Figure 5.21
$N(t) = N_0(0.5)^{t/14}$

After 50 days, approximately 8% of the original phosphorus (P^{32}) remains.

■ *Try Exercise* **16**, *page 295*.

The following tables and Figure 5.21 show the percent of phosphorus(P^{32}) that remains after 0, 14, 28, 42, 56, and 70 days.

Days t	0	14	28	42	56	70
Percent (P^{32}) remaining $N(t)$	100	50	25	12.5	6.25	3.125

Carbon Dating

The bone tissue in all living animals contains both carbon-12, which is nonradioactive, and carbon-14, which is radioactive with a half-life of approximately 5730 years. As long as the animal is alive, the ratio of carbon-14 to carbon-12 remains constant. When the animal dies ($t = 0$), the carbon-14 begins to decay. Thus a bone that has a smaller ratio of carbon-14 to carbon-12 is older than a bone that has a larger ratio. The amount of carbon-14 present at time t is

$$N(t) = N_0(0.5)^{t/5730}$$

where N_0 is the amount of carbon-14 present in the bone at time $t = 0$.

EXAMPLE 7   **Application**

Determine the age of a bone if it now contains 85 percent of the carbon-14 it had when $t = 0$.

Solution   Let t be the time at which $N(t) = 0.85N_0(0.5)^{t/5730}$.

$$0.85N_0 = N_0(0.5)^{t/5730}$$

$$0.85 = (0.5)^{t/5730} \quad \text{Divide each side by } N_0.$$

Taking the natural logarithm of each side produces

$$\ln 0.85 = \ln (0.5)^{t/5730}$$

$$\ln 0.85 = \frac{t}{5730} \ln 0.5 \quad \text{Power property}$$

$$\frac{\ln 0.85}{\ln 0.5} = \frac{t}{5730}$$

$$5730 \frac{\ln 0.85}{\ln 0.5} = t$$

$$1340 \approx t \qquad \text{To nearest 10 years}$$

The bone is about 1340 years old.

■ *Try Exercise* **18**, *page 295*.

The Decibel Scale

The range of sound intensities that the human ear can detect is so large that a special *decibel scale* (named after the inventor of the telephone, Alexander Graham Bell) is used to measure and compare different sound intensities. Specifically, the *intensity level N* of sound measured in decibels is directly proportional to the *power I* of the sound measured in watts per square centimeter. That is,

$$N(I) = 10 \log\left(\frac{I}{I_0}\right)$$

where I_0 is the power of sound that is barely audible to the human ear. By international agreement, I_0 is the constant 10^{-16} watts per square centimeter.

EXAMPLE 8   Solve an Application

The power of normal conversation is 10^{-10} watts per square centimeter. What is the intensity level N in decibels of normal conversation?

Solution   Evaluate $N(10^{-10})$.

$$N(I) = 10 \log\left(\frac{I}{10^{-16}}\right)$$

$$N(10^{-10}) = 10 \log\left(\frac{10^{-10}}{10^{-16}}\right) \quad \text{Substitute } 10^{-10} \text{ for } I.$$

$$= 10 \log(10^6) \quad \text{Since } \frac{10^{-10}}{10^{-16}} = 10^{-10-(-16)} = 10^6$$

$$= 10(6) = 60$$

The intensity level of normal conversation is 60 decibels.

■ *Try Exercise* **22,** *page 295.*

You might be tempted to conclude that three people each conversing at the same time, each with the power of normal conversation, would produce a combined intensity level of $3 \cdot 60$ decibels $= 180$ decibels; however, since the decibel scale is a logarithmic scale, this is not the case. In fact, three people conversing at the same time will produce a combined conversation with power of $3 \cdot (10^{-10})$ watts per square centimeter. To find the intensity level in decibels of their combined conversation, substitute $3 \cdot (10^{-10})$ for I in the intensity function.

$$N(I) = 10 \log\left(\frac{I}{10^{-16}}\right)$$

$$N(3 \cdot 10^{-10}) = 10 \log\left(\frac{3 \cdot 10^{-10}}{10^{-16}}\right) = 10 \log(3 \cdot 10^6) \approx 65$$

Thus the intensity of sound is not *multiplicative*. That is, although the power of the conversation of the three people is three times larger than the power of the conversation of a single person, the intensity of sound in decibels of the three people conversing (65 decibels) is not three times larger than the decibel rating of a single normal conversation (60 decibels).

EXERCISE SET 5.5

1. If $8000 is invested at an annual interest rate of 5 percent and compounded annually, find the balance after **a.** 4 years, **b.** 7 years.

2. If $22,000 is invested at an annual interest rate of 4.5 percent and compounded annually, find the balance after **a.** 2 years, **b.** 10 years.

3. If $38,000 is invested at an annual interest rate of 6.5 percent for 4 years, find the balance if the interest is compounded **a.** annually, **b.** daily, **c.** hourly.

4. If $12,500 is invested at an annual interest rate of 8 percent for 10 years, find the balance if the interest is compounded **a.** annually, **b.** daily, **c.** hourly.

5. Find the balance if $15,000 is invested at an annual rate of 10 percent for 5 years, compounded continuously.

6. Find the balance if $32,000 is invested at an annual rate of 8 percent for 3 years, compounded continuously.

7. The number of bacteria $N(t)$ present in a culture at time t hours is given by $N(t) = 2200(2)^t$. Find the number of bacteria present when **a.** $t = 0$ hours, **b.** $t = 3$ hours.

8. The population of a town grows exponentially according to the function

$$f(t) = 12,400(1.14)^t \quad \text{for } 0 \le t \le 5$$

Find the population of the town when t is **a.** 3 years, **b.** 4.25 years.

9. Find the growth function for a town whose population was 22,600 in 1980 and 24,200 in 1985. Use $t = 0$ to represent the year 1980.

10. Find the growth function for a town whose population was 53,700 in 1982 and 58,100 in 1988. Use $t = 0$ to represent the year 1982.

11. The function $P(t) = 9700(e^{0.08t})$ estimates the population of a city at time t years after 1985.

    a. Estimate the population of the city in 1993.

    b. Estimate the year the population will be double its 1985 population.

12. The function $P(t) = 15,600(e^{0.09t})$ estimates the population of a city at time t years after 1984.

    a. Estimate the population of the city in 1994.

    b. Estimate the year the population will be double its 1984 population.

13. Radium (Ra^{226}) has a half-life of 1660 years. Find the decay function for the percent of radium (Ra^{226}) that remains in a sample after t years.

14. Polonium (Po^{210}) has a half-life of 138 days. Find the decay function for the percent of polonium (Po^{210}) that remains in a sample after t days.

15. Use $N(t) = N_0(0.5)^{t/1660}$ to estimate the percentage of radium (Ra^{226}) that remains in a sample after 2250 years.

16. Use $N(t) = N_0(0.5)^{t/138}$ to estimate the percentage of polonium (Po^{210}) that remains in a sample after 2 years.

17. Determine the age of a bone if it now contains 77 percent of its original amount of carbon-14.

18. Determine the age of a bone if it now contains 65 percent of its original amount of carbon-14.

19. Newton's Law of Cooling states that if an object at temperature T_0 is placed into an environment at constant temperature A, then the temperature of the object will be $T(t)$ after t minutes according to the function given by $T(t) = A + (T_0 - A)e^{-kt}$, where k is a constant that depends on the object.

    a. Determine the constant k (to the nearest thousandth) for a canned soda drink that takes 5 minutes to cool from 75°F to 65°F after being placed in a refrigerator that maintains a constant temperature of 34°F.

    b. What will be the temperature (to the nearest degree) of the soda drink after 30 minutes?

    c. When (to the nearest minute) will the temperature of the soda drink be 36°F?

    d. When will the temperature of the soda drink be exactly 34°F?

20. Solve the sound intensity equation $N = 10 \log\left(\dfrac{I}{I_0}\right)$ for I.

21. How much more powerful is a sound that measures 120 decibels than a sound (at the same frequency) that measures 110 decibels?

22. The power of a band is 3.4×10^{-5} watts per square centimeter. What is the band's intensity level N in decibels?

23. If the power of a sound is doubled, what is the increase in the intensity level. (*Hint:* Find $N(2I) - N(I)$.)

24. According to a software company, the users of their typing tutorial can expect to type $N(t)$ words per minute

after t hours of practice with their product according to the function $N(t) = 100(1.04 - 0.99^t)$.

a. How many words per minute can a student expect to type after 2 hours of practice?

b. How many words per minute can a student expect to type after 40 hours of practice?

c. According to the function N, how many hours (to the nearest 1 hour) of practice will be required before a student can expect to type 60 words per minute?

25. In the city of Whispering Palms, the number of people $P(t)$ exposed to a rumor in t hours is given by the function $P(t) = 80,000(1 - e^{-0.0005t})$.

a. Find the number of hours until 10 percent of the population have heard the rumor.

b. Find the number of hours until 50 percent of the population have heard the rumor.

26. A lawyer has determined that the number of people $P(t)$ who have been exposed to a news item after t days is given by the function $P(t) = 1,200,000(1 - e^{-0.03t})$.

a. Find the number of days after a major crime has been reported that 40 percent of the population have heard of the crime.

b. A defense lawyer knows that it will be very difficult to pick an unbiased jury after 80 percent of the population have heard of the crime. How many days are required until 80 percent of the population will have heard of the crime?

27. An automobile depreciates according to the function $V(t) = V_0(1 - r)^t$, where $V(t)$ is the value in dollars after t years, V_0 is the original value, and r is the yearly depreciation rate. If a car has a yearly depreciation rate of 20 percent, determine in how many years the car will depreciate to half its original value.

28. The current $I(t)$ (measured in amperes) of a circuit is given by the function $I(t) = 6(1 - e^{-2.5t})$, where t is the number of seconds after the switch is closed.

a. Find the current when $t = 0$.

b. Find the current when $t = 0.5$.

c. Solve the equation for t.

Supplemental Exercises

29. The Prime Number Theorem states that the number of prime numbers $P(n)$ less than a number n can be approximated by the function

$$P(n) = \frac{n}{\ln n}.$$

a. The actual number of prime numbers less than 100 is 25. Compute $P(100)$ and $P(100)/25$.

b. The actual number of prime numbers less than 10,000 is 1229. Compute $P(10,000)$ and $P(10,000)/1229$.

c. The actual number of prime numbers less than 1,000,000 is 78,498. Compute $P(1,000,000)$, and then compute the ratio $P(1,000,000)/78,498$.

30. The number $n!$ (read "n factorial") is defined as

$$n! = n(n - 1)(n - 2) \cdots 1$$

for all positive integers n. Thus, $4! = 4 \cdot 3 \cdot 2 \cdot 1 = 12$. *Stirling's Formula* (after James Stirling, 1692–1770)

$$n! \approx \left(\frac{n}{e}\right)^n \sqrt{2\pi n}$$

is often used to approximate very large factorials. Use Stirling's Formula to approximate $10!$, and then compute the ratio of Stirling's approximation of $10!$ divided by the actual value of $10!$, which is 3,628,800.

31. A population that grows or decays according to the function $P(t) = P_0 e^{kt}$ is called a **Malthusian model.** This formula models the growth or decay of many populations with unlimited resources. However, if such factors as limited food supply and other limited resources affect the growth of the population, then it may be necessary to model the population growth using the following function, which is known as the **logistic law:**

$$P(t) = \frac{mP_0}{P_0 + (m - P_0)e^{-kt}}$$

where m is the maximum possible population and k is a positive constant. Notice that the graph of P as shown in the figure starts out with an exponential type of growth

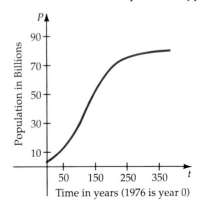

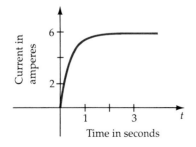

but eventually levels off and approaches the graph of $y = m$ asymptotically. Assume that the world's population growth satisfies the logistics law with $m = 80$ billion. If P was 4 billion in 1976 (think of this as the year $t = 0$) and 5 billion in 1986 ($t = 10$), find the

a. constant k;

b. world's predicted population (according to the logistic law) for the year 2000, and

c. world's predicted population for the year 3000.

32. The world's population reached 3 billion in 1961. How does this compare with the result obtained by using the logistic law and the value of k obtained in Exercise 31?

33. The population of squirrels in a nature reserve satisfies the logistics law, with $P_0 = 1500$, $k = 0.29$, and $P(2) = 2500$.

a. What is the maximum number of squirrels (to the nearest thousand) that the reserve can support? (*Hint:* compute m.)

b. Find the number of squirrels when $t = 10$.

34. The population of walruses in a colony satisfies the logistics law, with $P_0 = 800$, $P(1) = 900$, and $k = 0.14$.

a. What is the maximum number of walruses (to the nearest hundred) that the colony can support? (*Hint:* compute m.)

b. Find the number of walruses when $t = 5$ years.

35. Solve the logistic law in Exercise 31 for t.

36. A farmer knows that planting the same crop in the same field year after year reduces the yield. If the yield on each succeeding years crop is 90 percent of the preceding years yield, then the yield $Y(t)$ at any time t is given by the function $Y(t) = Y_0(0.90)^t$, where Y_0 is the yield when $t = 0$. In how many years (to the nearest year) will the yield be 60 percent of Y_0?

37. Crude oil leaks from a tank at a rate that depends on the amount of oil that remains in the tank. Since 1/8 of the oil in the tank leaks out every 2 hours, the volume of oil $V(t)$ in the tank at time t is given by the function $V(t) = V_0(0.875)^{t/2}$, where $V_0 = 350,000$ gallons is the number of gallons in the tank at the time the tank started to leak ($t = 0$).

a. How many gallons does the tank hold after 3 hours?

b. How many gallons does the tank hold after 5 hours?

c. How long will it take until 90 percent of the oil has leaked from the tank?

38. How many times stronger is an earthquake that measures 6 on the Richter scale than one that measures 3 on the Richter scale?

39. How many times stronger was the Chile earthquake of 1960, which measured 9.5 on the Richter scale, than the San Francisco earthquake of 1906, which measured 8.3 on the Richter scale?

40. How many years will it take the price of goods to double if the annual rate of inflation is 5 percent per year? Use continuous compounding.

41. The current rate of inflation will cause the price of goods to double in the next 10 years. Determine the current rate of inflation. Use continuous compounding.

42. The height h in feet of any point P on the cable shown is a function of the horizontal distance in feet from point P to the origin given by the function

$$h(x) = \frac{20}{2}(e^{x/20} + e^{-x/20}) \quad -40 \le x \le 40.$$

a. What is the height of the cable at point P if P is directly above the origin?

b. What is the height of the cable at point P if P is 25 feet to the right of the origin?

c. How far to the right or left of the origin is the cable 30 feet in height? (*Hint:* Use the method developed in Example 5, Section 4.)

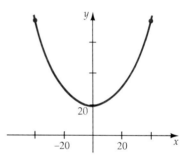

43. Logarithms and a function called the integer function (denoted by INT) can be used to determine the number of digits in a number written in exponential notation. The INT function is illustrated in the following examples:

$$INT(8.75) = 8 \qquad INT(102.003) = 102$$
$$INT(55) = 55 \qquad INT(e) = 2$$

Note that the INT function removes the decimal portion of a real number and returns the integer part of the real number as its output. The number of digits $N(x)$ in the number b^x, with $0 < b < 10$ and x a positive integer, is given by the function $N(x) = INT(x \log b) + 1$.

a. Find the number of digits in 3^{200}.

b. Find the number of digits in 7^{4005}.

c. The largest known prime number in 1980 was the number $2^{44,497} - 1$. Find the number of digits in this prime number.

d. The largest known prime number as of 1983 was $2^{132,049} - 1$. Find the number of digits in this prime number.

Chapter 5 Review

5.1   **Exponential Functions and Their Graphs**

For all positive real numbers b ($b \neq 1$), the exponential function $f(x) = b^x$ has the following properties:

- f has the set of real numbers as its domain.
- f has the set of positive real numbers as its range.
- f has a graph with a y-intercept of $(0, 1)$.
- f has a graph asymptotic to the x-axis.
- f is a one-to-one function.

   The exponential function $f(x) = b^x$ is

- an increasing function if $b > 1$.
- a decreasing function if $0 < b < 1$.

As n increases without bound, $(1 + 1/n)^n$ approaches an irrational number denoted by e. The value of e accurate to eight decimal places is 2.71828183. The function $f(x) = e^x$ is called the natural exponential function.

5.2   **Logarithms and Logarithmic Properties**

Definition of a Logarithm
If $x > 0$ and b is a positive constant ($b \neq 1$), then

$$y = \log_b x \quad \text{if and only if} \quad b^y = x.$$

In the equation $y = \log_b x$, y is referred to as the logarithm, b is the base, and x is the argument.
   The properties of logarithms are used to simplify logarithmic expressions and to solve both exponential and logarithmic equations. Logarithms with a base of 10 are called common logarithms. It is customary to write $\log_{10} x$ as $\log x$. Logarithms with a base of e are called natural logarithms. It is customary to write $\log_e x$ as $\ln x$.

Change-of-Base Formula   If $x > 0$, $a > 0$, $b > 0$, and neither a nor b equals 1, then

$$\log_b x = \frac{\log_a x}{\log_a b}.$$

5.3   **Logarithmic Functions and Their Graphs**

For all positive real numbers b, $b \neq 1$, the function $f(x) = \log_b x$ has the following properties:

- f has the set of positive real numbers as its domain.
- f has the set of real numbers as its range.
- f has a graph with an x-intercept of $(1, 0)$.

- f has a graph asymptotic to the y-axis.
- f is a one-to-one function.

    The logarithmic function $f(x) = \log_b x$ is:

- an increasing function if $b > 1$.
- a decreasing function if $0 < b < 1$.

5.4 Exponential and Logarithmic Equations

To solve exponential equations and logarithmic equations, we use the properties of exponents, the properties of logarithms, the Equality of Exponents Theorem, and the one-to-one property of logarithms.

Equality of Exponents Theorem   If b is a positive real number ($b \neq 1$) such that $b^x = b^y$, then $x = y$.

If M, N, and b are positive real numbers with $b \neq 1$, then $\log_b M = \log_b N$ if and only if $M = N$.

5.5 Applications of Exponential and Logarithmic Functions

The function $N(t) = N_0 e^{kt}$ is called an exponential growth function if k is positive, and it is called an exponential decay function if k is negative.

The Compound Interest Formula
A principal P invested at an annual interest rate r compounded n times per year, for t years produces the balance

$$B = P\left(1 + \frac{r}{n}\right)^{nt}.$$

Continuous Compounding Interest Formula
If an account with principal P and annual interest rate r is compounded continuously for t years, then the balance is $B = Pe^{rt}$.

CHALLENGE EXERCISES

In Exercises 1 to 14, answer true or false. If the statement is false, give an example.

1. If $7^x = 40$, then $\log_7 40 = x$.

2. If $\log_4 x = 3.1$, then $4^{3.1} = x$.

3. If $f(x) = \log x$ and $g(x) = 10^x$, then $f[g(x)] = x$ for all real numbers x.

4. If $f(x) = \log x$ and $g(x) = 10^x$, then $g[f(x)] = x$ for all real numbers x.

5. The exponential function $h(x) = b^x$ is an increasing function.

6. The logarithmic function $j(x) = \log_b x$ is an increasing function.

7. The exponential function $h(x) = b^x$ is a one-to-one function.

8. The logarithmic function $j(x) = \log_b x$ is a one-to-one function.

9. The graph of

$$f(x) = \frac{2^x + 2^{-x}}{2}$$

is symmetric with respect to the y-axis.

10. The graph of

$$f(x) = \frac{2^x - 2^{-x}}{2}$$

is symmetric with respect to the origin.

11. If $x > 0$ and $y > 0$, $\log(x + y) = \log x + \log y$.

12. If $x > 0$, $\log x^2 = 2 \log x$.

13. If M and N are positive real numbers, then

$$\ln\left(\frac{M}{N}\right) = \ln M - \ln N.$$

14. For all $p > 0$, $e^{\ln p} = p$.

REVIEW EXERCISES

In Exercises 1 to 12, solve each equation. Do not use a calculator.

1. $\log_5 25 = x$

2. $\log_3 81 = x$

3. $\ln e^3 = x$

4. $\ln e^\pi = x$

5. $3^{2x+7} = 27$

6. $5^{x-4} = 625$

7. $2^x = \dfrac{1}{8}$

8. $27(3^x) = 3^{-1}$

9. $\log x^2 = 6$

10. $\dfrac{1}{2} \log|x| = 5$

11. $10^{\log 2x} = 14$

12. $e^{\ln x^2} = 64$

In Exercises 13 to 18, use a calculator to evaluate each power. Give your answers accurate to six significant digits.

13. $7^{\sqrt{2}}$

14. $3^{\sqrt{5}}$

15. $e^{1.7}$

16. $e^{-2.2}$

17. $10^{1.135}$

18. $10^{-\sqrt{10}}$

In Exercises 19 to 32, sketch the graph of each function.

19. $f(x) = (2.5)^x$

20. $f(x) = \left(\dfrac{1}{4}\right)^x$

21. $f(x) = 3^{|x|}$

22. $f(x) = 4^{-|x|}$

23. $f(x) = 2^x - 3$

24. $f(x) = 2^{(x-3)}$

25. $f(x) = \dfrac{4^x + 4^{-x}}{2}$

26. $f(x) = \dfrac{3^x - 3^{-x}}{2}$

27. $f(x) = \dfrac{1}{3} \log x$

28. $f(x) = 3 \log x^{1/3}$

29. $f(x) = -x + \log x$

30. $f(x) = 2^{-x} \log x$

31. $f(x) = -\dfrac{1}{2} \ln x$

32. $f(x) = -\ln|x|$

In Exercises 33 to 36, change each logarithmic equation to its exponential form.

33. $\log_4 64 = 3$

34. $\log_{1/2} 8 = -3$

35. $\log_{\sqrt{2}} 4 = 4$

36. $\ln 1 = 0$

In Exercises 37 to 40, change each exponential equation to its logarithmic form.

37. $5^3 = 125$

38. $2^{10} = 1024$

39. $10^0 = 1$

40. $8^{1/2} = 2\sqrt{2}$

In Exercises 41 to 44, write the given logarithm in terms of logarithms of x, y, and z.

41. $\log_b \dfrac{x^2 y^3}{z}$

42. $\log_b \dfrac{\sqrt{x}}{y^2 z}$

43. $\ln xy^3$

44. $\ln \dfrac{\sqrt{xy}}{z^4}$

In Exercises 45 to 48, write each logarithmic expression as a single logarithm.

45. $2 \log x + \dfrac{1}{3} \log(x + 1)$

46. $5 \log x - 2 \log(x + 5)$

47. $\dfrac{1}{2} \ln 2xy - 3 \ln z$

48. $\ln x - (\ln y - \ln z)$

In Exercises 49 to 52, use the change-of-base formula and a calculator to approximate each logarithm accurate to six significant digits.

49. $\log_5 101$

50. $\log_3 40$

51. $\log_4 0.85$

52. $\log_8 0.3$

In Exercises 53 to 56, use a calculator to approximate N to three significant digits.

53. $\log N = 247$

54. $\log N = -0.48$

55. $\ln N = 51$

56. $\ln N = -0.09$

In Exercises 57 to 72, solve each equation for x. Give exact answers. Do not use a calculator.

57. $4^x = 30$

58. $5^{x+1} = 41$

59. $\ln 3x - \ln(x - 1) = \ln 4$

60. $\ln 3x + \ln 2 = 1$

61. $e^{\ln(x+2)} = 6$

62. $10^{\log(2x+1)} = 31$

63. $\dfrac{4^x + 4^{-x}}{4^x - 4^{-x}} = 2$

64. $\dfrac{5^x + 5^{-x}}{2} = 8$

65. $\log(\log x) = 3$

66. $\ln(\ln x) = 2$

67. $\log\sqrt{x - 5} = 3$

68. $\log x + \log(x - 15) = 1$

69. $\log_4(\log_3 x) = 1$

70. $\log_7(\log_5 x^2) = 0$

71. $\log_5 x^3 = \log_5 16x$

72. $25 = 16^{\log_4 x}$

73. Find the pH of tomatoes that have a hydronium-ion concentration of 6.28×10^{-5}.

74. Find the hydronium-ion concentration of rainwater that has a pH of 5.4.

75. Find the balance if $16,000 is invested at an annual rate of 8 percent for 3 years if the interest is compounded **a.** monthly, **b.** continuously.

76. Find the balance if $19,000 is invested at an annual rate of 6 percent for 5 years if the interest is compounded **a.** daily, **b.** continuously.

77. The scrap value S of a product with an expected life span of n years is given by $S(n) = P(1 - r)^n$, where P is the original purchase price of the product and r is the annual rate of depreciation. A taxicab is purchased for $12,400 and is expected to last 3 years. What is its scrap value if it depreciates at a rate of 29 percent per year?

78. A skin wound heals according to the function given by $N(t) = N_0 e^{-0.12t}$, where N is the number of square centimeters of unhealed skin t days after the injury, and N_0 is the number of square centimeters covered by the original wound.

　　a. What percentage of the wound will be healed after 10 days?

　　b. How many days will it take for 50 percent of the wound to heal?

　　c. How long will it take for 90 percent of the wound to heal?

In Exercises 79 to 82, find the exponential growth/decay function $N(t) = N_0 e^{kt}$ that satisfies the given conditions.

79. $N(0) = 1$,　$N(2) = 5$

80. $N(0) = 2$,　$N(3) = 11$

81. $N(1) = 4$,　$N(5) = 5$

82. $N(-1) = 2$,　$N(0) = 1$

6

Trigonometric Functions

A mathematician, like a painter or a poet, is a maker of patterns. If his patterns are more permanent than theirs, it is because they are made of ideas.

G. H. HARDY (1877–1947)

TSUNAMIS AND EARTHQUAKES

Imagine an undersea earthquake. The energy from the earthquake would be translated to the water as water waves. These water waves are called *tsunamis* or *tidal waves*. Although the phrase tidal waves is still used to describe these waves, tsunami is the preferred description because the waves have nothing to do with the tides.

In the open ocean, the distance between crests of a tsunami may be as great as 60 miles and the height of the wave no more than 2 feet. As the depth of the ocean decreases however, the water wave slows down. As it slows, the height of the wave increases. When a tsunami reaches the shore, the wave height can be quite high with crests 100 feet above the normal tide level.

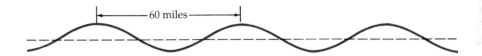

The earthquake that generated the tsunamis also creates waves within the earth. Two of the wave types that are created are the *primary* or *P wave* and the *secondary* or *S wave*. These two waves are quite different. The P wave is very much like a sound wave. It alternately compresses and dilates the substances within the earth. These waves can travel through solid rock and water.

S waves are slower than P waves and are more like water waves. As an S wave travels through the earth, it shears the rock sideways at right angles to the direction of travel. The S wave causes much of the structural damage from an earthquake. Wave phenomena exhibited by tsunami and earthquake waves can be described by trigonometric functions, the subject of this chapter.

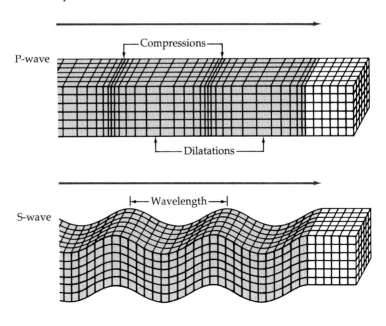

6.1

Measuring Angles

Figure 6.1

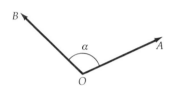

Figure 6.2

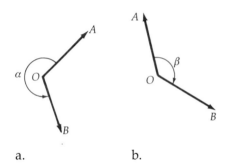

a.   b.

Figure 6.3
a. Positive angle;
b. Negative angle.

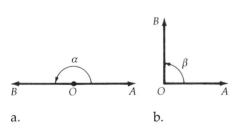

a.   b.

Figure 6.4
a. Straight angle (180°);
b. Right angle (90°).

Early Babylonians noticed that the seasons repeated about every 360 days. Thinking that the earth was at the center of the universe, they assumed the universe made one complete revolution in 360 days. Our concept of measuring angles in degrees is an outgrowth of those early beliefs.

A ray is a part of a line originating at a point and extending infinitely, as Figure 6.1 shows. An **angle** is formed by rotating a ray about its endpoint. The initial position of the ray is called the **initial side** of the angle. The position of the ray after it has been rotated is called the **terminal side** of the angle. The endpoint of the ray is called the vertex of the angle. Figure 6.2 shows an angle with vertex O.

In Figure 6.3 the ray OA is the **initial side** of the angle and OB is the **terminal side** of the angle. Angles formed by a counterclockwise rotation are considered **positive** angles; angles formed by a clockwise rotation are considered **negative** angles.

The measure of an angle is the amount of rotation of the ray. One unit of angle measurement is the degree.

Definition of Degree

> An angle formed by rotating a ray $\frac{1}{360}$ of a complete revolution has a measure of one **degree**. The symbol for degree is °.

Using this definition, an angle formed by rotating a ray one complete revolution has a measure of 360°. A **straight angle** is formed by rotating a ray one-half a complete revolution. The measure of a straight angle is 180°. A **right angle** is formed by rotating a ray one-quarter of a complete revolution. The measure of a right angle is 90°. See Figure 6.4.

An angle is **acute** if its measure is between 0° and 90°. An angle is **obtuse** if its measure is between 90° and 180°. See Figure 6.5.

Two angles are **complementary angles** if the sum of the measures of the angles is 90°. In this case, one angle is the *complement* of the other angle. Two angles are **supplementary angles** if the sum of the measures of the two angles is 180°. One angle is the *supplement* of the other angle. See Figure 6.6.

We also can measure angles greater than 360° (1 revolution). An angle of 720° is 2 revolutions counterclockwise. An angle of 450° (360° + 90°) is $1\frac{1}{4}$ revolutions counterclockwise. The angle −990° [−720° + (−270°)] is $2\frac{3}{4}$ revolutions clockwise.

Degrees may be subdivided into smaller units by using decimal degrees or by using the degree, minute, second system (DMS). In the decimal system,

8.53° means 8° plus 53 hundredths of 1°.

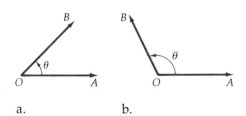

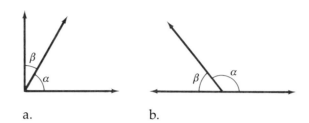

Figure 6.5
a. Acute angle: $0° < \theta < 90°$;
b. Obtuse angle: $90° < \theta < 180°$.

Figure 6.6
a. Complementary angles: $\alpha + \beta = 90°$;
b. Supplementary angles: $\alpha + \beta = 180°$.

2 Revolutions   1¼ Revolutions   2¾ Revolutions

Figure 6.7

In the DMS system, the degree is subdivided into smaller units of minutes, which are further divided into seconds.

$$\text{One minute } (1') = \left(\frac{1}{60}\right)° \quad \text{One-sixtieth of a degree}$$

$$\text{One second } (1'') = \left(\frac{1}{60}\right)' \quad \text{One-sixtieth of a minute}$$

Thus we can write $60' = 1°$ and $60'' = 1'$.

Therefore we know that $1° = 3600''$. Why?[1]

EXAMPLE 1 Convert from DMS to Decimal Degrees

Convert $20°14'$ to decimal degrees. Round to the nearest thousandth of a degree.

Solution We use the conversion factor $1° = 60'$ to convert minutes to decimal degrees.

$$20°14' = 20° + 14' = 20° + 14'\left(\frac{1°}{60'}\right) \quad 1° = 60', \text{ therefore } 1 = \frac{1°}{60'}.$$

$$\approx 20° + 0.233° = 20.233°$$

■ *Try Exercise **22**, page 312.*

---

[1] $1° = 60' = 60' \cdot \dfrac{60''}{1'} = 3600''$.

EXAMPLE 2 **Convert from Decimal Degrees to DMS**

Convert 42.82° to the DMS system of measurement.

Solution

$$42.82° = 42° + 0.82°$$

$$= 42° + 0.82°\left(\frac{60'}{1°}\right) \qquad 1 = \frac{60'}{1°}.$$

$$= 42° + 49.2'$$

$$= 42° + 49' + 0.2'$$

$$= 42° + 49' + 0.2'\left(\frac{60''}{1'}\right) \qquad 1 = \frac{60''}{1'}.$$

$$= 42° + 49' + 12''$$

$$= 42°49'12''$$

■ *Try Exercise* **30**, *page 312.*

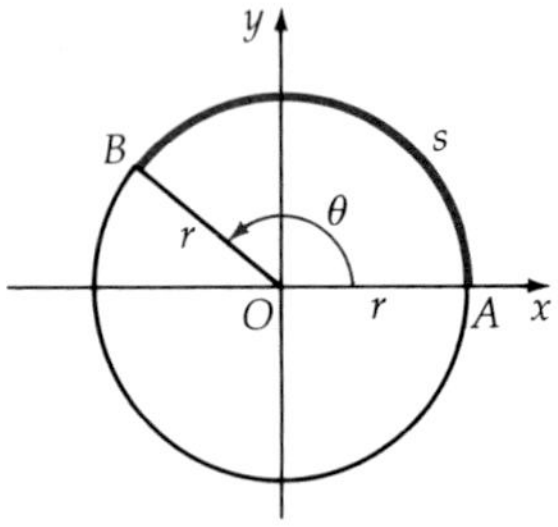

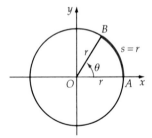

Figure 6.8

Another common angle measurement is the *radian*. To define a radian, first we consider a circle of radius r and two radii OA and OB. The angle θ formed by the two radii is a **central angle**. The portion of the circle between A and B is an **arc** of the circle and is written $\overset{\frown}{AB}$. We say that $\overset{\frown}{AB}$ *subtends* the angle θ. The length of $\overset{\frown}{AB}$ is s (see Figure 6.8).

Definition of Radian

> One **radian** is the measure of the central angle subtended by an arc of length r on a circle of radius r.

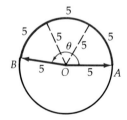

Figure 6.9

For example, an arc length of 15 centimeters on a circle with a radius of 5 centimeters will subtend an angle of 3 radians, as shown in Figure 6.10. We can obtain the same result by dividing 15 centimeters by 5 centimeters. To find the number of radians in *any central angle* θ, divide the length s of the arc that subtends θ by the radius of the circle.

Radian Measure

> Given an arc of length s on a circle of radius r, the radian measure of the angle subtended by the arc is $\theta = s/r$.

Figure 6.11 shows the measure of the central angles for different arc lengths. For $\frac{1}{2}$ revolution we see that the central angle measure is π radians, and for a complete revolution the central angle measure is 2π radians.

Figure 6.10

a. 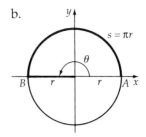   b. 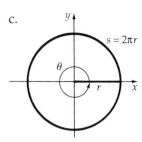   c.

Figure 6.11
a. $\theta = s/r = 2r/r = 2$ radians;
b. $\theta = s/r = \pi r/r = \pi$ radians;
c. $\theta = s/r = 2\pi r/r = 2\pi$ radians.

EXAMPLE 3   Find the Measure of a Central Angle

An arc of length 10 centimeters on a circle with a radius of 4 centimeters subtends an angle θ. Find the measure of θ in radians.

Solution   Use the formula $\theta = \dfrac{s}{r}$ to find the central angle.

$$\theta = \frac{s}{r} = \frac{10\,\text{cm}}{4\,\text{cm}} = 2.5\,\text{rad}$$

■ *Try Exercise* **74,** *page 312.*

Remark   From Example 3, note that the magnitude of a radian is the ratio of two length quantities (centimeters, in this case). Thus a radian is a dimensionless quantity, a fact that is used in application problems.

EXAMPLE 4   Convert Revolutions to Radian Measure

Find the measure in radians of an angle formed by rotating the initial side $1\dfrac{1}{3}$ revolutions in a counterclockwise direction.

Solution   From Figure 6.11, one complete counterclockwise revolution corresponds to 2π radians. Thus to find the number of radians in $1\dfrac{1}{3}$ counterclockwise revolutions, multiply $1\dfrac{1}{3}$ by 2π.

$$\theta = 1\frac{1}{3} \cdot 2\pi = \frac{8}{3}\pi$$

■ *Try Exercise* **82,** *page 312.*

Referring again to Figure 6.11, one complete revolution has a degree measure of $360°$, or a radian measure of 2π. Thus $360° = 2\pi$ radians, and

$180° = \pi$ radians. From the last equation, we can derive the following conversion factors to convert degrees to radians or radians to degrees:

Radian/Degree Conversion Factors

$$1 \text{ radian} = \frac{180°}{\pi} \qquad 1° = \frac{\pi}{180} \text{ radians}$$

Remark   Using a calculator, we have

$$1 \text{ radian} \approx 57.29577951° \quad \text{and} \quad 1° \approx 0.017453292 \text{ radian.}$$

EXAMPLE 5   **Convert Degree Measure to Radian Measure**

Convert 240° to radians.

Solution   We use the conversion factor π rad/180° to convert degrees to radians. The exact answer can be given in terms of π. A decimal approximation is obtained when an approximate value of π is used.

$$240° = 240°\left(\frac{\pi}{180°} \text{ radians}\right)$$

$$= \frac{4}{3}\pi \text{ radians} \qquad \text{Exact answer}$$

$$\approx 4.19 \text{ radians} \qquad \text{Approximate answer}$$

■ *Try Exercise* **42**, *page 312.*

Remark   In most cases the radian measure of angles will be stated in terms of π and the decimal approximation will not be used.

EXAMPLE 6   **Convert Radians to Degrees**

Convert $\dfrac{11\pi}{12}$ radians to degrees.

Solution   We use the conversion factor $\dfrac{180°}{\pi}$ radians to convert radians to degrees.

$$\frac{11\pi}{12} \text{ radians} = \frac{11\pi}{12} \cdot \frac{180°}{\pi} = 165°$$

■ *Try Exercise* **52**, *page 312.*

The following table lists the degree and radian measure of selected angles. Figure 6.12 illustrates the relationships.

Degrees	Radians
0	0
30	$\pi/6$
45	$\pi/4$
60	$\pi/3$
90	$\pi/2$
120	$2\pi/3$
135	$3\pi/4$
150	$5\pi/6$
180	π
210	$7\pi/6$
225	$5\pi/4$
240	$4\pi/3$
270	$3\pi/2$
300	$5\pi/3$
315	$7\pi/4$
330	$11\pi/6$
360	2π

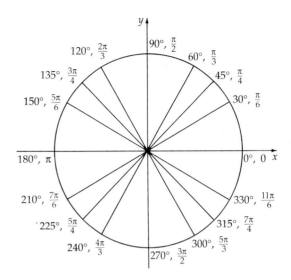

Figure 6.12
Degree and radian measures of selected angles.

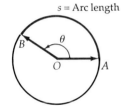

Figure 6.13

Consider a circle of radius r. The use of radians is helpful for finding the measure of the length of an arc s. By solving the formula $\theta = s/r$ for s, we have an equation that gives the length of the arc of a circle.

Arc Length of a Circle

Let r be the radius of a circle C and θ the radian measure of a central angle of C. Then the length of the arc s that subtends θ is $s = r\theta$.

Remark When the circle C has a radius $r = 1$, $s = \theta$. This circle, called the unit circle, will be particularly useful in later sections.

EXAMPLE 7 Find Arc Length of a Circle

Find the length of an arc that subtends a central angle of $120°$ in a circle of radius 10 centimeters.

Solution The formula for the length of a circular arc requires that the angle be measured in radians. Convert $120°$ to radians and use the formula $s = r\theta$ to find the length of the arc.

$$\theta = 120° = 120 \cdot \frac{\pi}{180°} \text{ radians} = \frac{2\pi}{3} \text{ radian}$$

$$s = 10 \text{ cm}\left(\frac{2\pi}{3}\right) = \frac{20\pi}{3} \text{ cm}$$

■ *Try Exercise **80**, page 312.*

EXAMPLE 8   **Solve an Application Involving Radians**

A pulley with a radius of 10 inches uses a belt to drive a pulley with a radius of 6 inches. The 10-inch pulley turns through an angle of 2π radians. Find the angle through which the 6-inch pulley turns (see Figure 6.14).

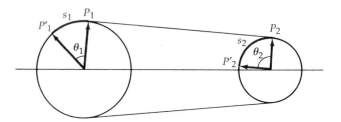

Figure 6.14

Solution

A point P_1 on the 10-inch pulley moves: $s_1 = r_1\theta_1$

A point P_2 on the 6-inch pulley moves: $s_2 = r_2\theta_2$

Since one point on the belt moves the same distance as any other point on the belt, the point P_1 moves through the same distance as P_2. Thus $s_1 = s_2$ and $r_1\theta_1 = r_2\theta_2$. Solve for θ_2. Substitute the given values for r_1, r_2, and θ_1 and simplify.

$$\theta_2 = \left(\frac{r_1}{r_2}\right)\theta_1 = \frac{10}{6}(2\pi) = \frac{10}{3}\pi \quad r_1 = 10,\ r_2 = 6,\ \theta_1 = 2\pi$$

■ *Try Exercise **84**, page 312.*

A car traveling at a speed of 55 miles per hour travels 55 miles in 1 hour, 110 miles in 2 hours. **Linear velocity** is defined as the distance traveled per unit time, or, in equation form, $v = d/t$, where v is the velocity, d the distance, and t the time. The units may be expressed in feet per second, miles per hour, meters per second, or in any unit of length per unit of time.

Remark   Velocity actually has two characteristics: speed and direction. The distinction between speed and velocity is discussed when we discuss vectors.

The floppy disk in a computer revolving at 300 rpm makes 300 revolutions in 1 minute. A radius OP of the disk would move through or generate an angle θ in time t. **Angular velocity** is defined as the angle generated per unit of time, or $\omega = \theta/t$, where ω is the angular velocity in units of revolutions per second, revolutions per minute, radians per second, or in any unit of angle measure per unit of time. The angle generated is θ and t is the time.

EXAMPLE 9   Solve an Application Involving Angular Velocity

A hard disk in a computer rotates at 3600 revolutions per minute (rpm). Convert the angular velocity to radians per second. Round to the nearest tenth.

Solution   Since we must find the angular velocity in radians per second, first we find the total number of radians.

$$3600(2\pi) = 7200\pi$$

Use the formula $\omega = \theta/t$ to find the angular velocity in radians per second.

$$\omega = \frac{\theta}{t} = \frac{7200\pi}{60} = 377.0 \text{ rad/s}$$

■ *Try Exercise 86, page 312.*

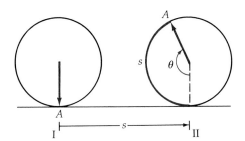

Figure 6.15

A rotating circle may have both linear and angular velocity related by a simple equation. Assume that a wheel is rolling without slipping. Point A on the wheel in position I is in contact with the ground. As the wheel moves a distance s, the point A moves through an angle θ. The arc length subtending angle θ in position II is equal to s. (See Figure 6.15.)

The linear velocity of the wheel is given by $v = s/t$
Substitute $s = r\theta$ for s $= r\theta/t$
Substitute ω for θ/t $= r\omega$

This expression $v = r\omega$ gives the velocity of a point on a rotating body at a distance r from the axis of rotation.

EXAMPLE 10   Solve an Application Involving Linear Velocity

The tires of an automobile are rotating at a rate of 600 revolutions per minute. The radius of the tire is 14 inches. Find the speed of the automobile in miles per hour. Round to the nearest mile per hour.

Solution   First convert 600 revolutions per minute to 36,000 revolutions per hour by multiplying by 60. Next multiply by 2π to get $\omega = 72000\pi$. Now use $v = r\omega$ to find the velocity.

$$v = \frac{14}{63,360} \cdot 72000\pi \approx 50 \text{ mph}   \text{There are 63,368 inches per mile.}$$

■ *Try Exercise 88, page 312.*

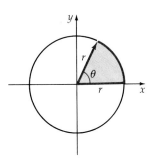

Figure 6.16

A **sector** of a circle is the figure bounded by two radii and the intercepted arc as Figure 6.16 shows. It can be shown that the ratio of the area of the sector (A) to the area of the circle (πr^2) is equal to the ratio of the *central angle* (θ) of the sector to one complete revolution of the circle (2π).

$$\frac{A}{\pi r^2} = \frac{\theta}{2\pi}$$

Solving for A, we have $A = \frac{1}{2}r^2\theta .$

EXERCISE SET 6.1

In Exercises 1 to 9, find the complement of the given angle.

1. 15°   **2.** 72°   **3.** 86.3°

4. 66.14°   **5.** 77.55°   **6.** 7°4′

7. 46°9′   **8.** 54°36′14″   **9.** 10°55′35″

In Exercises 10 to 18, find the supplement of the given angle.

10. 37°   **11.** 67°   **12.** 101.5°

13. 172.34°   **14.** 34.56°   **15.** 123.8°

16. 102.4°   **17.** 45.78°   **18.** 147.66°

In Exercises 19 to 27, convert the DMS measure of each angle to decimal degree measure to the nearest thousandth of a degree.

19. 78°8′   **20.** 5°39′   **21.** 16°44″

22. 35°42″   **23.** 47°20′   **24.** 20°4′45″

25. 165°36′54″   **26.** 68°16′50″   **27.** 95°28′8″

In Exercises 28 to 36, convert the decimal degree measure of each angle to the DMS system to the nearest second.

28. 110.4°   **29.** 36.6°   **30.** 55.44°

31. 66.72°   **32.** 7.05°   **33.** 6.19°

34. 12.28°   **35.** 132.58°   **36.** 102.76°

In Exercises 37 to 42, convert the measure of the angle to exact radian measure.

37. 15°   **38.** 165°   **39.** 315°

40. 420°   **41.** 630°   **42.** 585°

In Exercises 43 to 48, convert the degree measure of the angle to radian measure to the nearest hundredth of a radian.

43. 29°   **44.** 148°   **45.** 166°

46. 434°   **47.** 610°   **48.** 295°

In Exercises 49 to 54, convert the radian measure of the angles to exact degree measure.

49. $\pi/6$ **50.** $\pi/9$ **51.** $3\pi/8$

52. $11\pi/18$ **53.** $11\pi/3$ **54.** $6\pi/5$

In Exercises 55 to 60, convert the radian measure of the angles to degree measure to the nearest hundredth of a degree.

55. 1.2   **56.** 0.35   **57.** 0.64

58. 5.66   **59.** 4.38   **60.** 8

In Exercises 61 to 66, find the complement of each angle measured in radians. Express your answer in terms of π.

61. $\pi/6$ **62.** $\pi/4$ **63.** $5\pi/12$

64. 0.75   **65.** 1.22   **66.** 0.30

In Exercises 67 to 72, find the supplement of each angle measured in radians. Express your answer in terms of π.

67. $3\pi/4$ **68.** $7\pi/12$ **69.** $7\pi/8$

70. 2.5   **71.** 1.76   **72.** 0.55

In Exercises 73 to 76, find the measure in radians and in degrees of the central angle of a circle with the given radius and arc length.

73. $r = 2$ in, $s = 8$ in   **74.** $r = 7$ ft, $s = 4$ ft

75. $r = 5.2$ cm, $s = 12.4$ cm   **76.** $r = 35.8$ m, $s = 84.3$ m

In Exercises 77 to 80, find the measure of the intercepted arc of a circle with the given radius and central angle.

77. $r = 8$ in., $\phi = \pi/4$ radians

78. $r = 3$ ft, $\theta = 7\pi/2$ radians

79. $r = 25$ cm, $\phi = 42°$ **80.** $r = 5$ m, $\theta = 144°$

81. Find the equivalent number of radians in $1\frac{1}{2}$ revolutions.

82. Find the equivalent number of radians in $\frac{3}{8}$ revolution.

83. A pulley with a radius of 14 inches uses a belt to drive a pulley with a radius of 28 inches. The 14-inch pulley turns through an angle of 150°. Find the angle through which the 28-inch pulley turns.

84. A pulley with a diameter of 1.2 meters uses a belt to drive a pulley with a diameter of 0.8 meter. The 1.2-meter pulley turns through an angle of 240°. Find the angle through which the 0.8-meter pulley turns.

85. Find the angular velocity of the second hand of a watch.

86. Find the angular velocity of a point on the equator in radians per second. The radius of the earth is 3960 miles.

87. A wheel is rotating at 50 revolutions per second. Find the angular velocity in radians per second.

88. A wheel is rotating at 200 revolutions per minute. Find the angular velocity in radians per second.

89. A car with a 14-inch wheel is moving with a velocity of 55 mph. Find the angular velocity of the wheel in radians per second.

90. The 15-inch tires of an automobile are rotating at a rate of 450 revolutions per minute. Find the velocity of the automobile in miles per hour.

91. The 18-inch tires of a truck are rotating at a rate of 500 revolutions per minute. Find the velocity of the truck in miles per hour.

Supplemental Exercises

In Exercises 92 to 97 find the area of the sector of a circle with the given radius and central angle.

92. $r = 5$ in, $\theta = \pi/3$ rad   **93.** $r = 2.8$ ft, $\theta = 5\pi/2$ rad

94. $r = 120$ cm, $\theta = 0.65$ rad   **95.** $r = 30$ ft, $\theta = 62°$

96. $r = 20$ m, $\theta = 125°$ **97.** $r = 25$ cm, $\theta = 220°$

98. The minute hand on the clock atop city hall measures 6 ft 3 inches from the tip to its axle.

   a. Through what angle does the minute hand pass between 9:12 A.M. and 9:48 A.M.?

   b. What distance does the tip of the minute hand travel during this period?

99. At a time when the earth was 93,000,000 miles from the sun, using a transit you observed through a properly smoked glass that the diameter of the sun occupied an arc of 31′. Calculate the approximate diameter of the sun to 2 significant digits.

100. A merry-go-round horse is 11.6 meters from the center. The merry-go-round makes $14\frac{1}{4}$ revolutions per ride. How many meters does the horse travel? How fast is it moving in meters per second?

101. a. A car with 13-inch tires makes an 8-mile trip. Find the number of revolutions the tire makes on the 8-mile trip.

   b. A car with 15-inch tires makes an 8-mile trip. Find the number of revolutions the tire makes on the 8-mile trip.

102. A water wheel has a 10-foot radius. When the wheel makes 18 revolutions per minute, what is the speed of the river?

103. A pulley with a 50-centimeter diameter drives a pulley with a 20-centimeter diameter. The larger pulley makes 30 revolutions per minute. What is the linear speed of a point on the surface of the smaller pulley?

104. Find the area of the shaded portion of the graph shown. The radius of the circle is 9 inches.

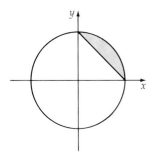

105. Latitude describes the position of a point on the earth's surface in relation to the equator. A point on the equator has a latitude of 0°. The north pole has a latitude of 90°. The radius of the earth is approximately 3960 miles. Assuming that the earth is a perfect sphere, find the distance along the earth's surface that subtends a central angle of latitude (a) 1°, (b) 1′, and (c) 1″. Express your answer to 3 significant digits.

6.2

Trigonometric Functions of Acute Angles

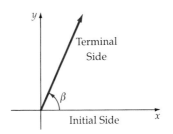

Figure 6.17

The study of trigonometry, which means "triangle measurement," began more than 2000 years ago, partially as a means to solving surveying problems. Early trigonometry used the length of a chord of a circle as the value of a *trigonometric function*. In the sixteenth century, right triangles were used to define a trigonometric function. We will use a modifed right triangle approach to define the trigonometric functions by placing one of the acute angles of a right triangle on a coordinate grid.

An angle is in **standard position** when the initial side coincides with the x-axis and the vertex is at the origin of the coordinate axes. The angle β in Figure 6.17 is in standard position.

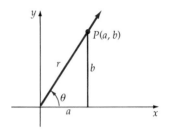

Figure 6.18

Now consider angle θ in Figure 6.18 in standard position and a point $P(a, b)$ on the terminal side of the angle. A line perpendicular to the x-axis has been drawn. The values a, b and r represent the measure of the legs and hypotenuse of the right triangle formed. Six possible ratios can be formed:

$$\frac{a}{r}, \frac{b}{r}, \frac{r}{a}, \frac{r}{b}, \frac{a}{b}, \frac{b}{a}.$$

Each ratio defines a value of a trigonometric function; the functions are the sine (sin), cosine (cos), tangent (tan), cosecant (csc), secant (sec), and cotangent (cot).

Trigonometric Functions of an Acute Angle

Let θ be an acute angle in standard position and $P(a, b)$ a point on the terminal side of the angle. The six trigonometric functions of θ are

$$\sin \theta = \frac{b}{r} \qquad \cos \theta = \frac{a}{r} \qquad \tan \theta = \frac{b}{a}$$

$$\csc \theta = \frac{r}{b} \qquad \sec \theta = \frac{r}{a} \qquad \cot \theta = \frac{a}{b}$$

where $r = \sqrt{a^2 + b^2}$.

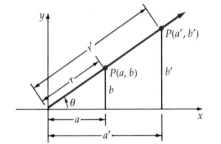

Figure 6.19

Consider any two points on the terminal side of the angle in standard position and drawing the perpendicular lines from the two points to the x-axis. The first triangle formed has legs of lengths a and b, and the second triangle formed has legs of lengths a' and b'. The right triangles formed are similar triangles, and the corresponding sides are proportional. Thus,

$$\frac{b}{r} = \frac{b'}{r'} \quad \text{and} \quad \sin \theta = \frac{b}{r} = \frac{b'}{r'}.$$

Thus the value of the sine function is independent of the length of the sides of a triangle. The values of the remaining trigonometric functions can also be shown to be independent of the lengths of the sides.

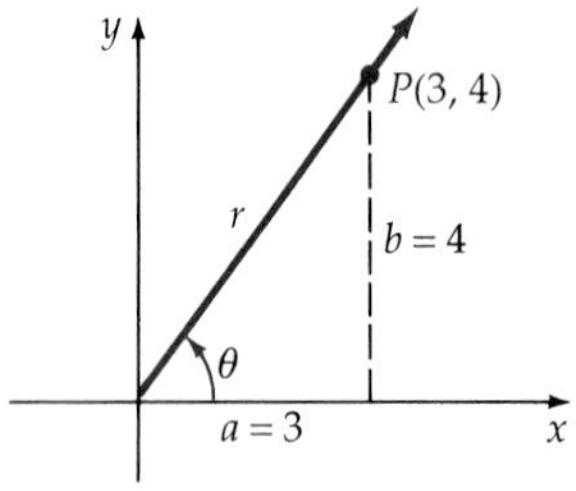

EXAMPLE 1 Evaluate Trigonometric Functions

Find the values of the six trigonometric functions of the acute angle in standard position with the terminal side passing through the point $P(3, 4)$.

Solution Sketch the angle θ on the coordinate axes. (See Figure 6.20.) Find the length of r by using the Pythagorean Theorem.

$$r = \sqrt{3^2 + 4^2} = 5$$

Figure 6.20

Use the definitions of the trigonometric functions to evaluate the trigonometric functions.

$$\sin \theta = \frac{b}{r} = \frac{4}{5} \qquad \csc \theta = \frac{r}{b} = \frac{5}{4} \qquad a = 3, b = 4, r = 5$$

$$\cos \theta = \frac{a}{r} = \frac{3}{5} \qquad \sec \theta = \frac{r}{a} = \frac{5}{3}$$

$$\tan \theta = \frac{b}{a} = \frac{4}{3} \qquad \cot \theta = \frac{a}{b} = \frac{3}{4}$$

■ *Try Exercise* **8,** *page 319.*

EXAMPLE 2 **Evaluate Trigonometric Functions**

If ϕ is an acute angle in standard position and $\sin \phi = \frac{5}{13}$, find the values of the other five trigonometric functions of ϕ.

Solution   Sketch the angle in standard position (see Figure 6.21). Because $\sin \phi = \frac{b}{r} = \frac{5}{13}$, we use $r = 13$, $b = 5$.

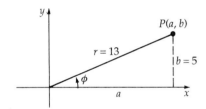

$$r^2 = a^2 + b^2 \quad \text{Use the Pythagorean Theorem to find } a.$$

$$13^2 = a^2 + 5^2$$

$$a = 12$$

$$\cos \phi = \frac{a}{r} = \frac{12}{13} \qquad \tan \phi = \frac{b}{a} = \frac{5}{12}$$

$$\csc \phi = \frac{r}{b} = \frac{13}{5} \qquad \sec \phi = \frac{r}{a} = \frac{13}{12} \qquad \cot \phi = \frac{a}{b} = \frac{12}{5}$$

■ *Try Exercise* **18,** *page 319.*

Figure 6.21

For most angles, advanced mathematical methods are required to find the value of a trigonometric function. However, the value of a trigonometric function for some *special angles* can be found by geometric methods. These special angles are 0° (0), 30° ($\pi/6$), 45° ($\pi/4$), 60° ($\pi/3$), and 90° ($\pi/2$).

First we will find the six trigonometric functions of 45°. (The discussion is based on angles measured in degrees. However, we could have used radian measure without changing the results.) Figure 6.22 shows a 45° angle in standard position with a perpendicular line drawn from $P(a, b)$ to the x-axis. The measure of the legs of the triangle are equal. Let the length of each leg be equal to a. By the Pythagorean Theorem,

$$r^2 = a^2 + a^2 = 2a^2$$

$$r = a\sqrt{2}.$$

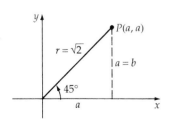

Figure 6.22

The values of the six trigonometric functions of 45° are

$$\sin 45° = \frac{a}{a\sqrt{2}} = \frac{1}{\sqrt{2}} = \frac{\sqrt{2}}{2} \qquad \csc 45° = \frac{a\sqrt{2}}{a} = \sqrt{2}$$

$$\cos 45° = \frac{a}{a\sqrt{2}} = \frac{1}{\sqrt{2}} = \frac{\sqrt{2}}{2} \qquad \sec 45° = \frac{a\sqrt{2}}{a} = \sqrt{2}$$

$$\tan 45° = \frac{a}{a} = 1 \qquad\qquad \cot 45° = \frac{a}{a} = 1$$

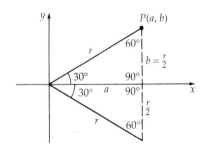

Figure 6.23

The trigonometric values of the special angles 30° and 60° can be found by drawing an equilateral triangle and bisecting one of the angles, as Figure 6.23 shows. The angle bisector also bisects one of the sides. Thus $\alpha = 30°$ and the measure of the side opposite the 30° angle is one-half the hypotenuse.

Let r be the measure of the hypotenuse. Then the measure of the side opposite 30° is $r/2$. The length of the side adjacent to the 30° angle (a) is found by using the Pythagorean Theorem.

$$r^2 = \left(\frac{r}{2}\right)^2 + a^2$$

$$\frac{3r^2}{4} = a^2$$

$$\frac{\sqrt{3}r}{2} = a$$

Note that in a 30°–60° right triangle, the side opposite the 30° angle is one-half the hypotenuse and the side opposite the 60° angle is $\sqrt{3}/2$ times the hypotenuse. The point $P(\sqrt{3}r/2, r/2)$ is on the terminal side of a 30° angle. (See Figure 6.24.) The values of the six trigonometric functions of 30° are

$$\sin 30° = \frac{r/2}{r} = \frac{1}{2} \qquad\qquad \cos 30° = \frac{\sqrt{3}r/2}{r} = \frac{\sqrt{3}}{2}$$

$$\tan 30° = \frac{r/2}{\sqrt{3}r/2} = \frac{\sqrt{3}}{3} \qquad \csc 30° = \frac{r}{r/2} = 2$$

$$\sec 30° = \frac{r}{\sqrt{3}r/2} = \frac{2\sqrt{3}}{3} \qquad \cot 30° = \frac{\sqrt{3}r/2}{r/2} = \sqrt{3}$$

The trigonometric functions of 60° can be found by placing a 60° angle in standard position and using the information in Figure 6.25. The mea-

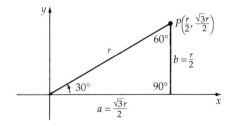

Figure 6.24

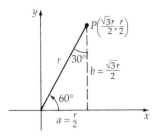

Figure 6.25

sure of the hypotenuse is r. The measure of the side *opposite* the 60° angle is $\sqrt{3}r/2$, and the measure of the side *adjacent* to the 60° angle is $r/2$. The point $P(r/2, \sqrt{3}r/2)$ is on the terminal side of the 60° angle. The values of the six trigonometric functions of 60° are

$$\sin 60° = \frac{\sqrt{3}r/2}{r} = \frac{\sqrt{3}}{2} \qquad \cos 60° = \frac{r/2}{r} = \frac{1}{2}$$

$$\tan 60° = \frac{\sqrt{3}r/2}{r/2} = \sqrt{3} \qquad \csc 60° = \frac{r}{\sqrt{3}r/2} = \frac{2\sqrt{3}}{3}$$

$$\sec 60° = \frac{r}{r/2} = 2 \qquad \cot 60° = \frac{r/2}{\sqrt{3}r/2} = \frac{\sqrt{3}}{3}$$

Table 6.1 summarizes the values of the trigonometric functions for the special angles 30° ($\pi/6$), 45° ($\pi/4$), and 60° ($\pi/3$).

TABLE 6.1   Values of Trigonometric Functions for 30°, 45° and 60°

	$\sin \phi$	$\cos \phi$	$\tan \phi$	$\csc \phi$	$\sec \phi$	$\cot \phi$
$\phi = 30° = \dfrac{\pi}{6}$	$\dfrac{1}{2}$	$\dfrac{\sqrt{3}}{2}$	$\dfrac{\sqrt{3}}{3}$	2	$\dfrac{2\sqrt{3}}{3}$	$\sqrt{3}$
$\phi = 45° = \dfrac{\pi}{4}$	$\dfrac{\sqrt{2}}{2}$	$\dfrac{\sqrt{2}}{2}$	1	$\sqrt{2}$	$\sqrt{2}$	1
$\phi = 60° = \dfrac{\pi}{3}$	$\dfrac{\sqrt{3}}{2}$	$\dfrac{1}{2}$	$\sqrt{3}$	$\dfrac{2\sqrt{3}}{3}$	2	$\dfrac{\sqrt{3}}{3}$

EXAMPLE 3   **Evaluate Trigonometric Expressions with Special Angles**

Find the exact value of $\sin^2 \pi/6 + \cos^2 \pi/6$.

Solution   Substitute the values for $\sin \pi/6$ and $\cos \pi/6$ into the expression and simplify.

$$\sin^2 \frac{\pi}{6} + \cos^2 \frac{\pi}{6} = \left(\frac{1}{2}\right)^2 + \left(\frac{\sqrt{3}}{2}\right)^2 = \frac{1}{4} + \frac{3}{4} = 1 \qquad \begin{array}{l} \sin \pi/6 = 1/2 \\ \cos \pi/6 = \sqrt{3}/2 \end{array}$$

■ *Try Exercise **70**, page 320.*

Appendix III includes a table of values of the trigonometric functions to four decimal places in increments of 10′. For more accuracy, a calculator or interpolation can be used. (Interpolation is discussed in Appendix I.)

To find the values of trigonometric functions using a scientific calculator, first notice that the calculator has only the function keys sin, cos, and tan; the values of the cosecant, secant, and cotangent are found by using *reciprocal* keys.

From the definitions of the sine and the cosecant functions,

$$(\sin \theta)(\csc \theta) = \frac{b}{r} \cdot \frac{r}{b} = 1$$

Thus

$$(\sin \theta)(\csc \theta) = 1.$$

By rewriting the above equation, the sine and cosecant functions can be written in the following forms:

$$\sin \theta = \frac{1}{\csc \theta} \quad \text{or} \quad \csc \theta = \frac{1}{\sin \theta}$$

The sine and cosecant functions are called **reciprocal functions.**

The secant is the reciprocal of the cosine and the cotangent is the reciprocal of the tangent. Why?[2]

Table 6.2 shows each trigonometric function and its reciprocal. These relationships are true for all values of the variable θ for which the functions are defined.

TABLE 6.2   Trigonometric Functions and Their Reciprocals.

$$\sin \theta = \frac{1}{\csc \theta} \qquad \cos \theta = \frac{1}{\sec \theta} \qquad \tan \theta = \frac{1}{\cot \theta}$$

$$\csc \theta = \frac{1}{\sin \theta} \qquad \sec \theta = \frac{1}{\cos \theta} \qquad \cot \theta = \frac{1}{\tan \theta}$$

For example, to calculate cos 33°, cos 58°, and sec 58°, perform the following sequence of keystrokes. (Be sure the calculator is in *degree* mode.)

Trigonometric Function	Key Sequence	Calculator Display
cos 33°	33 [cos]	0.83867057
cos 58°	58 [cos]	0.52991926
sec 58°	58 [cos] [1/x]	1.88707992

Remark   The [1/x] key is used to calculate the reciprocal of the number in the display. By pressing this key when the value of cos 58° is shown in the display, we are performing the mathematical equivalent of

$$\frac{1}{\cos 58°} = \sec 58°.$$

As a second *important* point, many calculator errors are a result of not placing the calculator in the correct degree or radian mode before beginning a calculation. Always ensure that the calculator is in the correct mode before starting a calculation.

---

[2] $\sec \theta = r/a$ and $\cos \theta = a/r$, therefore $(\sec \theta)(\cos \theta) = 1$; $\cot \theta = a/b$, $\tan \theta = b/a$, therefore $(\cot \theta)(\tan \theta) = 1$.

EXAMPLE 4   **Evaluate Trigonometric Functions Using a Calculator**

Use a scientific calculator to find:

a. $\sin 23.33°$ b. $\csc 79.35°$ c. $\tan \dfrac{\pi}{12}$

Solution

Trigonometric Function	Key Sequence	Calculator Display
a. $\sin 23.33°$	23.33 $\boxed{\sin}$	$\boxed{0.39602635}$
b. $\csc 79.35°$	79.35 $\boxed{\sin}$ $\boxed{1/x}$	$\boxed{1.01752747}$
c. $\tan \dfrac{\pi}{12}$	π $\boxed{\div}$ 12 $\boxed{=}$ $\boxed{\tan}$	$\boxed{0.26794919}$

■ *Try Exercise* **48**, *page 319.*

EXERCISE SET 6.2 _____

In Exercises 1 to 12, find the six trigonometric function values with the given points on the terminal side of the angle in standard position.

1. $P(1, 3)$ **2.** $P(1, 7)$ **3.** $P(2, 2)$
4. $P(5, 2)$ **5.** $P(5, 5)$ **6.** $P(3, 5)$
7. $P(3, 2)$ **8.** $P(10, 6)$ **9.** $P(8, 4)$
10. $P(\sqrt{3}/2, 1)$ **11.** $P(\sqrt{2}, \sqrt{2})$ **12.** $P(2, \sqrt{3})$

In Exercises 13 to 15, if θ is an acute angle and $\sin \theta = 3/5$, use the definitions of the trigonometric functions to find the following:

13. $\tan \theta$ **14.** $\sec \theta$ **15.** $\cos \theta$

In Exercises 16 to 18, if θ is an acute angle and $\tan \theta = 4/3$, use the definitions of the trigonometric functions to find the following:

16. $\sin \theta$ **17.** $\cot \theta$ **18.** $\sec \theta$

In Exercises 19 to 21, if β is an acute angle and $\sec \beta = 13/12$, use the definitions of the trigonometric functions to find the following:

19. $\cos \beta$ **20.** $\cot \beta$ **21.** $\csc \beta$

In Exercises 22 to 24, if θ is an acute angle and $\cot \theta = 3$, use the definitions of the trigonometric functions to find the following:

22. $\sin \theta$ **23.** $\sec \theta$ **24.** $\tan \theta$

In Exercises 25 to 36, find the exact values of the trigonometric functions.

25. $\tan 45°$ **26.** $\sin 60°$ **27.** $\csc 30°$ **28.** $\cos 45°$
29. $\cot 30°$ **30.** $\sec 60°$ **31.** $\sin \pi/4$ **32.** $\cot \pi/6$
33. $\cos \pi/3$ **34.** $\sec \pi/6$ **35.** $\tan \pi/3$ **36.** $\csc \pi/4$

In Exercises 37 to 54, find the values of the trigonometric functions to 4 decimal places.

37. $\sin 12°$ **38.** $\cos 49°$ **39.** $\tan 32°$
40. $\sec 88°$ **41.** $\csc 63°20'$ **42.** $\cot 55°50'$
43. $\cos 34.7°$ **44.** $\tan 81.3°$ **45.** $\sec 5.9°$
46. $\sin \pi/5$ **47.** $\tan \pi/7$ **48.** $\sec 3\pi/8$
49. $\csc 1.2$ **50.** $\sin 0.45$ **51.** $\cos 1.25$
52. $\tan 3/4$ **53.** $\sec 5/8$ **54.** $\cot 3/5$

In Exercises 55 to 74, find the exact value of each expression.

55. $\sin 45° + \cos 45°$ **56.** $\csc 45° - \sec 45°$
57. $\sin 30° - \cos 60°$ **58.** $\tan 45° - \cot 45°$
59. $\sin 30° \cos 60° - \tan 45°$
60. $\csc 60° \sec 30° + \cot 45°$
61. $\cos 30° - \tan 30°$
62. $\sin 60° - \tan 60° \cos 0°$
63. $\sin 30° \cos 60° + \tan 45°$

64. $\sec 30° \cos 30° - \tan 60° \cot 60°$

65. $2 \sin 60° - \sec 45° \tan 60°$

66. $\sec 45° \cot 30° + 3 \cos 60°$

67. $\sin \dfrac{\pi}{3} + \cos \dfrac{\pi}{6}$ **68.** $\csc \dfrac{\pi}{6} - \sec \dfrac{\pi}{3}$

69. $\sin \dfrac{\pi}{4} + \tan \dfrac{\pi}{6}$ **70.** $\sin \dfrac{\pi}{3} \cos \dfrac{\pi}{4} - \tan \dfrac{\pi}{4}$

71. $\sec \dfrac{\pi}{3} \cos \dfrac{\pi}{3} - \tan \dfrac{\pi}{6}$ **72.** $\cos \dfrac{\pi}{4} \tan \dfrac{\pi}{6} + 2 \tan \dfrac{\pi}{3}$

73. $2 \csc \dfrac{\pi}{4} - \sec \dfrac{\pi}{3} \cos \dfrac{\pi}{6}$ **74.** $3 \tan \dfrac{\pi}{4} + \sec \dfrac{\pi}{6} \sin \dfrac{\pi}{3}$

Supplemental Exercises

In Exercises 75 to 86, find the value of the acute angle β in degrees and radians without using a calculator.

75. $\sin \beta = 1/2$ **76.** $\cos \beta = \sqrt{3}/2$

77. $\tan \beta = \sqrt{3}/3$ **78.** $\sec \beta = 2$

79. $\csc \beta = \sqrt{2}$ **80.** $\cot \beta = \sqrt{3}$

81. $\cos \beta = \sqrt{2}/2$ **82.** $\tan \beta = 1$

83. $\sec \beta = 2\sqrt{3}/3$ **84.** $\sin \beta = \sqrt{2}/2$

85. $\csc \beta = 2\sqrt{3}/3$ **86.** $\cot \beta = 1$

In Exercises 87 to 92, find the values of the trigonometric functions to 4 decimal places.

87. $\sin 36°23'4''$ **88.** $\tan 67°38'26''$ **89.** $\sec 5°45'34''$

90. $\csc 34°49'17''$ **91.** $\cos 50°45'9''$ **92.** $\cot 55°22'12''$

In Exercises 93 to 96, show that the equations are true.

93. $\sin(90° - 30°) = \cos 30°$

94. $\tan(90° - 60°) = \cot 60°$

95. $\cos(\pi/2 - \pi/3) = \sin \pi/3$

96. $\sec(\pi/2) - \pi/3) = \csc \pi/3$

In Exercises 97 to 100, use the definitions of the trigonometric functions to verify the equations.

97. $\sin^2 \theta + \cos^2 \theta = 1$ **98.** $1 + \tan^2 \phi = \sec^2 \phi$

99. $\tan \beta = \dfrac{\sin \beta}{\cos \beta}$ **100.** $\cot^2 \phi + 1 = \csc^2 \phi$

In Exercises 101 to 108, evaluate the given expressions.

101. $\sin^2 30° - 2 \cos^2 45°$

102. $\cos 45° \tan^2 60° + 3 \cot^2 45°$

103. $\csc^2 \dfrac{\pi}{3} \sec^2 \dfrac{\pi}{6} + 2 \cot^2 \dfrac{\pi}{4}$

104. $\sin^3 \dfrac{\pi}{6} \csc^3 \dfrac{\pi}{6} + \cos^2 \dfrac{\pi}{4}$

105. $\sec 45° - \csc 60° \sin 60°$

106. $\cot 30° \tan 60° - \tan 60°$

107. $\sec \dfrac{\pi}{3} \sin \dfrac{\pi}{6} - \tan \dfrac{\pi}{4}$

108. $\sin^2 \dfrac{\pi}{6} - \tan^2 \dfrac{\pi}{3} \cos \dfrac{\pi}{6}$

In Exercises 109 to 112, determine which of the following equations are true.

109. $\tan \phi \cos \phi = \sin \phi$ **110.** $\cot \phi \sin \phi = \cos \phi$

111. $\sec \phi \sin \phi = \tan \phi$ **112.** $\csc \phi \cos \phi = \cot \phi$

6.3
Trigonometric Functions of Any Angle

The application of trigonometry would be quite limited if all angles had to be acute. Fortunately, this is not the case. In this section we extend the definition of a trigonometric function to include any angle.

We begin by considering an angle θ in standard position. Let $P(a, b)$ be any point along the terminal side of the angle θ. We then define the trigonometric functions according to the following definitions.

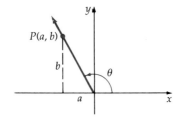

Figure 6.26

Definition of the Trigonometric Functions of Any Angle

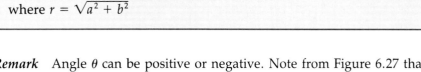

Let $P(a, b)$ be any point, except the origin, on the terminal side of an angle θ in standard position. Let $r = d(O, P)$, the distance from the origin to P. The six trigonometric functions of θ are

$$\sin \theta = \frac{b}{r} \qquad \cos \theta = \frac{a}{r} \qquad \tan \theta = \frac{b}{a}, \quad a \neq 0$$

$$\csc \theta = \frac{r}{b}, \quad b \neq 0 \qquad \sec \theta = \frac{r}{a}, \quad a \neq 0 \qquad \cot \theta = \frac{a}{b}, \quad b \neq 0$$

where $r = \sqrt{a^2 + b^2}$

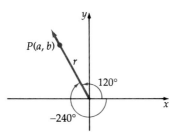

Figure 6.27

Remark Angle θ can be positive or negative. Note from Figure 6.27 that $\sin 120° = \sin(-240°)$, because point $P(a, b)$ is on the terminal side of the 120° angle and the −240° angle.

The advantage of our extended definition is that we can compute the trigonometric functions of any angle. The numbers a and b can be positive, negative, or zero, depending on where the terminal side of the angle is located.

Any point in a rectangular coordinate system can determine an angle in standard position. For example, the point $P(-4, 3)$ in the second quadrant (Figure 6.28) determines an angle θ in standard position with $r = \sqrt{(-4)^2 + 3^2} = 5$. The values of the trigonometric functions of θ are

$$\sin \theta = \frac{3}{5} \qquad \cos \theta = \frac{-4}{5} = -\frac{4}{5} \qquad \tan \theta = \frac{3}{-4} = -\frac{3}{4}$$

$$\csc \theta = \frac{5}{3} \qquad \sec \theta = \frac{5}{-4} = -\frac{5}{4} \qquad \cot \theta = \frac{-4}{3} = -\frac{4}{3}$$

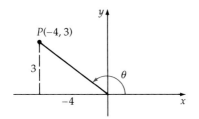

Figure 6.28

As this example shows, the sign of a trigonometric function depends on the quadrant in which the terminal side of the angle lies. For example, if θ is in the third quadrant and $P(a, b)$ is on the terminal side of angle θ, both a and b are negative. Thus only the tangent and cotangent functions are positive in the third quadrant. Since a and b are both negative, the quotient of b/a or a/b will be positive.

Table 6.3 lists the sign of the six trigonometric functions in each quadrant. Figure 6.29 indicates the quadrant in which the functions are positive.

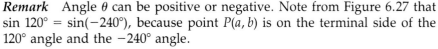

Table 6.3

	θ in Quadrant			
	I	II	III	IV
$\sin \theta$ and $\csc \theta$	Positive	Positive	Negative	Negative
$\cos \theta$ and $\sec \theta$	Positive	Negative	Negative	Positive
$\tan \theta$ and $\cot \theta$	Positive	Negative	Positive	Negative

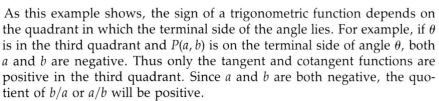

Figure 6.29

EXAMPLE 1   Evaluate the Trigonometric Functions of an Angle in the Third Quadrant

Find the six trigonometric functions of an angle whose terminal side contains the point $(-3, -2)$.

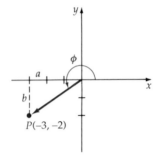

Figure 6.30

Solution   Sketch the angle (Figure 6.30). Find r by using the Pythagorean Theorem.

$$r = \sqrt{(-3)^2 + (-2)^2} = \sqrt{9 + 4} = \sqrt{13} \quad a = -3, b = -2$$

$$\sin \phi = \frac{-2}{\sqrt{13}} = -\frac{2\sqrt{13}}{13} \qquad \cos \phi = \frac{-3}{\sqrt{13}} = -\frac{3\sqrt{13}}{13} \qquad \tan \phi = \frac{-2}{-3} = \frac{2}{3}$$

$$\csc \phi = \frac{\sqrt{13}}{-2} = -\frac{\sqrt{13}}{2} \qquad \sec \phi = \frac{\sqrt{13}}{-3} = -\frac{\sqrt{13}}{3} \qquad \cot \phi = \frac{-3}{-2} = \frac{3}{2}$$

■ *Try Exercise 6, page 326.*

A **quadrantal angle** is an angle of $0°$ (0), $90°$ $(\pi/2)$, $180°$ (π), or $270°$ $(3\pi/2)$. The terminal side of a quadrantal angle coincides with the x- or y-axis. The trigonometric function value of a quadrantal angle can be found by choosing any point on the terminal side of the quandrantal angle and then applying the definition of the trigonometric function.

The terminal side of $0°$ coincides with the positive x-axes. Let $P(a, 0)$ be any point on the x-axis. Then $b = 0$ and $r = a$. The values of the six trigonometric functions at $0°$ are given by

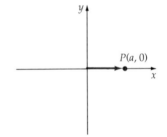

Figure 6.31

$$\sin 0° = \frac{0}{r} = 0 \qquad \cos 0° = \frac{a}{r} = \frac{a}{a} = 1 \qquad \tan 0° = \frac{0}{a} = 0$$

$$\csc 0° \text{ is undefined} \qquad \sec 0° = \frac{r}{a} = \frac{a}{a} = 1 \qquad \cot 0° \text{ is undefined}$$

Remark   For the point $(a, 0)$, $\csc 0° = r/0$, which is undefined. A similar statement is true for $\cot 0°$.

In like manner, the trigonometric functions of $90°$, $180°$, and $270°$ can be found by using a point on the terminal side of each angle and using the definitions of the trigonometric functions. The results are shown in Table 6.4.

The values of the trigonometric functions in the table in Appendix III are for angles between $0°$ and $90°$. To find the value of a trigonometric function for some other angle, a reference angle is used.

TABLE 6.4

θ	$\sin \theta$	$\cos \theta$	$\tan \theta$	$\csc \theta$	$\sec \theta$	$\cot \theta$
0°	0	1	0	Undefined	1	Undefined
90°	1	0	Undefined	1	Undefined	0
180°	0	−1	0	Undefined	−1	Undefined
270°	−1	0	Undefined	−1	Undefined	0

Reference Angle

> The **reference angle** for the angle α is the positive acute angle formed by the terminal side of α and the x-axis.

Figure 6.32 shows the reference angle β for an angle α whose terminal side lies in the second, third, or fourth quadrants. Quadrantal angles do not have a reference angle. The reference angle for an angle in the first quadrant is the given angle.

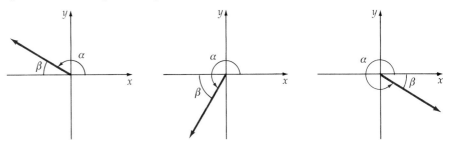

Figure 6.32

Figure 6.33 illustrates three examples of finding a reference angle:

1. The terminal side of a 224° angle lies in the third quadrant. The reference angle is the angle formed by the terminal side of the angle and the negative x-axis. The measurement of the reference angle is 44°.

2. The terminal side of a −47° angle lies in the fourth quadrant. The reference angle is the angle formed by the terminal side of the angle and the positive x-axis. The measurement of the reference angle is 47°.

3. The terminal side of an angle with measure $-8\pi/3$ radians lies in the third quadrant. The reference angle is the angle formed by the terminal side of the angle and the negative x-axis. The measurement of the reference angle is $\pi/3$.

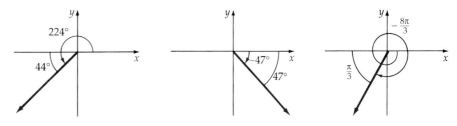

Figure 6.33

Value of a Trigonometric Function for Any Angle

> Let θ be any angle and β be the reference angle for θ. The trigonometric function of θ is the trigonometric function of β with the appropriate sign depending on the quadrant where the terminal side lies.

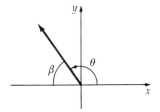

Figure 6.34

EXAMPLE 2   Evaluate the Six Trigonometric Functions of an Angle By Using the Reference Angle

Find the exact value of the six trigonometric functions of 150°.

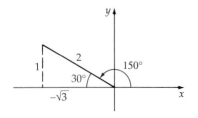

Figure 6.35

Solution   Sketch a 150° angle in standard position. The measure of the reference angle is 180° − 150° = 30°. Use the values of the trigonometric functions for a 30° angle and Table 6.3 to attach the appropriate sign.

$$\sin 150° = \frac{1}{2} \qquad \cos 150° = -\frac{\sqrt{3}}{2} \qquad \tan 150° = -\frac{\sqrt{3}}{3}$$

$$\csc 150° = 2 \qquad \sec 150° = -\frac{2\sqrt{3}}{3} \qquad \cot 150° = -\sqrt{3}$$

■ *Try Exercise* **10**, *page 326.*

EXAMPLE 3   Evaluate Trigonometric Functions

Given that $\cos \phi = \dfrac{-\sqrt{3}}{2}$ for an angle ϕ in the third quadrant. Find the exact values of the other five trigonomeric functions.

Solution   Sketch a right triangle with the angle ϕ in standard position and b the unknown side.

$$r^2 = a^2 + b^2 \qquad \text{Use the Pythagorean Theorem to find } b.$$
$$2^2 = (-\sqrt{3})^2 + b^2$$
$$b = \pm 1$$
$$ = -1 \qquad \text{Because the terminal side of the angle is in the third quadrant, } b \text{ is a negative number.}$$

Use the definition of the trigonometric functions to find their exact values. We have $a = -\sqrt{3}$, $b = -1$, and $r = 2$ (Figure 6.36).

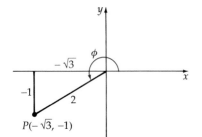

Figure 6.36

$$\sin \phi = \frac{b}{r} = -\frac{1}{2} \qquad\qquad \tan \phi = \frac{b}{a} = \frac{-1}{-\sqrt{3}} = \frac{\sqrt{3}}{3}$$

$$\csc \phi = \frac{r}{b} = \frac{2}{-1} = -2 \qquad \sec \phi = \frac{r}{a} = \frac{2}{-\sqrt{3}} = -\frac{2\sqrt{3}}{3}$$

$$\cot \phi = \frac{a}{b} = \frac{-\sqrt{3}}{-1} = \sqrt{3}$$

■ *Try Exercise* **40**, *page 327.*

EXAMPLE 4   Evaluate a Trigonometric Expression of Special Angles

Find the exact value of $\sin \dfrac{\pi}{2} \cos \dfrac{5\pi}{6} - \tan \dfrac{2\pi}{3}$.

Solution   Sketch the angles in standard form and find the reference angles. Substitute the function values into the trigonometric expression and simplify.

$$\sin \frac{\pi}{2} = 1$$

$$\cos \frac{5\pi}{6} = -\cos \frac{\pi}{6} = -\frac{\sqrt{3}}{2}$$

$$\tan \frac{2\pi}{3} = -\tan \frac{\pi}{3} = -\sqrt{3}$$

$$\sin \frac{\pi}{2} \cos \frac{5\pi}{6} - \tan \frac{2\pi}{3} = (1)\left(-\frac{\sqrt{3}}{2}\right) - (-\sqrt{3}) = -\frac{\sqrt{3}}{2} + \sqrt{3} = \frac{\sqrt{3}}{2}$$

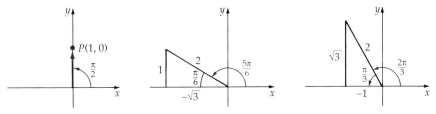

Figure 6.37

■ *Try Exercise **26**, page 327.*

When tables are used to find the value of a trigonometric function, the proper sign is attached to the value of the function depending on the quadrant in which the angle lies. A calculator, on the other hand, will correctly evaluate a trigonometric function, including the appropriate sign.

EXAMPLE 5   **Evaluate a Trigonometric Expression By Using a Calculator**

Find   a.   csc 322.3°   b. $\tan \frac{7\pi}{12}$ c.   sin 4.34.

Solution   Use a calculator and follow the key sequence shown.

Function	Key Sequence	Calculator Display
a. csc 322.3°	322.3 [sin] [1/x]	-1.635250666
b. $\tan \dfrac{7\pi}{12}$	7 [×] [π] [÷] 12 [=] [tan]	-3.732050808
c. sin 4.34	4.34 [sin]	-0.931460793

■ *Try Exercises **54**, page 327.*

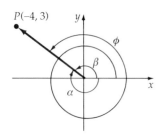

Figure 6.38

Caution   In b and c there are no units on the argument of the function. When this occurs, the argument of the function is in radians.

Figure 6.38 shows the point $P(-4, 3)$ on the terminal side of an angle β in standard position. The same point is also on the terminal side of the negative angle α. Note also that $\phi = \beta + 360$. Angles that have the same initial and terminal sides are called **coterminal** angles. The trigonometric function values of coterminal angles are equal. Therefore $\sin \beta = \sin \alpha$, and $\sin \beta = \sin \phi$.

To find the reference angle for an angle greater than 360°, first find the coterminal angle less than 360° and then find the reference angle. The terminal side of 660° lies in the fourth quadrant ($660° - 360° = 300°$). Thus the reference angle of 660° is the same as the reference angle of 300°. The reference angle is $360° - 300° = 60°$.

EXAMPLE 6   **Evaluate Trigonometric Functions By Using a Coterminal Angle**

Find the exact value of the six trigonometric functions of 585°.

Solution   First we find the angle $0 \le \phi < 360°$ that is coterminal with 585°. Since 585° is less than two complete rotations, we subtract 360° from 585°: $585° - 360° = 225°$. Now we determine the reference angle of 225°. Since 225° is in the third quadrant, the reference angle is formed by the terminal side of ϕ and the negative x-axis. Thus the reference angle is $225° - 180° = 45°$.

Evaluate the trigonometric functions of 45° and attach the signs associated with an angle in the third quadrant.

$$\sin 585° = -\sin 45° = -\frac{\sqrt{2}}{2} \qquad \cos 585° = -\cos 45° = -\frac{\sqrt{2}}{2}$$

$$\tan 585° = \tan 45° = 1 \qquad \csc 585° = -\csc 45° = -\sqrt{2}$$

$$\sec 585° = -\sec 45° = -\sqrt{2} \qquad \cot 585° = \cot 45° = 1$$

■ *Try Exercise **56**, page 327.*

EXERCISE SET 6.3

In Exercises 1 to 8, find the six trigonometric function values with the given points on the terminal side of the angle.

1. $P(2, 3)$ **2.** $P(3, 7)$ **3.** $P(-2, 3)$ **4.** $P(-3, 5)$
5. $P(-8, -5)$ **6.** $P(-6, -9)$ **7.** $P(-5, 0)$ **8.** $P(0, 2)$

In Exercises 9 to 16, find the exact values of the six trigonometric functions of the given angles. Do not use a calculator.

9. 330°   **10.** 225°   **11.** 210°   **12.** 315°
13. $\pi/3$ **14.** $7\pi/6$ **15.** $11\pi/6$ **16.** $3\pi/4$

In Exercises 17 to 28, find the exact value of each expression.

17. $\cos 0° - \sin 30° \tan 45°$

18. $\sin 90° \cos 60° + \cos 60°$

19. $\sin 210° - \cos 330° \tan 330°$

20. $\tan 225° + \sin 240° \cos 60°$

21. $\sin^2 30° + \cos^2 30°$

22. $\tan^2 60° - 2 \sin 30°$

23. $\cos \pi \sin(7\pi/4) - \tan(11\pi/6)$

24. $\cos(7\pi/6)\tan(3\pi/4) - \sin(\pi/6)$

25. $\sin(3\pi/2)\tan(\pi/4) - \cos(\pi/3)$

26. $\cos(7\pi/4)\tan(4\pi/3) + \cos(7\pi/6)$

27. $\sin^2(5\pi/4) + \cos^2(5\pi/4)$

28. $\tan^2(7\pi/4) - \sec^2(7\pi/4)$

In Exercises 29 to 34, let ϕ be an angle in standard position. State the quadrant in which the terminal side of ϕ lies.

29. $\sin\phi > 0, \quad \cos\phi > 0$

30. $\tan\phi < 0, \quad \sin\phi < 0$

31. $\cos\phi > 0, \quad \tan\phi < 0$

32. $\sin\phi < 0, \quad \cos\phi > 0$

33. $\sin\phi < 0, \quad \cos\phi < 0$

34. $\tan\phi < 0, \quad \cos\phi < 0$

In Exercises 35 to 48, find the value of each expression.

35. $\sin\phi = -1/2$; ϕ in quadrant III, find $\tan\phi$.

36. $\tan\phi = -\sqrt{3}$; ϕ in quadrant IV, find $\sin\phi$.

37. $\cos\phi = -\sqrt{3}/2$; ϕ in quadrant III, find $\sin\phi$.

38. $\cot\phi = -1$; ϕ in quadrant II, find $\cos\phi$.

39. $\csc\phi = \sqrt{2}$; ϕ in quadrant II, find $\cot\phi$.

40. $\sec\phi = 2\sqrt{3}/3$; ϕ in quadrant IV, find $\sin\phi$.

41. $\sin\phi = -1/2$ and $\cos\phi > 0$, find $\tan\phi$.

42. $\tan\phi = 1$ and $\sin < 0$, find $\cos\phi$.

43. $\cos\phi = 1/2$ and $\tan\phi = \sqrt{3}$, find $\csc\phi$.

44. $\tan\phi = 1$ and $\sin\phi = -\sqrt{2}/2$, find $\sec\phi$.

45. $\cos\phi = -1/2$ and $\sin\phi = \sqrt{3}/2$, find $\cot\phi$.

46. $\cot\phi = 1$ and $\csc\phi = -\sqrt{2}$, find $\cos\phi$.

47. $\sec\phi = 2\sqrt{3}/3$ and $\sin\phi = -1/2$, find $\cot\phi$.

48. $\sin\phi = -\sqrt{2}/2$ and $\sec\phi = -\sqrt{2}$, find $\cot\phi$.

Calculator Exercises

In Exercises 49 to 64, use a calculator to evaluate the following functions to 4 decimal places.

49. $\sin 127°$ 50. $\sin 257°$ 51. $\cos 116°$ 52. $\cos 355°$

53. $\tan 548°$ 54. $\sin 398°$ 55. $\cos 578°$ 56. $\sin 740°$

57. $\sin \pi/5$ 58. $\cos 3\pi/7$ 59. $\cos 9\pi/5$ 60. $\tan 11\pi/8$

61. $\sin 4.12$ 62. $\sin 6.98$ 63. $\cos 4.45$ 64. $\cos 0.34$

Supplemental Exercises

In Exercises 65 to 72, find the exact values of the angle ϕ, $0 \le \phi \le 360°$.

65. $\sin\phi = 1/2$

66. $\tan\phi = -\sqrt{3}$

67. $\cos\phi = -\sqrt{3}/2$

68. $\tan\phi = 1$

69. $\csc\phi = -\sqrt{2}$

70. $\cot\phi = -1$

71. $\csc\phi = -2\sqrt{3}/3$

72. $\sin\phi = \sqrt{3}/2$

In Exercises 73 to 80, find the exact values of the angle ϕ, $0 \le \phi < 2\pi$.

73. $\tan\phi = -1$

74. $\cos\phi = 1/2$

75. $\tan\phi = -\sqrt{3}/3$

76. $\sec\phi = -2\sqrt{3}/3$

77. $\sin\phi = \sqrt{3}/2$

78. $\cos\phi = -1/2$

79. $\cot\phi = \sqrt{3}$

80. $\sin\phi = \sqrt{2}/2$

In Exercises 81 to 94, verify the given equation.

81. $\sin^2 30° + \cos^2 30° = 1$

82. $1 + \tan^2 60° = \sec^2 60°$

83. $\sin(180° + 30°) = -\sin 30°$

84. $\sin(90° + 60°) = \cos 60°$

85. $\cos 90° = \cos 60° \cos 30° - \sin 60° \sin 30°$

86. $\tan(90° + 30°) = -\cot 30°$

87. $\sin 30° \tan 30° = \sec 30° - \cos 30°$

88. $\sin 120° \sec 120° = \tan 120°$

89. $\dfrac{\sin 45° + \cos 45°}{\cos 45°} = \tan 45° + 1$

90. $\tan^2 60° \cos^2 60° = \sin^2 60°$

91. $\cot^2 210° \sin^2 210° = \cos^2 210°$

92. $\dfrac{\sin 30° + \cos 30°}{\cos 30°} = \tan 30° + 1$

93. $1 + \tan^2 135° = \dfrac{1}{\cos^2 135°}$

94. $\dfrac{1}{\sin 60° \cos 60°} = \csc 60° \sec 60°$

In Exercises 95 to 98, use a calculator to find the trigonometric functions to 4 decimal places.

95. $\sin 34°23'12''$

96. $\cos 123°14'56''$

97. $\csc 235°48'5''$

98. $\cot 145°34'51''$

In Exercises 99 to 102, verify the equation by using the definitions of the trigonometric functions.

99. $\sin(-\phi) = -\sin\phi$

100. $\tan(-\phi) = -\tan\phi$

101. $\cos(-\phi) = \cos\phi$

102. $\sin(180° + \phi) = -\sin\phi$

103. The sine function can be approximated by the following polynomial function, where x is measured in radians.

$$\sin x \approx x - \frac{x^3}{6} + \frac{x^5}{120} - \frac{x^7}{5040}$$

a. Use the polynomial approximation of the sine function to evaluate sin 0.5. Compare with the calculator value of the function.

b. Use the polynomial approximation of the sine function to evaluate sin $\pi/3$. Compare with the calculator value of the function.

6.4

Circular Functions

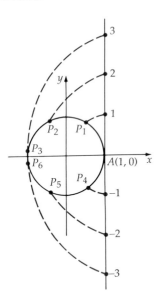

Figure 6.39

During the seventeenth century, applications of trigonometry were broadened to problems in physics and engineering. These kinds of problems required trigonometric functions whose domains of the functions were sets of real numbers rather than sets of angles. The definitions of trigonometric functions were extended by using a correspondence between a number and an angle.

One correspondence involves a circle with a radius of 1, called the **unit circle.** Consider a vertical number line tangent to a unit circle at the point $A(1, 0)$. Wrapping the line around the circle creates a one-to-one correspondence between a point t on the line and a point $P(x, y)$ on the unit circle. By wrapping the line *counterclockwise*, a positive real number is paired with a point on the unit circle. A *clockwise* wrapping of the line pairs a negative real number with a point on the unit circle.

Each real number t defines an arc $\overset{\frown}{AP}$ of measure t (see Figure 6.40). The arc $\overset{\frown}{AP}$ subtends an angle θ in standard position.

Recall that an arc length t on a circle of radius r subtends an angle θ such that $t = r\theta$ or $t = \theta$ for a unit circle, $r = 1$. Thus, on the unit circle, *the measure of a central angle and the length of an arc can be represented by the same real number t.*

As an example, consider a unit circle and the point $\pi/3$ on the number line tangent to the circle at $A(1, 0)$ paired with a point $P(x, y)$ on the circle. The length of the arc $t = \pi/3$. Since the measure of an angle and the length of the arc in a unit circle can be represented by the same number, $\theta = \pi/3$. From the definitions of the sine and cosine functions, we can find the coordinates of the point $P(x, y)$.

$$\cos \theta = \cos \frac{\pi}{3} = \frac{1}{2} = \frac{x}{r} = x \qquad r = 1$$

$$\sin \theta = \sin \frac{\pi}{3} = \frac{\sqrt{3}}{2} = \frac{y}{r} = y$$

Thus the coordinates of P are $P(\frac{1}{2}, \frac{\sqrt{3}}{2})$.

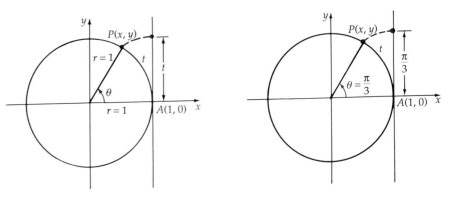

Figure 6.40   **Figure 6.41**

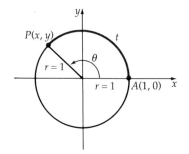

Figure 6.42

Remark The cosine of the real number $\pi/3$ is the x-coordinate and the sine of a real number is the y-coordinate.

To review, a point t on a line corresponds (by wrapping the line around the unit circle) to a point $P(x, y)$ on the unit circle, which in turn corresponds to an angle. With this correspondence, we can define the circular functions of a real number t.

Circular Functions

Let t be a real number and $P(x, y)$ be a point on a unit circle corresponding to t. Then
$$\cos t = x \qquad \sin t = y \qquad \tan t = y/x, \; x \neq 0$$
$$\sec t = 1/x, \; x \neq 0 \qquad \csc t = 1/y, \; y \neq 0 \qquad \cot t = x/y, \; y \neq 0$$

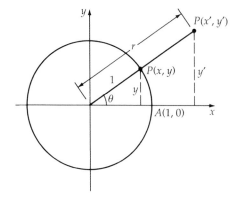

Figure 6.43

Remark The circular functions are defined for real numbers t, whereas the trigonometric functions are defined for angles. However, there is a relationship between the two.

Consider an angle θ (in radians) in standard position. Let $P(x, y)$ and $P'(x', y')$ be two points on the terminal side of the angle, with $x^2 + y^2 = 1$ and $(x')^2 + (y')^2 = r^2$. Let t be the length of the arc from $(1, 0)$ to (x, y). Then

$$\sin \theta = \frac{y}{1} = \frac{y'}{r} = \sin t.$$

Thus the value of the sine function of an angle measured in radians is equal to the value of the sine function of the real number t. This is true for each of the trigonometric functions.

The domain and range of the circular functions can be found from the definition of these functions. If t is any real number and $P(x, y)$ is the point on the unit circle corresponding to t, then by definition $\cos t = x$ and $\sin t = y$. Because t takes on all values of the real numbers, $\cos t$ and $\sin t$ are defined for all real numbers t. Thus the domain of the sine and cosine functions is the set of real numbers.

Because the radius of the unit circle is 1, we have $-1 \leq x \leq 1$ and $-1 \leq y \leq 1$. Therefore, with $x = \cos t$ and $y = \sin t$, we have

$$-1 \leq \cos t \leq 1 \quad \text{and} \quad -1 \leq \sin t \leq 1.$$

Thus, the range of the sine and cosine functions is $[-1, 1]$.

Domain and Range of the Sine and Cosine Functions

The domain of the sine and cosine functions is the set of real numbers. The range of the sine and cosine functions is the set of numbers $[-1, 1]$.

Using the definition of tangent and secant, we have

$$\tan t = \frac{y}{x} \quad \text{and} \quad \sec t = \frac{1}{x}.$$

Note the tangent and secant functions are undefined when $x = 0$. The domain of the tangent and secant functions is all real numbers t except those for which the x-coordinate of $P(x, y)$ is zero. The x-coordinate is zero when $t = \pm\frac{\pi}{2}, \pm\frac{3\pi}{2}, \pm\frac{5\pi}{2}$, and in general $\pm(2n + 1)\frac{\pi}{2}$, where n is a whole number. Thus the domain of the tangent and the secant function is all real numbers t except $t = \pm(2n + 1)\frac{\pi}{2}$, where n is a whole number.

Using the definition of cotangent and cosecant, we have

$$\cot t = \frac{x}{y} \quad \text{and} \quad \csc t = \frac{1}{y}.$$

The domain of the cotangent and cosecant functions is all real numbers t except those for which the y-coordinate of $P(x, y)$ is zero. The y-coordinate is zero when $t = 0, \pm\pi, \pm2\pi, \pm3\pi$, and in general $\pm n\pi$, where n is a whole number. Thus the domain of the cotangent and the cosecant function is all real numbers t except $t = \pm n\pi$, where n is a whole number.

We can use the same techniques we used before to evaluate a circular function of a real number t. Consider t an angle in standard position of radian measure t. Evaluate the trigonometric function of the angle by using a calculator (in radian mode) or the tables in Appendix III.

For example,

Function	Key Sequence	Calculator Display
sin(0.5)	0.5 $\boxed{\text{sin}}$	$\boxed{0.47942554}$
cot 20	20 $\boxed{\text{tan}}$ $\boxed{1/x}$	$\boxed{0.44699511}$
sec(−2)	2 $\boxed{+/-}$ $\boxed{\text{cos}}$ $\boxed{1/x}$	$\boxed{-2.40299796}$

Period of the Sine and Cosine Functions

Because the circumference of the unit circle is 2π (Why?)[3], the point $P(x, y)$ on the circle that corresponds to, say, 2 on the number line will also correspond to $2 + 2\pi$, $2 + 4\pi$, $2 + 6\pi$, and in general $2 + n(2\pi)$, where n is a positive integer. Furthermore, it is also true that the point $P(x, y)$ that corresponds to 2 also corresponds to $2 - 2\pi$, $2 - 4\pi$, and in general $2 - n(2\pi)$, where n is a natural number.

For any real number t, the point $P(x, y)$ on the unit circle that corresponds to t also corresponds to $t + 2\pi$, $t + 4\pi$, $t + 6\pi$, and in general, $t + n(2\pi)$, where n is an integer. Thus $\cos t$ and $\cos(t + n \cdot 2\pi)$ correspond to the same x-coordinate. Similarly, $\sin t$ and $\sin(t + n \cdot 2\pi)$ correspond to the same y-coordinate. Therefore,

$$\cos t = \cos(t + n \cdot 2\pi) \quad \text{and} \quad \sin t = \sin(t + n \cdot 2\pi)$$

for all real numbers t and integers n. These equations state that the values of the cosine and sine functions are repeated in every interval of length 2π.

A function that repeats itself is said to be *periodic*.

---

[3] Since the radius is 1, $C = 2\pi r = 2\pi(1) = 2\pi$.

Period of a Function

> Let p be a constant. If $f(t) = f(t + p)$ for all t in the domain of f, then f is a **periodic function.** The **period** of f is the smallest positive value of p for which $f(t) = f(t + p)$.

The sine and cosine functions are periodic. The period of each of the functions is 2π.

EXAMPLE 1 **Use the Periodic Property of the Sine Function**

Use the periodic property of the sine function to evaluate $\sin \dfrac{20\pi}{3}$.

Solution

$$\sin \frac{20\pi}{3} = \sin\left(\frac{2\pi}{3} + \frac{18\pi}{3}\right) = \sin\left(\frac{2\pi}{3} + 6\pi\right)$$

$$= \sin\left(\frac{2\pi}{3} + 2(3\pi)\right)$$

$$= \sin \frac{2\pi}{3} \qquad\qquad \text{Sin has period } 2\pi.$$

$$= \frac{\sqrt{3}}{2}$$

■ *Try Exercise* **48,** *page 335*

EXAMPLE 2 **Use the Unit Circle to Verify an Equation**

Use the unit circle to show that $\sin(t + \pi) = -\sin t$.

Solution Sketch a unit circle, and let P be the point on the unit circle corresponding to t. Draw a line from P through the origin. Label the point Q. Because PQ is a diameter, the length of $\overset{\frown}{PQ}$ is π. Thus the arc is $t + \pi$. For any line through the origin, if (a, b) is a point on the line, then $(-a, -b)$ is also a point on the line. Thus if P has coordinates (a, b), then Q has coordinates $(-a, -b)$. From the definition of the sine function, we obtain

$$\sin t = a \qquad \text{and} \qquad \sin(t + \pi) = -a.$$

Therefore, $\sin(t + \pi) = -\sin t$

■ *Try Exercise* **58,** *page 335.*

Certain important relationships exist among the trigonometric functions. Some of these, the reciprocal functions, we discussed previously. We now derive additional relationships.

Consider the two numbers t and $-t$ represented by the points P_1 and P_2, as shown on the unit circle in Figure 6.45. From the definition of the circular functions:

$$\sin t = y \text{ and } \sin(-t) = -y, \qquad \text{and} \qquad \cos t = x \text{ and } \cos(-t) = x.$$

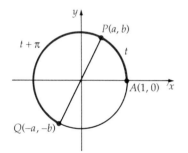

Figure 6.44

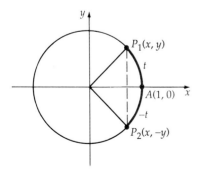

Figure 6.45

Substituting $\sin t$ for y and $\cos t$ for x, we have

$$\sin(-t) = -\sin t \text{ and } \cos(-t) = \cos t.$$

Thus the sine function is an odd function and the cosine function is an even function.

Recall that an equation that is true for every number in the domain of the variable is an identity. The trigonometric reciprocal functions defined earlier are examples of trigonometric identities. The statement

$$\sin \theta = \frac{1}{\csc \theta} \quad \csc \theta \neq 0$$

is a **trigonometric identity** because the two expressions produce the same result for all values of θ for which the functions are defined. By using the definitions of the trigonometric functions, we can prove other trigonometric identities.

The **ratio identities** are obtained by writing the tangent and cotangent functions in terms of the sine and cosine functions. Let $P(x, y)$ be the point on a unit circle corresponding to θ. Recall that $x = \cos \theta$, and $y = \sin \theta$. Substitute for x and y in the equations. $\tan \theta = y/x$ and $\cot \theta = x/y$. Substitute $x = \cos \theta$ and $y = \sin \theta$. The resulting ratio identities are

$$\tan \theta = \frac{y}{x} = \frac{\sin \theta}{\cos \theta}$$

and

$$\cot \theta = \frac{x}{y} = \frac{\cos \theta}{\sin \theta}.$$

The **Pythagorean identities** are based on the fact that $\cos \theta$ and $\sin \theta$ are, respectively, the x- and y-coordinates of a point on a unit circle. Thus

$$x^2 + y^2 = 1 \quad \text{Equation of a unit circle}$$

$$\cos^2\theta + \sin^2\theta = 1 \quad \text{Substitute } x = \cos \theta \text{ and } y = \sin \theta.$$

This is the first Pythagorean identity.

Dividing each term of the first Pythagorean identity by $\cos^2\theta$, we have

$$\frac{\cos^2\theta}{\cos^2\theta} + \frac{\sin^2\theta}{\cos^2\theta} = \frac{1}{\cos^2\theta} \quad \cos \theta \neq 0$$

$$1 + \tan^2\theta = \sec^2\theta \quad \frac{\sin \theta}{\cos \theta} = \tan \theta$$

This is the second Pythagorean identity.

Dividing each term of the first Pythagorean identity by $\sin^2\theta$, we have

$$\frac{\cos^2\theta}{\sin^2\theta} + \frac{\sin^2\theta}{\sin^2\theta} = \frac{1}{\sin^2\theta} \quad \sin \theta \neq 0$$

$$\cot^2\theta + 1 = \csc^2\theta \quad \frac{\sin \theta}{\cos \theta} = \tan \theta$$

This is the third Pythagorean identity.

The Fundamental Trigonometric Identities

The reciprocal, ratio, and Pythagorean identities are known as the eight **fundamental trigonometric identities** and are used to simplify and rewrite trigonometric expressions.

Reciprocal identities $\qquad \sin \theta = \dfrac{1}{\csc \theta} \qquad \cos \theta = \dfrac{1}{\sec \theta} \qquad \tan \theta = \dfrac{1}{\cot \theta}$

Ratio identities $\qquad \tan \theta = \dfrac{\sin \theta}{\cos \theta} \qquad \cot \theta = \dfrac{\cos \theta}{\sin \theta}$

Pythagorean identities $\qquad \sin^2\theta + \cos^2\theta = 1 \qquad \tan^2\theta + 1 = \sec^2\theta$

$$1 + \cot^2\theta = \csc^2\theta$$

EXAMPLE 3 **Evaluate Trigonometric Functions Using a Pythagorean Identity**

Given that $\sin \theta = \dfrac{1}{2}$ for $\dfrac{\pi}{2} < \theta < \pi$, find $\cos \theta$.

Solution Use the Pythagorean identity $\sin^2\theta + \cos^2\theta = 1$.

$$\sin^2\theta + \cos^2\theta = 1$$

$$\left(\frac{1}{2}\right)^2 + \cos^2\theta = 1$$

$$\cos^2\theta = \frac{3}{4}$$

$$\cos \theta = -\frac{\sqrt{3}}{2} \qquad \cos \theta \text{ is negative when } \pi/2 < \theta < \pi.$$

■ *Try Exercise* **96,** *page 335.*

EXAMPLE 4 **Express a Trigonometric Expression in Terms of the Sine Function**

Express the function $\cot^2 x + 1$ as an expression containing the sine function.

Solution $\cot^2 x + 1 = \dfrac{\cos^2 x}{\sin^2 x} + 1 = \dfrac{\cos^2 x}{\sin^2 x} + \dfrac{\sin^2 x}{\sin^2 x} \qquad \cot x = \dfrac{\cos x}{\sin x}$

$$= \dfrac{\cos^2 x + \sin^2 x}{\sin^2 x} = \dfrac{1}{\sin^2 x} \qquad \sin^2 x + \cos^2 x = 1$$

■ *Try Exercise* **80,** *page 335.*

EXAMPLE 5   **Express the Cotangent Function in Terms of the Cosine Function**

Express $\cot \beta$ in terms of $\cos \beta$, where the terminal side of β is in the third quadrant.

Solution

$$\cot \beta = \frac{\cos \beta}{\sin \beta} \qquad \text{Ratio identity}$$

$$= -\frac{\cos \beta}{\sqrt{1 - \cos^2 \beta}} \quad \begin{array}{l} \sin \beta = -\sqrt{1 - \cos^2 \beta}. \text{ Use the negative} \\ \text{because } \sin \beta < 0 \text{ in quadrant III.} \end{array}$$

■ *Try Exercise* **106**, *page 335.*

EXAMPLE 6   **Evaluate a Trigonometric Function By Using Identities**

Given that $\tan \beta = -\frac{1}{2}$, where β is an angle whose terminal side is in the fourth quadrant. Find $\sec \beta$ and $\cos \beta$.

Solution $1 + \tan^2 \beta = \sec^2 \beta$ \qquad Pythagorean identity

$$1 + \left(-\frac{1}{2}\right)^2 = \sec^2 \beta \qquad \tan \beta = -1/2$$

$$\frac{5}{4} = \sec^2 \beta$$

$$\frac{\sqrt{5}}{2} = \sec \beta \qquad \begin{array}{l} \sec \beta > 0 \text{ in} \\ \text{quadrant IV} \end{array}$$

$$\cos \beta = \frac{1}{\sec \beta} = \frac{1}{\sqrt{5}/2} = \frac{2}{\sqrt{5}} = \frac{2\sqrt{5}}{5}$$

■ *Try Exercise* **98**, *page 335.*

EXERCISE SET 6.4 _____

In Exercises 1 to 6, find the arc length (t) intercepted by the given angle (θ) in a unit circle.

1. $\theta = \pi/3$ radians   **2.** $\theta = 2\pi/3$ radians
3. $\theta = 7\pi/4$ radians   **4.** $\theta = 45°$
5. $\theta = 120°$ **6.** $\theta = 225°$

In Exercises 7 to 12, find the angle (in radians) subtended by the given arc (t) in a unit circle.

7. $t = \pi/4$ **8.** $t = 2\pi/3$ **9.** $t = 7\pi/4$

10. $t = 5\pi/4$ **11.** $t = 2.4$ **12.** $t = 3.4$

In Exercises 13 to 18, find the angle (in degrees) subtended by the given arc (t) in a unit circle.

13. $t = \pi/6$ **14.** $t = 5\pi/3$ **15.** $t = 11\pi/6$
16. $t = 5\pi/4$ **17.** $t = 2.2$ **18.** $t = 1.76$

In Exercises 19 to 34, find the coordinates of the point $P(x, y)$ on a unit circle that corresponds to the given real number t.

19. $t = \pi/3$ **20.** $t = \pi/4$ **21.** $t = 3\pi/4$
22. $t = 7\pi/4$ **23.** $t = 7\pi/6$ **24.** $t = 4\pi/3$
25. $t = 5\pi/3$ **26.** $t = \pi/6$ **27.** $t = 11\pi/6$
28. $t = 0$ **29.** $t = \pi$ **30.** $t = 9\pi/4$
31. $t = -\pi/3$ **32.** $t = -7\pi/4$
33. $t = -2\pi/3$ **34.** $t = -\pi$

In Exercises 35 to 46, evaluate the circular functions.

35. $\sin 1.22$ **36.** $\cos 4.22$
37. $\tan 5$ **38.** $\sec 3.5$
39. $\csc(-1.05)$ **40.** $\sin(-0.55)$
41. $\tan 11\pi/12$ **42.** $\cot 2\pi/5$
43. $\cos(-\pi/5)$ **44.** $\csc 8.2$
45. $\sec 1.55$ **46.** $\cot 2.11$

In Exercises 47 to 56, make use of the period to evaluate each function.

47. $\sin 17\pi/4$ **48.** $\sin 37\pi/4$
49. $\cos 13\pi/6$ **50.** $\cos 31\pi/6$
51. $\sin 29\pi/3$ **52.** $\sin 43\pi/6$
53. $\cos 43\pi/3$ **54.** $\cos 65\pi/6$
55. $\sin 41\pi/3$ **56.** $\cos 29\pi/6$

In Exercises 57 to 64, use the unit circle to show that the following expressions are true.

57. $\cos(-\phi) = \cos \phi$ **58.** $\tan(\phi - 180°) = \tan \phi$
59. $\cos \phi = -\cos(\phi + 180°)$ **60.** $\sin(-\phi) = -\sin \phi$
61. $\sin(\phi - 180°) = -\sin \phi$ **62.** $\sec(-\phi) = \sec \phi$
63. $\csc(\alpha) = -\csc \alpha$ **64.** $\tan(-\alpha) = -\tan \alpha$

In Exercises 65 to 85, use the fundamental identities to transform each expression in terms of the sine and cosine and simplify.

65. $\tan \phi \cos \phi$ **66.** $\cot \phi \sin \phi$

67. $\dfrac{\csc \phi}{\cot \phi}$ **68.** $\dfrac{\sec \phi}{\tan \phi}$

69. $1 + \tan^2\phi$ **70.** $1 + \cot^2\phi$

71. $\dfrac{\tan \phi}{\sec \phi}$ **72.** $\dfrac{\cot \phi}{\csc \phi}$

73. $\tan \phi + \cot \phi$ **74.** $\sec \phi + \csc \phi$

75. $1 - \sec^2\phi$ **76.** $1 - \csc^2\phi$

77. $\tan \phi - \dfrac{\sec^2\phi}{\tan \phi}$ **78.** $\dfrac{\csc^2\phi}{\cot \phi} - \cot \phi$

79. $\sec^2\phi + \csc^2\phi$ **80.** $\cos \phi \sec^2\phi - \dfrac{\cos \phi}{\cot^2\phi}$

81. $\dfrac{1 - \cos^2\phi}{\tan^2\phi}$ **82.** $\dfrac{1 - \sin^2\phi}{\cot^2\phi}$

83. $\sec \phi - \tan \phi \sin \phi$ **84.** $\dfrac{1}{1 - \cos \phi} + \dfrac{1}{1 + \cos \phi}$

85. $\dfrac{1}{1 - \sin \phi} + \dfrac{1}{1 + \sin \phi}$

In Exercises 86 to 95, simplify the first expression to the second expression.

86. $\dfrac{\sec^2\phi - \tan^2\phi}{\sec^2\phi}$; $\cos^2\phi$ **87.** $\dfrac{1 - \cos^2\phi}{\cos^2\phi}$; $\tan^2\phi$

88. $\dfrac{\tan \phi + \cot \phi}{\tan \phi}$; $\csc^2\phi$ **89.** $\dfrac{\csc \phi - \sin \phi}{\csc \phi}$; $\cos^2\phi$

90. $\dfrac{1 - \sin^2\phi}{1 - \cos^2\phi}$; $\cot^2\phi$ **91.** $\dfrac{1 + \tan^2\phi}{1 + \cot^2\phi}$; $\tan^2\phi$

92. $\sin^2\phi(1 + \cot^2\phi)$; 1 **93.** $\cos^2\phi(1 + \tan^2\phi)$; 1

94. $1 + \cot^2\phi$; $\csc^2\phi$ **95.** $\dfrac{1}{\tan \phi \cot \phi}$; 1

In Exercises 96 to 107, use the fundamental identities to find the other circular functions of ϕ.

96. $\csc \phi = \sqrt{2}$; ϕ in quadrant I, find $\sin \phi$.
97. $\cos \phi = 1/2$; ϕ in quadrant IV, find $\sin \phi$.
98. $\sin \phi = 1/2$; ϕ in quadrant II, find $\tan \phi$.
99. $\cot \phi = \sqrt{3}/3$; ϕ in quadrant III, find $\cos \phi$.
100. $\sec \phi = -2$; ϕ in quadrant III, find $\tan \phi$.
101. $\cot \phi = -1$; ϕ in quadrant IV, find $\tan \phi$.
102. Write $\sin \phi$ in terms of $\cos \phi$.
103. Write $\tan \phi$ in terms of $\sec \phi$.
104. Write $\csc \phi$ in terms of $\cot \phi$.
105. Write $\sec \phi$ in terms of $\tan \phi$.
106. Write $\tan \phi$ in terms of $\sin \phi$.
107. Write $\cot \phi$ in terms of $\csc \phi$.

Supplemental Exercises

108. Use the unit circle and the triangles in the figure to write each function in terms of the length of a line segment: **a.** $\sin \phi$;   **b.** $\cos \phi$;   **c.** $\tan \phi$.

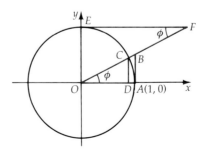

109. Use the unit circle and the triangles in the previous figure to write each function in terms of the length of a line segment: **a.** csc ϕ; **b.** sec ϕ;   **c.** cot ϕ.

In Exercises 110 to 119, simplify the first expression to the second expression.

110. $\dfrac{\sin^2\phi + \cos^2\phi}{\sin^2\phi}$; csc$^2\phi$

111. $\dfrac{\sin^2\phi + \cos^2\phi}{\cos^2\phi}$; sec$^2\phi$

112. $(\cos\phi - 1)(\cos\phi + 1)$; $-\sin^2\phi$

113. $(\sec\phi - 1)(\sec\phi + 1)$; $\tan^2\phi$

114. $\sec^2\beta - 1$; $\tan^2\beta$

115. $1 - \sin^2\beta$; $\cos^2\beta$

116. $\dfrac{\sec\phi}{\cos\phi} + \dfrac{\csc\phi}{\sin\phi}$; $\sec^2\phi\,\csc^2\phi$

117. $\dfrac{1}{\sin\phi\,\cos\phi}$; $\tan\phi + \cot\phi$

118. $1 + \tan^2\phi$; $\dfrac{1}{\cos^2\phi}$

119. $1 + \cot^2\phi$; $\csc^2\phi$

6.5

Graphs of the Sine and Cosine Functions

Graphing the Sine Function

The trigonometric functions of real numbers can be graphed on a rectangular coordinate system by plotting the points whose coordinates satisfy the function. We begin with the sine function.

Table 6.5 lists some of the ordered pairs (x, y), where $y = \sin x$ between 0 and 2π. Plot the ordered pairs and draw a smooth curve through the points to obtain the graph of $y = \sin x$.

TABLE 6.5

x	0	$\pi/4$	$\pi/2$	$3\pi/4$	π	$5\pi/4$	$3\pi/2$	$7\pi/4$	2π
$\sin x$	0	0.707	1	0.707	0	-0.707	-1	-0.707	0
(x, y)	$(0, 0)$	$\left(\dfrac{\pi}{4}, 0.707\right)$	$\left(\dfrac{\pi}{2}, 1\right)$	$\left(\dfrac{3\pi}{4}, 0.707\right)$	$(\pi, 0)$	$\left(\dfrac{5\pi}{4}, -0.707\right)$	$\left(\dfrac{3\pi}{2}, -1\right)$	$\left(\dfrac{7\pi}{4}, -0.707\right)$	$(2\pi, 0)$

Recall that the period of the sine function is 2π and the domain is the set of real numbers. Thus the graph of the sine function on the interval $[0, 2\pi]$ duplicates itself every 2π. Thus the graph in Figure 6.46 can be ex-

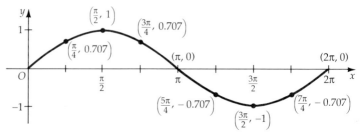

Figure 6.46
$y = \sin x,\ 0 \le x \le 2\pi$

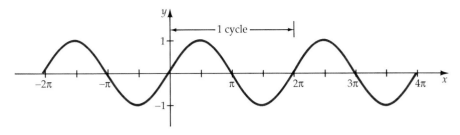

Figure 6.47
$y = \sin x, \ -2\pi \le x \le 4\pi$

tended indefinitely in both directions along the x-axes. The part of the graph corresponding to an interval of one period (2π) is referred to as one **cycle** of the graph.

Properties of the Sine Function

1. The domain of sin x is the set of real numbers.
2. The range of sin x is $[-1, 1]$.
3. The function sin x is periodic and the period is 2π.
4. The function sin x is an odd function. The graph of the function is symmetric with respect to the origin.

One period of the sine function can be graphed by using five key domain values; which include the maximum and minimum values of the sine as well as the zeros of the function (see Figure 6.48):

1. Beginning point of one cycle
2. Quarter point of one cycle
3. Middle point of one cycle
4. Three-quarter point of one cycle
5. Endpoint of one cycle

The maximum value attained by the function $f(x) = \sin x$ is 1 and the minimum value is -1. The **amplitude** of a sine or cosine function is defined as one-half the difference between the maximum and minimum val-

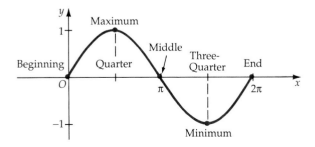

Figure 6.48

ues of the function. For example, if the maximum value of the function is M and the minimum value is m, then

$$\text{Amplitude} = \frac{1}{2}(M - m).$$

For $y = \sin x$, the amplitude is 1:

$$\text{Amplitude} = \frac{1}{2}(1 - (-1)) = 1.$$

Figure 6.49 shows the graph of $f(x) = 3 \sin x$. The graph can be drawn by plotting key points.

1. Beginning point: $f(0) = 3 \sin 0 = 0$.
2. Quarter point: $f(\pi/2) = 3 \sin \pi/2 = 3$ (maximum value).
3. Middle point $f(\pi) = 3 \sin \pi = 0$.
4. Three-quarter point: $f(3\pi/2) = 3 \sin 3\pi/2 = -3$ (minimum value).
5. Endpoint: $f(2\pi) = 3 \sin 2\pi = 0$.

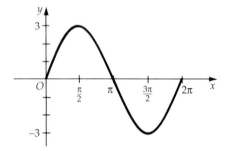

Figure 6.49

x	0	$\pi/2$	π	$3\pi/2$	2π
$\sin x$	0	1	0	-1	0
$3 \sin x$	0	3	0	-3	0
(x, y)	$(0, 0)$	$\left(\dfrac{\pi}{2}, 3\right)$	$(\pi, 0)$	$\left(\dfrac{3\pi}{2}, -3\right)$	$(2\pi, 0)$

Remark   Note that the value of the function $f(x) = 3 \sin x$ at the beginning, middle, and end of the period are zeros of the function. The y-value of the quarter point is the maximum of the function, and the y-value of the three-quarter point is the minimum of the function.

The amplitude of $f(x) = 3 \sin x$ is 3 because

$$\frac{1}{2}(M - m) = \frac{1}{2}(3 - (-3)) = 3.$$

From the graphs in Figures 6.48 and 6.49, the function $\sin x$ has an amplitude of 1, and the amplitude of $f(x) = 3 \sin x$ is 3. This suggests the following theorem.

Amplitude of $f(x) = a \sin x$

> Amplitude of function $f(x) = a \sin x$ is $|a|$.

EXAMPLE 1   Graph a Sine Function With an Amplitude of 2

Graph the sine function $f(x) = -2 \sin x$.

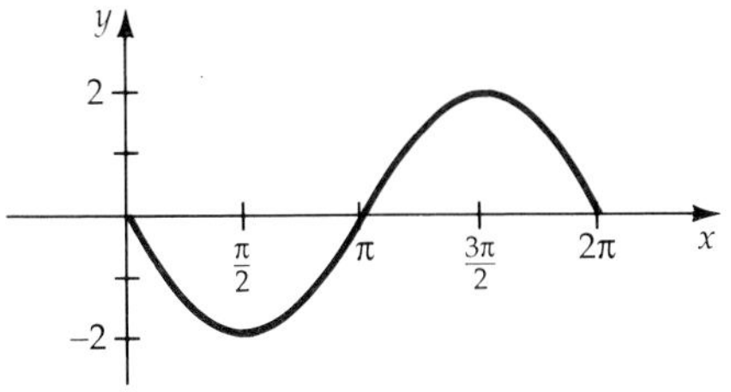

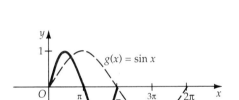

Figure 6.50
$f(x) = -2 \sin x$

Solution The amplitude of the function $f(x) = -2 \sin x$ is 2.

1. Beginning point: $f(0) = -2 \sin 0 = 0$.
2. Quarter point: $f(\pi/2) = -2 \sin \pi/2 = -2$ (minimum value)
3. Middle point: $f(\pi) = -2 \sin \pi = 0$.
4. Three-quarter point: $f(3\pi/2) = -2 \sin 3\pi/2 = 2$ (maximum value).
5. Endpoint: $f(2\pi) = -2 \sin 2\pi = 0$.

x	0	$\pi/2$	π	$3\pi/2$	2π
$\sin x$	0	1	0	-1	0
$-2 \sin x$	0	-2	0	2	0
(x, y)	$(0, 0)$	$\left(\dfrac{\pi}{2}, -2\right)$	$(\pi, 0)$	$\left(\dfrac{3\pi}{2}, 2\right)$	$(2\pi, 0)$

■ *Try Exercise* **20**, *page 345.*

Remark Note that the graph of $f(x) = -2 \sin x$ is a *reflection* across the x-axis of $2 \sin x$.

Figure 6.51 is the graph of the function $f(x) = \sin 2x$. The dashed graph is that of $g(x) = \sin x$. Because one cycle of the graph of $f(x) = \sin 2x$ is completed in an interval of length π, the period of f is π. Algebraically, one cycle of the function $f(x) = \sin 2x$ is completed as $2x$ varies from 0 to 2π. Thus

$$0 \le 2x \le 2\pi \quad \text{or} \quad 0 \le x \le \pi \quad \text{The period of } f(x) = \sin 2x \text{ is } \pi.$$

1. Beginning point: $f(0) = \sin 0 = 0$.
2. Quarter point: $f(\pi/4) = \sin \pi/2 = 1$.
3. Middle point: $f(\pi/2) = \sin \pi = 0$.
4. Three-quarter point: $f(3\pi/4) = \sin(3\pi/2) = -1$.
5. Endpoint: $f(\pi) = \sin 2\pi = 0$.

Figure 6.51

x	0	$\pi/4$	$\pi/2$	$3\pi/4$	π
$2x$	0	$\dfrac{\pi}{2}$	π	$\dfrac{3\pi}{2}$	2π
$\sin 2x$	0	1	0	-1	0
(x, y)	$(0, 0)$	$\left(\dfrac{\pi}{4}, 1\right)$	$\left(\dfrac{\pi}{2}, 0\right)$	$\left(\dfrac{3\pi}{4}, -1\right)$	$(\pi, 0)$

Figure 6.52 is the graph of the function $f(x) = \sin(x/2)$. Because one cycle of the graph of $f(x) = \sin(x/2)$ is completed in an interval of length 4π, the period of f is 4π. Algebraically, one cycle of the function $f(x) = \sin(x/2)$ is completed as $x/2$ varies from 0 to 2π. Thus

$$0 \le x/2 \le 2\pi \quad \text{or} \quad 0 \le x \le 4\pi \quad \text{The period of } f(x) = \sin x/2 \text{ is } 4\pi.$$

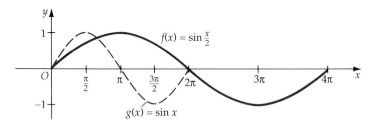

Figure 6.52

1. Beginning point: $f(0) = \sin 0 = 0$.
2. Quarter point: $f(\pi) = \sin \pi/2 = 1$.
3. Middle point: $f(2\pi) = \sin \pi = 0$.
4. Three-quarter point: $f(3\pi) = \sin(3\pi/2) = -1$.
5. Endpoint: $f(4\pi) = \sin 2\pi = 0$.

x	0	π	2π	3π	4π
$\dfrac{x}{2}$	0	$\dfrac{\pi}{2}$	π	$\dfrac{3\pi}{2}$	2π
$\sin\dfrac{x}{2}$	0	1	0	-1	0
(x, y)	$(0, 0)$	$(\pi, 1)$	$(2\pi, 0)$	$(3\pi, -1)$	$(4\pi, 0)$

The previous examples suggest that one cycle of the function $f(x) = \sin bx$, where $b > 0$, is completed as bx varies from 0 to 2π. Algebraically,

$$0 \leq bx \leq 2\pi$$

$$0 \leq x \leq \frac{2\pi}{b}. \quad \text{Divide by } b.$$

The length of the interval, $2\pi/b$, is the period of $f(x) = \sin bx$.

Remark   If $b > 0$, then $f(x) = \sin(-bx) = -\sin bx$. Therefore, the period is $2\pi/b$. Table 6.6 shows the amplitude and period for several sine functions.

TABLE 6.6

Function	Amplitude	Period
$f(x) = a \sin bx$	$\lvert a \rvert$	$\dfrac{2\pi}{b}$
$f(x) = 3 \sin(-2x)$	$\lvert 3 \rvert = 3$	$\dfrac{2\pi}{2} = \pi$
$f(x) = -\sin\dfrac{x}{3}$	$\lvert -1 \rvert = 1$	$\dfrac{2\pi}{1/3} = 6\pi$
$f(x) = -2 \sin\dfrac{3x}{4}$	$\lvert -2 \rvert = 2$	$\dfrac{2\pi}{3/4} = \dfrac{8\pi}{3}$

Here is a review of important properties used to graph the sine function $f(x) = a \sin bx$ where $0 \le x \le 2\pi/b$.

1. The amplitude of the function is $|a|$.
2. The period of the function is $2\pi/b$.
3. The zeros of the function are 0, π/b, and $2\pi/b$.

EXAMPLE 2   Graph a Sine Function

Graph one cycle of the function $f(x) = \dfrac{1}{2} \sin \dfrac{2\pi}{3} x$.

Solution   Amplitude: $\left|\dfrac{1}{2}\right| = \dfrac{1}{2}$,   Period: $\dfrac{2\pi}{2\pi/3} = 3$

Use the five key values to graph the function for values of x such that $0 \le x \le 3$.

1. Beginning point $f(0) = \frac{1}{2} \sin 0 = 0$.
2. Quarter point: $f(3/4) = \frac{1}{2} \sin \pi/2 = \frac{1}{2}$.
3. Middle point: $f(3/2) = \frac{1}{2} \sin \pi = 0$.
4. Three-quarter point: $f(9/4) = \frac{1}{2} \sin 3\pi/2 = -\frac{1}{2}$.
5. Endpoint: $f(3) = \frac{1}{2} \sin 2\pi = 0$.

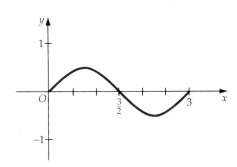

Figure 6.53

$$f(x) = \dfrac{1}{2} \sin \dfrac{2\pi}{3} x$$

x	0	$3/4$	$3/2$	$9/4$	3
$\dfrac{1}{2} \sin \dfrac{2\pi}{3} x$	0	$\dfrac{1}{2}$	0	$-\dfrac{1}{2}$	0
(x, y)	$(0, 0)$	$\left(\dfrac{3}{4}, \dfrac{1}{2}\right)$	$\left(\dfrac{3}{2}, 0\right)$	$\left(\dfrac{9}{4}, -\dfrac{1}{2}\right)$	$(3, 0)$

■ *Try Exercise **32**, page 345.*

The Cosine Function

Figure 6.54 shows the graph of the ordered pairs (x, y) where $y = \cos x$.

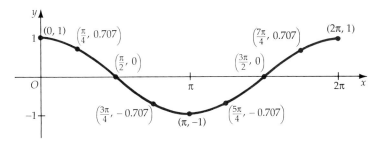

Figure 6.54
$y = \cos x$, $0 \le x \le 2\pi$

Table 6.7 lists the ordered pairs (x, y), where $y = \cos x$ between 0 and 2π.

TABLE 6.7

x	0	$\pi/4$	$\pi/2$	$3\pi/4$	π	$5\pi/4$	$3\pi/2$	$7\pi/4$	2π
$\cos x$	1	0.707	0	-0.707	-1	-0.707	0	0.707	1
(x, y)	$(0, 1)$	$\left(\dfrac{\pi}{4}, 0.707\right)$	$\left(\dfrac{\pi}{2}, 0\right)$	$\left(\dfrac{3\pi}{4}, -0.707\right)$	$(\pi, -1)$	$\left(\dfrac{5\pi}{4}, -0.707\right)$	$\left(\dfrac{3\pi}{2}, 0\right)$	$\left(\dfrac{7\pi}{4}, 0.707\right)$	$(2\pi, 1)$

Recall that the cosine function is periodic with period 2π. Figure 6.55 shows the graph of the cosine function along the x-axis for $-2\pi \leq x \leq 4\pi$. The graph of one cycle is repeated every 2π units along the horizontal axes.

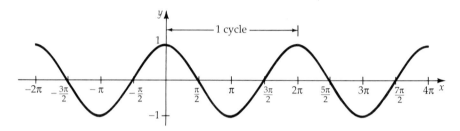

Figure 6.55
$f(x) = \cos x, \ -2\pi \leq x \leq 4\pi$

Properties of the Cosine Function

1. The domain of the function $\cos x$ is the set of real numbers.
2. The range of the function $\cos x$ is $[-1, 1]$.
3. The function $\cos x$ is periodic and the period is 2π.
4. The function $\cos x$ is an even function. The graph of $\cos x$ is symmetric to the y-axis.

The key points we used to graph the sine function can be used to graph the cosine function. Figure 6.56 shows the graph of $f(x) = 2 \cos x$.

1. Beginning point: $f(0) = 2 \cos 0 = 2$.
2. Quarter point: $f(\pi/2) = 2 \cos \pi/2 = 0$.
3. Middle point: $f(\pi) = 2 \cos \pi = -2$.
4. Three-quarter point: $f(3\pi/2) = 2 \cos 3\pi/2 = 0$.
5. Endpoint: $f(2\pi) = 2 \cos 2\pi = 2$.

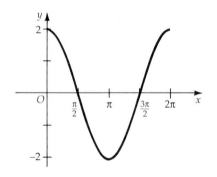

Figure 6.56
$f(x) = 2 \cos x$

x	0	$\pi/2$	π	$3\pi/2$	2π
$\cos x$	1	0	-1	0	1
$2 \cos x$	2	0	-2	0	2
(x, y)	$(0, 2)$	$\left(\dfrac{\pi}{2}, 0\right)$	$(\pi, -2)$	$\left(\dfrac{3\pi}{2}, 0\right)$	$(2\pi, 2)$

It is clear from the previous graphs that the amplitude of $f(x) = \cos x$ is 1 and the amplitude of $f(x) = 2 \cos x$ is 2.

Amplitude of $f(x) = a \cos x$

> Amplitude of the function $f(x) = a \cos x$ is $|a|$.

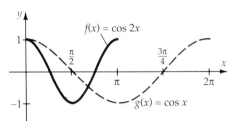

Figure 6.57

The five key points have been used in Figure 6.57 to graph $f(x) = \cos 2x$. (The dashed graph is that of $g(x) = \cos x$.) The period of $f(x) = \cos 2x$ is π.

1. Beginning point: $f(0) = \cos 0 = 1$.
2. Quarter point: $f(\pi/4) = \cos \pi/2 = 0$.
3. Middle point: $f(\pi/2) = \cos \pi = -1$.
4. Three-quarter point: $f(3\pi/4) = \cos 3\pi/2 = 0$.
5. Endpoint: $f(\pi) = \cos 2\pi = 1$.

x	0	$\pi/4$	$\pi/2$	$3\pi/4$	π
$\cos 2x$	1	0	-1	0	1
(x, y)	$(0, 1)$	$\left(\dfrac{\pi}{4}, 0\right)$	$\left(\dfrac{\pi}{2}, -1\right)$	$\left(\dfrac{3\pi}{4}, 0\right)$	$(\pi, 1)$

As with the sine function, the period of $f(x) = \cos bx$, where $b > 0$, is $2\pi/b$.

Remark If $b > 0$, $f(x) = \cos(-bx) = \cos bx$. Therefore, f has period $2\pi/b$.

Table 6.8 shows the amplitude and period for several cosine functions.

TABLE 6.8

Function	Amplitude	Period
$a \cos bx$	$|a|$	$\dfrac{2\pi}{b}$
$2 \cos 3x$	$|2|$	$\dfrac{2\pi}{3}$
$-3 \cos \dfrac{2x}{3}$	$|-3| = 3$	$\dfrac{2\pi}{2/3} = 3\pi$

Here is a review of important properties used to graph $f(x) = a\cos bx$ where $0 \le x \le 2\pi/b$.

1. The amplitude of $a\cos bx$ is $|a|$.
2. The period of $a \cos bx$ is $2\pi/b$.
3. The zeros of $a \cos bx$ are $\pi/2b$ and $3\pi/2b$.

EXAMPLE 3   **Graph a Cosine Function**

Graph one cycle of the function $f(x) = \dfrac{1}{2}\cos 3x$.

Solution   Amplitude: $\left|\dfrac{1}{2}\right| = \dfrac{1}{2}$ Period: $\dfrac{2\pi}{3}$

Use the five key values to graph the function for values of x for $0 \le x \le 2\pi/3$.

1. Beginning point: $f(0) = 1/2 \cos 0 = 1/2$.
2. Quarter point: $f(\pi/6) = 1/2 \cos \pi/2 = 0$.
3. Middle point: $f(\pi/3) = 1/2 \cos \pi = -1/2$.
4. Three-quarter point: $f(\pi/2) = 1/2 \cos 3\pi/2 = 0$.
5. Endpoint: $f(2\pi/3) = 1/2 \cos 2\pi = 1/2$.

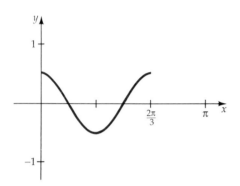

Figure 6.58

$f(x) = \dfrac{1}{2}\cos 3x$

x	0	$\pi/6$	$\pi/3$	$\pi/2$	$2\pi/3$
$\dfrac{1}{2}\cos 3x$	$1/2$	0	$-1/2$	0	$1/2$
(x, y)	$\left(0, \dfrac{1}{2}\right)$	$\left(\dfrac{\pi}{6}, 0\right)$	$\left(\dfrac{\pi}{3}, -\dfrac{1}{2}\right)$	$\left(\dfrac{\pi}{2}, 0\right)$	$\left(\dfrac{2\pi}{3}, \dfrac{1}{2}\right)$

■ *Try Exercise* **28,** *page 345.*

EXAMPLE 4   **Graph a Cosine Function**

Graph one cycle of the function $f(x) = -2 \cos \dfrac{\pi x}{2}$.

Solution   Amplitude: $|-2| = 2$,   Period: $\dfrac{2\pi}{\pi/2} = 4$

Use the five key values to graph the function for the values of x for $0 \le x \le 4$.

1. Beginning point: $f(0) = -2 \cos 0 = -2$.
2. Quarter point: $f(1) = -2 \cos \pi/2 = 0$.
3. Middle point: $f(2) = -2 \cos \pi = 2$.
4. Three-quarter point: $f(3) = -2 \cos 3\pi/2 = 0$.
5. Endpoint: $f(4) = -2 \cos 2\pi = -2$.

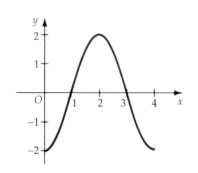

Figure 6.59

$f(x) = -2 \cos \dfrac{\pi x}{2}$

x	0	1	2	3	4
$-2 \cos \dfrac{\pi x}{2}$	-2	0	2	0	-2
(x, y)	$(0, -2)$	$(1, 0)$	$(2, 2)$	$(3, 0)$	$(4, -2)$

■ *Try Exercise* **34,** *page 345.*

EXAMPLE 5   **Graph the Absolute Value of the Cosine Function**

Graph the function $f(x) = |\cos x|$ from $0 \le x \le 2\pi$.

Solution   Since $|\cos x| \ge 0$, the graph of $f(x) = |\cos x|$ can be drawn by reflecting the negative portions of the graph $y = \cos x$ across the x-axes.

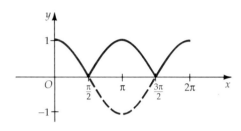

Figure 6.60
$f(x) = |\cos x|$

■ *Try Exercise* **40,** *page 345.*

EXERCISE SET 6.5

In Exercises 1 to 16, state the amplitude and period of the trigonometric function.

1. $f(x) = 2 \sin x$

2. $f(x) = -\dfrac{1}{2} \sin x$

3. $f(x) = \sin 2x$

4. $f(x) = \sin \dfrac{2x}{3}$

5. $f(x) = \dfrac{1}{2} \sin 2x$

6. $f(x) = 2 \sin \dfrac{x}{3}$

7. $f(x) = -2 \sin \dfrac{x}{2}$

8. $f(x) = -\dfrac{1}{2} \sin \dfrac{x}{2}$

9. $f(x) = \dfrac{1}{2} \cos x$

10. $f(x) = -3 \cos x$

11. $f(x) = \cos \dfrac{x}{4}$

12. $f(x) = \cos 3x$

13. $f(x) = 2 \cos \dfrac{x}{3}$

14. $f(x) = \dfrac{1}{2} \cos 2x$

15. $f(x) = -3 \cos \dfrac{2x}{3}$

16. $f(x) = \dfrac{3}{4} \cos 4x$

In Exercises 17 to 42, graph the trigonometric functions.

17. $f(x) = \dfrac{1}{2} \sin x$

18. $f(x) = \dfrac{3}{2} \cos x$

19. $f(x) = 3 \cos x$

20. $f(x) = \dfrac{3}{2} \sin x$

21. $f(x) = 4 \cos \dfrac{x}{2}$

22. $f(x) = 2 \cos \dfrac{3x}{4}$

23. $f(x) = -2 \cos \dfrac{x}{3}$

24. $f(x) = -\dfrac{4}{3} \cos 3x$

25. $f(x) = 2 \sin \pi x$

26. $f(x) = \dfrac{1}{2} \sin \dfrac{\pi x}{3}$

27. $f(x) = \dfrac{3}{2} \cos \dfrac{\pi x}{2}$

28. $f(x) = 3 \cos \dfrac{\pi x}{3}$

29. $f(x) = -4 \sin \dfrac{2\pi x}{3}$

30. $f(x) = 3 \cos \dfrac{3\pi x}{2}$

31. $f(x) = 2 \cos 2x$

32. $f(x) = \dfrac{1}{2} \sin 2.5x$

33. $f(x) = -2 \sin 1.5x$

34. $f(x) = -\dfrac{3}{4} \cos 5x$

35. $f(x) = \left| 2 \sin \dfrac{x}{2} \right|$

36. $f(x) = \left| \dfrac{1}{2} \sin 3x \right|$

37. $f(x) = |-2 \cos 3x|$

38. $f(x) = \left| -\dfrac{1}{2} \cos \dfrac{x}{2} \right|$

39. $f(x) = -\left| 2 \sin \dfrac{x}{3} \right|$

40. $f(x) = -\left| 3 \sin \dfrac{2x}{3} \right|$

41. $f(x) = -|3 \cos \pi x|$

42. $f(x) = -\left| 2 \cos \dfrac{\pi x}{2} \right|$

In Exercises 43 to 48, find an equation of each function shown in the accompanying graphs.

43.

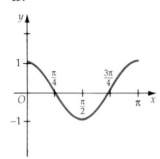

44.

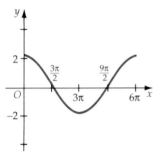

45.

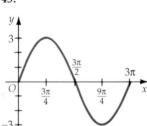

46.

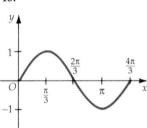

47.

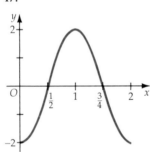

48.

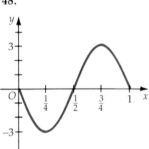

49. Sketch the graph of

$$f(x) = 2 \sin \frac{2x}{3} \text{ from } -3\pi \text{ to } 6\pi.$$

50. Sketch the graph of

$$f(x) = -3 \cos \frac{3x}{4} \text{ from } -2\pi \text{ to } 4\pi.$$

51. Sketch the graphs of

$$f(x) = 2 \cos \frac{x}{2} \text{ and } g(x) = 2 \cos x$$

on the same set of axes from -2π to 4π.

52. Sketch the graphs of

$$f(x) = \sin \pi x \text{ and } g(x) = \sin \frac{\pi x}{3}$$

on the same set of axes from -2 to 4.

Supplemental Exercises

53. Graph $f(x) = \sin^2 x$.

54. Graph $f(x) = 3^{\cos^2 x} \cdot 3^{\sin^2 x}$.

55. Graph $f(x) = \sin^2 x + \cos^2 x$.

56. Graph $f(x) = \cos |x|$.

In Exercises 57 and 58, discuss the symmetry of the trigonometric functions.

57. $f(x) = \sin x \cos x$

58. $f(x) = \dfrac{\sin x}{x}$

In Exercises 59 to 64, write an equation for a sine function with the given information.

59. Amplitude: 2, period: 3π

60. Amplitude: 5, period: $2\pi/3$

61. Amplitude: 0.5, period: $5\pi/4$

62. Amplitude: 1.5, period: 5π

63. Amplitude: 4, period: 2

64. Amplitude: 2.5, period: 3.2

In Exercises 65 to 70, write an equation for a cosine function with the given information.

65. Amplitude: 3, period: $\pi/2$

66. Amplitude: 0.8, period: 4π

67. Amplitude: 1.8, period: $3\pi/2$

68. Amplitude: 5, period: $7\pi/3$

69. Amplitude: 3, period: 2.5

70. Amplitude: 4.2, period: 1

71. A tidal wave that is caused by an earthquake under the sea is called a **tsunami wave**. These waves can be described by the formula $f(t) = A \cos Bt$. Find the equation of a tsunami wave that has a height of 60 feet, a period of 20 seconds, and travels at 120 feet per second. Graph two cycles of the equation and find the wavelength of one wave or the interval of one cycle.

6.6

Graphs of the Other Trigonometric Functions

The tangent function can be graphed by the same method of plotting points. Figure 6.61 shows the graph of the tangent function for the interval $-\pi/3 \le x \le \pi/3$.

TABLE 6.9

x	$-\pi/3$	$-\pi/4$	$-\pi/6$	0	$\pi/6$	$\pi/4$	$\pi/3$
$\tan x$	-1.732	-1	-0.577	0	0.577	1	1.732
(x, y)	$\left(-\dfrac{\pi}{3}, -1.732\right)$	$\left(-\dfrac{\pi}{4}, -1\right)$	$\left(-\dfrac{\pi}{6}, -0.577\right)$	$(0, 0)$	$\left(\dfrac{\pi}{6}, 0.577\right)$	$\left(\dfrac{\pi}{4}, 1\right)$	$\left(\dfrac{\pi}{3}, 1.732\right)$

Because $\cos \pi/2 = 0$,

$$\tan x = \frac{\sin x}{\cos x}$$

is not defined at $\pi/2$. Table 6.10 shows that as x approaches $\pi/2$ from the left, $\tan x$ becomes larger and larger (increases without bound). We say that the vertical line $x = \pi/2$ is an asymptote of the tangent function. Vertical asymptotes of certain trigonometric functions will occur at the values of the domain for which the function is undefined. Figure 6.62 shows the graph of the tangent function for the interval $-\pi/2 < x < \pi/2$.

TABLE 6.10

x	1.5	1.52	1.55	1.57	1.5707
$\tan x$	14.1	19.66	48.08	1255.8	10,381.3
(x, y)	$(1.5, 14.1)$	$(1.52, 19.66)$	$(1.55, 48.08)$	$(1.57, 1255.8)$	$(1.5707, 10,381.3)$

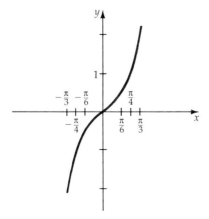

Figure 6.61
$f(x) = \tan x, \; -\pi/3 \le x \le \pi/3$

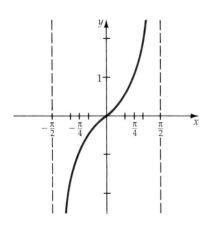

Figure 6.62
$f(x) = \tan x, \; -\pi/2 < x < \pi/2$

Remark   The asymptotes of $f(x) = \tan x$ will occur at $x = \pi/2 + k\pi$, where k is an integer since

$$\tan x = \frac{\sin x}{\cos x}$$

asymptotes will occur where the tangent function is undefined. The tangent function is undefined where the cosine function is zero, or where $x = \pi/2 + k\pi$, k is an integer.

Figure 6.63 shows the result of graphing the tangent function for values of x between $-3\pi/2$ and $3\pi/2$.

Important relationships of the function $f(x) = \tan x$ can be observed from the graph of the tangent function:

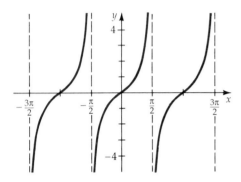

Figure 6.63
$f(x) = \tan x$, $-3\pi/2 < x < 3\pi/2$

1. The domain of $f(x) = \tan x$ is the set of real number except the values $x = \pi/2 + k\pi$. Vertical asymptotes occur at the points $x = \pi/2 + k\pi$, where k is an integer.

2. The range of $f(x) = \tan x$ is the set of real numbers.

3. The function $f(x) = \tan x$ is periodic, and the period is π.

4. Since $\tan(-x) = -\tan x$, $\tan x$ is an odd function. Why?[4] The graph of the function is symmetric with respect to the origin.

Key values also can be used in graphing one period of the tangent function $f(x) = a \tan bx$. Since the tangent function is not bounded, the tangent does not have an amplitude. The value a simply changes the rate of increase or decrease of the tangent function. The constant b determines the period of the function.

Since the period of $\tan x$ is π, the period of the tangent function $\tan bx$ is π/b. If $b < 0$, we use the identity $\tan(-bx) = -\tan bx$ to rewrite the function with $b > 0$. Normally we will show the graph of $f(x) = a \tan bx$ on the interval $-\pi/2b < x < \pi/2b$. Figure 6.64 shows the graph of the function $f(x) = 2 \tan 2x$.

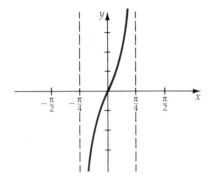

Figure 6.64
$f(x) = 2 \tan 2x$

Period: $\dfrac{\pi}{2}$

1. Beginning point: $f(-\pi/4) = 2 \tan(-\pi/2)$ (undefined).
2. Quarter point: $f(-\pi/8) = 2 \tan(-\pi/4) = -2$.
3. Middle point: $f(0) = 2 \tan 0 = 0$.
4. Three-quarter point: $f(\pi/8) = 2 \tan \pi/4 = 2$.
5. End point: $f(\pi/4) = 2 \tan \pi/2$ (undefined).

x	$-\pi/4$	$-\pi/8$	0	$\pi/8$	$\pi/4$
$2x$	$-\dfrac{\pi}{2}$	$-\dfrac{\pi}{4}$	0	$\dfrac{\pi}{4}$	$\dfrac{\pi}{2}$
$2 \tan 2x$	Undefined	-2	0	2	Undefined
(x, y)	—	$\left(-\dfrac{\pi}{8}, -2\right)$	$(0, 0)$	$\left(\dfrac{\pi}{8}, 2\right)$	—

[4] Since $\tan(-x) = \dfrac{\sin(-x)}{\cos(-x)} = \dfrac{-\sin x}{\cos x} = -\tan x$, $\tan x$ is an odd function.

We can also use the five-point method to graph tangent functions. The vertical asymptotes of the tangent functions occur where the function is undefined.

EXAMPLE 1   Graph a Tangent Function

Graph the tangent function $f(x) = 2 \tan \dfrac{x}{2}$ from $-\pi$ to 3π.

Solution   Period: $\dfrac{\pi}{1/2} = 2\pi$

Use the five-point method to graph one period of the function for values of x such that $-\pi < x < \pi$; then duplicate the graph on the interval $\pi < x < 3\pi$.

1. Beginning point: $f(-\pi) = 2 \tan(-\pi/2)$ (undefined).
2. Quarter point: $f(-\pi/2) = 2 \tan(-\pi/4) = -2$.
3. Middle point: $f(0) = 2 \tan 0 = 0$.
4. Three-quarter point: $f(\pi/2) = 2 \tan \pi/4 = 2$.
5. Endpoint: $f(\pi) = 2 \tan \pi/2$ (undefined).

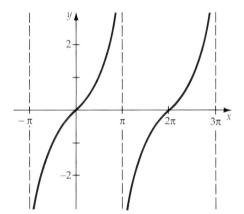

Figure 6.65

$f(x) = 2 \tan \dfrac{x}{2}$

x	$-\pi$	$-\pi/2$	0	$\pi/2$	π
$2 \tan \dfrac{x}{2}$	Undefined	-2	0	2	Undefined
(x, y)	—	$\left(-\dfrac{\pi}{2}, -2\right)$	$(0, 0)$	$\left(\dfrac{\pi}{2}, 2\right)$	—

■ *Try Exercise **30**, page 353.*

The Cotangent Function

Because the period of $\tan x$ is π and $\cot x = 1/\tan x$, the period of $\cot x$ is π. Since

$$\cot x = \frac{\cos x}{\sin x},$$

the cotangent function has vertical asymptotes where $\sin x = 0$; that is, when $x = k\pi$, where k is an integer.

Vertical asymptotes of $f(x) = \cot x$ will occur at $x = 0$ and $x = \pi$. Figure 6.66 shows the graph of one complete cycle of the cotangent function. The graph was achieved by using the five key values.

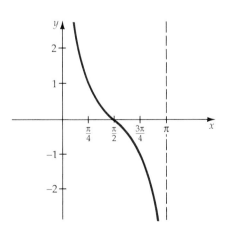

Figure 6.66
$f(x) = \cot x$

1. Beginning point: $f(0) = \cot 0$ (undefined).
2. Quarter point: $f(\pi/4) = \cot \pi/4 = 1$.
3. Middle point: $f(\pi/2) = \cot \pi/2 = 0$.
4. Three-quarter point: $f(3\pi/4) = \cot 3\pi/4 = -1$.
5. Endpoint: $f(x) = \cot \pi$ (undefined).

x	0	$\pi/4$	$\pi/2$	$3\pi/4$	π
$\cot x$	Undefined	1	0	-1	Undefined
(x, y)	—	$\left(\dfrac{\pi}{4}, 1\right)$	$\left(\dfrac{\pi}{2}, 0\right)$	$\left(\dfrac{3\pi}{4}, -1\right)$	—

Four important characteristics of the cotangent function $\cot x$ are now summarized:

1. The domain of $\cot x$ is the set of real numbers x except $x = k\pi$, k an integer. Vertical asymptotes occur, when $x = k\pi$, where k is an integer.
2. The range of $\cot x$ is the set of real numbers.
3. The function $\cot x$ is periodic, and the period is π.
4. The function $\cot x$ is an odd function. The function is symmetric to the origin.

Just as with the tangent function, the constant a in the function $f(x) = a \cot bx$ changes the rate of increase or decrease of the function. Since the period of $\cot x$ is π, the period of $a \cot bx$ is π/b. If $b > 0$, then $f(x) = \cot(-bx) = -\cot bx$. Therefore, the period of f is π/b. The interval of one period of the cotangent function is $0 \le x \le \pi/b$.

EXAMPLE 2   Graph a Cotangent Function

Graph one cycle of the function $f(x) = 3 \cot \dfrac{x}{2}$.

Solution   Period: $\dfrac{\pi}{1/2} = 2\pi$

1. Beginning point: $f(0) = 3 \cot 0$ (undefined).
2. Quarter point: $f(\pi/2) = 3 \cot \pi/4 = 3$.
3. Middle point: $f(\pi) = 3 \cot \pi/2 = 0$.
4. Three-quarter point: $f(3\pi/2) = 3 \cot 3\pi/4 = -3$.
5. Endpoint: $f(2\pi) = 3 \cot \pi$ (undefined).

x	0	$\pi/2$	π	$3\pi/2$	2π
$3 \cot \dfrac{x}{2}$	Undefined	3	0	-3	Undefined
(x, y)	—	$\left(\dfrac{\pi}{2}, 3\right)$	$(\pi, 0)$	$\left(\dfrac{3\pi}{2}, -3\right)$	—

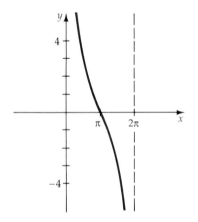

Figure 6.67

$f(x) = 3 \cot \dfrac{x}{2}$

■ *Try Exercise 32, page 353.*

The Cosecant Function

The cosecant function $\csc x$ is the reciprocal of the sine function, that is $\csc x = 1/\sin x$. Therefore $\csc x$ is undefined at those values for which $\sin x = 0$. Vertical asymptotes of the graph of $y = \csc x$ will occur at those values. To find the value of $\csc x$, we take the reciprocal of the values of sin x. In Table 6.11 we have calculated csc x using this procedure for selected values of x.

TABLE 6.11

x	0	$\pi/4$	$\pi/2$	$3\pi/4$	π	$5\pi/4$	$3\pi/2$	$7\pi/4$	2π
$\sin x$	0	0.707	1	0.707	0	−0.707	−1	−0.707	0
csc x	Undefined	1.414	1	1.414	Undefined	−1.414	−1	−1.414	Undefined
(x, y)	—	$\left(\dfrac{\pi}{4}, 1.414\right)$	$\left(\dfrac{\pi}{2}, 1\right)$	$\left(\dfrac{3\pi}{4}, 1.414\right)$	—	$\left(\dfrac{5\pi}{4}, -1.414\right)$	$\left(\dfrac{3\pi}{2}, -1\right)$	$\left(\dfrac{7\pi}{4}, -1.414\right)$	—

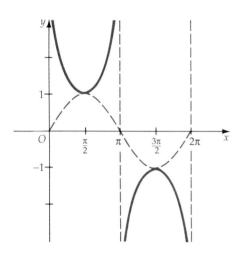

Figure 6.68
$f(x) = \csc x$

Remark The sine function has been sketched as a dashed line in Figure 6.68. Note the relationships among the zeros of the sine function and the asymptotes of the cosecant function. Note also that the turning points of the cosecant function will occur at the maximum and minimum points of the sine function.

Four important characteristics of the function $f(x) = \csc x$ are now summarized:

1. The domain of csc x is the set of real numbers, except at $x = k\pi$ where k is an integer. Vertical asymptotes occur at $x = k\pi$.
2. The range of csc x is the set of real numbers $\{y \mid y \geq 1, y \leq -1\}$.
3. The cosecant function has a period of 2π.
4. The function csc x is an odd function. The graph is symmetric with respect to the origin.

EXAMPLE 3 **Graph a Cosecant Function**

Graph one cycle of the function $f(x) = \csc 2x$.

> *Solution*   Period: $\dfrac{2\pi}{2} = \pi$

Use the five key values to graph the function for $0 < x < \pi$. Note that the graph of $g(x) = \sin 2x$ is helpful in sketching the graph.

1. Beginning point: $f(0) = \csc 0$ (undefined).
2. Quarter point: $f(\pi/4) = \csc \pi/2 = 1$.
3. Middle point: $f(\pi/2) = \csc \pi$ (undefined).
4. Three-quarter point: $f(3\pi/4) = \csc 3\pi/2 = -1$.
5. Endpoint: $f(\pi) = \csc 2\pi$ (undefined).

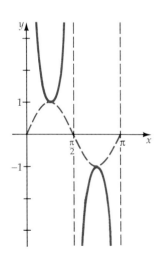

Figure 6.69
$f(x) = \csc 2x$

x	0	$\pi/4$	$\pi/2$	$3\pi/4$	π
csc $2x$	Undefined	1	Undefined	−1	Undefined
(x, y)	—	$\left(\dfrac{\pi}{4}, 1\right)$	—	$\left(\dfrac{3\pi}{4}, -1\right)$	—

■ *Try Exercise* **38**, *page 353.*

The Secant Function

The secant function is the reciprocal of the cosine function: $\sec x = 1/\cos x$. The secant function is undefined at those points for which $\cos x = 0$. Vertical asymptotes of the secant will occur at those values. The values of the secant function are reciprocals of the values of the cosine function. Table 6.12 contains values of $\sec x$ for selected values of x.

TABLE 6.12

x	0	$\pi/4$	$\pi/2$	$3\pi/4$	π	$5\pi/4$	$3\pi/2$	$7\pi/4$	2π
$\cos x$	1	0.707	0	-0.707	-1	-0.707	0	0.707	1
$\sec x$	1	1.414	Undefined	-1.414	-1	-1.414	Undefined	1.414	1
(x, y)	$(0, 1)$	$\left(\dfrac{\pi}{4}, 1.414\right)$	—	$\left(\dfrac{3\pi}{4}, -1.414\right)$	$(\pi, -1)$	$\left(\dfrac{5\pi}{4}, -1.414\right)$	—	$\left(\dfrac{7\pi}{4}, 1.414\right)$	$(2\pi, 1)$

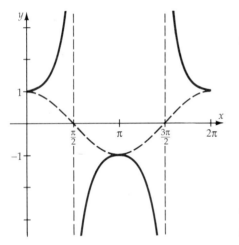

Figure 6.70
$f(x) = \sec x$

Four important relationships of the function $f(x) = \sec x$ are now summarized.

1. The domain of $\sec x$ is the set of real numbers x except $x = \pi/2 + k\pi$. Vertical asymptotes occur at $x = \pi/2 + k\pi$, where k is an integer.
2. The range of $\sec x$ is the set of real numbers $\{y \mid y \geq 1, y \leq -1\}$.
3. Since the cosine function has a period of 2π, the secant function has a period of 2π.
4. The function $\sec x$ is an even function. The graph of $y = \sec x$ is symmetric with respect to the y-axis.

EXAMPLE 4   Graph a Secant Function

Graph the function $f(x) = \sec 2x$ from 0 to 2π.

Solution   Period: $\dfrac{2\pi}{2} = \pi$

Use the five key values to graph the function for values of x for $0 \leq x \leq \pi$, then duplicate the graph on the interval $\pi \leq x \leq 2\pi$. Note that the graph of the corresponding cosine curve $g(x) = \cos 2x$ is helpful in sketching the graph.

1. Beginning point: $f(0) = \sec 0 = 1$.
2. Quarter point: $f(\pi/4) = \sec \pi/2$ (undefined).
3. Middle point: $f(\pi/2) = \sec \pi = -1$.
4. Three-quarter point: $f(3\pi/4) = \sec(3\pi/2)$ (undefined).
5. Endpoint: $f(\pi) = \sec 2\pi = 1$.

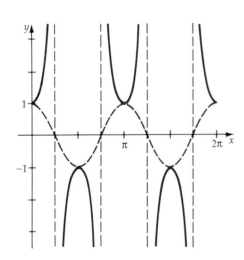

Figure 6.71
$f(x) = \sec 2x$

x	0	$\pi/4$	$\pi/2$	$3\pi/4$	π
$\sec 2x$	1	Undefined	-1	Undefined	1
(x, y)	$(0, 1)$	—	$\left(\dfrac{\pi}{2}, -1\right)$	—	$(\pi, 1)$

■ *Try Exercise* **42**, *page 353.*

EXERCISE SET 6.6

1. For what values of x is $y = \tan x$ undefined?

2. For what values of x is $y = \cot x$ undefined?

3. For what values of x is $y = \sec x$ undefined?

4. For what values of x is $y = \csc x$ undefined?

In Exercises 5 to 20, state the period of the given function.

5. $f(x) = \sec x$ **6.** $f(x) = \cot x$ **7.** $f(x) = \tan x$

8. $f(x) = \csc x$ **9.** $f(x) = 2 \tan \dfrac{x}{2}$ **10.** $f(x) = \dfrac{1}{2} \cot 2x$

11. $f(x) = \csc 3x$ **12.** $f(x) = \csc \dfrac{x}{2}$

13. $f(x) = -\tan 3x$ **14.** $f(x) = -3 \cot \dfrac{2x}{3}$

15. $f(x) = -3 \sec \dfrac{x}{4}$ **16.** $f(x) = -\dfrac{1}{2} \csc 2x$

17. $f(x) = \cot \pi x$ **18.** $f(x) = \cot \dfrac{\pi x}{3}$

19. $f(x) = 2 \csc \dfrac{\pi x}{2}$ **20.** $f(x) = -3 \cot \pi x$

In Exercises 21 to 40, sketch the graph of the given function.

21. $f(x) = 3 \tan x$ **22.** $f(x) = \dfrac{1}{3} \tan x$

23. $f(x) = \dfrac{3}{2} \cot x$ **24.** $f(x) = 4 \cot x$

25. $f(x) = 2 \sec x$ **26.** $f(x) = \dfrac{3}{4} \sec x$

27. $f(x) = \dfrac{1}{2} \csc x$ **28.** $f(x) = 2 \csc x$

29. $f(x) = 2 \tan \dfrac{x}{2}$ **30.** $f(x) = -3 \tan 3x$

31. $f(x) = -3 \cot \dfrac{x}{2}$ **32.** $f(x) = \dfrac{1}{2} \cot 2x$

33. $f(x) = -2 \csc \dfrac{x}{3}$ **34.** $f(x) = \dfrac{3}{2} \csc 3x$

35. $f(x) = \dfrac{1}{2} \sec 2x$ **36.** $f(x) = -3 \sec \dfrac{2x}{3}$

37. $f(x) = -2 \sec \pi x$ **38.** $f(x) = 3 \csc \dfrac{\pi x}{2}$

39. $f(x) = 3 \tan 2\pi x$ **40.** $f(x) = -\dfrac{1}{2} \cot \dfrac{\pi x}{2}$

41. Graph $f(x) = 2 \csc 3x$ from -2π to 2π

42. Graph $f(x) = \sec \dfrac{x}{2}$ from -4π to 4π

43. Graph $f(x) = 3 \sec \pi x$ from -2 to 4

44. Graph $f(x) = \csc \dfrac{\pi x}{2}$ from -4 to 4

45. Graph $f(x) = 2 \cot 2x$ from $-\pi$ to π

46. Graph $f(x) = \dfrac{1}{2} \tan \dfrac{x}{2}$ from -4π to 4π

47. Graph $f(x) = 3 \tan \pi x$ from -2 to 2

48. Graph $f(x) = \cot \dfrac{\pi x}{2}$ from -4 to 4

In Exercises 49 to 54, find an equation of each function.

49.

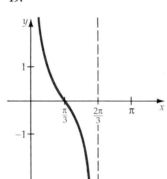

50.

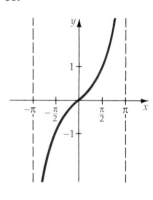

51.

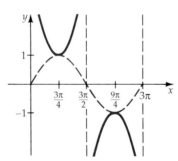

52.

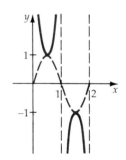

53.

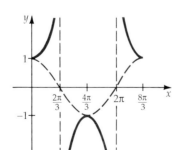

54.

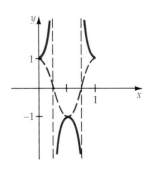

Supplemental Exercises

In Exercises 55 to 62, write an equation of the form $y = \tan bx$, $y = \cot bx$, $y = \sec bx$ or $y = \csc bx$ that satisifes the given conditions.

55. Tangent, period: $\pi/3$

56. Cotangent, period: $\pi/2$

57. Secant, period: $3\pi/4$

58. Cosecant, period: $5\pi/2$

59. Cotangent, period: 2

60. Tangent, period: 0.5

61. Cosecant, period: 1.5

62. Secant, period: 3

In Exercises 63 to 66, sketch the graph of the given function.

63. $f(x) = \tan |x|$

64. $f(x) = \sec |x|$

65. $f(x) = |\csc x|$

66. $f(x) = |\cot x|$

67. Graph $y = \tan x$ and $x = \tan y$ on the same coordinate axes.

68. Graph $y = \sin x$ and $x = \sin y$ on the same coordinate axes.

6.7

Phase Shift and Addition of Ordinates

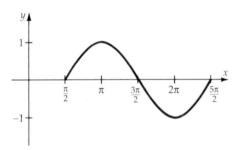

Figure 6.72
$f(x) = \sin(x - \pi/2)$

Recall that for $c > 0$, the graph of $y = f(x - c)$ is the graph of $y = f(x)$ shifted c units to the right on the x-axis and $y = f(x + c)$ is the graph of $y = f(x)$ shifted c units to the left on the x-axes. This property of functions that shifts a function horizontally is used to graph the functions of the form $f(x) = a \sin(bx \pm c)$ and $f(x) = a \cos(bx \pm c)$.

Figure 6.72 shows the graph of the function $f(x) = \sin(x - \pi/2)$. Each point on the curve $g(x) = \sin x$ is translated $\pi/2$ units to the right. This is called the **phase shift** of the graph.

We are now ready to graph sine or cosine functions of the form $y = a \cos(bx \pm c)$ or $y = a \sin(bx \pm c)$. The amplitude of each function is $|a|$ and the period is $2\pi/b$, $b > 0$. One cycle of the function can be determined by solving an inequality for x:

$$0 \le bx + c \le 2\pi$$

$$-\frac{c}{b} \le x \le -\frac{c}{b} + \frac{2\pi}{b}$$

The number $-\dfrac{c}{b}$ is the phase shift of the graph.

Properties of $y = a\,\sin(bx + c)$ and $y = a\,\cos(bx + c)$

> The graphs of
>
> $$y = a\,\sin(bx + c) \quad \text{and} \quad y = a\,\cos(bx + c)$$
>
> have the following properties:
>
> $$\text{amplitude} = |a|, \quad \text{period} = \frac{2\pi}{b}, \quad \text{phase shift} = -\frac{c}{b}.$$
>
> The interval containing one cycle can be found by solving the inequality $0 \le bx + c \le 2\pi$.

EXAMPLE 1   **Graph a Cosine Function with a Phase Shift**

Graph one cycle of the function $f(x) = \cos\left(2x + \dfrac{\pi}{2}\right)$.

   Solution   To determine the interval of one cycle, we consider the inequality

$$0 \leq 2x + \frac{\pi}{2} \leq 2\pi$$

$$-\frac{\pi}{2} \leq 2x \leq \frac{3\pi}{2}$$

$$-\frac{\pi}{4} \leq x \leq \frac{3\pi}{4}$$

Amplitude $= |a| = 1$,   Period $= \dfrac{2\pi}{2} = \pi$,   Phase Shift $= \dfrac{-\pi/2}{2} = -\dfrac{\pi}{4}$.

The period of the function is equal to $\dfrac{2\pi}{b} = \dfrac{2\pi}{2} = \pi$.

   Use the five key values starting at $-\dfrac{\pi}{4}$ and ending at $\dfrac{3\pi}{4}$.

x	$-\pi/4$	0	$\pi/4$	$\pi/2$	$3\pi/4$
$\cos\left(2x + \dfrac{\pi}{2}\right)$	1	0	-1	0	1
(x, y)	$\left(-\dfrac{\pi}{4}, 1\right)$	$(0, 0)$	$\left(\dfrac{\pi}{4}, -1\right)$	$\left(\dfrac{\pi}{2}, 0\right)$	$\left(\dfrac{3\pi}{4}, 1\right)$

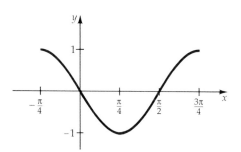

Figure 6.73

$f(x) = \cos\left(2x + \dfrac{\pi}{2}\right)$

■ *Try Exercise* **20**, *page 359.*

The Phase Shift for the Tangent and Cotangent Functions

The phase shift of the graphs

$$f(x) = a \tan(bx \pm c) \quad \text{or} \quad f(x) = a \cot(bx \pm c)$$

can be determined by using a technique similar to the one we used for $a \sin(bx + c)$.

Properties of $y = a \tan(bx + c)$ and $y = a \cot(bx + c)$

The graphs of

$$y = a \tan(bx + c) \quad \text{and} \quad y = a \cot(bx + c)$$

have the following properties:

$$\text{period} = \frac{\pi}{b}, \quad \text{phase shift} = -\frac{c}{b}.$$

Successive vertical asymptotes for $y = a \tan(bx + c)$ are the vertical lines $x = -c/b \pm \pi/2b$.
Successive vertical asymptotes for $y = a \cot(bx + c)$ are the vertical lines $x = -c/b$ and $x = -c/b + \pi$.

Remark   The interval containing one cycle of $y = a\tan(bx + c)$ can be found by solving the inequality $-\pi/2 < bx + c < \pi/2$. The corresponding interval for $y = a\cot(bx + c)$ can be found by solving the inequality $0 < bx + c < \pi$.

EXAMPLE 2   Graph a Tangent Function with a Phase Shift

Graph one cycle of the function $f(x) = 2\tan\left(x - \dfrac{\pi}{4}\right)$.

Solution   To determine the interval of one cycle of the graph solve the inequality.

$$-\frac{\pi}{2} < x - \frac{\pi}{4} < \frac{\pi}{2}$$

$$-\frac{\pi}{4} < x < \frac{3\pi}{4}$$

$$\text{period} = \pi, \qquad \text{phase shift} = \frac{\pi}{4}.$$

Use the five key values to graph one cycle of the function for values of x for $-\dfrac{\pi}{4} < x < \dfrac{3\pi}{4}$.

x	$-\pi/4$	0	$\pi/4$	$\pi/2$	$3\pi/4$
$2\tan\left(x - \dfrac{\pi}{4}\right)$	Undefined	-2	0	2	Undefined
(x, y)	—	$(0, -2)$	$\left(\dfrac{\pi}{4}, 0\right)$	$\left(\dfrac{\pi}{2}, 2\right)$	—

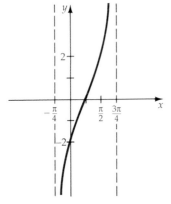

Figure 6.74

$f(x) = 2\tan\left(x - \dfrac{\pi}{4}\right)$

■ *Try Exercise* **22**, *page 359.*

The Graph of the Cosecant and Secant Functions

The graphs of the functions $y = a\csc(bx \pm c)$ or $y = a\sec(bx \pm c)$ can be found by finding the reciprocals of the corresponding sine or cosine functions and graphing point by point.

EXAMPLE 3   Graph a Cosecant Function

Graph one cycle of the function $f(x) = 2\csc(2x - \pi)$.

Solution   First, sketch $y = 2\sin(2x - \pi)$ as a dashed graph.

$$\text{Amplitude} = 2, \qquad \text{Period} = \pi, \qquad \text{Phase Shift} = \frac{\pi}{2}.$$

Because $\csc x = \dfrac{1}{\sin x}$, we use the reciprocal of the y-coordinates of $y = 2\sin(2x - \pi)$ to produce the graph of $y = 2\csc(2x - \pi)$.

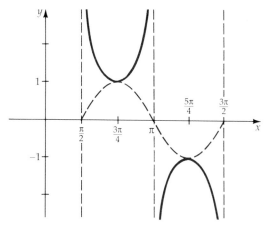

Figure 6.75
$$f(x) = 2 \csc(2x - \pi)$$

■ *Try Exercise* **28**, *page 360.*

Vertical Translations of Circular Functions

Recall that for $c > 0$ the function $y = f(x) + c$ is the graph of $y = f(x)$ shifted vertically c units. For example, the graph of $f(x) = \sin x + 1$ is the graph of $g(x) = \sin x$ shifted upward one unit, as shown in Figure 6.76.

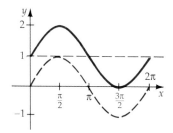

Figure 6.76

Remark Observe from the graph that a vertical shift has no effect on the amplitude, period, or phase shift.

EXAMPLE 4 **Graph a Sine Function with a Vertical Translation**

Graph one cycle of the function $f(x) = 2 \sin \left(2x - \dfrac{\pi}{2} \right) - 2$.

Solution First sketch $y = 2 \sin \left(2x - \dfrac{\pi}{2} \right)$.

$$\text{Amplitude} = 2, \qquad \text{Period} = \pi, \qquad \text{Phase Shift} = \dfrac{\pi}{4}$$

The graph of $f(x) = 2 \sin \left(2x - \dfrac{\pi}{2} \right) - 2$ is the graph of $y = 2 \sin \left(2x - \dfrac{\pi}{2} \right)$ shifted 2 units downward.

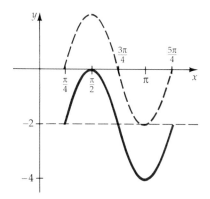

Figure 6.77

$$f(x) = 2 \sin \left(2x - \dfrac{\pi}{2} \right) - 2$$

■ *Try Exercise* **40**, *page 360.*

EXAMPLE 5 **Graph a Cosecant Function with a Vertical Translation**

Graph one cycle of the function $f(x) = \csc x + 1$.

Solution First sketch the graph of $y = \csc x$ as a dashed graph. (See Figure 6.78.) The graph of $f(x) = \csc x + 1$ is the graph of $y = \csc x$

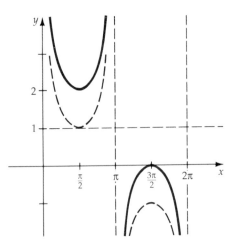

Figure 6.78
$f(x) = \csc x + 1$

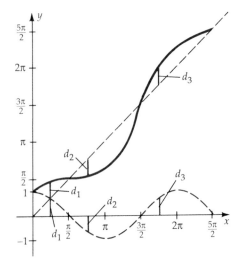

Figure 6.79
$f(x) = x + \cos x$

shifted one unit upward.

■ *Try Exercise* **44**, *page 360.*

Addition of Ordinates

Given two functions $g(x)$ and $h(x)$, the sum of the two functions is the function $f(x)$ defined by $f(x) = g(x) + h(x)$. The graph of the sum $f(x)$ can be obtained by graphing $g(x)$ and $h(x)$ separately and then geometrically adding the y coordinates of each function for a given x value. It is convenient when adding functions to pick zeros or maximum or minimum points of the function. Examples 6 and 7 illustrate this procedure.

EXAMPLE 6   Graph by Addition of Ordinates

Graph the function $f(x) = x + \cos x$ from 0 to 2π.

Solution   Graph the functions $g(x) = x$ and $h(x) = \cos x$ on the same coordinate system. Then add the y-coordinates geometrically point by point. The graph in Figure 6.79 shows the results of adding, by using a ruler, the y-coordinates of the two functions for selected values of x.

■ *Try Exercise* **50**, *page 360.*

EXAMPLE 7   Graph by Addition of Ordinates

Graph the function $f(x) = \sin x - \cos x$ from 0 to 2π.

Solution   The graph of $f(x) = \sin x - \cos x$ can be accomplished by sketching the graph of $y_1 = \sin x$ and the graph of $y_2 = -\cos x$ and adding the y-coordinates of selected values of x. A smooth curve is then drawn through the points.

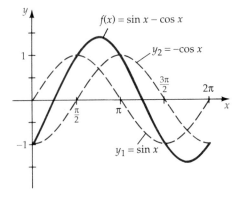

Figure 6.80
$f(x) = \sin x - \cos x$

■ *Try Exercise* **52**, *page 360.*

The points on the graph of the product of two functions can be found by multiplying the values of the y-coordinates at each given x-coordinate. It is convenient to choose values of x for which one of the functions is 0 or 1.

EXAMPLE 8   **Graph the Product of Functions**

Graph the function $f(x) = x \sin x$.

Solution   This is the product of $y_1 = x$ and $y_2 = \sin x$. The graph in Figure 6.81 was obtained by selecting values for which y_1 or y_2 is 0 or 1.

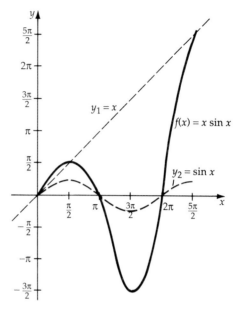

Figure 6.81
$f(x) = x \sin x$

■ *Try Exercise* **58,** *page* 360.

EXERCISE SET 6.7

In Exercises 1 to 8, find the amplitude, phase shift, and period for the given functions.

1. $f(x) = 2 \sin\left(x - \dfrac{\pi}{2}\right)$ **2.** $f(x) = -3 \sin(x + \pi)$

3. $f(x) = \cos\left(2x - \dfrac{\pi}{4}\right)$ **4.** $f(x) = \dfrac{3}{4} \cos\left(\dfrac{x}{2} + \dfrac{\pi}{3}\right)$

5. $f(x) = -4 \sin\left(\dfrac{2x}{3} + \dfrac{\pi}{6}\right)$ **6.** $f(x) = \dfrac{3}{2} \sin\left(\dfrac{x}{4} - \dfrac{3\pi}{4}\right)$

7. $f(x) = \dfrac{5}{4} \cos(3x - 2\pi)$ **8.** $f(x) = 6 \cos\left(\dfrac{x}{3} - \dfrac{\pi}{6}\right)$

In Exercises 9 to 16, find the phase shift and the period for the given functions.

9. $f(x) = 2 \tan\left(2x - \dfrac{\pi}{4}\right)$ **10.** $f(x) = \dfrac{1}{2} \tan\left(\dfrac{x}{2} - \pi\right)$

11. $f(x) = -3 \csc\left(\dfrac{x}{3} + \pi\right)$ **12.** $f(x) = -4 \csc\left(3x - \dfrac{\pi}{6}\right)$

13. $f(x) = 2 \sec\left(2x - \dfrac{\pi}{8}\right)$ **14.** $f(x) = 3 \sec\left(\dfrac{x}{4} - \dfrac{\pi}{2}\right)$

15. $f(x) = -3 \cot\left(\dfrac{x}{4} + 3\pi\right)$ **16.** $f(x) = \dfrac{3}{2} \cot\left(2x - \dfrac{\pi}{4}\right)$

In Exercises 17 to 32, graph one cycle of the given function.

17. $f(x) = \sin\left(x - \dfrac{\pi}{2}\right)$ **18.** $f(x) = \sin\left(x + \dfrac{\pi}{6}\right)$

19. $f(x) = \cos\left(\dfrac{x}{2} + \dfrac{\pi}{3}\right)$ **20.** $f(x) = \cos\left(2x - \dfrac{\pi}{3}\right)$

21. $f(x) = \tan\left(x + \dfrac{\pi}{4}\right)$ **22.** $f(x) = \tan(x - \pi)$

23. $f(x) = 2 \cot\left(\dfrac{x}{2} - \dfrac{\pi}{8}\right)$ **24.** $f(x) = \dfrac{3}{2} \cot\left(3x + \dfrac{\pi}{4}\right)$

25. $f(x) = \sec\left(x + \dfrac{\pi}{4}\right)$ **26.** $f(x) = \csc(2x + \pi)$

27. $f(x) = \csc\left(\dfrac{x}{3} - \dfrac{\pi}{2}\right)$ **28.** $f(x) = \sec\left(2x + \dfrac{\pi}{6}\right)$

29. $f(x) = -2\sin\left(\dfrac{x}{3} - \dfrac{2\pi}{3}\right)$ **30.** $f(x) = -\dfrac{3}{2}\sin\left(2x + \dfrac{\pi}{4}\right)$

31. $f(x) = -3\cos\left(3x + \dfrac{\pi}{4}\right)$ **32.** $f(x) = -4\cos\left(\dfrac{3x}{2} + 2\pi\right)$

In Exercises 33 to 46, graph the functions by using vertical translations.

33. $f(x) = \sin x - 1$ **34.** $f(x) = -\sin x + 1$

35. $f(x) = -\cos x - 2$ **36.** $f(x) = 2\sin x + 3$

37. $f(x) = \sin 2x - 2$ **38.** $f(x) = -\cos\dfrac{x}{2} + 2$

39. $f(x) = \sin\left(x - \dfrac{\pi}{2}\right) - \dfrac{1}{2}$

40. $f(x) = -2\cos\left(x + \dfrac{\pi}{3}\right) + 3$

41. $f(x) = \tan\dfrac{x}{2} - 4$ **42.** $f(x) = \cot 2x + 3$

43. $f(x) = \sec 2x - 2$ **44.** $f(x) = \csc\dfrac{x}{3} + 4$

45. $f(x) = \csc\dfrac{x}{2} - 1$ **46.** $f(x) = \sec\left(x - \dfrac{\pi}{2}\right) + 1$

In Exercises 47 to 56, graph the given functions by using the addition of ordinates.

47. $f(x) = x - \sin x$ **48.** $f(x) = \dfrac{x}{2} + \cos x$

49. $f(x) = x + \sin 2x$ **50.** $f(x) = \dfrac{2x}{3} - \sin\dfrac{x}{2}$

51. $f(x) = \sin x + \cos x$ **52.** $f(x) = -\sin x + \cos x$

53. $f(x) = \sin x - \cos\dfrac{x}{2}$ **54.** $f(x) = 2\sin 2x - \cos x$

55. $f(x) = 2\cos x + \sin\dfrac{x}{2}$ **56.** $f(x) = -\dfrac{1}{2}\cos 2x + \sin\dfrac{x}{2}$

In Exercises 57 to 62, graph the following functions.

57. $f(x) = \dfrac{x}{2}\sin x$ **58.** $f(x) = x\cos x$

59. $f(x) = x\sin\dfrac{x}{2}$ **60.** $f(x) = \dfrac{x}{2}\cos\dfrac{x}{2}$

61. $f(x) = x\sin\left(x + \dfrac{\pi}{2}\right)$ **62.** $f(x) = x\cos\left(x - \dfrac{\pi}{2}\right)$

In Exercises 63 to 68, find an equation of the trigonometric function from the accompanying graph.

63.

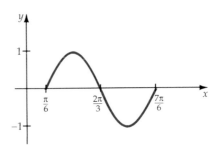

64.

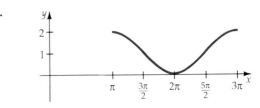

65.   **66.**

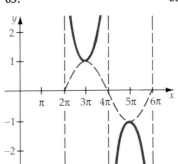
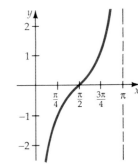

67.

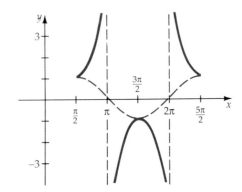

68.

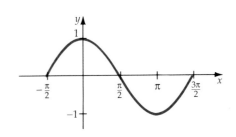

Supplemental Exercises

69. Find an equation of the sine function with amplitude 2, period π, and phase shift $\pi/3$.

70. Find an equation of the cosine function with amplitude 3, period 3π, and phase shift $-\pi/4$.

71. Find an equation of the tangent function with period 2π and phase shift $\pi/2$.

72. Find an equation of the cotangent function with period $\pi/2$ and phase shift $-\pi/4$.

73. Find an equation of the secant function with period 4π and phase shift $3\pi/4$.

74. Find an equation of the cosecant function with period $3\pi/2$ and phase shift $\pi/4$.

75. If $f(x) = \sin^2 x$ and $g(x) = \cos^2 x$, find $f(x) + g(x)$.

76. If $f(x) = 2 \sin x - 3$ and $g(x) = 4 \cos x + 2$, find the sum $f(x) + g(x)$.

77. If $f(x) = x^2 + 2$ and $g(x) = \cos x$, find $f[g(x)]$.

78. If $f(x) = \sin x$ and $g(x) = x^2 + 2x + 1$, find $g[f(x)]$.

In Exercises 79 to 82, sketch the graph of the given function.

79. $f(x) = \dfrac{\sin x}{x}$

80. $f(x) = 2 + \sec \dfrac{x}{2}$

81. $f(x) = |x| \sin x$

82. $f(x) = |x| \cos x$

83. The average depth at a pier in a small port is 9 feet. During a 24-hour day, the tides raise and lower the depth of water at the pier as shown in the figure. Write an equation in the form $f(t) = A \cos Bt + k$ and find the depth of the water at 6 P.M.

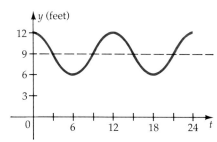

t is the number of hours from 6 AM

84. During a summer day, the ground temperature at a desert location was recorded and graphed as a function of time as shown in the figure. The graph can be approximated by the equation $f(x) = A \cos(bx + c) + k$. Find the equation and find the approximate temperature at 1:00 P.M.

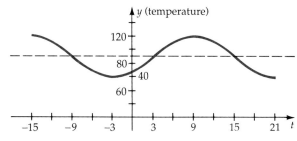

t is the number of hours from 6 AM

6.8

Simple Harmonic Motion—An Application of the Sine and Cosine Functions

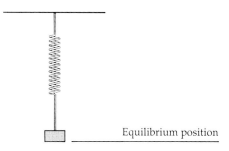

Figure 6.82

Many phenomena occur in nature that can be modeled by periodic functions, including vibrations in buildings, sound waves, electromagnetic waves, and vibrations of a swing or in a spring. These phenomena can be described by the *sinusoidal* functions, which are the sine and cosine functions or the sum of these two functions.

We will consider a mass on a spring to illustrate vibratory motion. Assume that we have placed a mass on a spring and allowed the spring to come to rest, as shown in Figure 6.82. The system is said to be in equilibrium when the mass is at rest. The point of rest is called the origin of the system. We consider the distance above the equilibrium point as positive and the distance below the equilibrium point as negative.

If the mass is now lifted a distance a and released, the mass will oscillate up and down in periodic motion, which means that the motion repeats itself in a certain period of time. The distance a is called the displacement from the origin. The number of times the mass oscillates in 1 second is called the frequency of the motion and the time for one oscillation is the period of the motion. For small oscillations, this period is a constant and the motion is referred to as simple harmonic motion. Figure 6.83 shows the position y of the mass for one oscillation for $t = 0$, $p/4$, $p/2$, $3p/4$, and p when the period is p.

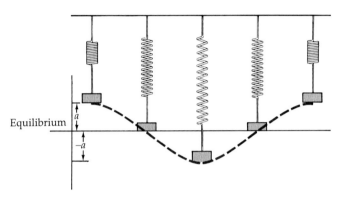

Figure 6.83

Remark   Note that if we were to graph the displacement y as a function of t and draw a smooth line through the points, we would have a cosine curve.

There is a relationship between the frequency and the period. Assume that we have a mass that will make two oscillations (an oscillation is a back-and-forth motion) in 1 second. The time for one oscillation is $\frac{1}{2}$ second. Thus the period is $\frac{1}{2}$ second. The frequency and the period are related by the formula $f = 1/\text{period}$.

The maximum displacement from the equilibrium position is called the amplitude of the motion. Vibratory motion can be quite complicated. However, the motion that we have described with the mass on the spring is called simple harmonic motion and can be described by the following equation.

Definition of Simple Harmonic Motion

> Simple harmonic motion is motion that can be modeled by one of the following equations:
>
> $$y = a \cos 2\pi ft \qquad \text{or} \qquad y = a \sin 2\pi ft$$
>
> where a is the amplitude (maximum displacement), f is the frequency, and $1/f$ is the period, y is the displacement and t is the time.

Remark   We have been given two equations of simple harmonic motion. The cosine function is used if the displacement from the origin is at a maximum at time $t = 0$. The sine function is used if the displacement at time $t = 0$ is zero.

EXAMPLE 1   Find the Equation of Motion of a Mass on a Spring

A mass on a spring has been displaced 4 centimeters above the equilibrium point and released. The mass is vibrating with a frequency of $\frac{1}{2}$ cycles per second. Write the equation of motion and graph three cycles of the displacement as a function of time.

Solution   Since the maximum displacement is 4 centimeters when $t = 0$, use $y = a \cos 2\pi ft$.

$$y = a \cos 2\pi ft \qquad \text{Equation for simple harmonic motion}$$

$$= 4 \cos 2\pi\left(\frac{1}{2}\right)t \quad a = 4, f = \tfrac{1}{2}.$$

$$= 4 \cos \pi t$$

■ *Try Exercise **20**, page 365.*

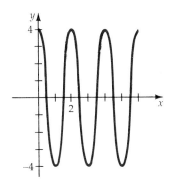

Figure 6.84
$y = 4 \cos \pi t$

From physical laws determined by experiment, the frequency of oscillation of a mass on a spring is given by

$$f = \frac{1}{2\pi}\sqrt{\frac{k}{m}} \quad \begin{array}{l} k \text{ is a spring constant determined} \\ \text{by experiment and } m \text{ is the mass.} \end{array}$$

The motion of the mass on the spring can then be described by

$$y = a \cos 2\pi ft = a \cos 2\pi\left(\frac{1}{2\pi}\sqrt{\frac{k}{m}}\right)t$$

$$= a \cos \sqrt{\frac{k}{m}}t.$$

The equation of motion for zero displacement at $t = 0$ is

$$y = a \sin \sqrt{\frac{k}{m}}t.$$

EXAMPLE 2   Find the Equation of Motion of a Mass on a Spring

A mass of 2 units is in equilibrium suspended from a spring. The mass is pulled down 0.5 units and released. Find the period, frequency, and amplitude of the resulting motion. Write the equation of the motion if $k = 18$, and graph two cycles of the displacement as a function of time.

Solution   At the start of the motion, the displacement is at a maximum but in the negative direction. The resulting motion is described by

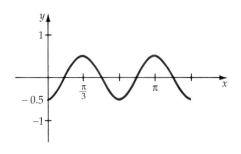

Figure 6.85
$y = -0.5 \cos 3t$

the cosine function. We know that $a = -0.5$, $k = 18$, $m = 2$.

$$y = a \cos \sqrt{\frac{k}{m}}\, t = -0.5 \cos \sqrt{\frac{18}{2}}\, t \quad \text{Substitute for } a,\, k,\, \text{and } m.$$

$$= -0.5 \cos 3t \qquad\qquad\qquad \text{Equation of motion}$$

Period: $\dfrac{2\pi}{3}$,   Frequency: $\dfrac{3}{2\pi}$,   Frequency = 1/period.

Amplitude: 0.5

■ *Try Exercise **28**, page 365.*

A simple pendulum is a system that exhibits approximate simple harmonic motion. A simple pendulum consists of a mass suspended from a string that is attached to a fixed point. If the mass is displaced through an angle θ and released, the pendulum will oscillate back and fourth in a plane. It can be shown from physical laws that if the angle θ is small, the equation of motion is approximated by the equation

$$y = a \cos 2\pi ft \qquad \text{or} \qquad y = a \sin 2\pi ft$$

where y is the displacement at time t, a the maximum displacement, and f the frequency of motion. The period of the motion is $1/f$. For the pendulum, we will consider all displacements positive.

From measurements taken from experiments on a simple pendulum,

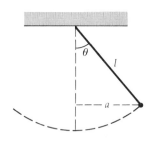

Figure 6.86

$$f = \frac{1}{2\pi} \sqrt{\frac{g}{l}}$$

where g is the gravitational constant 32 feet per second squared, and l is the length of the pendulum in feet.

Thus the equation for the motion of a pendulum is given by

$$y = a \cos \sqrt{\frac{g}{l}}\, t \qquad \text{or} \qquad y = a \sin \sqrt{\frac{g}{l}}\, t.$$

EXAMPLE 3   **Find the Period and Frequency of the Motion of the Pendulum**

Find the period and frequency of a pendulum with a length of 8 feet. Graph two cycles of the motion if the maximum displacement is 1.5 feet and the displacement is zero at $t = 0$.

Solution   Use the formula to find the period and the frequency.

Period: $\dfrac{2\pi}{\sqrt{g/l}} = \dfrac{2\pi}{\sqrt{32/8}} = \dfrac{2\pi}{2} = \pi$,   Frequency: 1/period $= 1/\pi$

To graph the motion, find the equation of motion. We use the equation

$$y = a \sin \sqrt{\frac{g}{l}}\, t \text{ because the displacement is zero at } t = 0.$$

$$y = 1.5 \sin 2t \quad a = 1.5, \text{ period} = \pi.$$

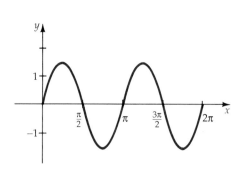

Figure 6.87
$y = 1.5 \sin 2t$

■ *Try Exercise **30**, page 365.*

EXERCISE SET 6.8

In Exercises 1 to 8, find the amplitude, period, and frequency of the harmonic motion.

1. $y = 2 \sin 2t$

2. $y = \dfrac{2}{3} \cos \dfrac{t}{3}$

3. $y = 3 \cos \dfrac{2t}{3}$

4. $y = 4 \sin 3t$

5. $y = 4 \cos \pi t$

6. $y = 2 \sin \dfrac{\pi t}{3}$

7. $y = \dfrac{3}{4} \sin \dfrac{\pi t}{2}$

8. $y = 5 \cos 2\pi t$

In Exercises 9 to 12, write the equation of motion and graph the amplitude as a function of time for the following given values. Assume that the motion is harmonic motion and maximum displacement occurs at $t = 0$.

9. frequency = 1.5 cycles per second, $a = 4$ inches

10. frequency = 0.8 cycle per second, $a = 4$ centimeters

11. period = 1.5 seconds, $a = \frac{3}{2}$ feet

12. period = 0.6 second, $a = 1$ meter

In Exercises 13 to 18, find the equation of simple harmonic motion with the given conditions. Assume zero displacement at $t = 0$.

13. Amplitude 2 centimeters, period π seconds

14. Amplitude 4 inches, period $\pi/2$ seconds

15. Amplitude 1 inch, period 2 seconds

16. Amplitude 3 centimeters, period 1 second

17. Amplitude 2 centimeters, frequency 1 second

18. Amplitude 4 inches, frequency 4 seconds

In Exercises 19 to 26, write an equation for simple harmonic motion. Assume that the maximum displacement occurs when $t = 0$.

19. Amplitude $\frac{1}{2}$ centimeters, frequency $2/\pi$ cycles per second

20. Amplitude 3 inches, frequency $1/\pi$ cycles per second

21. Amplitude 2.5 inches, frequency 0.5 cycles per second

22. Amplitude 5 inches, frequency $\frac{1}{8}$ cycles per second

23. Amplitude $\frac{1}{2}$ inch, period 3 seconds

24. Amplitude 5 centimeters, period 5 seconds

25. Amplitude 4 inches, period $\pi/2$ seconds

26. Amplitude 2 centimeters, period π seconds

27. A mass of 32 units is in equilibrium suspended from a spring. The mass is pulled down 2 feet and released. Find the period, frequency, and amplitude of the resulting motion. Write the equation of motion. Let $k = 8$.

28. A mass of 27 units is in equilibrium suspended from a spring. The mass is pulled down 1.5 feet and released.

Find the period, frequency, and amplitude of the resulting motion. Write the equation of motion. Let $k = 3$.

29. A pendulum 6 feet long is displaced a distance of 1 foot and released. Write the equation for the displacement as a function of time. Find the period and frequency of the motion.

30. A pendulum 20 feet long is displaced a distance of 4 feet and released. Write the equation for the displacement as a function of time. Find the period and frequency of the motion.

31. A mass of 5 units is suspended from a spring. The spring is compressed 6 inches and released. Find the period, frequency, and amplitude of the resulting motion. The constant $k = 2$. Write the equation of motion.

32. A pendulum of 3 feet long is displaced a distance of 6 inches and released. Find the period and frequency of the motion and write the equation of motion.

Supplemental Exercises

33. A pendulum with a length of 4 feet is released with an initial displacement of 6 inches. Write the equation for the displacement and find the frequency of the motion.

34. A mass of 0.5 unit is suspended from a spring with a constant of 32. The mass is displaced a distance of 10 inches and released. Write the equation for the motion and find the frequency of the motion.

35. A weight on a spring is displaced 6 inches from its equilibrium position and then released. The weight oscillated with a frequency of 1.5 seconds. Find the period and the equation of the motion.

36. A weight on a spring is displaced 9 inches from its equilibrium position and then released. The weight oscillated with a frequency of 2 seconds. Find the period and the equation of the motion.

37. What effect does doubling the length of a pendulum have on the period of the pendulum?

38. The length of a pendulum is changed until the period is cut in half. Find the length of the pendulum as compared to the original length.

Chapter 6 Review

6.1   **Measuring Angles**

An angle is formed by rotating a ray about a point 0.

An angle in standard position has the initial side along the positive x-axis and the vertex at the origin of the coordinate axes.

Coterminal angles have the same initial and terminal sides.

Degrees and radians are used for the measure of angles.

Arc length of a circle is the product of the radius r and central angle θ in radians is $s = r\theta$.

The area of a sector of radius r and central angle θ is given by the formula $A = \frac{1}{2}r^2\theta$.

Angular velocity is given by the formula: $\omega = \theta/t$.

The velocity of a point on a rotating body is given by the formula $V = r\omega$.

6.2   **Trigonometric Functions of Acute Angles**

The six trigonometric functions of an angle θ in standard position are defined by

$$\sin \theta = \frac{b}{r} \qquad \cos \theta = \frac{a}{r} \qquad \tan \theta = \frac{b}{a}, a \neq 0$$

$$\csc \theta = \frac{r}{b}, b \neq 0 \qquad \sec \theta = \frac{r}{a}, a \neq 0 \qquad \cot \theta = \frac{a}{b}, b \neq 0$$

6.3   **Trigonometric Functions of Any Angle**

Quadrantal angles are angles in which the terminal side coincides with the x- or y-axis.

A reference angle is the positive acute angle made by the terminal side of an angle in standard position and the x-axis.

6.4   **Circular Functions**

A circle of radius one is a unit circle.

The circular functions are defined in terms of the unit circle and the domain of the circular functions is the set of real numbers. If $P(x, y)$ is a point on a unit circle corresponding to the real number t, then

$$\cos t = x \qquad \sin t = y \qquad \tan t = \frac{y}{x}, x \neq 0$$

$$\sec t = \frac{1}{x}, x \neq 0 \qquad \csc t = \frac{1}{y}, y \neq 0 \qquad \cot t = \frac{x}{y}, y \neq 0$$

The reciprocal identities are:

$$\sin \theta = \frac{1}{\csc \theta} \qquad \cos \theta = \frac{1}{\sec \theta} \qquad \tan \theta = \frac{1}{\cot \theta}$$

The ratio identities are

$$\tan \theta = \frac{\sin \theta}{\cos \theta} \qquad \cos \theta = \frac{\cos \theta}{\sin \theta}$$

The Pythagorean identities are

$$\cos^2 \theta + \sin^2 \theta = 1 \qquad 1 + \tan^2 \theta = \sec^2 \theta \qquad 1 + \cot^2 \theta = \csc^2 \theta$$

6.5   Graphs of the Sine and Cosine Functions

The amplitude of $y = a \sin bx$ and $y = a \cos bx$ is $|a|$. The period of each function is given by $2\pi/b$ ($b > 0$).

6.6   Graphs of the Other Trigonometric Functions

The period of $y = a \tan bx$ and $y = a \cot bx$ is π/b. The graphs of $y = a \sec bx$ and $y = a \csc bx$ are obtained by using the reciprocals of the y-coordinates of the graphs of $y = a \cos bx$ and $y = a \sin bx$, respectively.

6.7   Phase Shift and Addition of Ordinates

The phase shift of $y = a \sin(bx + c)$ and $y = a \cos(bx + c)$, where $b > 0$, is $-c/b$.
    The phase shift of $y = a \tan(bx + c)$ and $y = \cot(bx + c)$ is $-c/b$.
    The graph of the sum of two trigonometric functions can be obtained by the method of addition of ordinates.

6.8   Simple Harmonic Motion

The equations of simple harmonic motion are $y = a \cos 2\pi f t$ or $y = a \sin 2\pi f t$, where a is the amplitude and f is the frequency.

CHALLENGE EXERCISES

In Exercises 1 to 14, answer true or false. If the statement is false, give an example.

1. An angle is in standard position when the vertex is at the origin of a coordinate system.

2. The angle θ is in radians in standard position with the terminal side in the second quadrant. The reference angle of θ is $\pi - \theta$.

3. In the formula $s = r\theta$, the angle θ must be measured in radians.

4. If $\tan \theta < 0$ and $\cos \theta > 0$, then the terminal side θ is in the third quadrant.

5. $\sec^2 \theta + \tan^2 \theta = 1$ is an identity.

6. The amplitude of the function $f(x) = 2 \tan x$ is 2.

7. The period of the function $\cos \theta$ is π.

8. The graph of the function $\sin \theta$ is symmetric to the origin.

9. For any acute angle θ, $\sin \theta + \cos(90° - \theta) = 1$.

10. $\sin(x + y) = \sin x + \sin y$.

11. $\sin^2 x = \sin x^2$.

12. The phase shift of the function $f(x) = 2 \sin\left(2x - \dfrac{\pi}{3}\right)$ is $\dfrac{\pi}{3}$.

13. One radian has approximately the same measure as one degree.

14. The measure of one radian differs depending on the radius of the circle used.

REVIEW EXERCISES

1. Sketch the angles in standard position: a. 120°; b. −135°.

2. Convert the angular measure to decimal form: a. 37° 34′; b. −142° 46′ 8″.

3. Convert the angular measurement to DMS: a. 114.8°; b. −38.38°.

4. Convert the radian measure to degree measure: a. $7\pi/4$ rad;   b. 2 rad.

5. Convert the angle measurement to radian measure: a. 315°;   b. 97.4°.

6. Find the arc length subtended by an angle with a measure of 75° in a circle with a 3-meter radius.

7. Find the arc length subtended by an angle with a measure of 42° in a circle with a 5-inch radius.

8. Find the radian measure of a central angle subtended by an arc length of 12 centimeters in a circle with a 40-centimeter radius.

9. Find the degree measure of a central angle subtended by an arc length of 5 inches in a circle with a 20-inch radius.

10. Find the area of a sector of a circle with a 16-inch radius. The central angle of the sector is 40°.

11. Find the area of a sector of a circle with a 4-meter radius. The central angle of the sector is 25°.

12. A car with a 16-inch wheel is moving with a velocity of 50 miles per hour. Find the angular velocity of the wheel in radians per second.

13. A truck with an 18-inch wheel is moving with a velocity of 55 miles per hour. Find the angular velocity of the wheel in radians per second.

14. Find the six trigonometric function values of an angle in standard position with the point $P(-2, 5)$ on the terminal side of the angle.

15. Find the six trigonometric function values of an angle in standard position with the point $P(1, -3)$ on the terminal side of the angle.

In Exercises 16 to 19, $\csc \theta = 3/2$, in the first quadrant. Evaluate the following trigonometric functions.

16. $\cos \theta$ 17. $\cot \theta$ 18. $\sin \theta$ 19. $\sec \theta$

20. Find the exact values of: a. $\sec 150°$;   b. $\tan(-3\pi/4)$.

21. Find the exact values of: a. $\cot(-225°)$;   b. $\cos 2\pi/3$.

22. Find the value of: a. $\cos 123°$;   b. $\cot 4.22$.

23. Find the value of: a. $\sec 612°$;   b. $\tan 7\pi$.

24. $\cos \phi = -\sqrt{3}/2$, ϕ in quadrant three, find the exact values of: a. $\sin \phi$;   b. $\tan \phi$.

25. $\tan \phi = -\sqrt{3}/3$, ϕ in quadrant two, find the exact values of: a. $\sec \phi$;   b. $\cot \phi$.

26. $\tan \phi = 1$, ϕ in quadrant three, find the exact values of: a. $\csc \phi$;   b. $\cot \phi$.

27. $\sin \phi = -\sqrt{2}/2$, ϕ in quadrant four, find the exact values of: a. $\sec \phi$;   b. $\cot \phi$.

In Exercises 28 to 31, find the value of each expression.

28. $\sin 120° \tan 45° - \cos 315°$.

29. $\tan \dfrac{3\pi}{4} - \sin \dfrac{\pi}{3} \cos \dfrac{3\pi}{4}$

30. $\tan 225° - \cot 30° \sin 150°$.

31. $\cos \dfrac{5\pi}{3} \cot \dfrac{\pi}{6} + \csc \dfrac{7\pi}{4}$

In Exercises 32 to 35, find the exact value of the indicated function.

32. $\sec \phi = -\sqrt{2}$, ϕ in quadrant II, find $\tan \phi$.

33. $\cot \phi = \sqrt{3}$, ϕ in quadrant III, find $\sin \phi$.

34. $\sin \phi = -1/2$, ϕ in quadrant IV, find $\cot \phi$.

35. $\tan \phi = 1$, ϕ in quadrant III, find $\cos \phi$.

In Exercises 36 to 39, find the arc length intercepted by the given angle in a unit circle.

36. $2\pi/5$ radians   37. 112°

38. 0.6 radians   39. 2°

In Exercises 40 to 43, find the angle (in radians) subtended by the given arc in a unit circle.

40. $3\pi/4$ 41. 2.6   42. 2   43. 0.85

44. Use the unit circle to show that the following expressions are true:

a. $\cos(180° + \phi) = -\cos \phi$; b. $\tan(-\phi) = -\tan \phi$.

In Exercises 45 to 50, simplify the first expression to show that it is equivalent to the second expression.

45. $1 + \dfrac{\sin^2\phi}{\cos^2\phi}$: $\sec^2\phi$

46. $\dfrac{\tan\phi + 1}{\cot\phi + 1}$: $\tan\phi$

47. $\dfrac{\cos^2\phi + \sin^2\phi}{\csc\phi}$: $\sin\phi$

48. $\sin^2\phi(\tan^2\phi + 1)$: $\tan^2\phi$

49. $1 + \dfrac{1}{\tan^2\phi}$: $\csc^2\phi$

50. $\dfrac{\cos^2\phi}{1 - \sin\phi}$: $1 + \sin\phi$

In Exercises 51 to 64, graph the given functions.

51. $f(x) = 2\cos\pi x$

52. $f(x) = -\sin\dfrac{2x}{3}$

53. $f(x) = 2\sin\dfrac{3x}{2}$

54. $f(x) = \cos\left(x - \dfrac{\pi}{2}\right)$

55. $f(x) = \dfrac{1}{2}\sin\left(2x + \dfrac{\pi}{4}\right)$

56. $f(x) = 3\cos 3(x - \pi)$

57. $f(x) = -\tan\dfrac{x}{2}$

58. $f(x) = 2\cot 2x$

59. $f(x) = \tan\left(x - \dfrac{\pi}{2}\right)$

60. $f(x) = -\cot\left(2x + \dfrac{\pi}{4}\right)$

61. $f(x) = -2\csc\left(2x - \dfrac{\pi}{3}\right)$

62. $f(x) = 3\sec\left(x + \dfrac{\pi}{4}\right)$

63. $f(x) = 2 - \sin 2x$

64. $f(x) = \sin x - \sqrt{3}\cos x$

65. Find the amplitude, period, and frequency of the harmonic motion given by the equation $y = 2.5\sin 50t$.

66. A pendulum 5 feet long is displaced a distance of 9 inches and released. Write the equation for the displacement as a function of time. Find the period and frequency of the motion.

67. A mass of 5 kilograms is in equilibrium suspended from a spring. The mass is pulled down 0.5 feet and released. Find the period, frequency, and amplitude of the resulting motion. Write the equation of motion. (Let $k = 20$.)

7

Trigonometric Identities and Equations

In most sciences, one generation tears down what another has built, and what one has established, the next undoes. In mathematics alone, each generation builds a new story to the old structure.

HERMANN HANKEL (1839–1873)

RESONANCE PHENOMENA IN A PHYSICAL SYSTEM

Many vibrating systems can be described by simple harmonic motion. Simple harmonic motion can be described by equations of the form $y = a \cos 2\pi ft$. In pure simple harmonic motion, the amplitude of the vibrations remain constant. In an actual physical system, energy will be lost, and the amplitude of the vibrations will gradually decrease. When energy is put into a vibrating system, the amplitude of the vibrations may increase. For example, the amplitude of a swing will increase if impulses of energy are applied to the system at the proper times. *Resonance* is the building up of larger amplitudes of a vibrating system when small impulses are applied at the proper time.

An equation for a motion with resonance is $y = at \cos 2\pi ft$. The graph shows how the vibrations in a resonance system would increase with time.

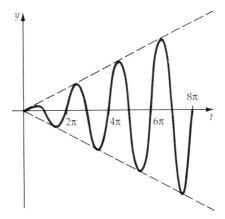

Resonance phenomena can occur in almost any physical system. It is possible for soldiers marching across a bridge to cause large vibrations, if the frequency of march is the same as the natural frequency of the bridge. Another example of resonance is the Tacoma Narrows bridge collapse, which occurred on July 2, 1940. A wind coming down a canyon created a

The Tacoma Narrows Bridge at Puget Sound, Washington. Wide World.

vibration in the bridge that continued to increase as the day passed. Vibrations in the concrete and steel structure reached an amplitude of 8 feet and the bridge collapsed.

7.1

The Fundamental Trigonometric Identities

An equation that is true for all replacements of the variables for which all terms of the equation are defined is called an **identity**. Trigonometric identities are used to simplify a trigonometric expression or to write a trigonometric expression in an equivalent form.

As for algebraic equations, the domain of a trigonometric identity will be all values of the variable for which the expression is meaningful. For example, the identity

$$\frac{\sin x \cos x}{\sin x} = \cos x$$

is true for all real numbers x such that $\sin x \neq 0$. Because $\sin x = 0$ when $x = n\pi$ and n is an integer, the domain of the equation must exclude all integral multiples of π. To verify an identity means to show that each side of the equation represents the same expression. For example, to verify the identity above, simplify the left side of the equation. We now have $\cos x = \cos x$. The left and right sides of the equation are the same expression.

There is no one method that can be used to verify identities. Generally we work only on one side of the equation. The list below provides some suggestions for verifying trigonometric identities.

1. If one side of the identity is more complex than the other, simplify the more complex side.

2. Perform algebraic operations in the identity such as
   a. Squaring
   b. Factoring
   c. Adding fractions
   d. Multiplying the numerator and denominator by a nonzero factor

3. Rewrite the identity in terms of sine and cosine functions.

4. Rewrite one side of the identity in terms of a single function.

TABLE 7.1   Fundamental Trigonometric Identities

Reciprocal identities	$\sin x = \dfrac{1}{\csc x}$	$\cos x = \dfrac{1}{\sec x}$	$\tan x = \dfrac{1}{\cot x}$
Ratio identities	$\tan x = \dfrac{\sin x}{\cos x}$	$\cot x = \dfrac{\cos x}{\sin x}$	
Pythagorean identities	$\sin^2 x + \cos^2 x = 1$	$\tan^2 x + 1 = \sec^2 x$	$1 + \cot^2 x = \csc^2 x$

Table 7.1 lists the eight fundamental identities established in the previous chapter. These identities will be valuable when we verify other trigonometric identities.

EXAMPLE 1   Verify a Trigonometric Identity by Converting to Sines and Cosines

Verify the identity $\sec x - \cos x = \sin x \tan x$.

Solution   Simplify the left side of the equation.

$$\sec x - \cos x = \frac{1}{\cos x} - \cos x = \frac{1 - \cos^2 x}{\cos x} \qquad \text{Write as a single fraction with a common denominator}$$

$$= \frac{\sin^2 x}{\cos x} = \sin x \cdot \frac{\sin x}{\cos x} \qquad 1 - \cos^2 x = \sin^2 x$$

$$= \sin x \tan x \qquad \frac{\sin x}{\cos x} = \tan x$$

Since the left side of the identity has been rewritten to be the right, we have verified the identity.

■ *Try Exercise* **30,** *page 375.*

EXAMPLE 2   Verify a Trigonometric Identity by Using a Pythagorean Identity

Verify the identity $1 - 2 \sin^2 x = 2 \cos^2 x - 1$.

Solution   The Pythagorean identity $\sin^2 x + \cos^2 x = 1$ implies that $\cos^2 x = 1 - \sin^2 x$. Use this identity; rewrite the right side of the equation.

$$2 \cos^2 x - 1 = 2(1 - \sin^2 x) - 1 \qquad \cos^2 x = 1 - \sin^2 x$$
$$= 2 - 2 \sin^2 x - 1$$
$$= 1 - 2 \sin^2 x$$

■ *Try Exercise* **40,** *page 375.*

EXAMPLE 3   Verify a Trigonometric Identity by Factoring

Verify the identity $\csc^2 x - \cos^2 x \csc^2 x = 1$.

Solution

$$\csc^2 x - \cos^2 x \csc^2 x = \csc^2 x (1 - \cos^2 x) \qquad \text{Factor out } \csc^2 x.$$

$$= \csc^2 x \sin^2 x$$

$$= \frac{1}{\sin^2 x} \cdot \sin^2 x = 1 \qquad \csc^2 x = \frac{1}{\sin^2 x}$$

■ *Try Exercise* **52,** *page 375.*

EXAMPLE 4   **Verify a Trigonometric Identity by Using a Conjugate**

Verify the identity $\dfrac{\sin x}{1 + \cos x} = \dfrac{1 - \cos x}{\sin x}$.

Solution   Multiply the numerator and denominator of the left side of the identity by the conjugate of $1 + \cos x$.

$$\frac{\sin x}{1 + \cos x} = \frac{\sin x}{1 + \cos x} \cdot \frac{1 - \cos x}{1 - \cos x} = \frac{\sin x(1 - \cos x)}{1 - \cos^2 x}$$

$$= \frac{\sin x(1 - \cos x)}{\sin^2 x} = \frac{1 - \cos x}{\sin x}$$

■ *Try Exercise* **62**, *page 375.*

EXAMPLE 5   **Verify a Trigonometric Identity**

Verify the identity $\dfrac{\sin x + \tan x}{1 + \cos x} = \tan x$.

Solution   Rewrite the left side of the identity in terms of sines and cosines.

$$\frac{\sin x + \tan x}{1 + \cos x} = \frac{\sin x + \dfrac{\sin x}{\cos x}}{1 + \cos x} \qquad \tan x = \frac{\sin x}{\cos x}$$

$$= \frac{\dfrac{\sin x \cos x + \sin x}{\cos x}}{1 + \cos x} \qquad \text{Write the numerator with a common denominator.}$$

$$= \frac{\sin x \cos x + \sin x}{\cos x(1 + \cos x)} \qquad \text{Simplify.}$$

$$= \frac{\sin x \,\cancel{(1 + \cos x)}}{\cos x \,\cancel{(1 + \cos x)}}$$

$$= \tan x$$

■ *Try Exercise* **72**, *page 376.*

EXERCISE SET 7.1

In Exercises 1 to 10, find the product.

1. $(\cos x - 2)(\cos x - 3)$

2. $(\sin x - 2)(\sin x + 7)$

3. $(\sin^2 x - 5)(\sin^2 x - 12)$

4. $(\tan^2 x + 5)(\tan^2 x + 3)$

5. $(3 \sin x - 2)(4 \sin x + 5)$

6. $(2 \cos x + 5)(4 \cos x + 1)$

7. $(\cos x - 2)(\sin x + 7)$

8. $(\tan x + 5)(\cot x - 3)$

9. $(\sin x - 2 \cos x)(3 \sin x + 4 \cos x)$

10. $(2 \tan x - 3 \cot x)(5 \tan x - \cot x)$

In Exercises 11 to 20, factor completely.

11. $\sin^2 x - 2 \sin x + 1$

12. $\tan^2 x + 2 \tan x + 1$

13. $8 \tan^2 x - 26x + 15$

14. $6 \sin^2 x + \sin x - 15$

15. $4 \sin^2 x - 1$

16. $\tan^2 x - 64$

17. $\sin^3 x - 27$

18. $1 + \tan^2 x$

19. $\sin x \cos x + \sin x + \cos x + 1$

20. $\tan x \cot x + \tan x - 2 \cot x - 2$

In Exercises 21 to 28, simplify by performing the indicated operation.

21. $\dfrac{1}{\sin x} + \dfrac{3}{\cos x}$

22. $\dfrac{2}{\cos x} - \dfrac{3}{\sin x}$

23. $\dfrac{2 \sin x}{\cos x} - \dfrac{\cos x}{\sin x}$

24. $\dfrac{4 \tan x}{\cot x} + \dfrac{5}{\tan x}$

25. $\dfrac{3}{3 \cos x + 1} + \dfrac{2}{\cos x}$

26. $\dfrac{2}{\sin x - 2} - \dfrac{4}{\sin x}$

27. $\dfrac{1 + \dfrac{1}{\sin x}}{1 - \dfrac{1}{\sin x}}$

28. $\dfrac{\sin x - 2 + \dfrac{1}{\sin x}}{\sin x - \dfrac{1}{\sin x}}$

In Exercises 29 to 82, verify the identities.

29. $\tan x \csc x \cos x = 1$

30. $\sin x \cot x \sec x = 1$

31. $\dfrac{4 \sin^2 x - 1}{2 \sin x + 1} = 2 \sin x - 1$

32. $\dfrac{\sin^2 x - 2 \sin x + 1}{\sin x - 1} = \sin x - 1$

33. $(\sin x - \cos x)(\sin x + \cos x) = 1 - 2 \cos^2 x$

34. $\tan x(1 - \cot x) = \tan x - 1$

35. $\dfrac{1}{\sin x} - \dfrac{1}{\cos x} = \dfrac{\cos x - \sin x}{\sin x \cos x}$

36. $\dfrac{1}{\sin x} + \dfrac{3}{\cos x} = \dfrac{\cos x + 3 \sin x}{\sin x \cos x}$

37. $\dfrac{\cos x}{1 - \sin x} = \sec x + \tan x$

38. $\dfrac{\sin x}{1 - \cos x} = \csc x + \cot x$

39. $\dfrac{1 - \tan^4 x}{\sec^2 x} = 1 - \tan^2 x$

40. $\sin^4 x - \cos^4 x = \sin^2 x - \cos^2 x$

41. $\dfrac{1 + \tan^3 x}{1 + \tan x} = 1 - \tan x + \tan^2 x$

42. $\dfrac{\cos x \tan x - \sin x}{\cot x} = 0$

43. $\dfrac{\sin x - 2 + \dfrac{1}{\sin x}}{\sin x - \dfrac{1}{\sin x}} = \dfrac{\sin x - 1}{\sin x + 1}$

44. $\dfrac{\sin x}{1 - \cos x} - \dfrac{\sin x}{1 + \cos x} = 2 \cot x$

45. $(\sin x + \cos x)^2 = 1 + 2 \sin x \cos x$

46. $(\tan x + 1)^2 = \sec^2 x + 2 \tan x$

47. $\dfrac{\cos x}{1 + \sin x} = \sec x - \tan x$

48. $\dfrac{\sin x}{1 + \cos x} = \csc x - \cot x$

49. $\csc x = \dfrac{\cot x + \tan x}{\sec x}$

50. $\sec x = \dfrac{\cot x + \tan x}{\csc x}$

51. $\dfrac{\cos x \tan x + 2 \cos x - \tan x - 2}{\tan x + 2} = \cos x - 1$

52. $\dfrac{2 \sin x \cot x + \sin x - 4 \cot x - 2}{2 \cot x + 1} = \sin x - 2$

53. $\sec x - \tan x = \dfrac{1 - \sin x}{\cos x}$

54. $\cot x - \csc x = \dfrac{\cos x - 1}{\sin x}$

55. $\sin^2 x - \cos^2 x = 2 \sin^2 x - 1$

56. $\sin^2 x - \cos^2 x = 1 - 2 \cos^2 x$

57. $\dfrac{1}{\sin^2 x} + \dfrac{1}{\cos^2 x} = \csc^2 x \sec^2 x$

58. $\dfrac{1}{\tan^2 x} - \dfrac{1}{\cot^2 x} = \csc^2 x - \sec^2 x$

59. $\sec x - \cos x = \sin x \tan x$

60. $\tan x + \cot x = \sec x \csc x$

61. $\dfrac{\dfrac{1}{\sin x} + 1}{\dfrac{1}{\sin x} - 1} = \tan^2 x + 2 \tan x \sec x + \sec^2 x$

62. $\dfrac{\dfrac{1}{\sin x} + \dfrac{1}{\cos x}}{\dfrac{1}{\sin x} - \dfrac{1}{\cos x}} = \dfrac{\cos^2 x - \sin^2 x}{1 - 2 \cos x \sin x}$

63. $\sin^4 x - \cos^4 x = 2 \sin^2 x - 1$

64. $\sin^6 x + \cos^6 x = \sin^4 x - \sin^2 x \cos^2 x + \cos^4 x$

65. $\dfrac{1}{1 - \cos x} = \dfrac{1 + \cos x}{\sin^2 x}$

66. $1 + \sin x = \dfrac{\cos^2 x}{1 - \sin x}$

67. $\dfrac{\sin x}{1 - \sin x} - \dfrac{\cos x}{1 - \sin x} = \dfrac{1 - \cot x}{\csc x - 1}$

68. $\dfrac{\tan x}{1 + \tan x} - \dfrac{\cot x}{1 + \tan x} = 1 - \cot x$

69. $\dfrac{1}{1 + \cos x} - \dfrac{1}{1 - \cos x} = -2 \cot x \csc x$

70. $\dfrac{1}{1 - \sin x} - \dfrac{1}{1 + \sin x} = 2 \tan x \sec x$

71. $\dfrac{\dfrac{1}{\sin x} + \csc x}{\dfrac{1}{\sin x} - \sin x} = \dfrac{2}{\cos^2 x}$

72. $\dfrac{\dfrac{1}{\tan x} + \cot x}{\dfrac{1}{\tan x} + \tan x} = \dfrac{2}{\sec^2 x}$

73. $\sqrt{\dfrac{1 + \sin x}{1 - \sin x}} = \dfrac{1 + \sin x}{\cos x}, \quad \cos x > 0$

74. $\dfrac{\cos x + \cot x \sin x}{\cot x} = 2 \sin x$

75. $\dfrac{\sin^3 x + \cos^3 x}{\sin x + \cos x} = 1 - \sin x \cos x$

76. $\dfrac{1 - \sin x}{1 + \sin x} - \dfrac{1 + \sin x}{1 - \sin x} = -4 \sec x \tan x$

77. $\dfrac{\sec x - 1}{\sec x + 1} - \dfrac{\sec x + 1}{\sec x - 1} = -4 \csc x \cot x$

78. $\dfrac{1}{1 - \cos x} - \dfrac{\cos x}{1 + \cos x} = 2 \csc^2 x - 1$

79. $\dfrac{1 + \sin x}{\cos x} - \dfrac{\cos x}{1 - \sin x} = 0$

80. $(\sin x + \cos x + 1)^2 = 2(\sin x + 1)(\cos x + 1)$

81. $\dfrac{\sec x + \tan x}{\sec x - \tan x} = \dfrac{(\sin x + 1)^2}{\cos^2 x}$

82. $\dfrac{\sin^3 x - \cos^3 x}{\sin x + \cos x} = \dfrac{\csc^2 x - \cot x - 2 \cos^2 x}{1 - \cot^2 x}$

Supplemental Exercises

83. Express $\cos x$ in terms of $\sin x$.

84. Express $\tan x$ in terms of $\cos x$.

85. Express $\sec x$ in terms of $\sin x$.

86. Express $\csc x$ in terms of $\sec x$.

In Exercises 87 to 92, verify the identity.

87. $\dfrac{1 - \sin x + \cos x}{1 + \sin x + \cos x} = \dfrac{\cos x}{\sin x + 1}$

88. $\dfrac{1 - \tan x + \sec x}{1 + \tan x - \sec x} = \dfrac{1 + \sec x}{\tan x}$

89. $\dfrac{2 \sin^4 x + 2 \sin^2 x \cos^2 x - 3 \sin^2 x - 3 \cos^2 x}{2 \sin^2 x}$

$$= 1 - \dfrac{3}{2} \csc^2 x$$

90. $\dfrac{4 \tan x \sec^2 x - 4 \tan x - \sec^2 x + 1}{4 \tan^3 x - \tan^2 x} = 1$

91. $\dfrac{\sin x (\tan x + 1) - 2 \tan x \cos x}{\sin x - \cos x} = \tan x$

92. $\dfrac{\sin^2 x \cos x + \cos^3 x - \sin^3 x \cos x - \sin x \cos^3 x}{1 - \sin^2 x}$

$$= \dfrac{\cos x}{1 + \sin x}$$

93. Verify the identity $\sin^4 x + \cos^4 x = 1 - 2 \sin^2 x \cos^2 x$ by completing the square of the left side of the identity.

94. Verify the identity $\tan^4 x + \sec^4 x = 1 + 2 \tan^2 x \sec^2 x$ by completing the square of the left side of the identity.

7.2
Sum and Difference Identities

There are several useful identities relating the sum and difference of two angles $(\alpha \pm \beta)$. We begin by finding an identity for $\cos(\alpha - \beta)$.

In Figure 7.1, angles α and β are drawn in standard position, with OA and OB as the terminal sides of α and β, respectively. The coordinates of A are $(\cos \alpha, \sin \alpha)$, and the coordinates of B are $(\cos \beta, \sin \beta)$. The angle $(\alpha - \beta)$ is formed by the terminal sides of the angles α and β (angle AOB).

An angle equal in measure to angle $(\alpha - \beta)$ is placed in standard position in the same figure. The chords AB and CB are equal, because there is a theorem from geometry that states that if two central angles of a circle have the same measure, then the measure of their chords are also equal. Using the distance formula, we can calculate the lengths of the chords AB and CD.

$$d(A, B) = \sqrt{(\sin \alpha - \sin \beta)^2 + (\cos \alpha - \cos \beta)^2}$$

$$d(C, D) = \sqrt{[\cos(\alpha - \beta) - 1]^2 + [\sin(\alpha - \beta) - 0]^2}$$

Since $d(A, B) = d(C, D)$, we have

$$\sqrt{(\sin \alpha - \sin \beta)^2 + (\cos \alpha - \cos \beta)^2}$$
$$= \sqrt{[\cos(\alpha - \beta) - 1]^2 + [\sin(\alpha - \beta) - 0]^2}$$

Squaring each side of the equation and simplifying, we obtain

$$(\sin \alpha - \sin \beta)^2 + (\cos \alpha - \cos \beta)^2 = [\cos(\alpha - \beta) - 1]^2 + [\sin(\alpha - \beta) - 0]^2$$

$$\sin^2\alpha - 2 \sin \alpha \sin \beta + \sin^2\beta + \cos^2\alpha - 2 \cos \alpha \cos \beta + \cos^2\beta$$
$$= \cos^2(\alpha - \beta) - 2 \cos(\alpha - \beta) + 1 + \sin^2(\alpha - \beta)$$

$$\sin^2\alpha + \cos^2\alpha + \sin^2\beta + \cos^2\beta - 2 \sin \alpha \sin \beta - 2 \cos \alpha \cos \beta$$
$$= \sin^2(\alpha - \beta) + \cos^2(\alpha - \beta) + 1 - 2 \cos(\alpha - \beta)$$

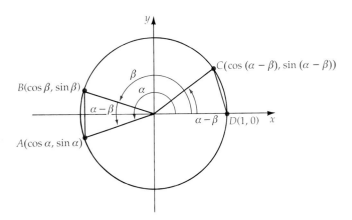

Figure 7.1

Simplifying by using the Pythagorean identity $\sin^2\theta + \cos^2\theta = 1$, we have

$$2 - 2\sin\alpha\sin\beta - 2\cos\alpha\cos\beta = 2 - 2\cos(\alpha - \beta).$$

Solving for $\cos(\alpha - \beta)$ gives us

$$\cos(\alpha - \beta) = \cos\alpha\cos\beta + \sin\alpha\sin\beta.$$

To derive the identity for the cosine of the sum of two angles, substitute $-\beta$ for β in $\cos(\alpha - \beta)$.

$$\cos(\alpha - \beta) = \cos[\alpha - (-\beta)] = \cos\alpha\cos(-\beta) + \sin\alpha\sin(-\beta)$$

Recall that $\cos(-\beta) = \cos\beta$ and $\sin(-\beta) = -\sin\beta$. Substituting into the previous equation, we obtain the identity

$$\cos(\alpha + \beta) = \cos\alpha\cos\beta - \sin\alpha\sin\beta.$$

EXAMPLE 1   Evaluate a Trigonometric Function

Evaluate $\cos\dfrac{5\pi}{12}$.

Solution   Since $\dfrac{5\pi}{12} = \dfrac{\pi}{4} + \dfrac{\pi}{6}$, we can substitute $\alpha = \dfrac{\pi}{4}$ and $\beta = \dfrac{\pi}{6}$ in the identity for the cosine of the sum of two angles.

$$\cos\frac{5\pi}{12} = \cos\left(\frac{\pi}{4} + \frac{\pi}{6}\right) = \cos\frac{\pi}{4}\cos\frac{\pi}{6} - \sin\frac{\pi}{4}\sin\frac{\pi}{6}$$

$$= \frac{\sqrt{2}}{2}\cdot\frac{\sqrt{3}}{2} - \frac{\sqrt{2}}{2}\cdot\frac{1}{2} = \frac{\sqrt{6}}{4} - \frac{\sqrt{2}}{4} = \frac{\sqrt{6} - \sqrt{2}}{4}$$

■ *Try Exercise* **22**, *page 382.*

EXAMPLE 2   Verify an Identity

Verify the identity $\cos(\pi - \theta) = -\cos\theta$.

Solution   Use the identity for the cosine of the difference of two angles. Recall that $\cos\pi = -1$ and $\sin\pi = 0$.

$$\cos(\pi - \theta) = \cos\pi\cos\theta + \sin\pi\sin\theta = -1\cdot\cos\theta + 0\cdot\sin\theta = -\cos\theta$$

■ *Try Exercise* **70**, *page 383.*

EXAMPLE 3   Evaluate a Trigonometric Function

Given $\sin\alpha = \dfrac{1}{2}$ for α in quadrant II and $\cos\beta = \dfrac{\sqrt{3}}{2}$ for β in quadrant IV, find $\cos(\alpha + \beta)$.

Solution   We sketch the diagrams of angles α and β with the lengths

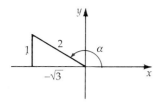

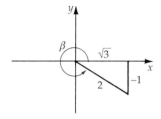

Figure 7.2

of the sides indicated. Use the Pythagorean Theorem to evaluate $\cos \alpha$ and $\sin \beta$. Use the identity of the cosine of the sum of two angles.

$$\cos(\alpha + \beta) = \cos \alpha \cos \beta - \sin \alpha \sin \beta$$
$$= \frac{-\sqrt{3}}{2} \cdot \frac{\sqrt{3}}{2} - \frac{1}{2} \cdot \frac{-1}{2} = \frac{-3}{4} + \frac{1}{4} = -\frac{1}{2}$$

■ *Try Exercise **62**, page 383.*

EXAMPLE 4   **Verify an Identity**

Verify the identity $\dfrac{\cos 4\theta}{\sin \theta} - \dfrac{\sin 4\theta}{\cos \theta} = \dfrac{\cos 5\theta}{\sin \theta \cos \theta}$.

Solution   Subtract the fractions on the left side of the equation.

$$\frac{\cos 4\theta}{\sin \theta} - \frac{\sin 4\theta}{\cos \theta} = \frac{\cos 4\theta \cos \theta - \sin 4\theta \sin\theta}{\sin \theta \cos \theta}$$

$$= \frac{\cos(4\theta + \theta)}{\sin \theta \cos \theta} = \frac{\cos 5\theta}{\sin \theta \cos \theta} \quad \begin{array}{l}\text{Use the identity} \\ \text{for } \cos(\alpha + \beta).\end{array}$$

■ *Try Exercise **82**, page 383.*

If we apply the difference identity for the cosine function to the expression $\cos(90° - \alpha)$, we obtain the following result:

$$\cos(90° - \alpha) = \cos 90° \cos \alpha + \sin 90° \sin \alpha = 0 \cdot \cos \alpha + 1 \cdot \sin \alpha$$

$$\cos(90° - \alpha) = \sin \alpha$$

Thus the sine of an angle is equal to the cosine of its complement. Similarly, we have

$$\cos \alpha = \cos[90° - (90° - \alpha)]$$
$$= \cos 90° \cos(90° - \alpha) + \sin 90° \sin(90° - \alpha)$$
$$= 0 \cdot \cos(90° - \alpha) + 1 \cdot \sin(90° - \alpha)$$

which gives us

$$\cos \alpha = \sin(90° - \alpha)$$

Thus the cosine of an angle is equal to the sine of its complement. Any pair of functions with this property are said to be **cofunctions**.

We can use the ratio identity to show that the tangent and cotangent functions are cofunctions. (The secant and cosecant are also cofunctions.)

$$\tan(90° - \theta) = \frac{\sin(90° - \theta)}{\cos(90° - \theta)} \qquad \cot(90° - \theta) = \frac{\cos(90° - \theta)}{\sin(90° - \theta)}$$

$$= \frac{\cos \theta}{\sin \theta} = \cot \theta \qquad\qquad = \frac{\sin \theta}{\cos \theta} = \tan \theta$$

To derive the identity for the sine of the sum of two angles, substitute $\theta = \alpha + \beta$ in the cofunction identity $\sin \theta = \cos(90° - \theta)$.

$$\sin \theta = \cos(90° - \theta)$$

$$\sin(\alpha + \beta) = \cos[90° - (\alpha + \beta)]$$

$$= \cos[(90° - \alpha) - \beta] \quad \text{Rewrite as the difference of two angles.}$$

$$= \cos(90° - \alpha) \cos \beta + \sin(90° - \alpha) \sin \beta$$

$$\sin(\alpha + \beta) = \sin \alpha \cos \beta + \cos \alpha \sin \beta$$

We also can derive the difference identity for the sine by substituting $-\beta$ for β in the sum identity for the $\sin(\alpha + \beta)$.

$$\sin(\alpha - \beta) = \sin[\alpha + (-\beta)]$$

$$= \sin \alpha \cos(-\beta) + \cos \alpha \sin(-\beta) \quad \cos(-\beta) = \cos \beta,$$
$$\sin(-\beta) = -\sin \beta.$$

$$\sin(\alpha - \beta) = \sin \alpha \cos \beta - \cos \alpha \sin \beta$$

EXAMPLE 5   Evaluate a Trigonometric Function

Find the exact value of $\sin 105°$.

Solution Since $105°$ is the sum of two of the special angles, $60°$ and $45°$, use the sum identity for the sine.

$$\sin 105° = \sin(60° + 45°) = \sin 60° \cos 45° + \cos 60° \sin 45°$$

$$= \frac{\sqrt{3}}{2} \cdot \frac{\sqrt{2}}{2} + \frac{1}{2} \cdot \frac{\sqrt{2}}{2} = \frac{\sqrt{6} + \sqrt{2}}{4}$$

■ *Try Exercise **20**, page 382.*

EXAMPLE 6   Verify an Identity

Verify the identity $\sin(\alpha + \beta) \cdot \sin(\alpha - \beta) = \sin^2\alpha - \sin^2\beta$.

Solution Work on the left side of the identity.

$$\sin(\alpha + \beta) \cdot \sin(\alpha - \beta)$$

$$= (\sin \alpha \cos \beta + \cos \alpha \sin \beta)(\sin \alpha \cos \beta - \cos \alpha \sin \beta)$$

$$= \sin^2\alpha \cos^2\beta - \cos^2\alpha \sin^2\beta$$

$$= \sin^2\alpha(1 - \sin^2\beta) - (1 - \sin^2\alpha)(\sin^2\beta)$$

$$= \sin^2\alpha - \sin^2\alpha \sin^2\beta - \sin^2\beta + \sin^2\alpha \sin^2\beta$$

$$= \sin^2\alpha - \sin^2\beta$$

■ *Try Exercise **86**, page 383.*

The sum identity for the tangent function is a direct result of the ratio identity $\tan \theta = \sin \theta/\cos \theta$ and the sum identity of the sine and cosine:

$$\tan(\alpha + \beta) = \frac{\sin(\alpha + \beta)}{\cos(\alpha + \beta)} = \frac{\sin \alpha \cos \beta + \cos \alpha \sin \beta}{\cos \alpha \cos \beta - \sin \alpha \sin \beta}$$

$$= \frac{\dfrac{\sin \alpha \cos \beta}{\cos \alpha \cos \beta} + \dfrac{\cos \alpha \sin \beta}{\cos \alpha \cos \beta}}{\dfrac{\cos \alpha \cos \beta}{\cos \alpha \cos \beta} - \dfrac{\sin \alpha \sin \beta}{\cos \alpha \cos \beta}}$$

Divide each term by $\cos \alpha \cos \beta$ and simplify.

$$\tan(\alpha + \beta) = \frac{\tan \alpha + \tan \beta}{1 - \tan \alpha \tan \beta}$$

The tangent function is an odd function; thus $\tan(-\theta) = -\tan \theta$. Substituting $-\beta$ for β in $\tan(\alpha + \beta)$, we derive the difference identity for $\tan(\alpha - \beta)$.

$$\tan[\alpha + (-\beta)] = \frac{\tan \alpha + \tan(-\beta)}{1 - \tan \alpha \tan(-\beta)}$$

$$\tan(\alpha - \beta) = \frac{\tan \alpha - \tan \beta}{1 + \tan \alpha \tan \beta}$$

EXAMPLE 7 Evaluate a Trigonometric Function

Find the exact value of $\tan 75°$.

Solution   Since $75° = 45° + 30°$, use the identity for the tangent of the sum of two angles.

$$\tan 75° = \tan(45° + 30°) = \frac{\tan 45° + \tan 30°}{1 - \tan 45° \tan 30°}$$

$$= \frac{1 + \dfrac{\sqrt{3}}{3}}{1 - (1)\dfrac{\sqrt{3}}{3}} = \frac{3 + \sqrt{3}}{3 - \sqrt{3}}$$

$$= \frac{(3 + \sqrt{3})(3 + \sqrt{3})}{(3 - \sqrt{3})(3 + \sqrt{3})}$$

$$= \frac{9 + 6\sqrt{3} + 3}{9 - 3} = 2 + \sqrt{3}$$

■ *Try Exercise* **24,** *page 382.*

The identities we developed in this section are used frequently in problems involving trigonometric functions. Following is a list of the identities for convenient reference.

Sum or Difference of Two Angle Identities

$$\cos(\alpha - \beta) = \cos \alpha \cos \beta + \sin \alpha \sin \beta$$

$$\cos(\alpha + \beta) = \cos \alpha \cos \beta - \sin \alpha \sin \beta$$

$$\sin(\alpha - \beta) = \sin \alpha \cos \beta - \cos \alpha \sin \beta$$

$$\sin(\alpha + \beta) = \sin \alpha \cos \beta + \cos \alpha \sin \beta$$

$$\tan(\alpha + \beta) = \frac{\tan \alpha + \tan \beta}{1 - \tan \alpha \tan \beta}$$

$$\tan(\alpha - \beta) = \frac{\tan \alpha - \tan \beta}{1 + \tan \alpha \tan \beta}$$

Cofunction Identities

$$\sin(90° - \theta) = \cos \theta \quad \cos(90° - \theta) = \sin \theta \quad \tan(90° - \theta) = \cot \theta$$

$$\csc(90° - \theta) = \sec \theta \quad \sec(90° - \theta) = \csc \theta \quad \cot(90° - \theta) = \tan \theta$$

The cofunction identities are also true when 90° is replaced by $\pi/2$ radians.

EXERCISE SET 7.2

In Exercises 1 to 34, find the exact value of the given function or expression.

1. $\sin(45° + 30°)$
2. $\sin(330° + 45°)$
3. $\cos(45° - 30°)$
4. $\cos(120° - 45°)$
5. $\tan(45° + 135°)$
6. $\tan(240° - 45°)$
7. $\sin\left(\dfrac{5\pi}{4} - \dfrac{\pi}{6}\right)$
8. $\sin\left(\dfrac{4\pi}{3} + \dfrac{\pi}{4}\right)$
9. $\cos\left(\dfrac{3\pi}{4} + \dfrac{\pi}{6}\right)$
10. $\cos\left(\dfrac{\pi}{4} - \dfrac{\pi}{3}\right)$
11. $\tan\left(\dfrac{\pi}{6} + \dfrac{\pi}{4}\right)$
12. $\tan\left(\dfrac{11\pi}{6} - \dfrac{\pi}{4}\right)$

13. $\sin 15°$
14. $\sin 285°$
15. $\cos 195°$
16. $\cos 165°$
17. $\tan 75°$
18. $\tan 285°$
19. $\sin \dfrac{5\pi}{12}$
20. $\sin \dfrac{11\pi}{12}$
21. $\cos \dfrac{13\pi}{12}$
22. $\cos \dfrac{\pi}{12}$
23. $\tan \dfrac{19\pi}{12}$
24. $\tan \dfrac{7\pi}{12}$

25. $\cos 212° \cos 122° + \sin 212° \sin 122°$
26. $\cos 82° \cos 37° + \sin 82° \sin 37°$
27. $\sin 167° \cos 107° - \cos 167° \sin 107°$
28. $\sin 178° \cos 58° - \cos 178° \sin 58°$
29. $\sin \dfrac{5\pi}{12} \cos \dfrac{\pi}{4} - \cos \dfrac{5\pi}{12} \sin \dfrac{\pi}{4}$
30. $\cos \dfrac{7\pi}{12} \cos \dfrac{\pi}{4} + \sin \dfrac{7\pi}{12} \sin \dfrac{\pi}{4}$
31. $\cos \dfrac{\pi}{12} \cos \dfrac{\pi}{4} - \sin \dfrac{\pi}{12} \sin \dfrac{\pi}{4}$
32. $\sin \dfrac{\pi}{3} \cos \dfrac{\pi}{6} + \cos \dfrac{\pi}{3} \sin \dfrac{\pi}{6}$
33. $\dfrac{\tan 7\pi/12 - \tan \pi/4}{1 + \tan 7\pi/12 \tan \pi/4}$
34. $\dfrac{\tan \pi/6 + \tan \pi/3}{1 - \tan \pi/6 \tan \pi/3}$

In Exercises 35 to 44, evaluate each expression.

35. $\cos 30° \cos 40° + \sin 30° \sin 40°$

36. $\cos 125° \cos 65° - \sin 125° \sin 65°$

37. $\sin 8° \cos 48° - \cos 8° \sin 48°$

38. $\sin 230° \cos 88° + \cos 230° \sin 88°$

39. $\sin \dfrac{\pi}{12} \cos \dfrac{\pi}{3} + \cos \dfrac{\pi}{12} \sin \dfrac{\pi}{3}$

40. $\sin \dfrac{11\pi}{6} \cos \dfrac{\pi}{4} - \cos \dfrac{11\pi}{6} \sin \dfrac{\pi}{4}$

41. $\cos \dfrac{\pi}{4} \cos \dfrac{\pi}{6} - \sin \dfrac{\pi}{4} \sin \dfrac{\pi}{6}$

42. $\cos \dfrac{\pi}{3} \cos \dfrac{\pi}{4} + \sin \dfrac{\pi}{3} \sin \dfrac{\pi}{4}$

43. $\dfrac{\tan \pi/3 + \tan \pi/4}{1 - \tan \pi/3 \tan \pi/4}$

44. $\dfrac{\tan 11\pi/6 - \tan \pi/4}{1 + \tan 11\pi/6 \tan \pi/4}$

In Exercises 45 to 56, write each expression in terms of a single trigonometric function.

45. $\sin 7x \cos 2x - \cos 7x \sin 2x$

46. $\sin x \cos 3x + \cos 3x \sin x$

47. $\cos x \cos 2x + \sin x \sin 2x$

48. $\cos 4x \cos 2x - \sin 4x \sin 2x$

49. $\sin 7x \cos 3x - \cos 7x \sin 3x$

50. $\cos x \cos 5x - \sin x \sin 5x$

51. $\cos 4x \cos(-2x) - \sin 4x \sin(-2x)$

52. $\sin(-x) \cos 3x - \cos(-x) \sin 3x$

53. $\sin \dfrac{x}{3} \cos \dfrac{2x}{3} + \cos \dfrac{2x}{3} \sin \dfrac{x}{3}$

54. $\cos \dfrac{3x}{4} \cos \dfrac{x}{4} + \sin \dfrac{3x}{4} \sin \dfrac{x}{4}$

55. $\dfrac{\tan 3x + \tan 4x}{1 - \tan 3x \tan 4x}$

56. $\dfrac{\tan 2x - \tan 3x}{1 + \tan 2x \tan 3x}$

In Exercises 57 to 60, evaluate each expression. ·

57. $\sin 2 \cos 3 - \cos 2 \sin 3$

58. $\cos 4 \cos 2 - \sin 4 \sin 2$

59. $\cos 5 \cos 8 + \sin 5 \sin 8$

60. $\sin 4 \cos 6 + \cos 4 \sin 6$

In Exercises 61 to 68, find the exact value of the given functions.

61. Given $\sin \alpha = \frac{3}{5}$ in quadrant I and $\cos \beta = -\frac{5}{13}$ in quadrant II, find **a.** $\sin(\alpha - \beta)$ and **b.** $\cos(\alpha + \beta)$.

62. Given $\sin \alpha = \frac{24}{25}$ in quadrant II and $\cos \beta = -\frac{4}{5}$ in quadrant III, find **a.** $\cos(\beta - \alpha)$ and **b.** $\sin(\alpha + \beta)$.

63. Given $\sin \alpha = -\frac{4}{5}$ in quadrant III and $\cos \beta = -\frac{12}{13}$ in quadrant II, find **a.** $\sin(\alpha - \beta)$, **b.** $\cos(\alpha + \beta)$, and **c.** $\tan(\alpha + \beta)$.

64. Given $\sin \alpha = -\frac{7}{25}$ in quadrant IV and $\cos \beta = \frac{8}{17}$ in quadrant IV, find **a.** $\sin(\alpha + \beta)$, **b.** $\cos(\alpha - \beta)$, and **c.** $\tan(\alpha + \beta)$.

65. Given $\cos \alpha = \frac{15}{17}$ in quadrant I and $\sin \beta = -\frac{3}{5}$ in quadrant III, find **a.** $\sin(\alpha + \beta)$, **b.** $\cos(\alpha - \beta)$, and **c.** $\tan(\alpha - \beta)$.

66. Given $\cos \alpha = -\frac{7}{25}$ in quadrant II and $\sin \beta = -\frac{12}{13}$ in quadrant IV, find **a.** $\sin(\alpha + \beta)$, **b.** $\cos(\alpha + \beta)$, and **c.** $\tan(\alpha - \beta)$.

67. Given $\cos \alpha = -\frac{3}{5}$ in quadrant III and $\sin \beta = \frac{5}{13}$ in quadrant I, find **a.** $\sin(\alpha - \beta)$, **b.** $\cos(\alpha + \beta)$, and **c.** $\tan(\alpha + \beta)$.

68. Given $\cos \alpha = \frac{8}{17}$ in quadrant IV and $\sin \beta = -\frac{24}{25}$ in quadrant III, find **a.** $\sin(\alpha - \beta)$, **b.** $\cos(\alpha + \beta)$, and **c.** $\tan(\alpha + \beta)$.

In Exercises 69 to 92, verify the identities.

69. $\cos\left(\dfrac{\pi}{2} - \theta\right) = \sin \theta$

70. $\cos(\theta + \pi) = -\cos \theta$

71. $\sin\left(\theta + \dfrac{\pi}{2}\right) = \cos \theta$

72. $\sin(\theta + \pi) = -\sin \theta$

73. $\tan\left(\theta + \dfrac{\pi}{4}\right) = \dfrac{\tan \theta + 1}{1 - \tan \theta}$

74. $\tan 2\theta = \dfrac{2 \tan \theta}{1 - \tan^2\theta}$

75. $\cos\left(\dfrac{3\pi}{2} - \theta\right) = -\sin \theta$

76. $\sin\left(\dfrac{3\pi}{2} + \theta\right) = -\cos \theta$

77. $\cot\left(\dfrac{\pi}{2} - \theta\right) = \tan \theta$

78. $\cot(\pi + \theta) = \cot \theta$

79. $\csc(\pi - \theta) = \csc \theta$

80. $\sec\left(\dfrac{\pi}{2} - \theta\right) = \csc \theta$

81. $\sin 6x \cos 2x - \cos 6x \sin 2x = 2 \sin 2x \cos 2x$

82. $\cos 5x \cos 3x + \sin 5x \sin 3x = \cos^2x - \sin^2x$

83. $\cos(\alpha + \beta) + \cos(\alpha - \beta) = 2 \cos \alpha \cos \beta$

84. $\cos(\alpha - \beta) - \cos(\alpha + \beta) = 2 \sin \alpha \sin \beta$

85. $\sin(\alpha + \beta) + \sin(\alpha - \beta) = 2 \sin \alpha \cos \beta$

86. $\sin(\alpha - \beta) - \sin(\alpha + \beta) = -2 \cos \alpha \sin \beta$

87. $\dfrac{\cos(\alpha - \beta)}{\sin(\alpha + \beta)} = \dfrac{\cot \alpha + \tan \beta}{1 + \cot \alpha \tan \beta}$

88. $\dfrac{\sin(\alpha + \beta)}{\sin(\alpha - \beta)} = \dfrac{1 + \cot \alpha \tan \beta}{1 - \cot \alpha \tan \beta}$

89. $\sin(\pi/2 + \alpha - \beta) = \cos \alpha \cos \beta + \sin \alpha \sin \beta$

90. $\cos(\pi/2 + \alpha + \beta) = -(\sin \alpha \cos \beta + \cos \alpha \sin \beta)$

91. $\sin 3x = 3 \sin x - 4 \sin^3 x$

92. $\cos 3x = 4 \cos^3 x - 3 \cos x$

Supplemental Exercises

In Exercises 93 to 99, verify the identities.

93. $\sin(x - y) \cdot \sin(x + y) = \sin^2 x \cos^2 y - \cos^2 x \sin^2 y$

94. $\sin(x + y + z) = \sin x \cos y \cos z + \cos x \sin y \cos z$
$+ \cos x \cos y \sin z - \sin x \sin y \sin z$

95. $\cos(x + y + z) = \cos x \cos y \cos z - \sin x \sin y \cos z$
$- \sin x \cos y \sin z - \cos x \sin y \sin z$

96. $\dfrac{\sin(x + y)}{\sin x \sin y} = \cot x + \cot y$

97. $\dfrac{\cos(x - y)}{\cos x \sin y} = \cot y + \tan x$

98. $\dfrac{\sin(x + h) - \sin x}{h} = \cos x \dfrac{\sin h}{h} + \sin x \dfrac{\cos h - 1}{h}$

99. $\dfrac{\cos(x + h) - \cos x}{h} = \cos x \dfrac{\cos h - 1}{h} - \sin x \dfrac{\sin h}{h}$

7.3

Double- and Half-Angle Identities

By using the sum identities, we can derive identities for twice the argument of a trigonometic function. These are called the *double-angle* identities. To find the sine of a double angle, substitute α for β in the identity for $\sin(\alpha + \beta)$ to obtain

$$\sin(\alpha + \beta) = \sin \alpha \cos \beta + \cos \alpha \sin \beta$$

$$\sin(\alpha + \alpha) = \sin \alpha \cos \alpha + \cos \alpha \sin \alpha \quad \text{Let } \alpha = \beta.$$

$$\sin 2\alpha = 2 \sin \alpha \cos \alpha.$$

The double-angle identity for cosine is derived in a similar manner.

$$\cos(\alpha + \beta) = \cos \alpha \cos \beta - \sin \alpha \sin \beta$$

$$\cos(\alpha + \alpha) = \cos \alpha \cos \alpha - \sin \alpha \sin \alpha \quad \text{Let } \alpha = \beta.$$

$$\cos 2\alpha = \cos^2 \alpha - \sin^2 \alpha$$

There are two alternative forms of the double-angle identity for the cosine. Using a Pythagorean identity, we can rewrite the identity for $\cos 2\alpha$ as follows:

$$\cos 2\alpha = \cos^2 \alpha - \sin^2 \alpha$$

$$\cos 2\alpha = (1 - \sin^2 \alpha) - \sin^2 \alpha \quad \cos^2 \alpha = 1 - \sin^2 \alpha$$

$$\cos 2\alpha = 1 - 2 \sin^2 \alpha$$

In addition, we can rewrite $\cos 2\alpha$ as

$$\cos 2\alpha = \cos^2\alpha - \sin^2\alpha$$

$$\cos 2\alpha = \cos^2\alpha - (1 - \cos^2\alpha) \quad \sin^2\alpha = 1 - \cos^2\alpha$$

$$\cos 2\alpha = 2\cos^2\alpha - 1$$

The double-angle identity for the tangent function is derived from the identity for the tangent of the sum of two angles.

$$\tan(\alpha + \beta) = \frac{\tan \alpha + \tan \beta}{1 - \tan \alpha \tan \beta}$$

$$\tan(\alpha + \alpha) = \frac{\tan \alpha + \tan \alpha}{1 - \tan \alpha \tan \alpha} \quad \text{Let } \alpha = \beta.$$

$$\tan 2\alpha = \frac{2 \tan \alpha}{1 - \tan^2\alpha}$$

EXAMPLE 1   Evaluate a Trigonometric Function

For an angle α in quadrant I, $\sin \alpha = \dfrac{4}{5}$. Find $\sin 2\alpha$.

Solution   We will use $\sin 2\alpha = 2 \sin \alpha \cos \alpha$. Find $\cos \alpha$ by substituting for $\sin \alpha$ in $\sin^2\alpha + \cos^2\alpha = 1$ and solve for $\cos \alpha$.

$$\cos \alpha = \sqrt{1 - \sin^2 \alpha} = \sqrt{1 - \left(\frac{4}{5}\right)^2} = \frac{3}{5} \quad \cos \alpha > 0 \text{ in quadrant I.}$$

Substitute the values of $\sin \alpha$ and $\cos \alpha$ in the double angle-formula for the sine function.

$$\sin 2\alpha = 2 \sin \alpha \cos \alpha = 2\left(\frac{4}{5}\right)\left(\frac{3}{5}\right) = \frac{24}{25}$$

■ *Try Exercise **26**, page 389.*

EXAMPLE 2   Verify a Double-Angle Identity

Verify the identity $\csc 2\alpha = \dfrac{1}{2}(\tan \alpha + \cot \alpha)$.

Solution   Work on the right-hand side of the equation.

$$\frac{1}{2}(\tan \alpha + \cot \alpha) = \frac{1}{2}\left(\frac{\sin \alpha}{\cos \alpha} + \frac{\cos \alpha}{\sin \alpha}\right) = \frac{1}{2}\left(\frac{\sin^2\alpha + \cos^2\alpha}{\cos \alpha \sin \alpha}\right)$$

$$= \frac{1}{2 \cos \alpha \sin \alpha} = \frac{1}{\sin 2\alpha} = \csc 2\alpha$$

■ *Try Exercise **54**, page 390.*

EXAMPLE 3   **Verify a Double-Angle Identity**

Verify the identity $\sin^2 x = \dfrac{1}{2}(1 - \cos 2x)$.

Solution   Work on the right side of the equation.

$$\frac{1}{2}(1 - \cos 2x) = \frac{1}{2}[1 - (1 - 2\sin^2 x)] = \frac{1}{2}(1 - 1 + 2\sin^2 x) = \sin^2 x$$

■ *Try Exercise* **62**, *page 390.*

EXAMPLE 4   **Verify a Double-Angle Identity**

Verify the identity $\tan 2x = \dfrac{2}{\cot x - \tan x}$.

Solution   Work on the right side of the equation.

$$\frac{2}{\cot x - \tan x} = \frac{2}{\dfrac{1}{\tan x} - \tan x} = \frac{2}{\dfrac{1}{\tan x} - \tan x} \cdot \frac{\tan x}{\tan x}$$

$$= \frac{2\tan x}{1 - \tan^2 x} = \tan 2x$$

■ *Try Exercise* **58**, *page 390.*

EXAMPLE 5   **Verify an Identity**

Verify the identity $\sin 3\theta = 3\sin\theta - 4\sin^3\theta$.

Solution   Work on the left side of the identity.

$$\sin 3\theta = \sin(2\theta + \theta) = \sin 2\theta \cos\theta + \cos 2\theta \sin\theta$$
$$= (2\sin\theta\cos\theta)\cos\theta + (1 - 2\sin^2\theta)\sin\theta$$
$$= 2\sin\theta\cos^2\theta + \sin\theta - 2\sin^3\theta$$
$$= 2\sin\theta(1 - \sin^2\theta) + \sin\theta - 2\sin^3\theta$$
$$= 2\sin\theta - 2\sin^3\theta + \sin\theta - 2\sin^3\theta$$
$$= 3\sin\theta - 4\sin^3\theta$$

■ *Try Exercise* **68**, *page 390.*

The double-angle identity for the cosine function can be used to derive the *half-angle* identities. To derive an identity for $\sin \alpha/2$, we solve for $\sin^2\theta$ in the double-angle identity for $\cos\theta$.

$$\cos 2\theta = 1 - 2\sin^2\theta$$

$$\sin^2\theta = \frac{1 - \cos 2\theta}{2}$$

Substitute $\alpha/2$ for θ and take square root of both sides of the equation

$$\sin^2 \frac{\alpha}{2} = \frac{1 - \cos 2(\alpha/2)}{2}$$

$$\sin \frac{\alpha}{2} = \pm \sqrt{\frac{1 - \cos \alpha}{2}}$$

The sign of the radical is determined by the quadrant in which the terminal side of angle $\alpha/2$ lies.

In a similar manner, we derive an identity for $\cos \alpha/2$.

$$\cos 2\theta = 2 \cos^2\theta - 1$$

$$\cos^2\theta = \frac{1 + \cos 2\theta}{2}$$

Substitute $\alpha/2$ for θ and take the square root of both sides of the equation.

$$\cos^2 \frac{\alpha}{2} = \frac{1 + \cos 2(\alpha/2)}{2}$$

$$\cos \frac{\alpha}{2} = \pm \sqrt{\frac{1 + \cos \alpha}{2}}$$

A half-angle identity for the tangent is derived from the ratio identity for the tangent function. Two different forms of the half-angle identity for the tangent function are possible.

$$\tan \alpha/2 = \frac{\sin \alpha/2}{\cos \alpha/2} = \frac{\sin \alpha/2}{\cos \alpha/2} \cdot \frac{2 \cos \alpha/2}{2 \cos \alpha/2}$$

$$= \frac{2 \sin \alpha/2 \cos \alpha/2}{2 \cos^2\alpha/2}$$

$$= \frac{\sin 2(\alpha/2)}{1 + \cos 2(\alpha/2)} = \frac{\sin \alpha}{1 + \cos \alpha}$$

$$\tan \frac{\alpha}{2} = \frac{\sin \alpha}{1 + \cos \alpha}.$$

To obtain an equivalent identity for $\tan \alpha/2$, multiply by the conjugate of the denominator.

$$\tan \frac{\alpha}{2} = \frac{\sin \alpha}{1 + \cos \alpha} \cdot \frac{1 - \cos \alpha}{1 - \cos \alpha}$$

$$= \frac{\sin \alpha(1 - \cos \alpha)}{1 - \cos^2\alpha}$$

$$= \frac{\sin \alpha(1 - \cos \alpha)}{\sin^2\alpha}$$

$$\tan \frac{\alpha}{2} = \frac{1 - \cos \alpha}{\sin \alpha}$$

EXAMPLE 6   Evaluate a Trigonometric Function

Find sin 15° by using the half-angle identity.

Solution   Use the half-angle for the sine function in the first quadrant. Since $15° = \dfrac{30°}{2}$, let $\alpha = 30°$.

$$\sin \frac{\alpha}{2} = \sqrt{\frac{1 - \cos \alpha}{2}}$$

$$\sin \frac{30°}{2} = \sqrt{\frac{1 - \cos 30°}{2}}$$

$$\sin 15° = \sqrt{\frac{1 - \sqrt{3}/2}{2}} = \sqrt{\frac{2 - \sqrt{3}}{4}} = \frac{\sqrt{2 - \sqrt{3}}}{2}$$

■ *Try Exercise* **38**, *page 390.*

EXAMPLE 7   Verify a Half-Angle Identity

Verify the identity $2 \csc x \cos^2 \dfrac{x}{2} = \dfrac{\sin x}{1 - \cos x}$.

Solution   Work on the left side of the identity.

$$2 \csc x \cos^2 \frac{x}{2} = 2 \csc x \frac{1 + \cos x}{2} \qquad \cos^2 \frac{x}{2} = \frac{1 + \cos x}{2}$$

$$= \frac{1 + \cos x}{\sin x} \qquad\qquad \csc x = \frac{1}{\sin x}$$

$$= \frac{1 - \cos x}{1 - \cos x} \cdot \frac{1 + \cos x}{\sin x} \qquad \text{Multiply the numerator and}$$
$$\text{denominator by the conjugate}$$
$$\text{of the numerator.}$$

$$= \frac{1 - \cos^2 x}{(1 - \cos x) \sin x}$$

$$= \frac{\sin^2 x}{(1 - \cos x)(\sin x)} \qquad 1 - \cos^2 x = \sin^2 x$$

$$= \frac{\sin x}{1 - \cos x}$$

■ *Try Exercise* **72**, *page 390.*

EXAMPLE 8   Verify a Half-Angle Identity

Verify the identity $\tan \dfrac{\alpha}{2} = \sin \alpha + \cos \alpha \cot \alpha - \cot \alpha$.

Solution Work on the left side of the identity.

$$\tan \frac{\alpha}{2} = \frac{1 - \cos \alpha}{\sin \alpha} = \frac{\sin^2\alpha + \cos^2\alpha - \cos \alpha}{\sin \alpha} \qquad 1 = \sin^2\alpha + \cos^2\alpha$$

$$= \frac{\sin^2\alpha}{\sin \alpha} + \frac{\cos^2\alpha}{\sin \alpha} - \frac{\cos \alpha}{\sin \alpha} \qquad \text{Divide each term by } \sin \alpha.$$

$$= \sin \alpha + \cos \alpha \cot \alpha - \cot \alpha$$

■ *Try Exercise* **78,** *page 390.*

Here is a summary of the double- and half-angle identities:

Double-Angle Identities

$$\sin 2\alpha = 2 \sin \alpha \cos \alpha$$

$$\cos 2\alpha = \cos^2\alpha - \sin^2\alpha = 1 - 2 \sin^2\alpha = 2 \cos^2\alpha - 1$$

$$\tan 2\alpha = \frac{2 \tan \alpha}{1 - \tan^2\alpha}$$

Half-Angle Identities

$$\sin \frac{\alpha}{2} = \pm\sqrt{\frac{1 - \cos \alpha}{2}} \qquad \tan \frac{\alpha}{2} = \frac{\sin \alpha}{1 + \cos \alpha} = \frac{1 - \cos \alpha}{\sin \alpha}$$

$$\cos \frac{\alpha}{2} = \pm\sqrt{\frac{1 + \cos \alpha}{2}}$$

EXERCISE SET 7.3

In Exercises 1 to 8, write the trigonometric expressions in terms of a single trigonometric function.

1. $2 \sin 2\alpha \cos 2\alpha$

2. $2 \sin 3\theta \cos 3\theta$

3. $1 - 2 \sin^2 5\beta$

4. $2 \cos^2 2\beta - 1$

5. $\cos^2 3\alpha - \sin^2 3\alpha$

6. $\cos^2 6\alpha - \sin^2 6\alpha$

7. $\dfrac{2 \tan 3\alpha}{1 - \tan^2 3\alpha}$

8. $\dfrac{2 \tan 4\theta}{1 - \tan^2 4\theta}$

In Exercises 9 to 24, use the half-angle identities to evaluate the trigonometric expressions.

9. $\sin 75°$

10. $\cos 105°$

11. $\tan 67.5°$

12. $\tan 165°$

13. $\cos 157.5°$

14. $\sin 112.5°$

15. $\sin 22.5°$

16. $\cos 67.5°$

17. $\sin \dfrac{7\pi}{8}$

18. $\cos \dfrac{5\pi}{8}$

19. $\cos \dfrac{5\pi}{12}$

20. $\sin \dfrac{3\pi}{8}$

21. $\tan \dfrac{7\pi}{12}$

22. $\tan \dfrac{3\pi}{8}$

23. $\cos \dfrac{\pi}{12}$

24. $\sin \dfrac{\pi}{8}$

In Exercises 25 to 36, find the exact value of $\sin 2\theta$, $\cos 2\theta$ and $\tan 2\theta$ given the following information.

25. $\cos \theta = -\dfrac{4}{5}$, θ is in quadrant II.

26. $\cos \theta = \dfrac{24}{25}$, θ is in quadrant IV.

27. $\sin \theta = \dfrac{8}{17}$, θ is in quadrant II.

28. $\sin \theta = -\dfrac{9}{41}$, θ is in quadrant III.

29. $\tan \theta = -\dfrac{24}{7}$, θ is in quadrant IV.

30. $\tan \theta = \dfrac{4}{3}$, θ is in quadrant I.

31. $\sin \theta = \dfrac{15}{17}$, θ is in quadrant I.

32. $\sin \theta = -\dfrac{3}{5}$, θ is in quadrant III.

33. $\cos \theta = \dfrac{40}{41}$, θ is in quadrant IV.

34. $\cos \theta = \dfrac{4}{5}$, θ is in quadrant IV.

35. $\tan \theta = \dfrac{15}{8}$, θ is in quadrant III.

36. $\tan \theta = -\dfrac{40}{9}$, θ is in quadrant II.

In Exercises 37 to 48, find the value of the sine, cosine, and tangent of $\alpha/2$ given the following information.

37. $\sin \alpha = \dfrac{5}{13}$, α is in quadrant II.

38. $\sin \alpha = -\dfrac{7}{25}$, α is in quadrant III.

39. $\cos \alpha = -\dfrac{8}{17}$, α is in quadrant III.

40. $\cos \alpha = \dfrac{12}{13}$, α is in quadrant I.

41. $\tan \alpha = \dfrac{4}{3}$, α is in quadrant I.

42. $\tan \alpha = -\dfrac{8}{15}$, α is in quadrant II.

43. $\cos \alpha = \dfrac{24}{25}$, α is in quadrant IV.

44. $\sin \alpha = -\dfrac{9}{41}$, α is in quadrant IV.

45. $\sec \alpha = \dfrac{17}{15}$, α is in quadrant I.

46. $\csc \alpha = -\dfrac{5}{3}$, α is in quadrant IV.

47. $\cot \alpha = \dfrac{8}{15}$, α is in quadrant III.

48. $\sec \alpha = -\dfrac{13}{5}$, α is in quadrant II.

In Exercises 49 to 94, verify the given identity.

49. $\sin 3x \cos 3x = \dfrac{1}{2} \sin 6x$

50. $\cos 8x = \cos^2 4x - \sin^2 4x$

51. $\sin^2 x + \cos 2x = \cos^2 x$

52. $\dfrac{\cos 2x}{\sin^2 x} = \cot^2 x - 1$

53. $\dfrac{1 + \cos 2x}{\sin 2x} = \cot x$

54. $\dfrac{1}{1 - \cos 2x} = \dfrac{1}{2} \csc^2 x$

55. $\dfrac{\sin 2x}{1 - \sin^2 x} = 2 \tan x$

56. $\dfrac{\cos^2 x - \sin^2 x}{2 \sin x \cos x} = \cot 2x$

57. $1 - \tan^2 x = \dfrac{\cos 2x}{\cos^2 x}$

58. $\tan 2x = \dfrac{2 \sin x \cos x}{\cos^2 x - \sin^2 x}$

59. $\sin 2x - \tan x = \tan x \cos 2x$

60. $\sin 2x - \cot x = -\cot x \cos 2x$

61. $\cos^4 x - \sin^4 x = \cos 2x$

62. $\sin 4x = 4 \sin x \cos^3 x - 4 \cos x \sin^3 x$

63. $\cos^2 x - 2 \sin^2 x \cos^2 x - \sin^2 x + 2 \sin^4 x = \cos^2 2x$

64. $2 \cos^4 x - \cos^2 x - 2 \sin^2 x \cos^2 x + \sin^2 x = \cos^2 2x$

65. $\cos 4x = 1 - 8 \cos^2 x + 8 \cos^4 x$

66. $\sin 4x = 4 \sin x \cos x - 8 \cos x \sin^3 x$

67. $\cos 3x - \cos x = 4 \cos^3 x - 4 \cos x$

68. $\sin 3x + \sin x = 4 \sin x - 4 \sin^3 x$

69. $\sin^3 x + \cos^3 x = (\sin x + \cos x)\left(1 - \dfrac{1}{2} \sin 2x\right)$

70. $\cos^3 x - \sin^3 x = (\cos x - \sin x)\left(1 + \dfrac{1}{2} \sin 2x\right)$

71. $\sin^2 \dfrac{x}{2} = \dfrac{\sec x - 1}{2 \sec x}$

72. $\cos^2 \dfrac{x}{2} = \dfrac{\sec x + 1}{2 \sec x}$

73. $\tan \dfrac{x}{2} = \csc x - \cot x$

74. $\tan \dfrac{x}{2} = \dfrac{\tan x}{\sec x + 1}$

75. $2 \sin \dfrac{x}{2} \cos \dfrac{x}{2} = \sin x$

76. $\cos^2 \dfrac{x}{2} - \sin^2 \dfrac{x}{2} = \cos x$

77. $\left(\cos \dfrac{x}{2} + \sin \dfrac{x}{2}\right)^2 = 1 + \sin x$

78. $\tan^2 \dfrac{x}{2} = \dfrac{\sec x - 1}{\sec x + 1}$

79. $\sin^2 \dfrac{x}{2} \sec x = \dfrac{1}{2}(\sec x - 1)$

80. $\cos^2 \dfrac{x}{2} \sec x = \dfrac{1}{2}(\sec x + 1)$

81. $\cos^2 \dfrac{x}{2} - \cos x = \sin^2 \dfrac{x}{2}$

82. $\sin^2 \dfrac{x}{2} + \cos x = \cos^2 \dfrac{x}{2}$

83. $\sin^2 \dfrac{x}{2} - \cos^2 \dfrac{x}{2} = -\cos x$

84. $\cos^2 \dfrac{x}{2} - \sin^2 \dfrac{x}{2} = \dfrac{1}{2} \csc x \sin 2x$

85. $\sin 2x - \cos x = \cos x(2 \sin x - 1)$

86. $\dfrac{\cos 2x}{\sin^2 x} = \csc^2 x - 2$

87. $\tan 2x = \dfrac{2}{\cot x - \tan x}$

88. $\dfrac{2 \cos 2x}{\sin 2x} = \cot x - \tan x$

89. $2 \tan \dfrac{x}{2} = \dfrac{\sin^2 x + 1 - \cos^2 x}{\sin x(1 + \cos x)}$

90. $\dfrac{1}{2} \csc^2 \dfrac{x}{2} = \csc^2 x + \cot x \csc x$

91. $\csc 2x = \dfrac{1}{2} \csc x \sec x$ **92.** $\sec 2x = \dfrac{\sec^2 x}{2 - \sec^2 x}$

93. $\cos \dfrac{x}{5} = 1 - 2 \sin^2 \dfrac{x}{10}$ **94.** $\sec^2 \dfrac{x}{2} = \dfrac{2}{1 + \cos x}$

Supplemental Exercises

In Exercises 95 to 98, verify the identities.

95. $\dfrac{\sin^3 x + \cos^3 x}{\sin x + \cos x} = 1 - \dfrac{1}{2} \sin 2x$

96. $\cos^4 x = \dfrac{1}{8} \cos 4x + \dfrac{1}{2} \cos 2x + \dfrac{3}{8}$

97. $\sin \dfrac{x}{2} - \cos \dfrac{x}{2} = \sqrt{1 - \sin x}, \, 0 \le x \le 90°$

98. $\dfrac{\sin x - \sin 2x}{\cos x + \cos 2x} = -\tan \dfrac{x}{2}$

99. If $x + y = 90°$; verify $\sin(x - y) = -\cos 2x$.

100. If $x + y = 90°$; verify $\sin(x - y) = \cos 2y$.

101. If $x + y = 180°$; verify $\sin(x - y) = -\sin 2x$.

102. If $x + y = 180°$; verify $\cos(x - y) = -\cos 2x$.

7.4

Identities Involving the Sum of Trigonometric Functions

Some applications require that a product of trigonometric functions be written as a sum or difference of these functions. Other applications require that the sum or difference of trigonometric functions be represented as a product of these functions. The product-to-sum identities are particularly useful to these types of problems.

The Product-to-Sum Identities

The product-to-sum identities can be derived by using the sum and difference identities. Adding the identities for $\sin(\alpha + \beta)$ and $\sin(\alpha - \beta)$, we have

$$\sin(\alpha + \beta) = \sin \alpha \cos \beta + \cos \alpha \sin \beta$$

$$\underline{\sin(\alpha - \beta) = \sin \alpha \cos \beta - \cos \alpha \sin \beta}$$

$$\sin(\alpha + \beta) + \sin(\alpha - \beta) = 2 \sin \alpha \cos \beta$$

Solving for $\sin \alpha \cos \beta$, we obtain the first product-to-sum identity:

$$\sin \alpha \cos \beta = \frac{1}{2}[\sin(\alpha + \beta) + \sin(\alpha - \beta)].$$

The identity for $\cos \alpha \sin \beta$ is obtained when $\sin(\alpha - \beta)$ is subtracted from $\sin(\alpha + \beta)$. The result is

$$\cos \alpha \sin \beta = \frac{1}{2}[\sin(\alpha + \beta) - \sin(\alpha - \beta)].$$

In like manner, the identities for $\cos(\alpha + \beta)$ and $\cos(\alpha - \beta)$ are used to derive the identities for $\cos \alpha \cos \beta$ and $\sin \alpha \sin \beta$.

$$\cos \alpha \cos \beta = \frac{1}{2}[\cos(\alpha + \beta) + \cos(\alpha - \beta)]$$

$$\sin \alpha \sin \beta = \frac{1}{2}[\cos(\alpha - \beta) - \cos(\alpha + \beta)]$$

EXAMPLE 1   Use the Product-to-Sum Identity to Evaluate an Expression

Use a product-to-sum identity to evaluate $\sin 75° \cos 15°$.

Solution $\sin 75° \cos 15° = \frac{1}{2}(\sin[75° + 15°] + \sin[75° - 15°])$

$$= \frac{1}{2}(\sin 90° + \sin 60°)$$

$$= \frac{1}{2}\left(1 + \frac{\sqrt{3}}{2}\right) = \frac{1}{2} + \frac{\sqrt{3}}{4} = \frac{2 + \sqrt{3}}{4}$$

■ *Try Exercise* **12,** *page 397.*

EXAMPLE 2   Verify an Identity By Using the Product-to-Sum Identity

Verify the identity $\cos 2x \sin 5x = \frac{1}{2}(\sin 7x + \sin 3x)$.

Solution

$\cos 2x \sin 5x = \frac{1}{2}[\sin(2x + 5x) - \sin(2x - 5x)]$ Use the product-to-sum identity: $\cos \alpha \sin \beta$.

$$= \frac{1}{2}[\sin 7x - \sin(-3x)]$$

$$= \frac{1}{2}(\sin 7x + \sin 3x)$$ $\sin(-3x) = -\sin 3x$

■ *Try Exercise* **36,** *page 397.*

The Sum-to-Product Identities

The sum-to-product identities can be derived from the product-to-sum identities. To derive the sum-to-product identity for $\sin x + \sin y$ first we let $x = \alpha + \beta$ and $y = \alpha - \beta$. Then

$$x + y = \alpha + \beta + \alpha - \beta \qquad \text{and} \qquad x - y = \alpha + \beta - (\alpha - \beta)$$

$$x + y = 2\alpha \qquad\qquad\qquad\qquad x - y = 2\beta$$

$$\alpha = \frac{x + y}{2} \qquad\qquad\qquad\qquad \beta = \frac{x - y}{2} \qquad\qquad (1)$$

Substituting α and β in the product-to-sum identity

$$\frac{1}{2}[\sin(\alpha + \beta) + \sin(\alpha - \beta)] = \sin \alpha \cos \beta,$$

we have

$$\sin\left(\frac{x + y}{2} + \frac{x - y}{2}\right) + \sin\left(\frac{x + y}{2} - \frac{x - y}{2}\right) = 2 \sin \frac{x + y}{2} \cos \frac{x - y}{2}.$$

Simplifying the left side, we have a sum-to-product identity.

$$\sin x + \sin y = 2 \sin \frac{x + y}{2} \cos \frac{x - y}{2}$$

To derive the sum to product identity for $\cos x - \cos y$, substitute the expressions for α and β from equation (1) in the product identity $\frac{1}{2}[\cos(\alpha - \beta) - \cos(\alpha + \beta)] = \sin \alpha \sin \beta$ to get

$$\cos\left(\frac{x + y}{2} - \frac{x - y}{2}\right) - \cos\left(\frac{x + y}{2} + \frac{x - y}{2}\right) = 2 \sin \frac{x + y}{2} \sin \frac{x - y}{2}.$$

Simplifying the left side, we have a sum-to-product identity.

$$\cos y - \cos x = 2 \sin \frac{x + y}{2} \sin \frac{x - y}{2}$$

or, multiplying by -1,

$$\cos x - \cos y = -2 \sin \frac{x + y}{2} \sin \frac{x - y}{2}$$

In like manner, two other sum-to-product identities can be derived from the other product-to-sum identities. The proof of the identities are left as exercises.

$$\sin x - \sin y = 2 \cos \frac{x + y}{2} \sin \frac{x - y}{2}$$

$$\cos x + \cos y = 2 \cos \frac{x + y}{2} \cos \frac{x - y}{2}$$

EXAMPLE 3   **Write the Difference of Trigonometric Expressions As a Product**

Write $\sin 142° - \sin 80°$ as the product of two functions.

Solution

$$\sin 142° - \sin 80° = 2 \cos \frac{142° + 80°}{2} \sin \frac{142° - 80°}{2} = 2 \cos 111° \sin 31°$$

■ *Try Exercise* **22**, *page 397.*

EXAMPLE 4   **Verify a Sum-to-Product Identity**

Verify the identity $\dfrac{\sin 6x + \sin 2x}{\sin 6x - \sin 2x} = \tan 4x \cot 2x.$

Solution

$$\frac{\sin 6x + \sin 2x}{\sin 6x - \sin 2x} = \frac{2 \sin \dfrac{6x + 2x}{2} \cos \dfrac{6x - 2x}{2}}{2 \cos \dfrac{6x + 2x}{2} \sin \dfrac{6x - 2x}{2}} = \frac{\sin 4x \cos 2x}{\cos 4x \sin 2x}$$

$$= \tan 4x \cot 2x$$

■ *Try Exercise* **44**, *page 397.*

Functions of the Form $f(x) = a \sin x + b \cos x$

The function $f(x) = a \sin x + b \cos x$ can be written as $f(x) = k \sin(x + \alpha)$. This form of the function is useful in graphing and engineering applications because the amplitude, period, and phase shift can be readily calculated.

Let $P(a, b)$ be a point on a coordinate plane and let α represent an angle in standard position, as shown in Figure 7.3. To rewrite $y = a \sin x + b \cos x$, multiply and divide the expression $a \sin x + b \cos x$ by $\sqrt{a^2 + b^2}$.

$$a \sin x + b \cos x = \frac{\sqrt{a^2 + b^2}}{\sqrt{a^2 + b^2}} (a \sin x + b \cos x)$$

$$= \sqrt{(a^2 + b^2)} \left(\frac{a}{\sqrt{a^2 + b^2}} \sin x + \frac{b}{\sqrt{a^2 + b^2}} \cos x \right) \quad (1)$$

From the definition of the sine and cosine of an angle in standard position, let

$$k = \sqrt{a^2 + b^2}, \quad \cos \alpha = \frac{a}{\sqrt{a^2 + b^2}}, \quad \text{and} \quad \sin \alpha = \frac{b}{\sqrt{a^2 + b^2}}.$$

Substitute these expressions into Equation (1). We have

$$a \sin x + b \cos x = k(\cos \alpha \sin x + \sin \alpha \cos x).$$

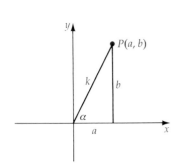

Figure 7.3

Now using the identity for the sum of two angles, we have

$$a \sin x + b \cos x = k \sin(x + \alpha).$$

Thus $a \sin x + b \cos x = k \sin(x + \alpha)$, where $k = \sqrt{a^2 + b^2}$ and α is the angle for which $\sin \alpha = b/\sqrt{a^2 + b^2}$ and $\cos \alpha = a/\sqrt{a^2 + b^2}$.

EXAMPLE 5   Write the Sum of the Sine and Cosine in Terms of the Sine

Write $f(x) = \sin x + \cos x$ in terms of $f(x) = k \sin(x + \alpha)$.

Solution   Since $a = 1$, $b = 1$, then we have $k = \sqrt{1^2 + 1^2} = \sqrt{2}$, $\sin \alpha = \dfrac{1}{\sqrt{2}}$, and $\cos \alpha = \dfrac{1}{\sqrt{2}}$. Thus, $\alpha = 45°$.

$$\sin x + \cos x = k \sin(x + \alpha) = \sqrt{2} \sin(x + 45°)$$

■ *Try Exercise **62**, page 397.*

The expression $y = a \sin x + b \cos x$ can be graphed by the addition of ordinates. However, it is easier to graph the function as a sine function with a phase shift.

EXAMPLE 6   Use an Identity to Graph a Trigonometric Function

Graph the function $f(x) = -\sin x + \sqrt{3} \cos x$.

Solution   First, we write $f(x)$ as $k \sin(x + \alpha)$. Let $a = -1$, $b = \sqrt{3}$, then $k = \sqrt{(-1)^2 + (\sqrt{3})^2} = 2$. The point $P(-1, \sqrt{3})$ is the second quadrant. Find the reference angle β.

$$\sin \beta = \frac{\sqrt{3}}{2} \qquad \beta \text{ is the reference angle of } \alpha.$$

$$\beta = \frac{\pi}{3}$$

$$\alpha = \pi - \frac{\pi}{3} = \frac{2\pi}{3}$$

Substituting the values of k and α in the formula $y = k \sin(x + \alpha)$, we have

$$y = 2 \sin\left(x + \frac{2\pi}{3}\right).$$

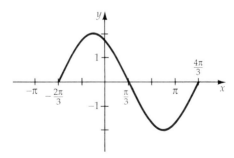

Figure 7.4
$f(x) = -\sin x + \sqrt{3} \cos x$

Graph by using the key values:

Amplitude = 2   Period = 2π Phase Shift = $-\dfrac{2\pi}{3}$

1. Beginning point: $f(-2\pi/3) = 2 \sin 0 = 0$.
2. Quarter point: $f(-\pi/6) = 2 \sin \pi/2 = 2$.
3. Middle point: $f(\pi/3) = 2 \sin \pi = 0$.

4. Three-quarter point: $f(5\pi/6) = 2 \sin 3\pi/2 = -2$.
5. End point: $f(4/3) = 2 \sin 2\pi = 0$.

x	$-2\pi/3$	$-\pi/6$	$\pi/3$	$5\pi/6$	$4\pi/3$
$2 \sin\left(x + \dfrac{2\pi}{3}\right)$	0	2	0	-2	0
(x, y)	$\left(-\dfrac{2\pi}{3}, 0\right)$	$\left(-\dfrac{\pi}{6}, 2\right)$	$\left(\dfrac{\pi}{3}, 0\right)$	$\left(\dfrac{5\pi}{6}, -2\right)$	$\left(\dfrac{4\pi}{3}, 0\right)$

■ *Try Exercise* **70,** *page 397.*

We now list all the product-to-sum and sum-to-product identities.

Product-to-Sum Identities

$$\sin \alpha \cos \beta = \frac{1}{2}[\sin(\alpha + \beta) + \sin(\alpha - \beta)]$$

$$\cos \alpha \sin \beta = \frac{1}{2}[\sin(\alpha + \beta) - \sin(\alpha - \beta)]$$

$$\cos \alpha \cos \beta = \frac{1}{2}[\cos(\alpha + \beta) + \cos(\alpha - \beta)]$$

$$\sin \alpha \sin \beta = \frac{1}{2}[\cos(\alpha - \beta) - \cos(\alpha + \beta)]$$

Sum-to-Product Identities

$$\sin x + \sin y = 2 \sin \frac{x + y}{2} \cos \frac{x - y}{2}$$

$$\cos x - \cos y = -2 \sin \frac{x + y}{2} \sin \frac{x - y}{2}$$

$$\sin x - \sin y = 2 \cos \frac{x + y}{2} \sin \frac{x - y}{2}$$

$$\cos x + \cos y = 2 \cos \frac{x + y}{2} \cos \frac{x - y}{2}$$

Sums of the Form $a \sin x + b \cos x$

$$a \sin x + b \cos x = k \sin(x + \alpha),$$

where $k = \sqrt{a^2 + b^2}$, $\sin \alpha = \dfrac{b}{\sqrt{a^2 + b^2}}$, and $\cos \alpha = \dfrac{a}{\sqrt{a^2 + b^2}}$.

EXERCISE SET 7.4

In Exercises 1 to 8, write each expression as the sum or difference of two functions.

1. $2 \sin x \cos 2x$

2. $2 \sin 4x \sin 2x$

3. $\cos 6x \sin 2x$

4. $\cos 3x \cos 5x$

5. $2 \sin 5x \cos 3x$

6. $2 \sin 2x \cos 6x$

7. $\sin x \sin 5x$

8. $\cos 3x \sin x$

In Exercises 9 to 16, evaluate each of the expressions. Do not use a calculator or tables.

9. $\cos 75° \cos 15°$

10. $\sin 105° \cos 15°$

11. $\cos 157.5° \sin 22.5°$

12. $\sin 195° \cos 15°$

13. $\sin \dfrac{13\pi}{12} \cos \dfrac{\pi}{12}$

14. $\sin \dfrac{11\pi}{12} \sin \dfrac{7\pi}{12}$

15. $\sin \dfrac{\pi}{12} \cos \dfrac{7\pi}{12}$

16. $\cos \dfrac{17\pi}{12} \sin \dfrac{7\pi}{12}$

In Exercises 17 to 24, write each expression as the product of two functions.

17. $\sin 4\theta + \sin 2\theta$

18. $\cos 5\theta - \cos 3\theta$

19. $\cos 3\theta + \cos \theta$

20. $\sin 7\theta - \sin 3\theta$

21. $\cos 6\theta - \cos 2\theta$

22. $\cos 3\theta + \cos 5\theta$

23. $\cos \theta + \cos 3\theta$

24. $\sin 3\theta + \sin 7\theta$

In Exercises 25 to 32, evaluate each of the expressions. Do not use a calculator or tables.

25. $\sin 75° + \sin 15°$

26. $\cos 105° + \cos 165°$

27. $\cos 105° - \cos 15°$

28. $\sin 255° - \sin 165°$

29. $\sin \dfrac{13\pi}{12} + \sin \dfrac{5\pi}{12}$

30. $\sin \dfrac{11\pi}{12} - \sin \dfrac{5\pi}{12}$

31. $\cos \dfrac{\pi}{12} + \cos \dfrac{5\pi}{12}$

32. $\cos \dfrac{7\pi}{8} + \cos \dfrac{\pi}{8}$

In Exercises 33 to 48, verify the identities.

33. $2 \cos \alpha \cos \beta = \cos(\alpha + \beta) + \cos(\alpha - \beta)$

34. $2 \sin \alpha \sin \beta = \cos(\alpha - \beta) - \cos(\alpha + \beta)$

35. $2 \cos 3x \sin x = 2 \sin x \cos x - 8 \cos x \sin^3 x$

36. $\sin 5x \cos 3x = \sin 4x \cos 4x + \sin x \cos x$

37. $2 \cos 5x \cos 7x = \cos^2 6x - \sin^2 6x + 2 \cos^2 x - 1$

38. $\sin 3x \cos x = \sin x \cos x(3 - 4 \sin^2 x)$

39. $\sin 3x - \sin x = 2 \sin x - 4 \sin^3 x$

40. $\cos 5x - \cos 3x = -8 \sin^2 x(2 \cos^3 x - \cos x)$

41. $\sin 2x + \sin 4x = 2 \sin x \cos x(4 \cos^2 x - 1)$

42. $\cos 3x + \cos x = 4 \cos^3 x - 2 \cos x$

43. $\dfrac{\sin 3x - \sin x}{\cos 3x - \cos x} = -\cot 2x$

44. $\dfrac{\cos 5x - \cos 3x}{\sin 5x + \sin 3x} = -\tan x$

45. $\dfrac{\sin 5x + \sin 3x}{4 \sin x \cos^3 x - 4 \sin^3 x \cos x} = 2 \cos x$

46. $\dfrac{\cos 4x - \cos 2x}{\sin 2x - \sin 4x} = \tan 3x$

47. $\sin(x + y) \cos(x - y) = \sin x \cos x + \sin y \cos y$

48. $\sin(x + y) \sin(x - y) = \sin^2 x - \sin^2 y$

In Exercises 49 to 58, write the given equation in the form $y = k \sin(x + \alpha)$, where the measure of α is in degrees.

49. $y = -\sin x - \cos x$

50. $y = \sqrt{3} \sin x - \cos x$

51. $y = \dfrac{1}{2} \sin x - \dfrac{\sqrt{3}}{2} \cos x$

52. $y = \dfrac{\sqrt{3}}{2} \sin x - \dfrac{1}{2} \cos x$

53. $y = \dfrac{1}{2} \sin x - \dfrac{1}{2} \cos x$

54. $y = -\dfrac{\sqrt{3}}{2} \sin x - \dfrac{1}{2} \cos x$

55. $y = 8 \sin x + 15 \cos x$

56. $y = -7 \sin x + 24 \cos x$

57. $y = 8 \sin x - 3 \cos x$

58. $y = -4 \sin x + 7 \cos x$

In Exercises 59 to 66, write the given equations in the form $k \sin(x + \alpha)$ where the measure of α is in radians.

59. $y = -\sin x + \cos x$

60. $y = -\sqrt{3} \sin x - \cos x$

61. $y = \dfrac{\sqrt{3}}{2} \sin x + \dfrac{1}{2} \cos x$

62. $y = \sin x + \sqrt{3} \cos x$

63. $y = -4 \sin x + 9 \cos x$

64. $y = -3 \sin x + 5 \cos x$

65. $y = -5 \sin x + 5 \cos x$

66. $y = 3 \sin x - 3 \cos x$

In Exercises 67 to 76, graph one cycle of the following equations.

67. $y = -\sin x - \sqrt{3} \cos x$

68. $y = -\sqrt{3} \sin x + \cos x$

69. $y = 2 \sin x + 2 \cos x$

70. $y = \sin x + \sqrt{3} \cos x$

71. $y = -\sqrt{3} \sin x - \cos x$

72. $y = -\sin x + \cos x$

73. $y = 2 \sin x - 5 \cos x$

74. $y = -6 \sin x - 10 \cos x$

75. $y = 3 \sin x - 4 \cos x$

76. $y = -5 \sin x + 9 \cos x$

Supplemental Exercises

77. Derive the sum-to-product identity:

$$\cos x + \cos y = 2 \cos \frac{x+y}{2} \cos \frac{x-y}{2}$$

78. Derive the product-to-sum identity:

$$\sin x \sin y = \frac{1}{2}[\cos(x-y) - \cos(x+y)]$$

79. If $x + y = 180°$; show that $\sin x + \sin y = 2 \sin x$.

80. If $x + y = 360°$; show that $\cos x + \cos y = 2 \cos x$.

In Exercises 81 to 86, verify the identities.

81. $\sin 2x + \sin 4x + \sin 6x = 4 \sin 3x \cos 2x \cos x$

82. $\sin 4x - \sin 2x + \sin 6x = 4 \cos 3x \sin 2x \cos x$

83. $\dfrac{\cos 10x + \cos 8x}{\sin 10x - \sin 8x} = \cot x$

84. $\dfrac{\sin 10x + \sin 2x}{\cos 10x + \cos 2x} = \dfrac{2 \tan 3x}{1 - \tan^2 3x}$

85. $\dfrac{\sin 2x + \sin 4x + \sin 6x}{\cos 2x + \cos 4x + \cos 6x} = \tan 4x$

86. $\dfrac{\sin 2x + \sin 6x}{\cos 6x - \cos 2x} = -\cot 2x$

87. Verify $\cos^2 x - \sin^2 x = \cos 2x$ by using a product-to-sum identity.

88. Verify $2 \sin x \cos x = \sin 2x$ by using a product-to-sum identity.

89. Verify that $a \sin x + b \cos x = k \cos(x - \alpha)$, where $k = \sqrt{a^2 + b^2}$ and $\tan \alpha = a/b$.

90. Verify that $a \sin cx + b \cos cx = k \sin(cx + \alpha)$, where $k = \sqrt{a^2 + b^2}$ and $\tan \alpha = b/a$.

In Exercises 91 to 96, find the amplitude, phase shift, and period and then graph the function.

91. $y = \sin \dfrac{x}{2} - \cos \dfrac{x}{2}$

92. $y = -\sqrt{3} \sin \dfrac{x}{2} + \cos \dfrac{x}{2}$

93. $y = \sqrt{3} \sin 2x - \cos 2x$

94. $y = -\sin 2x + \cos 2x$

95. $y = \sin \pi x + \sqrt{3} \cos \pi x$

96. $y = 3 \sin 2\pi x - 4 \cos 2\pi x$

97. Two nonvertical lines intersect in a plane. The slope of l_1 is m_1 and the slope of l_2 is m_2. Show that the tangent of the angle θ formed by the two lines is given by the expression:

$$\tan \theta = \frac{m_1 - m_2}{1 + m_1 m_2}.$$

98. Use the equation from Exercise 97 to find the angle from the line $y = x + 5$ to the line $y = 3x - 4$.

99. Use the equation from Exercise 97 to find the angle from the line $y = -3x/2 - 4$ to the line $y = 2x/3 + 3$.

7.5

Inverse Trigonometric Functions

Because the sine function is not a one-to-one function it does not have an inverse function. Figure 7.5 shows the graphs of the function $f(x) = \sin x$ and the equation $x = \sin y$. The sine function is one-to-one on the interval $-\pi/2 \le x \le \pi/2$. Thus the sine function, with domain restricted to that interval, does have an inverse function. Using this restricted domain, the graphs of the restricted sine function and the inverse sine function are shown in Figure 7.6. The range values of the inverse sine function are called the **principal values** of the function.

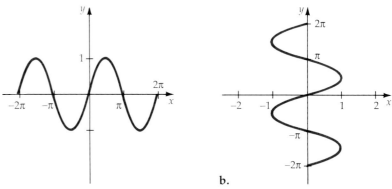

Figure 7.5
a. $y = \sin x$; b. $x = \sin y$

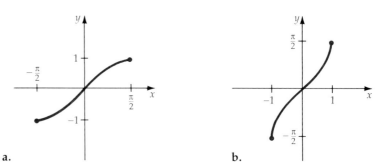

Figure 7.6
a. $y = \sin x$: $-\pi/2 \leq x \leq \pi/2$;
b. $x = \sin y$: $-\pi/2 \leq y \leq \pi/2$.

The graph shows that the domain of the sine function $y = \sin x$ with $-\pi/2 \leq x \leq \pi/2$ is the range of the inverse sine function $x = \sin y$. We can attempt to find an equation for the inverse sine function via the method we used for linear functions. We begin with

$$y = \sin x$$

Interchange the x and y. $x = \sin y$

Unfortunately, there is no algebraic solution for y. Thus we establish new notation and write

$$y = \sin^{-1} x$$

which is read "y is the inverse sine of x." Some textbooks use the notation $y = \arcsin x$ for the inverse sine function.

Caution   The -1 in $\sin^{-1}x$ is not an exponent. The -1 is used to denote the inverse function. To use -1 as an exponent for the sine function, enclose the function with parentheses.

$$(\sin x)^{-1} = \frac{1}{\sin x} \qquad \sin^{-1} x \neq \frac{1}{\sin x}$$

This convention is used for other inverse trigonometric functions presented in this text.

From our discussion, we can write the following definition of the inverse sine function:

Definition of the Inverse Sine Function

$$y = \sin^{-1}x \text{ if and only if } x = \sin y$$

The domain of $y = \sin^{-1}x$ is $\{x \mid -1 \le x \le 1\}$. The range is $\{y \mid -\pi/2 \le y \le \pi/2\}$.

Recall that a function f and its inverse function f^{-1} have the property that $f[f^{-1}(x)] = x$ and $f^{-1}[f(x)] = x$. Thus, if $f(x) = \sin x$ and $f^{-1}(x) = \sin^{-1}x$, then

$$f[f^{-1}(x)] = \sin(\sin^{-1}x) = x \quad \text{where } -1 \le x \le 1,$$

$$f^{-1}[f(x)] = \sin^{-1}(\sin x) = x \quad \text{where } -\frac{\pi}{2} \le x \le \frac{\pi}{2}$$

It is convenient to think of the value of an inverse trigonometric function as an angle. For example, $\sin^{-1}\frac{1}{2}$ is an angle whose sine is $\frac{1}{2}$. There is an infinite number of angles whose sign is $\frac{1}{2}$. (Why?[1]) However, the function values of the inverse sine function are restricted to angles whose measure y is $-\pi/2 \le y \le \pi/2$. Thus $\sin^{-1}\frac{1}{2} = \pi/6$.

EXAMPLE 1 Evaluate the Inverse Sine Function

Find the exact value of $y = \sin^{-1}\left(-\dfrac{\sqrt{3}}{2}\right)$, where y is a real number.

Solution The meaning of $\sin^{-1}\left(-\dfrac{\sqrt{3}}{2}\right)$ is the angle whose sine is $-\dfrac{\sqrt{3}}{2}$. Since the range of the inverse sine function is $-\dfrac{\pi}{2} \le y \le \dfrac{\pi}{2}$, the angle is in the fourth quadrant because the inverse sine function is negative.

$$y = \sin^{-1}\left(-\frac{\sqrt{3}}{2}\right)$$

$$\sin y = -\frac{\sqrt{3}}{2} \qquad \text{y is the angle whose sine is $-\sqrt{3}/2$,}$$
$$\text{and } -\pi/2 \le y \le \pi/2.$$

$$y = -\frac{\pi}{3}$$

■ *Try Exercise **2**, page 409.*

---

[1] The sine function is a periodic function. Thus, $\sin\frac{1}{2} = \sin(\frac{1}{2} + 2k\pi)$, where k is an integer. There is an infinite number of values because the set of integers is infinite.

EXAMPLE 2   **Evaluate the Inverse Sine Function**

Evaluate $\sin^{-1}\left(\sin\dfrac{2\pi}{3}\right)$.

> *Solution*   Recall that $\sin^{-1}(\sin x) = x$ when $-\dfrac{\pi}{2} \le x \le \dfrac{\pi}{2}$. Because $\dfrac{2\pi}{3}$ is not in this interval, we must first evaluate $\sin\dfrac{2\pi}{3}$.
>
> $$y = \sin^{-1}\left(\sin\frac{2\pi}{3}\right) = \sin^{-1}\left(\frac{\sqrt{3}}{2}\right) \quad \sin 2\pi/3 = \sqrt{3}/2.$$
>
> $$\sin y = \frac{\sqrt{3}}{2} \qquad -\frac{\pi}{2} \le y \le \frac{\pi}{2} \qquad \text{By the definition of } y = \sin^{-1}x$$
>
> $$y = \frac{\pi}{3}$$

■ *Try Exercise 54, page 410.*

Caution   In this case, $\sin^{-1}(\sin x) \ne x$. The range of the inverse sine function is $\{y \mid -\pi/2 \le y \le \pi/2\}$ and $2\pi/3$ is not in the range of the function.

EXAMPLE 3   **Evaluate Inverse Sine Functions with a Calculator**

Use a calculator to evaluate the following expressions:

a. $\sin^{-1} 0.8$ b. $\sin^{-1}(\sin 120°)$ c. $\sin^{-1}(\sin 1.2)$ d. $\sin(\sin^{-1} 2)$

> *Solution*

Function	Key Sequence	Calculator Display
a. $\sin^{-1} 0.8$	0.8 $\boxed{\text{INV}}$ $\boxed{\text{sin}}$	$\boxed{\texttt{0.92729522}}$
b. $\sin^{-1}(\sin 120°)$	120 $\boxed{\text{sin}}$ $\boxed{\text{INV}}$ $\boxed{\text{sin}}$	$\boxed{\texttt{60}}$
c. $\sin^{-1}(\sin 1.2)$	1.2 $\boxed{\text{sin}}$ $\boxed{\text{INV}}$ $\boxed{\text{sin}}$	$\boxed{\texttt{1.2}}$
d. $\sin(\sin^{-1} 2)$	2 $\boxed{\text{INV}}$ $\boxed{\text{sin}}$	$\boxed{\text{Error}}$

■ *Try Exercise 30, page 409.*

Remark   In part d, $\sin^{-1} 2$ is undefined. (Why?[2]) Therefore, $\sin(\sin^{-1} 2)$ results in an error message.

The cosine function is not a one-to-one function so it does not have an inverse function. Figure 7.7 shows the graphs of the function $y = \cos x$ and the equation $x = \cos y$.

---

[2] If $x = \sin^{-1} 2$, then $\sin x = 2$. However, the range of $\sin x$ is $-1 \le \sin x \le 1$. Therefore, there is no value of x for $\sin x = 2$.

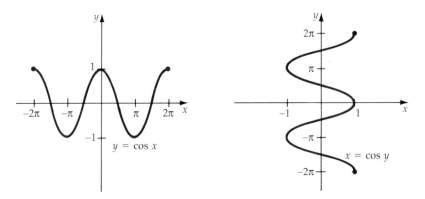

Figure 7.7

The cosine function is one-to-one in the interval $0 \le x \le \pi$. Thus the cosine function, with the domain restricted to that interval, has an inverse function. Using this restricted domain, the graphs of the restricted cosine function and the inverse cosine function are shown in Figure 7.8.

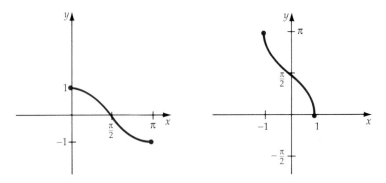

Figure 7.8
$y = \cos x$: $0 \le x \le \pi$; $x = \cos y$: $0 \le y \le \pi$.

Definition of the Inverse Cosine Function

$$y = \cos^{-1} x \text{ if and only if } x = \cos y$$

The domain of $y = \cos^{-1} x$ is $\{x \mid -1 \le x \le 1\}$. The range is $\{y \mid 0 \le y \le \pi\}$

A function f and its inverse f^{-1} have the property that $f[f^{-1}(x)] = x$ and $f^{-1}[f(x)] = x$. Thus, if $f(x) = \cos x$ and $f^{-1}(x) = \cos^{-1} x$, then

$$f[f^{-1}(x)] = \cos(\cos^{-1} x) = x \quad \text{where } -1 \le x \le 1$$

$$f^{-1}[f(x)] = \cos^{-1}(\cos x) = x \quad \text{where } 0 \le x \le \pi$$

EXAMPLE 4   Evaluate the Inverse Cosine Function with a Calculator

Use a calculator to evaluate the following expressions:

a. $\cos^{-1}(-0.2388)$ b. $\cos(\cos^{-1} 2)$ c. $\cos^{-1}(\cos 5)$

Solution

Function	Key Sequence	Calculator Display
a. $\cos^{-1}(-0.2388)$	-0.2388 $\boxed{\text{INV}}$ $\boxed{\cos}$	1.811926238
b. $\cos(\cos^{-1} 2)$	2 $\boxed{\text{INV}}$ $\boxed{\cos}$	Error
c. $\cos^{-1}(\cos 5)$	5 $\boxed{\cos}$ $\boxed{\text{INV}}$ $\boxed{\cos}$	1.283185307

■ *Try Exercise* **58,** *page 410.*

The domains of the other trigonometric functions can be restricted so that each has an inverse function. Table 7.2 shows the restricted function and the inverse function for tan x, csc x, sec x, and cot x. Pay close attention to the asymptotes of the function and the inverse function.

TABLE 7.2

	$y = \tan x$	$y = \tan^{-1}x$	$y = \csc x$	$y = \csc^{-1}x$
Domain	$-\dfrac{\pi}{2} < x < \dfrac{\pi}{2}$	$-\infty < x < \infty$	$-\dfrac{\pi}{2} \le x \le \dfrac{\pi}{2},$ $x \ne 0$	$x \le -1$ or $x \ge 1$
Range	$-\infty < y < \infty$	$-\dfrac{\pi}{2} < y < \dfrac{\pi}{2}$	$x \le -1, x \ge 1$	$-\dfrac{\pi}{2} \le y \le \dfrac{\pi}{2},$ $y \ne 0$
Asymptotes	$x = -\dfrac{\pi}{2}, x = \dfrac{\pi}{2}$	$y = -\dfrac{\pi}{2}, y = \dfrac{\pi}{2}$	$x = 0$	$y = 0$

(Continued)

TABLE 7.2 (*continued*)

	$y = \sec x$	$y = \sec^{-1}x$	$y = \cot x$	$y = \cot^{-1}x$
Domain	$0 \le x \le \pi,$ $x \ne \dfrac{\pi}{2}$	$x \le -1 \text{ or } x \ge 1$	$0 < x < \pi$	$-\infty < x < \infty$
Range	$y \le -1 \text{ or } y \ge 1$	$0 \le y \le \pi,$ $y \ne \dfrac{\pi}{2}$	$-\infty < y < \infty$	$0 < y < \pi$
Asymptotes	$x = \dfrac{\pi}{2}$	$y = \dfrac{\pi}{2}$	$x = 0, x = \pi$	$y = 0, y = \pi$

EXAMPLE 5   **Evaluate the Inverse Tangent Function**

Find the exact value of $y = \tan^{-1}\left(-\dfrac{\sqrt{3}}{3}\right)$.

Solution   We know that y is the real number whose tangent is $-\dfrac{\sqrt{3}}{3}$.

The range of the inverse tangent function is $-\pi/2 < y < \pi/2$. Thus y must be in the interval $-\pi/2 < y < 0$ because it is a negative number.

$$y = \tan^{-1}\left(-\frac{\sqrt{3}}{3}\right)$$

$$\tan y = -\frac{\sqrt{3}}{3}$$

$$y = -\frac{\pi}{6}$$

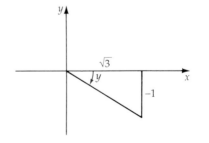

Figure 7.9

■ *Try Exercise* **18**, *page 409.*

A calculator may not have keys for the inverse secant, cosecant, and cotangent functions. The following procedure shows an identity for the inverse cosecant function in terms of the inverse sine function.

If we need to determine y, which is the angle (or number) whose cosecant is x, we can rewrite $y = \csc^{-1} x$ as follows.

$$y = \csc^{-1} x \qquad \text{Domain: } x \le -1 \text{ or } x \ge 1$$
$$\text{Range: } -\pi/2 \le y \le \pi/2, \ y \ne 0.$$

$$\csc y = x \qquad \text{Definition of inverse function}$$

$$\frac{1}{\sin y} = x \qquad \text{Substitute } 1/\sin y \text{ for } \csc y.$$

$$\sin y = \frac{1}{x} \qquad \text{Solve for } \sin y.$$

$$y = \sin^{-1} \frac{1}{x}$$

$$\csc^{-1} x = \sin^{-1} \frac{1}{x}$$

Thus the inverse cosecant of x is the same as the inverse sine of $1/x$.

There are similar expressions for the inverse secant and inverse cotangent.

$$\sec^{-1} x = \cos^{-1} 1/x \qquad \cot^{-1} x = \tan^{-1} 1/x$$

EXAMPLE 6   Find the Inverse Cosecant Function with a Calculator

Find $\csc^{-1} 4.9873$ using a calculator.

Solution   The inverse cosecant of an argument is the inverse sine of the reciprocal of the argument. Be sure your calculator is in the radian mode.

Function	Key Sequence	Calculator Display
$\csc^{-1} 4.9873$	4.9873 $\boxed{1/x}$ $\boxed{\text{INV}}$ $\boxed{\sin}$	$\boxed{0.20187774}$

■ *Try Exercise* **40,** *page 409.*

EXAMPLE 7   Evaluate the Inverse Secant Function

Find the exact value of $y = \sec^{-1} 2$.

Solution   Use the identity $\sec^{-1} x = \cos^{-1} \dfrac{1}{x}$.

$$\sec^{-1} 2 = \cos^{-1} \frac{1}{2} = \frac{\pi}{3}$$

■ *Try Exercise* **10,** *page 409.*

EXAMPLE 8   **Evaluate an Inverse Trigonometric Expression**

Evaluate $\sin\left[\cos^{-1}\left(-\dfrac{3}{5}\right)\right]$.

> *Solution*   Let $y = \cos^{-1}\left(-\dfrac{3}{5}\right)$, then $\cos y = -\dfrac{3}{5}$.
>
> $$\sin^2 y + \cos^2 y = 1$$
>
> $$\sin^2 y = 1 - \cos^2 y = 1 - \left(-\frac{3}{5}\right)^2 = \frac{16}{25}$$

Because $\cos y = -\dfrac{3}{5}$, we know $\dfrac{\pi}{2} < y < \pi$ and thus $\sin y > 0$. Therefore,

$$\sin y = \frac{4}{5}$$

The same problem worked with a calculator is as follows:

Function	Key Sequence	Calculator Display
$\sin\left(\cos^{-1}\left(-\dfrac{3}{5}\right)\right)$	3/5 $\boxed{+/-}$ $\boxed{\text{INV}}$ $\boxed{\cos}$ $\boxed{\sin}$	$\boxed{\text{0.8}}$

■ *Try Exercise* **70**, *page 410.*

EXAMPLE 9   **Evaluate an Inverse Trigonometric Expression**

Evaluate $\sin\left(\sin^{-1}\dfrac{3}{5} + \cos^{-1}\dfrac{5}{13}\right)$.

> *Solution*   Let $x = \sin^{-1}\dfrac{3}{5}$ and let $y = \cos^{-1}\dfrac{5}{13}$, then $\sin x = \dfrac{3}{5}$ and $\cos y = \dfrac{5}{13}$ with $0 < x < \dfrac{\pi}{2}$ and $0 < y < \dfrac{\pi}{2}$. Therefore,
>
> $$\cos x = \sqrt{1 - \sin^2 x} = \sqrt{1 - \left(\frac{3}{5}\right)^2} = \frac{4}{5}$$
>
> $$\sin y = \sqrt{1 - \cos^2 y} = \sqrt{1 - \left(\frac{5}{13}\right)^2} = \frac{12}{13}$$

$$\sin\left(\sin^{-1}\frac{3}{5} + \cos^{-1}\frac{5}{13}\right) = \sin(x + y) = \sin x \cos y + \cos x \sin y$$

$$= \frac{3}{5} \cdot \frac{5}{13} + \frac{4}{5} \cdot \frac{12}{13}$$

$$= \frac{15}{65} + \frac{48}{65} = \frac{63}{65}$$

■ *Try Exercise* **84**, *page 410.*

EXAMPLE 10   **Solve an Inverse Trigonometric Equation**

Solve the inverse trigonometric equation $\sin^{-1}x + \cos^{-1}\dfrac{3}{5} = \pi$.

Solution   Solve for $\sin^{-1}x$, then take the sine of both sides of the equation.

$$\sin^{-1}x + \cos^{-1}\frac{3}{5} = \pi$$

$$\sin^{-1}x = \pi - \cos^{-1}\frac{3}{5}$$

$$\sin(\sin^{-1}x) = \sin\left(\pi - \cos^{-1}\frac{3}{5}\right)$$

$$x = \sin(\pi - \alpha) \qquad \text{Let } \alpha = \cos^{-1}3/5.$$

$$= \sin\pi\cos\alpha - \cos\pi\sin\alpha$$

$$= (0)\cos\alpha - (-1)\sin\alpha$$

$$= \sin\alpha \qquad \begin{array}{l}\sin\alpha = 4/5 \text{ (see}\\ \text{Figure 7.10)}\end{array}$$

$$x = \frac{4}{5}$$

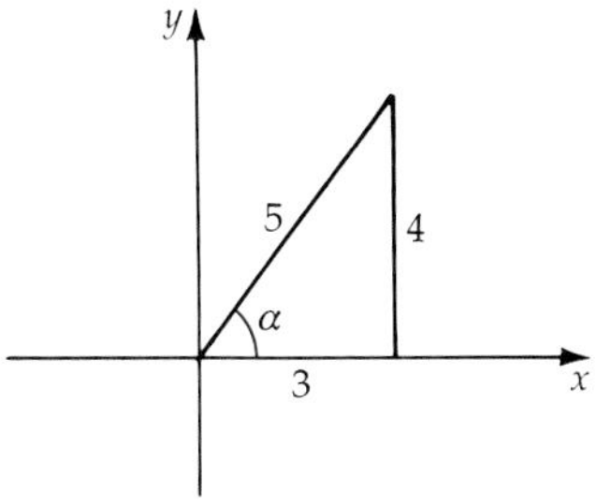

Figure 7.10

■ *Try Exercise **94**, page 410.*

EXAMPLE 11   **Verify an Inverse Trigonometric Identity**

Verify the identity $\sin^{-1}x + \cos^{-1}x = \dfrac{\pi}{2}$.

Solution   Let $\alpha = \sin^{-1}x$ and $\beta = \cos^{-1}x$ which implies that $\sin\alpha = x$ and $\cos\beta = x$. Then $\cos\alpha = \sqrt{1 - \sin^2\alpha} = \sqrt{1 - x^2}$ and in $\beta = \sqrt{1 - \cos^2\beta} = \sqrt{1 - x^2}$. Working with the left side of the identity produces

$$\sin^{-1}x + \cos^{-1}x = \alpha + \beta$$

$$= \sin^{-1}[\sin(\alpha + \beta)]$$

$$= \sin^{-1}[\sin\alpha\cos\beta + \cos\alpha\sin\beta]$$

$$= \sin^{-1}[x \cdot x + \sqrt{1 - x^2}\sqrt{1 - x^2}]$$

$$= \sin^{-1}[x^2 + 1 - x^2]$$

$$= \sin^{-1}1$$

$$= \frac{\pi}{2}$$

■ *Try Exercise **102**, page 410.*

The inverse trigonometric functions can be graphed by finding the upper and lower values of the range and then finding as many other points as needed. To find points on the graph, it is sometimes easier to solve for the independent variable x and assigning values to the dependent variable y. This procedure is shown by graphing the function $y = 2 \sin^{-1} 2x$.

$$y = 2 \sin^{-1} 2x$$

$$y/2 = \sin^{-1} 2x \quad \text{By dividing each side by 2}$$

We know that the range of the inverse sine function is from $-\pi/2$ to $\pi/2$. Thus we have

$$-\frac{\pi}{2} \le \sin^{-1} 2x \le \frac{\pi}{2},$$

and we can substitute $y/2$ for the inverse sine

$$-\frac{\pi}{2} \le \frac{y}{2} \le \frac{\pi}{2}$$

$$-\pi \le y \le \pi \quad \text{Multiplying by 2}$$

The range of $y = 2 \sin^{-1} 2x$ is $-\pi \le y \le \pi$. Thus the upper value of the range is π and the lower value is $-\pi$.

Now solve for x.

$$y = 2 \sin^{-1} 2x$$

$$\frac{y}{2} = \sin^{-1} 2x$$

$$\sin \frac{y}{2} = 2x$$

$$x = \frac{1}{2} \sin \frac{y}{2}$$

We have found the range of the function and solved the equation for x. Choose values from the range for y and calculate x to find points on the graph of the function. Assign values to y between $-\pi$ and π.

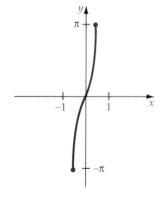

Figure 7.11
$y = 2 \sin^{-1} 2x$

	y	$\sin \dfrac{y}{2}$	$x = \dfrac{1}{2} \sin \dfrac{y}{2}$	(x, y)
Lower value	$-\pi$	-1	$-\dfrac{1}{2}$	$\left(-\dfrac{1}{2}, -\pi\right)$
Middle value	0	0	0	$(0, 0)$
Upper value	π	1	$\dfrac{1}{2}$	$\left(\dfrac{1}{2}, \pi\right)$

EXAMPLE 12 **Graph the Inverse Cosine Function**

Graph the function $y = 0.5 \cos^{-1}(x + 2)$.

Solution The graph of $y = 0.5 \cos^{-1}(x + 2)$ is the graph of $y = \cos^{-1} x$ shifted left horizontally 2 units and shrunk towards the *x*-axis, so that each *y*-coordinate is $\frac{1}{2}$ of its previous value. See Figure 7.12.

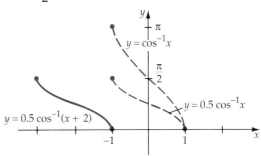

Figure 7.12
$y = 0.5 \cos^{-1}(x + 2)$

■ *Try Exercise* **106**, *page 410.*

EXERCISE SET 7.5

In Exercises 1 to 12, find the exact radian value for the given inverse functions.

1. $\sin^{-1} 1$

2. $\sin^{-1} \dfrac{\sqrt{2}}{2}$

3. $\cos^{-1}\left(-\dfrac{\sqrt{3}}{2}\right)$

4. $\cos^{-1}\left(-\dfrac{1}{2}\right)$

5. $\tan^{-1}(-1)$

6. $\tan^{-1}\sqrt{3}$

7. $\cot^{-1}\dfrac{\sqrt{3}}{3}$

8. $\cot^{-1} 1$

9. $\sec^{-1} 2$

10. $\sec^{-1}\dfrac{2\sqrt{3}}{3}$

11. $\csc^{-1}(-\sqrt{2})$

12. $\csc^{-1}(-2)$

In Exercises 13 to 20, find the exact degree value of the given inverse functions.

13. $\sin^{-1}\left(-\dfrac{\sqrt{3}}{2}\right)$

14. $\sin^{-1}\dfrac{1}{2}$

15. $\cos^{-1}\left(-\dfrac{1}{2}\right)$

16. $\cos^{-1}\dfrac{\sqrt{3}}{2}$

17. $\tan^{-1}\dfrac{\sqrt{3}}{3}$

18. $\tan^{-1} 1$

19. $\cot^{-1}\sqrt{3}$

20. $\cot^{-1}(-1)$

In Exercises 21 to 28, find the approximate radian value for the given functions to 4 significant digits.

21. $\sin^{-1} 0.4555$

22. $\sin^{-1} 0.8700$

23. $\cos^{-1}(-0.2357)$

24. $\cos^{-1}(-0.1298)$

25. $\tan^{-1}(-1.4344)$

26. $\tan^{-1}(-5.2691)$

27. $\cot^{-1} 0.9823$

28. $\cot^{-1} 4.2317$

In Exercises 29 to 40, find the approximate degree value for the given inverse functions to the nearest tenth of a degree.

29. $\sin^{-1}(-0.2781)$

30. $\sin^{-1}(-0.9650)$

31. $\cos^{-1} 0.5555$

32. $\cos^{-1} 0.1598$

33. $\tan^{-1}(-2.0440)$

34. $\tan^{-1} 10.0050$

35. $\cot^{-1}(-0.9752)$

36. $\cot^{-1} 1.0578$

37. $\sec^{-1}(-3.4785)$

38. $\sec^{-1} 9.4455$

39. $\csc^{-1} 1.0056$

40. $\csc^{-1}(-10.9856)$

In Exercises 41 to 60, evaluate the given expression.

41. $\cos\left(\cos^{-1}\dfrac{1}{2}\right)$

42. $\cos(\cos^{-1} 2)$

43. $\tan(\tan^{-1} 2)$

44. $\tan\left(\tan^{-1}\dfrac{1}{2}\right)$

45. $\sin\left(\tan^{-1}\dfrac{3}{4}\right)$

46. $\cos\left(\sin^{-1}\dfrac{5}{13}\right)$

47. $\tan\left(\sin^{-1}\dfrac{\sqrt{2}}{2}\right)$

48. $\sin\left[\cos^{-1}\left(-\dfrac{\sqrt{3}}{2}\right)\right]$

49. $\cos[\sec^{-1}(2)]$

50. $\sin^{-1}(\sin 2)$

51. $\sin^{-1}\left(\sin \dfrac{\pi}{6}\right)$

52. $\sin^{-1}\left(\sin \dfrac{5\pi}{6}\right)$

53. $\cos^{-1}\left(\sin \dfrac{\pi}{4}\right)$

54. $\sin^{-1}\left(\cos \dfrac{7\pi}{6}\right)$

55. $\sin^{-1}\left(\tan \dfrac{\pi}{3}\right)$

56. $\cos^{-1}\left(\tan \dfrac{2\pi}{3}\right)$

57. $\tan^{-1}\left(\sin \dfrac{\pi}{6}\right)$

58. $\cot^{-1}\left(\cos \dfrac{2\pi}{3}\right)$

59. $\sin^{-1}\left[\cos\left(-\dfrac{2\pi}{3}\right)\right]$

60. $\cos^{-1}\left[\tan\left(-\dfrac{\pi}{3}\right)\right]$

In Exercises 61 to 100, solve the following inverse trigonometric equations.

61. $y = \sin^{-1}\left(-\dfrac{\sqrt{3}}{2}\right)$

62. $y = \tan^{-1}\dfrac{\sqrt{3}}{3}$

63. $y = \cos^{-1}(-0.5669)$

64. $y = \csc^{-1}(2.3033)$

65. $y = \cot^{-1}(-1.0886)$

66. $y = \sec^{-1}(-2.9071)$

67. $y = \cos^{-1}\dfrac{\pi}{4}$

68. $y = \tan^{-1}\left(-\dfrac{\pi}{3}\right)$

69. $y = \cos\left(\sin^{-1}\dfrac{7}{25}\right)$

70. $y = \tan\left(\cos^{-1}\dfrac{3}{5}\right)$

71. $y = \sec\left(\tan^{-1}\dfrac{12}{5}\right)$

72. $y = \csc\left(\sin^{-1}\dfrac{12}{13}\right)$

73. $y = \sin^{-1}\left(\sin \dfrac{2\pi}{3}\right)$

74. $y = \tan^{-1}\left(\tan \dfrac{5\pi}{4}\right)$

75. $y = \cos^{-1}\left[\cos\left(-\dfrac{\pi}{6}\right)\right]$

76. $y = \sin^{-1}\left(\sin \dfrac{5\pi}{3}\right)$

77. $y = \tan^{-1}\left(\tan \dfrac{3\pi}{4}\right)$

78. $y = \cos^{-1}\left(\cos \dfrac{5\pi}{6}\right)$

79. $y = \cos\left(2 \sin^{-1}\dfrac{\sqrt{2}}{2}\right)$

80. $y = \tan\left(2 \sin^{-1}\dfrac{\sqrt{3}}{2}\right)$

81. $y = \sin\left(2 \sin^{-1}\dfrac{4}{5}\right)$

82. $y = \cos(2 \tan^{-1}1)$

83. $\sin^{-1}x = \cos^{-1}\dfrac{5}{13}$

84. $\tan^{-1}x = \sin^{-1}\dfrac{24}{25}$

85. $\sin^{-1}(y - 1) = \dfrac{\pi}{2}$

86. $\cos^{-1}\left(y - \dfrac{1}{2}\right) = \dfrac{\pi}{3}$

87. $\tan^{-1}\left(y + \dfrac{\sqrt{2}}{2}\right) = \dfrac{\pi}{4}$

88. $\sin^{-1}(y - 2) = -\dfrac{\pi}{6}$

89. $y = \sin\left(\sin^{-1}\dfrac{2}{3} + \cos^{-1}\dfrac{1}{2}\right)$

90. $y = \cos\left(\sin^{-1}\dfrac{3}{4} + \cos^{-1}\dfrac{5}{13}\right)$

91. $y = \tan\left(\cos^{-1}\dfrac{1}{2} - \sin^{-1}\dfrac{3}{4}\right)$

92. $y = \sec\left(\cos^{-1}\dfrac{2}{3} + \sin^{-1}\dfrac{2}{3}\right)$

93. $\sin^{-1}\dfrac{3}{5} + \cos^{-1}x = \dfrac{\pi}{4}$

94. $\sin^{-1}x + \cos^{-1}\dfrac{4}{5} = \dfrac{\pi}{6}$

95. $\sin^{-1}x - \cos^{-1}\dfrac{\sqrt{2}}{2} = \dfrac{2\pi}{3}$

96. $\cos^{-1}x + \sin^{-1}\dfrac{\sqrt{3}}{2} = \dfrac{\pi}{4}$

In Exercises 97 to 100, evaluate each expression.

97. $y = \cos(\sin^{-1}x)$

98. $y = \tan(\cos^{-1}x)$

99. $y = \sin(\sec^{-1}x)$

100. $y = \sec(\sin^{-1}x)$

In Exercises 101 to 104, verify the identity.

101. $\sin^{-1}x + \sin^{-1}(-x) = 0$

102. $\cos^{-1}x + \cos^{-1}(-x) = \pi$

103. $\tan^{-1}x + \tan^{-1}\dfrac{1}{x} = \dfrac{\pi}{2}$

104. $\sec^{-1}\dfrac{1}{x} + \csc^{-1}\dfrac{1}{x} = \dfrac{\pi}{2}$

In Exercises 105 to 116, graph the given inverse functions.

105. $y = 2 \sin^{-1}x$

106. $y = \dfrac{1}{2} \sin^{-1}x$

107. $y = 2 \cos^{-1}\dfrac{x}{2}$

108. $y = \dfrac{1}{2} \cos^{-1}2x$

109. $y = 2 \sin^{-1}(x - 2)$

110. $y = 3 \sin^{-1}(x + 1)$

111. $y = 2 \cos^{-1}(x + 3)$

112. $y = \dfrac{1}{3} \cos^{-1}(x - 2)$

113. $y = \csc^{-1}2x$

114. $y = 0.5 \sec^{-1}\dfrac{x}{2}$

115. $y = \sec^{-1}(x - 1)$

116. $y = \sec^{-1}(x + \pi)$

Supplemental Exercises

In Exercises 117 to 120, show that the identities are true.

117. $\cos(\sin^{-1}x) = \sqrt{1 - x^2}$

118. $\sec(\sin^{-1}x) = \dfrac{\sqrt{1 - x^2}}{1 - x^2}$

119. $\tan(\csc^{-1}x) = \dfrac{\sqrt{x^2 - 1}}{x^2 - 1}$

120. $\sin(\cot^{-1}x) = \dfrac{\sqrt{x^2 - 1}}{x^2 + 1}$

In Exercises 121 to 124, solve for y in terms of x.

121. $5x = \tan^{-1}3y$

122. $2x = \dfrac{1}{2} \sin^{-1}2y$

123. $x - \dfrac{\pi}{3} = \cos^{-1}(y - 3)$ **124.** $x + \dfrac{\pi}{2} = \tan^{-1}(2y - 1)$

In Exercises 125 to 128, graph the given inverse function.

125. $y = 2 \tan^{-1} 2x$ **126.** $y = \tan^{-1}(x - 1)$

127. $y = \cot^{-1} \dfrac{x}{3}$ **128.** $y = 2 \cot^{-1}(x - 1)$

In Exercises 129 and 130, use the following formula. In dot-matrix printing, the *blank-area factor* is the ratio of the blank area (unprinted area) to the total area of the line. If circular dots are used to print, then the blank-area factor is given by

$$\frac{A}{SD} = 1 - \frac{1}{2}\left[1 - \left(\frac{S}{D}\right)^2 + \frac{D}{S}\sin^{-1}\left(\frac{S}{D}\right)\right]$$

where $A = A_1 + A_2$, with A_1 and A_2 shown in the figure, S equals the distance between centers of overlapping dots, and D is the diameter of a dot.

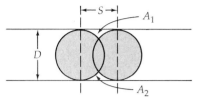

129. Calculate the blank-area factor where $D = 0.2$ millimeters and $S = 0.1$ millimeters.

130. Calculate the blank-area factor where $D = 0.16$ millimeters and $S = 0.1$ millimeters.

7.6

Trigonometric Equations

Consider the equation $\sin x = 1/2$. The graph of $y = \sin x$ along with the line $y = 1/2$ is shown in Figure 7.13. The intersections of the two graphs are the solutions of $\sin x = 1/2$. The solutions in the interval $0 \le x < 2\pi$ are $x = \pi/6$ and $5\pi/6$.

If we remove the restriction $0 \le x < 2\pi$, there are many more possible solutions. Because the sine function is periodic with a period of 2π, other solutions are obtained by adding $2k\pi$, k an integer, to either of the above solutions. Thus, the solutions of $\sin x = 1/2$ are

$$x = \frac{\pi}{6} + 2k\pi, \quad k \text{ is an integer,}$$

$$x = \frac{5\pi}{6} + 2k\pi, \quad k \text{ is an integer.}$$

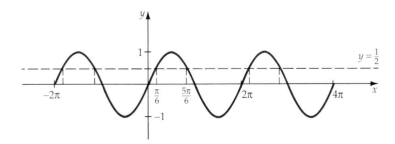

Figure 7.13

The identity for the sine of the sum of two angles can be used to check the above solutions.

$$\sin\left(\frac{\pi}{6} + 2k\pi\right) = \sin\frac{\pi}{6}\cos 2k\pi + \cos\frac{\pi}{6}\sin 2k\pi$$

$$= \sin\frac{\pi}{6}(1) + \cos\frac{\pi}{6}(0) = \sin\frac{\pi}{6} = \frac{1}{2}$$

The same check will show that $x = 5\pi/6 + 2k\pi$ are also solutions.

Algebraic methods and trigonometric identities are used frequently to find the solutions of trigonometric equations. Algebraic methods frequently used are solving by factoring, solving by using the quadratic formula, and squaring each side of the equation.

Remark   Squaring both sides of an equation may not produce an equivalent equation. Thus, when this method is used, the proposed solutions must be checked to eliminate the extraneous solutions.

EXAMPLE 1   Solve a Trigonometric Equation by Factoring

Solve the equation $2\sin^2 x\cos x - \cos x = 0$, $0 \le x < 2\pi$.

Solution   Factor the left side of the equation and set each factor equal to zero.

$$2\sin^2 x\cos x - \cos x = 0$$

$$\cos x(2\sin^2 x - 1) = 0$$

$$\cos x = 0 \quad \text{or} \quad 2\sin^2 x - 1 = 0$$

$$x = \frac{\pi}{2}, \frac{3\pi}{2} \qquad \sin^2 x = \frac{1}{2}$$

$$\sin x = \pm\frac{\sqrt{2}}{2}$$

$$x = \frac{\pi}{4}, \frac{3\pi}{4}, \frac{5\pi}{4}, \frac{7\pi}{4}$$

The solutions in the interval $0 \le x < 2\pi$ are $\dfrac{\pi}{4}, \dfrac{\pi}{2}, \dfrac{3\pi}{4}, \dfrac{5\pi}{4}, \dfrac{3\pi}{2}$, and $\dfrac{7\pi}{4}$.

■ *Try Exercise **44**, page 416.*

EXAMPLE 2   Solve a Trigonometric Equation

Solve the equation $\sin x + \cos x = 1$, $0 \le x < 2\pi$.

Solution   Solve for $\sin x$, then square each side of the equation.

$$\sin x + \cos x = 1$$

$$\sin x = 1 - \cos x$$

$$\sin^2 x = (1 - \cos x)^2 = 1 - 2 \cos x + \cos^2 x$$

$$1 - \cos^2 x = 1 - 2 \cos x + \cos^2 x$$

$$2 \cos^2 x - 2 \cos x = 0$$

Factor the left side and set each factor equal to zero.

$$2 \cos x(\cos x - 1) = 0$$

$$2 \cos x = 0 \quad \text{or} \quad \cos x = 1$$

$$x = \frac{\pi}{2}, \frac{3\pi}{2} \qquad x = 0$$

Squaring each side of an equation may introduce an extraneous root. Therefore, we must check the solution. A check will show that 0 and $\dfrac{\pi}{2}$ are solutions, but $\dfrac{3\pi}{2}$ is not a solution.

■ *Try Exercise* **52,** *page 416.*

EXAMPLE 3 **Solve a Trigonometric Equation by Using the Quadratic Formula**

Solve the equation $3 \cos^2 x - 5 \cos x - 4 = 0$, $0° \le x < 360°$.

Solution   This equation is quadratic in form and cannot be factored. However, we can use the quadratic formula to solve for $\cos x$.

$$3 \cos^2 x - 5 \cos x - 4 = 0 \qquad a = 3, b = -5, c = -4.$$

$$\cos x = \frac{-(-5) \pm \sqrt{(-5)^2 - 4(3)(-4)}}{(2)(3)}$$

$$= \frac{5 \pm \sqrt{73}}{6}$$

The equation $\cos x = \dfrac{5 - \sqrt{73}}{6}$ has no solution. Why?[3] Thus

$$\cos x = \frac{5 + \sqrt{73}}{6}$$

$$x \approx 126.2° \text{ or } 233.8°$$

■ *Try Exercise* **56,** *page 416.*

Caution   The condition $0° \le x < 360°$ requires that you list both 126.2° and 233.8° as solutions of the equation.

---

[3] The range of the cosine function is $-1 \le \cos x \le 1$. The value $(5 + \sqrt{73})/6$ is not in the range of the function.

When solving equations containing multiple angles, use care to make sure all the solutions of the equation are found for the given interval. Consider the equation $\sin 2x = \frac{1}{2}$. We first solve for $2x$.

$$\sin 2x = \frac{1}{2}$$

$$2x = \frac{\pi}{6} + 2k\pi \quad \text{or} \quad 2x = \frac{5\pi}{6} + 2k\pi \quad k \text{ is an integer.}$$

Solving for x, we have $x = \pi/12 + k\pi$ or $x = 5\pi/12 + k\pi$. Substituting integers for k, we obtain

$$k = 0 \qquad x = \frac{\pi}{12} \quad \text{or} \quad x = \frac{5\pi}{12}$$

$$k = 1 \qquad x = \frac{13\pi}{12} \quad \text{or} \quad x = \frac{17\pi}{12}$$

$$k = 2 \qquad x = \frac{25\pi}{12} \quad \text{or} \quad x = \frac{29\pi}{12}$$

Note that for $k > 1$, $x > 2\pi$ and the solutions to $\sin 2x = \frac{1}{2}$ are not in the interval $0 \le x < 2\pi$. Thus for $0 \le x < 2\pi$ the solutions are $\pi/12$, $5\pi/12$, $13\pi/12$, and $17\pi/12$.

EXAMPLE 4   Solve a Trigonometric Equation

Solve the equation $\sin 3x = 1$.

Solution   The equation

$$\sin 3x = 1$$

implies

$$3x = \frac{\pi}{2} + 2k\pi, \quad k \text{ an integer}$$

$$x = \frac{\pi}{6} + \frac{2k\pi}{3}, \quad k \text{ an integer}$$

■ *Try Exercise **66**, page 416.*

EXAMPLE 5   Solve a Trigonometric Equation

Solve the equation $\sin^2 2x - \dfrac{\sqrt{3}}{2} \sin 2x + \sin 2x - \dfrac{\sqrt{3}}{2} = 0$, $0° \le x < 360°$.

Solution   The required solutions for x are in the interval $0° \le x < 360°$. The interval for $2x$ is two times as long.

$$0° \le x < 360°$$

$$0° \le 2x < 720°$$

Now factor the left side of the equation by grouping and then set each factor equal to zero.

$$\sin^2 2x - \frac{\sqrt{3}}{2}\sin 2x + \sin 2x - \frac{\sqrt{3}}{2} = 0$$

$$\sin 2x\left(\sin 2x - \frac{\sqrt{3}}{2}\right) + \left(\sin 2x - \frac{\sqrt{3}}{2}\right) = 0$$

$$(\sin 2x + 1)\left(\sin 2x - \frac{\sqrt{3}}{2}\right) = 0$$

$$\sin 2x + 1 = 0 \quad \text{or} \quad \sin 2x - \frac{\sqrt{3}}{2} = 0$$

$$\sin 2x = -1 \qquad \sin 2x = \frac{\sqrt{3}}{2}$$

The equation $\sin 2x = -1$ implies that $2x = 270° + 360° \cdot k$, k an integer. Thus $x = 135° + 180° \cdot k$. The solutions of this equation with $0 \le x < 360°$ are $135°$ and $315°$. Similarly, the solutions of the equation $\sin 2x = \frac{\sqrt{3}}{2}$ with $0 \le x < 360°$ are

$$2x = 60°, 120°, 420°, 480°$$

$$x = 30°, 60°, 210°, 240°$$

The solutions are $30°$, $60°$, $135°$, $210°$, $240°$, $315°$.

■ *Try Exercise* **84**, *page 416.*

EXERCISE SET 7.6

In Exercises 1 to 22, solve the equation for all values in the interval $0 \le x < 2\pi$.

1. $\sec x - \sqrt{2} = 0$

2. $2 \sin x = \sqrt{3}$

3. $\tan x - \sqrt{3} = 0$

4. $\cos x - 1 = 0$

5. $2 \sin x \cos x = \sqrt{2} \cos x$

6. $2 \sin x \cos x = \sqrt{3} \sin x$

7. $\sin^2 x - 1 = 0$

8. $\cos^2 x - 1 = 0$

9. $4 \sin x \cos x - 2\sqrt{3} \sin x - 2\sqrt{2} \cos x + \sqrt{6} = 0$

10. $\sec^2 x + \sqrt{3} \sec x - \sqrt{2} \sec x - \sqrt{6} = 0$

11. $\csc x - \sqrt{2} = 0$

12. $3 \cot x + \sqrt{3} = 0$

13. $2 \sin^2 x + 1 = 3 \sin x$

14. $2 \cos^2 x + 1 = -3 \cos x$

15. $4 \cos^2 x - 3 = 0$

16. $2 \sin^2 x - 1 = 0$

17. $2 \sin^3 x = \sin x$

18. $4 \cos^3 x = 3 \cos x$

19. $4 \sin^2 x + 2\sqrt{3} \sin x - \sqrt{3} = 2 \sin x$

20. $\tan^2 x + \tan x - \sqrt{3} = \sqrt{3} \tan x$

21. $\sin^4 x = \sin^2 x$

22. $\cos^4 x = \cos^2 x$

In Exercises 23 to 60, solve the following equations, where $0 \le x < 360°$. Round to the nearest tenth of a degree.

23. $\cos x - 0.75 = 0$

24. $\sin x + 0.432 = 0$

25. $3 \sin x - 5 = 0$

26. $4 \cos x - 1 = 0$

27. $3 \sec x - 8 = 0$

28. $4 \csc x + 9 = 0$

29. $\cos x + 3 = 0$

30. $\sin x - 4 = 0$

31. $3 - 5 \sin x = 4 \sin x + 1$

32. $4 \cos x - 5 = \cos x - 3$

33. $\frac{1}{2} \sin x + \frac{2}{3} = \frac{3}{4} \sin x + \frac{3}{5}$

34. $\frac{2}{5} \cos x - \frac{1}{2} = \frac{1}{3} - \frac{1}{2} \cos x$

35. $3 \tan^2 x - 2 \tan x = 0$

36. $4 \cot^2 x + 3 \cot x = 0$

37. $3 \cos x + \sec x = 0$

38. $5 \sin x - \csc x = 0$

39. $\tan^2 x = 3 \sec^2 x - 2$

40. $\csc^2 x - 1 = 3 \cot^2 x + 2$

41. $2 \sin^2 x = 1 - \cos x$

42. $\cos^2 x + 4 = 2 \sin x - 3$

43. $3 \cos^2 x + 5 \cos x - 2 = 0$

44. $2 \sin^2 x + 5 \sin x + 3 = 0$

45. $2 \tan^2 x - \tan x - 10 = 0$

46. $2 \cot^2 x - 7 \cot x + 3 = 0$

47. $3 \sin x \cos x - \cos x = 0$

48. $\tan x \sin x - \sin x = 0$

49. $2 \sin x \cos x - \sin x - 2 \cos x + 1 = 0$

50. $6 \cos x \sin x - 3 \cos x - 4 \sin x + 2 = 0$

51. $2 \sin x - \cos x = 1$

52. $\sin x + 2 \cos x = 1$

53. $2 \sin x - 3 \cos x = 1$

54. $\sqrt{3} \sin x + \cos x = 1$

55. $3 \sin^2 x - \sin x - 1 = 0$

56. $2 \cos^2 x - 5 \cos x - 5 = 0$

57. $2 \cos x - 1 + 3 \sec x = 0$

58. $3 \sin x - 5 + \csc x = 0$

59. $\cos^2 x - 3 \sin x + 2 \sin^2 x = 0$

60. $\sin^2 x = 2 \cos x + 3 \cos^2 x$

In Exercises 61 to 70, solve the trigonometric equations.

61. $\tan 2x - 1 = 0$

62. $\sec 3x - \dfrac{2\sqrt{3}}{3} = 0$

63. $\sin 5x = 1$

64. $\cos 4x = -\dfrac{\sqrt{2}}{2}$

65. $\sin 2x - \sin x = 0$

66. $\cos 2x = -\dfrac{\sqrt{3}}{2}$

67. $\sin\left(2x + \dfrac{\pi}{6}\right) = -\dfrac{1}{2}$

68. $\cos\left(2x - \dfrac{\pi}{4}\right) = -\dfrac{\sqrt{2}}{2}$

69. $\sin^2 \dfrac{x}{2} + \cos x = 1$

70. $\cos^2 \dfrac{x}{2} - \cos x = 1$

In Exercises 71 to 84, solve each equation where $0 \le x < 2\pi$.

71. $\cos 2x = 1 - 3 \sin x$

72. $\cos 2x = 2 \cos x - 1$

73. $\sin 4x - \sin 2x = 0$

74. $\sin 4x - \cos 2x = 0$

75. $\tan \dfrac{x}{2} = \sin x$

76. $\tan \dfrac{x}{2} = 1 - \cos x$

77. $\sin 2x \cos x + \cos 2x \sin x = 0$

78. $\cos 2x \cos x - \sin 2x \sin x = 0$

79. $\sin x \cos 2x - \cos x \sin 2x = \dfrac{\sqrt{3}}{2}$

80. $\cos 2x \cos x + \sin 2x \sin x = -1$

81. $\sin 3x - \sin x = 0$

82. $\cos 3x + \cos x = 0$

83. $2 \sin x \cos x + 2 \sin x - \cos x - 1 = 0$

84. $2 \sin x \cos x - 2\sqrt{2} \sin x - \sqrt{3} \cos x + \sqrt{6} = 0$

Supplemental Exercises

In Exercises 85 to 94, solve the trigonometric equations for $0 \le x < 2\pi$.

85. $\sqrt{3} \sin x + \cos x = \sqrt{3}/2$

86. $\sin x - \cos x = 1$

87. $-\sin x + \sqrt{3} \cos x = \sqrt{3}$

88. $-\sqrt{3} \sin x - \cos x = 1$

89. $\cos 5x - \cos 3x = 0$

90. $\cos 5x - \cos x - \sin 3x = 0$

91. $\sin 3x + \sin x = 0$

92. $\sin 3x + \sin x - \sin 2x = 0$

93. $\cos 4x + \cos 2x = 0$

94. $\cos 4x + \cos 2x - \cos 3x = 0$

95. Find the area of the sector in terms of r and θ in the accompanying figure.

96. Find the area of triangle OAB in terms of r and θ in the figure.

97. Find the area of the shaded part of the figure. Find that area in terms of r and θ.

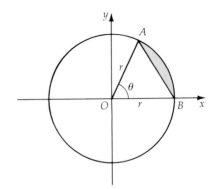

Chapter 7 Review

7.1 The Fundamental Trigonometric Identities

Trigonometric identities are verified by using algebraic methods and previously proven identities.

$$\sin x = \frac{1}{\csc x} \qquad \cos x = \frac{1}{\sec x} \qquad \tan x = \frac{1}{\cot x}$$

$$\tan x = \frac{\sin x}{\cos x} \qquad \cot x = \frac{\cos x}{\sin x}$$

$$\sin^2 x + \cos^2 x = 1 \qquad \tan^2 x + 1 = \sec^2 x \qquad 1 + \cot^2 x = \csc^2 x$$

7.2 Sum and Difference Identities

Sum and difference identities for the cosine function are

$$\cos(\alpha - \beta) = \cos \alpha \cos \beta + \sin \alpha \sin \beta$$

$$\cos(\alpha + \beta) = \cos \alpha \cos \beta - \sin \alpha \sin \beta$$

Sum and difference identities for the sine function are

$$\sin(\alpha - \beta) = \sin \alpha \cos \beta - \cos \alpha \sin \beta$$

$$\sin(\alpha + \beta) = \sin \alpha \cos \beta + \cos \alpha \sin \beta$$

Sum and difference identities for the tangent function are

$$\tan(\alpha + \beta) = \frac{\tan \alpha + \tan \beta}{1 - \tan \alpha \tan \beta}$$

$$\tan(\alpha - \beta) = \frac{\tan \alpha - \tan \beta}{1 + \tan \alpha \tan \beta}$$

7.3 Double- and Half-Angle Identities

The double-angle identities are

$$\sin 2\alpha = 2 \sin \alpha \cos \alpha$$

$$\cos 2\alpha = \cos^2 \alpha - \sin^2 \alpha$$

$$= 1 - 2 \sin^2 \alpha$$

$$= 2 \cos^2 \alpha - 1$$

$$\tan 2\alpha = \frac{2 \tan \alpha}{1 - \tan^2 \alpha}$$

The half-angle identities are

$$\sin \frac{\alpha}{2} = \pm\sqrt{\frac{1 - \cos \alpha}{2}} \qquad \tan \frac{\alpha}{2} = \frac{\sin \alpha}{1 + \cos \alpha}$$

$$\cos \frac{\alpha}{2} = \pm\sqrt{\frac{1 + \cos \alpha}{2}} \qquad \qquad = \frac{1 - \cos \alpha}{\sin \alpha}$$

7.4 Identities Involving the Sums of Trigonometric Functions

The product-to-sum identities are

$$\sin \alpha \cos \beta = \frac{1}{2}[\sin(\alpha + \beta) + \sin(\alpha - \beta)]$$

$$\cos \alpha \sin \beta = \frac{1}{2}[\sin(\alpha + \beta) - \sin(\alpha - \beta)]$$

$$\cos \alpha \cos \beta = \frac{1}{2}[\cos(\alpha + \beta) + \cos(\alpha - \beta)]$$

$$\sin \alpha \sin \beta = \frac{1}{2}[\cos(\alpha - \beta) - \cos(\alpha + \beta)]$$

The sum-to-product identities are

$$\sin x + \sin y = 2 \sin \frac{x + y}{2} \cos \frac{x - y}{2}$$

$$\cos x - \cos y = -2 \sin \frac{x + y}{2} \sin \frac{x - y}{2}$$

$$\sin x - \sin y = 2 \cos \frac{x + y}{2} \sin \frac{x - y}{2}$$

$$\cos x + \cos y = 2 \cos \frac{x + y}{2} \cos \frac{x - y}{2}$$

7.5 Inverse Trigonometric Functions

The inverse of $y = \sin t$ is $y = \sin^{-1}t$, with $-1 \le t \le 1$ and $-\pi/2 \le y \le \pi/2$.

The inverse of $y = \cos t$ is $y = \cos^{-1}t$, with $-1 \le t \le 1$ and $0 \le y \le \pi$.

The inverse of $y = \tan t$ is $y = \tan^{-1}t$, with $-\infty < t < \infty$ and $-\pi/2 < y < \pi/2$.

The inverse of $y = \cot t$ is $y = \cot^{-1}t$, with $-\infty < t < \infty$ and $0 < y < \pi$.

The inverse of $y = \csc t$ is $y = \csc^{-1}t$, with $t \le -1$ or $t \ge 1$ and $-\pi/2 \le y \le \pi/2$, $y \ne 0$.

The inverse of $y = \sec t$ is $y = \sec^{-1}t$, with $t \le -1$ or $t \ge 1$ and $0 \le y \le \pi$, $y \ne \pi/2$.

7.6   Trigonometric Equations

Algebraic methods and the use of identities are used to solve trigonometric equations. Since the trigonometric functions are periodic, there may be an infinite number of solutions.

CHALLENGE EXERCISES

In Exercises 1 to 12, answer true or false. If the statement is false, give an example.

1. $\dfrac{\tan \alpha}{\tan \beta} = \dfrac{\alpha}{\beta}$

2. $\dfrac{\sin x}{\cos y} = \tan \dfrac{x}{y}$

3. $\sin^{-1} x = \csc x$

4. $\sin 2\alpha = 2 \sin \alpha$ for all α

5. $\sin(\alpha + \beta) = \sin \alpha + \sin \beta$

6. An equation that has an infinite number of solutions is an identity.

7. If $\tan \alpha = \tan \beta$, then $\alpha = \beta$.

8. $\cos^{-1}(\cos x) = x$

9. $\cos(\cos^{-1}x) = x$

10. $\csc^{-1} \dfrac{1}{\alpha} = \dfrac{1}{\csc \alpha}$

11. If $0° \le \theta \le 90°$, then $\cos \theta = \sin(180° - \theta)$

12. $\sin^2 \theta = \sin \theta^2$

REVIEW EXERCISES

In Exercises 1 to 6, find the exact value of the given function.

1. $\cos(45° + 30°)$

2. $\tan(210° - 45°)$

3. $\sin\left(\dfrac{2\pi}{3} + \dfrac{\pi}{4}\right)$

4. $\sec\left(\dfrac{4\pi}{3} - \dfrac{\pi}{4}\right)$

5. $\sin(60° - 135°)$

6. $\cos\left(\dfrac{5\pi}{3} - \dfrac{7\pi}{4}\right)$

In Exercises 7 to 10, evaluate the functions by using the half-angle identities.

7. $\sin\left(22\dfrac{1}{2}\right)°$

8. $\cos 105°$

9. $\tan\left(67\dfrac{1}{2}\right)°$

10. $\sin 112.5°$

In Exercises 11 to 14, find the exact value of the given functions.

11. Given $\sin \alpha = \frac{1}{2}$, α in quadrant I and $\cos \beta = \frac{1}{2}$, β in quadrant IV, find a. $\cos(\alpha - \beta)$, b. $\tan 2\alpha$, c. $\sin \beta/2$.

12. Given $\sin \alpha = \frac{\sqrt{3}}{2}$, α in quadrant II and $\cos \beta = -\frac{1}{2}$, β in quadrant III, find a. $\sin(\alpha + \beta)$, b. $\sec 2\beta$ c. $\cos \frac{\alpha}{2}$.

13. Given $\sin \alpha = -\frac{1}{2}$, α in quadrant IV and $\cos \beta = -\frac{\sqrt{3}}{2}$, β in quadrant III, find a. $\sin(\alpha - \beta)$, b. $\tan 2\alpha$, c. $\cos \beta/2$.

14. Given $\sin \alpha = \frac{\sqrt{2}}{2}$, α in quadrant I and $\cos \beta = \frac{\sqrt{3}}{2}$, β in quadrant IV, find a. $\cos(\alpha - \beta)$, b. $\tan 2\beta$, c. $\sin 2\alpha$.

In Exercises 15 to 20, write the given expression as a single trigonometric function.

15. $2 \sin 3x \cos 3x$

16. $\dfrac{\tan 2x + \tan x}{1 - \tan 2x \tan x}$

17. $\sin 4x \cos x - \cos 4x \sin x$

18. $\cos^2 2\theta - \sin^2 2\theta$

19. $1 - 2 \sin^2 \dfrac{\beta}{2}$

20. $\pm\sqrt{\dfrac{1 - \cos 4\theta}{2}}$

In Exercises 21 to 24, evaluate each expression.

21. $\sin 47° \sin 22°$

22. $\cos 14° \cos 92°$

23. $2 \sin \dfrac{\pi}{3} \cos \dfrac{2\pi}{3}$

24. $2 \cos \dfrac{\pi}{4} \cos \dfrac{3\pi}{2}$

In Exercises 25 to 28, write each expression as the product of two functions.

25. $\cos 2\theta - \cos 4\theta$

26. $\sin 3\theta - \sin 5\theta$

27. $\sin 6\theta + \sin 2\theta$

28. $\sin 5\theta - \sin \theta$

In Exercises 29 to 46, verify the identities.

29. $\dfrac{1}{\sin x - 1} + \dfrac{1}{\sin x + 1} = -2 \tan x \sec x$

30. $\dfrac{\sin x}{1 - \cos x} = \csc x + \cot x, \quad 0 < x < \dfrac{\pi}{2}$

31. $\dfrac{1 + \sin x}{\cos^2 x} = \tan^2 x + 1 + \tan x \sec x$

32. $\cos^2 x - \sin^2 x - \sin 2x = \dfrac{\cos^2 2x - \sin^2 2x}{\cos 2x + \sin 2x}$

33. $\dfrac{1}{\cos x} - \cos x = \tan x \sec x$

34. $\sin(270° - \theta) - \cos(270° - \theta) = \sin \theta - \cos \theta$

35. $\sin\left(\dfrac{\pi}{4} - \alpha\right) = \dfrac{\sqrt{2}}{2}(\cos \alpha - \sin \alpha)$

36. $\sin(180° - \alpha + \beta) = \sin \alpha \cos \beta - \cos \alpha \sin \beta$

37. $\dfrac{\sin 4x - \sin 2x}{\cos 4x - \cos 2x} = -\cot 3x$

38. $2 \sin x \sin 3x = (1 - \cos 2x)(1 + 2 \cos 2x)$

39. $\sin x - \cos 2x = (2 \sin x - 1)(\sin x + 1)$

40. $\cos 4x = 1 - 8 \sin^2 x + 8 \sin^4 x$

41. $\tan 4x = \dfrac{4 \tan x - 4 \tan^3 x}{1 - 6 \tan^2 x + \tan^4 x}$

42. $\dfrac{\sin 2x - \sin x}{\cos 2x + \cos x} = \dfrac{1 - \cos x}{\sin x}$

43. $2 \cos 4x \sin 2x = 2 \sin 3x \cos 3x - 2 \sin x \cos x$

44. $2 \sin x \sin 2x = 4 \cos x \sin^2 x$

45. $\cos(x + y) \cos(x - y) = \cos^2 x + \cos^2 y - 1$

46. $\cos(x + y) \sin(x - y) = \sin x \cos x - \sin y \cos y$

In Exercises 47 to 52, solve the equation.

47. $y = \sec\left(\sin^{-1} \dfrac{12}{13}\right)$

48. $y = \cos\left(\sin^{-1} \dfrac{3}{5}\right)$

49. $2 \sin^{-1}(x - 1) = \dfrac{\pi}{3}$

50. $y = \cos\left(\sin^{-1}\left(-\dfrac{3}{5}\right)\right) + \cos^{-1} \dfrac{5}{13}$

51. $\sin^{-1} x - \cos^{-1} \dfrac{4}{5} = \dfrac{\pi}{2}$

52. $y = \cos\left[2 \sin^{-1}\left(\dfrac{3}{5}\right)\right]$

In Exercises 53 to 58, solve the equations for $0° \le x < 360°$.

53. $4 \sin^2 x + 2\sqrt{3} \sin x - 2 \sin x - \sqrt{3} = 0$

54. $2 \sin x \cos x - \sqrt{2} \cos x - 2 \sin x + \sqrt{2} = 0$

In Exercises 55 and 56, solve the trigonometric equation.

55. $3 \cos^2 x + \sin x = 1$

56. $\tan^2 x - 2 \tan x - 3 = 0$

In Exercises 57 and 58, solve the equation for $0 \le x < 2\pi$.

57. $\sin 3x \cos x - \cos 3x \sin x = \dfrac{1}{2}$

58. $\cos\left(2x - \dfrac{\pi}{3}\right) = -\dfrac{\sqrt{3}}{2}$

In Exercises 59 to 62, find the amplitude and phase shift of each function. Graph each function.

59. $f(x) = \sqrt{3} \sin x + \cos x$

60. $f(x) = -2 \sin x - 2 \cos x$

61. $f(x) = -\sin x - \sqrt{3} \cos x$

62. $f(x) = \dfrac{\sqrt{3}}{2} \sin x - \dfrac{1}{2} \cos x$

In Exercises 63 to 66, graph each function.

63. $f(x) = 2 \cos^{-1} x$

64. $f(x) = \sin^{-1}(x - 1)$

65. $f(x) = \sin^{-1} \dfrac{x}{2}$

66. $f(x) = \sec^{-1} 2x$

8

Applications of Trigonometry

When you can measure what you are talking about and express it in numbers, you know something about it.

LORD KELVIN (1824–1907)

THE ELECTROMAGNETIC SPECTRUM

Electromagnetic waves are generated by oscillations of an electrically charged particle. All electromagnetic waves travel through space at approximately 186,000 miles per second or 300,000 kilometers per second. This speed is frequently referred to as the speed of light.

Visible light however, comprises only a small portion of what is called the *electromagnetic spectrum*. The electromagnetic spectrum consists of bands of electromagnetic waves of different wavelengths. The major types of waves are gamma rays, x-rays, ultraviolet light, visible light, infrared rays, microwaves, and radio waves. Gamma rays have the shortest wavelength and radio waves have the longest wavelength.

The electromagnetic spectrum is shown below. The lengths of the waves are given in meters, and the frequency of the wave is given in terms of hertz. One hertz is 1 wave cycle per second. No sharp line can be drawn between various portions of the spectrum. The only difference between the waves is the wavelength or frequency of the wave.

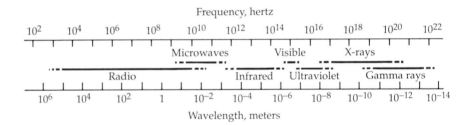

The Electromagnetic Spectrum

Notice that the wavelength of visible light is somewhere between 10^{-7} and 10^{-6} meters. The spectrum of visible light ranges over wavelengths of 4×10^{-7} to 7×10^{-7} meters.

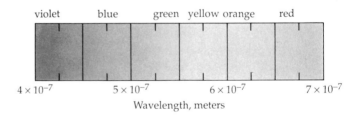

The Visible Light Spectrum

The frequencies of the waves that are broadcast on the AM (Amplitude Modulated) band of a radio range approximately from 530,000 hertz to 1,600,000 hertz. Setting the radio dial to 760 means that you are tuning your radio to receive radio waves that are broadcast at 760,000 hertz. The frequencies in the FM (Frequency Modulated) band range from 88,000,000 hertz to 108,000,000 hertz.

8.1

Right Triangles

A right triangle is a triangle with a right angle and two acute angles that are complementary. The sides opposite the acute angles are called the legs and the side opposite the right angle is called the hypotenuse. The angles in a right triangle are usually labeled with the capital letters A, B, and C with C being reserved for the right angle. The side opposite angle A is a, b is opposite angle B, and c is the hypotenuse.

Solving a triangle involves finding the lengths of all sides and the measures of all angles in a triangle. A right triangle can be solved when an acute angle and one side of the triangle are known, or when two sides are known. The Pythagorean Theorem and the trigonometric functions are used to solve a right triangle.

The right triangle in Figure 8.1 has an acute angle of 42.3° and the side opposite the 42.3° angle is 12.5 centimeters. To solve the triangle, find the unknown parts: angle B, the hypotenuse (c) and the unknown leg (b). Angle A and angle B are complementary, that is their sum is 90°. Thus, we have

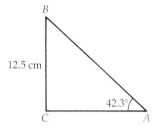

Figure 8.1

$$42.3° + B = 90°$$

$$B = 47.7°$$

Since angle A and the side opposite A are known, we can use the sine function to find the length of the hypotenuse:

$$\sin 42.3° = \frac{12.5}{c}$$

$$c = \frac{12.5}{\sin 42.3°} \approx 18.6$$

To find the length of side b, we use the tangent function.

$$\tan 42.3° = \frac{12.5}{b}$$

$$b = \frac{12.5}{\tan 42.3°} \approx 13.7$$

Angle $B = 47.7°$, hypotenuse $c \approx 18.6$ cm, and side $b \approx 13.7$ cm.

If a calculator were used for the calculations above, it would show that the calculator displays more digits than the problem warrants. The precision of the calculator is misleading given that the original values were measured in tenths of a centimeter and tenths of a degree. The following rounding convention will be used when solving triangles.

Angle Measure to the Nearest	Significant Digits of the Lengths
Degree	Two
Tenths of a degree	Three
Hundredths of a degree	Four

For example, because the angle of 42.3° in Figure 8.1 is measured to the nearest tenth of a degree, c and b are rounded to three significant digits. To minimize the error in rounding, the intermediate steps in a problem will not be rounded. Only round the final result using the rounding convention.

EXAMPLE 1   Solve a Right Triangle

Solve the right triangle with an acute angle of 14.28° and a hypotenuse of 146.2 feet.

Solution   Since A and B are complementary angles,

$$A = 90.00° - 14.28° = 75.72°$$

$$\sin 14.28° = \frac{b}{146.2} \quad \text{The sine function is used to find side } b.$$

$$b = 146.2(\sin 14.28°)$$

$$\approx 36.06 \text{ ft}$$

Now find a.

$$\cos 14.28° = \frac{a}{146.2}$$

$$a = 146.2(\cos 14.28°) \approx 141.7 \text{ ft}$$

Angle $B = 75.72°$, $b \approx 36.06$ ft, and $a \approx 141.7$ ft.

■ *Try Exercise* **14,** *page 426.*

In some application problems, a horizontal line is used as a reference. An angle measured above the line to a line of sight is called an **angle of elevation,** and the angle measured below the line to a line of sight is called an **angle of depression.** See Figure 8.2.

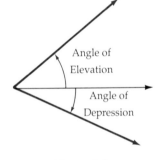

Figure 8.2

EXAMPLE 2   Solve an Application Involving Angle of Elevation

The measure of the angle of elevation from a position 62 feet from the base of the flagpole to the top of the flagpole is 34°. Find the height of the flagpole.

Solution   Sketch the figure representing the given conditions. Because the length of the side opposite the angle is unknown and the measure of the adjacent angle is known, the tangent function is used.

$$\tan 34° = \frac{h}{62}$$

$$h = 62 \tan 34° \approx 42 \text{ ft.}$$

■ *Try Exercise* **28,** *page 426.*

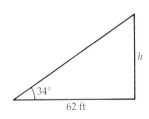

Figure 8.3

Two terms are used in navigation and surveying to represent direction: heading and bearing. **Heading** is the measure of an angle clockwise from north. Figure 8.4 shows a heading of 45° and a heading of 225°. **Bearing** is the measure of the acute angle formed by a north-south line and the line of direction. Figure 8.5 shows a bearing of N38°W and a bearing of S15°E.

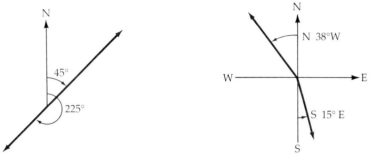

Figure 8.4   Figure 8.5

EXAMPLE 3   Solve an Application Involving Bearing

A motorist drove 40 miles at a bearing of S25°E. The motorist then drove 32 miles at a bearing S65°W. Find the distance of the motorist from the starting point.

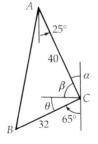

Figure 8.6

Solution   Sketch a diagram showing the information given. From the figure, we see that angle α is 25° because alternate interior angles of parallel lines are equal. The angles α and β are complementary; thus β is 65°. Angle θ is complementary to 65°. Thus $\theta = 25°$.

$\beta + \theta = 65° + 25° = 90°$. Thus, angle ACB is a right angle. Use the Pythagorean Theorem to solve the right triangle for the hypotenuse.

$$c = \sqrt{a^2 + b^2} = \sqrt{32^2 + 40^2} \quad a = 32, \ b = 40$$

$$= \sqrt{1024 + 1600}$$

$$= \sqrt{2624} \approx 51 \text{ mi} \qquad \text{Round to two significant digits.}$$

■ *Try Exercise **32**, page 426.*

EXAMPLE 4   Solve an Application Involving Heading

A car travels at 60 mph for 1 hour at a heading of 315°. The car then travels at 45 mph for 2 hours at a heading of 225°. At the end of 3 hours, how far is the car from its starting point?

Solution   Sketch a diagram. Triangle ABC can be shown to be a right triangle. Angle $\theta = 45°$ by subtracting 315° from 360°. Angle $\alpha = 45°$ by alternate interior angles of parallel lines. Angle $\beta = 45°$ by subtracting 180°

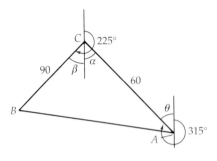

Figure 8.7

from 225°. Therefore angle C is 90° and triangle ABC is a right triangle.

$b = 60$ Traveling 60 mph for 1 hour

$a = 90$ Traveling 45 mph for 2 hours

Use the Pythagorean Theorem to find c.

$$c = \sqrt{60^2 + 90^2} = \sqrt{3600 + 8100}$$

$$= \sqrt{11,700} \approx 110 \text{ mi}$$ Round to two significant digits.

■ *Try Exercise **34**, page 426.*

EXERCISE SET 8.1

In Exercises 1 to 8, find the indicated part of the right triangles.

1. $a = 12$, $b = 15$; find angle B
2. $a = 40$, $b = 15$; find angle A
3. $a = 12$, $c = 30$; find angle A
4. $b = 25$, $c = 40$; find angle A
5. $b = 623$, $B = 23.4°$; find c
6. $a = 185$, $B = 72.7°$; find b
7. $a = 4.08$, $B = 40.4°$; find b
8. $b = 10.2$, $A = 50.5°$; find c

In Exercises 9 to 16, solve the given right triangle.

9. $A = 56°$, $a = 120$
10. $a = 340$, $b = 520$
11. $A = 23.8°$, $b = 125$
12. $B = 78.4°$, $a = 12.8$
13. $B = 23.47°$, $c = 1248$
14. $A = 45.89°$, $b = 1.228$
15. $a = 32.35$, $b = 12.67$
16. $a = 12.38$, $b = 4.367$

17. Find the leg opposite the 60° angle in a 30°-60° right triangle with a hypotenuse of 16 meters.
18. Find the leg of a 45°-45° right triangle with a hypotenuse of 60 centimeters.
19. The legs of a right triangle are $a = 4.0$ and $b = 8.0$. Find the six trigonometric functions of angle A.
20. The hypotenuse of a right triangle is 80, and one leg is 50. Find the six trigonometric functions of the angle opposite the given leg.
21. The sine of angle A of a right triangle is $\frac{3}{5}$. Find the cosine of angle B.
22. The sine of angle A of a right triangle is $\frac{2}{3}$. Find the cosecant of angle A.
23. The cotangent of angle B of a right triangle is $\frac{5}{8}$. Find the tangent of angle A.
24. The cosecant of angle A of a right triangle is 2. Find the sine of angle A.

25. An angle of a right triangle is 42° and the length of the side opposite the angle is 10 inches. Solve the triangle.
26. The length of the two sides of a right triangle are 8.0 inches and 12 inches. Solve the triangle.
27. The length of the hypotenuse of a right triangle is 6.0 feet, and one leg is 4.5 feet. Solve the triangle.
28. A telephone pole casts a shadow of 5.2 meters. Find the height of the telephone pole if the angle of elevation from the tip of the shadow to the top of the pole is 70°.
29. A flagpole casts a shadow of 18 feet. Find the height of the flagpole if the angle of elevation from the tip of the shadow to the top of the pole is 44°.
30. The angle of depression from a hotel room 82 feet from the ground to the base of a nearby building is 17°. Find the distance between the two buildings.
31. The angle of depression of one side of a lake measured from a balloon 2500 feet high is 43°. The angle of depression to the opposite side of the lake is 27°. Find the width of the lake.
32. A ship is 453 yards from a lighthouse and has a bearing of S33.8°W. A second ship is 1520 yards from the lighthouse at a bearing of S56.2°E. Find the distance between the two ships.
33. Two lookout stations are located 5.25 miles apart on an east-west line. A fire was observed from one lookout at a heading of 63.4° and from the other lookout at a heading of 296.6°. Find the distance of the fire from each lookout station.
34. A plane leaves an airport traveling at 215 mph at a heading of 65.4° at the same time another plane leaves the airport traveling at 480 mph at a heading of 335.4°. Find the distance between the two planes at the end of 2 hours.
35. A car goes 34 miles at a bearing of S70°E and then turns and travels 18 miles at a bearing of S20°W. Find the distance from the starting point.

Supplemental Exercises

36. Two buildings are 240 feet apart. The angle of elevation from the top of the shorter building to the top of the other building is 22°. If the shorter building is 80 feet high, how high is the taller building?

37. A circle is inscribed in a regular hexagon with a side of 6.0 meters. Find the radius of the circle.

38. The angle of elevation to the top of a radio antenna on the top of a building is 53.4°. After moving 200 feet closer to the building, the angle of elevation is 64.3°. Find the height of the building if the height of the antenna is 180 feet.

39. The angle of elevation to the top of a building is 32°. After moving 50 feet closer to the building, the angle of elevation is 53°. Find the height of the building.

40. An airplane is flying at an altitude of 30,000 feet. The pilot sees one side of a canyon at an angle of depression of 33.5°. The other side of the canyon is at an angle of depression of 27.4°. Find the width of the canyon accurate to three significant digits.

41. An airplane traveling 240 mph descends 42.0 feet in 1 second. Find the angle of descent.

42. An airplane flies directly overhead at 32,000 feet. Twenty seconds later the plane is observed at an angle of elevation of 67°. Find the velocity of the plane in miles per hour.

43. A submarine traveling at 9.0 mph is diving at an angle of depression of 5°. How long does it take the submarine to reach a depth of 80 feet?

44. Let $y = mx + b$, $m \neq 0$, be the equation of a line. Show that $\tan \alpha = m$, where m is the slope of the straight line and α is the angle made by the line and the positive x-axis.

8.2

The Law of Sines

An oblique triangle is a triangle that does not have a right angle. The *Law of Sines* can be used to solve oblique triangles when the following information is given:

1. Two angles and a side
2. Two sides and an angle opposite one of the given sides

In Figure 8.8, the altitude *CD* is drawn from *C* perpendicular to the opposite side. The length of the altitude is *h*. Triangles *ACD* and *BCD* are right triangles.
The sines of the angles *A* and *B* are

$$\sin A = \frac{h}{b} \qquad \sin B = \frac{h}{a}$$

$$h = b \sin A \qquad h = a \sin B.$$

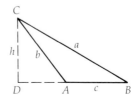 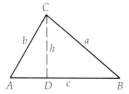

Figure 8.8

Thus $b \sin A = a \sin B$,

and by dividing each side of the equation by $\sin A \sin B$, we obtain

$$\frac{b}{\sin B} = \frac{a}{\sin A}.$$

Similarly, when the perpendicular is drawn to another side, the following formulas result:

$$\frac{c}{\sin C} = \frac{b}{\sin B} \quad \text{and} \quad \frac{c}{\sin C} = \frac{a}{\sin A}.$$

The Law of Sines

> If A, B, and C are the measures of the angles of a triangle and a, b, and c are the lengths of the sides opposite these angles, then
>
> $$\frac{a}{\sin A} = \frac{b}{\sin B} = \frac{c}{\sin C}.$$

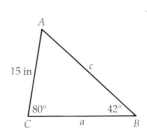

Figure 8.9

EXAMPLE 1   **Solve a Triangle Using the Law of Sines**

Solve triangle ABC if $B = 42°$, $C = 80°$, and $b = 15$ inches.

Solution   The sum of the angles in a triangle equals $180°$.

$$A + B + C = 180°$$

$$A + 80° + 42° = 180°$$

$$A = 58°$$

Since we know A, B, and b, we can use the Law of Sines to find side a.

$$\frac{b}{\sin B} = \frac{a}{\sin A}$$

$$\frac{15}{\sin 42°} = \frac{a}{\sin 58°}$$

$$a = \frac{15 \sin 58°}{\sin 42°} \approx 19 \text{ in.}$$

Now we use the Law of Sines to find side c.

$$\frac{b}{\sin B} = \frac{c}{\sin C}$$

$$\frac{15}{\sin 42°} = \frac{c}{\sin 80°}$$

$$c = \frac{15 \sin 80°}{\sin 42°} \approx 22 \text{ in.}$$

$A = 58°$, $a \approx 19$ in., $c \approx 22$ in.

■ *Try Exercise 4, page 431.*

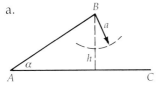

Figure 8.10

EXAMPLE 2   Solve an Application Using the Law of Sines

A ship with a heading of 330° first sighted a lighthouse at a heading of 65°. After traveling 8.5 miles, the ship observed the lighthouse at a heading of 130°. Find the distance to the lighthouse from the ship when the first sighting was made.

Solution   From Figure 8.10 we see that angle $A = 65° + 30° = 95°$; angle $C = 180° - 30° - 130° = 20°$; angle $B = 180° - 95° - 20° = 65°$. Use the Law of Sines to find c.

$$\frac{b}{\sin B} = \frac{c}{\sin C}$$

$$\frac{8.5}{\sin 65°} = \frac{c}{\sin 20°}$$

$$c = \frac{8.5 \sin 20°}{\sin 65°} \approx 3.2$$

The lighthouse was 3.2 miles, to the nearest tenth of a mile, from the ship when the first sighting was made.

■ *Try Exercise* **20,** *page 432.*

Not all triangles with two sides and an angle opposite one of the given sides are unique. Some information may result in two triangles, and some may result in no triangle at all. The case of two sides and an angle opposite one of them is called the ambiguous case of the Law of Sines.

Suppose we are given sides a and c and the nonincluded angle A and asked to solve triangle ABC. First we determine h, the height of the triangle, which is found by dropping a perpendicular from B to side b. Note that since $\sin A = h/c$, we have $h = c \sin A$. Now we examine what happens for various values of a. We consider two cases:

Case 1 A is an acute angle. In Figure 8.11a, $a < h$, no triangle. In 8.11b, $a = h$; one triangle. In 8.11c, $h \leq a \leq c$; two triangles. In 8.11d, $a > c$; one triangle.

Case 2 A is an obtuse angle. In Figure 8.11e, $a < c$; no triangle. In 8.11f, $a > c$; one triangle.

a. 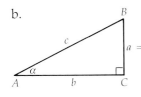   b. 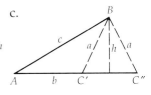   c. 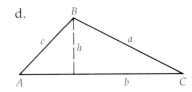   d.

e. 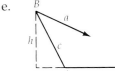   f.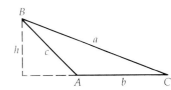

Figure 8.11

Remark   It is not necessary to remember all these cases. In solving such problems, the values in the problem will dictate the outcome that is possible in each case.

To find the possible solutions of the triangle in Figure 8.12 where the measure of $A = 57°$ and sides are $a = 15$ feet and $c = 20$ feet, we must first find h.

$$h = 20 \sin 57° \approx 17$$

Since $a < h$, no triangle is formed, and thus there is no solution.

Note the result when the Law of Sines is used to find angle C.

$$\frac{a}{\sin A} = \frac{c}{\sin C}$$

$$\frac{15}{\sin 57°} = \frac{20}{\sin C}$$

$$\sin C = \frac{20 \sin 57°}{15}$$

$$\approx 1.1182$$

Because 1.1182 is not in the domain of the sine function, there is no triangle for these values of A, a, and c.

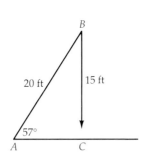

Figure 8.12

EXAMPLE 3   Solve the Ambiguous Case of the Law of Sines

Given the triangle ABC with $B = 32°$, $a = 42$, and $b = 30$, find angle A.

Solution   Find h to determine if this is the ambiguous case.

$$h = 42 \sin 32° = 22$$

We have $h < b < a$; thus we know there are two solutions. Use the Law of Sines to find the two possible values of angle A.

$$\frac{b}{\sin B} = \frac{a}{\sin A}$$

$$\frac{30}{\sin 32°} = \frac{42}{\sin A}$$

$$\sin A = \frac{42 \sin 32°}{30°}$$

$$\approx 0.7419$$

$A \approx 48°$ The two angles that have a sine of 0.719

$A \approx 132°$ are approximately 48° and 132°.

Angle $A \approx 48°$ or $A \approx 132°$.

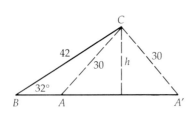

Figure 8.13

■ *Try Exercise* **14,** *page 431.*

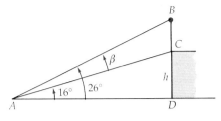

Figure 8.14

EXAMPLE 4 **Solve an Application Using the Law of Sines**

An 80-foot high radio antenna is located on top of an office building. At a distance *d* from the base of the building, the angle of elevation to the top of the antenna is 26°, and the angle of elevation to the bottom of the antenna is 16°. Find the height of the building.

Solution Sketch the diagram. Find angles *B* and *β*.

$$B = 90° - 26° = 64°$$

$$\beta = 26° - 16° = 10°$$

Since we know the length BC and the measure of *β*, we can use the Law of Sines to find length *AC*.

$$\frac{BC}{\sin \beta} = \frac{AC}{\sin B}$$

$$\frac{80}{\sin 10°} = \frac{AC}{\sin 64°}$$

$$AC = \frac{80 \sin 64°}{\sin 10°}$$

Having found *AC*, we can now find the height of the building.

$$\sin 16° = \frac{h}{AC}$$

$$h = AC \sin 16°$$

$$= \frac{80 \sin 64°}{\sin 10°} \sin 16° \approx 114 \text{ ft} \text{Substitute for } AC.$$

Using the rounding convention, the height of the building to two significant digits is 110 feet.

■ *Try Exercise **24**, page 432.*

EXERCISE SET 8.2

In Exercises 1 to 8, solve the triangles.

1. $A = 42°$, $B = 61°$, $a = 12$

2. $B = 25°$, $C = 125°$, $b = 5.0$

3. $A = 110°$, $C = 32°$, $b = 12$

4. $B = 28°$, $C = 78°$, $c = 44$

5. $A = 82.0°$, $B = 65.4°$, $b = 36.5$

6. $B = 54.8°$, $C = 72.6°$, $a = 14.4$

7. $A = 33.8°$, $C = 98.5°$, $c = 102$

8. $B = 36.9°$, $C = 69.2°$, $a = 166$

In Exercises 9 to 18, solve the triangles that exist.

9. $A = 37°$, $c = 40$, $a = 28$

10. $B = 32°$, $c = 14$, $b = 9$

11. $C = 65°$, $b = 10$, $c = 8.0$

12. $A = 42°$, $a = 12$, $c = 18$

13. $A = 30°$, $a = 1.0$, $b = 2.4$

14. $B = 22.6°$, $b = 5.55$, $a = 13.8$

15. $A = 14.8°$, $c = 6.35$, $a = 4.80$

16. $C = 37.9°$, $b = 3.50$, $c = 2.84$

17. $C = 47.2°$, $a = 8.25$, $c = 5.80$

18. $B = 52.7°$, $b = 12.3$, $c = 16.3$

Application Exercises

19. A navigator on a ship sights a lighthouse at a bearing of N36°E. After traveling 8.0 miles at a bearing of N28°W, the ship sights the lighthouse at a bearing of S82°E. How far is the ship from the lighthouse at the second sighting?

20. Two fire lookouts are located on mountains 20 miles apart. Lookout B is at a bearing of S65°E from A. A fire was sighted at a bearing of N50°E from A and N8°E from B. Find the distance of the fire from lookout A.

21. The navigator on a ship traveling at 8 mph due east sights a lighthouse at a heading of 125°. One hour later the lighthouse is sighted at a heading of 205°. Find the closest the ship came to the lighthouse.

22. Two observers are directly in line with a balloon and 220 feet apart. The angle of elevation of the balloon from one observer is 67°, and the angle of elevation of the balloon from the other observer is 31°. How far is the balloon from the observer who sees the balloon at an angle of 67°?

23. To find the distance across a canyon, a surveying team locates points A and B on one side of the canyon and point C on the other side of the canyon. The distance between A and B is 85 yards. The angle CAB is 68°, and the angle CBA is 88°. Find the distance across the canyon.

24. The longer side of a parallelogram is 6.0 meters. Angle A is 56°, and angle α is 35°. Find the length of the longer diagonal.

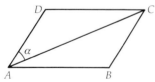

25. Two observers, in a line directly under a kite and a distance of 30 feet apart, observe the kite at an angle of elevation of 62° and 78°, respectively. Find the height of the kite.

26. A 35-foot high telephone pole is situated on an 11 percent slope from A. The angle of elevation from point A to the top of the pole is 32°. Find the length of the guy wire AC.

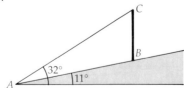

27. A surveying team determines the height of a hill by placing a 12-foot pole at the top of the hill and measuring the angles of elevation to the bottom and the top of the pole. Find the height of the hill.

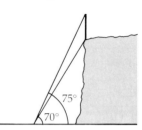

28. Three roads intersect in such a way to form a triangular piece of land. Find the lengths of the other two sides of the land.

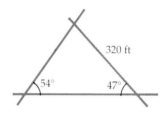

29. An airplane flew 450 miles at a heading of 65° from airport A to airport B. The pilot then flew at a heading of 142° to airport C. Find the distance from A to C if the heading from airport A to airport C is 120°.

Supplemental Exercises

30. A house B is located at a heading of 67° from house A. A house C is 300 meters at a heading of 112° from house A. House B is located at a heading of 349° from house C. Find the distance from house A to house B.

31. For any triangle ABC, show that

$$\frac{a - b}{b} = \frac{\sin A - \sin B}{\sin B}.$$

32. For any triangle ABC, show that

$$\frac{a + b}{b} = \frac{\sin A + \sin B}{\sin B}.$$

33. For any triangle ABC, show that

$$\frac{a - b}{a + b} = \frac{\sin A - \sin B}{\sin A + \sin B}.$$

8.3

The Law of Cosines and Area

The *Law of Cosines* can be used to solve triangles in which two sides and the included angle (angle between the two sides) are known or in which three sides are known. Consider the triangle in Figure 8.15. The altitude BD is drawn from B perpendicular to the x-axis. The triangle BDA is a right triangle, and the coordinates of B are $(a \cos C, a \sin C)$. The coordinates of A are $(b, 0)$. Using the distance formula, we can find the distance c.

$$c = \sqrt{(a \cos C - b)^2 + (a \sin C - 0)^2}$$

$$c^2 = a^2 \cos^2 C - 2ab \cos C + b^2 + a^2 \sin^2 C$$

$$c^2 = a^2(\cos^2 C + \sin^2 C) + b^2 - 2ab \cos C$$

$$c^2 = a^2 + b^2 - 2ab \cos C$$

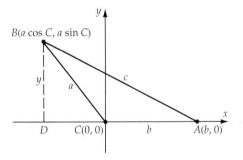

Figure 8.15

The Law of Cosines

> If A, B, and C are the angles of an oblique triangle and a, b, and c are the lengths of sides opposite these angles,
>
> $$c^2 = a^2 + b^2 - 2ab \cos C,$$
>
> $$a^2 = b^2 + c^2 - 2bc \cos A,$$
>
> $$b^2 = a^2 + c^2 - 2ac \cos B.$$

EXAMPLE 1   **Solve a Triangle Using the Law of Cosines**

In triangle ABC, angle $B = 110.0°$, side $a = 10.0$ centimeters, and side $c = 15.0$ centimeters. Find side b.

Solution   The Law of Cosines can be used because two sides and the

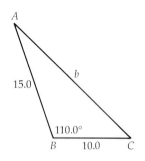

Figure 8.16

included angle are known.

$$b^2 = a^2 + c^2 - 2ac \cos B$$
$$= 10.0^2 + 15.0^2 - 2(10.0)(15.0) \cos 110.0°$$
$$\approx 100 + 225 + 103 = 428$$
$$b \approx 20.7 \text{ cm}$$

■ *Try Exercise* **12**, *page 438.*

EXAMPLE 2   Solve a Triangle Using the Law of Cosines

In triangle ABC, $a = 32$ feet, $b = 20$ feet, and $c = 40$ feet. Find angle B.

Solution $b^2 = a^2 + c^2 - 2ac \cos B$

$$\cos B = \frac{a^2 + c^2 - b^2}{2ac} \qquad \text{Solve for } \cos B.$$

$$= \frac{32^2 + 40^2 - 20^2}{2(32)(40)} \approx 0.8688 \quad \begin{array}{l}\text{Substitute the values} \\ \text{in the formula and} \\ \text{solve for angle } B.\end{array}$$

$$B \approx 30°$$

■ *Try Exercise* **18**, *page 438.*

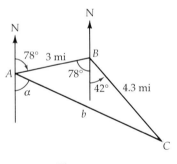

Figure 8.17

EXAMPLE 3   Solve an Application Using the Law of Cosines

A car traveled 3.0 miles at a bearing of N78°E. The road turned and another 4.3 miles was traveled at a bearing of S42°E. Find the distance and the bearing of the car from the starting point.

Solution   Sketch a diagram. First find the measure of angle B. Then use the Law of Cosines to find b.

$$B = 78° + 42° = 120°$$
$$b^2 = a^2 + c^2 - 2ac \cos B$$
$$= 3.0^2 + 4.3^2 - 2(3.0)(4.3) \cos 120°$$
$$= 9 + 18.49 + 12.9 = 40.39$$
$$b \approx 6.4 \text{ mi}$$

Find angle A.

$$\cos A = \frac{b^2 + c^2 - a^2}{2bc} = \frac{6.4^2 + 3.0^2 - 4.3^2}{(2)(6.4)(3.0)}$$

$$= \frac{40.96 + 9.0 - 18.49}{38.4} \approx 0.8195$$

$$A \approx 35°$$

The bearing of the present position of the car from the starting point A can be determined by calculating the measure of angle α from Figure 8.17.

$$\alpha = 180° - (78° + 35°) = 67°$$

The distance is approximately 6.4 miles, and the bearing to the nearest degree is S67°E.

■ *Try Exercise* **48,** *page 439.*

Remark   The measure of A in Example 3 also can be determined by using the Law of Sines.

Area

We used the formula $A = \frac{1}{2}bh$ for the area of a triangle when the base and height were given. In this section, we will find the area of triangles when the height is not given. We will use K for the area of a triangle since A is often used to represent an angle.

Consider the areas of the acute and obtuse triangles in Figure 8.18.

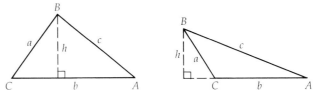

Figure 8.18
a. Acute triangle. b. Obtuse triangle.

Area of each triangle: $K = \dfrac{1}{2}bh$

Height of each triangle: $h = c \sin A$

Substitute for h: $K = \dfrac{1}{2}bc \sin A$

Thus we have established the following theorem.

Area of a Triangle

The area K of triangle ABC is one-half the product of the lengths of any two sides and the included angle. Thus

$$K = \frac{1}{2}bc \sin A,$$

$$K = \frac{1}{2}ab \sin C,$$

$$K = \frac{1}{2}ac \sin B.$$

EXAMPLE 4 Find the Area of a Triangle

Given Angle $A = 62°$, $b = 12$ meters, and $c = 5.0$ meters, find the area of triangle ABC.

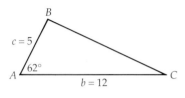

Figure 8.19

Solution Two sides and the included angle of the triangle are given. Using the formula for area, we have

$$K = \frac{1}{2}bc \sin A = \frac{1}{2}(12)(5.0)(\sin 62°) \approx 26 \text{ m}^2.$$

■ *Try Exercise **28**, page 438.*

When two angles and a side are given, the Law of Sines is used to derive a formula for the area of a triangle.

First, solve for c in the Law of Sines.

$$\frac{c}{\sin C} = \frac{b}{\sin B}$$

$$c = \frac{b \sin C}{\sin B}$$

Substitute for c in the first formula for area:

$$K = \frac{1}{2}b \cdot c \sin A$$

$$= \frac{1}{2}b \cdot \frac{b \sin C}{\sin B} \sin A$$

$$K = \frac{b^2 \sin C \sin A}{2 \sin B}$$

In like manner, the following two alternate formulas can be derived for the area of a triangle:

$$K = \frac{a^2 \sin B \sin C}{2 \sin A} \quad \text{and} \quad K = \frac{c^2 \sin A \sin B}{2 \sin C}$$

EXAMPLE 5 Find the Area of a Triangle

Given angle $A = 32°$, angle $C = 77°$, and side $a = 14$ inches. Find the area of triangle ABC.

Solution First we find the third angle.

$$B = 180° - 32° - 77° = 71°$$

Thus

$$K = \frac{a^2 \sin B \sin C}{2 \sin A} = \frac{14^2 \sin 71° \sin 77°}{2 \sin 32°} \approx 170 \text{ in}^2$$

■ *Try Exercise **26**, page 438.*

The Law of Cosines can be used to derive *Heron's formula* for the area of a triangle in which three sides of the triangle are given.

Heron's Formula for Finding the Area of a Triangle

If a, b, and c are the lengths of the sides of a triangle, then the area K of the triangle is

$$K = \sqrt{s(s - a)(s - b)(s - c)}, \quad \text{where } s = \frac{1}{2}(a + b + c)$$

EXAMPLE 6   **Find an Area by Heron's Formula**

Find the area of the triangle with $a = 7.0$ meters, $b = 15$ meters, and $c = 12$ meters.

Solution   Using Heron's formula, we have

$$s = \frac{a + b + c}{2} = \frac{7.0 + 15 + 12}{2} = 17$$

$$K = \sqrt{s(s - a)(s - b)(s - c)}$$

$$= \sqrt{17(17 - 7)(17 - 15)(17 - 12)}$$

$$= \sqrt{1700} \approx 41 \text{ m}^2$$

■ *Try Exercise 36, page 438.*

EXAMPLE 7   **Solve an Application Using Heron's Formula**

A commercial piece of real estate is priced at $6.50 per square foot. Find the cost of a triangular piece of commercial property measuring 200 feet by 350 feet by 400 feet.

Solution

$$s = \frac{a + b + c}{2} = \frac{200 + 350 + 400}{2} = 475$$

$$K = \sqrt{s(s - a)(s - b)(s - c)}$$

$$= \sqrt{475(475 - 200)(475 - 350)(475 - 400)}$$

$$= \sqrt{475(275)(125)(75)} = \sqrt{1224609375}$$

$$\approx 35,000$$

The area is approximately 35,000 square feet. Find the value of the property by multiplying the cost per foot by the number of square feet.

$$\text{Cost} = 6.50(35,000) = 227,500$$

The cost of the commercial property is approximately $227,500.

■ *Try Exercise 56, page 439.*

EXERCISE SET 8.3

In Exercises 1 to 14, find the third side of the triangle.

1. $a = 12$, $b = 18$, $C = 44°$
2. $b = 30$, $c = 24$, $A = 120°$
3. $a = 120$, $c = 180$, $B = 56°$
4. $a = 400$, $b = 620$, $C = 116°$
5. $b = 60$, $c = 84$, $A = 13°$
6. $a = 122$, $c = 144$, $B = 48°$
7. $a = 9.0$, $b = 7.0$, $C = 72°$
8. $b = 12$, $c = 22$, $A = 55°$
9. $a = 4.6$, $b = 7.2$, $C = 124°$
10. $b = 12.3$, $c = 14.5$, $A = 6.5°$
11. $a = 25.9$, $c = 33.4$, $B = 84°$
12. $a = 14.2$, $b = 9.30$, $C = 9.20°$
13. $a = 122$, $c = 55.9$, $B = 44.2°$
14. $b = 444.8$, $c = 389.6$, $A = 78.44°$

In Exercises 15 to 24, given three sides of a triangle, find the specified angle.

15. $a = 25$, $b = 32$, $c = 40$; find A.
16. $a = 60$, $b = 88$, $c = 120$; find B.
17. $a = 8.0$, $b = 9.0$, $c = 12$; find C.
18. $a = 108$, $b = 132$, $c = 160$; find A.
19. $a = 80.0$, $b = 92.0$, $c = 124$; find B.
20. $a = 166$, $b = 124$, $c = 139$; find B.
21. $a = 1025$, $b = 625.0$, $c = 1420$; find C.
22. $a = 4.7$, $b = 3.2$, $c = 5.9$; find A.
23. $a = 32.5$, $b = 40.1$, $c = 29.6$; find B.
24. $a = 112.4$, $b = 96.80$, $c = 129.2$; find C.

In Exercises 25 to 36, find the area of the given triangle.

25. $A = 105°$, $b = 12$, $c = 24$
26. $B = 127°$, $a = 32$, $c = 25$
27. $A = 42°$, $B = 76°$, $c = 12$
28. $B = 102°$, $C = 27°$, $a = 8.5$
29. $a = 10$, $b = 12$, $c = 14$
30. $a = 32$, $b = 24$, $c = 36$
31. $B = 54.3°$, $a = 22.4$, $b = 26.9$
32. $C = 18.2°$, $b = 13.4$, $a = 9.84$
33. $A = 116°$, $B = 34°$, $c = 8.5$
34. $B = 42.8°$, $C = 76.3°$, $c = 17.9$
35. $a = 3.6$, $b = 4.2$, $c = 4.8$
36. $a = 13.3$, $b = 15.4$, $c = 10.2$

Application Exercises

37. A plane leaves airport A and travels 560 miles to airport B at a heading of 32°. The plane leaves airport B and travels to airport C 320 miles away at a heading of 108°. Find the distance from airport A to airport C.

38. A developer has a triangular lot at the intersection of two streets. The streets meet at an angle of 72°, and the lot has 300 feet of frontage along one street and 416 feet of frontage along the other street. Find the length of the third side of the lot.

39. Two ships left a port at the same time. One ship traveled at a speed of 18 mph at a heading of 318°. The other ship traveled at a speed of 22 mph at a heading of 198°. Find the distance between the two ships after 10 hours of travel.

40. Find the distance across a lake using the measurements as shown in the figure.

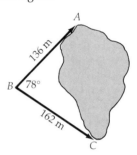

41. A regular hexagon is inscribed in a circle with a radius of 40 centimeters. Find the length of one side of the hexagon.

42. A regular pentagon is inscribed in a circle with a radius of 25 inches. Find the length of one side of the pentagon.

43. The length of the diagonals of a parallelogram are 20 inches and 32 inches. The diagonals intersect at an angle of 35°. Find the lengths of the sides of the parallelogram. (*Hint:* The diagonals of a parallelogram bisect one another.)

44. The sides of a parallelogram are 10 feet and 14 feet. The longer diagonal of the parallelogram is 18 feet. Find the length of the shorter diagonal of the parallelogram. (*Hint:* The diagonals of a parallelogram bisect one another.)

45. The sides of a parallelogram are 30 centimeters and 40 centimeters. The shorter diagonal of the parallelogram is 44 centimeters. Find the length of the longer diagonal of the parallelogram. (*Hint:* The diagonals of a parallelogram bisect one another.)

46. The sides of a triangular city lot have sides of 224 feet, 182 feet, and 165 feet. Find the angle between the longer two sides of the lot.

47. A plane traveling at 180 mph passes 400 feet directly overhead of an observer. The plane is traveling along a path with an angle of elevation of 14°. Find the distance of the plane from the observer 10 seconds after the plane has passed directly overhead.

48. A ship leaves a port at a speed of 16 mph at a bearing of N32°E. One hour later another ship leaves the port at a speed of 22 mph at a bearing of S74°W. Find the distance between the ships 4 hours after the first ship leaves the port.

49. Find the area of a triangular piece of land that is bounded by sides of 236 meters, 620 meters, and 814 meters.

50. Find the area of a parallelogram whose diagonals are 24 inches and 32 inches and the diagonals intersect at an angle of 40°.

51. Find the area of a parallelogram with sides of 12 meters and 18 meters and with one angle of 70°.

52. Find the area of a parallelogram with sides of 8 feet and 12 feet. The shorter diagonal is 10 feet.

53. Find the area of a square inscribed in a circle with a radius of 9 inches.

54. Find the area of a regular hexagon inscribed in a circle with a radius of 24 centimeters.

55. A commercial piece of real estate is priced at $2.20 per square foot. Find the cost of a triangular lot measuring 212 feet by 185 feet by 240 feet.

56. An industrial piece of real estate is priced at $4.15 per square foot. Find the cost of a triangular lot measuring 324 feet by 516 feet by 412 feet.

57. Find the number of acres in a pasture whose shape is a triangle measuring 800 feet by 1020 feet by 680 feet. (An acre is 43,560 square feet.)

58. Find the number of acres in a housing tract whose shape is a triangle measuring 420 yards by 540 yards by 500 yards. (An acre is 4840 square yards.)

Supplemental Exercises

59. Find the angle formed by the sides P_1P_2 and P_1P_3 of a triangle with the vertices at $P_1(-2, 4)$, $P_2(2, 1)$, and $P_3(4, -3)$.

60. The sides of a parallelogram are x and y, and the diagonals are w and z. Show that $w^2 + z^2 = 2x^2 + 2y^2$.

61. A rectangular box has dimensions of length 10 feet, width 4 feet, and height 3 feet. Find the angle between the diagonal of the bottom of the box and the diagonal of the end of the box.

62. A regular pentagon is inscribed in a circle with a radius of 4 inches. Find the perimeter of the pentagon.

63. An equilateral triangle is inscribed in a circle of radius 10 centimeters. Find the perimeter of the triangle.

64. Given a triangle ABC, prove that
$$a^2 = b^2 + c^2 - 2bc \cos A.$$

65. Use the Law of Cosines to show that
$$\cos A = \frac{(b + c - a)(b + c + a)}{2bc} - 1.$$

66. Prove that $K = xy \sin A$ for a parallelogram, where x and y are adjacent sides and A is the angle between x and y.

67. Show that the area of the parallelogram in the figure is $K = 2ab \sin C$.

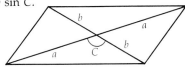

68. Given a regular hexagon inscribed in a circle of 10 inches, find the area of a sector (see the figure).

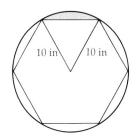

69. Find the volume of the pyramid-shaped piece of aluminum in the figure.

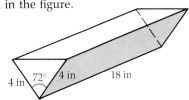

70. Show that the area of the circumscribed triangle in the figure is $K = rs$, where $s = \dfrac{a + b + c}{2}$.

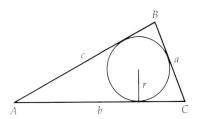

8.4

Vectors

In scientific applications, measurements are made that consist of a number quantity (magnitude) and a unit. Some of these measurements are temperature, force, displacement, and time. The magnitude and unit are sufficient to completely represent some of these measurements, but in other instances it is important to have a direction associated with the measurement. For example, a temperature of 37° does not need a direction to completely represent the quantity. However, a measurement, such as exerting 30 pounds of force on a box, must have the direction specified to completely represent the measurement. For example, is the force being used to lift the box up or push the box along the floor? Also, a displacement (change in location) of 2 miles east is different from a displacement of 2 miles north.

Physical measurements are divided into two classes: scalars and vectors. **Scalars** are quantities that are fully represented by a magnitude only. Temperature is an example of a scalar quantity. **Vector quantities** are represented by a magnitude and a direction. Force and velocity are examples of vector quantities. Force is specified by a magnitude and a direction. Velocity is a vector quantity with magnitude and direction. Speed is a scalar quantity; it is the magnitude part of velocity. For example, 55 meters per second east is a vector. The speed is 55 meters per second.

Definition of a Vector

> A **vector** is a directed line segment. The length of the vector is the magnitude of the vector, and the direction of the vector is measured by an angle.

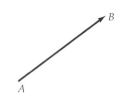

Figure 8.20

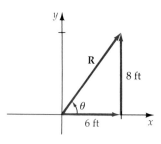

Figure 8.21

The point A of the vector in Figure 8.20 is called the initial point (or tail), and the point B is the terminal point (or head) of the vector. An arrow over the letters $\overrightarrow{AB}$, an arrow over a single letter $\vec{V}$, or boldface type **AB** or **V** is used to represent a vector. The magnitude of the vector is represented by $|\overrightarrow{AB}|$, $|\vec{V}|$, $|\mathbf{V}|$, or $|\mathbf{AB}|$.

The sum of two vectors is the single vector that will have the same effect as the two vectors. For example, the sum of a vector of magnitude 6 feet in the positive x direction and a vector with a magnitude of 8 feet in the positive y direction is equivalent to a vector of magnitude 10 feet at an angle of approximately 53° to the positive x-axis. This vector is called the **resultant** vector. The magnitude of the vector is

$$|\mathbf{V}| = \sqrt{6^2 + 8^2} = 10 \text{ ft.}$$

To find the direction of the vector from the positive x-axis, solve $\tan \theta = \frac{8}{6}$ for θ.

$$\tan \theta = \frac{8}{6}$$

$$\theta = \tan^{-1} \frac{8}{6} \approx 53°$$

The sum of two vectors is another vector that has an equivalent effect of the two vectors. In our example, walking a distance of 10 feet at an angle of 53° with the positive x-axis will put you in the same place as walking 6 feet east and 8 feet north.

One method of adding vectors graphically is to place the tail of one vector **U** at the head of the other vector **V** and complete the triangle, as Figure 8.22 shows. This is called the triangle method of adding vectors.

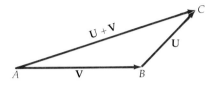

Figure 8.22

An alternate method of adding vectors graphically is to place the tails of the two vectors **U** and **V** together, as Figure 8.23 shows. Complete the parallelogram so that **U** and **V** are sides of the parallelogram. The diagonal of the parallelogram is the resultant.

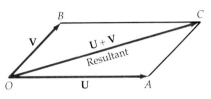

Figure 8.23

Multiplying a vector by a positive number other than 1 changes the length of the vector but does not affect the direction of the vector. If **V** is any vector, then **2V** denotes the vector that has the same direction as **V** but twice its magnitude. We say that **V** has been multiplied by the scalar 2. Multiplying a vector by a negative number reverses the direction of the vector.

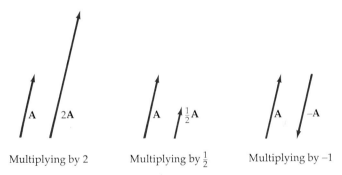

Multiplying by 2   Multiplying by $\frac{1}{2}$ Multiplying by -1

Figure 8.24

To substract **U** from **V** geometrically, add the opposite of **U** to **V**, as Figure 8.25 shows.

$$\mathbf{V} - \mathbf{U} = \mathbf{V} + (-\mathbf{U})$$

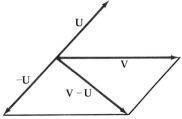

Figure 8.25

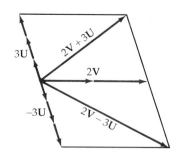

Figure 8.26

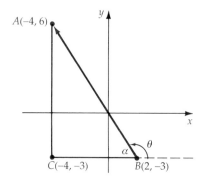

Figure 8.27

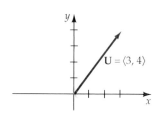

Figure 8.28

Assume that the two vectors **U** and **V** are given as shown in Figure 8.26. The sum $2\mathbf{V} + 3\mathbf{U}$ and the difference $2\mathbf{V} - 3\mathbf{U}$ are shown graphically.

The distance formula and the trigonometric functions can be used to find the magnitude and direction of a vector in the coordinate plane. The direction of a vector is the measure of the angle made by the vector and the positive x-axis.

EXAMPLE 1 Find the Direction and Magnitude of a Vector

Find the magnitude and direction of a vector that has its tail at the point $B(2, -3)$ and its head at the point $A(-4, 6)$.

Solution The vector is sketched as **BA** in Figure 8.27. We use the Pythagorean Theorem to find the magnitude of the vector.

$$|\mathbf{BA}| = \sqrt{[2 - (-4)]^2 + (-3 - 6)^2}$$

$$= \sqrt{36 + 81} = \sqrt{117}$$

Sketch the lines BC and AC to form a right triangle (see Figure 8.27). Find the lengths of BC and AC and then use the tangent function to find the reference angle for the direction angle.

$$a = |-4 - 2| = 6 \quad \text{and} \quad b = |6 - (-3)| = 9$$

$$\tan \alpha = \frac{b}{a} = \frac{9}{6}$$

$\alpha \approx 56°$ The reference angle for the vector is 56°.

$\theta = 180° - 56° = 124°$ The direction angle is the angle made by the vector and the positive x-axis.

The magnitude of the vector is $\sqrt{117}$ and the direction angle is 124°.

■ *Try Exercise **2**, page 450.*

The notation $\mathbf{U} = \langle 3, 4 \rangle$ is used for a vector with its tail at the origin and its head at the point $(3, 4)$. The distance 3 along the x-axis is called the magnitude of the x-component of the vector, and the distance 4 along the y-axis is called the magnitude of the y-component of the vector. The vector $\langle 0, 0 \rangle$ is called the zero (**0**) vector. For any vector $\mathbf{V} + (-\mathbf{V}) = \mathbf{0}$.

Caution The notation $\langle a, b \rangle$ is used for a vector, whereas (a, b) is used for a point. The vector $\langle a, b \rangle$ has its tail at $(0, 0)$ and its head at (a, b).

EXAMPLE 2 Graph Vectors on the Coordinate Plane

Graph each vector.   a. $\mathbf{U} = \langle -2, 4 \rangle$ b. $\mathbf{V} = \langle 3, -1 \rangle$

Solution See Figure 8.29

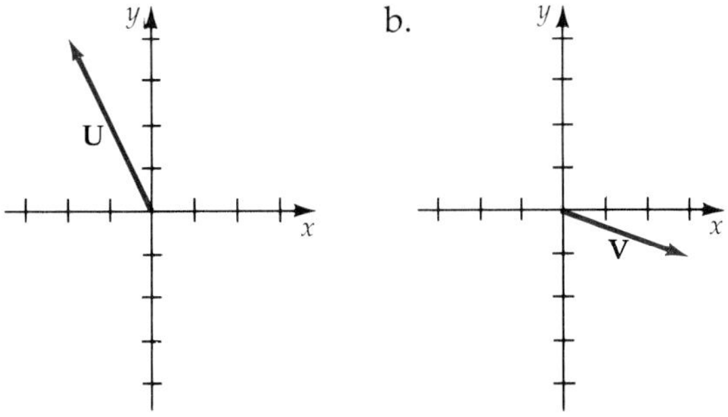

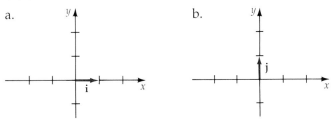

Figure 8.29

■ *Try Exercise* **8**, *page 450.*

The vector $\langle 1, 0 \rangle$ has a magnitude of 1, and its direction is in the positive *x*-direction. It is called the **unit vector i.** The vector $\langle 0, 1 \rangle$ has a magnitude of 1 and a direction in the positive *y*-direction. It is called the **unit vector j.** See Figure 8.30.

Figure 8.30
a. Unit vector **i**; b. Unit vector **j**.

Any vector in the plane can be written in terms of the unit vectors **i** and **j**. The vector $\mathbf{U} = \langle 3, 4 \rangle$ can be written $\mathbf{U} = 3\mathbf{i} + 4\mathbf{j}$. The vector $3\mathbf{i}$ is called the *x*-**component** of the vector **U**, and the vector $4\mathbf{j}$ is called the *y*-**component** of the vector **U**. The magnitude of the vector can be found by the Pythagorean Theorem, and the direction angle can be found by using the tangent function.

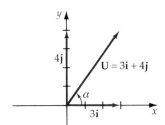

Figure 8.31

$$|\mathbf{U}| = \sqrt{3^2 + 4^2} = 5 \qquad \tan \alpha = \frac{4}{3}$$

$$\alpha \approx 53°$$

The vector $\mathbf{U} = 3\mathbf{i} + 4\mathbf{j}$ has a magnitude of 5 and a direction angle of 53°.

EXAMPLE 3   Find the Magnitude and Direction of Vectors

Find the magnitude and direction angle of each vector.

a. $\mathbf{U} = 5\mathbf{i} - 2\mathbf{j}$ b. $\mathbf{V} = \langle -2, -3 \rangle$

Solution   Use the Pythagorean Theorem to find the magnitude of the

a.

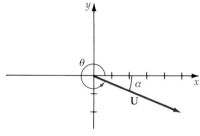

b.

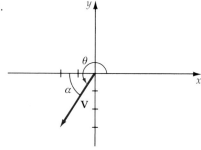

Figure 8.32

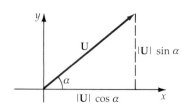

Figure 8.33

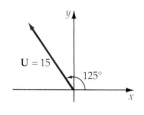

Figure 8.34

vector and the tangent function to find the reference angle for the direction. For each vector, α is the reference angle and θ is the direction angle.

a. $|\mathbf{U}| = \sqrt{5^2 + (-2)^2} \approx 5.4$ $\tan \alpha = \left| -\dfrac{2}{5} \right|$

$$\alpha \approx 22°$$

$$\theta = 360° - 22° = 338°$$

The magnitude is 5.4 and the direction is 338°.

b. $|\mathbf{V}| = \sqrt{(-2)^2 + (-3)^2} \approx 3.6$ $\tan = \left| \dfrac{-3}{-2} \right|$

$$\alpha \approx 56°$$

$$\theta = 180° + 56° = 236°$$

The magnitude is 3.6 and the direction angle is 236°.

■ *Try Exercises* **22**, *page 450.*

Any vector $\mathbf{U}$ can be written in the form $\mathbf{U} = x\mathbf{i} + y\mathbf{j}$. Figure 8.33 shows that the magnitude of the x- and y-components of vector $\mathbf{U} = x\mathbf{i} + y\mathbf{j}$ are given by $x = |\mathbf{U}| \cos \alpha$ and $y = |\mathbf{U}| \sin \alpha$. Thus

$$\mathbf{U} = |\mathbf{U}|(\cos \alpha)\mathbf{i} + |\mathbf{U}|(\sin \alpha)\mathbf{j}.$$

EXAMPLE 4   **Find the Components of a Vector**

Find the x- and y-components of a vector $\mathbf{U}$ with a magnitude of 15.0 and direction of 125°. Write $\mathbf{U}$ in terms of unit vectors. (See Figure 8.34.)

Solution   Find the magnitude of the x- and y-components.

$$x = 15 \cos 125° \approx -8.60 \qquad y = 15 \sin 125° \approx 12.3$$

Writing in terms of unit vectors, we have $\mathbf{U} \approx -8.60\mathbf{i} + 12.3\mathbf{j}$.

■ *Try Exercise* **28**, *page 450.*

Vectors can be added algebraically by adding their x- and y-components.

Definition of the Addition of Vectors

If $\mathbf{U} = a\mathbf{i} + b\mathbf{j}$ and $\mathbf{V} = c\mathbf{i} + d\mathbf{j}$, then

$$\mathbf{U} + \mathbf{V} = (a\mathbf{i} + b\mathbf{j}) + (c\mathbf{i} + d\mathbf{j}) = (a + c)\mathbf{i} + (b + d)\mathbf{j} = \langle a + c, b + d \rangle.$$

EXAMPLE 5   **Add Vectors Algebraically**

Add the following vectors:

a. $\mathbf{U} = 4\mathbf{i} + 2\mathbf{j}$ and $\mathbf{V} = -3\mathbf{i} + 2\mathbf{j}$ b. $\mathbf{W} = \langle -4, 5 \rangle$ and $\mathbf{Z} = \langle 2, 3 \rangle$

Solution

a. $\mathbf{U} + \mathbf{V} = (4\mathbf{i} + 2\mathbf{j}) + (-3\mathbf{i} + 2\mathbf{j}) = (4 - 3)\mathbf{i} + (2 + 2)\mathbf{j} = \mathbf{i} + 4\mathbf{j}$

b. $\mathbf{W} + \mathbf{Z} = \langle -4, 5 \rangle + \langle 2, 3 \rangle = \langle -4 + 2, 5 + 3 \rangle = \langle -2, 8 \rangle$

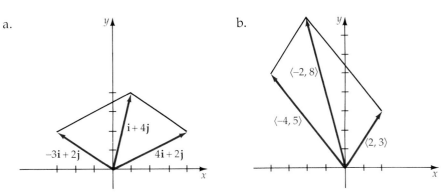

Figure 8.35

■ *Try Exercise* **30,** *page 450.*

To subtract $\mathbf{U}$ from $\mathbf{V}$, add $-\mathbf{U}$ to $\mathbf{V}$.

EXAMPLE 6   **Subtract Vectors Algebraically**

Subtract $\mathbf{U} = 2\mathbf{i} - 3\mathbf{j}$ from $\mathbf{V} = 4\mathbf{i} - \mathbf{j}$ and find the magnitude and direction of the new vector.

$$
\begin{aligned}
\textit{Solution} \quad \mathbf{V} - \mathbf{U} &= (4\mathbf{i} - \mathbf{j}) - (2\mathbf{i} - 3\mathbf{j}) \\
&= (4\mathbf{i} - \mathbf{j}) + (-2\mathbf{i} + 3\mathbf{j}) \\
&= (4 - 2)\mathbf{i} + (-1 + 3)\mathbf{j} \\
&= 2\mathbf{i} + 2\mathbf{j}
\end{aligned}
$$

Use the Pythagorean Theorem and the tangent function to find the magnitude and the direction.

$$|\mathbf{V} - \mathbf{U}| = \sqrt{2^2 + (2)^2} = 2\sqrt{2} \qquad \tan \theta = \frac{2}{2} = 1$$

$$\theta = 45°$$

The magnitude $\mathbf{V} - \mathbf{U}$ is $2\sqrt{2}$ and the direction angle is $45°$.

■ *Try Exercise* **32,** *page 450.*

To multiply a vector $x\mathbf{i} + y\mathbf{j}$ by a scalar a, multiply each component of the vector by a.

Definition of the Product of a Scalar and a Vector

If a is a scalar and $\mathbf{V}$ is the vector $x\mathbf{i} + y\mathbf{j} = \langle x, y \rangle$, then

$$a\mathbf{V} = a(x\mathbf{i} + y\mathbf{j}) = ax\mathbf{i} + ay\mathbf{j} = \langle ax, ay \rangle.$$

EXAMPLE 7   **Perform Vector Operations**

Let $\mathbf{U} = \mathbf{i} - 3\mathbf{j}$ and $\mathbf{V} = -2\mathbf{i} + 2\mathbf{j}$. Find $2\mathbf{U} - 3\mathbf{V}$.

Solution

$$2\mathbf{U} - 3\mathbf{V} = 2(\mathbf{i} - 3\mathbf{j}) - 3(-2\mathbf{i} + 2\mathbf{j}) = 2\mathbf{i} - 6\mathbf{j} + 6\mathbf{i} - 6\mathbf{j} = 8\mathbf{i} - 12\mathbf{j}$$

■ *Try Exercise* **40,** *page 450.*

Dot Product

The *dot product* of two vectors is one way to multiply a vector by a vector. The dot product is a scalar and is useful in certain types of physics and engineering problems involving forces and work.

Definition of the Dot Product

> Given $\mathbf{U} = a\mathbf{i} + b\mathbf{j}$ and $\mathbf{V} = c\mathbf{i} + d\mathbf{j}$, the dot product $\mathbf{U} \cdot \mathbf{V}$ is given by
>
> $$\mathbf{U} \cdot \mathbf{V} = (a\mathbf{i} + b\mathbf{j}) \cdot (c\mathbf{i} + d\mathbf{j}) = ac + bd.$$

EXAMPLE 8   **Find the Dot Product**

Find the dot product of $\mathbf{U} = -3\mathbf{i} + 4\mathbf{j}$ and $\mathbf{V} = 2\mathbf{i} + \mathbf{j}$.

Solution $\mathbf{U} \cdot \mathbf{V} = (-3\mathbf{i} + 4\mathbf{j}) \cdot (2\mathbf{i} + \mathbf{j})$ Use the definition of the dot product.

$$= (-3)(2) + (4)(1) = -2$$

■ *Try Exercise* **56,** *page 450.*

Caution   Remember that a dot product is a scalar.

The dot product and the Law of Cosines can be used to derive an alternate expression for the dot product. Given the vectors $\mathbf{U} = a\mathbf{i} + b\mathbf{j}$ and $\mathbf{V} = c\mathbf{i} + d\mathbf{j}$ shown in Figure 8.36, we can derive an expression for the cosine of the angle formed by the two vectors.

Using the Law of Cosines for the triangle *OAB*, we have

$$|\mathbf{AB}|^2 = |\mathbf{U}|^2 + |\mathbf{V}|^2 - 2|\mathbf{U}||\mathbf{V}| \cos \alpha.$$

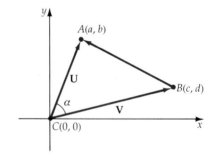

Figure. 8.36

By the distance formula, $|\mathbf{AB}|^2 = (a - c)^2 + (b - d)^2$, $|\mathbf{U}|^2 = a^2 + b^2$, and $|\mathbf{V}|^2 = c^2 + d^2$. Thus,

$$(a - c)^2 + (b - d)^2 = (a^2 + b^2) + (c^2 + d^2) - 2|\mathbf{U}||\mathbf{V}| \cos \alpha$$

$$a^2 - 2ac + c^2 + b^2 - 2bd + d^2 = a^2 + b^2 + c^2 + d^2 - 2|\mathbf{U}||\mathbf{V}| \cos \alpha$$

$$-2ac - 2bd = -2|\mathbf{U}||\mathbf{V}| \cos \alpha$$

$$ac + bd = |\mathbf{U}||\mathbf{V}| \cos \alpha$$

$$\mathbf{U} \cdot \mathbf{V} = |\mathbf{U}||\mathbf{V}| \cos \alpha \quad \text{Substitute } \mathbf{U} \cdot \mathbf{V} \text{ for } ac + bd$$

$$\cos \alpha = \frac{\mathbf{U} \cdot \mathbf{V}}{|\mathbf{U}||\mathbf{V}|}$$

Remark The equation for cos α can be used to find the angle between any two vectors.

The Angle Between Two Vectors

If α is the angle between two nonzero vectors **U** and **V**, $0 \le \alpha \le 180°$, then

$$\cos \alpha = \frac{\mathbf{U} \cdot \mathbf{V}}{|\mathbf{U}||\mathbf{V}|}.$$

EXAMPLE 9 **Find the Angle Between Two Vectors Using the Dot Product**

Find the angle between the vectors **U** = 3**i** + 2**j** and **V** = −5**i** + **j**.

Solution Use the equation

$$\cos \alpha = \frac{\mathbf{U} \cdot \mathbf{V}}{|\mathbf{U}||\mathbf{V}|}$$

$$= \frac{(3\mathbf{i} + 2\mathbf{j}) \cdot (-5\mathbf{i} + \mathbf{j})}{(\sqrt{3^2 + 2^2})(\sqrt{(-5)^2 + 1^2})}$$

$$= \frac{-15 + 2}{\sqrt{13}\sqrt{26}}$$

$$= \frac{-13}{\sqrt{338}}$$

$$\approx -0.7071$$

$$\alpha \approx 135°$$

The angle between the vectors is approximately 135°.

■ *Try Exercise* **62**, *page 450.*

Application Problems Using Vectors

We will consider an object on which two vectors are acting simultaneously. This occurs when a boat is moving in a current or an airplane is flying in a wind. We will refer to the airspeed of an airplane as if there were no wind. The actual velocity is the velocity relative to the earth. The magnitude of the actual velocity is called the ground speed.

Remark The bearing of an airplane is the direction in which it is pointed or the direction of the airspeed. The course is the actual direction the airplane is moving relative to the ground or the direction of the ground speed.

Consider an airplane flying due east at an airspeed of 525 mph. The wind is from the south at a speed of 40.0 mph. Suppose we are asked to find the ground speed and course of the airplane.

The airplane is headed east with a speed of 525 mph (airspeed). The airplane is also carried north by the wind. The actual velocity of the airplane (ground speed) is the sum of two vectors.

Graph the two vectors in a coordinate plane and find the resultant.

$$\tan \theta = \frac{40.0}{525}$$

$$\theta \approx 4.36°$$

$$\alpha = 90.00 - 4.36° = 85.64°$$

The sine function is used to find the ground speed $|\mathbf{V}|$.

$$\sin 4.36° = \frac{40.0}{|\mathbf{V}|}$$

$$|\mathbf{V}| = \frac{40.0}{\sin 4.36°} \approx 526$$

The airplane is on a course N85.6°E traveling at approximately 526 mph.

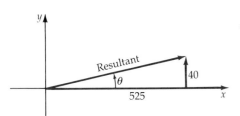

Figure 8.37

EXAMPLE 10   Solve an Application Involving Airspeed

An airplane traveling at an airspeed of 300 mph has a heading of 62.0°. The airplane is traveling through a 40.0-mph wind that has a heading of 125.0°. Find the ground speed and the course of the airplane. (Assume all lengths are accurate to three significant digits.)

Solution   Sketch a vector diagram showing the given information. The vector **AB** represents the heading and the airspeed of the airplane. **AD** represents the wind velocity, and **AC** represents the course and the ground speed of the airplane. The angle ϕ (direction of vector **AB**) is equal to 28.0°, and the angle β (direction of vector **AD**) is equal to −35.0°. The resultant vector **AC** can be found by vector addition or by using the Law of Cosines. We will use vector addition.

We find the resultant vector by rewriting the heading and airspeed of the airplane and the wind velocity into component vectors and adding.

$$\mathbf{AB} = 300 \cos 28.°\mathbf{i} + 300 \sin 28.°\mathbf{j}$$

and

$$\mathbf{AD} = 40 \cos(-35°)\mathbf{i} + 40 \sin(-35°).$$

Substitute for **AB** and **AD** in the equation **AC** = **AB** + **AD**.

$$\mathbf{AC} = (300 \cos 28.°\mathbf{i} + 300 \sin 28.°\mathbf{j}) + [40 \cos(-35°)\mathbf{i} + 40 \sin(-35°)\mathbf{j}]$$

$$\approx (264.9\mathbf{i} + 140.8\mathbf{j}) + (32.76\mathbf{i} - 22.94\mathbf{j})$$

$$= 297.7\mathbf{i} + 117.9\mathbf{j}$$

The magnitude of the resultant is found by using the Pythagorean Theorem.

$$|\mathbf{R}| = \sqrt{(297.7)^2 + (117.9)^2} \approx 320$$

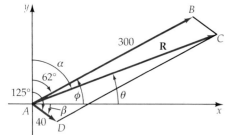

Figure 8.38

The angle that the resultant makes with the positive x-axis is found by using the tangent function. The course is found by subtracting this angle from 90°.

$$\tan \theta = \frac{117.9}{297.7}$$

$$\theta = 21.6°$$

$$\alpha = 90.0° - 21.6°$$

$$= 68.4°$$

The airplane is on a course N68.4°E traveling at approximately 320 mph.

■ *Try Exercise 68, page 451.*

EXAMPLE 11   Solve an Application Involving Force

A 100-pound box is on a 20° ramp. Find the magnitude of the component of the force vector parallel to the ramp.

Solution   We are finding the magnitude of the component of a force in a direction other than that of an axis. The force of gravity vector (100 pounds) can be written as the sum of two components, one parallel to the ramp and the other perpendicular to the ramp, as shown in Figure 8.39. The vector **AB** represents the force that pulls the box down the ramp. The angle φ is 20° because two angles are equal if they have mutually perpendicular sides; PQ is perpendicular to AD, and PR is perpendicular to AC. Triangle ADC is a right triangle. The magnitude of the component of the force vector is |**AB**|.

$$\sin 20° = \frac{|\mathbf{AB}|}{100}$$

$$|\mathbf{AB}| = 34$$

The magnitude of the component of the force vector parallel to the ramp is 34 pounds.

■ *Try Exercise 70, page 451.*

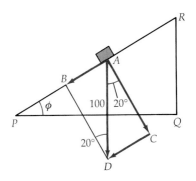

Figure 8.39

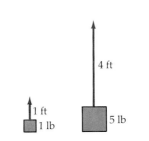

Figure 8.40

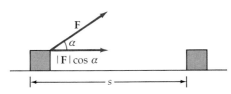

Figure 8.41

Work is done only if there is a displacement (or movement through a distance) of an object on which a force acts. **Work** is defined as a force exerted through a distance. Lifting 1 pound a distance of 1 foot is 1 foot pound (ft-lb) of work. Lifting 5 pounds a distance of 4 feet is $5 \cdot 4 = 20$ ft-lb of work. When the force is in the same direction as the displacement, the work is equal to the force times the displacement.

When the force is applied at a nonzero angle to the displacement, as shown in Figure 8.41, the work W is equal to the dot product of the force and the displacement vectors.

$$W = \mathbf{F} \cdot \mathbf{S}$$

$$= |\mathbf{F}||\mathbf{S}| \cos \alpha \qquad \mathbf{F} \cdot \mathbf{S} = |\mathbf{F}||\mathbf{S}| \cos \alpha.$$

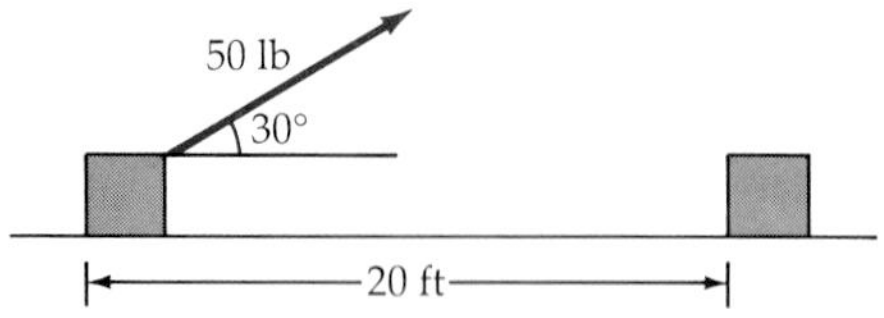

50 lb

30°

|←——20 ft——→|

Figure 8.42

EXAMPLE 12   **Find the Work Done by a Force**

Find the work done when a force of 50 pounds is used to pull a crate 20 feet along a level path if the force is at an angle of 30°. (See Figure 8.42.)

Solution   Use the definition for work.

$$W = \mathbf{F} \cdot \mathbf{S} = |\mathbf{F}||\mathbf{S}| \cos 30° = 50 \cdot 20 \cos 30° \approx 870$$

The work is 870 foot-pounds (to the nearest 10 ft-lbs).

■ *Try Exercise* **72**, *page 451.*

EXERCISE SET 8.4

In Exercises 1 to 6, find the magnitude and direction of the vectors with the tail at P_1 and the head at P_2.

1. $P_1(2, 3)$, $P_2(-1, 4)$

2. $P_1(-5, 6)$, $P_2(-4, -7)$

3. $P_1(-4, -6)$, $P_2(5, -6)$

4. $P_1(3, -8)$, $P_2(6, 3)$

5. $P_1(20, 30)$, $P_2(-10, 30)$

6. $P_1(-45, 10)$, $P_2(25, 25)$

In Exercises 7 to 10, graph each vector.

7. $\mathbf{U} = \langle -1, 4 \rangle$

8. $\mathbf{V} = \langle 3, -2 \rangle$

9. $\mathbf{V} = \langle -5, -3 \rangle$

10. $\mathbf{U} = \langle 3, 0 \rangle$

In Exercises 11 to 18, find the magnitude and direction of the given vector.

11. $\mathbf{U} = \langle -3, 4 \rangle$

12. $\mathbf{U} = \langle 6, 10 \rangle$

13. $\mathbf{V} = \langle 20, -40 \rangle$

14. $\mathbf{V} = \langle -50, 30 \rangle$

15. $\mathbf{V} = 2\mathbf{i} - 4\mathbf{j}$

16. $\mathbf{V} = -5\mathbf{i} + 6\mathbf{j}$

17. $\mathbf{U} = 42\mathbf{i} - 18\mathbf{j}$

18. $\mathbf{U} = -22\mathbf{i} - 32\mathbf{j}$

In Exercises 19 to 28, find the magnitude of the x- and y-components of each vector.

19. $\mathbf{U} = \langle -8, 5 \rangle$

20. $\mathbf{V} = \langle 4, -3 \rangle$

21. $\mathbf{U} = \langle -2, -6 \rangle$

22. $\mathbf{V} = \langle -3, 4 \rangle$

23. $\mathbf{V} = -3\mathbf{i} + 4\mathbf{j}$

24. $\mathbf{U} = 8\mathbf{i} + 5\mathbf{j}$

25. $\mathbf{V} = 6\mathbf{i} - 4\mathbf{j}$

26. $\mathbf{U} = 12\mathbf{i} + 9\mathbf{j}$

27. $|\mathbf{U}| = 6.0$, $\theta = 50°$

28. $|\mathbf{U}| = 15$, $\theta = 140°$

In Exercises 29 to 32, perform the indicated operation.

29. $(2\mathbf{i} + 3\mathbf{j}) + (4\mathbf{i} - 2\mathbf{j})$

30. $(-2\mathbf{i} + \mathbf{j}) + (-\mathbf{i} + 2\mathbf{j})$

31. $(-3\mathbf{i} + \mathbf{j}) - (2\mathbf{i} + 3\mathbf{j})$

32. $(\mathbf{i} - 4\mathbf{j}) - (-3\mathbf{i} - 2\mathbf{j})$

In Exercises 33 to 40, perform the indicated operation where $\mathbf{U} = 2\mathbf{i} + 3\mathbf{j}$, $\mathbf{V} = -\mathbf{i} + 2\mathbf{j}$.

33. $3\mathbf{U}$

34. $-4\mathbf{V}$

35. $2\mathbf{U} - \mathbf{V}$

36. $3\mathbf{U} + 2\mathbf{V}$

37. $5\mathbf{V} - \dfrac{1}{2}\mathbf{U}$

38. $\dfrac{3}{4}\mathbf{V} + 2\mathbf{U}$

39. $\dfrac{2}{3}\mathbf{U} + \dfrac{1}{2}\mathbf{V}$

40. $\dfrac{3}{4}\mathbf{U} - \dfrac{1}{2}\mathbf{V}$

In Exercises 41 to 48, write the vectors in the form $x\mathbf{i} + y\mathbf{j}$.

41. $\mathbf{U} = \langle -3, 4 \rangle$

42. $\mathbf{U} = \langle 7, -10 \rangle$

43. $\mathbf{V} = \langle 1, -3 \rangle$

44. $\mathbf{U} = \langle -2, -5 \rangle$

45. $|\mathbf{U}| = 5.0$, $\theta = 44°$

46. $|\mathbf{U}| = 60$, $\theta = 125°$

47. $|\mathbf{V}| = 8.0$, $\theta = 320°$

48. $|\mathbf{U}| = 12$, $\theta = 245°$

In Exercises 49 to 56, find the dot product of the vectors.

49. $\mathbf{U} = \langle 2, -3 \rangle$, $\mathbf{V} = \langle 1, 4 \rangle$

50. $\mathbf{U} = \langle -3, 2 \rangle$, $\mathbf{V} = \langle 5, 5 \rangle$

51. $\mathbf{U} = \langle 2, 0 \rangle$, $\mathbf{V} = \langle -5, 4 \rangle$

52. $\mathbf{U} = \langle -5, -2 \rangle$, $\mathbf{V} = \langle 7, -1 \rangle$

53. $\mathbf{U} = 2\mathbf{i} + 6\mathbf{j}$, $\mathbf{V} = -3\mathbf{i} + 5\mathbf{j}$

54. $\mathbf{U} = -\mathbf{i} + \mathbf{j}$, $\mathbf{V} = -4\mathbf{i} + 6\mathbf{j}$

55. $\mathbf{U} = 8\mathbf{i} - 13\mathbf{j}$, $\mathbf{V} = 7\mathbf{i} - 10\mathbf{j}$

56. $\mathbf{U} = 15 - 10\mathbf{j}$, $\mathbf{V} = 18\mathbf{i} + 16\mathbf{j}$

In Exercises 57 to 64, find the angle between the vectors.

57. $\mathbf{U} = \langle -3, 2 \rangle$, $\mathbf{V} = \langle 1, 4 \rangle$

58. $\mathbf{U} = \langle 5, 5 \rangle$, $\mathbf{V} = \langle -3, -3 \rangle$

59. $\mathbf{U} = \langle 5, 0 \rangle$, $\mathbf{V} = \langle 2, -4 \rangle$

60. $\mathbf{U} = \langle 5, -2 \rangle$, $\mathbf{V} = \langle -4, -1 \rangle$

61. $\mathbf{U} = 3\mathbf{i} - 2\mathbf{j}$, $\mathbf{V} = 4\mathbf{i} + \mathbf{j}$

62. $\mathbf{U} = -\mathbf{i} + 5\mathbf{j}$, $\mathbf{V} = 4\mathbf{i} + 7\mathbf{j}$

63. $\mathbf{U} = -2\mathbf{i} - 3\mathbf{j}$, $\mathbf{V} = 3\mathbf{i} + 5\mathbf{j}$

64. $\mathbf{U} = 5\mathbf{i} - 3\mathbf{j}$, $\mathbf{V} = 6\mathbf{i} + \mathbf{j}$

65. Two forces of 68.0 pounds and 120 pounds are acting on an object. The angle between the forces is 17.0°. Find the resultant and the angle that the resultant makes with the smaller force.

66. Two forces of 85 pounds and 110 pounds act on an object. The 85-pound force acts at an angle of 9.4° with the positive *x*-axis, and the 110-pound force acts at an angle of −12.0° with the positive *x*-axis. Find the resultant vector.

67. A plane is flying at an airspeed of 340 mph at a heading of S56°E. A wind of 45 mph is from the west. Find the ground speed of the plane.

68. A person who can row 2.6 mph in still water wants to row due east across a river. The river is flowing from the north at a rate of 0.8 mph. Determine the heading of the boat to travel due east across the river.

69. Find the magnitude of force necessary to keep a 3000-pound car from sliding down a ramp of 5.6°.

70. A 120-pound force keeps an 800-pound object from sliding down an inclined plane. Find the angle of the inclined plane.

71. A 200-pound object is dragged 18 feet along a level floor. Find the work done if the force used is 75 pounds at an angle of 28° above the horizontal.

72. An 800-pound force is pulling on a sled loaded with 2200 pounds of lumber. The force is at an angle of 35° above the horizontal. Find the work done if the sled is pulled 45 feet.

Supplemental Exercises

73. Show that if $\mathbf{U} \cdot \mathbf{V} = 0$, the vectors $\mathbf{U}$ and $\mathbf{V}$ are orthogonal, that is, the angle between the vectors is a right angle.

74. Show that if $\mathbf{U} \cdot \mathbf{V} = |\mathbf{U}||\mathbf{V}|$, then $\mathbf{U}$ and $\mathbf{V}$ are parallel, that is, the angle between the vectors is 0° or 180°.

75. Show that $\mathbf{i} \cdot \mathbf{i} = 1$.

76. Show that $\mathbf{i} \cdot \mathbf{j} = 0$.

77. Find the sum of the three vectors: $\mathbf{U} = 6\mathbf{i} + 3\mathbf{j}$, $\mathbf{V} = -2\mathbf{i} + 2\mathbf{j}$, $\mathbf{W} = -\mathbf{i} - 4\mathbf{j}$.

In Exercises 78 to 81, find the projection of the vector in the specified direction. The projection of $\mathbf{V}$ in the direction of $\mathbf{U}$ is given by $\dfrac{\mathbf{U} \cdot \mathbf{V}}{|\mathbf{U}|}\mathbf{U}$.

78. Projection of vector $\mathbf{V} = -2\mathbf{i} + 4\mathbf{j}$ in the direction of vector $\mathbf{U} = 3\mathbf{i} - 2\mathbf{j}$.

79. Projection of vector $\mathbf{V} = 4\mathbf{i} - 2\mathbf{j}$ in the direction of vector $\mathbf{U} = -3\mathbf{i} + \mathbf{j}$.

80. Projection of vector $\mathbf{V} = \mathbf{i}$ in the direction of vector $\mathbf{U} = \mathbf{j}$.

81. Projection of vector $\mathbf{V} = -\mathbf{i} + \mathbf{j}$ in the direction of vector $\mathbf{U} = \mathbf{i} - \mathbf{j}$.

8.5
Complex Numbers and DeMoivre's Theorem

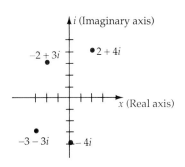

Figure 8.43

Real numbers are graphed as points on a number line. Complex numbers can be graphed on a coordinate grid called an **Argand diagram.** The horizontal axis of the coordinate plane is called the **real axis;** the vertical axis is called the **imaginary axis.** This coordinate system is called the **complex plane.**

A complex number written in the form $z = a + bi$ is written in **standard or rectangular form.** The graph of $a + bi$ is associated with the point $P(a, b)$ in the complex plane. Figure 8.43 shows the graphs of several complex numbers.

Figure 8.44 is the graph of the complex number $z = -3 + 4i$. The length of the line segment drawn from the origin to the point $P(-3, 4)$ on the complex plane is the *absolute value* of z. From the Pythagorean Theorem, the absolute value of $z = -3 + 4i$ is

$$\sqrt{(-3)^2 + 4^2} = \sqrt{25} = 5.$$

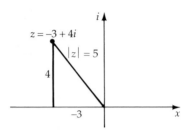

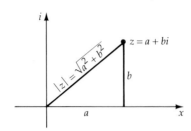

Figure 8.44   **Figure 8.45**

The absolute value of the complex number $z = a + bi$, denoted $|z|$ is

$$|z| = |a + bi| = \sqrt{a^2 + b^2}.$$

Thus $|z|$ is the distance from the origin to z (see Figure 8.45).

A complex number $z = a + bi$ can be written in terms of trigonometric functions. Consider the complex number graphed in Figure 8.46. We can write a and b in terms of the sine and the cosine.

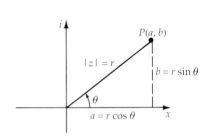

Figure 8.46

$$\cos \theta = \frac{a}{r} \qquad \sin \theta = \frac{b}{r} \qquad \text{where } r = |z| = \sqrt{a^2 + b^2}$$

$$a = r \cos \theta \qquad b = r \sin \theta$$

Substituting for a and b in $z = a + bi$, we obtain

$$z = r \cos \theta + i \sin \theta = r(\cos \theta + i \sin \theta).$$

The expression $z = r(\cos \theta + ir \sin \theta)$ is known as the trigonometric form of a complex number. The value of r is called the **modulus** of the complex number z, and the angle θ is called the **argument** of the complex number z.

The modulus r and the argument θ of a complex number $z = a + bi$ are given by

$$r = \sqrt{a^2 + b^2} \qquad \text{and} \qquad \cos \theta = \frac{a}{r}, \sin \theta = \frac{b}{r}, \qquad \text{where } 0 \le \theta < 360°.$$

We also can write $\alpha = \tan^{-1}\left|\dfrac{b}{a}\right|$ where α is the reference angle for θ. Because of the periodic nature of the sine and cosine functions, the trigonometric form of a complex number is not unique. Since $\cos \theta = \cos(\theta + 2k\pi)$ and $\sin \theta = \sin(\theta + 2k\pi)$, where k is an integer, the following complex numbers are equal:

$$r(\cos \theta + i \sin \theta) = r[\cos(\theta + 2k\pi) + i \sin(\theta + 2k\pi)] \quad \text{for } k \text{ an integer}.$$

For example, $2\left(\cos \dfrac{\pi}{6} + i \sin \dfrac{\pi}{6}\right) = 2\left[\cos\left(\dfrac{\pi}{6} + 2\pi\right) + i \sin\left(\dfrac{\pi}{6} + 2\pi\right)\right]$

EXAMPLE 1   Write a Complex Number in Trigonometric Form

Write $z = -2 - 2i$ in trigonometric form.

Solution   Find the modulus and the argument of z. Then substitute these values in the trigonometric form of z.

$$r = \sqrt{(-2)^2 + (-2)^2} = \sqrt{8} = 2\sqrt{2}$$

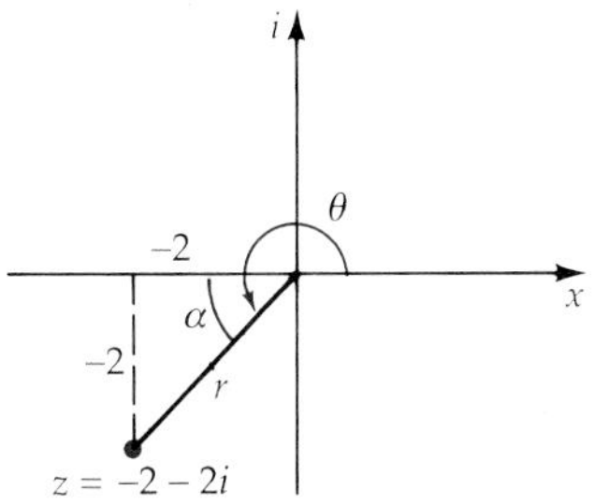

$$\alpha = \tan^{-1}\left|\frac{b}{a}\right|$$ α is the reference angle of angle θ.

$$\alpha = \tan^{-1}\left|\frac{-2}{-2}\right| = \tan^{-1}1$$

$$= 45°$$

$$\theta = 180° + 45° = 225°$$ Since z is in the third quadrant, $180° \le \theta \le 270°$.

$$z = 2\sqrt{2}\,(\cos 225° + i \sin 225°)$$

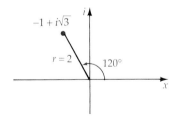

Figure 8.47

■ *Try Exercise* **12,** *page 459.*

Remark   In this case, we can find θ by looking at its graph.

EXAMPLE 2   **Write Complex Numbers in Standard Form**

Write $z = 2(\cos 120° + i \sin 120°)$ in standard form.

Solution   To change a complex number from its trigometric form to its standard form, evaluate $\cos \theta$ and $\sin \theta$.

$$z = 2(\cos 120° + i \sin 120°)$$

$$= 2\left(-\frac{1}{2} + \frac{\sqrt{3}}{2}i\right)$$

$$= -1 + i\sqrt{3}$$

Figure 8.48

■ *Try Exercise* **26,** *page 459.*

Let z_1 and z_2 be two complex numbers written in trogometric form. The product of z_1 and z_2 can be found by using several trigonometric identities.

Let $z_1 = r_1(\cos \theta_1 + i \sin \theta_1)$, and $z_2 = r_2(\cos \theta_2 + i \sin \theta_2)$. Then

$$z_1 z_2 = r_1(\cos \theta_1 + i \sin \theta_1) \cdot r_2(\cos \theta_2 + i \sin \theta_2)$$

$$= r_1 r_2(\cos \theta_1 \cos \theta_2 + i \cos \theta_1 \sin \theta_2 + i \sin \theta_1 \cos \theta_2 + i^2 \sin \theta_1 \sin \theta_2)$$

$$= r_1 r_2[(\cos \theta_1 \cos \theta_2 - \sin \theta_1 \sin \theta_2) + i(\cos \theta_1 \sin \theta_2 + \sin \theta_1 \cos \theta_2)]$$

Use the sum of two angle identities to simplify the product $z_1 z_2$.

$$z_1 z_2 = r_1 r_2[\cos(\theta_1 + \theta_2) + i \sin(\theta_1 + \theta_2)]$$

The modulus for the product of two complex numbers in trigonometric form is the product of the moduli of the two complex numbers, and the argument of the product is the sum of the arguments of the two numbers.

EXAMPLE 3 Find the Product of Two Complex Numbers in Trigonometric Form

Find the product of $z_1 = -1 + \sqrt{3}i$ and $z_2 = -\sqrt{3} + i$ by using the trigonometric form of the complex numbers. Write the answer in standard form.

Solution

$$z_1 = 2\left(\cos \frac{2\pi}{3} + i \sin \frac{2\pi}{3}\right)$$

$$z_2 = 2\left(\cos \frac{5\pi}{6} + i \sin \frac{5\pi}{6}\right) \qquad \text{See Figure 8.49.}$$

$$z_1 z_2 = 2\left(\cos \frac{2\pi}{3} + i \sin \frac{2\pi}{3}\right) \cdot 2\left(\cos \frac{5\pi}{6} + i \sin \frac{5\pi}{6}\right)$$

$$= 2 \cdot 2\left[\cos\left(\frac{2\pi}{3} + \frac{5\pi}{6}\right) + i \sin\left(\frac{2\pi}{3} + \frac{5\pi}{6}\right)\right]$$

$$= 4\left(\cos \frac{3\pi}{2} + i \sin \frac{3\pi}{2}\right)$$

$$= 4(0 - i) = -4i$$

■ *Try Exercise 40, page 459.*

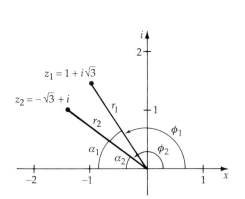

Figure 8.49

Let z_1 and z_2 be two complex numbers written in trigonometric form. The quotient of z_1 and z_2 can be found by using several trigonometric identities.

$$z_1 = r_1(\cos \theta_1 + i \sin \theta_1) \qquad z_2 = r_2(\cos \theta_2 + i \sin \theta_2)$$

$$\frac{z_1}{z_2} = \frac{r_1(\cos \theta_1 + i \sin \theta_1)}{r_2(\cos \theta_2 + i \sin \theta_2)}$$

$$= \frac{r_1(\cos \theta_1 + i \sin \theta_1)(\cos \theta_2 - i \sin \theta_2)}{r_2(\cos \theta_2 + i \sin \theta_2)(\cos \theta_2 - i \sin \theta_2)}$$

$$= \frac{r_1(\cos \theta_1 \cos \theta_2 - i \cos \theta_1 \sin \theta_2 + i \sin \theta_1 \cos \theta_2 - i^2 \sin \theta_1 \sin \theta_2)}{r_2(\cos^2\theta_2 - i^2 \sin^2\theta_2)}$$

$$= \frac{r_1[(\cos \theta_1 \cos \theta_2 + \sin \theta_1 \sin \theta_2) + i(\sin \theta_1 \cos_2 - \cos \theta_1 \sin \theta_2)]}{r_2(\cos^2\theta_2 + \sin^2\theta_2)}$$

$$\frac{z_1}{z_2} = \frac{r_1}{r_2}[(\cos(\theta_1 - \theta_2) + i \sin(\theta_1 - \theta_2)]$$

The modulus for the quotient of two complex numbers in trigonometric form is the quotient of the moduli of the two complex numbers, and the argument of the quotient is the difference of the arguments of the two numbers.

EXAMPLE 4   Divide Complex Numbers in Trigonometric Form

Use the trigonometric forms of the complex numbers $z_1 = -1 + i$ and $z_2 = \sqrt{3} - i$ to divide $\dfrac{z_1}{z_2}$. Write the answer in standard form.

Solution   From Figure 8.50, let $z_1 = r_1(\cos \theta_1 + i \sin \theta_1)$ and $z_2 = r_2(\cos \theta_2 + i \sin \theta_2)$, then

$$z_1 = \sqrt{2}\,(\cos 135° + i \sin 135°)$$

$$z_2 = 2(\cos 330° + i \sin 330°)$$

$$\frac{z_1}{z_2} = \frac{\sqrt{2}\,(\cos 135° + i \sin 135°)}{2(\cos 330° + i \sin 330°)}$$

$$= \frac{\sqrt{2}}{2}[\cos(135° - 330°) + i \sin(135° - 330°)]$$

$$= \frac{\sqrt{2}}{2}[\cos(-195°) + i \sin(-195°)]$$

$$= \frac{\sqrt{2}}{2}(\cos 195° - i \sin 195°)$$

$$\frac{z_1}{z_2} \approx \frac{\sqrt{2}}{2}(-0.9659 + 0.2588i) \approx -0.6830 + 0.1830i$$

■ *Try Exercise* **52**, *page 459.*

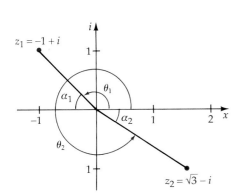

Figure 8.50

DeMoivre's Theorem is a procedure for finding powers and roots of complex numbers when the complex numbers are expressed in trigonometric form. This theorem can be illustrated by repeated multiplication of a complex number.

Let $z = r(\cos \theta + i \sin \theta)$. Then z^2 can be written as

$$z \cdot z = r(\cos \theta + i \sin \theta) \cdot r(\cos \theta + i \sin \theta)$$

$$z^2 = r^2(\cos 2\theta + i \sin 2\theta)$$

The product of $z^2 \cdot z$ is

$$z^2 \cdot z = r^2(\cos 2\theta + i \sin 2\theta) \cdot r(\cos \theta + i \sin \theta)$$

$$z^3 = r^3(\cos 3\theta + i \sin 3\theta)$$

If we continue this process, the results suggest a formula for the nth power of a complex number that is known as DeMoivre's Theorem.

DeMoivre's Theorem

If $z = r(\cos \theta + i \sin \theta)$ is a complex number and n is a positive integer, then

$$z^n = r^n(\cos n\theta + i \sin n\theta)$$

EXAMPLE 5   Find the Power of a Complex Number Using DeMoivre's Theorem

Find $(1 + i)^8$ using DeMoivre's Theorem. Write the answer in standard form.

Solution   Convert $1 + i$ to trigonometric form and then use DeMoivre's Theorem.

$$1 + i = \sqrt{2}\,(\cos 45° + i \sin 45°)$$

$$(1 + i)^8 = (\sqrt{2})^8 [\cos 8(45°) + i \sin 8(45°) \qquad \text{DeMoivre's Theorem}$$

$$= 16(\cos 360° + i \sin 360°) = 16(1 + 0i) = 16$$

■ *Try Exercise* **70,** *page 460.*

DeMoivre's Theorem can be extended to include finding the nth roots of any number. Recall that if $w^n = z$, then w is the nth root of z. All the nth roots of a complex number can be found by using DeMoivre's Theorem.

Let $w^n = z$, for $z = r(\cos \theta + i \sin \theta)$, and $w = R(\cos \alpha + i \sin \alpha)$. Then by DeMoivre's Theorem,

$$w^n = R^n(\cos n\alpha + i \sin n\alpha).$$

Recall that the sine and cosine functions are periodic functions with a period of 2π or $360°$. Therefore, for k an integer,

$$z = r(\cos \theta + i \sin \theta) = r[\cos(\theta + 360°k) + i \sin(\theta + 360°k)$$

Substituting for w^n and z in the equation $w^n = z$,

$$R^n(\cos n\alpha + i \sin \alpha) = r[\cos(\theta + 360°k) + i \sin(\theta + 360°k)].$$

Two complex numbers written in trigonometric form are equal if and only if the moduli are equal and their arguments are equal. Thus,

$$R^n = r \qquad \text{and} \qquad n\alpha = \theta + 360°k$$

$$R = r^{1/n} \qquad\qquad \alpha = \frac{\theta + 360°k}{n}$$

Since $w = R(\cos \alpha + i \sin \alpha)$, by substituting $r^{1/n}$ for R and $\left(\dfrac{\theta + 360°k}{n}\right)$ for α, we have

$$w = r^{1/n}\left[\cos\left(\frac{\theta + 360°k}{n}\right) + i \sin\left(\frac{\theta + 360°k}{n}\right)\right]$$

DeMoivre's Theorem for Finding Roots

If $z = r(\cos \theta + i \sin \theta)$ is a complex number, then there are n distinct nth roots of z given by the formula

$$w = r^{1/n}\left[\cos\left(\frac{\theta + 360°k}{n}\right) + i \sin\left(\frac{\theta + 360°k}{n}\right)\right]$$

for $k = 0, 1, 2, \ldots, n - 1$.

EXAMPLE 6   **Find Cube Roots by DeMoivre's Theorem**

Find the three cube roots of 27.

Solution   Write 27 in trigonometric form: $27 = 27(\cos 0° + i \sin 0°)$. Then from DeMoivre's Theorem it follows that the cube roots of 27 are

$$= 27^{1/3}\left[\cos\left(\frac{0° + 360°k}{3}\right) + i\,\sin\left(\frac{0° + 360°k}{3}\right)\right] \quad \text{for } k = 0, 1, 2.$$

Find the values of the arguments of the roots for which $k = 0, 1, 2$.

$$k = 0 \qquad \frac{0°}{3} = 0$$

$$k = 1 \qquad \frac{0° + 360°}{3} = 120°$$

$$k = 2 \qquad \frac{0° + 720°}{3} = 240°$$

For $k = 3$, $\dfrac{0° + 1080°}{3} = 360°$. The angles start repeating; thus there are only three cube roots of 27. These roots are shown below and also graphed in Figure 8.51.

$$w_1 = 3(\cos 0° + i \sin 0°) = 3$$

$$w_2 = 3(\cos 120° + i \sin 120°)$$

$$= -\frac{3}{2} + \frac{3\sqrt{3}}{2}i$$

$$w_3 = 3(\cos 240° + i \sin 240°)$$

$$= -\frac{3}{2} - \frac{3\sqrt{3}}{2}i$$

Check

$$3^3 = 27$$

$$[3(\cos 120° + i \sin 120°)]^3 = 27(\cos 360° + 1 \sin 360°)$$

$$= 27(1 + 0) = 27$$

$$[3(\cos 240° + i \sin 240°)]^3 = 27(\cos 720° + i \sin 720°)$$

$$= 27(1 + 0) = 27$$

Note from Figure 8.51 that the arguments of the cube roots of 27 are 0°, 120° and 240° and that $|w| = 3$. This means that the complex numbers representing the cube roots of 27 are equally spaced around a circle of radius 3.

■ *Try Exercise* **78**, *page 460.*

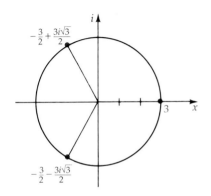

Figure 8.51

EXAMPLE 7   **Find the Fifth Roots of a Complex Number**

Find the fifth roots of $z = 1 + i\sqrt{3}$.

Solution   Write z in trigonometric form $z = r(\cos\theta + i\sin\theta)$.

$$r = \sqrt{1^2 + (\sqrt{3})^2} = 2 \qquad \alpha = \tan^{-1}\left|\frac{b}{a}\right| \quad \alpha \text{ is the reference angle for } \theta.$$

$$= \tan^{-1}\frac{\sqrt{3}}{1} \quad \begin{array}{l}\alpha \text{ is in the first quadrant.}\\ \text{Therefore } \alpha = \theta.\end{array}$$

$$\theta = 60°$$

$$z = 2(\cos 60° + i\sin 60°)$$

From DeMoivre's Theorem, the modulus of each root is $\sqrt[5]{2}$, and the arguments are determined by $\dfrac{60° + 360°k}{5}$, $k = 0, 1, 2, 3, 4$.

$$w = \sqrt[5]{2}\left(\cos\frac{60 + 360°k}{5} + i\sin\frac{60° + 360°k}{5}\right) \quad k = 0, 1, 2, 3, 4$$

$$w = \sqrt[5]{2}[\cos(12° + 72°k) + i\sin(12° + 72°k)] \quad k = 0, 1, 2, 3, 4$$

Substitute for k to find the five roots of z:

$$k = 0 \quad w_1 = \sqrt[5]{2}(\cos 12° + i\sin 12°)$$
$$k = 1 \quad w_2 = \sqrt[5]{2}(\cos 84° + i\sin 84°)$$
$$k = 2 \quad w_3 = \sqrt[5]{2}(\cos 156° + i\sin 156°)$$
$$k = 3 \quad w_4 = \sqrt[5]{2}(\cos 228° + i\sin 228°)$$
$$k = 4 \quad w_5 = \sqrt[5]{2}(\cos 300° + i\sin 300°)$$

■ *Try Exercise* **84,** *page 460.*

Remark   The five roots are graphed in Figure 8.52. The radius of the circle is $\sqrt[5]{2} \approx 1.22$, and the complex numbers are spaced equally around the circle.

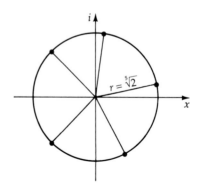

Figure 8.52

EXERCISE SET 8.5

In Exercises 1 to 8, graph the complex numbers. Find the absolute value of each complex number.

1. $z = -2 - 2i$ **2.** $z = 4 - 4i$ **3.** $z = \sqrt{3} - i$

4. $z = 1 + i\sqrt{3}$ **5.** $z = -2i$ **6.** $z = -5$

7. $z = 3 - 5i$ **8.** $z = -5 - 4i$

In Exercises 9 to 16, write the complex number in trigonometric form.

9. $z = 1 - i$ **10.** $z = -4 - 4i$ **11.** $z = \sqrt{3} - i$

12. $z = 1 + \sqrt{3}$ **13.** $z = 3i$ **14.** $z = -2i$

15. $z = -5$ **16.** $z = 3$

In Exercises 17 to 38, write the complex number in standard form.

17. $z = 2(\cos 45° + i \sin 45°)$

18. $z = 3(\cos 240° + i \sin 240°)$

19. $z = \cos 315° + i \sin 315°$

20. $z = 5(\cos 120° + i \sin 120°)$

21. $z = 6(\cos 140° + i \sin 140°)$

22. $z = \cos 305° + i \sin 305°$

23. $z = 8(\cos 0° + i \sin 0°)$

24. $z = 5(\cos 90° + i \sin 90°)$

25. $z = 2\left(\cos \dfrac{5\pi}{6} + i \sin \dfrac{5\pi}{6}\right)$

26. $z = 4\left(\cos \dfrac{5\pi}{3} + i \sin \dfrac{5\pi}{3}\right)$

27. $z = 3\left(\cos \dfrac{3\pi}{2} + i \sin \dfrac{3\pi}{2}\right)$

28. $z = 5(\cos \pi + i \sin \pi)$

29. $z = 8\left(\cos \dfrac{3\pi}{4} + i \sin \dfrac{3\pi}{4}\right)$

30. $z = 9\left(\cos \dfrac{4\pi}{3} + i \sin \dfrac{4\pi}{3}\right)$

31. $z = 9\left(\cos \dfrac{11\pi}{6} + i \sin \dfrac{11\pi}{6}\right)$

32. $z = \cos \dfrac{3\pi}{2} + i \sin \dfrac{3\pi}{2}$

33. $z = 2(\cos 2 + i \sin 2)$

34. $z = 5(\cos 4 + i \sin 4)$

35. $z = 3(\cos 300° + i \sin 300°)$

36. $z = 2(\cos 225° + i \sin 225°)$

37. $z = 8(\cos 210° + i \sin 210°)$

38. $z = 4(\cos 45° + i \sin 45°)$

In Exercises 39 to 48, multiply the complex numbers. Leave the answer in trigonometric form.

39. $2(\cos 30° + i \sin 30°) \cdot 3(\cos 225° + i \sin 225°)$

40. $4(\cos 120° + i \sin 120°) \cdot 6(\cos 315° + i \sin 315°)$

41. $3(\cos 122° + i \sin 122°) \cdot 4(\cos 213° + i \sin 213°)$

42. $5\left(\cos \dfrac{2\pi}{3} + i \sin \dfrac{2\pi}{3}\right) \cdot 2\left(\cos \dfrac{2\pi}{5} + i \sin \dfrac{2\pi}{5}\right)$

43. $2(\cos 49° + i \sin 49°) \cdot 5(\cos 135° + i \sin 135°)$

44. $4(\cos 97° + i \sin 97°) \cdot 7(\cos 279° + i \sin 279°)$

45. $5\left(\cos \dfrac{11\pi}{12} + i \sin \dfrac{11\pi}{12}\right) \cdot 3\left(\cos \dfrac{4\pi}{3} + i \sin \dfrac{4\pi}{3}\right)$

46. $4(\cos 2.4 + i \sin 2.4) \cdot 6(\cos 4.1 + i \sin 4.1)$

47. $3(\cos 45° + i \sin 45°) \cdot 5(\cos 120° + i \sin 120°)$

48. $2(\cos 60° + i \sin 60°) \cdot 6(\cos 150° + i \sin 150°)$

In Exercises 49 to 56, divide the complex numbers. Convert the answers to standard form.

49. $\dfrac{32(\cos 30° + i \sin 30°)}{4(\cos 150° + i \sin 150°)}$ **50.** $\dfrac{15(\cos 240° + i \sin 240°)}{3(\cos 135° + i \sin 135°)}$

51. $\dfrac{12(\cos 2\pi/3 + i \sin 2\pi/3)}{4(\cos 11\pi/6 + i \sin 11\pi/6)}$

52. $\dfrac{10(\cos \pi/3 + i \sin \pi/3)}{5(\cos \pi/4 + i \sin \pi/4)}$

53. $\dfrac{27(\cos 315° + i \sin 315°)}{9(\cos 225° + i \sin 225°)}$ **54.** $\dfrac{9(\cos 25° + i \sin 25°)}{3(\cos 175° + i \sin 175°)}$

55. $\dfrac{25(\cos 189° + i \sin 189°)}{5(\cos 69° + i \sin 69°)}$ **56.** $\dfrac{18 (\cos 420° + i \sin 420°)}{6(\cos 150° + i \sin 150°)}$

In Exercises 57 to 60, perform the indicated operation in trigonometric form. Convert the solution to standard form.

57. $\dfrac{1 + i\sqrt{3}}{1 - i\sqrt{3}}$ **58.** $\dfrac{1 + i}{1 - i}$

59. $\dfrac{\sqrt{3} + i\sqrt{3}}{(1 - i\sqrt{3})(2 - 2i)}$ **60.** $\dfrac{(2 - 2i\sqrt{3})(1 - i\sqrt{3})}{\sqrt{3} + i}$

In Exercises 61 to 74, find the indicated power. Leave the answers in trigonometric form.

61. $[2(\cos 30° + i \sin 30°)]^8$

62. $(\cos 240° + i \sin 240°)^{12}$

63. $[2(\cos 240° + i \sin 240°)]^5$

64. $[2(\cos 45° + i \sin 45°)]^{10}$

65. $[2(\cos 225° + i \sin 225°)]^5$

66. $[3(\cos 330° + i \sin 330°)]^4$

67. $[2(\cos 120° + i \sin 120°)]^6$

68. $[4(\cos 150° + i \sin 150°)]^3$

69. $(1 - i)^{10}$

70. $(1 + i\sqrt{3})^8$

71. $(2 + 2i)^7$

72. $(2\sqrt{3} - 2i)^5$

73. $\left(\dfrac{\sqrt{2}}{2} + i\dfrac{\sqrt{2}}{2}\right)^6$

74. $\left(-\dfrac{\sqrt{2}}{2} + i\dfrac{\sqrt{2}}{2}\right)^{12}$

In Exercises 75 to 88, find the indicated root. Leave all answers in standard form.

75. Square root of $9(\cos 135° + i \sin 135°)$

76. Square root of $16(\cos 45° + i \sin 45°)$

77. Sixth root of $64(\cos 120° + i \sin 120°)$

78. Fifth root of $\cos 315° + i \sin 315°$

79. Fifth root of $\cos 90° + i \sin 90°$

80. Fourth root of $\cos 0° + i \sin 0°$

81. Cube root of 1

82. Cube root of i

83. Fourth root of $1 + i$

84. Fifth root of $-1 + i$

85. Cube root of $2 - 2i\sqrt{3}$

86. Cube root of $-2 + 2i\sqrt{3}$

87. Square root of $-16 + 16i\sqrt{3}$

88. Square root of $-1 + \sqrt{3}i$

In Exercises 89 to 94, find all roots of the equations. Leave the answers in trigonometric form.

89. $x^4 + i = 0$

90. $x^3 - 2i = 0$

91. $x^3 - 27 = 0$

92. $x^5 + 32i = 0$

93. $x^4 - (1 - \sqrt{3}i) = 0$

94. $x^3 + (2\sqrt{3} - 2i) = 0$

Supplemental Exercises

95. Show that the conjugate of $z = r(\cos \theta + i \sin \theta)$ is equal to $\bar{z} = r(\cos \theta - i \sin \theta)$.

96. If $z = r(\cos \theta + i \sin \theta)$, show that
$$z^{-1} = r^{-1}(\cos \theta - i \sin \theta).$$

97. If $z = r(\cos \theta + i \sin \theta)$, show that
$$z^{-2} = r^{-2}(\cos 2\theta - i \sin 2\theta).$$

(Note that Exercises 96 and 97 suggest that the general expression $z^{-n} = r^{-n}(\cos n\theta - i \sin n\theta)$.

98. Use the results of Exercise 97 to find z^{-4} for $z = 1 - i\sqrt{3}$.

99. Raise $(\cos \theta + i \sin \theta)$ to the second power by using DeMoivre's Theorem. Now square $(\cos \theta + i \sin \theta)$ as a binomial. Equate the real and imaginary parts of the two complex numbers and show that:
   a. $\cos 2\theta = \cos^2 \theta - \sin^2 \theta$ **b.** $\sin 2\theta = 2 \sin \theta \cos \theta$.

100. Raise $(\cos \theta + i \sin \theta)$ to the fourth power by using DeMoivre's Theorem. Now find the fourth power of the binomial $(\cos \theta + i \sin \theta)$. Equate the real and imaginary parts of the two complex numbers and show that:
   a. $\cos 4\theta = \cos^2 2\theta - 2 \sin^2 2\theta$,
   b. $\sin 4\theta = 4 \cos^3 \theta \sin \theta - 4 \cos \sin^3 \theta$.

Chapter 8 Review

8.1   Right Triangles

To solve a right triangle means to find the lengths of all the sides and the measures of all the angles of the triangle.

8.2   The Law of Sines

The Law of Sines is used to solve general triangles when two angles and the included side are given or in triangles in which two sides and an angle opposite one of them are given.

$$\frac{a}{\sin A} = \frac{b}{\sin B} = \frac{c}{\sin C}$$

8.3 The Law of Cosines and Area

The Law of Cosines $a^2 = b^2 + c^2 - 2bc \cos A$ is used to solve general triangles when two sides and the included angle or three sides of the triangle are given.

Area K of a triangle ABC is

$$K = \frac{1}{2}bc \sin A = \frac{b^2 \sin C \sin A}{2 \sin B}.$$

Area for a triangle in which three sides are given (Heron's formula):

$$K = \sqrt{s(s-a)(s-b)(s-c)}, \quad \text{where } s = \frac{1}{2}(a + b + c).$$

8.4 Vectors

A vector is a quantity with magnitude and direction. Two vectors are equal if they have the same magnitude and direction. The resultant of two or more vectors is the single vector that will have the equivalent effect of the vectors.

Vectors can be added by parallelogram addition, triangle addition, or by adding the x- and y-components.

Multiplication of a vector by a scalar is given by

$$a(x\mathbf{i} + y\mathbf{j}) = ax\mathbf{i} + ay\mathbf{j}.$$

The dot product of two vectors $\mathbf{U} = a\mathbf{i} + b\mathbf{j}$ and $\mathbf{V} = c\mathbf{i} + d\mathbf{j}$ is given by

$$\mathbf{U} \cdot \mathbf{V} = (a\mathbf{i} + b\mathbf{j}) \cdot (c\mathbf{i} + d\mathbf{j}) = ac + bd$$

The cosine of the angle α between two vectors is given by

$$\cos \alpha = \frac{\mathbf{U} \cdot \mathbf{V}}{|\mathbf{U}||\mathbf{V}|}$$

8.5 Complex Numbers and DeMoivre's Theorem

The standard form of a complex number is $z = a + bi$.

The trigonometric form of a complex number is $z = r(\cos \theta + i \sin \theta)$.

The product of two complex numbers in trigonometric form is given by:

$$z_1 z_2 = r_1 r_2 [\cos(\theta_1 + \theta_2) + i \sin(\theta_1 + \theta_2)].$$

The quotient of two complex numbers $z_1 = r_1(\cos \theta_1 + i \sin \theta_1)$ and $z_2 = r_2(\cos \theta_2 + i \sin \theta_2)$ in trigonometric form is given by:

$$\frac{z_1}{z_2} = \frac{r_1}{r_2}[\cos \theta_1 - \theta_2) + i \sin(\theta_1 - \theta_2)].$$

DeMoivre's Theorem

If $z = r(\cos \theta + i \sin \theta)$ is a complex number and n is a positive integer, then

$$z^n = r^n(\cos n\theta + i \sin n\theta).$$

If $w^n = z$, then

$$w = r^{1/n}\left[\cos\left(\frac{\theta + 360°k}{n}\right) + i \sin\left(\frac{\theta + 360°k}{n}\right)\right], \quad k = 0, 1, 2, \ldots, n - 1.$$

CHALLENGE EXERCISES

For Exercises 1 to 16, answer true or false. If the statement is false, give an example.

1. The Law of Cosines can be used to solve any triangle given two sides and an angle.

2. The law of sines can be used to solve any triangle given two angles and any side.

3. In any triangle, the largest side is opposite the largest angle.

4. If two vectors have the same magnitude, then they are equal.

5. It is possible for the sum of two nonzero vectors to equal zero.

6. The expression $a^2 = b^2 + c^2 + 2bc \cos D$ is true for triangle ABC in which angle D is the supplement of angle A.

7. The measure of angle α formed by two vectors is greater than or equal to $0°$ and less than or equal to $180°$.

8. If A, B, and C are the angles of a triangle, then
$$\sin(A + B + C) = 0.$$

9. Real numbers are complex numbers.

10. Let $\mathbf{V} = a\mathbf{i} + b\mathbf{j}$; then $\mathbf{V} \cdot \mathbf{V} = a^2\mathbf{i} + b^2\mathbf{j}$.

11. If $\mathbf{V}$ and $\mathbf{W}$ are vectors with $\mathbf{V} \cdot \mathbf{W} = 0$, then $\mathbf{V} = 0$ or $\mathbf{W} = 0$.

12. The n roots of a complex number can be graphed on a circle and are equally spaced around the circle.

13. Let $z = r(\cos \theta + i \sin \theta)$, then $z^2 = r^2(\cos^2\theta + i \sin^2\theta)$.

14. $|a + bi| = \sqrt{a^2 + b^2}$.

15. $i = \cos \pi + i \sin \pi$.

16. $z = \cos 45° + i \sin 45°$ is a square root of i.

REVIEW EXERCISES

In Exercises 1 to 10, solve the triangles.

1. $A = 37°$, $b = 14$, $C = 90°$
2. $B = 77.4°$, $c = 11.8$, $C = 90°$
3. $a = 12$, $b = 15$, $c = 20$
4. $a = 24$, $b = 32$, $c = 28$
5. $a = 18$, $b = 22$, $C = 35°$
6. $b = 102$, $c = 150$, $A = 82°$
7. $A = 105°$, $a = 8$, $c = 10$
8. $C = 55°$, $c = 80$, $b = 110$
9. $A = 55°$, $B = 80°$, $c = 25$
10. $B = 25°$, $C = 40°$, $c = 40$

In Exercises 11 to 18, find the area of each triangle.

11. $a = 24$, $b = 30$, $c = 36$
12. $a = 9.0$, $b = 7.0$, $c = 12$
13. $a = 60$, $b = 44$, $C = 44°$
14. $b = 8.0$, $c = 12$, $A = 75°$
15. $b = 50$, $c = 75$, $C = 15°$
16. $b = 18$, $a = 25$, $A = 68°$
17. $A = 110°$, $a = 32$, $b = 15$
18. $C = 45°$, $c = 22$, $b = 18$

In Exercises 19 to 26, find the magnitude and direction of the given vector.

19. $\mathbf{U} = \langle 4, -2 \rangle$
20. $\mathbf{U} = \langle -2, -5 \rangle$
21. $\mathbf{V} = \langle -4, 2 \rangle$
22. $\mathbf{V} = \langle 6, -3 \rangle$
23. $\mathbf{U} = -2\mathbf{i} + 3\mathbf{j}$
24. $\mathbf{U} = -4\mathbf{i} - 7\mathbf{j}$

25. $V = 5i + j$ **26.** $V = 3i - 5j$

In Exercises 27 to 30, find the magnitude of the x- and y-components of the following vectors.

27. $U = \langle -8, 5 \rangle$ **28.** $U = \langle 7, -12 \rangle$

29. $V = 10i + 6j$ **30.** $V = 8i - 5j$

In Exercises 31 to 34, perform the indicated operation. $U = 3i + 2j$, $V = -4i - j$.

31. $V - U$ **32.** $2U - 3V$

33. $-U + \dfrac{1}{2}V$ **34.** $\dfrac{2}{3}V - \dfrac{3}{4}U$

In Exercises 35 to 38, find the dot product of the vectors.

35. $U = \langle 3, -2 \rangle$, $V = \langle -1, 3 \rangle$

36. $V = \langle -8, 5 \rangle$, $U = \langle 2, -1 \rangle$

37. $V = -4i - j$, $U = 2i + j$

38. $U = -3i + 7j$, $V = -2i + 2j$

In Exercises 39 to 42, find the angle between the vectors.

39. $U = \langle 7, -4 \rangle$, $V = \langle 2, 3 \rangle$

40. $V = \langle -5, 2 \rangle$, $U = \langle 2, -4 \rangle$

41. $V = 6i - 11j$, $U = 2i + 4j$

42. $U = i - 5j$, $V = i + 5j$

In Exercises 43 and 44, find the modulus and the argument and graph the complex numbers.

43. $z = 2 - 3i$ **44.** $z = -5 + i\sqrt{3}$

In Exercises 45 and 46, write the complex numbers in trigonometric form.

45. $z = 2 - 2i$ **46.** $z = -\sqrt{3} + 3i$

In Exercises 47 and 48, write the complex number in standard form.

47. $z = 5(\cos 315° + i \sin 315°)$

48. $z = 6\left(\cos \dfrac{4\pi}{3} + i \sin \dfrac{4\pi}{3} \right)$

In Exercises 49 to 51, multiply the complex numbers. Convert the answers to standard form.

49. $5(\cos 162° + i \sin 162°) \cdot 2(\cos 63° + i \sin 63°)$

50. $3(\cos 12° + i \sin 12°) \cdot 4(\cos 126° + i \sin 126°)$

51. $3(\cos 1.8 + i \sin 1.8) \cdot 5(\cos 2.5 + i \sin 2.5)$

In Exercises 52 to 55, divide the complex numbers. Leave the answers in trigonometric form.

52. $\dfrac{6(\cos 50° + i \sin 50°)}{2(\cos 150° + i \sin 150°)}$ **53.** $\dfrac{30(\cos 165° + i \sin 165°)}{10(\cos 55° + i \sin 55°)}$

54. $\dfrac{40(\cos 66° + i \sin 66°)}{8(\cos 125° + i \sin 125°)}$ **55.** $\dfrac{\sqrt{3} - i}{1 + i}$

In Exercises 56 to 59, find the indicated power. Leave the answers in standard form.

56. $[3(\cos 45° + i \sin 45°)]^5$ **57** $\left(\cos \dfrac{11\pi}{6} + i \sin \dfrac{11\pi}{6} \right)^8$

58. $(1 - i\sqrt{3})^7$ **59.** $(-2 - 2i)^{10}$

In Exercises 60 to 63, find the indicated roots. Leave the answers in trigonometric form.

60. Cube roots of $27i$ **61.** Fourth roots of $8i$

62. Fourth roots of $256(\cos 120° + i \sin 120°)$

63. Fifth roots of $-1 - i$

9

Topics in Analytic Geometry

My thesis, then, is that the essence of analytical geometry is the study of loci by means of their equations, and that this was known to the Greeks and was the basis of their study in conic sections.

J. L. COOLIDGE

KIDNEY STONES AND ELLIPSES

The conic sections, some of the curves we will study in this chapter, were studied by the ancient Greeks. The names of the curves are derived by looking at various sections of a right circular cone.

One of the conic sections is an ellipse. Besides applications of the ellipse to astronomy and optics, the ellipse now has an application to medicine. The ellipse is used in a nonsurgical treatment of kidney stones. To understand this application, we will introduce some properties of an ellipse.

An ellipse is an oval shaped curve. Inside the ellipse there are two points called foci of the ellipse. Sound or light waves emitted from one focus are reflected off the surface of the ellipse to the other focus. In an analagous way, think of a billiard table shaped like an ellipse. A ball struck from one focus would pass through the other focus no matter the direction of the initial shot.

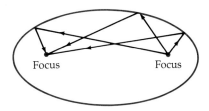

The treatment of kidney stones is based on this reflective property of an ellipse. An electrode is placed at one focus of an ellipse and the patient is placed at the other focus. When the electrode is discharged, ultrasound waves are produced. These waves hit the walls of the ellipse and are reflected to the kidney stone. Because of the reflective property of the ellipse, there is very little energy loss. As a result, it is as if the electrode actually discharges at the kidney stone. The energy of the discharge pulverizes the kidney stone thereby allowing the fragments to be passed through the system.

Research evaluation of this method has been encouraging. Hospitals are reporting success rates of 90% to 95%. Besides successfully treating the condition, the patient spends less time in the hospital and the complete recovery time is shorter.

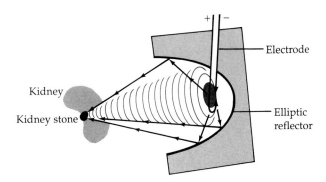

9.1

Conic Sections

The graph of a parabola, circle, ellipse, or a hyperbola can be formed by the intersection of a plane and a double cone. Hence these figures are referred to as **conic sections.**

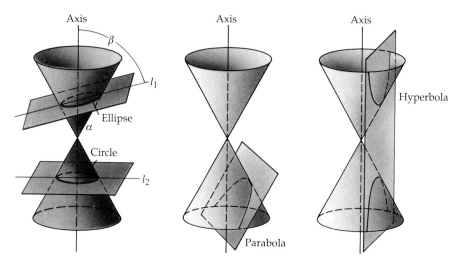

Figure 9.1
Cones intersected by planes.

A plane perpendicular to the axis of the cone intersects the cone in a circle (plane C). The plane E, tilted so that it is not perpendicular to the axis, intersects the cone in an ellipse. When the plane is parallel to a line on the surface of the cone, the plane intersects the cone in a parabola. When the plane intersects both parts of the cone, a hyperbola is formed.

Parabolas with Vertex at $(0, 0)$

Besides the geometric description of a conic section just given, a conic can be defined as a set of points. This method uses some specified conditions about the curve to determine which points in a coordinate system are points of the graph. For example, a parabola can be defined by the following set of points.

Definition of a Parabola

> A **parabola** is the set of points in the plane that are equidistance from a fixed line (the **directrix**) and a fixed point (the **focus**) not on the directrix.

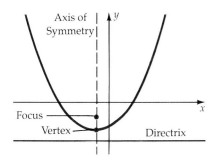

Figure 9.2

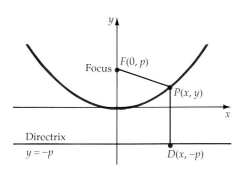

Figure 9.3

The line that passes through the focus and is perpendicular to the directrix is called the **axis of symmetry** of the parabola. The midpoint of the line segment between the focus and directrix on the axis of symmetry is the **vertex** of the parabola as shown in Figure 9.2.

Using this definition of a parabola, we can determine an equation of a parabola. Suppose that the coordinates of the vertex of a parabola are $(0, 0)$ and the axis of symmetry is the y-axis. The equation of the directrix is $y = -p$, $p > 0$. The focus lies on the axis of symmetry and is the same distance from the vertex as the vertex is from the directrix. Thus the coordinates of the focus are $(0, p)$ as shown in Figure 9.3.

Let $P(x, y)$ be any point on the parabola. Then, using the distance formula and the fact that the distance between any point on the parabola and the focus is equal to the distance from the point P to the directrix (why?[1]), we can write the equation

$$d(P, F) = d(P, D).$$

By the distance formula,

$$\sqrt{(x - 0)^2 + (y - p)^2} = y + p$$

Now squaring each side and simplifying,

$$(\sqrt{(x - 0)^2 + (y - p)^2})^2 = (y + p)^2$$

$$x^2 + y^2 - 2py + p^2 = y^2 + 2py + p^2$$

$$x^2 = 4py.$$

This is an equation of a parabola with a vertex at the origin and a vertical axis of symmetry. The equation of a parabola with a horizontal axis of symmetry is derived in a similar manner.

Standard Form of the Equation of a Parabola with Vertex at the Origin

Vertical Axis of Symmetry
The standard form of the equation of a parabola with vertex $(0, 0)$ and vertical axis of symmetry is $x^2 = 4py$. The focus is $(0, p)$, and the equation of the directrix is $y = -p$.

Horizontal Axis of Symmetry
The standard form of the equation of a parabola with vertex $(0, 0)$ and horizontal axis of symmetry is $y^2 = 4px$. The focus is $(p, 0)$, and the equation of the directrix is $x = -p$.

Remark   In the equation $x^2 = 4py$, $x^2 \geq 0$. Therefore $4py \geq 0$. Thus if $p > 0$, then $y \geq 0$ and the parabola opens up. If $p < 0$, then $y \leq 0$ and the parabola opens down. A similar analysis shows that for $y^2 = 4px$, the parabola opens to the right when $p > 0$ and opens to the left when $p < 0$.

---

[1] From the definition of a parabola, the distance from a point on the parabola to the focus is the same as the distance from the point to the directrix.

EXAMPLE 1   **Find the Focus and Directrix of a Parabola**

Find the focus and directrix of the parabola given by the equation $y = -\dfrac{1}{2}x^2$.

Solution   Because the x term is squared, the standard form of the equation is $x^2 = 4py$. Write the given equation in standard form:

$$y = -\frac{1}{2}x^2$$

$$x^2 = -2y.$$

Comparing this equation with the equation in standard form gives

$$4p = -2$$

or
$$p = -\frac{1}{2}.$$

Because p is negative, the parabola will open down and the focus will be below the vertex $(0, 0)$.

The coordinates of the focus are $\left(0, -\dfrac{1}{2}\right)$. The equation of the directrix is $y = \dfrac{1}{2}$.

■ *Try Exercise **4**, page 471.*

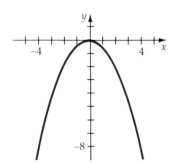

Figure 9.4
$y = -\dfrac{1}{2}x^2$

EXAMPLE 2   **Find the Equation of a Parabola in Standard Form**

Find the equation of the parabola in standard form with vertex at the origin and focus at $(-2, 0)$.

Solution   Because the vertex is at $(0, 0)$ and the focus is at $(-2, 0)$, $p = -2$. The graph of the parabola opens toward the focus and thus, in this case, the parabola is opening to the left. The equation of the parabola in standard form that opens to the left is $y^2 = 4px$. Substitute -2 for p in this equation and simplify:

$$y^2 = 4(-2)x = -8x$$

The equation is $y^2 = -8x$.

■ *Try Exercise **28**, page 471.*

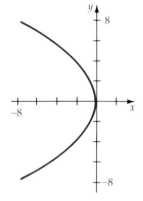

Figure 9.5
$y^2 = -8x$

Parabolas with the Vertex at (h, k)

The equation of a parabola with vertex at a point (h, k) can be found by using the translations discussed previously. Consider a coordinate system with coordinate axes labeled x' and y' placed so that its origin is at (h, k) of the xy-coordinate system.

The relationship between an ordered pair in the $x'y'$-coordinate sys-

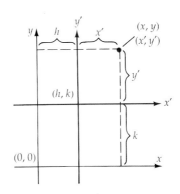

Figure 9.6
$x = x' + h, \; y = y' + k$

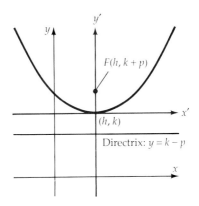

Figure 9.7

tem and the xy-coordinate system is given by the transformation equations

$$x' = x - h$$
$$y' = y - k \tag{1}$$

Now consider a parabola that opens up with vertex at (h, k). Place a new coordinate system labeled x' and y' with its origin at (h, k). The equation of a parabola in the $x'y'$-coordinate system is

$$(x')^2 = 4py', \quad p > 0. \tag{2}$$

Using the transformation equations (1), we can substitute the expressions for x' and y' into Equation (2). The standard form of the equation of the parabola with vertex (h, k) and a vertical axis of symmetry is

$$(x - h)^2 = 4p(y - k)$$

Similarly, we can derive the standard form of the equation of the parabola with vertex (h, k) and a horizontal axis of symmetry.

Standard Form of the Equation of a Parabola with Vertex at (h, k)

Vertical Axis of Symmetry
The standard form of the equation of the parabola with vertex (h, k) and vertical axis of symmetry is

$$(x - h)^2 = 4p(y - k)$$

The focus is $(h, k + p)$, and the equation of the directrix is $y = k - p$.

Horizontal Axis of Symmetry
The standard form of the equation of the parabola with vertex (h, k) and horizontal axis of symmetry is

$$(y - k)^2 = 4p(x - h)$$

The focus is $(h + p, k)$, and the equation of the directrix is $x = h - p$.

EXAMPLE 3 **Find the Focus and Directrix of a Parabola**

Find the equation of the directrix and the coordinates of the vertex and focus of the parabola given by the equation $3x + 2y^2 + 8y - 4 = 0$.

Solution   Rewrite the equation and then complete the square.

$$3x + 2y^2 + 8y - 4 = 0$$
$$2y^2 + 8y = -3x + 4$$
$$2(y^2 + 4y) = -3x + 4$$
$$2(y^2 + 4y + 4) = -3x + 4 + 8 \quad \text{Complete the square. Note that}$$
$$2(y + 2)^2 = -3(x - 4) \qquad \quad 2 \cdot 4 = 8 \text{ is added to each side.}$$

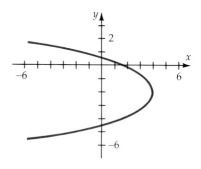

Figure 9.8
$(y + 2)^2 = -\dfrac{3}{2}(x - 4)$

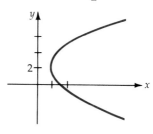

Figure 9.9
$(y - 2)^2 = 8(x - 1)$

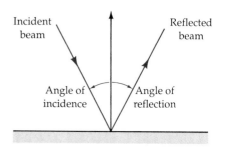

Figure 9.10

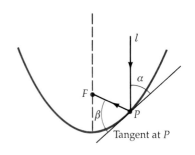

Figure 9.11

Write the equation in standard form.

$$(y + 2)^2 = -\dfrac{3}{2}(x - 4)$$

Comparing this equation to $(y - k)^2 = 4p(x - h)$, we have a parabola that opens to the left with vertex $(4, -2)$ and $4p = -\dfrac{3}{2}$. Thus $p = -\dfrac{3}{8}$.

The coordinates of the focus are $\left(4 + \left(-\dfrac{3}{8}\right), -2\right) = \left(\dfrac{29}{8}, -2\right)$. The equation of the directrix is $x = 4 - \left(-\dfrac{3}{8}\right) = \dfrac{35}{8}$.

Choosing some values for y and finding the corresponding values for x, we plot a few points. We use the fact that the line $y = -2$ is the axis of symmetry. Thus, for a point on one side of the axis of symmetry, there is a corresponding point on the other side. Two points are $(-2, 1)$ and $(-2, -5)$.

■ *Try Exercise 20, page 471.*

EXAMPLE 4   **Find the Equation of a Parabola in Standard Form**

Find the equation in standard form of the parabola with directrix $x = -1$ and focus $(3, 2)$.

Solution   The vertex is the midpoint of the line segment joining $(3, 2)$ and the point $(-1, 2)$ on the directrix.

$$(h, k) = \left(\dfrac{-1 + 3}{2}, \dfrac{2 + 2}{2}\right) = (1, 2).$$

The standard form of the equation will be of the form $(y - k)^2 = 4p(x - h)$. The distance from the vertex to the focus is 2. Thus $4p = 4(2) = 8$ and the equation of the parabola is

$$(y - 2)^2 = 8(x - 1).$$

■ *Try Exercise 30, page 471.*

A principle of physics states that when light is reflected from a point P on a surface, the angle of incidence (incoming ray) equals the angle of reflection (outgoing ray). This principle applied to parabolas has some useful consequences.

Optical Property of a Parabola

The line tangent to a parabola at a point P makes equal angles with the line through P and parallel to the axis of symmetry and the line through P and the focus of the parabola (see Figure 9.11).

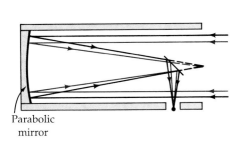

Parabolic
mirror

Figure 9.12   **Figure 9.13**

The reflecting mirror of a telescope is designed in the shape of a parabola. All the incoming light is reflected from the surface of the mirror and to the focus. See Figure 9.12.

Flashlights and car headlights also make use of this property. The light bulb is positioned at the focus of the parabolic reflector, causing the reflected light to be reflected outward in parallel rays. See Figure 9.13.

EXERCISE SET 9.1

In Exercises 1 to 26, find the vertex, focus, and directrix of the parabola given by each equation. Sketch the graph.

1. $x^2 = -4y$

2. $2y^2 = x$

3. $y^2 = \dfrac{1}{3}x$

4. $x^2 = -\dfrac{1}{4}y$

5. $(x - 2)^2 = 8(y + 3)$

6. $(y + 1)^2 = 6(x - 1)$

7. $(y + 4)^2 = -4(x - 2)$

8. $(x - 3)^2 = -(y + 2)$

9. $(y - 1)^2 = 2x + 8$

10. $(x + 2)^2 = 3y - 6$

11. $(2x - 4)^2 = 8y - 16$

12. $(3x + 6)^2 = 18y - 36$

13. $x^2 + 8x - y + 6 = 0$

14. $x^2 - 6x + y + 10 = 0$

15. $x + y^2 - 3y + 4 = 0$

16. $x - y^2 - 4y + 9 = 0$

17. $2x - y^2 - 6y + 1 = 0$

18. $3x + y^2 + 8y + 4 = 0$

19. $x^2 + 3x + 3y - 1 = 0$

20. $x^2 + 5x - 4y - 1 = 0$

21. $2x^2 - 8x - 4y + 3 = 0$

22. $6x - 3y^2 - 12y + 4 = 0$

23. $2x + 4y^2 + 8y - 5 = 0$

24. $4x^2 - 12x + 12y + 7 = 0$

25. $3x^2 - 6x - 9y + 4 = 0$

26. $2x - 3y^2 + 9y + 5 = 0$

27. Find the equation in standard form of the parabola with vertex at the origin and focus $(0, -4)$.

28. Find the equation in standard form of the parabola with vertex at the origin and focus $(5, 0)$.

29. Find the equation in standard form of the parabola with vertex at $(-1, 2)$ and focus $(-1, 3)$.

30. Find the equation in standard form of the parabola with vertex at $(2, -3)$ and focus $(0, -3)$.

31. Find the equation in standard form of the parabola with focus $(3, -3)$ and directrix $y = -5$.

32. Find the equation in standard form of the parabola with focus $(-2, 4)$ and directrix $x = 4$.

33. Find the equation in standard form of the parabola with vertex $(-4, 1)$, axis of symmetry parallel to the y-axis, and passing through the point $(-2, 2)$.

34. Find the equation in standard form of the parabola with vertex $(3, -5)$, axis of symmetry parallel to the x-axis, and passing through the point $(4, 3)$.

Supplemental Exercises

In Exercises 35 to 37, use the following definition of latus rectum: the line segment with endpoints on the parabola, through the focus of a parabola and perpendicular to the axis of symmetry is called the **latus rectum** of the parabola.

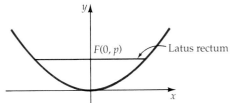

35. Find the length of the latus rectum for the parabola $x^2 = 4y$.

36. Find the length of the latus rectum for the parabola $y^2 = -8x$.

37. Find the length of the latus rectum for any parabola in terms of $|p|$, the distance from the vertex of the parabola to the focus.

The result of Exercise 37 can be stated as the following theorem. **Theorem** Two points on a parabola will be $2|p|$ units on each side of the axis of symmetry on the line through the focus and perpendicular to that axis.

38. Use the theorem to sketch a graph of the parabola given by the equation $(x - 3)^2 = 2(y + 1)$.

39. Use the theorem to sketch a graph of the parabola given by the equation $(y + 4)^2 = -(x - 1)$.

40. Use the theorem to sketch a graph of the parabola given by the equation $4x - y^2 + 8y = 0$.

41. Show that the point on the parabola closest to the focus is the vertex. (*Hint:* Consider the parabola $x^2 = 4py$ and a point on the parabola (a, b). Find the square of the distance between the point (a, b) and the focus. You may want to review the technique of minimizing a quadratic expression.)

42. By using the definition for a parabola, find the equation in standard form of the parabola with $V(0, 0)$, $F(-c, 0)$, and directrix $x = c$.

43. Sketch a graph of $4(y - 2) = (x|x| - 1)$.

44. Find the equation of the directrix of the parabola with vertex at the origin and whose focus is the point $(1, 1)$.

45. Find the equation of the parabola with vertex at the origin and focus at the point $(1, 1)$. (*Hint:* You will need the answer to Exercise 44 and the definition of a parabola.)

9.2

Ellipses

An ellipse is another of the conic sections formed when a plane intersects a right circular cone. If β is the angle at which the plane intersects the axis of the cone and α is the angle shown in Figure 9.14, an ellipse is formed when $\alpha < \beta < 90°$. If $\beta = 90°$, then a circle is formed.

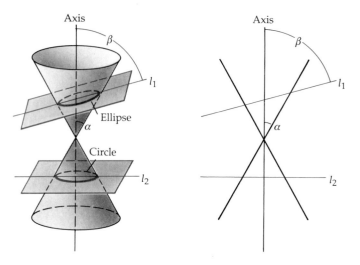

Figure 9.14

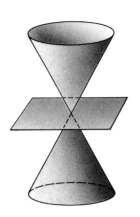

Figure 9.15

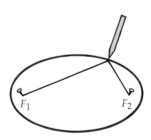

Figure 9.16

Remark If the plane were to intersect the cone at the vertex of the cone so that the resulting figure is a point, the point is a **degenerate ellipse.**

As was the case for a parabola, there is a definition for an ellipse in terms of a certain set of points in the plane.

Definition of an Ellipse

> An **ellipse** is the set of all points in the plane, the sum of whose distances from two fixed points (foci) is a positive constant.

This definition can be used to draw an ellipse using a piece of string and two tacks (see Figure 9.16). Tack the ends of the string to the foci, and trace a curve with a pencil held tight against the string. The resulting curve is an ellipse. The positive constant is the length of the string.

Ellipses with Center at $(0, 0)$

The graph of an ellipse is oval-shaped, with two axes of symmetry (see Figure 9.17). The longer axis is called the **major axis.** The foci of the ellipse are on the major axis. The shorter axis is called the **minor axis.** It is customary to denote the length of the major axis as $2a$ and the length of the minor axis as $2b$. The length of the **semiaxes** are one-half the axes. Thus the length of the semimajor axis is denoted by a and the length of the semiminor axis by b. The **center** of the ellipse is the midpoint of the major axis. The endpoints of the major axis are the **vertices** (plural of vertex) of the ellipse.

Consider the point $(a, 0)$, which is one vertex of an ellipse, and the point $(c, 0)$ and $(-c, 0)$, which are the foci of the ellipse shown in Figure 9.18. The distance from $(a, 0)$ to $(c, 0)$ is $a - c$. Similarly, the distance from $(a, 0)$ to $(-c, 0)$ is $a + c$. From the definition of an ellipse, the sum of distances from any point on the ellipse to the foci is a constant. By adding the expressions $a - c$ and $a + c$, we have

$$(a - c) + (a + c) = 2a$$

Thus the constant is precisely the length of the major axis.

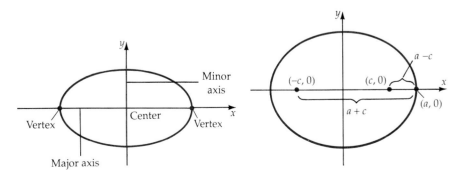

Figure 9.17   **Figure 9.18**

Now let $P(x, y)$ be any point on the ellipse. By using the definition of an ellipse, we have

$$d(P, F_1) + d(P, F_2) = 2a$$
$$\sqrt{(x + c)^2 + y^2} + \sqrt{(x - c)^2 + y^2} = 2a$$

Subtract the second radical from each side of the equation and then square each side.

$$[\sqrt{(x + c)^2 + y^2}]^2 = [2a - \sqrt{(x - c)^2 + y^2}]^2$$
$$(x + c)^2 + y^2 = 4a^2 - 4a\sqrt{(x - c)^2 + y^2} + (x - c)^2 + y^2$$
$$x^2 + 2cx + c^2 + y^2 = 4a^2 - 4a\sqrt{(x - c)^2 + y^2} + x^2 - 2cx + c^2 + y^2$$
$$4cx - 4a^2 = -4a\sqrt{(x - c)^2 + y^2}$$

Divide each side by -4 and then square each side again.

$$[-cx + a^2]^2 = [a\sqrt{(x - c)^2 + y^2}]^2$$
$$c^2x^2 - 2cxa^2 + a^4 = a^2x^2 - 2cxa^2 + a^2c^2 + a^2y^2$$

Simplify and then rewrite with x and y terms on the left side.

$$-a^2x^2 + c^2x^2 - a^2y^2 = -a^4 + a^2c^2$$
$$-(a^2 - c^2)x^2 - a^2y^2 = -a^2(a^2 - c^2) \quad \text{Factor and let } b^2 = a^2 - c^2.$$
$$-b^2x^2 - a^2y^2 = -a^2b^2 \qquad \text{Divide each side by } -a^2b^2.$$
$$\frac{x^2}{a^2} + \frac{y^2}{b^2} = 1 \qquad \text{An equation of an ellipse}$$

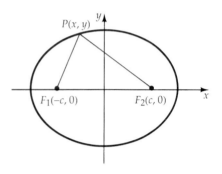

Figure 9.19

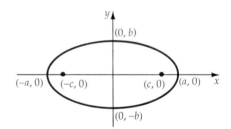

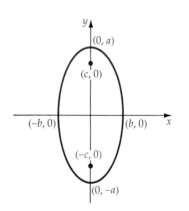

Figure 9.20

Standard Forms of the Equation of an Ellipse with Center at the Origin

Major Axis on the x-axis
The standard form of the equation of an ellipse with the center at the origin and major axis on the x-axis is given by

$$\frac{x^2}{a^2} + \frac{y^2}{b^2} = 1 \quad a > b.$$

The coordinates of the vertices are $(a, 0)$ and $(-a, 0)$, and the coordinates of the foci are $(c, 0)$ and $(-c, 0)$, where $c^2 = a^2 - b^2$.

Major Axis on the y-axis
The standard form of the equation of an ellipse with the center at the origin and major axis on the y-axis is given by

$$\frac{x^2}{b^2} + \frac{y^2}{a^2} = 1 \quad a > b.$$

The coordinates of the vertices are $(0, a)$ and $(0, -a)$, and the coordinates of the foci are $(0, c)$ and $(0, -c)$, where $c^2 = a^2 - b^2$.

Remark By looking at the standard form of the equations of an ellipse and noting that $a > b$, observe that the orientation of the major axis is determined by the larger denominator. When the x^2 term has the larger denominator, the major axis is on the x-axis. When the y^2 term has the larger denominator, the major axis is on the y-axis.

EXAMPLE 1 **Find the Vertices and Foci of an Ellipse**

Find the vertices and foci of the ellipse given by the equation $\dfrac{x^2}{25} + \dfrac{y^2}{49} = 1$. Sketch the graph.

Solution Because the y^2 term has the larger denominator, the major axis is on the y-axis.

$$a^2 = 49 \qquad b^2 = 25 \qquad c^2 = a^2 - b^2$$
$$a = 7 \qquad b = 5 \qquad = 49 - 25 = 24$$
$$c = \sqrt{24} = 2\sqrt{6}$$

The vertices are $(0, 7)$ and $(0, -7)$. The foci are $(0, 2\sqrt{6})$ and $(0, -2\sqrt{6})$.

■ *Try Exercise 8, page 479.*

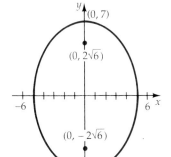

Figure 9.21
$$\frac{x^2}{25} + \frac{y^2}{49} = 1$$

EXAMPLE 2 **Find the Equation of an Ellipse**

Find the equation in standard form of the ellipse with foci $(3, 0)$ and $(-3, 0)$ and major axis of length 10. Sketch the graph.

Solution Because the foci are on the major axis, the major axis is on the x-axis. The length of the major axis is $2a$.

Thus $2a = 10$. Solving for a, we have $a = 5$.

Because the foci are $(3, 0)$ and $(-3, 0)$ and the center of the ellipse is the midpoint between the two foci, the distance from the center of the ellipse to a focus is 3. Therefore, $c = 3$.

To find b^2, use the equation $c^2 = a^2 - b^2$.

$$9 = 25 - b^2$$
$$b^2 = 16$$

The equation of the ellipse is $\dfrac{x^2}{25} + \dfrac{y^2}{16} = 1$.

■ *Try Exercise 22, page 480.*

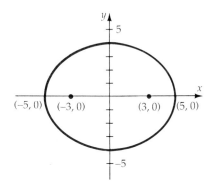

Figure 9.22
$$\frac{x^2}{25} + \frac{y^2}{16} = 1$$

Ellipses with the Center at (h, k)

The equation of an ellipse with center (h, k) and with horizontal or vertical major axes can be found by using a translation of coordinates. Given a

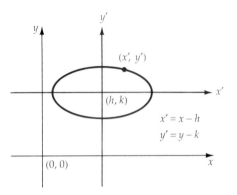

Figure 9.23

coordinate system with axes labeled x' and y', the standard form of the equation of an ellipse with center at the origin is

$$\frac{(x')^2}{a^2} + \frac{(y')^2}{b^2} = 1.$$

Now place the origin of the $x'y'$-coordinate system at (h, k) in an xy-coordinate system.

The relationship between an ordered pair in the $x'y'$-coordinate system and the xy-coordinate system is given by transformation equations

$$x' = x - h$$

$$y' = y - k$$

Substitute the expressions for x' and y' into the equation of an ellipse. The equation of the ellipse with center at (h, k) is

$$\frac{(x - h)^2}{a^2} + \frac{(y - k)^2}{b^2} = 1.$$

Standard Form of Ellipses with Center at (h, k)

Major Axis Parallel to the x-axis
The standard form of the equation of an ellipse with the center at (h, k) and major axis parallel to the x-axis is given by

$$\frac{(x - h)^2}{a^2} + \frac{(y - k)^2}{b^2} = 1 \quad a > b.$$

The coordinates of the vertices are $(h + a, k)$ and $(h - a, k)$, and the coordinates of the foci are $(h + c, k)$ and $(h - c, k)$, where $c^2 = a^2 - b^2$.

Major Axis Parallel to the y-axis
The standard form of the equation of an ellipse with the center at (h, k) and major axis parallel to the y-axis is given by

$$\frac{(x - h)^2}{b^2} + \frac{(y - k)^2}{a^2} = 1 \quad a > b.$$

The coordinates of the vertices are $(h, k + a)$ and $(h, k - a)$, and the coordinates of the foci are $(h, k + c)$ and $(h, k - c)$, where $c^2 = a^2 - b^2$.

EXAMPLE 3   Find the Vertices and Foci of an Ellipse

Find the vertices and foci of the ellipse $4x^2 + 9y^2 - 8x + 36y + 4 = 0$. Sketch the graph.

Solution   Write the equation of the ellipse in standard form by completing the square.

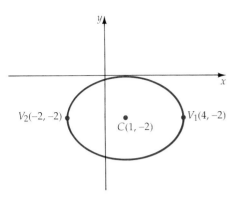

Figure 9.24
$$\frac{(x - 1)^2}{9} + \frac{(y + 2)^2}{4} = 1$$

$$4x^2 + 9y^2 - 8x + 36y + 4 = 0$$

$4x^2 - 8x + 9y^2 + 36y = -4$ Rearrange terms.

$4(x^2 - 2x) + 9(y^2 + 4y) = -4$ Factor.

$4(x^2 - 2x + 1) + 9(y^2 + 4y + 4) = -4 + 4 + 36$ Complete the square

$4(x - 1)^2 + 9(y + 2)^2 = 36$ Factor.

$$\frac{(x - 1)^2}{9} + \frac{(y + 2)^2}{4} = 1$$ Divide by 36.

From the equation of the ellipse in standard form, we see that the coordinates of the center of the ellipse are $(1, -2)$. Because the larger denominator is 9, the major axis is parallel to the x-axis and $a^2 = 9$. Thus $a = 3$. The vertices are $(4, -2)$ and $(-2, -2)$.

To find the coordinates of the foci, we find c.

$$c^2 = a^2 - b^2 = 9 - 4 = 5$$

thus $$c = \sqrt{5}.$$

The foci are $(1 + \sqrt{5}, -2)$ and $(1 - \sqrt{5}, -2)$.

■ *Try Exercise* **14**, *page 480.*

EXAMPLE 4   Find the Equation of an Ellipse

Find the standard form of the equation of the ellipse with center at $(4, -2)$, foci $(4, 1)$ and $(4, -5)$, and minor axis of length 10.

Solution Because the foci are on the major axis, the major axis is parallel to the y-axis. The distance from the center of the ellipse to a focus is c. The distance between $(4, -2)$ and $(4, 1)$ is 3. Therefore $c = 3$.

Recall that the length of the minor axis is $2b$. Thus $2b = 10$. Solving for b, we have $b = 5$.

To find a^2, use the equation $c^2 = a^2 - b^2$.

$$9 = a^2 - 25$$

$$a^2 = 34$$

Thus, the equation is

$$\frac{(x - 4)^2}{25} + \frac{(y + 2)^2}{34} = 1.$$

■ *Try Exercise* **30**, *page 480.*

Eccentricity of an Ellipse

The graph of an ellipse can be very long and thin, or it can be much like a circle. The **eccentricity** of an ellipse is a measure of its "roundness."

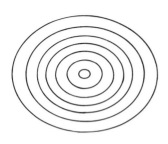

Figure 9.25

Figure 9.26

Eccentricity (*e*) of an Ellipse

The eccentricity *e* of an ellipse is the ratio of *c* to *a*, where *c* is the distance from the center to the focus and *a* is the length of the semimajor axis. That is,

$$e = \frac{c}{a}.$$

Because $c < a$, for an ellipse, $0 < e < 1$. If $c \approx 0$, then $e \approx 0$ and the graph will be almost like a circle.

If $c \approx a$, then $e \approx 1$ and the graph will be long and thin.

EXAMPLE 5   **Find the Eccentricity of an Ellipse**

Find the eccentricity of the ellipse $8x^2 + 9y^2 = 18$.

Solution   First, write the equation of the ellipse in standard form. Divide each side of the equation by 18.

$$\frac{8x^2}{18} + \frac{9y^2}{18} = 1$$

$$\frac{4x^2}{9} + \frac{y^2}{2} = 1$$

$$\frac{x^2}{9/4} + \frac{y^2}{2} = 1$$

The last step is necessary because the standard form of the equation has coefficients of 1 in the numerator. Thus we have

$$a^2 = \frac{9}{4}$$

$$a = \frac{3}{2}$$

Use the equation $c^2 = a^2 - b^2$ to find *c*.

$$c^2 = \frac{9}{4} - 2 = \frac{1}{4}$$

$$c = \sqrt{\frac{1}{4}} = \frac{1}{2}.$$

Now we can find the eccentricity.

$$e = \frac{c}{a} = \frac{1/2}{3/2} = \frac{1}{3}$$

The eccentricity is $\frac{1}{3}$.

■ *Try Exercise* **36,** *page 480.*

The planets revolve around the sun in elliptical orbits. The eccentricities of planets in our solar system are given below.
Which planet has the most circular orbit? Why?[2]

Planet	Eccentricity
Mercury	0.206
Venus	0.007
Earth	0.017
Mars	0.093
Jupiter	0.049
Saturn	0.051
Uranus	0.046
Neptune	0.005
Pluto	0.250

Acoustic Property of an Ellipse

Sound waves, although different from light waves, have a similiar reflective property. When sound is reflected from a point P on a surface, the angle of incidence equals the angle of reflection. Applying this principle to an ellipse results in what are called "whispering galleries." These galleries are based on the following theorem.

The Reflective Property of an Ellipse

> The lines from the foci to a point on the ellipse make equal angles with the tangent line at that point.

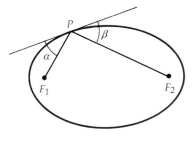

Figure 9.27
$\angle \alpha = \angle \beta$.

The Rotunda of the Capitol Building in Washington, D.C., is a whispering gallery. Two people standing at the foci of the elliptical ceiling can whisper and yet hear each other even though they are some distance apart. The whisper from one person is reflected to the person standing at the other focus.

[2] Neptune has the smallest eccentricity, it is therefore the planet with the most circular orbit.

EXERCISE SET 9.2

In Exercises 1 to 20, find the vertices and foci of the ellipse given by each equation. Sketch the graph.

1. $\dfrac{x^2}{16} + \dfrac{y^2}{25} = 1$

2. $\dfrac{x^2}{49} + \dfrac{y^2}{36} = 1$

3. $\dfrac{x^2}{9} + \dfrac{y^2}{4} = 1$

4. $\dfrac{x^2}{64} + \dfrac{y^2}{25} = 1$

5. $3x^2 + 4y^2 = 12$

6. $5x^2 + 4y^2 = 20$

7. $25x^2 + 16y^2 = 400$

8. $25x^2 + 12y^2 = 300$

9. $64x^2 + 25y^2 = 400$

10. $9x^2 + 64y^2 = 144$

11. $4x^2 + y^2 - 24x - 8y + 48 = 0$

12. $x^2 + 9y^2 + 6x - 36y + 36 = 0$

13. $5x^2 + 9y^2 - 20x + 54y + 56 = 0$

14. $9x^2 + 16y^2 + 36x - 16y - 104 = 0$

15. $16x^2 + 9y^2 - 64x - 80 = 0$

16. $16x^2 + 9y^2 + 36y - 108 = 0$

17. $25x^2 + 16y^2 + 50x - 32y - 359 = 0$

18. $16x^2 + 9y^2 - 64x - 54y + 1 = 0$

19. $8x^2 + 25y^2 - 48x + 50y + 47 = 0$

20. $4x^2 + 9y^2 + 24x + 18y + 44 = 0$

In Exercises 21 to 32, find the equation in standard form of each ellipse, given the information provided.

21. Center $(0, 0)$, major axis of length 10, foci at $(4, 0)$ and $(-4, 0)$

22. Center $(0, 0)$, minor axis of length 6, foci at $(0, 4)$ and $(0, -4)$

23. Vertices $(6, 0)$, $(-6, 0)$; ellipse passes through $(0, -4)$, and $(0, 4)$

24. Vertices $(5, 0)$, $(-5, 0)$; ellipse passes through $(0, 7)$, and $(0, -7)$

25. Major axis of length 12 on the x-axis, center at $(0, 0)$, and passing through $(2, -3)$

26. Minor axis of length 8, center at $(0, 0)$, and passing through $(-2, 2)$

27. Center $(-2, 4)$, vertices $(-6, 4)$ and $(2, 4)$, foci $(-5, 4)$ and $(1, 4)$

28. Center $(0, 3)$, minor axis of length 4, foci $(0, 0)$ and $(0, 6)$

29. Center $(2, 4)$, major axis parallel to the y-axis and of length 10, the ellipse passes through the point $(3, 3)$

30. Center $(-4, 1)$, minor axis parallel to the y-axis of length 8, and the ellipse passes through the point $(0, 4)$

31. Vertices $(5, 6)$ and $(5, -4)$, foci $(5, 4)$ and $(5, -2)$

32. Vertices $(-7, -1)$ and $(5, -1)$, foci $(-5, -1)$ and $(3, -1)$

In Exercises 33 to 40, use the eccentricity of the ellipse to find the equation in standard form of each of the following ellipses.

33. Eccentricity $2/5$, major axis on the x-axis of length 10, and center at $(0, 0)$

34. Eccentricity $3/4$, foci at $(9, 0)$ and $(-9, 0)$

35. Foci at $(0, -4)$ and $(0, 4)$, eccentricity $2/3$

36. Foci at $(0, -3)$ and $(0, 3)$, eccentricity $1/4$

37. Eccentricity $2/5$, foci $(-1, 3)$ and $(3, 3)$

38. Eccentricity $1/4$, foci $(-2, 4)$ and $(-2, -2)$

39. Eccentricity $2/3$, major axis of length 24 on the y-axis, centered at $(0, 0)$

40. Eccentricity $3/5$, major axis of length 15 on the x-axis, center at $(0, 0)$

Supplemental Exercises

41. Explain why the graph of the equation $4x^2 + 9y^2 - 8x + 36y + 76 = 0$ is or is not an ellipse.

42. Explain why the graph of the equation $4x^2 + 9y - 16x - 2 = 0$ is or is not an ellipse. Sketch the graph of this equation. See section 9.1 for assistance.

In Exercises 43 to 46, find the equation in standard form of an ellipse by using the definition of an ellipse.

43. Find the equation of the ellipse with foci at $(-3, 0)$ and $(3, 0)$ that passes through the point $(3, 9/2)$.

44. Find the equation of the ellipse with foci at $(0, 4)$ and $(0, -4)$ that passes through the point $(9/5, 4)$.

45. Find the equation of the ellipse with foci at $(-1, 2)$ and $(3, 2)$ that passes through the point $(3, 5)$.

46. Find the equation of the ellipse with foci at $(-1, 1)$ and $(-1, 7)$ that passes through the point $(3/4, 1)$.

In Exercises 47 and 48, find the latus rectum of the given ellipse. The line segment with endpoints on the ellipse that is perpendicular to the major axis and passes through the focus is the **latus rectum** of the ellipse.

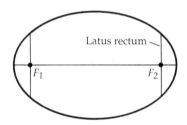

47. Find the length of the latus rectum of the ellipse given by

$$\frac{(x-1)^2}{9} + \frac{(y+1)^2}{16} = 1.$$

48. Find the length of the latus rectum of the ellipse given by

$$9x^2 + 16y^2 - 36x + 96y + 36 = 0.$$

49. Show that for any ellipse, the length of the latus rectum is $\dfrac{2b^2}{a}$.

50. Use the definition of an ellipse to find the equation of an ellipse with center at $(0, 0)$ and foci $(0, c)$ and $(0, -c)$.

Recall that a parabola has a directrix that is a line perpendicular to the axis of symmetry. An ellipse has two directrixes, both of which are perpendicular to the major axis and outside the ellipse. For an ellipse with center at the origin and whose major axis is the x-axis, the equations of the directrixes are $x = a^2/c$ and $x = -a^2/c$.

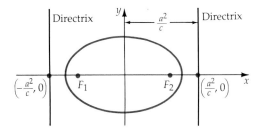

51. Find the directrix of the ellipse in Exercise 3.

52. Find the directrix of the ellipse in Exercise 4.

53. Let $P(x, y)$ be a point on the ellipse $\dfrac{x^2}{12} + \dfrac{y^2}{8} = 1$. Show that the distance from the point P to the focus $(2, 0)$ divided by the distance from the point P to the directrix $x = 6$ equals the eccentricity. (*Hint:* Solve the equation

of the ellipse for y^2. Substitute this value for y^2 after applying the distance formula.)

54. Let $P(x, y)$ be a point on the ellipse $\dfrac{x^2}{25} + \dfrac{y^2}{16} = 1$. Show that the distance from the point P to the focus $(3, 0)$ divided by the distance from the point to the directrix $x = 25/3$ equals the eccentricity. (*Hint:* Solve the equation of the ellipse for y^2. Substitute this value for y^2 after applying the distance formula.)

55. Generalize the results of Exercises 53 and 54. That is, show that if $P(x, y)$ is a point on the ellipse $\dfrac{x^2}{a^2} + \dfrac{y^2}{b^2} = 1$, where $F(c, 0)$ is the focus and $x = a^2/c$ is the directrix, then the following equation is true: $e = d(P, F)/d(P, D)$. (*Hint:* Solve the equation of the ellipse for y^2. Substitute this value for y^2 after applying this distance formula.)

9.3
Hyperbolas

The hyperbola is a conic section formed when a plane intersects a right circular cone at a certain angle. If β is the angle at which the plane intersects the axis of the cone and α is the angle shown in Figure 9.28, a hyperbola is formed when $0° < \beta < \alpha$ or when the plane is parallel to the axis of the cone.

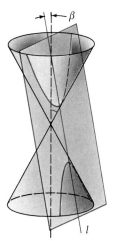
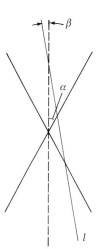

Figure 9.28

Figure 9.29

Remark If the plane intersects the cone along the axis of the cone, the resulting curve is two intersecting straight lines. This is the **degenerate** form of a hyperbola.

As with the other conic sections, there is a definition of a hyperbola in terms of a certain set of points in the plane.

Definition of a Hyperbola

> A **hyperbola** is the set of all points in the plane, the difference of whose distances from two fixed points (foci) is a positive constant.

Remark This definition differs from that of an ellipse in that the ellipse was defined in terms of the *sum* of the distances, whereas the hyperbola is defined in terms of the *difference* of two distances.

Hyperbolas with Center at $(0,0)$

The **transverse axis** is the line segment joining the intercepts through the foci of a hyperbola (see Figure 9.30). The midpoint of the transverse axis is called the **center** of the hyperbola. The **conjugate** axis passes through the center of the hyperbola and is perpendicular to the transverse axis.

The length of the transverse axis is customarily denoted $2a$, and the distance between the two foci is denoted $2c$. The length of the conjugate axis is denoted $2b$.

The **vertices** of a hyperbola are the points where the hyperbola intersects the transverse axis.

To determine the positive constant stated in the definition of a hyperbola, consider the point $V(a, 0)$, which is one vertex of a hyperbola, and the points $F_1(c, 0)$ and $F_2(-c, 0)$, which are the foci of the hyperbola (see Figure 9.31). The difference of the distance from V to F_1, $c - a$, and the distance from $V(a, 0)$ to $F_2(-c, 0)$, $a + c$, must be a constant. By subtracting these distances, we find

$$|(c - a) - (c + a)| = |-2a| = 2a$$

Thus the constant is $2a$ and is the length of the transverse axis. The absolute value was used to ensure that the distance is a positive number.

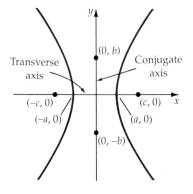

Figure 9.30

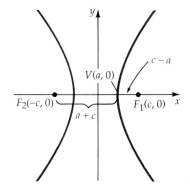

Figure 9.31

**Standard Forms of the Equation of a Hyperbola
with Center at the Origin**

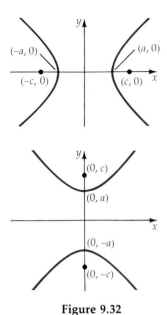

Figure 9.32

> *Transverse axis on the x-axis*
> The standard form of the equation of a hyperbola with the center at the origin and transverse axis on the x-axis is given by
> $$\frac{x^2}{a^2} - \frac{y^2}{b^2} = 1$$
>
> The coordinates of the vertices are $(a, 0)$ and $(-a, 0)$, and the coordinates of the foci are $(c, 0)$ and $(-c, 0)$, where $c^2 = a^2 + b^2$.
>
> *Transverse axis on the y-axis*
> The standard form of the equation of a hyperbola with the center at the origin and transverse axis on the y-axis is given by
> $$\frac{y^2}{a^2} - \frac{x^2}{b^2} = 1$$
>
> The coordinates of the vertices are $(0, a)$ and $(0, -a)$, and the coordinates of the foci are $(0, c)$ and $(0, -c)$, where $c^2 = a^2 + b^2$.

Remark By looking at the equations, note that it is possible to determine the transverse axis by finding which term in the equation is positive. If the x^2 term is positive, then the transverse axis is on the x-axis. When the y^2 term is positive, the transverse axis is on the y-axis.

Consider the hyperbola given by the equation $\frac{x^2}{16} - \frac{y^2}{9} = 1$. Because the x^2 term is positive, the transverse axis is on the x-axis, $a^2 = 16$, thus $a = 4$. The vertices are $(4, 0)$ and $(-4, 0)$. To find the foci, we determine c.

$$c^2 = a^2 + b^2 = 16 + 9 = 25$$
$$c = \sqrt{25} = 5.$$

The foci are $(5, 0)$ and $(-5, 0)$. The graph is shown in Figure 9.33.

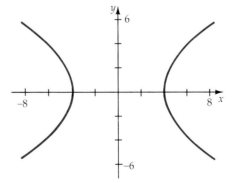

Figure 9.33
$$\frac{x^2}{16} - \frac{y^2}{9} = 1$$

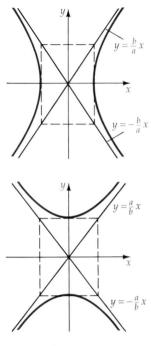

Figure 9.34

The asymptotes of the hyperbola are a useful guide to sketching the graph of the hyperbola. Each hyperbola has two asymptotes that pass through the center of the hyperbola.

Asymptotes of a Hyperbola with Center at the Origin

The **asymptotes** of the hyperbola $\dfrac{x^2}{a^2} - \dfrac{y^2}{b^2} = 1$ are given by the equations $y = \dfrac{b}{a}x$ and $y = -\dfrac{b}{a}x$. The asymptotes of the hyperbola $\dfrac{y^2}{a^2} - \dfrac{x^2}{b^2} = 1$ are given by the equations $y = \dfrac{a}{b}x$ and $y = -\dfrac{a}{b}x$.

We can outline a proof for the equations of the asymptotes by using the equation of a hyperbola in standard form.

$$\frac{x^2}{a^2} - \frac{y^2}{b^2} = 1$$

$$y^2 = b^2\left(\frac{x^2}{a^2} - 1\right) \qquad \text{Solve for } y^2.$$

$$= \frac{b^2}{a^2}(x^2 - a^2) \qquad \text{Factor out } 1/a^2.$$

$$= \frac{b^2}{a^2}x^2\left(1 - \frac{a^2}{x^2}\right) \qquad \text{Factor out } x^2.$$

$$y = \pm\frac{b}{a}x\sqrt{1 - \frac{a^2}{x^2}} \qquad \text{Take the square root of each side.}$$

As $|x|$ becomes larger and larger, $1 - \dfrac{a^2}{x^2}$ approaches 1. (Why?[3]) For large values of $|x|$, $y \approx \pm\dfrac{b}{a}x$, and thus $y = \pm\dfrac{b}{a}x$ are asymptotes for the hyperbola. A similar outline of a proof can be given for hyperbolas with the transverse axis on the y-axis.

Remark   One method for remembering the equations of the asymptotes is to write the equation of a hyperbola in standard form but replace 1 by 0 and then solve for y.

$$\frac{x^2}{a^2} - \frac{y^2}{b^2} = 0 \quad \text{thus } y^2 = \frac{b^2}{a^2}x^2 \text{ or } y = \pm\frac{b}{a}x$$

$$\frac{y^2}{a^2} - \frac{x^2}{b^2} = 0 \quad \text{thus } y^2 = \frac{a^2}{b^2}x^2 \text{ or } y = \pm\frac{a}{b}x$$

---

[3] Because a^2 is a constant, a^2/x^2 approaches 0 as $|x|$ becomes very large. Thus $1 - a^2/x^2$ approaches $1 - 0$ or 1.

EXAMPLE 1   **Find the Vertices, Foci, and Asymptotes of a Hyperbola**

Find the foci, vertices, and asymptotes of the hyperbola given by the equation $\dfrac{y^2}{9} - \dfrac{x^2}{4} = 1$. Sketch the graph.

Solution   Because the y^2 term is positive, the transverse axis is the y-axis. We know $a^2 = 9$; thus $a = 3$. The vertices are $V_1(0, 3)$ and $V_2(0, -3)$.

$$c^2 = a^2 + b^2 = 9 + 4$$

$$c = \sqrt{13}$$

The foci are $F_1(0, \sqrt{13})$ and $F_2(0, -\sqrt{13})$.

Because $a = 3$ and $b = 2$ ($b^2 = 4$), the equations of the asymptotes are $y = \dfrac{3}{2}x$ and $y = -\dfrac{3}{2}x$.

To sketch the graph, we draw a rectangle with its center at the origin that has dimensions equal to the lengths of the transverse and conjugate axes. The asymptotes are extensions of the diagonals of the rectangle.

■ *Try Exercise* **4,** *page 489.*

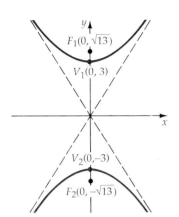

Figure 9.35
$$\dfrac{y^2}{9} - \dfrac{x^2}{4} = 1$$

Hyperbolas with the Center at the Point (h, k)

Using a translation of coordinates similar to that used for ellipses, we can write the equation of a hyperbola with its center at the point (h, k). Given coordinates axes labeled x' and y', an equation of a hyperbola with center at the origin is

$$\dfrac{(x')^2}{a^2} - \dfrac{(y')^2}{b^2} = 1. \tag{1}$$

Now place the origin of this coordinate system at the point (h, k) of the xy-coordinate system. The relationship between an ordered pair in

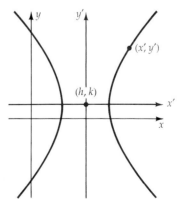

Figure 9.36

the $x'y'$-coordinate system and the xy-coordinate system is given by the transformation equations

$$x' = x - h$$

$$y' = y - k$$

Substitute the expressions for x' and y' into Equation (1). The equation of the hyperbola with center at (h, k) is

$$\frac{(x - h)^2}{a^2} - \frac{(y - k)^2}{b^2} = 1.$$

Standard Form of Hyperbolas with Center at (h, k)

> ***Transverse Axis Parallel to the x-axis***
> The standard form of the equation of a hyperbola with center (h, k) and transverse axis parallel to the x-axis is given by
>
> $$\frac{(x - h)^2}{a^2} - \frac{(y - k)^2}{b^2} = 1.$$
>
> The coordinates of the vertices are $V_1(h + a, k)$ and $V_2(h - a, k)$. The coordinates of the foci are $F_1(h + c, k)$ and $F_2(h - c, k)$. The equations of the asymptotes are $y - k = \frac{b}{a}(x - h)$ and $y - k = -\frac{b}{a}(x - h)$.
>
> ***Transverse Axis Parallel to the y-axis***
> The standard form of the equation of a hyperbola with center (h, k) and transverse axis parallel to the y-axis is given by
>
> $$\frac{(y - k)^2}{a^2} - \frac{(x - h)^2}{b^2} = 1.$$
>
> The coordinates of the vertices are $V_1(h, k + a)$ and $V_2(h, k - a)$. The coordinates of the foci are $F_1(h, k + c)$ and $F_2(h, k - c)$. The equations of the asymptotes are $y - k = \frac{a}{b}(x - h)$ and $y - k = -\frac{a}{b}(x - h)$.

EXAMPLE 2   **Find the Vertices, Foci, and Asymptotes of a Hyperbola**

Find the vertices, foci, and asymptotes of the hyperbola given by the equation $4x^2 - 9y^2 - 16x + 54y - 29 = 0$. Sketch the graph.

Solution   Write the equation of the hyperbola in standard form by completing the square.

$$4x^2 - 9y^2 - 16x + 54y - 29 = 0$$

$4x^2 - 16x - 9y^2 + 54y = 29$	Rearrange terms.
$4(x^2 - 4x) - 9(y^2 - 6y) = 29$	Factor.

$$4(x^2 - 4x + 4) - 9(y^2 - 6y + 9) = 29 + 16 - 81 \quad \text{Complete the square.}$$

$$4(x - 2)^2 - 9(y - 3)^2 = -36 \qquad \text{Factor.}$$

$$\frac{(y - 3)^2}{4} - \frac{(x - 2)^2}{9} = 1 \qquad \text{Divide by } -36.$$

The coordinates of the center are $(2, 3)$. Because the term containing $(y - 3)^2$ is positive, the transverse axis is parallel to the y-axis. We know $a^2 = 4$; thus $a = 2$. The vertices are $(2, 5)$ and $(2, 1)$.

$$c^2 = a^2 + b^2 = 4 + 9$$

$$c = \sqrt{13}.$$

The foci are $(2, 3 + \sqrt{13})$ and $(2, 3 - \sqrt{13})$. We know $b^2 = 9$; thus $b = 3$. The equations of the asymptotes are

$$y = \frac{2}{3}x + \frac{5}{3} \quad \text{and} \quad y = -\frac{2}{3}x + \frac{13}{3}.$$

■ *Try Exercise **18**, page 489.*

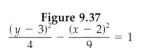

Figure 9.37
$$\frac{(y - 3)^2}{4} - \frac{(x - 2)^2}{9} = 1$$

Eccentricity of a Hyperbola

The graph of a hyperbola can be very wide or very narrow. The **eccentricity** of a hyperbola is a measure of its "wideness."

Eccentricity (*e*) of a Hyperbola

> The eccentricity e of a hyperbola is the ratio of c to a, where c is the distance from the center to a focus and a is the length of the semi-transverse axis.
>
> $$e = \frac{c}{a}$$

For a hyperbola, $c > a$ and therefore $e > 1$. As the eccentricity of the hyperbola increases, the graph becomes wider and wider.

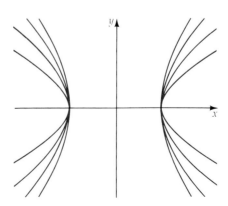

Figure 9.38

EXAMPLE 3   **Find the Equation of a Hyperbola Given Its Eccentricity**

Find the standard form of the equation of the hyperbola that has eccentricity 3/2, center at the origin, and a focus (6, 0).

Solution   Because the focus is located at (6, 0) and the center is at the origin, $c = 6$. An extension of the transverse axis contain the foci, and thus the transverse axis is on the x-axis.

$$e = \frac{3}{2} = \frac{c}{a}$$

$$\frac{3}{2} = \frac{6}{a} \quad \text{Substitute the value for } c.$$

$$a = 4 \quad \text{Solve for } a.$$

To find b, use the equation $c^2 = a^2 + b^2$ and the values for c and a.

$$c^2 = a^2 + b^2$$

$$36 = 16 + b^2$$

$$b^2 = 20$$

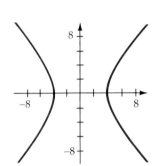

Figure 9.39

$$\frac{x^2}{16} - \frac{y^2}{20} = 1$$

The equation of the hyperbola is $\dfrac{x^2}{16} - \dfrac{y^2}{20} = 1$.

■ *Try Exercise* **40**, *page 490*.

Orbits of Comets

In Section 2 we noted that orbits of the planets are elliptical. Some comets have elliptical orbits also, the most notable being Halley's comet, whose eccentricity is 0.97.

Other comets have hyperbolic orbits with the sun at a focus. These comets pass by the sun only once. The velocity of a comet determines whether its orbit is elliptical or hyperbolic.

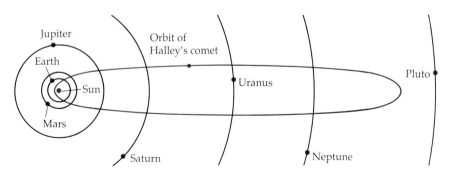

Figure 9.40

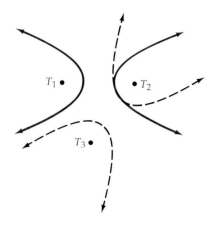

Figure 9.41

Hyperbolas as an Aid to Navigation

Consider two radio transmitters, T_1 and T_2, placed some distance apart. A ship with electronic equipment measures the difference in time it takes signals from the transmitters to reach the ship. Because the difference in time is proportional to the distance of the ship from the transmitter, the ship must be located on the hyperbola with foci at the two transmitters.

Using a third transmitter, T_3, a second hyperbola can be found with foci T_2 and T_3. The ship lies on the intersection of the two hyperbolas.

Optical Property of a Hyperbola

A ray of light directed toward one focus of a hyperbolic mirror is reflected toward the other focus. This property, along with the reflective property of an ellipse, is used in telescopes to focus light.

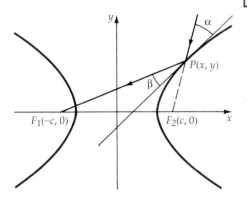

Figure 9.42

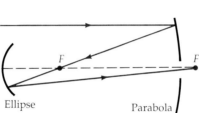

Figure 9.43

EXERCISE SET 9.3

In Exercises 1 to 24, find the center, vertices, foci, and asymptotes for the hyperbola given by each equation. Sketch the graph.

1. $\dfrac{x^2}{16} - \dfrac{y^2}{25} = 1$

2. $\dfrac{x^2}{16} - \dfrac{y^2}{9} = 1$

3. $\dfrac{y^2}{4} - \dfrac{x^2}{25} = 1$

4. $\dfrac{y^2}{25} - \dfrac{x^2}{36} = 1$

5. $\dfrac{(x-3)^2}{16} - \dfrac{(y+4)^2}{9} = 1$

6. $\dfrac{(x+3)^2}{25} - \dfrac{y^2}{4} = 1$

7. $\dfrac{(y+2)^2}{4} - \dfrac{(x-1)^2}{16} = 1$

8. $\dfrac{(y-2)^2}{36} - \dfrac{(x+1)^2}{49} = 1$

9. $x^2 - y^2 = 9$

10. $4x^2 - y^2 = 16$

11. $16y^2 - 9x^2 = 144$

12. $9y^2 - 25x^2 = 225$

13. $9y^2 - 36x^2 = 4$

14. $16x^2 - 25y^2 = 9$

15. $x^2 - y^2 - 6x + 8y - 3 = 0$

16. $4x^2 - 25y^2 + 16x + 50y - 109 = 0$

17. $9x^2 - 4y^2 + 36x - 8y + 68 = 0$

18. $16x^2 - 9y^2 - 32x - 54y + 79 = 0$

19. $4x^2 - y^2 + 32x + 6y + 39 = 0$

20. $x^2 - 16y^2 + 8x - 64y + 16 = 0$

21. $9x^2 - 16y^2 - 36x - 64y + 116 = 0$

22. $2x^2 - 9y^2 + 12x - 18y + 18 = 0$

23. $4x^2 - 9y^2 + 8x - 18y - 6 = 0$

24. $2x^2 - 9y^2 - 8x + 36y - 46 = 0$

In Exercises 25 to 38, find the equation in standard form of the hyperbola satisfying the stated conditions.

25. Vertices $(3, 0)$ and $(-3, 0)$, foci $(4, 0)$ and $(-4, 0)$

26. Vertices $(0, 2)$ and $(0, -2)$, foci $(0, 3)$ and $(0, -3)$

27. Foci $(0, 5)$ and $(0, -5)$, asymptotes $y = 2x$ and $y = -2x$

28. Foci $(4, 0)$ and $(-4, 0)$, asymptotes $y = x$ and $y = -x$

29. Vertices $(0, 3)$ and $(0, -3)$ and passing through $(2, 4)$

30. Vertices $(5, 0)$ and $(-5, 0)$ and passing through $(-1, 3)$

31. Asymptotes $y = \frac{1}{2}x$ and $y = -\frac{1}{2}x$, vertices $(0, 4)$ and $(0, -4)$

32. Asymptotes $y = \frac{2}{3}x$ and $y = -\frac{2}{3}x$, vertices $(6, 0)$ and $(-6, 0)$

33. Vertices $(6, 3)$ and $(2, 3)$, foci $(7, 3)$ and $(1, 3)$

34. Vertices $(-1, 5)$ and $(-1, -1)$, foci $(-1, 7)$ and $(-1, -3)$

35. Foci $(1, -2)$ and $(7, -2)$, slope of an asymptote $5/4$

36. Foci $(-3, -6)$ and $(-3, -2)$, slope of an asymptote 1

37. Passing through $(9, 4)$, slope of an asymptote $1/2$, center $(7, 2)$, transverse axis parallel to the y-axis

38. Passing through $(6, 1)$, slope of an asymptote 2, center $(3, 3)$, transverse axis parallel to the x-axis

In Exercises 39 to 44, use the eccentricity to find the equation in standard form of a hyperbola.

39. Vertices $(1, 6)$ and $(1, 8)$, eccentricity 2

40. Vertices $(2, 3)$ and $(-2, 3)$, eccentricity $5/2$

41. Eccentricity 2, foci $(4, 0)$ and $(-4, 0)$

42. Eccentricity $4/3$, foci $(0, 6)$ and $(0, -6)$

43. Center $(4, 1)$, conjugate axis length 4, eccentricity $4/3$

44. Center $(-3, -3)$, conjugate axis length 6, eccentricity 2

In Exercises 45 to 52, identify the graph of each equation as a parabola, ellipse, or hyperbola. Sketch the graph.

45. $4x^2 + 9y^2 - 16x - 36y + 16 = 0$

46. $2x^2 + 3y - 8x + 2 = 0$

47. $5x - 4y^2 + 24y - 11 = 0$

48. $9x^2 - 25y^2 - 18x + 50y = 0$

49. $x^2 + 2y - 8x = 0$

50. $9x^2 + 16y^2 + 36x - 64y - 44 = 0$

51. $25x^2 + 9y^2 - 50x - 72y - 56 = 0$

52. $(x - 3)^2 + (y - 4)^2 = (x + 1)^2$

53. Find the equation of the path of Halley's comet in astronomical units by letting the sun (one focus) be at the

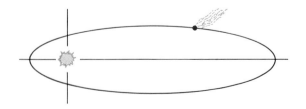

origin and the other focus on the positive x-axis. The length of the major axis of the orbit of Halley's comet is approximately 36 astronomical units (36 AU) and the length of the minor axis 9 AU wide (1 AU = 92,600,000 miles).

54. A foot suspension bridge is 100 feet long and supported by cables that hang in the shape of a parabola. Find the equation of the parabola if the positive x-axis is along the footpath, as in the figure.

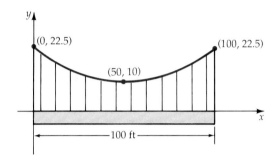

55. Two radio towers are positioned along the coast of California, 200 miles apart. A signal is sent simultaneously from each tower to a ship off the coast. The signal from tower B is received by the ship 500 microseconds after the signal sent by A. If the radio signal travels 0.2 miles per microsecond, find the equation of the hyperbola on which the ship is located.

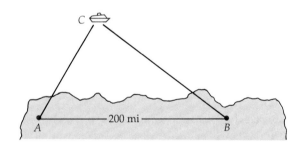

56. A softball player releases a softball at a height of 5 feet above the ground with a speed of 88 feet per second (60 mph). The initial trajectory of the ball is 45°. Neglecting air resistance, the path of the ball is a parabola given by $y = -0.004x^2 + x + 5$. How far will the ball travel before hitting the ground? *Hint:* Let $y = 0$ and solve for x.

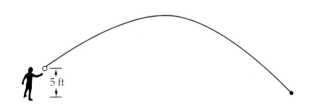

Supplemental Exercises

In Exercises 57 to 60, use the definition for a hyperbola to find the equation of the hyperbola in standard form.

57. Foci $(2, 0)$ and $(-2, 0)$ and passes through the point $(2, 3)$

58. Foci $(0, 3)$ and $(0, -3)$ and passes through the point $(5/2, 3)$

59. Foci $(0, 4)$ and $(0, -4)$ and passes through the point $(7/3, 4)$

60. Foci $(5, 0)$ and $(-5, 0)$ and passes through the point $(5, 9/4)$

Recall that an ellipse has two directrixes that are lines perpendicular to the line containing the foci. A hyperbola also has two directrixes that are perpendicular to the transverse axis and outside the hyperbola. For a hyperbola with center

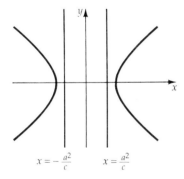

$$x = -\frac{a^2}{c} \qquad x = \frac{a^2}{c}$$

at the origin and transverse axis on the x-axis, the equations of the directrixes are $x = a^2/c$ and $x = -a^2/c$. In Exercises 61 to 65, use this information to solve each exercise.

61. Find the directrixes for the ellipse in Exercise 21.

62. Find the directrixes for the ellipse in Exercise 22.

63. Let $P(x, y)$ be a point on the hyperbola $\dfrac{x^2}{9} - \dfrac{y^2}{16} = 1$.

Show that the distance from the point P to the focus $(5, 0)$ divided by the distance from the point P to the directrix $x = 9/5$ equals the eccentricity.

64. Let $P(x, y)$ be a point on the hyperbola $\dfrac{x^2}{7} - \dfrac{y^2}{9} = 1$. Show that the distance from the point P to the focus $(4, 0)$ divided by the distance from the point to the directrix $x = 7/4$ equals the eccentricity.

65. Generalize the results of Exercises 63 and 64. That is, show that if $P(x, y)$ is a point on the hyperbola $\dfrac{x^2}{a^2} - \dfrac{y^2}{b^2} = 1$, $F(c, 0)$ is the focus, and $x = a^2/c$ is the directrix, then the following equation is true: $e = d(P, F)/d(P, D)$.

66. Derive the equation of a hyperbola with center at the origin, foci at $(0, c)$ and $(0, -c)$, and vertices $(0, a)$ and $(0, -a)$.

67. Sketch a graph of $\dfrac{x|x|}{16} - \dfrac{y|y|}{9} = 1$.

68. Sketch a graph of $\dfrac{x|x|}{16} + \dfrac{y|y|}{9} = 1$.

9.4

Introduction to Polar Coordinates

Until now, *rectangular coordinate systems* have been used to locate a point in the coordinate plane. An alternate method is to use a *polar coordinate system*. Using this method, a point is located in a manner similar to giving a distance and an angle from some fixed direction.

The Polar Coordinate System

A **polar coordinate system** is formed by drawing a horizontal ray. The ray is called the **polar axis,** and the beginning point is called the **pole.** A point $P(r, \theta)$ in the plane is located by specifying a distance r from the pole and

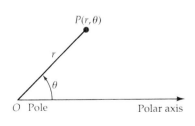

Figure 9.44

an angle θ measured from the polar axis to the line segment OP. The angle can be measured in degrees or radians.

The coordinates of the pole are $(0, \theta)$, where θ is an arbitrary angle. Positive angles are measured counterclockwise from the polar axis. Negative angles are measured clockwise from the axis. Positive values of r are measured along the ray that makes an angle θ from the polar axis.

Negative values of r are measured along the ray that makes an angle of $\theta + 180°$ from the polar axis.

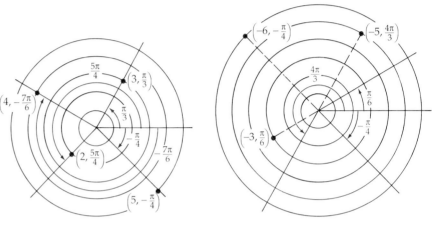

Figure 9.45   Figure 9.46

In a rectangular coordinate system, there is a one-to-one correspondence between the points in the plane and the ordered pairs (x, y). This is not true for a polar coordinate system. For polar coordinates, the relation is one-to-many. For each point $P(r, \theta)$ in a polar coordinate system there corresponds infinitely many ordered pair descriptions of that point.

For example, consider a point whose coordinates are $P(3, 45°)$. Because there are $360°$ in one complete revolution around a circle, the point P could also be written as $(3, 405°)$, $(3, 765°)$, $(3, 1125°)$, and generally as $(3, 45° + n \cdot 360°)$, where n is an integer. It is also possible to describe the point $P(3, 45°)$ by $(-3, 225°)$, $(-3, -135°)$, and $(3, -315°)$, to name just a few.

Remark   The relationship between an ordered pair and a point is not one-to-many. That is, given an ordered pair (r, θ), there is exactly one point in the plane that corresponds to that point.

Graphs of Equations in a Polar Coordinate System

A **polar equation** is an equation in r and θ. A **solution** to a polar equation is an ordered pair (r, θ) that satisfies the equation. The **graph** of a polar equation is the set of all points whose ordered pairs that are solutions of the equation.

Figure 9.47 is the graph of the polar equation $r = 2$. The graph is drawn on a polar coordinate grid that consists of concentric circles and rays beginning at the pole. Because r is independent of θ, r is 2 units from

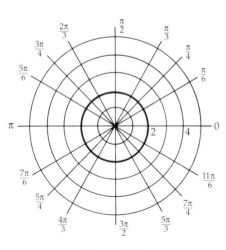

Figure 9.47
$r = 2$

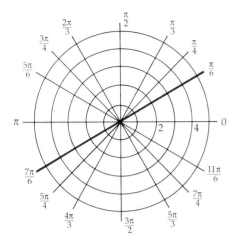

Figure 9.48

$$\theta = \frac{\pi}{6}$$

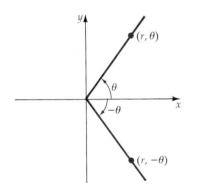

Figure 9.49

the pole for all values of θ. The graph is a circle of radius 2 with center at the pole.

The Graph of $r = a$

> The graph of $r = a$ is a circle with center at the pole and radius a.

The graph of the polar equation $\theta = \pi/6$ is a line. Because θ is independent of r, θ is $\pi/6$ radians from the polar axis for all values of r. The graph is a straight line that makes an angle of $\pi/6$ radians (30°) from the polar axis.

The Graph of $\theta = \alpha$

> The graph of $\theta = \alpha$ is a straight line through the pole at an angle of α from the polar axis.

As an aid to graphing other polar equations, it is useful to examine the equation for symmetries with respect to the pole, the polar axis, or the line $\theta = \pi/2$. By superimposing an xy-coordinate system on the polar coordinate system, this means finding symmetries with respect to the origin, the x-axis, or the y-axis.

Suppose that whenever the ordered pair (r, θ) lies on the graph of a polar equation, $(r, -\theta)$ also lies on the graph. From Figure 9.49, the graph will have symmetry with respect to the polar axis. Thus one test for symmetry is to replace θ by $-\theta$ in the polar equation. If the resulting equation is equivalent to the original equation, the graph is symmetric with respect to the polar axis.

EXAMPLE 1   Graph a Polar Equation That is Symmetric With Respect to the Polar Axis

Show that the graph of $r = 4 \cos \theta$ is symmetric to the polar axis. Graph the equation.

Solution   Test for symmetry with respect to the polar axis. Replace θ by $-\theta$.
$$r = 4 \cos(-\theta) = 4 \cos \theta \quad \cos(-\theta) = \cos \theta.$$

Because replacing θ by $-\theta$ results in the same equation, the graph is symmetric with respect to the polar axis.

To graph the equation, begin choosing various values of θ and finding the corresponding values of r. However, before doing so, two further observations will reduce the number of points we must choose.

First, because cosine is a periodic function with period 2π, it is only necessary to choose points between 0 and 2π (0 and 360°). Second, when $\frac{\pi f63}{2} < \theta < \frac{3\pi}{2}$, $\cos \theta$ is negative, which means that any θ between these values will produce a negative r. Thus the point will be in the first or

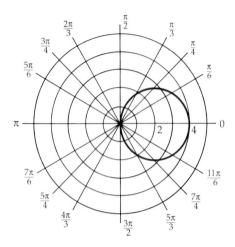

Figure 9.50
$r = 4 \cos \theta$

fourth quadrant. That is, we need consider only angles θ in the first or fourth quadrants. However, since the graph is symmetric with respect to the polar axis, it is only necessary to choose values of θ between 0 and $\pi/2$.

θ	0	$\pi/6$	$\pi/4$	$\pi/3$	$\pi/2$	$-\pi/6$	$-\pi/4$	$-\pi/3$	$-\pi/2$
r	4.0	3.5	2.8	2.0	0.0	3.5	2.8	2.0	0.0

The last four columns are labeled "By symmetry".

As we will show later, the graph of $r = 4 \cos \theta$ is a circle with a center at $(2, 0)$. See Figure 9.50.

■ *Try Exercise 14, page 501.*

The following table shows the types of symmetry and their associated tests. For each type, if the recommended substitution results in an equivalent equation, the graph will have the indicated symmetry.

Tests for Symmetry

Substitution	Symmetry with respect to
$-\theta$ for θ	The line $\theta = 0$
$\pi - \theta$ for θ, $-r$ for r	The line $\theta = 0$
$\pi - \theta$ for θ	The line $\theta = \pi/2$
$-\theta$ for θ, $-r$ for r	The line $\theta = \pi/2$
$-r$ for r	The pole (origin)
$\pi + \theta$ for θ	The pole (origin)

Caution   The graph of a polar equation may have a symmetry even though a test for that symmetry fails. For example, as you will see later, the graph of $r = \sin 2\theta$ is symmetric with respect to the line $\theta = 0$. However, using the symmetry test of substituting $-\theta$ for θ, we have

$$\sin 2(-\theta) = -\sin 2\theta \neq r.$$

Thus this test fails to show symmetry with respect to the line $\theta = 0$. The symmetry test of substituting $\pi - \theta$ for θ and $-r$ for r will establish symmetry with respect to the line $\theta = 0$. Why?[4]

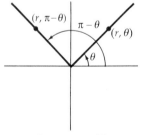
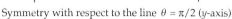
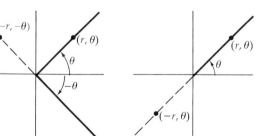
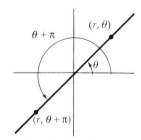

Symmetry with respect to the line $\theta = \pi/2$ (y-axis)   Symmetry with respect to the pole (origin)

Figure 9.51

[4] $\sin(2(\pi - \theta)) = \sin(2\pi - 2\theta) = \sin(-2\theta) = -\sin 2\theta = -r.$

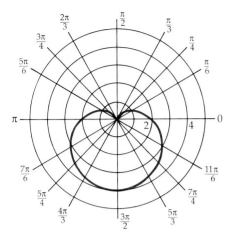

Figure 9.52
$r = 2 - 2 \sin \theta$

EXAMPLE 2   **Sketch the Graph of a Cardioid**

Sketch a graph of $r = 2 - 2 \sin \theta$.

Solution   Try each of the tests for symmetry (here we will show only the first and third tests). Substitute $-\theta$ for θ.

$$2 - 2 \sin(-\theta) = 2 + 2 \sin \theta \neq 2 - 2 \sin \theta \quad \sin(-\theta) = -\sin \theta$$

This test fails to show symmetry with respect to the line $\theta = 0$.
Substitute $\pi - \theta$ for θ.

$$2 - 2 \sin(\pi - \theta) = 2 - 2 \sin \theta \quad \sin(\pi - \theta) = \sin \theta$$

The graph is symmetric with respect to the line $\theta = \pi/2$. A check of the other tests for symmetry will show that the graph has no other symmetries.

Choose some values of θ and find the corresponding values for r. Plot these points and use symmetry to sketch the graph.

θ	$-\pi/2$	$-\pi/3$	$-\pi/4$	$-\pi/6$	0	$\pi/6$	$\pi/4$	$\pi/3$	$\pi/2$
r	4.0	3.7	3.4	3.0	2.0	1.0	0.6	0.3	0.0

The graph is called a **cardioid**. See Figure 9.52

■ *Try Exercise* **18,** *page 501.*

EXAMPLE 3   **Sketch the Graph of a Cardioid**

Sketch a graph of $r = 3 - 2 \cos \theta$.

Solution   Try each of the tests for symmetry (here we will show only the first and sixth tests). Substitute $-\theta$ for θ.

$$3 - 2 \cos(-\theta) = 3 - 2 \cos \theta \quad \cos(-\theta) = \cos \theta$$

The graph is symmetric with respect to the line $\theta = 0$.
Substitute $\pi + \theta$ for θ.

$$3 - 2 \cos(\pi + \theta) = 3 + 2 \cos \theta \neq 3 - 2 \cos \theta \quad \cos(\pi + \theta) = -\cos \theta$$

This test fails to show symmetry with respect to the line $\theta = \pi/2$. This graph does not have any other symmetries.

Choose some values of θ and find the corresponding values of r. Plot these points and use symmetry to draw the graph.

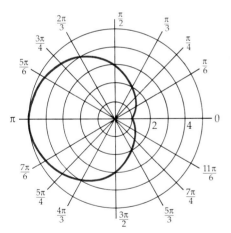

Figure 9.53
$r = 3 - 2 \cos \theta$

θ	0	$\pi/6$	$\pi/4$	$\pi/3$	$\pi/2$	$2\pi/3$	$3\pi/4$	$5\pi/6$	π
r	1.0	1.3	1.6	2.0	3.0	4.0	4.4	4.7	5.0

This is also a cardioid. See Figure 9.53.

■ *Try Exercise* **20,** *page 501.*

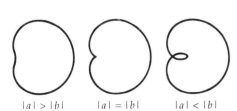

$|a| > |b|$ $|a| = |b|$ $|a| < |b|$

Figure 9.54

Examples 2 and 3 illustrate two possible graphs of a cardioid. There is also a third possibility, shown in Figure 9.54.

General Equations of a Cardioid

> The graph of the equation $r = a + b \cos \theta$ is a cardioid that is symmetric with respect to the line $\theta = 0$. The graph of the equation $r = a + b \sin \theta$ is a cardioid that is symmetric with respect to the line $\theta = \frac{\pi}{2}$.

The graphs illustrate the three forms of a cardioid with symmetry with respect to the line $\theta = 0$. The magnitudes of the absolute values of a and b determine the shape of the graph. All cardioids are reflections or rotations of the three basic graphs.

EXAMPLE 4   Sketch the Graph of a Three-Leaf Rose

Sketch a graph of $r = 3 \sin 3\theta$.

Solution   Try each of the tests for symmetry (here we will show only the second and fourth tests).

$-r = 3 \sin 3(\pi - \theta)$ Substitute $\pi - \theta$ for θ and $-r$ for r.

$r = -3 \sin 3(\pi - \theta)$ Multiply by -1.

$= -3 \sin 3\theta \neq 3 \sin 3\theta$ $\sin(\pi - \theta) = \sin \theta$.

This test fails to show symmetry with respect to the line $\theta = 0$.

$-r = 3 \sin 3(-\theta)$ Substitute $-\theta$ for θ and $-r$ for r.

$r = -3 \sin 3(-\theta)$ Multiply by -1.

$= 3 \sin 3\theta$ $\sin(-\theta) = -\sin \theta$.

The graph is symmetric with the line $\theta = \dfrac{\pi}{2}$. The graph does not have any other symmetries.

Choose some values for θ and find the corresponding values of r. Use symmetry to sketch the graph.

θ	0	$\pi/18$	$\pi/6$	$5\pi/18$	$\pi/3$	$7\pi/18$	$\pi/2$
r	0.0	1.5	3.0	1.5	0.0	-1.5	-3.0

The graph is a **three-leaf rose**. See Figure 9.55.

■ *Try Exercise **28**, page 501.*

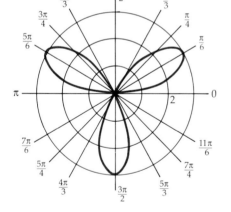

Figure 9.55
$r = 3 \sin 3\theta$

EXAMPLE 5   Sketch the Graph of a Four-Leaf Rose

Sketch a graph of $r = 4 \cos 2\theta$

Solution   Try each of the tests for symmetry (here we will show only the first and third tests). Substitute $-\theta$ for θ.

$$4 \cos(-2\theta) = 4 \cos 2\theta \quad \cos(-\theta) = \cos \theta$$

The graph is symmetric with respect to the line $\theta = 0$.

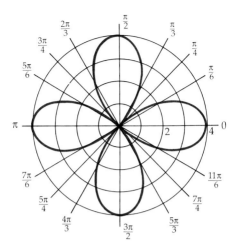

Figure 9.56
$r = 4 \cos 2\theta$

Substitute $\pi - \theta$ for θ.

$$4 \cos(2(\pi - \theta)) = 4 \cos(2\pi - 2\theta) = 4 \cos 2\theta \quad \cos(2\pi - \alpha) = \cos \alpha$$

The graph is symmetric with respect to the line $\theta = \pi/2$. The graph does not have any other symmetries.

Choose some values of θ and find the corresponding values of r. Use symmetry to sketch the graph. The graph is a **four-leaf rose.**

θ	0	$\pi/12$	$\pi/6$	$\pi/4$	$\pi/3$	$5\pi/12$	$\pi/2$
r	4	3.5	2	0	-2	-3.5	-4

■ *Try Exercise* **32**, *page 501.*

Rose curves have the form $r = a \cos n\theta$ or $r = a \sin n\theta$.

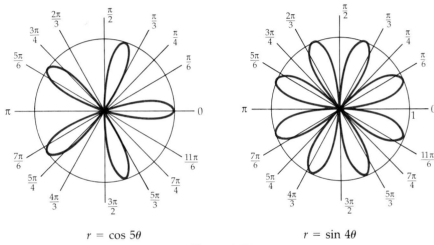

$r = \cos 5\theta$ $\qquad\qquad\qquad$ $r = \sin 4\theta$

Figure 9.57

General Equations of Rose Curves

> The graphs of the equation $r = a \cos n\theta$ and $r = a \sin n\theta$ are rose curves. When n is an even number, the number of petals is $2n$. When n is an odd number, the number of petals is n.

Transformations Between Rectangular and Polar Coordinates

A transformation between coordinate systems is a set of equations that relate the coordinates of one system with the coordinates in a second system. By superimposing a rectangular coordinate system on a polar system, we can derive the set of transformation equations.

Construct a polar coordinate system and a rectangular system so that the pole coincides with the origin and the polar axis coincides with the positive x-axis. Let a point P have coordinates (x, y) in one system and (r, θ) in the other $(r > 0)$.

From the definitions of the sine and cosine of an acute angle in a right triangle, we have

$$\frac{x}{r} = \cos\theta \qquad \text{or} \qquad x = r\cos\theta,$$

and

$$\frac{y}{r} = \sin\theta \qquad \text{or} \qquad y = r\sin\theta.$$

It can be shown that these equations are also true when $r < 0$.

Thus given the point (r, θ) in a polar coordinate system, the coordinates of the point in the xy-coordinate system are given by

$$x = r\cos\theta \qquad y = r\sin\theta$$

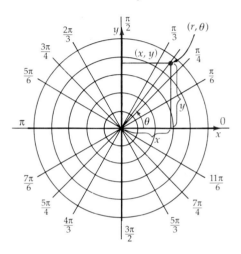

Figure 9.58

For example, to find the point in the xy-coordinate system that corresponds to the point $\left(4, \dfrac{2\pi}{3}\right)$ in the $r\theta$-coordinate system, substitute into the equations and solve for x and y.

$$x = 4\cos\left(\frac{2\pi}{3}\right) = 4\left(-\frac{1}{2}\right) = -2$$

$$y = 4\sin\left(\frac{2\pi}{3}\right) = 4\left(\frac{\sqrt{3}}{2}\right) = 2\sqrt{3}$$

The point is $(-2, 2\sqrt{3})$.

To find the polar coordinates of a given point in the xy-coordinate system, use the Pythagorean Theorem and the definition of the tangent function. Let $P(x, y)$ be a point in the plane and r the distance from the origin to the point P. Then

$$r^2 = x^2 + y^2 \qquad \text{or} \qquad r = \sqrt{x^2 + y^2}.$$

From the definition of the tangent function of an angle in a right triangle,

$$\tan\theta = \frac{y}{x}.$$

Thus θ is the angle whose tangent is y/x. The quadrant for θ is chosen according to the following:

- $x > 0$ and $y > 0$, θ is a first-quadrant angle.
- $x < 0$ and $y > 0$, θ is a second-quadrant angle.
- $x < 0$ and $y < 0$, θ is a third-quadrant angle.
- $x > 0$ and $y < 0$, θ is a fourth-quadrant angle.

The equations of transformations between a polar and rectangular coordinate system are now summarized.

Transformations between Polar and Rectangular Coordinates

Given the point (r, θ) in the polar coordinate system, the transformation equations to change from polar to rectangular coordinates are

$$x = r \cos \theta \qquad y = r \sin \theta$$

Given the point (x, y) in the rectangular coordinate system, the transformation equations to change from rectangular to polar coordinates are

$$r = \sqrt{x^2 + y^2} \qquad \tan \theta = \frac{y}{x}, \, x \neq 0$$

where $r \geq 0$, $0 \leq \theta < 2\pi$, and θ is chosen so that the point lies in the appropriate quadrant. If $x = 0$, then $\theta = \pi/2$ or $\theta = 3\pi/2$.

EXAMPLE 6 **Transform from Polar to Rectangular Coordinates**

Find the rectangular coordinates of the points whose polar coordinates are   a. $(6, 3\pi/4)$ b. $(-4, 30°)$.

Solution   Use the two transformation equations $x = r \cos \theta$ and $y = r \sin \theta$.

a. $x = 6 \cos\left(\dfrac{3\pi}{4}\right) = 6 \cdot -\dfrac{\sqrt{2}}{2} = -3\sqrt{2}$

$y = 6 \sin\left(\dfrac{3\pi}{4}\right) = 6 \cdot \dfrac{\sqrt{2}}{2} = 3\sqrt{2}$

The rectangular coordinates are $(-3\sqrt{2}, 3\sqrt{2})$.

b. $x = -4 \cos(30°) = -4 \cdot \dfrac{\sqrt{3}}{2} = -2\sqrt{3}$

$y = -4 \sin(30°) = -4 \cdot \dfrac{1}{2} = -2$

The rectangular coordinates are $(-2\sqrt{3}, -2)$.

■ *Try Exercise* **44,** *page 501.*

EXAMPLE 7   **Transform from Rectangular to Polar Coordinates**

Find the polar coordinates of the points whose rectangular coordinates are   a. $(-3, 4)$ b. $(-2, -2\sqrt{3})$.

Solution   Use the transformation equations $r = \sqrt{x^2 + y^2}$ and $\tan \theta = \dfrac{y}{x}$.

a.
$$r = \sqrt{(-3)^2 + 4^2} = \sqrt{9 + 16} = \sqrt{25} = 5$$

$$\tan \theta = \frac{4}{-3} = -\frac{4}{3}$$

From this and the fact that $(-3, 4)$ lies in the second quadrant, $\theta \approx 127°$. The approximate polar coordinates of the point are $(5, 127°)$.

b.
$$r = \sqrt{(-2)^2 + (-2\sqrt{3})^2} = \sqrt{4 + 12} = \sqrt{16} = 4$$

$$\tan \theta = \frac{-2\sqrt{3}}{-2} = \sqrt{3}$$

From this and the fact that $(-2, -2\sqrt{3})$ lies in the third quadrant, $\theta = 4\pi/3$. The polar coordinates of the point are $(4, 4\pi/3)$.

■ *Try Exercise* **48,** *page 501.*

Using the transformation equations, it is possible to write a polar equation in rectangular form or a rectangular coordinate equation in polar form.

EXAMPLE 8   **Write a Rectangular Coordinate Equation in Polar Form**

Find a polar form of the equation $x^2 + y^2 - 2x = 3$.

Solution
$$x^2 + y^2 - 2x = 3$$

$(r \cos \theta)^2 + (r \sin \theta)^2 - 2r \cos \theta = 3$ Use the transformation equations.

$r^2(\cos^2\theta + \sin^2\theta) - 2r \cos \theta = 3$ Factor.

$$r^2 - 2r \cos \theta = 3$$

A polar form of the equation is $r^2 - 2r \cos \theta = 3$.

■ *Try Exercise* **56,** *page 501.*

EXAMPLE 9   **Write a Polar Coordinate Equation in Rectangular Form.**

Find a rectangular form of the equation $r^2 \cos 2\theta = 3$.

Solution

$$r^2 \cos 2\theta = 3$$

$$r^2(1 - 2\sin^2\theta) = 3 \quad \cos 2\theta = 1 - 2\sin^2\theta$$

$$r^2 - 2r^2\sin^2\theta = 3$$

$$x^2 + y^2 - 2y^2 = 3 \quad \text{Use the transformation equations.}$$

$$x^2 - y^2 = 3$$

A rectangular form of the equation is $x^2 - y^2 = 3$.

■ *Try Exercise **58**, page 501.*

EXERCISE SET 9.4 _____

In Exercises 1 to 8, plot the point on a polar coordinate system.

1. $(2, 60°)$ **2.** $(3, -90°)$ **3.** $(1, 315°)$

4. $(2, 400°)$ **5.** $\left(-2, \dfrac{\pi}{4}\right)$ **6.** $\left(4, \dfrac{7\pi}{6}\right)$

7. $\left(-3, \dfrac{5\pi}{3}\right)$ **8.** $(-3, \pi)$

In Exercises 9 to 40, sketch the graphs of the polar equations.

9. $r = 3$ **10.** $r = 5$

11. $\theta = 2$ **12.** $\theta = -\dfrac{\pi}{3}$

13. $r = 6\cos\theta$ **14.** $r = 4\sin\theta$

15. $r = 3 + 3\cos\theta$ **16.** $r = 4 - 4\sin\theta$

17. $r = 2 - 3\sin\theta$ **18.** $r = 2 - 2\cos\theta$

19. $r = 4 + 3\sin\theta$ **20.** $r = 2 + 4\sin\theta$

21. $r = -2 + 2\cos\theta$ **22.** $r = -1 - \cos\theta$

23. $r = 2 - 4\sin\theta$ **24.** $r = 4(1 - \sin\theta)$

25. $r = 3\sin 2\theta$ **26.** $r = 2\cos 2\theta$

27. $r = 4\cos 3\theta$ **28.** $r = 5\sin 3\theta$

29. $r = 2\sin 5\theta$ **30.** $r = 3\cos 5\theta$

31. $r = 3\cos 4\theta$ **32.** $r = 6\sin 4\theta$

33. $r = 3\sec\theta$ **34.** $r = 4\csc\theta$

35. $r = 5\csc\theta$ **36.** $r = 2\sec\theta$

37. $r = \theta$ **38.** $r = -\theta$

39. $r = 2^\theta, \theta \geq 0$ **40.** $r = \dfrac{1}{\theta}, \theta > 0$

In Exercises 41 to 48, transform the given coordinates to the indicated ordered pair.

41. $(1, -\sqrt{3})$ to (r, θ) **42.** $(-2\sqrt{3}, 2)$ to (r, θ)

43. $\left(-3, \dfrac{2\pi}{3}\right)$ to (x, y) **44.** $\left(2, -\dfrac{\pi}{3}\right)$ to (x, y)

45. $\left(0, -\dfrac{\pi}{2}\right)$ to (x, y) **46.** $\left(3, \dfrac{5\pi}{6}\right)$ to (x, y)

47. $(3, 4)$ to (r, θ) **48.** $(12, -5)$ to (r, θ)

In Exercises 49 to 60, find an equation in x and y that has the same graph as the given polar equation.

49. $r = 3\cos\theta$ **50.** $r = 2\sin\theta$

51. $r = 3\sec\theta$ **52.** $r = 4\csc\theta$

53. $r = 4$ **54.** $\theta = \dfrac{\pi}{4}$

55. $r = \tan\theta$ **56.** $r = \cot\theta$

57. $r = \dfrac{2}{1 + \cos\theta}$ **58.** $r = \dfrac{2}{1 - \sin\theta}$

59. $r(\sin\theta - 2\cos\theta) = 6$ **60.** $r(2\cos\theta + \sin\theta) = 3$

In Exercises 61 to 68, find a polar equation that has the same graph as the given equation.

61. $y = 2$ **62.** $x = -4$

63. $x^2 + y^2 = 4$ **64.** $2x - 3y = 6$

65. $x^2 = 8y$ **66.** $y^2 = 4y$

67. $x^2 - y^2 = 25$ **68.** $x^2 + 4y^2 = 16$

Supplemental Exercises

For Exercises 69 to 76, sketch a graph of the polar equation.

69. $r^2 = 4\cos 2\theta$ (lemniscate)

70. $r^2 = -2\sin 2\theta$ (lemniscate)

71. $r = 2(1 + \sec\theta)$ (conchoid)

72. $r = 2\cos 2\theta \sec\theta$ (strophoid)

73. $r\theta = 2$ (spiral)

74. $r = 2 \sin \theta \cos^2 2\theta$ (bifolium)

75. $r = |\theta|$

76. $r = \ln \theta$

77. If $P_1(r_1, \theta_1)$ and $P_2(r_2, \theta_2)$ are two points in the $r\theta$-plane,

use the Law of Cosines to show that the distance between the two points $d(P_1, P_2)$ is given by

$$[d(P_1, P_2)]^2 = r_1^2 + r_2^2 - 2r_1 r_2 \cos(\theta_1 - \theta_2).$$

78. Prove that the graph of $r = a \sin \theta + b \cos \theta$, $ab \neq 0$, is a circle in polar coordinates.

Chapter 9 Review

9.1   **Conic Sections**

A parabola is the set of points in the plane that are equidistant from a fixed line (the directrix) and a fixed point (the focus) not on the directrix.

The equations of a parabola with vertex at (h, k) and axis of symmetry parallel to a coordinate axis are given by

$$(x - h)^2 = 4p(y - k) \quad \text{Focus: } (h, k + p), \text{ directrix: } y = k - p$$

$$(y - k)^2 = 4p(x - h) \quad \text{Focus: } (h + p, k), \text{ directrix: } x = h - p$$

9.2   **Ellipses**

An ellipse is the set of all points in the plane, the sum of whose distances from two fixed points (foci) is a positive constant.

The equations of an ellipse with center at (h, k) and major axis parallel to the coordinate axes are given by

$$\frac{(x - h)^2}{a^2} + \frac{(y - k)^2}{b^2} = 1 \quad \text{Foci: } (h \pm c, k), \text{ vertices: } (h \pm a, k)$$

$$\frac{(x - h)^2}{b^2} + \frac{(y - k)^2}{a^2} = 1 \quad \text{Foci: } (h, k \pm c), \text{ vertices: } (h, k \pm a)$$

For each equation, $a > b$ and $c^2 = a^2 - b^2$.

The eccentricity e of an ellipse is given by $e = c/a$.

9.3   **Hyperbolas**

A hyperbola is the set of all points in the plane, the difference of whose distances from two fixed points (foci) is a positive constant.

The equations of a hyperbola with center at (h, k) and transverse axis parallel to a coordinate axis are given by

$$\frac{(x - h)^2}{a^2} - \frac{(y - k)^2}{b^2} = 1 \quad \text{Foci: } (h \pm c, k), \text{ vertices: } (h \pm a, k)$$

$$\frac{(y - k)^2}{a^2} + \frac{(x - h)^2}{b^2} = 1 \quad \text{Foci: } (h, k \pm c), \text{ vertices: } (h, k \pm a)$$

For each equation, $c^2 = a^2 + b^2$.

The eccentricity e of a hyperbola is given by $e = c/a$.

9.4   Introduction to Polar Coordinates

A polar coordinate system is formed by drawing a ray (polar axis) and concentric circles with center at the beginning of the ray. The pole is the origin of a polar coordinate system.

A point is specified by coordinates (r, θ), where r is a directed distance from the pole and θ is an angle measured from the polar axis.

The transformation equations between a polar coordinate system and a rectangular coordinate system are

Polar to rectangular: $x = r \cos \theta$ Rectangular to polar: $r = \sqrt{x^2 + y^2}$

$$y = r \sin \theta \qquad\qquad \tan \theta = \frac{y}{x}$$

CHALLENGE EXERCISES

In Exercises 1 to 10, answer true or false. If the answer is false, give an example.

1. The graph of a parabola is the same shape as one branch of a hyperbola.

2. For the two axes of an ellipse, the major axis and the minor axis, the major axis is always the longer axis.

3. For the two axes of a hyperbola, the transverse axis and the conjugate axis, the transverse axis is always the longer axis.

4. If two ellipses have the same foci, they will have the same graph.

5. A hyperbola is similar to a parabola in that both curves have asymptotes.

6. If a hyperbola with center at the origin and a parabola with vertex at the origin have the same focus, $(0, c)$, the two graphs will always intersect.

7. The graphs of all the conic sections are not the graphs of a function.

8. If F_1 and F_2 are the two foci of an ellipse and P is a point on the ellipse, then $d(P, F_1) + d(P, F_2) = 2a$ where a is the length of the semimajor axis of the ellipse.

9. The eccentricity of a hyperbola is always greater than 1.

10. Each ordered pair (r, θ) in a polar coordinate system specifies exactly one point.

REVIEW EXERCISES

In Exercises 1 to 12, find the foci and vertices of each conic. If the conic is a hyperbola, find the asymptotes. Sketch the graph.

1. $x^2 - y^2 = 4$

2. $y^2 = 16x$

3. $x^2 + 4y^2 - 6x + 8y - 3 = 0$

4. $3x^2 - 4y^2 + 12x - 24y - 36 = 0$

5. $3x - 4y^2 + 8y + 2 = 0$

6. $3x + 2y^2 - 4y - 7 = 0$

7. $9x^2 + 4y^2 + 36x - 8y + 4 = 0$

8. $11x^2 - 25y^2 - 44x - 50y - 256 = 0$

9. $4x^2 - 9y^2 - 8x + 12y - 144 = 0$

10. $9x^2 + 16y^2 + 36x - 16y - 104 = 0$

11. $4x^2 + 28x + 32y + 81 = 0$

12. $x^2 - 6x - 9y + 27 = 0$

In Exercises 13 to 20, find the equation in standard form of the conic that satisfies the given conditions.

13. Ellipse with vertices at $(7, 3)$ and $(-3, 3)$; length of minor axis 8

14. Hyperbola with vertices at $(4, 1)$ and $(-2, 1)$; eccentricity $4/3$

15. Hyperbola with foci $(-5, 2)$ and $(1, 2)$; length of transverse axis 8

16. Parabola with focus $(2, -3)$ and directrix $x = 6$.

17. Parabola with vertex $(0, -2)$ and passing through the point $(3, 4)$

18. Ellipse with eccentricity $2/3$ and foci $(-4, -1)$ and $(0, -1)$

19. Hyperbola with vertices $(\pm 6, 0)$ and asymptotes whose equations are $y = \pm(1/9)x$

20. Parabola passing through the points $(1, 0)$, $(2, 1)$, and $(0, 1)$ with axis of symmetry parallel to the y-axis.

21. Find the equation of the parabola traced by a point $P(x, y)$ that moves so that the distance between $P(x, y)$ and the line $x = 2$ equals the distance between $P(x, y)$ and the point $(-2, 3)$.

22. Find the equation of the parabola traced by a point $P(x, y)$ that moves so that the distance between $P(x, y)$ and the line $y = 1$ equals the distance between $P(x, y)$ and the point $(-1, 2)$.

23. Find the equation of the ellipse traced by a point $P(x, y)$ that moves so that the sum of its distances to $(-3, 1)$ and $(5, 1)$ is 10.

24. Find the equation of the ellipse traced by a point $P(x, y)$ that moves so that the sum of its distances to $(3, 5)$ and $(3, -1)$ is 8.

In Exercises 25 to 34, sketch the graph of the polar equations.

25. $r = 4 \cos 3\theta$

26. $r = 1 + \cos \theta$

27. $r = 2(1 - 2 \sin \theta)$

28. $r = 4 \sin 4\theta$

29. $r = 5 \sin \theta$

30. $r = 3 \sec \theta$

31. $r = 4 \csc \theta$

32. $r = 4 \sin \theta$

33. $r = 3 + 2 \cos \theta$

34. $r = 4 + 2 \sin \theta$

In Exercises 35 to 38, change the equations to polar equations.

35. $y^2 = 16x$

36. $x^2 + y^2 + 4x + 3y = 0$

37. $3x - 2y = 6$

38. $xy = 4$

In Exercises 39 to 42, change the equations to rectangular equations.

39. $r = \dfrac{4}{1 - \cos \theta}$

40. $r = 3 \cos \theta - 4 \sin \theta$

41. $r^2 = \cos 2\theta$

42. $\theta = 1$

10

Systems of Equations

The greatest mathematicians, as Archimedes, Newton, and Gauss, always united theory and applications in equal measure.

FELIX KLEIN

BOARDWALK? PARK PLACE? HOW ABOUT NEW YORK?: A WINNING STRATEGY FOR MONOPOLY

During the Depression, an unemployed engineer named Charles Darrow invented Monopoly. In 1935, he sold the game to Parker Brothers and became a millionaire. Today, Monopoly is still a very popular board game. In fact, Parker Brothers has sponsored world championship Monopoly games in Atlantic City, home of Baltic and Mediterranean Avenues.

In the early 1980's, Stephen Heppe, at the time a student, became interested in winning Monopoly strategies. He wanted to know which properties on the Monopoly board paid a greater rate of return for each dollar invested. The answer to Heppe's question required solving a system of *linear equations*, the subject of this chapter. Heppe's system of equations contained 123 equations with 123 variables. It would be a prodigous effort to solve such a system without the aid of a computer. Fortunately, he did have one available. Some of the results of his calculations are

1. A player is less likely to land on Mediterranean Avenue during the course of the game than on any other property.
2. The chances that a player lands on Illinois Avenue is greater than that of any other property.
3. There is a 21% better chance of landing on Broadway than there is of landing on Park Place.

Besides knowing which properties have the greatest chance of being occupied, Heppe also wanted to know which properties paid the greatest return for each dollar invested in houses or hotels. Some of the conclusions reached by Heppe are

1. New York with a hotel has the highest rate of return.
2. The lowest rate of return for a property with a hotel is Mediterranean.
3. Assuming all the railroads are owned, the B&O railroad has the greatest rate of return of all the railroads. The reason is that it is more likely a player will land on this railroad than the other railroads.
4. With one house, Boardwalk has a greater return on investment than New York. With 4 houses, New York has a greater return on investment than Boardwalk.

For more information on the mathematics of monopoly, see "Matrix Mathematics: How to Win at Monopoly" by Dr. Crypton in the September 1985 issue of *Science Digest*.

10.1

Systems of Linear Equations in Two Variables

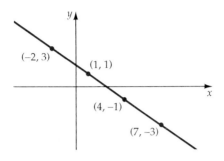

Figure 10.1
$2x + 3y = 5$

Recall that an equation of the form $Ax + By = C$ is a linear equation in two variables. A solution of a linear equation in two variables is an ordered pair (x, y), which makes the equation a true statement. For example, $(-2, 3)$ is a solution of the equation

$$2x + 3y = 5$$

since

$$2(-2) + 3(3) = 5.$$

The graph of a linear equation, a straight line, is the set of points whose ordered pairs satisfy the equation. Figure 10.1 is the graph of $2x + 3y = 5$.

A **system of equations** is two or more equations considered together. The following system of the equations is a **linear system of equations** in two variables.

$$\begin{cases} 2x + 3y = 4 \\ 3x - 2y = -7 \end{cases}$$

The **solution** of a system of equations is an ordered pair that is a solution of both equations.

In Figure 10.2, the graphs of the two equations in the system of equations above intersect at the point $(-1, 2)$. Since that point lies on both lines, $(-1, 2)$ is a solution of both equations and thus is a solution of the system of equations. The point $(5, -2)$ is a solution of the first equation but not the second equation. Therefore $(5, -2)$ is not a solution of the system of equations.

The graphs of two linear equations in two variables can intersect, be parallel, or be the same line. When the graphs intersect at a single point, the system is called a **consistent** system of equations. The system is called a **dependent** system of equations when the equations represent the same line. In this case, the system has an infinite number of solutions. When the graphs of the two lines are parallel, the system is called **inconsistent** and the system has no solution.

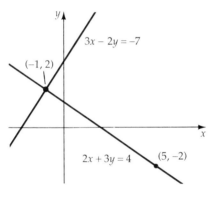

Figure 10.2

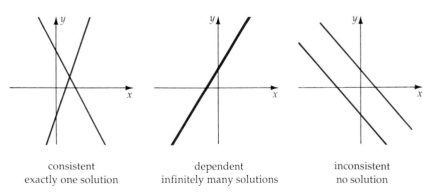

consistent
exactly one solution

dependent
infinitely many solutions

inconsistent
no solution

Figure 10.3

Substitution Method for Solving a System of Linear Equations

The **substitution method** is one procedure for solving a system of equations. This method is illustrated in Example 1.

EXAMPLE 1   Solve a System of Equations by the Substitution Method

Solve the system of equations

$$\begin{cases} 3x - 5y = 7 \\ \qquad y = 2x. \end{cases}$$

Solution   The solutions of the equation $y = 2x$ are the ordered pairs $(x, 2x)$. Since a solution of a system of equations must be a solution of each equation in the system, the ordered pair $(x, 2x)$ must also be a solution of the equation $3x - 5y = 7$.

Substitute $(x, 2x)$ into the first equation and solve for x. Think of this as *substituting $2x$ for y*.

$$3x - 5(2x) = 7$$
$$-7x = 7$$
$$x = -1$$

Substituting this value of x into the second equation and solving for y, we obtain

$$y = 2(-1) = -2.$$

The only solution of the system of equations is the ordered pair $(-1, -2)$. When a system of equations has a unique solution, the system is consistent.

■ *Try Exercise **6**, page 513.*

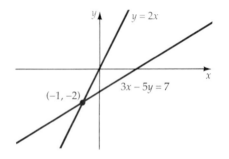

Figure 10.4

EXAMPLE 2   Solve a Dependent System of Equations

Solve the system of equations

$$\begin{cases} 4x - 8y = 16 & \text{(1)} \\ \ x - 2y = 4. & \text{(2)} \end{cases}$$

Solution   For convenience, we labeled the equations (1) and (2). Solve Equation (2) for x.

$$x - 2y = 4$$
$$x = 2y + 4$$

Each ordered pair solution of Equation (2) is of the form $(2y + 4, y)$. Since a solution of a system of equations must be a solution of each equation in the system, the ordered pair $(2y + 4, y)$ must also be a solution of Equation (1).

Substitute $(2y + 4, y)$ into Equation (1) and solve for y. Think of this as substituting $2y + 4$ for x.

$$4(2y + 4) - 8y = 16$$

$$16 = 16$$

Substituting the ordered pair $(2y + 4, y)$ into Equation (1) produced a true statement $16 = 16$. Thus the ordered pair is a solution of Equation (1). Since it is also a solution of Equation (2), the ordered pair is a solution of the system of equations.

By choosing any value of y, we will obtain a solution of the system of equations. Choose $y = -3$. Then $x = 2y + 4 = 2(-3) + 4 = -2$, so we get the solution $(-2, -3)$. By choosing $y = 2$, the solution $(8, 2)$ is obtained.

For the system of equations, if y is any real number c, then the ordered pair solutions are $(2c + 4, c)$. Because there are infinitely many real numbers c from which to choose, the system of equations has an infinite number of solutions. The graphs of the equations are the same line, and the system is dependent. See Figure 10.5.

■ *Try Exercise* **20,** *page 514.*

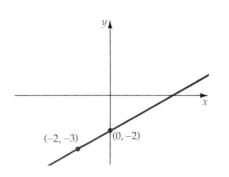

Figure 10.5
The graphs of the equations in the system are the same line.

Remark   Before leaving Example 2, note that there is more than one way to represent the ordered pair solutions. To illustrate this point, solve Equation (2) for y.

$$x - 2y = 4$$

$$y = \frac{1}{2}x - 2$$

If $x = -2$, then $y = -3$. One solution is $(-2, -3)$, which corresponds to one of the solutions found above. Since there are an infinite number of x's from which to choose, there are infinitely many solutions of the system. In general, if x is any real number b, then $(b, b/2 - 2)$ is a solution of the system. Either of the ordered pairs, $(2c + 4, c)$ or $(b, b/2 - 2)$, will generate all the solutions of the system.

EXAMPLE 3   Identify an Inconsistent System of Equations

Solve the following system of equations by substitution:

$$\begin{cases} 3x - y = 6 & \text{(1)} \\ 6x - 2y = 5 & \text{(2)} \end{cases}$$

Solution   Solve Equation (1) for y. Substitute into Equation (2) and solve for x.

$$3x - y = 6$$

$$y = 3x - 6$$

$$6x - 2(3x - 6) = 5 \qquad \text{Substitute } (x, 3x - 6) \text{ into Equation (2).}$$

$$12 = 5$$

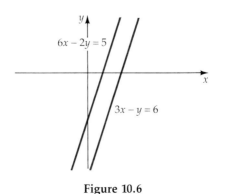

Figure 10.6

Because $12 = 5$ is a false equation, no ordered pairs of the form $(x, 3x - 6)$, which are the solutions of Equation (1), are also solutions of Equation (2). The system of equations has no ordered pairs in common and thus has no solution.

■ *Try Exercise* **18,** *page 514.*

Elimination Method of Solving a System of Linear Equations

A second strategy for solving a system of equations is similar to the strategy for solving first-degree equations in one variable. The system of equations is replaced by a series of equivalent systems until the solution is obvious.

Two systems of equations are **equivalent** if each system has exactly the same solutions. The systems

$$\begin{cases} 3x + 5y = 9 \\ 2x - 3y = -13 \end{cases} \quad \text{and} \quad \begin{cases} x = -2 \\ y = 3 \end{cases}$$

are equivalent systems of equations. Each system has solution $(-2, 3)$. Figure 10.7 shows the graphs of these systems.

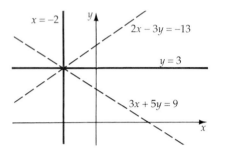

Figure 10.7

Operations That Produce an Equivalent System of Equations

> 1. Interchange any two equations.
> 2. Replace an equation with a nonzero multiple of that equation.
> 3. Replace an equation with the sum of an equation and a nonzero constant multiple of another equation in the system.

Since the order of the equations in a system does not affect the system, interchanging the equations does not affect the solution. The second operation restates the property that says that multiplying each side of an equation by the same nonzero constant does not change the solutions of the equation.

The third operation can be illustrated as follows. Consider the system of equations.

$$\begin{cases} 3x + 2y = 10 & \quad\quad (1) \\ 2x - 3y = -2 & \quad\quad (2) \end{cases}$$

Multiply each side of Equation (2) by 2. (Any nonzero number would work.) Add the resulting equation to Equation (1).

$$2(2x - 3y) = 2 \cdot (-2) \quad \text{or} \quad 4x - 6y = -4$$

$$\begin{aligned} 3x + 2y &= 10 \\ \underline{4x - 6y} &= \underline{-4} \\ 7x - 4y &= 6 \end{aligned}$$

Replace Equation (1) with the new equation to produce the following

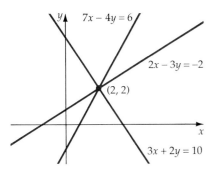

Figure 10.8

equivalent system.

$$\begin{cases} 7x - 4y = 6 \\ 2x - 3y = -2 \end{cases}$$

The third property says that the resulting system of equations has the same solution as the original system. This system is therefore equivalent to the first system. Figure 10.8 shows the graph of $7x - 4y = 6$. Note that the graph of the line intersects at the same point $(2, 2)$ as the graphs of the lines in the original system of equations.

The goal in solving a system of equations is to replace the given system with an equivalent system of the form

$$\begin{cases} x = a \\ y = b \end{cases}$$

The solution of the system is (a, b). You can do this by using the operations that produce an equivalent system of equations.

EXAMPLE 4   Solve a System of Equations by the Elimination Method

Solve the system of equations

$$\begin{cases} 3x - 4y = 10 & (1) \\ 2x + 5y = -1 & (2) \end{cases}$$

Solution   We will use the operations that produce an equivalent system to eliminate the x variable. Multiply Equation (1) by 2 and Equation (2) by -3. Add the equations and solve for y.

$$\begin{array}{r} 6x - 8y = 20 \\ -6x - 15y = 3 \\ \hline -23y = 23 \\ y = -1 \end{array}$$

The multipliers were chosen so that the coefficients of the variable we want to eliminate are additive inverses.

Replace one of the equations in the original system, we replaced Equation (2) by the equation $y = -1$.

$$\begin{cases} 3x - 4y = 10 \\ y = -1 \end{cases}$$ This system of equations is equivalent to the original system.

Solve Equation (1) by substituting -1 for y.

$$3x - 4(-1) = 10$$
$$x = 2$$

The solution of this system is $(2, -1)$.

■ *Try Exercise* **24**, *page 514.*

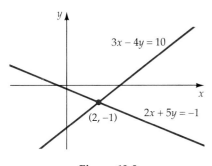

Figure 10.9

The method just described is called the **elimination method** for solving a system of equations since it involves eliminating a variable from one of the equations.

EXAMPLE 5   Solve a Dependent System of Equations

Solve the system of equations

$$\begin{cases} x - 2y = 4 & (1) \\ 3x - 6y = 12 & (2) \end{cases}$$

Solution   Eliminate x by multiplying Equation (2) by $-\dfrac{1}{3}$ and then adding the result to Equation (1).

$$\begin{aligned} x - 2y &= 4 \\ -x + 2y &= -4 \\ \hline 0 &= 0 \end{aligned}$$

Replace Equation (2) by $0 = 0$.

$$\begin{cases} x - 2y = 4 \\ \qquad\quad 0 = 0 \end{cases}$$ This system of equations is equivalent to the original system.

An ordered pair solution of a system of equations is a solution of both equations of the system. Any ordered pair that is a solution of $x - 2y = 4$ is also a solution of $0 = 0$ (this equation is an identity and therefore true for all values of x and y). Thus the solutions of the system are the solutions of $x - 2y = 4$. The graphs of the two lines will coincide.

The solutions of the system of equations are those ordered pairs (x, y) for which $x - 2y = 4$, or, solving for y, $y = x/2 - 2$. Thus, the ordered pairs can be written as $(x, x/2 - 2)$. Letting $x = 4$, $\sqrt{2}$, and -6 and substituting into $(x, x/2 - 2)$, we obtain the ordered pair solutions $(4, 0)$, $(\sqrt{2}, \sqrt{2}/2 - 2)$, and $(-6, -5)$.

■ *Try Exercise **28**, page 514.*

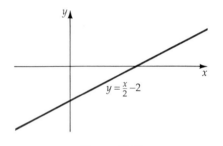

$y = \dfrac{x}{2} - 2$

Figure 10.10

If one equation of a system is replaced by a false equation, the system will have no solution. For example, the system

$$\begin{cases} x + y = 4 \\ \qquad\quad 0 = 5 \end{cases}$$

has no solution because the second equation has no ordered pair solutions.

EXAMPLE 6   Identify a System of Equations with no Solution

Solve the system of equations

$$\begin{cases} 6x - 9y = 14 & (1) \\ 4x - 6y = -5 & (2) \end{cases}$$

Solution   Multiply Equation (1) by 2 and Equation (2) by -3 and then add the equations.

$$\begin{aligned} 12x - 18y &= 28 \\ -12x + 18y &= 15 \\ \hline 0 &= 43 \end{aligned}$$

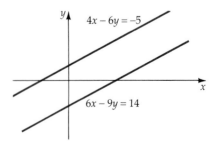

$4x - 6y = -5$

$6x - 9y = 14$

Figure 10.11

Because the equation $0 = 43$ has no solutions, the system of equations has no solution. The graphs of the lines are parallel. See Figure 10.11.

■ *Try Exercise 30, page 514.*

Applications of Systems of Equations

As application problems become more difficult, it becomes impossible to represent all unknowns in terms of a single variable. In such cases, a system of equations can be used.

EXAMPLE 7 **Solve an Application**

A rowing team rowing with the current traveled 18 miles in 2 hours. Against the current, the team rowed 10 miles in 2 hours. Find the rate of the rowing team in calm water and the rate of the current.

Solution Let r_1 represent the rate of the boat in calm water and r_2 represent the rate of the current.

The rate of the boat *with the current* is $r_1 + r_2$.
The rate of the boat *against the current* is $r_1 - r_2$.

Since the rowing team traveled 18 miles in 2 hours with the current, we use the equation $d = rt$ to write

$$d = r \cdot t$$

$$18 = (r_1 + r_2) \cdot 2 \quad d = 18, t = 2$$

$$9 = r_1 + r_2$$

Since the team rowed 10 miles in 2 hours against the current, we write

$$10 = (r_1 - r_2) \cdot 2 \quad d = 10, t = 2$$

$$5 = r_1 - r_2$$

Thus we have a system of two linear equations in the variables r_1 and r_2.

$$\begin{cases} 9 = r_1 + r_2 \\ 5 = r_1 - r_2 \end{cases}$$

Solving the system using the elimination method, we obtain r_1 is 7 mph and r_2 is 2 mph. Thus the rate of the boat in calm water is 7 mph and the rate of the current is 2 mph. You should verify these solutions.

■ *Try Exercise 44, page 514.*

EXERCISE SET 10.1

In Exercises 1 to 20, solve the following systems of equations by the substitution method.

1. $\begin{cases} 2x - 3y = 16 \\ \quad x = 2 \end{cases}$

2. $\begin{cases} 3x - 2y = -11 \\ \quad y = 1 \end{cases}$

3. $\begin{cases} 3x + 4y = 18 \\ \quad y = -2x + 3 \end{cases}$

4. $\begin{cases} 5x - 4y = -22 \\ \quad y = 5x - 2 \end{cases}$

5. $\begin{cases} -2x + 3y = 6 \\ \quad x = 2y - 5 \end{cases}$

6. $\begin{cases} 8x + 3y = -7 \\ \quad x = 3y + 15 \end{cases}$

7. $\begin{cases} 6x + 5y = 1 \\ x - 3y = 4 \end{cases}$

8. $\begin{cases} -3x + 7y = 14 \\ 2x - y = -13 \end{cases}$

9. $\begin{cases} 7x + 6y = -3 \\ y = \dfrac{2}{3}x - 6 \end{cases}$

10. $\begin{cases} 9x - 4y = 3 \\ x = \dfrac{4}{3}y + 3 \end{cases}$

11. $\begin{cases} y = 4x - 3 \\ y = 3x - 1 \end{cases}$

12. $\begin{cases} y = 5x + 1 \\ y = 4x - 2 \end{cases}$

13. $\begin{cases} y = 5x + 4 \\ x = -3y - 4 \end{cases}$

14. $\begin{cases} y = -2x - 6 \\ x = -2y - 2 \end{cases}$

15. $\begin{cases} 3x - 4y = 2 \\ 4x + 3y = 14 \end{cases}$

16. $\begin{cases} 6x + 7y = -4 \\ 2x + 5y = 4 \end{cases}$

17. $\begin{cases} 3x - 3y = 5 \\ 4x - 4y = 9 \end{cases}$

18. $\begin{cases} 3x - 4y = 8 \\ 6x - 8y = 9 \end{cases}$

19. $\begin{cases} 4x + 3y = 6 \\ y = -\dfrac{4}{3}x + 2 \end{cases}$

20. $\begin{cases} 5x + 2y = 2 \\ y = -\dfrac{5}{2}x + 1 \end{cases}$

In Exercises 21 to 40, solve the following systems of equations by the elimination method.

21. $\begin{cases} 3x - y = 10 \\ 4x + 3y = -4 \end{cases}$

22. $\begin{cases} 3x + 4y = -5 \\ x - 5y = -8 \end{cases}$

23. $\begin{cases} 4x + 7y = 21 \\ 5x - 4y = -12 \end{cases}$

24. $\begin{cases} 3x - 8y = -6 \\ -5x + 4y = 10 \end{cases}$

25. $\begin{cases} 5x - 3y = 0 \\ 10x - 6y = 0 \end{cases}$

26. $\begin{cases} 3x + 2y = 0 \\ 2x + 3y = 0 \end{cases}$

27. $\begin{cases} 6x + 6y = 1 \\ 4x + 9y = 4 \end{cases}$

28. $\begin{cases} 4x + 5y = 2 \\ 8x - 15y = 9 \end{cases}$

29. $\begin{cases} 3x + 6y = 11 \\ 2x + 4y = 9 \end{cases}$

30. $\begin{cases} 4x - 2y = 9 \\ 2x - y = 3 \end{cases}$

31. $\begin{cases} \dfrac{5}{6}x - \dfrac{1}{3}y = -6 \\ \dfrac{1}{6}x + \dfrac{2}{3}y = 1 \end{cases}$

32. $\begin{cases} \dfrac{3}{4}x + \dfrac{2}{5}y = 1 \\ \dfrac{1}{2}x - \dfrac{3}{5}y = -1 \end{cases}$

33. $\begin{cases} \dfrac{3}{4}x + \dfrac{1}{3}y = 1 \\ \dfrac{1}{2}x + \dfrac{2}{3}y = 0 \end{cases}$

34. $\begin{cases} \dfrac{3}{5}x - \dfrac{2}{3}y = 7 \\ \dfrac{2}{5}x - \dfrac{5}{6}y = 7 \end{cases}$

35. $\begin{cases} 2\sqrt{3}\,x - 3y = 3 \\ 3\sqrt{3}\,x + 2y = 24 \end{cases}$

36. $\begin{cases} 4x - 3\sqrt{5}\,y = -19 \\ 3x + 4\sqrt{5}\,y = 17 \end{cases}$

37. $\begin{cases} 3\pi x - 4y = 6 \\ 2\pi x + 3y = 5 \end{cases}$

38. $\begin{cases} 2x - 5\pi y = 3 \\ 3x + 4\pi y = 2 \end{cases}$

39. $\begin{cases} 3\sqrt{2}\,x - 4\sqrt{3}\,y = -6 \\ 2\sqrt{2}\,x + 3\sqrt{3}\,y = 13 \end{cases}$

40. $\begin{cases} 2\sqrt{2}\,x + 3\sqrt{5}\,y = 7 \\ 3\sqrt{2}\,x - \sqrt{5}\,y = -17 \end{cases}$

In Exercises 41 to 55, solve by using a system of equations.

41. Flying with the wind, a plane travels 450 miles in 3 hours. Flying against the wind, the plane traveled the same distance in 5 hours. Find the rate of the plane in calm air and the rate of the wind.

42. A plane flew 800 miles in 4 hours while flying with the wind. Against the wind, it took the plane 5 hours to travel the 800 miles. Find the rate of the plane in calm air and the rate of the wind.

43. A motorboat traveled a distance of 120 miles in 4 hours while traveling with the current. Against the current, the same trip took 6 hours. Find the rate of the boat in calm water and the rate of the current.

44. A canoeist can row 12 miles with the current in 2 hours. Rowing against the current, it takes the canoeist 4 hours to travel the same distance. Find the rate of the canoeist in calm water and the rate of the current.

45. A metallurgist made two purchases. The first purchase, which cost $1080, included 30 kilograms of an iron alloy and 45 kilograms of a lead alloy. The second purchase at the same prices cost $372 and included 15 kilograms of the iron alloy and 12 kilograms of the lead alloy. Find the cost per kilogram of the iron and lead alloys.

46. For $14.10, a chemist purchased 10 liters of hydrochloric acid and 15 liters of silver nitrate. A second purchase at the same prices cost $18.16 and included 12 liters of hydrochloric acid and 20 liters of silver nitrate. Find the cost per liter of the two chemicals.

47. A coin bank contains only nickels and dimes. The value of the coins is $1.30. If the nickels were dimes and dimes were nickels, the value of the coins would be $1.55. Find the original number of nickels and dimes in the bank.

48. The coin drawer of a cash register contains dimes and quarters. The value of the coins is $4.35. If the dimes were quarters and the quarters were dimes, the value of the coins would be $3.00. Find the original number of dimes and quarters in the cash register.

49. The sum of the digits of a two-digit number is 14. If the digits are reversed, the new number is 18 less than the original number. Find the original number.

50. The sum of the digits of a two-digit number is 11. If the digits are reversed, the new number is 63 more than the original number. Find the original number.

51. A broker invests $25,000 of a client's money in two different municipal bonds. The annual rate of return on one bond is 6 percent, and the annual rate of return on the second bond is 6.5 percent. If the investor receives a total annual interest payment from the two bonds of $1555, find the amount invested in each bond.

52. An investment of $3000 is placed in stocks and bonds.

The annual rate of return for the stocks is 4.5 percent, and the rate of return on the bonds is 8 percent. If the annual interest payment from the stocks and bonds is \$177, find the amount invested in bonds.

53. A goldsmith has two gold alloys. The first alloy is 40% gold; the second alloy is 60% gold. How many grams of each should be mixed to produce 20 grams of an alloy that is 52% gold?

54. One acetic acid solution contains 70% water and another contains 30% water. How many liters of each solution should be mixed to produce 20 liters of a solution that is 40% water?

55. A chemist wants to make 50 milliliters of a 16% acid solution. How many milliliters each of a 13% acid solution and an 18% acid solution should be mixed to produce the desired solution?

Supplemental Exercises

In Exercises 56 to 65, solve for x and y. Use the fact that if $z_1 = a_1 + b_1 i$ and $z_2 = a_2 + b_2 i$ are two complex numbers, then $z_1 = z_2$ if and only if $a_1 = a_2$ and $b_1 = b_2$.

56. $(2 + i)x + (3 - i)y = 7$

57. $(3 + 2i)x + (4 - 3i)y = 2 - 16i$

58. $(4 - 3i)x + (5 + 2i)y = 11 + 9i$

59. $(2 + 6i)x + (4 - 5i)y = -8 - 7i$

60. $(-3 - i)x - (4 + 2i)y = 1 - i$

61. $(5 - 2i)x + (-3 - 4i)y = 12 - 35i$

62. $\begin{cases} 2x + 5y = 11 + 3i \\ 3x + y = 10 - 2i \end{cases}$

63. $\begin{cases} 4x + 3y = 11 + 6i \\ 3x - 5y = 1 + 19i \end{cases}$

64. $\begin{cases} 2x + 3y = 11 + 5i \\ 3x - 3y = 9 - 15i \end{cases}$

65. $\begin{cases} 5x - 4y = 15 - 41i \\ 3x + 5y = 9 + 5i \end{cases}$

In Exercises 66 to 69, use the system of equations

$$\begin{cases} a_1 x + b_1 y = c_1 \\ a_2 x + b_2 y = c_2 \end{cases}$$

66. Assuming $a_1 b_2 - a_2 b_1 \neq 0$, solve the system for x and y.

67. Prove that the system of equations has a unique solution if and only if $a_1 b_2 - a_2 b_1 \neq 0$. (*Hint:* See the answer to Exercise 66.)

68. Prove that the system of equations has no solution if and only if

$$\frac{a_1}{a_2} = \frac{b_1}{b_2}, \quad \frac{a_1}{a_2} \neq \frac{c_1}{c_2}, \quad \frac{b_1}{b_2} \neq \frac{c_1}{c_2}.$$

69. Prove that the system of equations has infinitely many solutions if and only if

$$\frac{a_1}{a_2} = \frac{b_1}{b_2}, \quad \frac{a_1}{a_2} = \frac{c_1}{c_2}, \quad \frac{b_1}{b_2} = \frac{c_1}{c_2}.$$

10.2

Systems of Linear Equations in More Than Two Variables

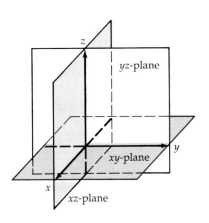

Figure 10.12

An equation of the form $ax + by + cz = d$, with a, b, and c not all zero, is a linear equation in three variables. A solution of an equation in three variables is an **ordered triple** (x, y, z).

The ordered triple $(2, -1, -3)$ is one of the solutions of the equation $2x - 3y + z = 4$. The ordered triple $(3, 1, 1)$ is another solution. In fact an infinite number of ordered triples are solutions of the equation.

Graphing an equation in three variables requires a third coordinate axis perpendicular to the xy-plane. This third axis is commonly called the **z-axis**. The result is a three-dimensional coordinate system called the xyz-coordinate system (Figure 10.12). To help visualize a three-dimensional coordinate system, think of a corner of a room: the floor is the xy-plane, one wall is the yz-plane, and the other wall is the xz-plane.

Graphing an ordered triple requires three moves, the first along the x-axis, the second along the y-axis, and the third along the z-axis, respectively. Figure 10.13 is the graph of the points $(-5, 4, 3)$ and $(4, 5, -2)$.

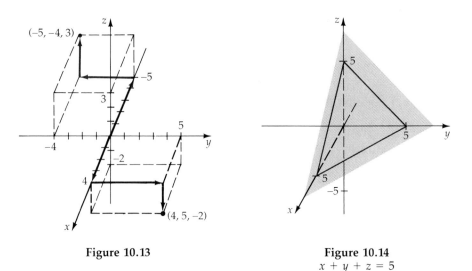

Figure 10.13

Figure 10.14
$x + y + z = 5$

The graph of a linear equation in three variables is a plane. That is, if all the solutions of a linear equation in three variables were plotted in an xyz-coordinate system, the graph would look like a large piece of paper extending infinitely. Figure 10.14 is the graph of $x + y + z = 5$.

There are different ways three planes can be oriented in an xyz-coordinate system. Figure 10.15 illustrates several ways.

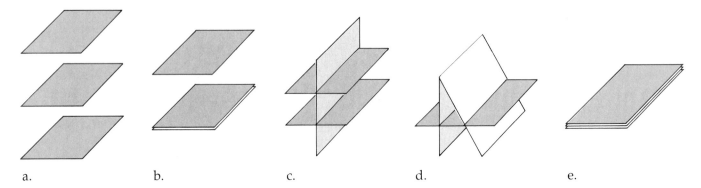

a.   b.   c.   d.   e.

Figure 10.15

For a linear system of three equations in three variables to have a solution, the graphs of the planes must intersect at a single point, along a common line, or all equations must have a graph that is the same plane. See Figure 10.16.

A system of equations in more than two variables can be solved by using the substitution or the elimination method. To illustrate the substitution method, consider the system of equations

$$\begin{cases} x - 2y + z = 7 & (1) \\ 2x + y - z = 0 & (2) \\ 3x + 2y - 2z = -2 & (3) \end{cases}$$

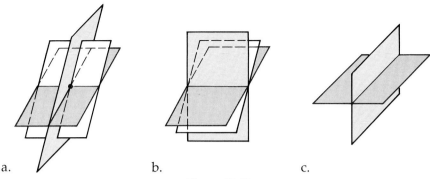

Figure 10.16

Solving Equation (1) for x and substituting the result into Equations (2) and (3), we have

$$x - 2y + z = 7 \quad \text{or} \quad x = 2y - z + 7 \tag{4}$$

$$2(2y - z + 7) + y - z = 0 \qquad \text{Substitute for } x \text{ in Equation (2) and simplify.}$$

$$4y - 2z + 14 + y - z = 0$$

$$5y - 3z = -14 \tag{5}$$

$$3(2y - z + 7) + 2y - 2z = -2 \qquad \text{Substitute for } x \text{ in Equation (3) and simplify.}$$

$$6y - 3z + 21 + 2y - 2z = -2$$

$$8y - 5z = -23 \tag{6}$$

Now we solve the system of equations formed from Equations (5) and (6).

$$\begin{cases} 5y - 3z = -14 \\ 8y - 5z = -23 \end{cases} \qquad \begin{aligned} 40y - 24z &= -112 \\ \underline{-40y + 25z} &= \underline{115} \\ z &= 3 \end{aligned}$$

Substitute 3 for z into Equation (5) and solve for y.

$$5y - 3z = -14$$
$$5y - 3(3) = -14$$
$$5y - 9 = -14$$
$$5y = -5$$
$$y = -1$$

Substitute -1 for y and 3 for z into Equation (4) and solve for x.

$$x = 2y - z + 7 = 2(-1) - 3 + 7 = 2$$

The ordered triple solution is $(2, -1, 3)$. The graphs of the three planes intersect at a single point.

In the remainder of this chapter we will use the elimination method for solving a system of equations. There are many approaches one could take to determine the solution by the elimination method. For consistency, we will always follow a plan that produces an equivalent system of equations in **triangular form**. Three examples of systems of equations in triangular form are

$$\begin{cases} 2x - 3y + z = -4 \\ 2y + 3z = 9 \\ -2z = -2 \end{cases} \qquad \begin{cases} w + 3x - 2y + 3z = 0 \\ 2x - y + 4z = 8 \\ -3y - 2z = -1 \\ 3z = 9 \end{cases} \qquad \begin{cases} 3x - 4y + z = 1 \\ 3y + 2z = 3 \end{cases}$$

Once a system of equations is written in this form, the solution can be found by *back substitution*, which is solving the last equation of the system and substituting back into the previous equation. This process is continued until the value of each variable has been found.

As an example of solving a triangular system by back substitution, consider the system

$$\begin{cases} 2x - 4y + z = -3 & (1) \\ 3y - 2z = 9 & (2) \\ 3z = -9 & (3) \end{cases}$$

We solve Equation (3) for z and substitute the value of z into Equation (2). We can now solve for y.

$$3z = -9 \qquad \text{or} \qquad z = -3$$

$$3y - 2z = 9$$

$$3y - 2(-3) = 9 \qquad \text{or} \qquad y = 1$$

Use the values of y and z to find x.

$$2x - 4y + z = -3$$

$$2x - 4(1) + (-3) = -3$$

$$2x - 7 = -3$$

$$x = 2$$

The solution is $(2, 1, -3)$.

EXAMPLE 1 Solve a System of Equations

Solve the system of equations by elimination

$$\begin{cases} x + 2y - z = 1 & (1) \\ 2x - y + z = 6 & (2) \\ x + 3y - 2z = -1 & (3) \end{cases}$$

Solution Eliminate x from Equation (2) by multiplying Equation (1) by -2 and adding it to Equation (2). Replace Equation (2) with the resulting equation.

$$\begin{cases} x + 2y - z = 1 \\ -5y + 3z = 4 \quad (4) \\ x + 3y - 2z = -1 \end{cases} \qquad \begin{array}{r} -2x - 4y + 2z = -2 \\ 2x - y + z = 6 \\ \hline -5y + 3z = 4 \end{array}$$

Eliminate x from Equation (3) by multiplying Equation (1) by -1 and adding it to Equation (3). This equation will replace Equation (3).

$$\begin{cases} x + 2y - \ z = 1 \\ \quad\ \ -5y + 3z = 4 \quad (4) \\ \qquad\quad\ y - \ z = -2 \quad (5) \end{cases} \qquad \begin{aligned} -x - 2y + \ z &= -1 \\ \underline{\quad x + 3y - 2z = -1} \\ y - \ z &= -2 \end{aligned}$$

Eliminate y from Equation (5) by multiplying Equation (5) by 5 and adding it to Equation (4). Replace Equation (5).

$$\begin{cases} x + 2y - \ z = 1 \\ \quad\ \ -5y + 3z = 4 \\ \qquad\qquad -2z = -6 \quad (6) \end{cases} \qquad \begin{aligned} -5y + 3z &= 4 \\ \underline{\quad 5y - 5z = -10} \\ -2z &= -6 \end{aligned}$$

The system of equations is now in triangular form. Solve by back substitution. Solve Equation (6) for z.

$$-2z = -6 \qquad \text{or} \qquad z = 3$$

Substitute 3 for z into Equation (4) and solve for y.

$$-5y + 3 \cdot 3 = 4 \qquad \text{or} \qquad y = 1$$

Now replace y and z in Equation (1) with their values and solve for x.

$$x + 2 \cdot 1 - 3 = 1 \qquad \text{or} \qquad x = 2$$

The solution is the ordered triple $(2, 1, 3)$. The graphs of the three planes of this system intersect at the single point $(2, 1, 3)$.

■ *Try Exercise* **12,** *page 524.*

EXAMPLE 2   **Solve a System of Equations**

Solve the system of equations

$$\begin{cases} x + \ 4y + 2z = 1 & (1) \\ 2x + \ 7y + 3z = 8 & (2) \\ 4x + 15y + 7z = 10 & (3) \end{cases}$$

Solution   Eliminate x from Equation (2). Multiply Equation (1) by -2 and add to Equation (2). Replace Equation (2).

$$\begin{cases} x + \ 4y + 2z = 1 \\ \qquad\ -y - \ z = 6 \\ 4x + 15y + 7z = 10 \end{cases} \qquad \begin{aligned} -2x - 8y - 4z &= -2 \\ \underline{\quad 2x + 7y + 3z = 8} \\ -y - \ z &= 6 \end{aligned}$$

Eliminate x from Equation (3). Multiply Equation (1) by -4 and add to Equation (3). Replace Equation (3).

$$\begin{cases} x + 4y + 2z = 1 \\ \quad\ -y - \ z = 6 \quad (4) \\ \quad\ -y - \ z = 6 \quad (5) \end{cases} \qquad \begin{aligned} -4x - 16y - 8z &= -4 \\ \underline{\quad 4x + 15y + 7z = 10} \\ -y - \ z &= 6 \end{aligned}$$

Multiply Equation (4) by -1 and add to Equation (5). Replace Equation (5).

The resulting equivalent system is

$$\begin{cases} x + 4y + 2z = 1 & (1) \\ \quad\;\; -y - z = 6 & (4) \\ \qquad\qquad\; 0 = 0 & (6) \end{cases} \qquad \begin{aligned} y + z &= -6 \\ \underline{-y - z} &= \underline{6} \\ 0 &= 0 \end{aligned}$$

Since any ordered triple (x, y, z) is a solution of Equation (6), the solutions of the system will be those ordered triples that are solutions to Equations (1) and (4).

Solve Equation (4) for z.

$$z = -y - 6$$

Substitute this expression into Equation (1) and solve for x.

$$x + 4y + 2(-y - 6) = 1$$
$$x + 4y - 2y - 12 = 1$$
$$x = -2y + 13$$

If y is any real number c, then the ordered triple solutions are

$$(-2c + 13, c, -c - 6).$$

■ *Try Exercise* **16,** *page 524.*

Since there are an infinite number of choices for c, there are an infinite number of ordered triple solutions of the system. The graphs of the equations of the system intersect along a line.

As in the case of a dependent system of equations in two variables, there is more than one way to represent the solutions of this dependent system. For example, letting $b = -2c + 13$, the x-component of the ordered triple $(-2c + 13, c, -c - 6)$, and then solving for c, we obtain

$$b = -2c + 13 \qquad \text{or} \qquad c = \frac{-b + 13}{2}.$$

Substituting this value of c into each component of the ordered triple solution $(-2c + 13, c, -c - 6)$, we have the three coordinates

$$-2\left(\frac{-b + 13}{2}\right) + 13 = b; \qquad \frac{-b + 13}{2}; \qquad -\left(\frac{-b + 13}{2}\right) - 6 = \frac{b - 25}{2}.$$

The solutions of the system are the ordered triple written as

$$\left(b, \frac{-b + 13}{2}, \frac{b - 25}{2}\right).$$

EXAMPLE 3   Identify an Inconsistent System of Equations

Solve the system of equations

$$\begin{cases} x + 2y + 3z = 4 & (1) \\ 2x - y - z = 3 & (2) \\ 3x + y + 2z = 5 & (3) \end{cases}$$

Solution   Eliminate x from Equation (2) by multiplying Equation (1) by -2 and adding to Equation (2). Replace Equation 2. Eliminate x from Equation (3) by multiplying Equation (1) by -3 and adding to Equation (3). Replace Equation (3). The equivalent system is

$$\begin{cases} x + 2y + 3z = 4 & \text{(1)} \\ \quad -5y - 7z = -5 & \text{(4)} \\ \quad -5y - 7z = -7 & \text{(5)} \end{cases}$$

Eliminate y from Equation (6) by multiplying Equation (5) by -1 and adding to Equation (6). Replace Equation (6). The equivalent system is

$$\begin{cases} x + 2y + 3z = 4 & \text{(1)} \\ \quad -5y - 7z = -5 & \text{(4)} \\ \qquad\qquad 0 = -2 & \text{(6)} \end{cases}$$

This system of equations contains a false equation. The system has no solutions.

■ *Try Exercise **18**, page 524.*

Although we will not discuss systems of equations with more than three variables in this chapter, such equations can be solved using similar methods. One application of a system of equations is "curve fitting." Given a set of points in the plane, try to find an equation whose graph passes through those points or "fits" those points.

EXAMPLE 4   Solve an Application of a System of Equations to Curve Fitting

Find an equation of the form $y = ax^2 + bx + c$ whose graph passes through the three points $(1, 4)$, $(-1, 6)$, and $(2, 9)$.

Solution   Substitute each of the given ordered pairs into the equation $y = ax^2 + bx + c$. Write the resulting system of equations.

$$\begin{cases} 4 = a(1)^2 + b(1) + c \\ 6 = a(-1)^2 + b(-1) + c \\ 9 = a(2)^2 + b(2) + c \end{cases} \quad \text{or} \quad \begin{cases} a + b + c = 4 & \text{(1)} \\ a - b + c = 6 & \text{(2)} \\ 4a + 2b + c = 9 & \text{(3)} \end{cases}$$

Solve the resulting system of equations for a, b, and c.

Eliminate a from Equation (2) by multiplying Equation (1) by -1 and adding to Equation (2). Now eliminate a from Equation (3) by multiplying Equation (1) by -4 and adding to Equation (3). The result is

$$\begin{cases} a + b + c = 4 \\ \quad -2b = 2 \\ \quad -2b - 3c = -7 \end{cases}$$

Although this system of equations is not in triangular form, we can solve the second equation for b and use this value to find a and c.

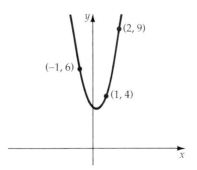

Figure 10.17
$y = 2x^2 - x + 3$

Solving by substitution, we obtain $a = 2$, $b = -1$, $c = 3$. The equation whose graph passes through the three points is $y = 2x^2 - x + 3$.

■ *Try Exercise* **34,** *page 524.*

Nonsquare Systems of Equations

The linear systems of equations that have been solved so far contain the same number of variables as equations. These are *square systems of equations*. If there are fewer equations than variables—a *nonsquare system of equations*—the system either will have no solution or an infinite number of solutions.

EXAMPLE 5   Solve a Nonsquare System of Equations

Solve the system of equations

$$\begin{cases} x - 2y + 2z = 3 & (1) \\ 2x - y - 2z = 15 & (2) \end{cases}$$

Solution   Eliminate x from Equation (2) by multiplying Equation (1) by -2 and adding to Equation (2). Replace Equation (2).

$$\begin{cases} x - 2y + 2z = 3 & (1) \\ 3y - 6z = 9 & (3) \end{cases} \qquad \begin{array}{r} -2x + 4y - 4z = -6 \\ \underline{2x - y - 2z = 15} \\ 3y - 6z = 9 \end{array}$$

Solve Equation (3) for y.

$$3y - 6z = 9$$
$$y = 2z + 3$$

Substitute $2z + 3$ for y into Equation (1) and solve for x.

$$x - 2y + 2z = 3$$
$$x - 2(2z + 3) + 2z = 3$$
$$x = 2z + 9$$

For each value of z selected, there corresponds values for x and y. If z is any real number a, then the solutions of the system are the ordered triples $(2a + 9, 2a + 3, a)$.

■ *Try Exercise* **20,** *page 524.*

Homogeneous Systems of Equations

A linear system of equations for which the constant term is zero for all equations is called a **homogeneous** system of equations. Two examples of homogeneous systems of equations are

$$\begin{cases} 3x + 4y = 0 \\ 2x + 3y = 0 \end{cases} \qquad \begin{cases} 2x - 3y + 5z = 0 \\ 3x + 2y + z = 0 \\ x - 4y + 5z = 0 \end{cases}$$

The solution $(0, 0)$ is always a solution of a homogeneous system of equations in two variables, and $(0, 0, 0)$ is always a solution of a homogeneous system of equations in three variables. This solution is called the **trivial** solution.

Sometimes a homogeneous system of equations may have solutions other than the trivial solution. For example, $(1, -1, -1)$ is a solution to the homogeneous system of three equations in three variables above.

If a homogeneous system of equations has a unique solution, the graphs intersect only at the origin. If the homogeneous system of equations has infinitely many solutions, the graphs intersect along a line or plane that passes through the origin.

Solutions to a homogeneous system of equations can be found by using the substitution or the elimination method.

EXAMPLE 6 **Solve a Homogeneous System of Equations**

Solve the system of equations.

$$\begin{cases} x + 2y - 3z = 0 & \quad (1) \\ 2x - y + z = 0 & \quad (2) \\ 3x + y - 2z = 0 & \quad (3) \end{cases}$$

Solution   Eliminate x from Equations (2) and (3).

$$\begin{cases} x + 2y - 3z = 0 & \quad (1) \\ -5y + 7z = 0 & \quad (4) \\ -5y + 7z = 0 & \quad (5) \end{cases}$$

Eliminate y from Equations (4) and (5). Replace Equation (5).

$$\begin{cases} x + 2y - 3z = 0 & \quad (1) \\ -5y + 7z = 0 & \quad (4) \\ 0 = 0 & \quad (6) \end{cases}$$

Since the last equation is an identity, the solutions of the system are the solutions of Equations (1) and (4).

Solve Equation (4) for y.

$$y = \frac{7}{5}z$$

Substitute the expression for y into Equation (1) and solve for x.

$$x + 2y - 3z = 0$$

$$x + 2\left(\frac{7}{5}z\right) - 3z = 0$$

$$x = \frac{1}{5}z$$

Letting z be any real number c, we find the solutions of the system are $\left(\frac{1}{5}c, \frac{7}{5}c, c\right)$.

■ *Try Exercise **32**, page 524.*

EXERCISE SET 10.2

In Exercises 1 to 24, solve each system of equations.

1. $\begin{cases} 2x - y + z = 8 \\ 2y - 3z = -11 \\ 3y + 2z = 3 \end{cases}$
2. $\begin{cases} 3x + y + 2z = -4 \\ - 3y - 2z = -5 \\ 2y + 5z = -4 \end{cases}$

3. $\begin{cases} x + 3y - 2z = 8 \\ 2x - y + z = 1 \\ 3x + 2y - 3z = 15 \end{cases}$
4. $\begin{cases} x - 2y + 3z = 5 \\ 3x - 3y + z = 9 \\ 5x + y - 3z = 3 \end{cases}$

5. $\begin{cases} 3x + 4y - z = -7 \\ x - 5y + 2z = 19 \\ 5x + y - 2z = 5 \end{cases}$
6. $\begin{cases} 2x - 3y - 2z = 12 \\ x + 4y + z = -9 \\ 4x + 2y - 3z = 6 \end{cases}$

7. $\begin{cases} 2x - 5y + 3z = -18 \\ 3x + 2y - z = -12 \\ x - 3y - 4z = -4 \end{cases}$
8. $\begin{cases} 4x - y + 2z = -1 \\ 2x + 3y - 3z = -13 \\ x + 5y + z = 7 \end{cases}$

9. $\begin{cases} x + 2y - 3z = -7 \\ 2x - y + 4z = 11 \\ 4x + 3y - 4z = -3 \end{cases}$
10. $\begin{cases} x - 3y + 2z = -11 \\ 3x + y + 4z = 4 \\ 5x - 5y + 8z = -18 \end{cases}$

11. $\begin{cases} 2x - 5y + 2z = -4 \\ 3x + 2y + 3z = 13 \\ 5x - 3y - 4z = -18 \end{cases}$
12. $\begin{cases} 3x + 2y - 5z = 6 \\ 5x - 4y + 3z = -12 \\ 4x + 5y - 2z = 15 \end{cases}$

13. $\begin{cases} 2x + y - z = -2 \\ 3x + 2y + 3z = 21 \\ 7x + 4y + z = 17 \end{cases}$
14. $\begin{cases} 3x + y + 2z = 2 \\ 4x - 2y + z = -4 \\ 11x - 3y + 4z = -6 \end{cases}$

15. $\begin{cases} 3x - 2y + 3z = 11 \\ 2x + 3y + z = 3 \\ 5x + 14y - z = 1 \end{cases}$
16. $\begin{cases} 2x + 3y + 2z = 14 \\ x - 3y + 4z = 4 \\ -x + 12y - 6z = 2 \end{cases}$

17. $\begin{cases} 2x - 3y + 6z = 3 \\ x + 2y - 4z = 5 \\ 3x + 4y - 8z = 7 \end{cases}$
18. $\begin{cases} 2x + 3y - 6z = 4 \\ 3x - 2y - 9z = -7 \\ 2x + 5y - 6z = 8 \end{cases}$

19. $\begin{cases} 2x - 3y + 5z = 14 \\ x + 4y - 3z = -2 \end{cases}$
20. $\begin{cases} x - 3y + 4z = 9 \\ 3x - 8y - 2z = 4 \end{cases}$

21. $\begin{cases} 6x - 9y + 6z = 7 \\ 4x - 6y + 4z = 9 \end{cases}$
22. $\begin{cases} 4x - 2y + 6z = 5 \\ 2x - y + 3z = 2 \end{cases}$

23. $\begin{cases} 5x + 3y + 2z = 10 \\ 3x - 4y - 4z = -5 \end{cases}$
24. $\begin{cases} 3x - 4y - 7z = -5 \\ 2x + 3y - 5z = 2 \end{cases}$

In Exercises 25 to 32, solve each homogeneous system of equations.

25. $\begin{cases} x + 3y - 4z = 0 \\ 2x + 7y + z = 0 \\ 3x - 5y - 2z = 0 \end{cases}$
26. $\begin{cases} x - 2y + 3z = 0 \\ 3x - 7y - 4z = 0 \\ 4x - 4y + z = 0 \end{cases}$

27. $\begin{cases} 2x - 3y + z = 0 \\ 2x + 4y - 3z = 0 \\ 6x - 2y - z = 0 \end{cases}$
28. $\begin{cases} 5x - 4y - 3z = 0 \\ 2x + y + 2z = 0 \\ x - 6y - 7z = 0 \end{cases}$

29. $\begin{cases} 3x - 5y + 3z = 0 \\ 2x - 3y + 4z = 0 \\ 7x - 11y + 11z = 0 \end{cases}$
30. $\begin{cases} 5x - 2y - 3z = 0 \\ 3x - y - 4z = 0 \\ 4x - y - 9z = 0 \end{cases}$

31. $\begin{cases} 4x - 7y - 2z = 0 \\ 2x + 4y + 3z = 0 \\ 3x - 2y - 5z = 0 \end{cases}$
32. $\begin{cases} 5x + 2y + 3z = 0 \\ 3x + y - 2z = 0 \\ 4x - 7y + 5z = 0 \end{cases}$

In Exercises 33 to 42, solve a system of equations.

33. Find an equation of the form $y = ax^2 + bx + c$ whose graph passes through the points $(2, 3)$, $(-2, 7)$, and $(1, -2)$.

34. Find an equation of the form $y = ax^2 + bx + c$ whose graph passes through the points $(1, -2)$, $(3, -4)$, and $(2, -2)$.

35. Find the equation of the circle whose graph passes through the points $(5, 3)$, $(-1, -5)$, and $(-2, 2)$. (*Hint:* Use the equation $x^2 + y^2 + ax + by + c = 0$.)

36. Find the equation of the circle whose graph passes through the points $(0, 6)$, $(1, 5)$, and $(-7, -1)$. (*Hint:* See Exercise 35.)

37. Find the center and radius of the circle whose graph passes through the points $(-2, 10)$, $(-12, -14)$, and $(5, 3)$. (*Hint:* See Exercise 35.)

38. Find the center and radius of the circle whose graph passes through the points $(2, 5)$, $(-4, -3)$, and $(3, 4)$. (*Hint:* See Exercise 35.)

39. A coin bank contains only nickels, dimes, and quarters. The value of the coins is $2. There are twice as many nickels as dimes and one more dime than quarters. Find the number of each coin in the bank.

40. A coin bank contains only nickels, dimes, and quarters. The value of the coins is $5.50. The number of nickels is six more than twice the number of quarters. The number of dimes is one-third the number of nickels. Find the number of each coin in the bank.

41. The sum of the digits of a positive three-digit number is 19. The tens digit is four less than twice the hundreds digit. The number is decreased by 99 when the digits are reversed. Find the number.

42. The sum of the digits of a positive three-digit number is 10. The hundreds digit is one less than twice the ones digit. The number is decreased by 198 when the digits are reversed. Find the number.

Supplemental Exercises

In Exercises 43 to 48, solve each system of equations.

43. $\begin{cases} 2x + y - 3z + 2w = -1 \\ 2y - 5z - 3w = 9 \\ 3y - 8z + w = -4 \\ 2y - 2z + 3w = -3 \end{cases}$

44. $\begin{cases} 3x - y + 2z - 3w = 5 \\ 2y - 5z + 2w = -7 \\ 4y - 9z + w = -19 \\ 3y + z - 2w = -12 \end{cases}$

45. $\begin{cases} x - 3y + 2z - w = 2 \\ 2x - 5y - 3z + 2w = 1 \\ 3x - 8y - 2z - 3w = 12 \\ -2x + 8y + z + 2w = -13 \end{cases}$

46. $\begin{cases} x - 2y + 3z + 2w = 8 \\ 3x - 7y - 2z + 3w = 18 \\ 2x - 5y + 2z - w = 19 \\ 4x - 8y + 3z + 2w = 29 \end{cases}$

47. $\begin{cases} x + 2y - 2z + 3w = 2 \\ 2x + 5y + 2z + 4w = 9 \\ 4x + 9y - 2z + 10w = 13 \\ -x - y + 8z - 5w = 3 \end{cases}$

48. $\begin{cases} x - 2y + 3z - 2w = -1 \\ 3x - 7y - 2z - 3w = -19 \\ 2x - 5y + 2z - w = -11 \\ -x + 3y - 2z - w = 3 \end{cases}$

In Exercises 49 and 50, use the system of equations

$$\begin{cases} x - 3y - 2z = A^2 \\ 2x - 5y + Az = 9 \\ 2x - 8y + z = 18 \end{cases}$$

49. Find all values of A for which the system has no solutions.

50. Find all values of A for which the system has a unique solution.

In Exercises 51 to 53, use the system of equations

$$\begin{cases} x + 2y + z = A^2 \\ -2x - 3y + Az = 1 \\ 7x + 12y + A^2 z = 4A^2 - 3 \end{cases}$$

51. Find all values of A for which the system has a unique solution.

52. Find all values of A for which the system has an infinite number of solutions.

53. Find all values of A for which the system has no solution.

54. Find an equation of a plane that contains the points $(2, 1, 1)$, $(-1, 2, 12)$, and $(3, 2, 0)$. (*Hint:* The equation of a plane can be written as $z = ax + by + c$.)

55. Find an equation of a plane that contains the points $(1, -1, 5)$, $(2, -2, 9)$, and $(-3, -1, -1)$. (*Hint:* The equation of a plane can be written as $z = ax + by + c$.)

10.3

Systems of Nonlinear Equations

A **nonlinear system of equations** is one in which one or more equations of the system is not a linear equation. Figure 10.18 shows examples of nonlinear systems of equations with their corresponding graphs of the equations.

Each point of intersection of the graphs is a solution of the system of equations. In the third example, the graphs do not intersect; therefore the system of equations has no real number solution.

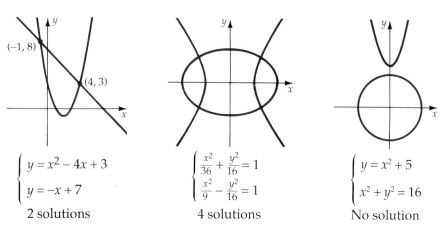

$$\begin{cases} y = x^2 - 4x + 3 \\ y = -x + 7 \end{cases}$$

2 solutions

$$\begin{cases} \dfrac{x^2}{36} + \dfrac{y^2}{16} = 1 \\ \dfrac{x^2}{9} - \dfrac{y^2}{16} = 1 \end{cases}$$

4 solutions

$$\begin{cases} y = x^2 + 5 \\ x^2 + y^2 = 16 \end{cases}$$

No solution

Figure 10.18

To solve a system of nonlinear equations, use the substitution or the elimination method. When solving a nonlinear system that contains a linear equation, the substitution method is usually easier.

EXAMPLE 1   **Solve a Nonlinear System by the Substitution Method**

Solve the nonlinear system of equations

$$\begin{cases} y = x^2 - x - 1 & (1) \\ 3x - y = 4 & (2) \end{cases}$$

Solution   We will use the substitution method. Using the equation $y = x^2 - x - 1$, substitute the expression for y into $3x - y = 4$.

$$3x - (x^2 - x - 1) = 4$$

$$-x^2 + 4x + 1 = 4$$

$$x^2 - 4x + 3 = 0 \quad \text{Write the quadratic equation in standard form.}$$

$$(x - 3)(x - 1) = 0 \quad \text{Solve for } x.$$

$$x - 3 = 0 \quad \text{or} \quad x - 1 = 0$$

$$x = 3 \quad \text{or} \quad x = 1$$

Substitute these values into the Equation (1) and solve for y.

$$y = 3^2 - 3 - 1 = 5 \qquad y = 1^2 - 1 - 1 = -1$$

The solutions are $(3, 5)$ and $(1, -1)$.

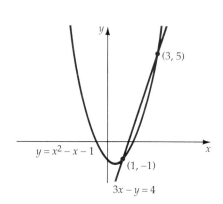

Figure 10.19

■ *Try Exercise 8, page 529.*

EXAMPLE 2   Solve a Nonlinear System by the Elimination Method

Solve the nonlinear system of equations

$$\begin{cases} 4x^2 + 3y^2 = 48 & (1) \\ 3x^2 + 2y^2 = 35 & (2) \end{cases}$$

Solution   We will eliminate the x^2 term. Multiply Equation (1) by -3 and Equation (2) by 4. Then add the two equations.

$$\begin{aligned} -12x^2 - 9y^2 &= -144 \\ \underline{12x^2 + 8y^2} &= \underline{140} \\ -y^2 &= -4 \\ y^2 &= 4 \,. \end{aligned}$$

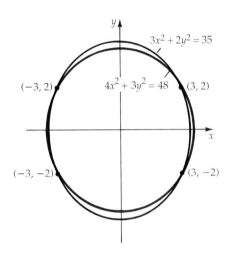

Thus $y = 2$ or $y = -2$. Substitute 2 and -2 for y into Equation (1) and solve for x

$$\begin{aligned} 4x^2 + 3(2)^2 &= 48 \\ 4x^2 &= 36 \\ x^2 &= 9 \end{aligned}$$

Thus $x = 3$ or $x = -3$. Because $(-2)^2 = 2^2$, by replacing y by -2 we will obtain the same values of x, $x = 3$ or $x = -3$. The solutions are $(3, 2)$, $(3, -2)$, $(-3, 2)$, and $(-3, -2)$.

■ *Try Exercise* **12,** *page 529.*

Figure 10.20

EXAMPLE 3   Identify a Nonlinear System of Equations with No Solution

Solve the nonlinear system of equations

$$\begin{cases} 4x^2 + 9y^2 = 36 & (1) \\ x^2 - y^2 = 25 & (2) \end{cases}$$

Solution   Using the elimination method, we will eliminate the x^2 term from each equation. Multiplying Equation (2) by -4 and then adding, we have

$$\begin{aligned} 4x^2 + 9y^2 &= 36 \\ \underline{-4x^2 + 4y^2} &= \underline{-100} \\ 13y^2 &= -64 \end{aligned}$$

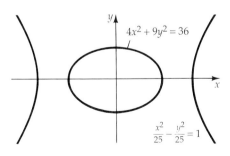

Because the equation $13y^2 = -64$ has no real solutions, the system of equations has no real solutions. The graphs of the equations do not intersect. See Figure 10.21.

■ *Try Exercise* **4,** *page 529.*

Figure 10.21

EXAMPLE 4   **Solve a System of Equations That Contains a Cubic Equation**

Solve the nonlinear system of equations

$$\begin{cases} y = x^3 + x^2 - x - 3 & (1) \\ y = 3x^2 - 5x + 5 & (2) \end{cases}$$

Solution   Because each equation is solved for y, set the expressions for y equal to each other and solve for x.

$$x^3 + x^2 - x - 3 = 3x^2 - 5x + 5$$

$$x^3 - 2x^2 + 4x - 8 = 0$$

$$x^2(x - 2) + 4(x - 2) = 0$$

$$(x^2 + 4)(x - 2) = 0$$

Since $x^2 + 4 = 0$ has no real number solutions, the only real solution is 2. Replace x by 2 in Equation (2) and solve for y.

$$y = 3 \cdot 2^2 - 5 \cdot 2 + 5.$$

Thus $y = 7$. The solution of the system is $(2, 7)$.

■ *Try Exercise* **10**, *page 529.*

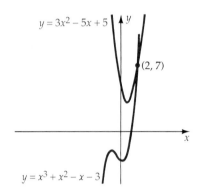

$y = 3x^2 - 5x + 5$

$(2, 7)$

$y = x^3 + x^2 - x - 3$

Figure 10.22

EXAMPLE 5   **Solve a Nonlinear System of Equations**

Solve the nonlinear system of equations

$$\begin{cases} (x + 3)^2 + (y - 4)^2 = 20 \\ (x + 4)^2 + (y - 3)^2 = 26 \end{cases}$$

Solution   Expand the binomials in each equation. Then subtract the two equations and simplify.

$$x^2 + 6x + 9 + y^2 - 8y + 16 = 20 \qquad (1)$$
$$\underline{x^2 + 8x + 16 + y^2 - 6y + 9 = 26} \qquad (2)$$
$$-2x - 7 -2y + 7 = -6$$
$$x + y = 3$$

Now solve the resulting equation for y.

$$y = -x + 3$$

Substitute $-x + 3$ for y into Equation (1) and solve for x.

$$x^2 + 6x + 9 + (-x + 3)^2 - 8(-x + 3) + 16 = 20$$

$$2(x^2 + 4x - 5) = 0$$

$$2(x + 5)(x - 1) = 0$$

$$x = -5 \qquad \text{or} \qquad x = 1$$

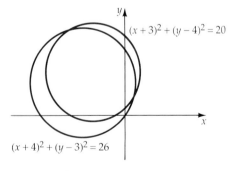

$(x + 3)^2 + (y - 4)^2 = 20$

$(x + 4)^2 + (y - 3)^2 = 26$

Figure 10.23

Substitute -5 and 1 for x into the equation $y = -x + 3$ and solve for y.

$$y = 8 \quad \text{or} \quad y = 2$$

The solutions of the system of equations are $(-5, 8)$ and $(1, 2)$.

■ *Try Exercise 20, page 529.*

EXERCISE SET 10.3

In Exercises 1 to 24, solve each system of equations.

1. $\begin{cases} y = x^2 - 2x + 3 \\ y = x^2 - x - 2 \end{cases}$

2. $\begin{cases} y = 2x^2 - x + 1 \\ y = x^2 + 2x + 5 \end{cases}$

3. $\begin{cases} x^2 - 2y^2 = 8 \\ x^2 + 3y^2 = 28 \end{cases}$

4. $\begin{cases} 2x^2 + 3y^2 = 5 \\ x^2 - 3y^2 = 4 \end{cases}$

5. $\begin{cases} x + y = 10 \\ xy = 24 \end{cases}$

6. $\begin{cases} x - 2y = 3 \\ xy = -1 \end{cases}$

7. $\begin{cases} 3x^2 - 2y^2 = 1 \\ y = 4x - 3 \end{cases}$

8. $\begin{cases} x^2 + 3y^2 = 7 \\ x + 4y = 6 \end{cases}$

9. $\begin{cases} y = x^3 + 4x^2 - 3x - 5 \\ y = 2x^2 - 2x - 3 \end{cases}$

10. $\begin{cases} y = x^3 - 2x^2 + 5x + 1 \\ y = x^2 + 7x - 5 \end{cases}$

11. $\begin{cases} 2x^2 + 4y^2 = 9 \\ 3x^2 + 8y^2 = 14 \end{cases}$

12. $\begin{cases} 2x^2 + 3y^2 = 11 \\ 3x^2 + 2y^2 = 14 \end{cases}$

13. $\begin{cases} x^2 - 2x + y^2 = 1 \\ 2x + y = 5 \end{cases}$

14. $\begin{cases} x^2 + y^2 + 3y = 22 \\ 2x + y = -1 \end{cases}$

15. $\begin{cases} (x - 3)^2 + (y + 1)^2 = 5 \\ x - 3y = 7 \end{cases}$

16. $\begin{cases} (x + 2)^2 + (y - 2)^2 = 13 \\ 2x + y = 6 \end{cases}$

17. $\begin{cases} x^2 - 3x + y^2 = 4 \\ 3x + y = 11 \end{cases}$

18. $\begin{cases} x^2 + y^2 - 4y = 4 \\ 5x - 2y = 2 \end{cases}$

19. $\begin{cases} (x - 1)^2 + (y + 2)^2 = 14 \\ (x + 2)^2 + (y - 1)^2 = 2 \end{cases}$

20. $\begin{cases} (x + 2)^2 + (y - 3)^2 = 10 \\ (x - 3)^2 + (y + 1)^2 = 13 \end{cases}$

21. $\begin{cases} (x + 3)^2 + (y - 2)^2 = 20 \\ (x - 2)^2 + (y - 3)^2 = 2 \end{cases}$

22. $\begin{cases} (x - 4)^2 + (y - 5)^2 = 8 \\ (x + 1)^2 + (y + 2)^2 = 34 \end{cases}$

23. $\begin{cases} (x - 1)^2 + (y + 1)^2 = 2 \\ (x + 2)^2 + (y - 3)^2 = 3 \end{cases}$

24. $\begin{cases} (x + 1)^2 + (y - 3)^2 = 4 \\ (x - 3)^2 + (y + 2)^2 = 2 \end{cases}$

Supplemental Exercises

In Exercises 25 to 30, solve the system of equations for rational number ordered pairs.

25. $\begin{cases} y = x^2 + 4 \\ x = y^2 - 24 \end{cases}$

26. $\begin{cases} y = x^2 - 5 \\ x = y^2 - 13 \end{cases}$

27. $\begin{cases} x^2 - 3xy + y^2 = 5 \\ x^2 - xy - 2y^2 = 0 \end{cases}$

(*Hint:* Factor the second equation. Now use the principle of zero products and the substitution principle.)

28. $\begin{cases} x^2 + 2xy - y^2 = 1 \\ x^2 + 3xy + 2y^2 = 0 \end{cases}$

(*Hint:* See Exercise 27.)

29. $\begin{cases} 2x^2 - 4xy - y^2 = 6 \\ 4x^2 - 3xy - y^2 = 6 \end{cases}$

(*Hint:* Subtract the two equations.)

30. $\begin{cases} 3x^2 + 2xy - 5y^2 = 11 \\ x^2 + 3xy + y^2 = 11 \end{cases}$

(*Hint:* Subtract the two equations.)

31. Show that the line $y = mx$ intersects the hyperbola $\dfrac{x^2}{a^2} - \dfrac{y^2}{b^2} = 1$ if and only if $|m| < \left| \dfrac{a}{b} \right|$.

32. Show that the line $y = mx$, $m \geq 0$, intersects the circle $(x - a)^2 + y^2 = a^2$, $a > 0$, at $(0, 0)$ and the point P whose coordinates are given by

$$P\left(\frac{2a}{1 + m^2}, \frac{2am}{1 + m^2} \right).$$

Now show that the slope of the line through P and $(2a, 0)$ is $-1/m$. What theorem from geometry does this prove?

33. In Example 5 of this section, the equation $x + y = 3$ was found by subtracting the original two equations of the system. Show that the graph of this line passes through the solutions of the system of equations. Do you understand that this would always happen? Think about operations that produce equivalent equations.

10.4

Partial Fractions

An algebraic application of systems of equations is to a technique known as *partial fractions*. In Chapter 1, we reviewed the problem of adding two rational expressions. For example,

$$\frac{5}{x-1} + \frac{1}{x+2} = \frac{6x+9}{(x-1)(x+2)}$$

Now we will take an opposite approach. That is, given a rational expression, find simpler rational expressions whose sum is the given expression. The method by which a more complicated rational expression is written as a sum of rational expressions is called **partial fraction decomposition**. This technique is based on the following theorem.

Partial Fraction Decomposition Theorem

If $$f(x) = \frac{p(x)}{q(x)}$$

is a rational expression in which the degree of the numerator is less than the degree of the denominator and $p(x)$ and $q(x)$ have no common factors, then $f(x)$ can be written as a partial fraction decomposition in the form

$$f(x) = f_1(x) + f_2(x) + \cdots + f_n(x)$$

where each $f_i(x)$ has one of the forms

$$\frac{A}{(px+q)^m} \quad \text{or} \quad \frac{Bx+C}{(ax^2+bx+c)^m}.$$

The procedure for finding a partial fraction decomposition of a rational expression depends on the factorization of the denominator of the rational expression. There are four cases.

Case 1   Nonrepeated Linear Factors

The partial fraction decomposition will contain an expression of the form $A/(x+a)$ for each nonrepeated linear factor of the denominator. Example:

$$\frac{3x-1}{x(3x+4)(x-2)} \quad \text{Each factor of the denominator occurs only once.}$$

Partial fraction decomposition:

$$\frac{3x-1}{x(3x+4)(x-2)} = \frac{A}{x} + \frac{B}{3x+4} + \frac{C}{x-2}$$

Case 2 Repeated Linear Factors

The partial fraction decomposition will contain an expression of the form

$$\frac{A_1}{(x + a)} + \frac{A_2}{(x + a)^2} + \cdots + \frac{A_m}{(x + a)^m}$$

for each repeated linear fractor. Example:

$$\frac{4x + 5}{(x - 2)^2(2x + 1)} \quad \begin{array}{l} (x - 2)^2 = (x - 2)(x - 2), \\ \text{a repeated linear factor.} \end{array}$$

Partial fraction decomposition:

$$\frac{4x + 5}{(x - 2)^2(2x + 1)} = \frac{A_1}{x - 2} + \frac{A_2}{(x - 2)^2} + \frac{B}{2x + 1}$$

Case 3 Nonrepeated Quadratic Factors

The partial fraction decomposition will contain an expression of the form

$$\frac{Ax + B}{ax^2 + bx + c}$$

for each quadratic factor irreducible over the real numbers. Example:

$$\frac{x - 4}{(x^2 + x + 1)(x - 4)} \quad \begin{array}{l} x^2 + x + 1 \text{ is irreducible} \\ \text{over the real numbers.} \end{array}$$

Partial fraction decomposition:

$$\frac{x - 4}{(x^2 + x + 1)(x - 4)} = \frac{Ax + B}{x^2 + x + 1} + \frac{C}{x - 4}$$

Case 4 Repeated Quadratic Factors

The partial fraction decomposition will contain an expression of the form

$$\frac{A_1x + B_1}{ax^2 + bx + c} + \frac{A_2x + B_2}{(ax^2 + bx + c)^2} + \cdots + \frac{A_mx + B_m}{(ax^2 + bx + c)^m}.$$

for each quadratic factor irreducible over the real numbers. Example:

$$\frac{2x}{(x - 2)(x^2 + 4)^2} \quad (x^2 + 4)^2 \text{ is a repeated quadratic factor.}$$

Partial fraction decomposition:

$$\frac{2x}{(x - 2)(x^2 + 4)^2} = \frac{A_1x + B_1}{x^2 + 4} + \frac{A_2x + B_2}{(x^2 + 4)^2} + \frac{C}{x - 2}$$

Remark All the denominators in these examples are shown in factored form. If the denominator is not in factored form, you must factor first before proceeding any further.

There are various methods by which the constants of a partial fraction decomposition can be found. One such method is based on a property of polynomials.

Equality of Polynomials

If the two polynomials $p(x) = a_n x^n + a_{n-1} x^{n-1} + \cdots + a_1 x + a_0$ and $r(x) = b_n x^n + b_{n-1} x^{n-1} + \cdots + b_1 x + b_0$ are of degree n, then $p(x) = r(x)$ if and only if $a_0 = b_0, a_1 = b_1, a_2 = b_2, \ldots, a_n = b_n$.

EXAMPLE 1 **Find a Partial Fraction Decomposition for Nonrepeated Factors**

Find a partial fraction decomposition of $\dfrac{x + 11}{x^2 - 2x - 15}$.

Solution   First factor the denominator.

$$x^2 - 2x - 15 = (x + 3)(x - 5)$$

The factors are nonrepeated linear factors. Therefore the partial fraction decomposition will have the form

$$\frac{x + 11}{(x + 3)(x - 5)} = \frac{A}{x + 3} + \frac{B}{x - 5}. \tag{1}$$

To solve for A and B, multiply each side of the equation by the least common multiple of the denominators, $(x + 3)(x - 5)$.

$$x + 11 = A(x - 5) + B(x + 3)$$

$$1x + 11 = (A + B)x + (-5A + 3B) \quad \text{Combine like terms.}$$

Using the equality of polynomials theorem, equate coefficients of like powers. The result will be the system of equations

$$\begin{cases} 1 = A + B \\ 11 = -5A + 3B \end{cases}$$

Solving the system of equations for A and B, we have $A = -1$ and $B = 2$. Substituting -1 for A and 2 for B into the form of the partial fraction decomposition (1), we obtain

$$\frac{x + 11}{(x + 3)(x - 5)} = \frac{-1}{x + 3} + \frac{2}{x - 5}.$$

You should add the two expressions and verify the equality.

■ *Try Exercise* **14,** *page 535.*

EXAMPLE 2 **Find a Partial Fraction Decomposition for Repeated Linear Factors**

Find the partial fraction decomposition of $\dfrac{x^2 + 2x + 7}{x(x - 1)^2}$.

Solution The denominator has one nonrepeated factor and one repeated factor. The partial fraction decomposition will have the form

$$\frac{x^2 + 2x + 7}{x(x-1)^2} = \frac{A}{x} + \frac{B}{x-1} + \frac{C}{(x-1)^2}.$$

Multiplying each side by the LCD $x(x-1)^2$, we have

$$x^2 + 2x + 7 = A(x-1)^2 + B(x-1)x + Cx$$

Expanding the right side and combining like terms gives

$$x^2 + 2x + 7 = (A+B)x^2 + (-2A - B + C)x + A.$$

Using the equality of polynomial theorem, equate coefficients of like powers. This will result in the system of equations

$$\begin{cases} 1 = A + B \\ 2 = -2A - B + C \\ 7 = A \end{cases}$$

The solution is $A = 7$, $B = -6$, and $C = 10$. Thus the partial fraction decomposition is

$$\frac{x^2 + 2x + 7}{x(x-1)^2} = \frac{7}{x} + \frac{-6}{x-1} + \frac{10}{(x-1)^2}.$$

■ *Try Exercise* **22,** *page 536.*

EXAMPLE 3   Find a Partial Fraction Decomposition with a Quadratic Factor

Find the partial fraction decomposition of $\dfrac{3x + 16}{(x-2)(x^2 + 7)}$.

Solution Because $(x-2)$ is a nonrepeated linear factor and $x^2 + 7$ is an irreducible quadratic over the real numbers, the partial fraction decomposition will have the form

$$\frac{3x + 16}{(x-2)(x^2+7)} = \frac{A}{x-2} + \frac{Bx + C}{x^2 + 7}.$$

Multiplying each side by the LCD $(x-2)(x^2+7)$, we obtain

$$3x + 16 = A(x^2 + 7) + (Bx + C)(x - 2).$$

Expanding the right side and combining like terms gives

$$0x^2 + 3x + 16 = (A+B)x^2 + (-2B + C)x + (7A - 2C)$$

Using the equality of polynomials theorem, equate coefficients of like powers. This will result in the system of equations

$$\begin{cases} 0 = A + B \\ 3 = -2B + C \\ 16 = 7A - 2C \end{cases}$$

The solution is $A = 2$, $B = -2$, and $C = -1$. Thus the partial fraction decomposition is

$$\frac{3x + 16}{(x - 2)(x^2 + 7)} = \frac{2}{x - 2} + \frac{-2x - 1}{x^2 + 7}.$$

■ *Try Exercise* **24**, *page 536.*

EXAMPLE 4   Find a Partial Fraction Decomposition with Repeated Quadratic Factors

Find the partial fraction decomposition of $\dfrac{4x^3 + 5x^2 + 7x - 1}{(x^2 + x + 1)^2}$.

Solution   The quadratic factor $(x^2 + x + 1)$ is irreducible over the real numbers and is a repeated factor. The partial fraction decomposition will be of the form

$$\frac{4x^3 + 5x^2 + 7x - 1}{(x^2 + x + 1)^2} = \frac{Ax + B}{x^2 + x + 1} + \frac{Cx + D}{(x^2 + x + 1)^2}.$$

Multiplying each side by the LCD $(x^2 + x + 1)^2$ and collecting like terms, we obtain

$$\begin{aligned}
4x^3 + 5x^2 + 7x - 1 &= (Ax + B)(x^2 + x + 1) + Cx + D \\
&= Ax^3 + Ax^2 + Ax + Bx^2 + Bx + B + Cx + D \\
&= Ax^3 + (A + B)x^2 + (A + B + C)x + (B + D).
\end{aligned}$$

Equating coefficients of like powers gives the system of equations

$$\begin{cases}
4 = A \\
5 = A + B \\
7 = A + B + C \\
-1 = B + D
\end{cases}$$

Solving this system, we have $A = 4$, $B = 1$, $C = 2$, and $D = -2$. Thus the partial fraction decomposition is

$$\frac{4x^3 + 5x^2 + 7x - 1}{(x^2 + x + 1)^2} = \frac{4x + 1}{x^2 + x + 1} + \frac{2x - 2}{(x^2 + x + 1)^2}.$$

■ *Try Exercise* **30**, *page 536.*

The partial fraction decomposition theorem requires that the degree of the numerator be less than the degree of the denominator. If this is *not* the case, use long division to first write the rational expression as a polynomial plus a remainder.

EXAMPLE 5   Find a Partial Fraction Decomposition When the Degree of the Numerator Exceeds the Denominator

Find the partial fraction decomposition of $F(x) = \dfrac{x^3 - 4x^2 - 19x - 35}{x^2 - 7x}$.

Solution   Because the degree of the denominator is less than the degree of the numerator, use long division to first obtain

$$F(x) = x + 3 + \frac{2x - 35}{x^2 - 7x}$$

The partial fraction decomposition of $\dfrac{2x - 35}{x^2 - 7x}$ will have the form

$$\frac{2x - 35}{x^2 - 7x} = \frac{2x - 35}{x(x - 7)} = \frac{A}{x} + \frac{B}{x - 7}.$$

Multiplying each side by $x(x - 7)$ and combining like terms gives

$$2x - 35 = (A + B)x + (-7A).$$

Equating coefficients, we have

$$\begin{cases} 2 = A + B \\ -35 = -7A \end{cases}$$

The solution of this system is $A = 5$ and $B = -3$. The partial fraction decomposition is

$$\frac{x^3 - 4x^2 - 19x - 35}{x(x - 7)} = x + 3 + \frac{5}{x} + \frac{-3}{x - 7}.$$

■ *Try Exercise* **20,** *page 535.*

EXERCISE SET 10.4

In Exercises 1 to 10, evaluate the constants A, B, C, and D.

1. $\dfrac{x + 15}{x(x - 5)} = \dfrac{A}{x} + \dfrac{B}{x - 5}$

2. $\dfrac{5x - 6}{x(x + 3)} = \dfrac{A}{x} + \dfrac{B}{x + 3}$

3. $\dfrac{1}{(2x + 3)(x - 1)} = \dfrac{A}{2x + 3} + \dfrac{B}{x - 1}$

4. $\dfrac{6x - 5}{(x + 4)(3x + 2)} = \dfrac{A}{x + 4} + \dfrac{B}{3x + 2}$

5. $\dfrac{x + 9}{x(x - 3)^2} = \dfrac{A}{x} + \dfrac{B}{(x - 3)} + \dfrac{C}{(x - 3)^2}$

6. $\dfrac{2x - 7}{(x + 1)(x - 2)^2} = \dfrac{A}{x + 1} + \dfrac{B}{x - 2} + \dfrac{C}{(x - 2)^2}$

7. $\dfrac{4x^2 + 3}{(x - 1)(x^2 + x + 5)} = \dfrac{A}{x - 1} + \dfrac{Bx + C}{x^2 + x + 5}$

8. $\dfrac{x^2 + x + 3}{(x^2 + 7)(x - 3)} = \dfrac{Ax + B}{x^2 + 7} + \dfrac{C}{x - 3}$

9. $\dfrac{x^3 + 2x}{(x^2 + 1)^2} = \dfrac{Ax + B}{x^2 + 1} + \dfrac{Cx + D}{(x^2 + 1)^2}$

10. $\dfrac{3x^3 + x^2 - x - 5}{(x^3 + 2x + 5)^2} = \dfrac{Ax + B}{x^2 + 2x + 5} + \dfrac{Cx + D}{(x^2 + 2x + 5)^2}$

In Exercises 11 to 36, find the partial fraction decomposition of the given rational expression.

11. $\dfrac{8x + 12}{x(x + 4)}$

12. $\dfrac{x - 14}{x(x - 7)}$

13. $\dfrac{3x + 50}{x^2 - 7x - 18}$

14. $\dfrac{7x + 44}{x^2 + 10x + 24}$

15. $\dfrac{16x + 34}{4x^2 + 16x + 15}$

16. $\dfrac{-15x + 37}{9x^2 - 12x - 5}$

17. $\dfrac{x - 5}{(3x + 5)(x - 2)}$

18. $\dfrac{1}{(x + 7)(2x - 5)}$

19. $\dfrac{x^3 + 3x^2 - 4x - 8}{x^2 - 4}$

20. $\dfrac{x^3 - 13x - 9}{x^2 - x - 12}$

21. $\dfrac{3x^2 + 49}{x(x + 7)^2}$

22. $\dfrac{x - 18}{x(x - 3)^2}$

23. $\dfrac{5x^2 - 7x + 2}{x^3 - 3x^2 + x}$

24. $\dfrac{9x^2 - 3x + 49}{x^3 - x^2 + 10x - 10}$

25. $\dfrac{2x^3 + 9x^2 + 26x + 41}{(x + 3)^2(x^2 + 1)}$

26. $\dfrac{12x^3 - 37x^2 + 48x - 36}{(x - 2)^2(x^2 + 4)}$

27. $\dfrac{3x - 7}{(x - 4)^2}$

28. $\dfrac{5x - 53}{(x - 11)^2}$

29. $\dfrac{3x^3 - x^2 + 34x - 10}{(x^2 + 10)^2}$

30. $\dfrac{2x^3 + 9x + 1}{x^4 + 14x^2 + 49}$

31. $\dfrac{1}{k^2 - x^2}$, where k is a constant

32. $\dfrac{1}{x(k + lx)}$, where k and l are constants

33. $\dfrac{x^3 - x^2 - x - 1}{x^2 - x}$

34. $\dfrac{2x^3 + 5x^2 + 3x - 8}{2x^2 + 3x - 2}$

35. $\dfrac{2x^3 - 4x^2 + 5}{x^2 - x - 1}$

36. $\dfrac{x^4 - 2x^3 - 2x^2 - x + 3}{x^2(x - 3)}$

Supplemental Exercises

In Exercises 37 to 42, find the partial fraction decomposition of the rational expressions.

37. $\dfrac{x^2 - 1}{(x - 1)(x + 2)(x - 3)}$

38. $\dfrac{x^2 + x}{x^2(x - 4)}$

39. $\dfrac{-x^4 - 4x^2 + 3x - 6}{x^4(x - 2)}$

40. $\dfrac{3x^2 - 2x - 1}{(x^2 - 1)^2}$

41. $\dfrac{2x^2 + 3x - 1}{x^3 - 1}$

42. $\dfrac{x^3 - 2x^2 + x - 2}{x^4 - x^3 + x - 1}$

Expressing a rational expression with real coefficients as a partial fraction decomposition requires factoring the denominator into the product of linear and quadratic factors. But is this always possible? For example, suppose the denominator of some rational expression was

$$x^4 + \pi x^3 - \sqrt{7}x^2 - 6x + \sqrt[3]{9}.$$

Can this be factored into a product of linear and quadratic factors?

The answer is yes to each question we asked, but it may be very difficult to find the factorization. Working Exercises 43 to 46 prove that any polynomial with real coefficients can always be factored into a product of linear and quadratic factors. You may want to review the material in Chapter 4 before trying to prove these exercises.

In Exercises 43 to 46, let $p(x)$ be a polynomial of degree n with real coefficients.

43. Prove that if c is a real number zero of $p(x)$, then $x - c$ is a linear factor of $p(x)$. (*Hint:* See the factor theorem in Chapter 4.)

44. Prove that if $c = a + bi$ is a complex zero of $p(x)$, then $a - bi$, the complex conjugate of c is also a complex zero of $p(x)$.

45. Use the result of Exercise 44 to show that if $a + bi$ is a complex zero of $p(x)$, then $x^2 - 2ax + (a^2 + b^2)$ is a quadratic factor of $p(x)$.

46. Combine the results of Exercises 43 and 45 to prove that any polynomial with real coefficients can be factored as the product of linear and quadratic factors.

10.5

Inequalities in Two Variables and Systems of Inequalities

Two examples of inequalities in two variables are

$$2x + 3y > 6 \quad \text{and} \quad xy \le 1.$$

A solution of an inequality in two variables is an ordered pair (x, y) that satisfies the inequality. For example, $(-2, 4)$ is a solution of the first inequality since $2(-2) + 3(4) > 6$. The ordered pair $(2, 1)$ is not a solution of the second inequality since $(2)(1) \nleq 1$.

The **solution set of an inequality** in two variables is the set of all ordered pairs that satisfy the inequality. The **graph** of an inequality is the graph of the solution set.

To sketch the graph of an inequality, first replace the inequality symbol by an equality sign and sketch the graph of the equation. Use a dashed graph for $<$ or $>$ to indicate that the curve is not part of the solution set. Use a solid graph for $\leq$ or $\geq$ to show that the graph is part of the solution set.

It is important to test a point in each region of the plane defined by the graph. If the point satisifies the inequality, shade that entire region. Do this for each region into which the graph divides the plane. For example, consider the inequality $xy \geq 2$. Figure 10.24 shows the three regions of the plane defined by this inequality. Because the inequality is $\geq$, a solid graph is used.

Choose a point in each of the three regions and determine if each point satisfies the inequality. In region I, choose $(-2, -4)$. Because $(-2)(-4) \geq 2$, shade region I. In region II, choose a point, say $(0, 0)$. Because $0 \cdot 0 \not\geq 2$, do not shade region II. In region III, choose $(4, 5)$. Because $4 \cdot 5 \geq 2$, shade region III.

You may choose any point not on the graph of the equation as a test point; $(0, 0)$ is usually a good choice.

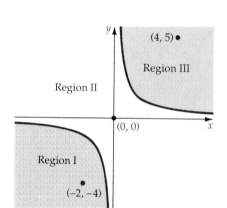

Figure 10.24
$xy \geq 2$

EXAMPLE 1   Graph a Linear Inequality

Sketch a graph of $3x + 4y > 12$.

Solution   Graph the line $3x + 4y = 12$ using a dashed line.

Test the point $(0, 0)$: $3(0) + 4(0) = 0 \not> 12$.

Since $(0, 0)$ does not satisfy the inequality, do not shade this region.

Test the point $(2, 3)$: $3(2) + 4(3) = 18 > 12$.

Since $(2, 3)$ satisfies the inequality, shade this region of the plane.

■ *Exercise* **6,** *page 541.*

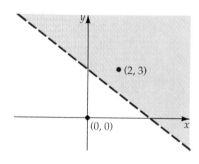

Figure 10.25
$3x + 4y > 12$

In general, the solution set of a *linear inequality in two variables* will be one of the regions of the plane separated by a line. Each region is called a **half-plane.**

EXAMPLE 2   Graph a Nonlinear Inequality

Sketch the graph of $y \leq x^2 + 2x - 3$.

Solution   Graph the parabola $y = x^2 + 2x - 3$ using a solid curve.

Test the point $(0, 0)$: $0 \not\leq 0^2 + 2(0) - 3$.

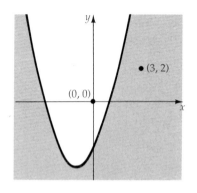

Figure 10.26
$y \leq x^2 + 2x - 3$

Since $(0, 0)$ does not satisfy the inequality, do not shade this region.

Test the point $(3, 2)$: $2 \leq 3^2 + 2(3) - 3$.

Since $(3, 2)$ satisfies the inequality, shade this region of the plane.

■ *Try Exercise* **12,** *page 541.*

EXAMPLE 3   Graph an Absolute Value Inequality

Sketch the graph of $y \geq |x| + 1$.

Solution   Graph the equation $y = |x| + 1$ using a solid line.

Test the point $(0, 0)$: $0 \not\geq |0| + 1$.

Since $0 \not\geq 1$, $(0, 0)$ does not belong to the solution set. Shade that portion of the plane that does not contain $(0, 0)$.

Test the point $(0, 4)$: $4 \geq |0| + 1$.

Since $(0, 4)$ satisfies the inequality, shade this region.

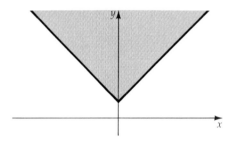

Figure 10.27
$y \geq |x| + 1$

■ *Try Exercise* **20,** *page 541.*

System of Inequalities in Two Variables

The solution set of a system of inequalities is the intersection of the solution sets of each inequality. To graph the solution set of a system of inequalities, first graph the solution set of each inequality. The solution set of the system of inequalities is the region of the plane represented by the intersection of the shaded regions.

EXAMPLE 4   Graph the Solution Set of a System of Linear Inequalities

Graph the solution set of the system of inequalities

$$\begin{cases} 3x - 2y > 6 \\ 2x - 5y \leq 10 \end{cases}$$

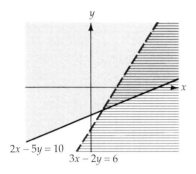

Figure 10.28

Solution   Graph the line $3x - 2y = 6$ using a dashed line. Test the point $(0, 0)$. Because $3(0) - 2(0) \not> 6$, shade the region below the line.

Graph the line $2x - 5y = 10$ using a solid line. Test the point $(0, 0)$. Because $2(0) - 5(0) \leq 10$, shade the region above the line.

The solution set is the region of the plane represented by the intersection of the solution sets of each inequality. See Figure 10.28.

■ *Try Exercise* **26**, *page 541.*

EXAMPLE 5   Graph the Solution Set of a Nonlinear System of Inequalities

Graph the solution set of the system of inequalities

$$\begin{cases} x^2 - y^2 \leq 9 \\ 2x + 3y > 12 \end{cases}$$

Solution   Graph the hyperbola $x^2 - y^2 = 9$ by using a solid graph. Test the point $(0, 0)$. Because $0^2 - 0^2 \leq 9$, shade the region containing the origin. By choosing points in the other two regions, you should show that those regions are not part of the solution set.

Graph the line $2x + 3y = 12$ by using a dashed graph. Test the point $(0, 0)$. Because $2(0) + 3(0) \not> 12$, shade the half-plane above the line.

The solution set is the region of the plane represented by the intersection of the solution sets of each inequality. See Figure 10.29.

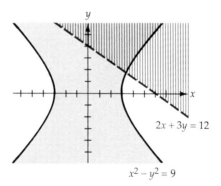

Figure 10.29

■ *Try Exercise* **36,** *page 541.*

EXAMPLE 6   Identify a System of Inequalities with No Solution

Graph the solution set of the system of inequalities

$$\begin{cases} x^2 + y^2 \leq 16 \\ x^2 - y^2 \geq 36 \end{cases}$$

Solution   Graph the circle $x^2 + y^2 = 16$ by using a solid graph. Test the point $(0, 0)$. Because $0^2 + 0^2 \leq 16$, shade the inside of the circle.

Graph the hyperbola $x^2 - y^2 = 36$ by using a solid graph. Test the point $(0, 0)$. Because $0^2 - 0^2 \not\geq 36$, shade inside the two branches of the hyperbola.

Because the solution sets of the inequalities do not intersect, the system has no solution. The solution set is the empty set.

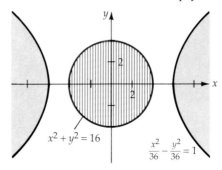

Figure 10.30

■ *Try Exercise* **40**, *page 541.*

EXAMPLE 7 Graph the Solution Set of a System of Four Inequalities

Graph the solution set of the system of inequalities

$$\begin{cases} 2x - 3y \leq 2 \\ 3x + 4y \geq 12 \\ x \geq -1, y \geq 2 \end{cases}$$

Solution First graph the inequalities $x \geq -1$ and $y \geq 2$. Because $x \geq -1$ and $y \geq 2$, the solution set for this system will be above the line $y = 2$ and to the right of the line $x = -1$. See Figure 10.31a.

Graph the solution set of $2x - 3y = 2$ by using a solid graph. Test the point $(0, 0)$. Because $2(0) - 3(0) \leq 2$, shade the region above the line.

Graph the solution set of $3x + 4y = 12$ by using a solid graph. Test the point $(0, 0)$. Because $3(0) + 4(0) \not\geq 12$, shade the region above the line.

The solution set of the system of equations is the region where the graphs of the solution sets of all four inequalities overlap (Figure 10.31b).

a.
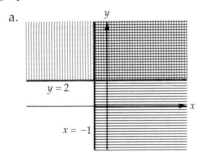

b.
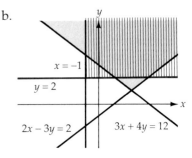

Figure 10.31

■ *Try Exercise* **42**, *page 541.*

EXERCISE SET 10.5

In Exercises 1 to 22, sketch the graph of each inequality.

1. $y \leq -2$ **2.** $x + y > -2$

3. $y \geq 2x + 3$ **4.** $y < -2x + 1$

5. $2x - 3y < 6$ **6.** $3x + 4y \leq 4$

7. $4x + 3y \leq 12$ **8.** $5x - 2y < 8$

9. $y < x^2$ **10.** $x > y^2$

11. $y \geq x^2 - 2x - 3$ **12.** $y < 2x^2 - x - 3$

13. $(x - 2)^2 + (y - 1)^2 < 16$

14. $(x + 2)^2 + (y - 3)^2 > 25$

15. $\dfrac{(x - 3)^2}{9} - \dfrac{(y + 1)^2}{16} > 1$ **16.** $\dfrac{(x + 1)^2}{25} - \dfrac{(y - 3)^2}{16} \leq 1$

17. $4x^2 + 9y^2 - 8x + 18y \geq 23$

18. $25x^2 - 16y^2 - 100x - 64y < 64$

19. $y < 2^{x-1}$ **20.** $y > \log_3(x)$

21. $y \leq \log_2(x - 1)$ **22.** $y > 3^x + 1$

40. $\begin{cases} \dfrac{(x + 1)^2}{36} + \dfrac{(y - 2)^2}{25} < 1 \\ \dfrac{(x + 1)^2}{25} + \dfrac{(y - 2)^2}{36} < 1 \end{cases}$

41. $\begin{cases} 2x - 3y \leq -5 \\ x + 2y \leq 7 \\ x \geq -1, y \geq 0 \end{cases}$ **42.** $\begin{cases} 5x + y \leq 9 \\ 2x + 3y \leq 14 \\ x \geq -2, y \geq 2 \end{cases}$

43. $\begin{cases} 3x + 2y \geq 14 \\ x + 3y \geq 14 \\ x \leq 10, y \leq 8 \end{cases}$ **44.** $\begin{cases} 4x + y \geq 13 \\ 3x + 2y \geq 16 \\ x \leq 15, y \leq 12 \end{cases}$

45. $\begin{cases} 3x + 4y \leq 12 \\ 2x + 5y \leq 10 \\ x \geq 0, y \geq 0 \end{cases}$ **46.** $\begin{cases} 5x + 3y \leq 15 \\ x + 4y \leq 8 \\ x \geq 0, y \geq 0 \end{cases}$

In Exercises 23 to 46, sketch the graph of the solution set of each system of inequalities.

23. $\begin{cases} 1 \leq x < 3 \\ -2 < y \leq 4 \end{cases}$ **24.** $\begin{cases} -2 < x < 4 \\ y \geq -1 \end{cases}$

25. $\begin{cases} 3x + 2y \geq 1 \\ x + 2y < -1 \end{cases}$ **26.** $\begin{cases} 2x - 5y < -6 \\ 3x + y < 8 \end{cases}$

27. $\begin{cases} 2x - y \geq -4 \\ 4x - 2y \leq -17 \end{cases}$ **28.** $\begin{cases} 4x + 2y > 5 \\ 6x + 3y > 10 \end{cases}$

29. $\begin{cases} 4x - 3y < 14 \\ 2x + 5y \leq -6 \end{cases}$ **30.** $\begin{cases} 3x + 5y \geq -8 \\ 2x - 3y \geq 1 \end{cases}$

31. $\begin{cases} y < 2x + 3 \\ y > 2x - 2 \end{cases}$ **32.** $\begin{cases} y > 3x + 1 \\ y < 3x - 2 \end{cases}$

33. $\begin{cases} y < 2x - 1 \\ y \geq x^2 + 3x - 7 \end{cases}$ **34.** $\begin{cases} y \leq 2x + 7 \\ y > x^2 + 3x + 1 \end{cases}$

35. $\begin{cases} x^2 + y^2 \leq 49 \\ 9x^2 + 4y^2 \geq 36 \end{cases}$ **36.** $\begin{cases} y < 2x - 1 \\ y > x^2 - 2x + 2 \end{cases}$

37. $\begin{cases} (x - 1)^2 + (y + 1)^2 \leq 16 \\ (x - 1)^2 + (y + 1)^2 \geq 4 \end{cases}$

38. $\begin{cases} (x + 2)^2 + (y - 3)^2 > 25 \\ (x + 2)^2 + (y - 3)^2 < 16 \end{cases}$

39. $\begin{cases} \dfrac{(x - 4)^2}{16} - \dfrac{(y + 2)^2}{9} > 1 \\ \dfrac{(x - 4)^2}{25} + \dfrac{(y + 2)^2}{9} < 1 \end{cases}$

Supplemental Exercises

In Exercises 47 to 58, sketch the graph of the inequality.

47. $y < |x|$ **48.** $y \geq |2x - 4|$

49. $|y| \geq |x|$ **50.** $|y| \leq |x - 1|$

51. $|x + y| \leq 1$ **52.** $|x - y| > 1$

53. $|x| + |y| \leq 1$ **54.** $|x| - |y| > 1$

55. $y > [\![x]\!]$, where $[\![x]\!]$ is the greatest integer function.

56. $y > x - [\![x]\!]$, where $[\![x]\!]$ is the greatest integer function.

57. Sketch the graphs of $xy > 1$ and $y > \dfrac{1}{x}$. Note that the two graphs are not the same, yet the second inequality can be derived from the first by dividing each side by x. Explain.

58. Sketch the graph of $\dfrac{x}{y} < 1$ and the graph of $x < y$. Note that the two graphs are not the same, yet the second inequality can be derived from the first by multiplying each side by y. Explain.

10.6

Linear Programming

Consider a business analyst who is trying to maximize the profit from the production of a product or an engineer who is trying to minimize the amount of energy an electrical circuit needs to operate. Generally, problems that seek to maximize or minimize a situation are called **optimization problems.** One strategy for solving certain of these problems was developed in the 1940s and is called **linear programming.**

A linear programming problem involves a **linear objective function,** which is the function that must be maximized or minimized. This objective function is subject to some **constraints,** which are inequalities or equations that restrict the values of the variables. To illustrate these concepts, suppose a manufacturer produces two types of computer monitors: monochrome and color. Past sales experience show that at least twice as many monochrome monitors are sold as color monitors. Suppose further that the manufacturing plant is capable of producing twelve monitors a day. Let x represent the number of monochrome monitors produced and y the number of color monitors produced. Then

$$\begin{cases} x \geq 2y \\ x + y \leq 12 \end{cases} \quad \text{These are the constraints.}$$

These two inequalities place a constraint or restriction on the manufacturer. For example, the manufacturer cannot produce five color monitors because that would require producing at least ten monochrome monitors, but $5 + 10 \nleq 12$.

The manufacturer will make a profit on the sale of the monitors. Suppose a profit of $50 is earned on each monochrome monitor sold and $75 is earned on each color monitor sold. Then the manufacturer's profit, P, is given by the equation

$$P = 50x + 75y. \quad \text{Objective function}$$

The equation $P = 50x + 75y$ defines the objective function. The goal of this linear programming problem is to determine how many of each monitor should be produced to maximize the manufacturer's profit and at the same time satisfy the constraints.

Because the manufacturer cannot produce less than zero units of either monitor, there are two other implied constraints, $x \geq 0$ and $y \geq 0$. Our linear programming problem now looks like

Objective function: $P = 50x + 75y$

Constraints: $\begin{cases} x \geq 0, y \geq 0 \\ x - 2y \geq 0 \\ x + \ y \leq 12 \end{cases}$

To solve this problem, graph the solution set of the constraints. The set of points in the solution set of the constraints is called the **set of feasible solutions.** Points in this set are used to evaluate the objective function to

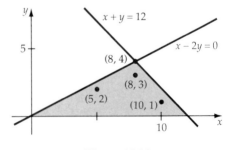

Figure 10.32

determine which point maximizes the profit. For example, $(5, 2)$, $(8, 3)$ and $(10, 1)$ are three points in the set. For these points, the profit would be

$$P = 50(5) + 75(2) = 400$$

$$P = 50(8) + 75(3) = 625$$

$$P = 50(10) + 75(1) = 575$$

It would be quite time consuming to check every point in the set of feasible solutions to find which point maximizes profit. Fortunately, we can find that point by solving the objective function $P = 50x + 75y$ for y.

$$y = -\frac{2}{3}x + \frac{P}{75}$$

In this form, the objective function is written as a linear equation with slope $-\frac{2}{3}$ and y-intercept $P/75$. If P is as large as possible (P a maximum), then the y-intercept will be as large as possible. Thus the maximum profit will occur on the line that has the largest y-intercept and intersects the set of feasible solutions.

From Figure 10.33, the largest y-intercept occurs when the line passes through the point $(8, 4)$. At this point, the profit is

$$P = 50(8) + 75(4) = 700.$$

The manufacturer will maximize profit by producing eight monochrome monitors and four color monitors each day. The profit will be $700 per day.

In general, the goal of any linear programming problem is to maximize or minimize the objective function subject to the constraints. Minimization problems occur, for example, when a manufacturer wants to minimize the cost of operations.

Suppose that a cost minimization problem results in the following constraints and objective function.

Objective function: $C = 3x + 4y$

Constraints:
$$\begin{cases} x \geq 0, y \geq 0 \\ x + y \geq 1 \\ 2x - y \leq 5 \\ x + 2y \leq 10 \end{cases}$$

Figure 10.34 is the graph of the solution set of the constraints. The task is to find the ordered pair that will give the smallest value of C. We again could solve the objective function for y and, since we want to minimize C, find the smallest y-intercept. However, a theorem from linear programming simplifies our task even more. The proof of this theorem, omitted here, is based on the techniques we used to solve our examples.

Fundamental Linear Programming Theorem

If an objective function has an optimal solution, then that solution will be at a vertex of the set of feasible solutions.

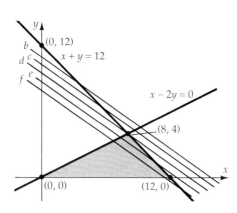

Figure 10.33

Figure 10.34

Following is a listing of the values of C at the vertices. The minimum value of the objective function occurs at the point $(1, 0)$.

$$(x, y) \quad C = 3x + 4y$$

$$(1, 0) \quad C = 3 \cdot 1 + 4 \cdot 0 = 3 \qquad \text{Minimum}$$

$$\left(\frac{5}{2}, 0\right) \quad C = 3 \cdot \frac{5}{2} + 4 \cdot 0 = 7.5$$

$$(4, 3) \quad C = 3 \cdot 4 + 4 \cdot 3 = 24 \qquad \text{Maximum}$$

$$(0, 5) \quad C = 3 \cdot 0 + 4 \cdot 5 = 20$$

$$(0, 1) \quad C = 3 \cdot 0 + 4 \cdot 1 = 4$$

The maximum value of the objective function can also be determined from the list. The maximum value occurs at the point $(4, 3)$.

It is important to realize that the maximum or minimum value of an objective function depends on the objective function and the set of feasible solutions. For example, using the same set of feasible solutions as in Figure 10.34 but changing the objective function to $C = 2x + 5y$ changes the maximum value of C to 25 at the ordered pair $(0, 5)$. You should verify this result by making a list similar to the one above.

EXAMPLE 1   Solve a Minimization Problem

Minimize the objective function $C = 4x + 7y$ with constraints:

$$\begin{cases} 3x + y \geq 6 \\ x + y \geq 4 \\ x + 3y \geq 6 \\ x \geq 0, y \geq 0 \end{cases}$$

Solution   Determine the set of feasible solutions by graphing the solution set of the inequalities. Note that the set of feasible solutions has no upper limit.

Find the vertices of the region by solving the following systems of equations. These systems are formed by the equations of the lines that intersect to form a vertex of the set of feasible solutions.

$$\begin{cases} 3x + y = 6 \\ x + y = 4 \end{cases} \qquad \begin{cases} x + 3y = 6 \\ x + y = 4 \end{cases}$$

The solutions of the two systems are $(1, 3)$ and $(3, 1)$, respectively. Also find the y- and x-intercepts of the set of feasible solutions. They are $(0, 6)$ and $(6, 0)$.

Evaluate the objective function at each of the four vertices of the feasible solutions.

$$(x, y) \quad C = 4x + 7y$$

$$(0, 6) \quad C = 4 \cdot 0 + 7 \cdot 6 = 42$$

$$(1, 3) \quad C = 4 \cdot 1 + 7 \cdot 3 = 25$$

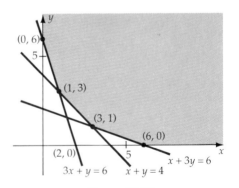

Figure 10.35

$$(3, 1) \quad C = 4 \cdot 3 + 7 \cdot 1 = 19$$

$$(6, 0) \quad C = 4 \cdot 6 + 7 \cdot 0 = 24$$

The minimum value of the objective function is 19 at $(3, 1)$.

■ *Try Exercise* **12,** *page 547.*

Linear programming can be used to determine the best allocation of the resources available to a company. In fact, the word "programming" refers to a "program to allocate resources."

EXAMPLE 2   **Solve an Applied Minimization Problem**

A manufacturer of animal food makes two grain mixtures, G_1 and G_2. Each kilogram of G_1 contains 300 grams of vitamins, 400 grams of protein, and 100 grams of carbohydrate. Each kilogram of G_2 contains 100 grams of vitamins, 300 grams of protein, and 200 grams of carbohydrate. Minimum nutritional guidelines require that a feed mixture made from these grains contain at least 900 grams of vitamins, 2200 grams of protein, and 800 grams of carbohydrate. If G_1 costs \$2.00 per kilogram to produce and G_2 costs \$1.25 per kilogram to produce, find the number of kilograms of each grain mixture that should be produced to minimize cost.

Solution   Let

$$x = \text{the number of kilograms of } G_1,$$

$$y = \text{the number of kilograms of } G_2.$$

The objective function is the cost function $C = 2x + 1.25y$.

Because x kilograms of G_1 require $300x$ grams of vitamins and y kilograms of G_2 require $100y$ grams of vitamins, the total amount of vitamins needed is $300x + 100y$. Since at least 900 grams of vitamins are necessary, $300x + 100y \geq 900$. Following similar reasoning, we have the constraints

$$\begin{cases} 300x + 100y \geq 900 \\ 400x + 300y \geq 2200 \\ 100x + 200y \geq 800 \\ x \geq 0, y \geq 0 \end{cases}$$

Two of the vertices of the set of feasible solutions can be found by solving two systems of equations. These systems are formed by the equations of the lines that intersect to form a vertex of the set of feasible solutions.

$$\begin{cases} 300x + 100y = 900 \\ 400x + 300y = 2200 \end{cases} \quad \text{The vertex is } (1, 6).$$

$$\begin{cases} 100x + 200y = 800 \\ 400x + 300y = 2200 \end{cases} \quad \text{The vertex is } (4, 2).$$

The vertices on the x- and y-axes are the x- and y-intercepts $(8, 0)$ and $(0, 9)$.

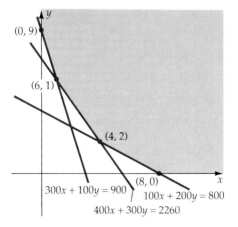

Figure 10.36

Substitute the coordinates of the vertices into the objective function.

$$(x, y) \quad C = 2x + 1.25y$$
$$(0, 9) \quad C = 2(0) + 1.25(9) = 11.25$$
$$(1, 6) \quad C = 2(1) + 1.25(6) = 9.50 \quad \text{Minimum}$$
$$(4, 2) \quad C = 2(4) + 1.25(2) = 10.50$$
$$(8, 0) \quad C = 2(8) + 1.25(0) = 16.00$$

The minimum value of the objective function is \$9.50 and occurs when the company produces a feed mixture that contains 1 kilogram of G_1 and 6 kilograms of G_2.

■ *Try Exercise **22**, page 548.*

EXAMPLE 3   Solve an Applied Maximization Problem

A chemical firm produces two types of industrial solvents: S_1 and S_2. Each solvent is a mixture of three chemicals. Each kiloliter of S_1 requires 12 liters of chemical 1, 9 liters of chemical 2, and 30 liters of chemical 3. Each kiloliter of S_2 requires 24 liters of chemical 1, 5 liters of chemical 2, and 30 liters of chemical 3. The profit per kiloliter of S_1 is \$100, and the profit per kiloliter of S_2 is \$85. The inventory of the company shows 480 liters of chemical 1, 180 liters of chemical 2, and 720 liters of chemical 3. Assuming the company can sell all the solvent it makes, find the number of kiloliters of each solvent the company should make to maximize profit.

Solution   Let

$$x = \text{the number of kiloliters of } S_1,$$
$$y = \text{the number of kiloliters of } S_2.$$

The objective function is the profit function $P = 100x + 85y$.

Because x kiloliters of S_1 requires $12x$ liters of chemical 1 and y kiloliters of S_2 require $24y$ liters of chemical 1, the total amount of chemical 1 needed is $12x + 24y$. Since there is 480 liters of chemical 1 in inventory, $12x + 24y \le 480$. Following similar reasoning, we have the constraints

$$\begin{cases} 12x + 24y \le 480 \\ 9x + 5y \le 180 \\ 30x + 30y \le 720 \end{cases}$$

There are two additional constraints, $x \ge 0$, $y \ge 0$, since the company will not make less than 0 kiloliters of either solvent.

Two of the vertices of the set of feasible solutions can be found by solving two systems of equations. These systems are formed by the equations of the lines that intersect to form a vertex of the set of feasible solutions.

$$\begin{cases} 12x + 24y = 480 \\ 30x + 30y = 720 \end{cases} \quad \text{The vertex is } (8, 16).$$

$$\begin{cases} 9x + 5y = 180 \\ 30x + 30y = 720 \end{cases} \quad \text{The vertex is } (15, 9).$$

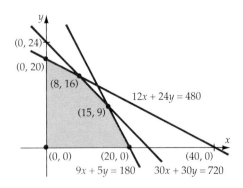

Figure 10.37

The vertices on the x- and y-axes are the x- and y-intercepts $(20, 0)$ and $(0, 20)$.

Substitute the coordinates of the vertices into the objective function.

$$(x, y) \quad P = 100x + 85y$$

$$(0, 20) \quad P = 100(0) + 85(20) = 1700$$

$$(8, 16) \quad P = 100(8) + 85(16) = 2160$$

$$(15, 9) \quad P = 100(15) + 85(9) = 2265 \quad \text{Maximum}$$

$$(20, 0) \quad P = 100(20) + 85(0) = 2000$$

The maximum value of the objective function is \$2265 and occurs when the company produces 15 kiloliters of S_1 and 9 kiloliters of S_2.

■ *Try Exercise* **24,** *page 548.*

EXERCISE SET 10.6

In Exercises 1 to 20, solve the linear programming problems. Assume $x \geq 0$ and $y \geq 0$.

1. Minimize $C = 4x + 2y$ with constraints:

$$\begin{cases} x + y \geq 7 \\ 4x + 3y \geq 24 \\ x \leq 10, \ y \leq 10 \end{cases}$$

2. Minimize $C = 5x + 4y$ with constraints:

$$\begin{cases} 3x + 4y \geq 32 \\ x + 4y \geq 24 \\ x \leq 12, \ y \leq 15 \end{cases}$$

3. Maximize $C = 6x + 7y$ with constraints:

$$\begin{cases} x + 2y \leq 16 \\ 5x + 3y \leq 45 \end{cases}$$

4. Maximize $C = 6x + 5y$ with constraints:

$$\begin{cases} 2x + 3y \leq 27 \\ 7x + 3y \leq 42 \end{cases}$$

5. Minimize $C = 5x + 6y$ with constraints:

$$\begin{cases} 4x - 3y \leq 2 \\ 2x + 3y \geq 10 \end{cases}$$

6. Maximize $C = 4x + 5y$ with constraints:

$$\begin{cases} 2x - y \leq 0 \\ 0 \leq y \leq 10 \\ 0 \leq x \leq 10 \end{cases}$$

7. Maximize $C = x + 6y$ with constraints:

$$\begin{cases} 5x + 8y \leq 120 \\ 7x + 16y \leq 192 \end{cases}$$

8. Minimize $C = 4x + 5y$ with constraints:

$$\begin{cases} x + 3y \geq 30 \\ 3x + 4y \geq 60 \end{cases}$$

9. Minimize $C = 4x + y$ with constraints:

$$\begin{cases} 3x + 5y \geq 120 \\ x + y \geq 32 \end{cases}$$

10. Maximize $C = 7x + 2y$ with constraints:

$$\begin{cases} x + 3y \leq 108 \\ 7x + 4y \leq 280 \end{cases}$$

11. Maximize $C = 2x + 7y$ with constraints:

$$\begin{cases} x + y \leq 10 \\ x + 2y \leq 16 \\ 2x + y \leq 16 \end{cases}$$

12. Minimize $C = 4x + 3y$ with constraints:

$$\begin{cases} 2x + y \geq 8 \\ 2x + 3y \geq 16 \\ x + 3y \geq 11 \\ x \leq 20, \ y \leq 20 \end{cases}$$

13. Minimize $C = 3x + 2y$ with constraints:

$$\begin{cases} 3x + y \geq 12 \\ 2x + 7y \geq 21 \\ x + y \geq 8 \end{cases}$$

14. Maximize $C = 2x + 6y$ with constraints:

$$\begin{cases} x + y \leq 12 \\ 3x + 4y \leq 40 \\ x + 2y \leq 18 \end{cases}$$

15. Maximize $C = 3x + 4y$ with constraints:

$$\begin{cases} 2x + y \le 10 \\ 2x + 3y \le 18 \\ x - y \le 2 \end{cases}$$

16. Minimize $C = 3x + 7y$ with constraints:

$$\begin{cases} x + y \ge 9 \\ 3x + 4y \ge 32 \\ x + 2y \ge 12 \end{cases}$$

17. Minimize $C = 3x + 2y$ with constraints:

$$\begin{cases} x + 2y \ge 8 \\ 3x + y \ge 9 \\ x + 4y \ge 12 \end{cases}$$

18. Maximize $C = 4x + 5y$ with constraints:

$$\begin{cases} 3x + 4y \le 250 \\ x + y \le 75 \\ 2x + 3y \le 180 \end{cases}$$

19. Maximize $C = 6x + 7y$ with constraints:

$$\begin{cases} x + 2y \le 900 \\ x + y \le 500 \\ 3x + 2y \le 1200 \end{cases}$$

20. Minimize $C = 11x + 16y$ with constraints:

$$\begin{cases} x + 2y \ge 45 \\ x + y \ge 40 \\ 2x + y \ge 45 \end{cases}$$

21. A farmer is planning to raise wheat and barley. Each acre of wheat yields a profit of $50, and each acre of barley yields a profit of $70. To sow the crop, two machines, a tractor and tiller, are rented. The tractor is available for 200 hours, and the tiller is available for 100 hours. Sowing an acre of barley requires 3 hours of tractor time and 2 hours of tilling. Sowing an acre of wheat requires 4 hours of tractor time and 1 hour of tilling. How many acres of each crop should be planted to maximize the farmer's profit?

22. An ice cream supplier has two machines that produce vanilla and chocolate ice cream. To meet one of its contractual obligations, the company must produce at least 60 gallons of vanilla ice cream and 100 gallons of chocolate ice cream. One machine makes 4 gallons of vanilla and 3 gallons of chocolate ice cream per hour. The second machine makes 5 gallons of vanilla and 10 gallons of chocolate ice cream per hour. If it cost $28 per hour to run machine 1 and $25 per hour to run machine 2, how many hours should each machine be operated to fulfill the contract at the least expense?

23. A manufacturer makes two types of golf clubs: a starter model and a professional model. The starter model requires 4 hours in the assembly room and 1 hour in the finishing room. The professional model requires 6 hours in the assembly room and 1 hour in the finishing room. The total number of hours available in the assembly room are 108. There are 24 hours available in the finishing room. The profit for each starter model is $35, and the profit for each professional model is $55. Assuming all the sets produced can be sold, how many of each set should be manufactured to maximize profit?

24. A company makes two types of telephone answering machines: the standard model and the deluxe model. Each machine passes through three processes: P_1, P_2, and P_3. One standard answering machine requires 1 hour in P_1, 1 hour in P_2, and 2 hours in P_3. One deluxe answering machine requires 3 hours in P_1, 1 hour in P_2, and 1 hour in P_3. Because of employee work schedules, P_1 is available for 24 hours, P_2 is available for 10 hours, and P_3 is available for 16 hours. If the profit is $25 for each standard model and $35 for each deluxe model, how many units of each type should be produced to maximize profit?

Supplemental Exercises

25. A dietitian formulates a special diet from two food groups: *A* and *B*. Each ounce of food group *A* contains 3 units of vitamin A, 1 unit of vitamin C, and 1 unit of vitamin D. Each unit of food group *B* contains 1 unit of vitamin A, 1 unit of vitamin C, and 3 units of vitamin D. Each ounce of food group *A* cost 40 cents, and each ounce of food group *B* costs 10 cents. The dietary constraints are such that at least 24 units of vitamin A, 16 units of vitamin C, and 30 units of vitamin D are required. Find the amount of each food group to be used to minimize the cost. What is the minimum cost?

26. Among the many products it produces, an oil refinery makes two specialized petroleum distillates: Pymex *A* and Pymex *B*. Each distillate passes through three stages S_1, S_2, and S_3. Each liter of Pymex *A* requires 1 hour in S_1, 3 hours in S_2, and 3 hours in S_3. Each liter of Pymex *B* requires 1 hour in S_1, 4 hours in S_2, and 2 hours in S_3. There are 10 hours available for S_1, 36 hours available in S_2, and 27 hours available in S_3. If the profit per liter of Pymex *A* is $12 and the profit per liter of Pymex *B* is $9, how many liters of each distillate should be produced to maximize profit? What is the maximum profit?

27. An engine reconditioning company works on 4- and 6-cylinder engines. Each 4-cylinder engine requires 1 hour for cleaning, 5 hours for overhauling, and 3 hours for testing. Each 6-cylinder engine requires 1 hour for cleaning, 10 hours for overhauling, and 2 hours for testing. The cleaning station is available for at most 9 hours, the overhauling equipment is available for at most 80 hours, and the testing equipment is available for at most 24 hours. For each reconditioned 4-cylinder engine, the company will make a profit of $150. A reconditioned 6-cylinder engine will give the company a profit of $250. Assuming the company can sell all the reconditioned engines it produces, how many of each type should be produced to maximize profit? What is the maximum profit?

28. A producer of animal feed makes two food products: F_1 and F_2. The products contain three major ingredients; M_1, M_2, and M_3. Each ton of F_1 requires 200 pounds of M_1, 100 pounds of M_2, and 100 pounds of M_3. Each ton of F_2 requires 100 pounds of M_1, 200 pounds of M_2, and 400 pounds of M_3. There are at least 5000 pounds of M_1 available, at least 7000 pounds of M_2 available, and at least 10,000 pounds of M_3 available. Each ton of F_1 costs $450 to make, and each ton of F_2 costs $300 to make. How many tons of each food product should be made to minimize cost? What is the minimum cost?

Chapter 10 Review

10.1 Systems of Linear Equations in Two Variables

A system of equations is two or more equations considered together. A solution of a system of equations in two variables is an ordered pair that satisfies each equation of the system. Equivalent systems of equations have the same solution set.

Operations that Produce Equivalent Systems of Equations

1. Interchange any two equations.
2. Replace an equation with a non-zero multiple of that equation.
3. Replace an equation with the sum of an equation and a nonzero constant multiple of another equation in the system.

A system of linear equations is consistent if it has exactly one solution. The linear system is dependent if it has infinitely many solutions. An inconsistent system of equations has no solution.

10.2 Systems of Linear Equations in More Than Two Variables

An equation of the form $ax + by + cz = d$, with a, b, and c not all zero, is a linear equation in three variables. The solution of a linear equation in three variables is an ordered triple (x, y, z). An ordered triple can be graphed in the *xyz*-coordinate system. The graph of a linear equation in three variables is a plane.

A linear system of equations for which the constant term is zero for all equations of the system is called a homogeneous system of equations.

10.3   Systems of Nonlinear Equations

A nonlinear system of equations is a system in which one or more of the equations is nonlinear.

10.4   Partial Fractions

A rational expression can be written as the sum of terms whose denominators are factors of the denominator of the rational expression. This is called a partial fraction decomposition.

10.5   Inequalities in Two Variables and Systems of Inequalitites

The graph of an inequality in two variables frequently separates the plane into two or more regions.

The solution set of a system of inequalities is the intersection of the solution sets of each inequality.

10.6   Linear Programming

A linear programming problem consists of a number of constraints that are linear equations or inequalities and a linear objective function. The goal of a linear programming problem is to maximize or minimize the objective function subject to the constraints.

The constraints of a linear programming problem provide a set of feasible solutions. The fundamental theorem of linear programming states that the objective function will attain its maximum and its minimum at one of the vertices of the set of feasible solutions.

CHALLENGE EXERCISES

In Exercises 1 to 10, answer true or false. If the statement is false, give an example.

1. A system of equations will always have a solution as long as the number of equations is equal to the number of variables.

2. A system of two different quadratic equations can have at most four solutions.

3. A homogeneous system of equations is one in which all the variables have the same exponent.

4. The intersection of two planes is a line in an *xyz*-coordinates system.

5. It is possible to find a partial fraction decomposition for any rational expression with real coefficients.

6. Two systems of equations with the same solution set have the same equations in their respective systems.

7. The systems of equations

$$\begin{cases} x = 0 \\ y = 0 \end{cases} \quad \text{and} \quad \begin{cases} y = x \\ y = -x \end{cases}$$

are equivalent systems of equations.

8. For a linear programming problem, one or more constraints are used to define the set of feasible solutions.

9. A system of three linear equations in three variables for which two of the planes are parallel and the third plane intersects the first two is a dependent system of equations.

10. The inequality $xy < 1$ and the inequality $y < 1/x$ are equivalent inequalities.

REVIEW EXERCISES

In Exercises 1 to 30, solve each system of equations.

1. $\begin{cases} 2x - 4y = -3 \\ 3x + 8y = -12 \end{cases}$

2. $\begin{cases} 4x - 3y = 15 \\ 2x + 5y = -12 \end{cases}$

3. $\begin{cases} 3x - 4y = -5 \\ y = \dfrac{2}{3}x + 1 \end{cases}$

4. $\begin{cases} 7x + 2y = -14 \\ y = -\dfrac{5}{2}x - 3 \end{cases}$

5. $\begin{cases} y = 2x - 5 \\ x = 4y - 1 \end{cases}$

6. $\begin{cases} y = 3x + 4 \\ x = 4y - 5 \end{cases}$

7. $\begin{cases} 6x + 9y = 15 \\ 10x + 15y = 25 \end{cases}$

8. $\begin{cases} 4x - 8y = 9 \\ 2x - 4y = 5 \end{cases}$

9. $\begin{cases} 2x - 3y + z = -9 \\ 2x + 5y - 2z = 18 \\ 4x - y + 3z = -4 \end{cases}$

10. $\begin{cases} x - 3y + 5z = 1 \\ 2x + 3y - 5z = 15 \\ 3x + 6y + 5z = 15 \end{cases}$

11. $\begin{cases} x + 3y - 5z = -12 \\ 3x - 2y + z = 7 \\ 5x + 4y - 9z = -17 \end{cases}$

12. $\begin{cases} 2x - y + 2z = 5 \\ x + 3y - 3z = 2 \\ 5x - 9y + 8z = 13 \end{cases}$

13. $\begin{cases} 3x + 4y - 6z = 10 \\ 2x + 2y - 3z = 6 \\ x - 6y + 9z = -4 \end{cases}$

14. $\begin{cases} x - 6y + 4z = 6 \\ 4x + 3y - 4z = 1 \\ 5x - 9y + 8z = 13 \end{cases}$

15. $\begin{cases} 2x + 3y - 2z = 0 \\ 3x - y - 4z = 0 \\ 5x + 13y - 4z = 0 \end{cases}$

16. $\begin{cases} 3x - 5y + z = 0 \\ x + 4y - 3z = 0 \\ 2x + y - 2z = 0 \end{cases}$

17. $\begin{cases} x - 2y + z = 1 \\ 3x + 2y - 3z = 1 \end{cases}$

18. $\begin{cases} 2x - 3y + z = 1 \\ 4x + 2y + 3z = 21 \end{cases}$

19. $\begin{cases} y = x^2 - 2x - 3 \\ y = 2x - 7 \end{cases}$

20. $\begin{cases} y = 2x^2 + x \\ y = 2x + 1 \end{cases}$

21. $\begin{cases} y = 3x^2 - x + 1 \\ y = x^2 + 2x - 1 \end{cases}$

22. $\begin{cases} y = 4x^2 - 2x - 3 \\ y = 2x^2 + 3x - 6 \end{cases}$

23. $\begin{cases} (x + 1)^2 + (y - 2)^2 = 4 \\ 2x + y = 4 \end{cases}$

24. $\begin{cases} (x - 1)^2 + (y + 1)^2 = 5 \\ y = 2x - 3 \end{cases}$

25. $\begin{cases} (x - 2)^2 + (y + 2)^2 = 4 \\ (x + 2)^2 + (y + 1)^2 = 17 \end{cases}$

26. $\begin{cases} (x + 1)^2 + (y - 2)^2 = 1 \\ (x - 2)^2 + (y + 2)^2 = 20 \end{cases}$

27. $\begin{cases} x^2 - 3xy + y^2 = -1 \\ 3x^2 - 5xy - 2y^2 = 0 \end{cases}$

28. $\begin{cases} 2x^2 + 2xy - y^2 = -1 \\ 6x^2 + xy - y^2 = 0 \end{cases}$

29. $\begin{cases} 2x^2 - 5xy + 2y^2 = 56 \\ 14x^2 - 3xy - 2y^2 = 56 \end{cases}$

30. $\begin{cases} 2x^2 + 7xy + 6y^2 = 1 \\ 6x^2 + 7xy + 2y^2 = 1 \end{cases}$

In Exercises 31 to 36, find the partial fraction decomposition.

31. $\dfrac{7x - 5}{x^2 - x - 2}$

32. $\dfrac{x + 1}{(x - 1)^2}$

33. $\dfrac{2x - 2}{(x^2 + 1)(x + 2)}$

34. $\dfrac{5x^2 - 10x + 9}{(x - 2)^2(x + 1)}$

35. $\dfrac{11x^2 - x - 2}{x^3 - x}$

36. $\dfrac{x^4 + x^3 + 4x^2 + x + 3}{(x^2 + 1)^2}$

In Exercises 37 to 48, graph the solution set of each inequality.

37. $4x - 5y < 20$

38. $2x + 7y \geq -14$

39. $y \geq 2x^2 - x - 1$

40. $y < x^2 - 5x - 6$

41. $(x - 2)^2 + (y - 1)^2 > 4$

42. $(x + 3)^2 + (y + 1)^2 \leq 9$

43. $\dfrac{(x - 3)^2}{16} - \dfrac{(y + 2)^2}{25} \leq 1$

44. $\dfrac{(x + 1)^2}{9} - \dfrac{(y - 3)^2}{4} < -1$

45. $(2x - y + 1)(x - 2y - 2) > 0$

46. $(2x - 3y - 6)(x + 2y - 4) < 0$

47. $x^2y^2 < 1$

48. $xy \geq 0$

In Exercises 49 to 60, graph the solution set of each system of inequalities.

49. $\begin{cases} 2x - 5y < 9 \\ 3x + 4y \geq 2 \end{cases}$

50. $\begin{cases} 3x + y > 7 \\ 2x + 5y < 9 \end{cases}$

51. $\begin{cases} 2x + 3y > 6 \\ 2x - y > -2 \\ x < 3 \end{cases}$

52. $\begin{cases} 2x + 5y > 10 \\ x - y > -2 \\ x \leq 4 \end{cases}$

53. $\begin{cases} 2x + 3y \leq 18 \\ x + y \leq 7 \\ x \geq 0, y \geq 0 \end{cases}$

54. $\begin{cases} 3x + 5y \geq 25 \\ 2x + 3y \geq 16 \\ x \geq 0, y \geq 0 \end{cases}$

55. $\begin{cases} 3x + y \geq 6 \\ x + 4y \geq 14 \\ 2x + 3y \geq 16 \\ x \geq 0, y \geq 0 \end{cases}$

56. $\begin{cases} 3x + 2y \geq 14 \\ x + y \geq 6 \\ 11x + 4y \leq 48 \\ x \geq 0, y \geq 0 \end{cases}$

57. $\begin{cases} y < x^2 - x - 2 \\ y \geq 2x - 4 \end{cases}$

58. $\begin{cases} y > 2x^2 + x - 1 \\ y > x + 3 \end{cases}$

59. $\begin{cases} x^2 + y^2 - 2x + 4y > 4 \\ y < 2x^2 - 1 \end{cases}$

60. $\begin{cases} x^2 - y^2 - 4x - 2y < -4 \\ x^2 + y^2 - 4x + 4y > 8 \end{cases}$

In Exercises 61 to 66, solve the linear programming problem. In each problem, assume $x \geq 0$ and $y \geq 0$.

61. Objective function: $P = 2x + 2y$
Constraints: $\begin{cases} x + 2y \leq 14 \\ 5x + 2y \leq 30 \end{cases}$
Maximize the objective function.

62. Objective function: $P = 4x + 5y$
Constraints: $\begin{cases} 2x + 3y \leq 24 \\ 4x + 3y \leq 36 \end{cases}$
Maximize the objective function.

63. Objective function: $P = 4x + y$
Constraints: $\begin{cases} 5x + 2y \geq 16 \\ x + 2y \geq 8 \\ x \leq 20, y \leq 20 \end{cases}$
Minimize the objective function.

64. Objective function: $P = 2x + 7y$
Constraints: $\begin{cases} 4x + 3y \geq 24 \\ 4x + 7y \geq 40 \\ x \leq 10, y \leq 10 \end{cases}$
Minimize the objective function.

65. Objective function: $P = 6x + 3y$
Constraints: $\begin{cases} 5x + 2y \geq 20 \\ x + y \geq 7 \\ x + 2y \geq 10 \\ x \leq 15, y \leq 15 \end{cases}$
Minimize the objective function.

66. Objective function: $P = 5x + 4y$
Constraints: $\begin{cases} x + y \leq 10 \\ 2x + y \leq 13 \\ 3x + y \leq 18 \end{cases}$
Maximize the objective function.

67. Find an equation of the form $y = ax^2 + bx + c$ whose graph passes through the points $(1, 0)$, $(-1, 5)$, and $(2, 3)$.

68. Find an equation of the circle that passes through the points $(4, 2)$, $(0, 1)$, and $(3, -1)$.

69. Find an equation of a plane that passes through the points $(2, 1, 2)$, $(3, 1, 0)$, and $(-2, -3, -2)$. Use the equation $z = ax + by + c$.

70. How many liters of a 20% acid solution should be mixed with 10 liters of a 10% acid solution so that the resulting solution is a 16% acid solution?

71. Flying with the wind, a small plane traveled 855 miles in 5 hours. Flying against the same wind, the plane traveled 575 miles in the same time. Find the rate of the wind and the rate of the plane in calm air.

72. A collection of ten coins has a value of $1.25. If the collection consists of only nickels, dimes, and quarters, how many of each coin is in the collection? (Hint: There is more than one solution?

73. Consider the ordered triple (a, b, c). Find all real number values for $a, b,$ and c so that the product of any two equals the remaining number.

11 Matrices and Determinants

Clouds are not spheres, mountains are not cones, coastlines are not circles, and bark is not smooth, nor does lightning travel in a straight line.

BENOIT MANDELBROT

THE MANDELBROT SET

The quote on the preceding page, from *The Fractal Geometry of Nature* by Benoit Mandelbrot, appears on the first page of his book and embodies the very essence of fractal geometry.

Mandelbrot says that he coined the word *fractal* from the Latin word *fragere* which means to break into irregular fragments. His studies in geometry have lead to *fractional dimensions* such as 1.26 or 1.67. According to fractal geometry, the snowflake curve shown in the introduction to the next chapter has dimension 1.26.

Mandelbrot also defined a set of complex numbers called the Mandelbrot set. The boundary of this set is a fractal curve. To construct the Mandelbrot set, begin with some complex number c. Substitute the value of c for z in the expression $z^2 + c$. The new number is $c_1 = c^2 + c$. Now substitute this number for z in $z^2 + c$. The new number is

$$c_2 = c_1^2 + c = (c^2 + c)^2 + c.$$

Repeat this process over and over. For example,

$$c_3 = c_2^2 + c = [(c^2 + c)^2 + c]^2 + c.$$

The Mandelbrot set is the set of all complex numbers c, for which $|c_n| < 2$ even after an infinite number of repetitions. The figure below represents a computer investigation into the Mandelbrot set. The white area is the set of complex numbers that belongs to the Mandelbrot set.

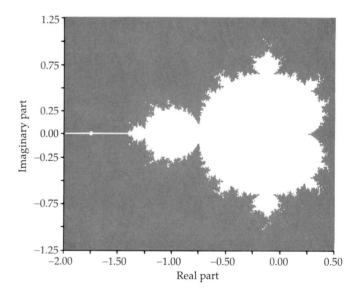

11.1

Gauss-Jordan Elimination Method

A **matrix** is a rectangular array of numbers. Each number in the matrix is called an **element** of the matrix. The matrix below, with three rows and four columns, is called a 3×4 (read 3 by 4) matrix.

$$\begin{bmatrix} 2 & 5 & -2 & 5 \\ -3 & 6 & 4 & 0 \\ 1 & 3 & 7 & 2 \end{bmatrix}$$

A matrix of m rows and n columns is said to be of **order** $m \times n$ or **dimension** $m \times n$. The matrix above has order 3×4. We will use the notation a_{ij} to refer to the element of a matrix in the ith row and jth column. For the matrix given above, $a_{23} = 4$, $a_{31} = 1$, and $a_{13} = -2$.

The elements $a_{11}, a_{22}, a_{33}, \ldots, a_{mm}$ form the **main diagonal** of a matrix. The elements 2, 6, and 7 form the main diagonal of the matrix shown above.

A matrix can be created from a system of linear equations. Consider the system of linear equations

$$\begin{cases} 2x - 3y + z = 2 \\ x - 3z = 4 \\ 4x - y + 4z = 3 \end{cases}$$

Using only the coefficients and constants of this system, we can write the 3×4 matrix

$$\begin{bmatrix} 2 & -3 & 1 & 2 \\ 1 & 0 & -3 & 4 \\ 4 & -1 & 4 & 3 \end{bmatrix}$$

This matrix is called the **augmented matrix** of the system of equations. The matrix formed by the coefficients of the system is the **coefficient matrix**. The matrix formed from the constants is the **constant matrix** for the system. The coefficient matrix and constant matrix for the given system are:

$$\text{Coefficient matrix:} \begin{bmatrix} 2 & -3 & 1 \\ 1 & 0 & -3 \\ 4 & -1 & 4 \end{bmatrix} \quad \text{Constant matrix:} \begin{bmatrix} 2 \\ 4 \\ 3 \end{bmatrix}$$

Remark When a term is missing from one of the equations of the system (as in the second equation), the coefficient of that term is 0 and a 0 is entered in the matrix.

We can write a system of equations from an augmented matrix:

$$\text{Augmented matrix:} \begin{bmatrix} 2 & -1 & 4 & 3 \\ 1 & 1 & 0 & 2 \\ 3 & -2 & -1 & 2 \end{bmatrix} \xrightarrow{\text{system}} \begin{cases} 2x - y + 4z = 3 \\ x + y = 2 \\ 3x - 2y - z = 2 \end{cases}$$

In certain cases, an augmented matrix represents a system of equations that we can solve by back substitution. Consider the following augmented matrix and the equivalent system of equations:

$$\begin{bmatrix} 1 & -3 & 4 & 5 \\ 0 & 1 & 2 & -4 \\ 0 & 0 & 1 & -1 \end{bmatrix} \xrightarrow{\text{equivalent system}} \begin{cases} x - 3y + 4z = 5 \\ y + 2z = -4 \\ z = -1 \end{cases}$$

Solving this system by using back substitution, we find that the solution is $(3, -2, -1)$. An augmented matrix is said to be in *echelon form* if the equivalent system of equations can be solved by back substitution.

Echelon Form

An augmented matrix is in echelon form if all the following conditions are satisfied:

1. The first nonzero number in any row is a 1.
2. Rows are arranged so that the column containing the first nonzero number is to the left of the column containing the first nonzero number of the next row.
3. All rows consisting entirely of zeros appear at the bottom of the matrix.

Following are three examples of matrices in echelon form:

$$\begin{bmatrix} 1 & -3 & 4 & 2 \\ 0 & 1 & -2 & -1 \\ 0 & 0 & 0 & 0 \end{bmatrix}, \quad \begin{bmatrix} 1 & 2 & -1 & 3 \\ 0 & 1 & 2 & -1 \end{bmatrix}, \quad \begin{bmatrix} 1 & -1 & 3 & 2 \\ 0 & 1 & 2 & 5 \\ 0 & 0 & 1 & -2 \end{bmatrix}.$$

We can write an augmented matrix in echelon form by using what are called **elementary row operations.** These operations are a rewording, using matrix terminology, of the operations that produce equivalent equations.

Elementary Row Operations

Given the augmented matrix for a system of linear equations, each of the following elementary row operations produces a matrix of an equivalent system of equations.

1. Interchanging two rows
2. Multiplying all the elements in a row by the same nonzero number
3. Replacing a row by the sum of that row and a nonzero multiple of any other row

It is convenient to specify each operation symbolically as follows:

1. Interchange the ith and jth rows: $R_i \longleftrightarrow R_j$.
2. Multiply the ith row by k, a nonzero constant: kR_i.
3. Replace the jth row by the sum of that row and a nonzero multiple of the ith row: $kR_i + R_j$.

To demonstrate these operations, we will use the 3×3 matrix

$$\begin{bmatrix} 2 & 1 & -2 \\ 3 & -2 & 2 \\ 1 & -2 & 3 \end{bmatrix}.$$

$$\begin{bmatrix} 2 & 1 & -2 \\ 3 & -2 & 2 \\ 1 & -2 & 3 \end{bmatrix} \xrightarrow{R_1 \longleftrightarrow R_3} \begin{bmatrix} 1 & -2 & 3 \\ 3 & -2 & 2 \\ 2 & 1 & -2 \end{bmatrix}$$
Interchange row 1 and row 3.

$$\begin{bmatrix} 2 & 1 & -2 \\ 3 & -2 & 2 \\ 1 & -2 & 3 \end{bmatrix} \xrightarrow{-3R_2} \begin{bmatrix} 2 & 1 & -2 \\ -9 & 6 & -6 \\ 1 & -2 & 3 \end{bmatrix}$$
Multiply row 2 by -3.

$$\begin{bmatrix} 2 & 1 & -2 \\ 3 & -2 & 2 \\ 1 & -2 & 3 \end{bmatrix} \xrightarrow{-2R_3 + R_1} \begin{bmatrix} 0 & 5 & -8 \\ 3 & -2 & 2 \\ 1 & -2 & 3 \end{bmatrix}$$
Multiply row 3 by -2 and add to row 1. Replace row 1 by the sum.

The **Gauss-Jordan** elimination method is an algorithm[1] that uses elementary row operations to solve a system of linear equations. The goal of this method is to rewrite an augmented matrix in echelon form.

We will now demonstrate how to solve a system of two equations in two variables by the Gauss-Jordan method. Consider the system of equations

$$\begin{cases} 2x + 5y = -1 \\ 3x - 2y = 8 \end{cases} \tag{1}$$

The augmented matrix for this system is

$$\begin{bmatrix} 2 & 5 & -1 \\ 3 & -2 & 8 \end{bmatrix}.$$

The goal of the Gauss-Jordan method is to rewrite the augmented matrix in echelon form by using elementary row operations. The row operations are chosen so that first, there is a 1 as a_{11}; second, there is a 0 as a_{21}; third, there is a 1 as a_{22}.

---

[1] An algorithm is a repetitive procedure used in calculations. The word is derived from Al-Khwarismi, the name of the author of an Arabic algebra book written around 825 A.D.

Begin by multiplying row 1 by $\frac{1}{2}$. The result is a 1 as a_{11}.

$$\begin{bmatrix} 2 & 5 & -1 \\ 3 & -2 & 8 \end{bmatrix} \xrightarrow{(1/2)R_1} \begin{bmatrix} 1 & \frac{5}{2} & -\frac{1}{2} \\ 3 & -2 & 8 \end{bmatrix}$$

Now multiply row 1 by -3 and add the result to row 2. Replace row 2. The result is a 0 as a_{21}.

$$\begin{bmatrix} 1 & \frac{5}{2} & -\frac{1}{2} \\ 3 & -2 & 8 \end{bmatrix} \xrightarrow{-3R_1 + R_2} \begin{bmatrix} 1 & \frac{5}{2} & -\frac{1}{2} \\ 0 & -\frac{19}{2} & \frac{19}{2} \end{bmatrix}$$

Now multiply row 2 by $-2/19$. The result is a 1 as a_{22}. The matrix is now in row echelon form.

$$\begin{bmatrix} 1 & \frac{5}{2} & -\frac{1}{2} \\ 0 & -\frac{19}{2} & \frac{19}{2} \end{bmatrix} \xrightarrow{(-2/19)R_2} \begin{bmatrix} 1 & \frac{5}{2} & -\frac{1}{2} \\ 0 & 1 & -1 \end{bmatrix}$$

The system of equations written from the echelon form of the matrix is

$$\begin{cases} x + \dfrac{5}{2}y = -\dfrac{1}{2} \\ \qquad\quad y = -1 \end{cases}$$

Solving by back substitution, replace y in the first equation by -1 and solve for x.

$$x + \left(\frac{5}{2}\right)(-1) = -\frac{1}{2}$$

$$x = 2$$

The solution of the system of equations (1) is $(2, -1)$.

Remark   The order in which a matrix is reduced to echelon form is important. Using elementary row operations, change a_{11} to a 1 and change the remaining elements in the first column to 0. Move to a_{22}. Change a_{22} to 1 and change the remaining elements below a_{22} in column 2 to 0. Move to the next column and repeat the procedure. Continue moving down the main diagonal, repeating the procedure until you reach a_{mm} or all remaining elements on the main diagonal are zero.

To conserve space, we will occasionally perform more than one elementary row operation in one step. For example, the notation

$$\begin{array}{c} 3R_1 + R_2 \\ -2R_1 + R_3 \\ \hline \longrightarrow \end{array}$$

means that two elementary row operations were performed. First multiply row 1 by 3 and add to row 2. Replace row 2. Now multiply row 1 by -2 and add to row 3. Replace row 3.

EXAMPLE 1 **Solve a System of Equations with Four Variables**

Solve the system of equations by using the Gauss-Jordan method.

$$\begin{cases} 3t - 8u + 8v + 7w = 41 \\ t - 2u + 2v + w = 9 \\ 2t - 2u + 6v - 4w = -1 \\ 2t - 2u + 3v - 3w = 3 \end{cases}$$

Solution Write the augmented matrix and then use elementary row operations to rewrite the matrix in echelon form.

$$\begin{bmatrix} 3 & -8 & 8 & 7 & 41 \\ 1 & -2 & 2 & 1 & 9 \\ 2 & -2 & 6 & -4 & -1 \\ 2 & -2 & 3 & -3 & 3 \end{bmatrix} \xrightarrow{R_1 \longleftrightarrow R_2} \begin{bmatrix} 1 & -2 & 2 & 1 & 9 \\ 3 & -8 & 8 & 7 & 41 \\ 2 & -2 & 6 & -4 & -1 \\ 2 & -2 & 3 & -3 & 3 \end{bmatrix}$$

$$\begin{matrix} -3R_1 + R_2 \\ -2R_1 + R_3 \\ -2R_1 + R_4 \\ \xrightarrow{} \end{matrix} \begin{bmatrix} 1 & -2 & 2 & 1 & 9 \\ 0 & -2 & 2 & 4 & 14 \\ 0 & 2 & 2 & -6 & -19 \\ 0 & 2 & -1 & -5 & -15 \end{bmatrix} \xrightarrow{(-1/2)R_2} \begin{bmatrix} 1 & -2 & 2 & 1 & 9 \\ 0 & 1 & -1 & -2 & -7 \\ 0 & 2 & 2 & -6 & -19 \\ 0 & 2 & -1 & -5 & -15 \end{bmatrix}$$

$$\begin{matrix} -2R_2 + R_3 \\ -2R_2 + R_4 \\ \xrightarrow{} \end{matrix} \begin{bmatrix} 1 & -2 & 2 & 1 & 9 \\ 0 & 1 & -1 & -2 & -7 \\ 0 & 0 & 4 & -2 & -5 \\ 0 & 0 & 1 & -1 & -1 \end{bmatrix} \xrightarrow{R_4 \longleftrightarrow R_3} \begin{bmatrix} 1 & -2 & 2 & 1 & 9 \\ 0 & 1 & -1 & -2 & -7 \\ 0 & 0 & 1 & -1 & -1 \\ 0 & 0 & 4 & -2 & -5 \end{bmatrix}$$

$$\xrightarrow{-4R_3 + R_4} \begin{bmatrix} 1 & -2 & 2 & 1 & 9 \\ 0 & 1 & -1 & -2 & -7 \\ 0 & 0 & 1 & -1 & -1 \\ 0 & 0 & 0 & 2 & -1 \end{bmatrix} \xrightarrow{(1/2)R_4} \begin{bmatrix} 1 & -2 & 2 & 1 & 9 \\ 0 & 1 & -1 & -2 & -7 \\ 0 & 0 & 1 & -1 & -1 \\ 0 & 0 & 0 & 1 & -\frac{1}{2} \end{bmatrix}$$

The last matrix is in echelon form. The system of equations written from the matrix is

$$\begin{cases} t - 2u + 2v + w = 9 \\ u - v - 2w = -7 \\ v - w = -1 \\ w = -\dfrac{1}{2} \end{cases}$$

Solve by back substitution. The solution is $\left(-\dfrac{13}{2}, -\dfrac{19}{2}, -\dfrac{3}{2}, -\dfrac{1}{2} \right)$.

■ *Try Exercise **14**, page 562.*

EXAMPLE 2　**Solve a System of Equations**

Solve the system of equations using the Gauss-Jordan method.

$$\begin{cases} x - 3y + z = 5 \\ 3x - 7y + 2z = 12 \\ 2x - 4y + z = 3 \end{cases}$$

Solution　Write the augmented matrix and then use elementary row operations to write the matrix in echelon form.

$$\begin{bmatrix} 1 & -3 & 1 & 5 \\ 3 & -7 & 2 & 12 \\ 2 & -4 & 1 & 3 \end{bmatrix} \xrightarrow[\substack{-3R_1 + R_2 \\ -2R_1 + R_3}]{} \begin{bmatrix} 1 & -3 & 1 & 5 \\ 0 & 2 & -1 & -3 \\ 0 & 2 & -1 & -7 \end{bmatrix}$$

$$\xrightarrow{(1/2)R_2} \begin{bmatrix} 1 & -3 & 1 & 5 \\ 0 & 1 & -\frac{1}{2} & -\frac{3}{2} \\ 0 & 2 & -1 & -7 \end{bmatrix} \xrightarrow{-2R_2 + R_3} \begin{bmatrix} 1 & -3 & 1 & 5 \\ 0 & 1 & -\frac{1}{2} & -\frac{3}{2} \\ 0 & 0 & 0 & -4 \end{bmatrix}$$

$$\begin{cases} x - 3y + z = 5 \\ \quad\quad y - \tfrac{1}{2}z = -\dfrac{3}{2} \quad \text{Equivalent system} \\ \quad\quad\quad\quad 0z = -4 \end{cases}$$

Because the equation $0z = -4$ has no solution, the system has no solution.

■ *Try Exercise* **20,** *page 562.*

EXAMPLE 3　**Solve a System of Equations with Infinitely Many Solutions**

Solve the system of equations using the Gauss-Jordan method.

$$\begin{cases} x - 3y + 4z = 1 \\ 2x - 5y + 3z = 6 \\ x - 2y - z = 5 \end{cases}$$

Solution　Write the augmented matrix and then use elementary row operations to reduce the matrix to echelon form.

$$\begin{bmatrix} 1 & -3 & 4 & 1 \\ 2 & -5 & 3 & 6 \\ 1 & -2 & -1 & 5 \end{bmatrix} \xrightarrow[\substack{-2R_1 + R_2 \\ -R_1 + R_3}]{} \begin{bmatrix} 1 & -3 & 4 & 1 \\ 0 & 1 & -5 & 4 \\ 0 & 1 & -5 & 4 \end{bmatrix}$$

$$\xrightarrow{-R_2 + R_3} \begin{bmatrix} 1 & -3 & 4 & 1 \\ 0 & 1 & -5 & 4 \\ 0 & 0 & 0 & 0 \end{bmatrix}$$

$$\begin{cases} x - 3y + 4z = 1 \\ \quad\quad y - 5z = 4 \end{cases} \quad \text{Equivalent system}$$

Any solution of the system of equations is a solution of $y - 5z = 4$. Solve this equation for y then substitute into the first equation.

$$y = 5z + 4$$
$$x - 3y + 4z = 1$$
$$x - 3(5z + 4) + 4z = 1$$

or

$$x = 11z + 13$$

Both x and y are expressed in terms of z. Let z be any real number c. The solutions of the system of equations are $(11c + 13, 5c + 4, c)$.

■ *Try Exercise **18**, page 562.*

EXAMPLE 4   Solve a System of Equations

Solve the system of equations using the Gauss-Jordan method.

$$\begin{cases} x_1 - 2x_2 - 3x_3 - 2x_4 = 1 \\ 2x_1 - 3x_2 - 4x_3 - 2x_4 = 3 \\ x_1 + x_2 + x_3 - 7x_4 = -7 \end{cases}$$

Solution   Write the augmented matrix and then use elementary row operations to reduce the matrix to echelon form.

$$\begin{bmatrix} 1 & -2 & -3 & -2 & 1 \\ 2 & -3 & -4 & -2 & 3 \\ 1 & 1 & 1 & -7 & -7 \end{bmatrix} \begin{array}{c} -2R_1 + R_2 \\ -1R_1 + R_3 \\ \xrightarrow{\hspace{1cm}} \end{array} \begin{bmatrix} 1 & -2 & -3 & -2 & 1 \\ 0 & 1 & 2 & 2 & 1 \\ 0 & 3 & 4 & -5 & -8 \end{bmatrix}$$

$$\begin{array}{c} -3R_2 - \frac{1}{2}R_3 \\ \xrightarrow{\hspace{1cm}} \end{array} \begin{bmatrix} 1 & -2 & -3 & -2 & 1 \\ 0 & 1 & 2 & 2 & 1 \\ 0 & 0 & -1 & -\frac{11}{2} & -\frac{11}{2} \end{bmatrix}$$

$$\begin{cases} x_1 - 2x_2 - 3x_3 - 2x_4 = 1 \\ x_2 + 2x_3 + 2x_4 = 1 \quad \text{Equivalent system} \\ x_3 + \frac{11}{2}x_4 = \frac{11}{2} \end{cases}$$

We now express each of the variables in terms of x_4. Solve the third equation for x_3.

$$x_3 = -\frac{11}{2}x_4 + \frac{11}{2}$$

Substitute this value into the second equation and solve for x_2.

$$x_2 + 2\left(-\frac{11}{2}x_4 + \frac{11}{2}\right) + 2x_4 = 1$$

Simplifying, we have $x_2 = 9x_4 - 10$.

Substitute the values for x_2 and x_3 into the first equation and solve for x_1.

$$x_1 - 2(9x_4 - 10) - 3\left(-\frac{11}{2}x_4 + \frac{11}{2}\right) - 2x_4 = 1$$

Simplifying, we have $x_1 = \frac{7}{2}x_4 - \frac{5}{2}$. If x_4 is any real number c, the solu-

tion is of the form $\left(\frac{7}{2}c - \frac{5}{2}, 9c - 10, -\frac{11}{2}c + \frac{11}{2}, c\right)$.

■ *Try Exercise* **36,** *page 563.*

EXERCISE SET 11.1 _____

In Exercises 1 to 4, write the augmented matrix, the coefficient matrix, and the constant matrix.

1. $\begin{cases} 2x - 3y + z = 1 \\ 3x - 2y + 3z = 0 \\ x \quad\quad + 5z = 4 \end{cases}$

2. $\begin{cases} -3y + 2z = 3 \\ 2x - y \quad\quad = -1 \\ 3x - 2y + 3z = 4 \end{cases}$

3. $\begin{cases} 2x - 3y - 4z + w = 2 \\ 2y + z \quad\quad = 2 \\ x - y + 2z \quad\quad = 4 \\ 3x - 3y - 2z \quad\quad = 1 \end{cases}$

4. $\begin{cases} x - y + 2z + 3w = -2 \\ 2x \quad\quad + z - 2w = 1 \\ 3x \quad\quad\quad - 2w = 3 \\ -x + 3y - z \quad\quad = 3 \end{cases}$

In Exercises 5 to 12, use elementary row operations to write the matrix in echelon form.

5. $\begin{bmatrix} 2 & -1 & 3 & -2 \\ 1 & -1 & 2 & 2 \\ 3 & 2 & -1 & 3 \end{bmatrix}$

6. $\begin{bmatrix} 1 & 2 & 4 & 1 \\ 2 & 2 & 7 & 3 \\ 3 & 6 & 8 & -1 \end{bmatrix}$

7. $\begin{bmatrix} 4 & -5 & -1 & 2 \\ 3 & -4 & 1 & -2 \\ 1 & -2 & -1 & 3 \end{bmatrix}$

8. $\begin{bmatrix} -2 & 1 & -1 & 3 \\ 2 & 2 & 4 & 6 \\ 3 & 1 & -1 & 2 \end{bmatrix}$

9. $\begin{bmatrix} 1 & -2 & 3 & -4 \\ 3 & -6 & 10 & -14 \\ 5 & -8 & 19 & -21 \\ 2 & -4 & 7 & -10 \end{bmatrix}$

10. $\begin{bmatrix} 2 & -1 & 3 & 2 \\ 1 & 2 & -1 & 3 \\ 3 & 5 & -2 & 2 \\ 4 & 3 & 1 & 8 \end{bmatrix}$

11. $\begin{bmatrix} 1 & -3 & 4 & 2 & 1 \\ 2 & -3 & 5 & -2 & -1 \\ -1 & 2 & -3 & 1 & 3 \end{bmatrix}$

12. $\begin{bmatrix} 2 & -1 & 3 & 2 & 2 \\ 1 & -2 & 2 & 1 & -1 \\ 3 & -5 & -1 & -2 & 3 \end{bmatrix}$

In Exercises 13 to 38, solve the system of equations by the Gauss-Jordon method.

13. $\begin{cases} x + 2y - 2z = -2 \\ 5x + 9y - 4z = -3 \\ 3x + 4y - 5z = -3 \end{cases}$

14. $\begin{cases} x - 3y + z = 8 \\ 2x - 5y - 3z = 2 \\ x + 4y + z = 1 \end{cases}$

15. $\begin{cases} 3x + 7y - 7z = -4 \\ x + 2y - 3z = 0 \\ 5x + 6y + z = -8 \end{cases}$

16. $\begin{cases} 2x - 3y + 2z = 13 \\ 3x - 4y - 3z = 1 \\ 3x + y - z = 2 \end{cases}$

17. $\begin{cases} x + 2y - 2z = 3 \\ 5x + 8y - 6z = 14 \\ 3x + 4y - 2z = 8 \end{cases}$

18. $\begin{cases} 3x - 5y + 2z = 4 \\ x - 3y + 2z = 4 \\ 5x - 11y + 6z = 12 \end{cases}$

19. $\begin{cases} 3x + 2y - z = 1 \\ 2x + 3y - z = 1 \\ x - y + 2z = 3 \end{cases}$

20. $\begin{cases} 2x + 5y + 2z = -1 \\ x + 2y - 3z = 5 \\ 5x + 12y + z = 10 \end{cases}$

21. $\begin{cases} x - 3y + 2z = 0 \\ 2x - 5y - 2z = 0 \\ 4x - 11y + 2z = 0 \end{cases}$

22. $\begin{cases} x + y - 2z = 0 \\ 3x + 4y - z = 0 \\ 5x + 6y - 5z = 0 \end{cases}$

23. $\begin{cases} 2x + y - 3z = 4 \\ 3x + 2y + z = 2 \end{cases}$

24. $\begin{cases} 3x - 6y + 2z = 2 \\ 2x + 5y - 3z = 2 \end{cases}$

25. $\begin{cases} 2x + 2y - 4z = 4 \\ 2x + 3y - 5z = 4 \\ 4x + 5y - 9z = 8 \end{cases}$

26. $\begin{cases} 3x - 10y + 2z = 34 \\ x - 4y + z = 13 \\ 5x - 2y + 7z = 31 \end{cases}$

27. $\begin{cases} x + 3y + 4z = 11 \\ 2x + 3y + 2z = 7 \\ 4x + 9y + 10z = 20 \\ 3x - 2y + z = 1 \end{cases}$

28. $\begin{cases} x - 4y + 3z = 4 \\ 3x - 10y + 3z = 4 \\ 5x - 18y + 9z = 10 \\ 2x + 2y - 3z = -11 \end{cases}$

29. $\begin{cases} t + 2u - 3v + w = -7 \\ 3t + 5u - 8v + 5w = -8 \\ 2t + 3u - 7v + 3w = -11 \\ 4t + 8u - 10v + 7w = -10 \end{cases}$

30. $\begin{cases} t + 4u + 2v - 3w = 11 \\ 2t + 10u + 3v - 5w = 17 \\ 4t + 16u + 7v - 9w = 34 \\ t + 4u + v - w = 4 \end{cases}$

31. $\begin{cases} 2t - u + 3v + 2w = 2 \\ t - u + 2v + w = 2 \\ 3t - 2v - 3w = 13 \\ 2t + 2u - 2w = 6 \end{cases}$

32. $\begin{cases} 4t + 7u - 10v + 3w = -29 \\ 3t + 5u - 7v + 2w = -20 \\ t + 2u - 3v + w = -9 \\ 2t - u + 2v - 4w = 15 \end{cases}$

33. $\begin{cases} 3t + 10u + 7v - 6w = 7 \\ 2t + 8u + 6v - 5w = 5 \\ t + 4u + 2v - 3w = 2 \\ 4t + 14u + 9v - 8w = 8 \end{cases}$

34. $\begin{cases} t - 3u + 2v + 4w = 13 \\ 3t - 8u + 4v + 13w = 35 \\ 2t - 7u + 8v + 5w = 28 \\ 4t - 11u + 6v + 17w = 56 \end{cases}$

35. $\begin{cases} t - u + 2v - 3w = 9 \\ 4t + 11v - 10w = 46 \\ 3t - u + 8v - 6w = 27 \end{cases}$

36. $\begin{cases} t - u + 3v - 5w = 10 \\ 2t - 3u + 4v + w = 7 \\ 3t + 4u - 2v - 2w = 6 \end{cases}$

37. $\begin{cases} 3t - 4u + v = 2 \\ t + u - 2v + 3w = 1 \end{cases}$

38. $\begin{cases} 2t + 3v - 4w = 2 \\ t + 2u - 4v + w = -3 \end{cases}$

Supplemental Exercises

In Exercises 39 to 44, solve by using the Gauss-Jordan method.

39. $\begin{cases} x_1 + 2x_2 - x_3 + 2x_4 + 3x_5 = 11 \\ x_1 - x_2 + 2x_3 - x_4 + 2x_5 = 0 \\ 2x_1 + x_2 - x_3 + 2x_4 - x_5 = 4 \\ 3x_1 + 2x_2 - x_3 + x_4 - 2x_5 = 2 \\ 2x_1 + x_2 - x_3 - 2x_4 + x_5 = 4 \end{cases}$

40. $\begin{cases} x_1 - 2x_2 + 2x_3 - 3x_4 + 2x_5 = 5 \\ x_1 - 3x_2 - x_3 + 2x_4 - x_5 = -4 \\ 3x_1 + x_2 - 2x_3 + x_4 + 3x_5 = 9 \\ 2x_1 - x_2 + 3x_3 - x_4 - 2x_5 = 2 \\ -x_1 + 2x_2 - 2x_3 + 3x_4 - x_5 = -4 \end{cases}$

41. $\begin{cases} x_1 + 2x_2 - 3x_3 - x_4 + 2x_5 = -10 \\ -x_1 - 3x_2 + x_3 + x_4 - x_5 = 4 \\ 2x_1 + 3x_2 - 5x_3 + 2x_4 + 3x_5 = -20 \\ 3x_1 + 4x_2 - 7x_3 + 3x_4 - 2x_5 = -16 \\ 2x_1 + x_2 - 6x_3 + 4x_4 - 3x_5 = -12 \end{cases}$

42. $\begin{cases} x_1 - 2x_2 + 2x_3 - 3x_4 + x_5 = 5 \\ 2x_1 - 3x_2 + 4x_3 - 5x_4 - x_5 = 13 \\ x_1 + x_2 - 2x_3 + 2x_4 + 2x_5 = -11 \\ 3x_1 - 2x_2 + 2x_3 - 2x_4 - 2x_5 = 7 \\ 4x_1 - 4x_2 + 4x_3 - 5x_4 - x_5 = 12 \end{cases}$

43. $\begin{cases} 2x_1 - 3x_2 - x_3 + 5x_4 - 3x_5 = 0 \\ x_1 - 2x_2 - x_3 + 2x_4 + 3x_5 = 3 \\ x_1 + 2x_2 - x_3 + 2x_4 + 3x_5 = 11 \end{cases}$

44. $\begin{cases} x_1 + 2x_2 + x_3 - 2x_4 - 3x_5 = -2 \\ 2x_1 + 3x_2 - x_3 + x_4 + 2x_5 = 14 \\ 4x_1 + 7x_2 - 2x_3 + 2x_4 - 3x_5 = 16 \end{cases}$

In Exercises 45 to 47, use the system of equations

$$\begin{cases} x + 3y - a^2z = a^2 \\ 2x + 3y + az = 2 \\ 3x + 4y + 2z = 3 \end{cases}$$

45. Find all values of a for which the system of equations has a unique solution.

46. Find all values of a for which the system of equations has infinitely many solutions.

47. Find all values of a for which the system of equations has no solutions.

48. Find an equation of the plane that passes through the points $(1, 2, 6)$, $(-1, 1, 7)$, and $(4, 2, 0)$. Use the equation $z = ax + by + c$.

49. Find an equation of the plane that passes through the points $(-1, 0, -4)$, $(2, 1, 5)$, and $(-1, 1, -1)$. Use the equation $z = ax + by + c$.

50. Find a cubic function whose graph passes through the points $(0, 2)$, $(1, 0)$, $(-2, -12)$, and $(3, 8)$. (*Hint:* Use the equation $f(x) = y = ax^3 + bx^2 + cx + d$.)

51. Find a cubic function whose graph passes through the points $(0, 0)$, $(1, 1)$, $(2, 6)$, and $(-1, 0)$. (*Hint:* Use the equation $f(x) = y = ax^3 + bx^2 + cx + d$.)

11.2
The Algebra of Matrices

Besides being convenient for solving a system of equations, matrices are appropriate for solving problems in business and science. To effectively apply matrices in these areas, we must develop some of the theory of matrices.

Throughout this chapter we denote a matrix by either a capital letter or by surrounding the corresponding lower-case letter with brackets. For example, a matrix could be denoted as

$$A \quad \text{or} \quad [a_{ij}].$$

Caution   Remember that $[a_{ij}]$ is a *matrix* and a_{ij} is the *element* in the *i*th row and *j*th column of the matrix.

An important concept involving matrices is the principle of equality.

Definition of Equality of Two Matrices

> Two matrices $A = [a_{ij}]$ and $B = [b_{ij}]$ are equal if and only if
> $$a_{ij} = b_{ij}$$
> for every *i* and *j*.

For example, the 2×3 matrices $\begin{bmatrix} a & -2 & b \\ 3 & c & 1 \end{bmatrix}$ and $\begin{bmatrix} 3 & x & -4 \\ 3 & -1 & y \end{bmatrix}$ are equal if and only if $a = 3$, $x = -2$, $b = -4$, $c = -1$, and $y = 1$.

Remark   The definition of equality implies that the two matrices have the same order. Why?[2]

Once equality is defined, we can begin the definitions of the operations on matrices.

Definition of Matrix Addition

> If A and B are matrices of order $m \times n$, then the sum of the matrices is the $m \times n$ matrix given by
> $$A + B = [a_{ij} + b_{ij}].$$

This definition states that the sum of two matrices is found by adding their corresponding elements. Note from the definition that both matrices have the same order. The sum of two matrices of different order is not defined.

---

[2] If the two matrices were of different order, there would be an element of one matrix for which there was no corresponding element of the second matrix.

Given $A = \begin{bmatrix} 2 & -2 & 3 \\ 1 & 3 & -4 \end{bmatrix}$ and $B = \begin{bmatrix} 5 & -2 & 6 \\ -2 & 3 & 5 \end{bmatrix}$, then

$$A + B = \begin{bmatrix} 2 & -2 & 3 \\ 1 & 3 & -4 \end{bmatrix} + \begin{bmatrix} 5 & -2 & 6 \\ -2 & 3 & 5 \end{bmatrix}$$

$$= \begin{bmatrix} 2 + 5 & -2 + (-2) & 3 + 6 \\ 1 + (-2) & 3 + 3 & -4 + 5 \end{bmatrix} = \begin{bmatrix} 7 & -4 & 9 \\ -1 & 6 & 1 \end{bmatrix}$$

Now let $A = \begin{bmatrix} 2 & -3 \\ 4 & 1 \end{bmatrix}$ and $B = \begin{bmatrix} 3 & 2 & 0 \\ 1 & -5 & 0 \end{bmatrix}$.

Then $A + B$ is not defined because the matrices do not have the same order.

To define the subtraction of two matrices, we first define the additive inverse of a matrix.

Additive Inverse of a Matrix

Given the matrix $A = [a_{ij}]$, the additive inverse of A is $-A = [-a_{ij}]$.

For example, if $A = \begin{bmatrix} -2 & 3 & -1 \\ 0 & -1 & 4 \end{bmatrix}$, then the additive inverse of A is

$$-A = -\begin{bmatrix} -2 & 3 & -1 \\ 0 & -1 & 4 \end{bmatrix} = \begin{bmatrix} 2 & -3 & 1 \\ 0 & 1 & -4 \end{bmatrix}.$$

Subtraction of two matrices is defined in terms of the additive inverse of a matrix.

Subtraction of Matrices

Given two matrices A and B of order $m \times n$, the subtraction of the two matrices $A - B$ is
$$A - B = A + (-B).$$

As an example, let $A = \begin{bmatrix} 2 & -3 \\ -1 & 2 \\ 2 & 4 \end{bmatrix}$ and $B = \begin{bmatrix} -1 & 2 \\ -4 & 1 \\ 3 & -2 \end{bmatrix}$;

then $A - B = \begin{bmatrix} 2 & -3 \\ -1 & 2 \\ 2 & 4 \end{bmatrix} - \begin{bmatrix} -1 & 2 \\ -4 & 1 \\ 3 & -2 \end{bmatrix}$

$$= \begin{bmatrix} 2 & -3 \\ -1 & 2 \\ 2 & 4 \end{bmatrix} + \begin{bmatrix} 1 & -2 \\ 4 & -1 \\ -3 & 2 \end{bmatrix} = \begin{bmatrix} 3 & -5 \\ 3 & 1 \\ -1 & 6 \end{bmatrix}$$

Of special importance is the *zero matrix*, which is the matrix that consists of all zeros. The zero matrix is the additive identity for matrices.

Definition of the Zero Matrix

The $m \times n$ **zero matrix,** denoted by $\mathbf{0}$, is the matrix whose elements are all zeros.

Three examples of zero matrices are

$$\begin{bmatrix} 0 & 0 & 0 \\ 0 & 0 & 0 \end{bmatrix} \qquad \begin{bmatrix} 0 & 0 & 0 & 0 \\ 0 & 0 & 0 & 0 \\ 0 & 0 & 0 & 0 \end{bmatrix} \qquad \begin{bmatrix} 0 & 0 \\ 0 & 0 \end{bmatrix}$$

Properties of Matrix Addition

Given matrices A, B, C and the zero matrix $\mathbf{0}$ each of order $m \times n$, then the following properties hold:

Commutative law $A + B = B + A$

Associative law $A + (B + C) = (A + B) + C$

Additive inverse $A + (-A) = \mathbf{0}$

Additive identity $A + \mathbf{0} = \mathbf{0} + A = A$

Two types of products involve matrices. The first product is the product of a real number and a matrix. The second product is the product of two matrices.

Definition of the Product of a Real Number and a Matrix

Given the $m \times n$ matrix $A = [a_{ij}]$ and a real number c, then $cA = [ca_{ij}]$.

Remark   The product of a real number and a matrix is referred to as **scalar multiplication.**

As an example of this definition, consider the matrix

$$A = \begin{bmatrix} 2 & -3 & 1 \\ 3 & 1 & -2 \\ 1 & -1 & 4 \end{bmatrix}$$

and the constant $c = -2$; then

$$-2A = -2 \begin{bmatrix} 2 & -3 & 1 \\ 3 & 1 & -2 \\ 1 & -1 & 4 \end{bmatrix} = \begin{bmatrix} -2(2) & -2(-3) & -2(1) \\ -2(3) & -2(1) & -2(-2) \\ -2(1) & -2(-1) & -2(4) \end{bmatrix} = \begin{bmatrix} -4 & 6 & -2 \\ -6 & -2 & 4 \\ -2 & 2 & -8 \end{bmatrix}$$

This definition is also used to factor a constant from a matrix.

$$\begin{bmatrix} \frac{3}{2} & -\frac{5}{4} & \frac{1}{4} \\ \frac{3}{4} & \frac{1}{2} & \frac{5}{2} \end{bmatrix} = \frac{1}{4}\begin{bmatrix} 6 & -5 & 1 \\ 3 & 2 & 10 \end{bmatrix}$$

Properties of Scalar Multiplication

Given real numbers a, b, and c and matrices $A = [a_{ij}]$ and $B = [b_{ij}]$ each of order $m \times n$, then

$$(b + c)A = bA + cA$$

$$c(A + B) = cA + cB$$

$$a(bA) = (ab)A$$

EXAMPLE 1   Find the Sum of Two Matrices

Given $A = \begin{bmatrix} -2 & 3 \\ 4 & -2 \\ 0 & 4 \end{bmatrix}$ and $B = \begin{bmatrix} 8 & -2 \\ -3 & 2 \\ -4 & 7 \end{bmatrix}$, find $2A + 5B$.

Solution $2A + 5B = 2\begin{bmatrix} -2 & 3 \\ 4 & -2 \\ 0 & 4 \end{bmatrix} + 5\begin{bmatrix} 8 & -2 \\ -3 & 2 \\ -4 & 7 \end{bmatrix}$

$$= \begin{bmatrix} -4 & 6 \\ 8 & -4 \\ 0 & 8 \end{bmatrix} + \begin{bmatrix} 40 & -10 \\ -15 & 10 \\ -20 & 35 \end{bmatrix} = \begin{bmatrix} 36 & -4 \\ -7 & 6 \\ -20 & 43 \end{bmatrix}$$

■ *Try Exercise 6, page 572.*

To define the multiplication of two matrices, we begin with a simple problem. Suppose that for a certain week an egg producer sold 200 dozen small eggs, 300 dozen medium eggs, 250 dozen large eggs, and 50 dozen extra large eggs. If the price per dozen was $.65 for small eggs, $.75 for medium eggs, $.90 for large eggs, and $1.05 for extra-large eggs, find the producer's total revenue from the sale of the eggs.

To solve this problem, represent the week's sale of eggs by the *row matrix*

[200   300   250   50]   A row matrix.

Represent the selling price per dozen by the *column matrix*

$$\begin{bmatrix} 0.65 \\ 0.75 \\ 0.90 \\ 1.05 \end{bmatrix}$$ A column matrix.

The total revenue from the sale of the eggs is the sum of the products of the number of dozen of each type sold and the price per dozen;

Total revenue $= 200 \cdot 0.65 + 300 \cdot 0.75 + 250 \cdot 0.90 + 50 \cdot 1.05 = 632.50$

In matrix terms, we can write

$$[200 \quad 300 \quad 250 \quad 50] \begin{bmatrix} 0.65 \\ 0.75 \\ 0.90 \\ 1.05 \end{bmatrix}$$

$$= 200 \cdot 0.65 + 300 \cdot 0.75 + 250 \cdot 0.90 + 50 \cdot 1.05 = 632.50$$

In general, if A is a row matrix of order $1 \times n$,

$$A = [a_1 \quad a_2 \quad a_3 \cdots a_n],$$

and B is a column matrix of order $n \times 1$,

$$B = \begin{bmatrix} b_2 \\ b_2 \\ b_3 \\ \cdot \\ \cdot \\ \cdot \\ b_n \end{bmatrix},$$

then the product of A and B, written AB, is

$$AB = [a_1 \quad a_2 \quad a_3 \cdots a_n] \begin{bmatrix} b_1 \\ b_2 \\ b_3 \\ \cdot \\ \cdot \\ \cdot \\ b_n \end{bmatrix} = a_1 b_1 + a_2 b_2 + a_3 b_3 + \cdots + a_n b_n$$

The definition of the product of two matrices is an extension of the definition of the product of a row matrix and column matrix.

Definition of the Product of Two Matrices

Let $A = [a_{ij}]$ be a matrix of order $m \times n$ and $B = [b_{ij}]$ be a matrix of order $n \times p$. Then the product AB is the matrix of order $m \times p$ given by $AB = [c_{ij}]$, where each element c_{ij} is

$$c_{ij} = [a_{i1} \quad a_{i2} \quad a_{i3} \cdots a_{in}] \begin{bmatrix} b_{1j} \\ b_{2j} \\ b_{3j} \\ \cdot \\ \cdot \\ \cdot \\ b_{nj} \end{bmatrix}$$

$$= a_{i1} b_{1j} + a_{i2} b_{2j} + a_{i3} b_{3j} + \cdots + a_{in} b_{nj}$$

Remark This definition may appear complicated, but basically, to multiply two matrices, multiply each row vector of the first matrix by each column vector of the second matrix.

For the product of two matrices to be possible, the number of columns of the first matrix must equal the number of rows of the second matrix.

$$
\begin{array}{ccccc}
A & \cdot & B & = & C \\
m \times n & & n \times p & & m \times p
\end{array}
$$

Must be equal
Order of product matrix

The product matrix has as many rows as the first matrix and as many columns as the second matrix. For example, let

$$
A = \begin{bmatrix} 2 & -3 & 0 \\ 1 & 4 & -1 \end{bmatrix} \quad \text{and} \quad B = \begin{bmatrix} 1 & 0 \\ 4 & -2 \\ 3 & 5 \end{bmatrix}.
$$

Then A has order 2×3 and B has order 3×2. Thus AB has order 2×2.

$$
\begin{bmatrix} 2 & -3 & 0 \\ 1 & 4 & -1 \end{bmatrix}_{2\times3} \begin{bmatrix} 1 & 0 \\ 4 & -2 \\ 3 & 5 \end{bmatrix}_{3\times2}
$$

$$
= \begin{bmatrix} [2 \quad -3 \quad 0]\begin{bmatrix}1\\4\\3\end{bmatrix} & [2 \quad -3 \quad 0]\begin{bmatrix}0\\-2\\5\end{bmatrix} \\ [1 \quad 4 \quad -1]\begin{bmatrix}1\\4\\3\end{bmatrix} & [1 \quad 4 \quad -1]\begin{bmatrix}0\\-2\\5\end{bmatrix} \end{bmatrix}_{2\times2}
$$

$$
= \begin{bmatrix} 2(1) + (-3)4 + 0(3) & 2(0) + (-3)(-2) + 0(5) \\ 1(1) + 4(4) + (-1)3 & 1(0) + 4(-2) + (-1)5 \end{bmatrix}_{2\times2}
$$

$$
= \begin{bmatrix} -10 & 6 \\ 14 & -13 \end{bmatrix}_{2\times2}
$$

EXAMPLE 2 Find the Product of Two Matrices

Find the following products:

a. $\begin{bmatrix} 2 & 3 \\ -3 & 1 \\ 1 & -3 \end{bmatrix} \begin{bmatrix} 1 & 2 & -2 & 3 \\ -1 & 0 & 3 & -4 \end{bmatrix}$

b.
$$\begin{bmatrix} 1 & -1 & 3 \\ 2 & 2 & -1 \\ 0 & -2 & 3 \end{bmatrix} \begin{bmatrix} 4 & -2 & 0 \\ -1 & 3 & 1 \\ 2 & -3 & 1 \end{bmatrix}$$

Solution

a.

$$\begin{bmatrix} 2 & 3 \\ -3 & 1 \\ 1 & -3 \end{bmatrix} \begin{bmatrix} 1 & 2 & -2 & 3 \\ -1 & 0 & 3 & -4 \end{bmatrix}$$

$$= \begin{bmatrix} 2(1) + 3(-1) & 2(2) + 3(0) & 2(-2) + 3(3) & 2(3) + 3(-4) \\ -3(1) + 1(-1) & -3(2) + 1(0) & (-3)(-2) + 1(3) & (-3)3 + 1(-4) \\ 1(1) + (-3)(-1) & 1(2) + (-3)0 & 1(-2) + (-3)3 & 1(3) + (-3)(-4) \end{bmatrix}$$

$$= \begin{bmatrix} -1 & 4 & 5 & -6 \\ -4 & -6 & 9 & -13 \\ 4 & 2 & -11 & 15 \end{bmatrix}$$

b.

$$\begin{bmatrix} 1 & -1 & 3 \\ 2 & 2 & -1 \\ 0 & -2 & 3 \end{bmatrix} \begin{bmatrix} 4 & -2 & 0 \\ -1 & 3 & 1 \\ 2 & -3 & 1 \end{bmatrix}$$

$$= \begin{bmatrix} 4 + 1 + 6 & -2 + (-3) + (-9) & 0 + (-1) + 3 \\ 8 + (-2) + (-2) & -4 + 6 + 3 & 0 + 2 + (-1) \\ 0 + 2 + 6 & 0 + (-6) + (-9) & 0 + (-2) + 3 \end{bmatrix}$$

$$= \begin{bmatrix} 11 & -14 & 2 \\ 4 & 5 & 1 \\ 8 & -15 & 1 \end{bmatrix}$$

■ *Try Exercise* **16**, *page 573.*

Generally, matrix multiplication is not commutative. That is, given two matrices A and B, $AB \neq BA$. In some cases, as in Example 2a, if the matrices were reversed, the product would not be defined.

$$\begin{bmatrix} 1 & 2 & -2 & 3 \\ -1 & 0 & 3 & -4 \end{bmatrix}_{2 \times 4} \begin{bmatrix} 2 & 3 \\ -3 & 1 \\ 1 & -3 \end{bmatrix}_{3 \times 2}$$

Columns $\neq$ Rows

But even in those cases where multiplication is defined, the product AB and BA may not equal. The product of part b of Example 2 with the matrices reversed illustrates this point.

$$\begin{bmatrix} 4 & -2 & 0 \\ -1 & 3 & 1 \\ 2 & -3 & 1 \end{bmatrix} \begin{bmatrix} 1 & -1 & 3 \\ 2 & 2 & -1 \\ 0 & -2 & 3 \end{bmatrix} = \begin{bmatrix} 0 & -8 & 14 \\ 5 & 5 & -3 \\ -4 & -10 & 12 \end{bmatrix} \neq \begin{bmatrix} 11 & -14 & 2 \\ 4 & 5 & 1 \\ 8 & -15 & 1 \end{bmatrix}$$

Although matrix multiplication is not commutative, the associative property of multiplication and distributive property do hold for matrices.

Properties of Matrix Multiplication

Associative property   Given matrices A, B, and C of orders $m \times n$, $n \times p$, and $p \times q$, respectively, then

$$A(BC) = (AB)C$$

Distributive property   Given matrices A_1 and A_2 of order $m \times n$ and matrices B_1 and B_2 of order $n \times p$, then

$$A_1(B_1 + B_2) = A_1 B_1 + A_1 B_2 \quad \text{Left distributive law}$$

and

$$(A_1 + A_2)B_1 = A_1 B_1 + A_2 B_1 \qquad \text{Right distributive law}$$

A system of equations can be expressed as the products of matrices. consider the matrix equation

$$\begin{bmatrix} 2 & -3 & 4 \\ 3 & 0 & 1 \\ 1 & -2 & -5 \end{bmatrix}_{3\times 3} \begin{bmatrix} x \\ y \\ z \end{bmatrix}_{3\times 1} = \begin{bmatrix} 9 \\ 4 \\ -2 \end{bmatrix}_{3\times 1}. \tag{1}$$

Multiplying the two matrices on the left side of the equation, we have

$$\begin{bmatrix} 2x - 3y + 4z \\ 3x \quad\;\; + z \\ x - 2y - 5z \end{bmatrix}_{3\times 1} = \begin{bmatrix} 9 \\ 4 \\ -2 \end{bmatrix}_{3\times 1}$$

Because we multiplied a 3×3 matrix by a 3×1 matrix, the result is a 3×1 matrix. Now using the definition of matrix equality, we have

$$\begin{cases} 2x - 3y + 4z = 9 \\ 3x \quad\;\; + z = 4 \\ x - 2y - 5z = -2 \end{cases}$$ The matrix equation (1) is equivalent to this system of equations.

As we shall see in the next section, writing a system of equations as a matrix equation is another method for solving a system of linear equations.

The square matrix that has a 1 for each element on the main diagonal and zeros elsewhere is called the *identity matrix*.

Identity Matrix

The **identity matrix** of order n, denoted I_n, is the $n \times n$ matrix

$$I_n = \begin{bmatrix} 1 & 0 & 0 & \cdots & 0 \\ 0 & 1 & 0 & \cdots & 0 \\ 0 & 0 & 1 & \cdots & 0 \\ \vdots & \vdots & \vdots & \cdots & \vdots \\ 0 & 0 & 0 & \cdots & 1 \end{bmatrix}.$$

The identity matrix has properties similar to the real number 1. For example, the product of a matrix A and I is A.

$$\begin{bmatrix} 2 & -3 & 0 \\ 4 & 7 & -5 \\ 9 & 8 & -6 \end{bmatrix} \begin{bmatrix} 1 & 0 & 0 \\ 0 & 1 & 0 \\ 0 & 0 & 1 \end{bmatrix} = \begin{bmatrix} 2 & -3 & 0 \\ 4 & 7 & -5 \\ 9 & 8 & -6 \end{bmatrix}.$$

If A is a square matrix of order $n \times n$ and I_n is the identity matrix of order n, then $AI_n = I_n A = A$.

EXERCISE SET 11.2

In Exercises 1 to 8, find **a.** $A + B$, **b.** $A - B$, **c.** $2B$, **d.** $2A - 3B$.

1. $A = \begin{bmatrix} 2 & -1 \\ 3 & 3 \end{bmatrix}$, $B = \begin{bmatrix} -1 & 3 \\ 2 & 1 \end{bmatrix}$

2. $A = \begin{bmatrix} 0 & -2 \\ 2 & 3 \end{bmatrix}$, $B = \begin{bmatrix} 5 & -1 \\ 3 & 0 \end{bmatrix}$

3. $A = \begin{bmatrix} 0 & -1 & 3 \\ 1 & 0 & -2 \end{bmatrix}$, $B = \begin{bmatrix} -3 & 1 & 2 \\ 2 & 5 & -3 \end{bmatrix}$

4. $A = \begin{bmatrix} 2 & -2 & 4 \\ 0 & -3 & -4 \end{bmatrix}$, $B = \begin{bmatrix} 1 & -5 & 6 \\ 4 & -2 & -3 \end{bmatrix}$

5. $A = \begin{bmatrix} -3 & 4 \\ 2 & -3 \\ -1 & 0 \end{bmatrix}$, $B = \begin{bmatrix} 4 & 1 \\ 1 & -2 \\ 3 & -4 \end{bmatrix}$

6. $A = \begin{bmatrix} 2 & -2 \\ 3 & 4 \\ 1 & 0 \end{bmatrix}$, $B = \begin{bmatrix} -1 & 8 \\ 2 & -2 \\ -4 & 3 \end{bmatrix}$

7. $A = \begin{bmatrix} -2 & 3 & -1 \\ 0 & -1 & 2 \\ -4 & 3 & 3 \end{bmatrix}$, $B = \begin{bmatrix} 1 & -2 & 0 \\ 2 & 3 & -1 \\ 3 & -1 & 2 \end{bmatrix}$

8. $A = \begin{bmatrix} 0 & 2 & 0 \\ 1 & -3 & 3 \\ 5 & 4 & -2 \end{bmatrix}$, $B = \begin{bmatrix} -1 & 2 & 4 \\ 3 & 3 & -2 \\ -4 & 4 & 3 \end{bmatrix}$

In Exercises 9 to 16, find AB and BA.

9. $A = \begin{bmatrix} 2 & -3 \\ 1 & 4 \end{bmatrix}$, $B = \begin{bmatrix} -2 & 4 \\ 2 & -3 \end{bmatrix}$

10. $A = \begin{bmatrix} 3 & -2 \\ 4 & 1 \end{bmatrix}$, $B = \begin{bmatrix} -1 & -1 \\ 0 & 4 \end{bmatrix}$

11. $A = \begin{bmatrix} 3 & -1 \\ 2 & 3 \end{bmatrix}$, $B = \begin{bmatrix} 4 & 1 \\ 2 & -3 \end{bmatrix}$

12. $A = \begin{bmatrix} -3 & 2 \\ 2 & -2 \end{bmatrix}$, $B = \begin{bmatrix} 0 & 2 \\ -2 & 4 \end{bmatrix}$

13. $A = \begin{bmatrix} 2 & -1 \\ 0 & 3 \\ 1 & -2 \end{bmatrix}$, $B = \begin{bmatrix} 1 & -2 & 3 \\ 2 & 0 & 1 \end{bmatrix}$

14. $A = \begin{bmatrix} -1 & 3 \\ 2 & 1 \\ -3 & -2 \end{bmatrix}$, $B = \begin{bmatrix} 0 & -1 & 2 \\ 1 & 2 & -4 \end{bmatrix}$

15. $A = \begin{bmatrix} 2 & -1 & 3 \\ 0 & 2 & -1 \\ 0 & 0 & 2 \end{bmatrix}$, $B = \begin{bmatrix} 2 & 0 & 0 \\ 1 & -1 & 0 \\ 2 & -1 & -2 \end{bmatrix}$

16. $A = \begin{bmatrix} -1 & 2 & 0 \\ 2 & -1 & 1 \\ -2 & 2 & -1 \end{bmatrix}$, $B = \begin{bmatrix} 2 & -1 & 0 \\ 1 & 5 & -1 \\ 0 & -1 & 3 \end{bmatrix}$

In Exercises 17 to 24, find AB, if possible.

17. $A = \begin{bmatrix} 1 & -2 & 3 \end{bmatrix}$, $B = \begin{bmatrix} 1 & 0 \\ 2 & -1 \\ 1 & 2 \end{bmatrix}$

18. $A = \begin{bmatrix} -2 & 3 \\ 1 & -2 \\ 0 & 2 \end{bmatrix}$, $B = \begin{bmatrix} 3 \\ -2 \end{bmatrix}$

19. $A = \begin{bmatrix} 2 & -1 \\ 3 & 3 \end{bmatrix}$, $B = \begin{bmatrix} 1 & -2 \\ 3 & 1 \\ 0 & -2 \end{bmatrix}$

20. $A = \begin{bmatrix} 2 & 0 & -1 \\ 3 & 4 & -3 \end{bmatrix}$, $B = \begin{bmatrix} 3 & -1 & 0 \\ 2 & 4 & 5 \end{bmatrix}$

21. $A = \begin{bmatrix} 2 & 3 \\ -4 & -6 \end{bmatrix}$, $B = \begin{bmatrix} 3 & 6 \\ -2 & -4 \end{bmatrix}$

22. $A = \begin{bmatrix} 2 & -1 & 3 \\ -1 & 2 & 1 \end{bmatrix}$, $B = \begin{bmatrix} 1 & 3 & 2 \\ 2 & -1 & 0 \\ 3 & 1 & 2 \end{bmatrix}$

23. $A = \begin{bmatrix} 1 & 2 & -2 & 3 \\ 0 & -2 & 1 & -3 \end{bmatrix}$, $B = \begin{bmatrix} -2 & 0 \\ 4 & -2 \end{bmatrix}$

24. $A = \begin{bmatrix} 2 & -2 & 4 \\ 1 & 0 & -1 \\ 2 & 1 & 3 \end{bmatrix}$, $B = \begin{bmatrix} 2 & 1 & -3 & 0 \\ 0 & -2 & 1 & -2 \\ 1 & -1 & 0 & 2 \end{bmatrix}$

In Exercises 25 to 28, given the matrices

$$A = \begin{bmatrix} -1 & 3 \\ 2 & -1 \\ 3 & 1 \end{bmatrix} \quad \text{and} \quad B = \begin{bmatrix} 0 & -2 \\ 1 & 3 \\ 4 & -3 \end{bmatrix},$$

find the 3×2 matrix X that is a solution of the equation.

25. $3X + A = B$ **26.** $2A - 3X = 5B$

27. $2X - A = X + B$ **28.** $3X + 2B = X - 2A$

In Exercises 29 to 32, use the matrices

$$A = \begin{bmatrix} 2 & -3 \\ 1 & -1 \end{bmatrix} \quad \text{and} \quad B = \begin{bmatrix} 3 & -1 & 0 \\ 2 & -2 & -1 \\ 1 & 0 & 2 \end{bmatrix}.$$

If A is a square matrix, then $A^n = A \cdot A \cdot A \cdots A$, where the matrix A is repeated n times.

29. Find A^2. **30.** Find A^3.

31. Find B^2. **32.** Find B^3.

In Exercises 33 to 38, find the system of equations that is equivalent to the given matrix equation.

33. $\begin{bmatrix} 3 & -8 \\ 4 & 3 \end{bmatrix} \begin{bmatrix} x \\ y \end{bmatrix} = \begin{bmatrix} 11 \\ 1 \end{bmatrix}$ **34.** $\begin{bmatrix} 2 & 7 \\ 3 & -4 \end{bmatrix} \begin{bmatrix} x \\ y \end{bmatrix} = \begin{bmatrix} 1 \\ 16 \end{bmatrix}$

35. $\begin{bmatrix} 1 & -3 & -2 \\ 3 & 1 & 0 \\ 2 & -4 & 5 \end{bmatrix} \begin{bmatrix} x \\ y \\ z \end{bmatrix} = \begin{bmatrix} 6 \\ 2 \\ 1 \end{bmatrix}$

36. $\begin{bmatrix} 2 & 0 & 5 \\ 3 & -5 & 1 \\ 4 & -7 & 6 \end{bmatrix} \begin{bmatrix} x \\ y \\ z \end{bmatrix} = \begin{bmatrix} 9 \\ 7 \\ 14 \end{bmatrix}$

37. $\begin{bmatrix} 2 & -1 & 0 & 2 \\ 4 & 1 & 2 & -3 \\ 6 & 0 & 1 & -2 \\ 5 & 2 & -1 & -4 \end{bmatrix} \begin{bmatrix} x_1 \\ x_2 \\ x_3 \\ x_4 \end{bmatrix} = \begin{bmatrix} 5 \\ 6 \\ 10 \\ 8 \end{bmatrix}$

38. $\begin{bmatrix} 5 & -1 & 2 & -3 \\ 4 & 0 & 2 & 0 \\ 2 & -2 & 5 & -4 \\ 3 & 1 & -3 & 4 \end{bmatrix} \begin{bmatrix} x_1 \\ x_2 \\ x_3 \\ x_4 \end{bmatrix} = \begin{bmatrix} -2 \\ 2 \\ -1 \\ 2 \end{bmatrix}$

Supplemental Exercises

The elements of a matrix can be complex numbers. In Exercises 39 to 48, let

$$A = \begin{bmatrix} 2 + 3i & 1 - 2i \\ 1 + i & 2 - i \end{bmatrix} \quad \text{and} \quad B = \begin{bmatrix} 1 - i & 2 + 3i \\ 3 + 2i & 4 - i \end{bmatrix}.$$

Perform the indicated operations.

39. $3A$ **40.** $-2B$ **41.** $2iB$ **42.** $3iA$

43. $A + B$ **44.** $A - B$ **45.** AB **46.** BA

47. A^2 **48.** B^2

Matrices with complex number elements play a role in the theory of the atom. The following three matrices, called Pauli spin matrices, were used by Linus Pauli in his early study of the electron. In Exercises 49 to 51, use the matrices.

$$\sigma_1 = \begin{bmatrix} 0 & 1 \\ 1 & 0 \end{bmatrix}, \quad \sigma_2 = \begin{bmatrix} 0 & -i \\ i & 0 \end{bmatrix}, \quad \sigma_3 = \begin{bmatrix} 1 & 0 \\ 0 & -1 \end{bmatrix}$$

49. Show that for $i = 1$, 2, and 3, $(\sigma_i)^2 = I_2$.

50. Show that $\sigma_1 \cdot \sigma_2 = i\sigma_3$.

51. Show that $\sigma_1 \cdot \sigma_2 + \sigma_2 \cdot \sigma_1 = 0$.

52. Given two real numbers a and b and a matrix A of order 2×2, prove that $(a + b)A = aA + bA$.

53. Given two real numbers a and b and a matrix A of order 2×2, prove that $a(bA) = (ab)A$.

11.3

The Inverse of a Matrix

Recall that the multiplicative inverse of a nonzero real number c is $1/c$, the number whose product with c is 1. For example, the multiplicative inverse of $\frac{2}{3}$ is $\frac{3}{2}$ because $\frac{2}{3} \cdot \frac{3}{2} = 1$.

For some square matrices we can define a multiplicative inverse.

Multiplicative Inverse of a Matrix

If A is a square matrix of order n, then the inverse of matrix A, denoted by A^{-1}, has the property that

$$A \cdot A^{-1} = A^{-1} \cdot A = I_n$$

where I_n is the identity matrix of order n.

Remark   As we will see shortly, not all square matrices have multiplicative inverses. Are there any real numbers that do not have multiplicative inverses?[3]

---

[3] The real number zero does not have a multiplicative inverse.

A procedure for finding the inverse (we will simply use inverse for multiplicative inverse) uses elementary row operations. The procedure will be illustrated by finding the inverse of a 2 × 2 matrix.

Let $A = \begin{bmatrix} 2 & 4 \\ 7 & 1 \end{bmatrix}$. To the matrix A we will merge the identity matrix I_2

to the right of A and denote this new matrix by $[A : I_2]$.

$$[A:I_2] = \begin{bmatrix} 2 & 7 & 1 & 0 \\ 1 & 4 & 0 & 1 \end{bmatrix}.$$

$$A \underline{\qquad} \uparrow \qquad \uparrow \underline{\qquad} I_2$$

Now we use elementary row operations in a manner similar to that of the Gauss-Jordan method. The goal is to produce

$$[I_2 : A^{-1}] = \begin{bmatrix} 1 & 0 & b_{11} & b_{12} \\ 0 & 1 & b_{21} & b_{22} \end{bmatrix}$$

$$I_2 \underline{\qquad} \uparrow \qquad \uparrow \underline{\qquad} A^{-1}$$

In this form, the inverse matrix is the matrix that is to the right of the identity matrix. That is,

$$A^{-1} = \begin{bmatrix} b_{11} & b_{12} \\ b_{21} & b_{22} \end{bmatrix}.$$

To find A^{-1}, we use a series of elementary row operations that will result in a 1 in the first row and first column.

$$\begin{bmatrix} 2 & 7 & 1 & 0 \\ 1 & 4 & 0 & 1 \end{bmatrix} \xrightarrow{(1/2)R_1} \begin{bmatrix} 1 & \frac{7}{2} & \frac{1}{2} & 0 \\ 1 & 4 & 0 & 1 \end{bmatrix}$$

$$\xrightarrow{-1R_1 + R_2} \begin{bmatrix} 1 & \frac{7}{2} & \frac{1}{2} & 0 \\ 0 & \frac{1}{2} & -\frac{1}{2} & 1 \end{bmatrix} \xrightarrow{2R_2} \begin{bmatrix} 1 & \frac{7}{2} & \frac{1}{2} & 0 \\ 0 & 1 & -1 & 2 \end{bmatrix}$$

$$\xrightarrow{(-7/2)R_2 + R_1} \begin{bmatrix} 1 & 0 & 4 & -7 \\ 0 & 1 & -1 & 2 \end{bmatrix}$$

Now that the original matrix has been transformed into the identity matrix, the inverse matrix is the matrix to the right of the identity matrix. Therefore,

$$A^{-1} = \begin{bmatrix} 4 & -7 \\ -1 & 2 \end{bmatrix}.$$

Each elementary row operation is chosen to advance the process of transforming the original matrix into the identity matrix.

EXAMPLE 1   Find the Inverse of a 3 × 3 Matrix

Find the inverse of the matrix $A = \begin{bmatrix} 1 & -1 & 2 \\ 2 & 0 & 6 \\ 3 & -5 & 7 \end{bmatrix}$.

Solution

$\begin{bmatrix} 1 & -1 & 2 & 1 & 0 & 0 \\ 2 & 0 & 6 & 0 & 1 & 0 \\ 3 & -5 & 7 & 0 & 0 & 1 \end{bmatrix}$ Merge the given matrix with the identity matrix I_3

$\begin{matrix} -2R_1 + R_2 \\ -3R_1 + R_3 \\ \longrightarrow \end{matrix} \begin{bmatrix} 1 & -1 & 2 & 1 & 0 & 0 \\ 0 & 2 & 2 & -2 & 1 & 0 \\ 0 & -2 & 1 & -3 & 0 & 1 \end{bmatrix}$ Since a_{11} is already 1, we next position zeros in a_{21} and a_{31}.

$\begin{matrix} (1/2)R_2 \\ \longrightarrow \end{matrix} \begin{bmatrix} 1 & -1 & 2 & 1 & 0 & 0 \\ 0 & 1 & 1 & -1 & \frac{1}{2} & 0 \\ 0 & -2 & 1 & -3 & 0 & 1 \end{bmatrix}$ Position a 1 in a_{22}.

$\begin{matrix} 2R_2 + R_3 \\ \longrightarrow \end{matrix} \begin{bmatrix} 1 & -1 & 2 & 1 & 0 & 0 \\ 0 & 1 & 1 & -1 & \frac{1}{2} & 0 \\ 0 & 0 & 3 & -5 & 1 & 1 \end{bmatrix}$ Position a 0 in a_{32}.

$\begin{matrix} (1/3)R_3 \\ \longrightarrow \end{matrix} \begin{bmatrix} 1 & -1 & 2 & 1 & 0 & 0 \\ 0 & 1 & 1 & -1 & \frac{1}{2} & 0 \\ 0 & 0 & 1 & -\frac{5}{3} & \frac{1}{3} & \frac{1}{3} \end{bmatrix}$ Position a 1 in a_{33}.

$\begin{matrix} -1R_3 + R_2 \\ -2R_3 + R_1 \\ \longrightarrow \end{matrix} \begin{bmatrix} 1 & -1 & 0 & \frac{13}{3} & -\frac{2}{3} & -\frac{2}{3} \\ 0 & 1 & 0 & \frac{2}{3} & \frac{1}{6} & -\frac{1}{3} \\ 0 & 0 & 1 & -\frac{5}{3} & \frac{1}{3} & \frac{1}{3} \end{bmatrix}$ Now work upward. Position a 0 in a_{23} and a_{13}.

$\begin{matrix} R_2 + R_1 \\ \longrightarrow \end{matrix} \begin{bmatrix} 1 & 0 & 0 & 5 & -\frac{1}{2} & -1 \\ 0 & 1 & 0 & \frac{2}{3} & \frac{1}{6} & -\frac{1}{3} \\ 0 & 0 & 1 & -\frac{5}{3} & \frac{1}{3} & \frac{1}{3} \end{bmatrix}$ Position a 0 in a_{12}.

The inverse matrix is $A^{-1} = \begin{bmatrix} 5 & -\frac{1}{2} & -1 \\ \frac{2}{3} & \frac{1}{6} & -\frac{1}{3} \\ -\frac{5}{3} & \frac{1}{3} & \frac{1}{3} \end{bmatrix}$.

You should verify that this matrix satisfies the condition of an inverse matrix. That is, show that $A^{-1} \cdot A = A \cdot A^{-1} = I_3$.

■ *Try Exercise* **6,** *page 581.*

A **singular matrix** is a matrix that does not have a multiplicative inverse. A matrix that has a multiplicative inverse is a **nonsingular matrix.** As you apply the procedure above to a singular matrix, there will come a point when there are zeros in a row of the original matrix. When that condition exists, the matrix does not have an inverse.

EXAMPLE 2 **Identify a Singular Matrix**

Show that the matrix $\begin{bmatrix} 1 & -1 & -1 \\ 2 & -3 & 0 \\ 1 & -2 & 1 \end{bmatrix}$ is a singular matrix.

Solution

$$\left[\begin{array}{ccc|ccc} 1 & -1 & -1 & 1 & 0 & 0 \\ 2 & -3 & 0 & 0 & 1 & 0 \\ 1 & -2 & 1 & 0 & 0 & 1 \end{array}\right] \xrightarrow[\;-1R_1 + R_3\;]{-2R_1 + R_2} \left[\begin{array}{ccc|ccc} 1 & -1 & -1 & 1 & 0 & 0 \\ 0 & -1 & 2 & -2 & 1 & 0 \\ 0 & -1 & 2 & -1 & 0 & 1 \end{array}\right]$$

$$\xrightarrow{-1 \cdot R_2} \left[\begin{array}{ccc|ccc} 1 & -1 & -1 & 1 & 0 & 0 \\ 0 & 1 & -2 & 2 & -1 & 0 \\ 0 & -1 & 2 & -1 & 0 & 1 \end{array}\right] \xrightarrow{R_2 + R_3} \left[\begin{array}{ccc|ccc} 1 & -1 & -1 & -1 & 0 & 0 \\ 0 & 1 & -2 & 2 & -1 & 0 \\ 0 & 0 & 0 & 1 & -1 & 1 \end{array}\right]$$

There are zeros in a row of the original matrix. The original matrix does not have an inverse.

■ *Try Exercise* **10,** *page 582.*

Systems of linear equations can be solved by finding the inverse of the coefficient matrix. Consider the system of equations

$$\begin{cases} 3x_1 + 4x_2 = -1 \\ 3x_1 + 5x_2 = 1 \end{cases} \tag{1}$$

Using matrix multiplication and the concept of equality of matrices, this system can be written as a matrix equation.

$$\begin{bmatrix} 3 & 4 \\ 3 & 5 \end{bmatrix} \begin{bmatrix} x_1 \\ x_2 \end{bmatrix} = \begin{bmatrix} -1 \\ 1 \end{bmatrix} \tag{2}$$

If we let

$$A = \begin{bmatrix} 3 & 4 \\ 3 & 5 \end{bmatrix}, \quad X = \begin{bmatrix} x_1 \\ x_2 \end{bmatrix}, \quad B = \begin{bmatrix} -1 \\ 1 \end{bmatrix}$$

then Equation (2) can be written as $AX = B$.

The inverse of the coefficient matrix A is $A^{-1} = \begin{bmatrix} \frac{5}{3} & -\frac{4}{3} \\ -1 & 1 \end{bmatrix}$.

To solve the system of equations, multiply each side of the equation $AX = B$ by the inverse A^{-1}.

$$\begin{bmatrix} \frac{5}{3} & -\frac{4}{3} \\ -1 & 1 \end{bmatrix}\begin{bmatrix} 3 & 4 \\ 3 & 5 \end{bmatrix}\begin{bmatrix} x_1 \\ x_2 \end{bmatrix} = \begin{bmatrix} \frac{5}{3} & -\frac{4}{3} \\ -1 & 1 \end{bmatrix}\begin{bmatrix} -1 \\ 1 \end{bmatrix}$$

$$\begin{bmatrix} x_1 \\ x_2 \end{bmatrix} = \begin{bmatrix} -3 \\ 2 \end{bmatrix}$$

Thus $x_1 = -3$ and $x_2 = 2$. The solution to (1) is $(-3, 2)$.

EXAMPLE 3   Solve a System of Equations By Using the Inverse of the Coefficient Matrix

Find the solution of the system of equations by using the inverse of the coefficient matrix.

$$\begin{cases} x_1 + \quad\quad 7x_3 = 20 \\ 2x_1 + x_2 - x_3 = -3 \\ 7x_1 + 3x_2 + x_3 = 2 \end{cases} \tag{1}$$

Solution   Write the system as a matrix equation.

$$\begin{bmatrix} 1 & 0 & 7 \\ 2 & 1 & -1 \\ 7 & 3 & 1 \end{bmatrix}\begin{bmatrix} x_1 \\ x_2 \\ x_3 \end{bmatrix} = \begin{bmatrix} 20 \\ -3 \\ 2 \end{bmatrix} \tag{2}$$

The inverse of the coefficient matrix is $\begin{bmatrix} -\frac{4}{3} & -7 & \frac{7}{3} \\ 3 & 16 & -5 \\ \frac{1}{3} & 1 & -\frac{1}{3} \end{bmatrix}$.

Multiplying each side of the matrix equation (2) by the inverse, we have

$$\begin{bmatrix} -\frac{4}{3} & -7 & \frac{7}{3} \\ 3 & 16 & -5 \\ \frac{1}{3} & 1 & -\frac{1}{3} \end{bmatrix}\begin{bmatrix} 1 & 0 & 7 \\ 2 & 1 & -1 \\ 7 & 3 & 1 \end{bmatrix}\begin{bmatrix} x_1 \\ x_2 \\ x_3 \end{bmatrix} = \begin{bmatrix} -\frac{4}{3} & -7 & \frac{7}{3} \\ 3 & 16 & -5 \\ \frac{1}{3} & 1 & -\frac{1}{3} \end{bmatrix}\begin{bmatrix} 20 \\ -3 \\ 2 \end{bmatrix}$$

$$\begin{bmatrix} x_1 \\ x_2 \\ x_3 \end{bmatrix} = \begin{bmatrix} -1 \\ 2 \\ 3 \end{bmatrix}$$

Thus $x_1 = -1$, $x_2 = 2$, and $x_3 = 3$. The solution to (1) is $(-1, 2, 3)$.

■ *Try Exercise 20, page 582.*

The advantage of using the inverse matrix to solve a system of equations is not apparent unless it is necessary to repeatedly solve a system of equations with the same coefficient matrix but different constant matrices. *Input-output analysis* is one such application of this method.

Input-Output Analysis

In an economy, some of the output of an industry is used by the industry to produce its product. For example, an electric company uses water and electricity to produce electricity and a water company uses water and electricity to produce drinking water. **Input-output analysis** attempts to determine the necessary output of industries to satisfy each other's demands plus the demands of consumers. Vassily Leontieff, a Harvard economist, was awarded the Nobel prize for his work in this field.

An **input-output matrix** is used to express the interdependence among industries in an economy. Each column of this matrix gives the dollar values of the inputs an industry needs to produce $1 worth of output.

To illustrate the concepts, we will assume an economy with only three industries: agriculture, transportation and oil. Suppose that to produce $1 of agricultural products requires $.05 of agriculture, $.02 of transportation, and $.05 of oil. To produce $1 of transportation requires $.10 of agriculture, $.08 of transportation, and $.10 of oil. To produce $1 of oil requires $.10 of agriculture, $.15 of transportation, and $.13 of oil. The input-output matrix A is

$$
\begin{array}{c}
\\
\text{from} \quad
\begin{array}{c}
\text{Agriculture} \\
\text{Transportation} \\
\text{Oil}
\end{array}
\end{array}
\overset{\begin{array}{c}\text{Input requirements of}\\ \text{Agriculture} \quad \text{Transportation} \quad \text{Oil}\end{array}}{
\begin{bmatrix}
.05 & .10 & .10 \\
.02 & .08 & .15 \\
.05 & .10 & .13
\end{bmatrix}}
$$

Consumers (other than the industries themselves) want to purchase some of the output from these industries. The amount of output that the consumer will want is called the **final demand** on the economy. This is represented by a column matrix.

Suppose in our example that the final demand is $3 billion of agriculture, $1 billion of transportation, and $2 billion of oil. The final demand matrix is

$$
\begin{bmatrix}
3 \\
1 \\
2
\end{bmatrix} = D.
$$

We represent the total output of each industry (in billions of dollars) as follows:

x = total output of agriculture.

y = total output of transportation.

z = total output of oil.

The object of input-output analysis is to determine the values of x, y, and z that will satisfy the amount the consumer demands. To find these values, consider agriculture. The amount of agriculture left for the consumer (demand d) is

$$d = x - \text{(amount agriculture used by industries)} \qquad (1)$$

To find the amount of agriculture used by the three industries in our economy, refer to the input-output matrix. Production of x billion dollars of agriculture takes $.05x$ of agriculture, production of y billion dollars of transportation takes $.10y$ of agriculture, and production of z billion dollars of oil takes $.10z$ of agriculture. Thus,

$$\text{Amount of agriculture used by industries} = .05x + .10y + .10z \qquad (2)$$

Combining Equations (1) and (2), we have

$$d = x - (.05x + .10y + .10z)$$

$$3 = .95x - .10y - .10z \qquad d \text{ is \$3 billion for agriculture.}$$

We could continue this way for each of the other industries. The result would be a system of equations. Instead, however, we will use a matrix approach.

If X = total output of the three industries of the economy, then

$$X = \begin{bmatrix} x \\ y \\ z \end{bmatrix}.$$

The product of A, the input-output matrix, and X is

$$AX = \begin{bmatrix} .05 & .10 & .10 \\ .02 & .08 & .15 \\ .05 & .10 & .13 \end{bmatrix} \begin{bmatrix} x \\ y \\ z \end{bmatrix}$$

This matrix represents the dollar amount of products used in the production for all three industries. Thus the amount available for consumer demand is $X - AX$. As a matrix equation, we can write

$$X - AX = D$$

Solving this equation for X, we determine the output necessary to meet the needs of our industries and the consumer.

$$IX - AX = D \qquad I \text{ is the identity matrix. Thus } IX = X.$$

$$(I - A)X = D$$

$$X = (I - A)^{-1}D \quad \text{Assuming the inverse of } (I - A) \text{ exists}$$

The last equation states that the solution to an input-output problem can be found by multiplying the demand matrix D by the inverse of

$(I - A)$. In our example, we have

$$I - A = \begin{bmatrix} 1 & 0 & 0 \\ 0 & 1 & 0 \\ 0 & 0 & 1 \end{bmatrix} - \begin{bmatrix} .05 & .10 & .10 \\ .02 & .08 & .15 \\ .05 & .10 & .13 \end{bmatrix} = \begin{bmatrix} .95 & -.10 & -.10 \\ -.02 & .92 & -.15 \\ -.05 & -.10 & .87 \end{bmatrix}$$

$$(I - A)^{-1} \approx \begin{bmatrix} 1.06 & .13 & .15 \\ .03 & 1.11 & .20 \\ .07 & .14 & 1.18 \end{bmatrix}.$$

The consumer demand is

$$X = (I - A)^{-1}D$$

$$X \approx \begin{bmatrix} 1.06 & .13 & .15 \\ .03 & 1.11 & .20 \\ .07 & .14 & 1.18 \end{bmatrix} \begin{bmatrix} 3 \\ 1 \\ 2 \end{bmatrix} = \begin{bmatrix} 3.61 \\ 1.60 \\ 2.71 \end{bmatrix}$$

This matrix indicates that $3.61 billion of agriculture, $1.60 billion of transportation, and $2.71 billion of oil must be produced by the industries to satisfy consumers' demands and the industries internal requirements.

If we change the final demand matrix,

$$D = \begin{bmatrix} 2 \\ 2 \\ 3 \end{bmatrix},$$

then the total output of the economy can be found as

$$X \approx \begin{bmatrix} 1.06 & .13 & .15 \\ .03 & 1.11 & .20 \\ .07 & .14 & 1.18 \end{bmatrix} \begin{bmatrix} 2 \\ 2 \\ 3 \end{bmatrix} = \begin{bmatrix} 2.83 \\ 2.88 \\ 3.96 \end{bmatrix}.$$

Thus, agriculture must produce $2.83 billion, transportation must produce $2.88 billion, and oil must produce $3.96 billion to satisfy the given consumer demand and the industries internal requirements.

EXERCISE SET 11.3

In Exercises 1 to 14, find the inverse of the given matrix.

1. $\begin{bmatrix} 1 & -3 \\ -2 & 5 \end{bmatrix}$

2. $\begin{bmatrix} 1 & 2 \\ -2 & -3 \end{bmatrix}$

3. $\begin{bmatrix} 1 & 4 \\ 2 & 10 \end{bmatrix}$

4. $\begin{bmatrix} -2 & 3 \\ -6 & -8 \end{bmatrix}$

5. $\begin{bmatrix} 1 & 2 & -1 \\ 2 & 5 & 1 \\ 3 & 6 & -2 \end{bmatrix}$

6. $\begin{bmatrix} 1 & 3 & -2 \\ -1 & -5 & 6 \\ 2 & 6 & -3 \end{bmatrix}$

7. $\begin{bmatrix} 1 & 2 & -1 \\ 2 & 6 & 1 \\ 3 & 6 & -4 \end{bmatrix}$

8. $\begin{bmatrix} 2 & 1 & -1 \\ 6 & 4 & -1 \\ 4 & 2 & -3 \end{bmatrix}$

9. $\begin{bmatrix} 2 & 4 & -4 \\ 1 & 3 & -4 \\ 2 & 4 & -3 \end{bmatrix}$ 10. $\begin{bmatrix} 1 & -2 & 2 \\ 2 & -3 & 1 \\ 3 & -6 & 6 \end{bmatrix}$

11. $\begin{bmatrix} 1 & -1 & 2 & 1 \\ 2 & -1 & 5 & 1 \\ 3 & -3 & 7 & 5 \\ -2 & 3 & -4 & -1 \end{bmatrix}$ 12. $\begin{bmatrix} 1 & 1 & -1 & 2 \\ 3 & 2 & -1 & 5 \\ 2 & 2 & -1 & 5 \\ 4 & 4 & -4 & 7 \end{bmatrix}$

13. $\begin{bmatrix} 1 & -1 & 1 & 3 \\ 2 & -1 & 4 & 8 \\ 1 & 1 & 6 & 10 \\ -1 & 5 & 5 & 4 \end{bmatrix}$ 14. $\begin{bmatrix} 1 & -1 & 1 & 2 \\ 2 & -1 & 6 & 6 \\ 3 & -1 & 12 & 12 \\ -2 & -1 & -14 & -10 \end{bmatrix}$

In Exercises 15 to 24, solve the system of equations by using inverse matrix methods.

15. $\begin{cases} x + 4y = 6 \\ 2x + 7y = 11 \end{cases}$ 16. $\begin{cases} 2x + 3y = 5 \\ x + 2y = 4 \end{cases}$

17. $\begin{cases} x - 2y = 8 \\ 3x + 2y = -1 \end{cases}$ 18. $\begin{cases} 3x - 5y = -18 \\ 2x - 3y = -11 \end{cases}$

19. $\begin{cases} x + y + 2z = 4 \\ 2x + 3y + 3z = 5 \\ 3x + 3y + 7z = 14 \end{cases}$ 20. $\begin{cases} x + 2y - z = 5 \\ 2x + 3y - z = 8 \\ 3x + 6y - 2z = 14 \end{cases}$

21. $\begin{cases} x + 2y + 2z = 5 \\ -2x - 5y - 2z = 8 \\ 2x + 4y + 7z = 19 \end{cases}$ 22. $\begin{cases} x - y + 3z = 5 \\ 3x - y + 10z = 16 \\ 2x - 2y + 5z = 9 \end{cases}$

23. $\begin{cases} w + 2x + z = 6 \\ 2w + 5x + y + 2z = 10 \\ 2w + 4x + y + z = 8 \\ 3w + 6x + 4z = 16 \end{cases}$

24. $\begin{cases} w - x + 2y = 5 \\ 2w - x + 6y + 2z = 16 \\ 3w - 2x + 9y + 4z = 28 \\ w - 2x - z = 2 \end{cases}$

25. A vacation resort offers a helicopter tour of an island. The price for an adult ticket is $20; the price for a children's ticket is $15. The records of the tour operator show that 100 people took the tour on Saturday and 120 people took the tour on Sunday. The total receipts for Saturday were $1900, and on Sunday the receipts were $2275. Find the number of adults and children who took the tour on Saturday and Sunday.

26. A company sells a standard and a deluxe model tape recorder. Each standard tape recorder costs $45 to manufacture, and each deluxe model costs $60 to manufac-

ture. The January manufacturing budget for 90 of these recorders was $4650; the February budget for 100 recorders was $5250. Find the number of each type of recorder manufactured in January and February.

27. The following table shows the active chemical content of three different soil additives.

	Grams per 100 grams		
Additive	Ammonium Nitrate	Phospherous	Iron
1	30	10	10
2	40	15	10
3	50	5	5

A soil chemist wants to prepare two chemical samples. The first sample contains 280 grams of ammonium nitrate, 95 grams of phospherous, and 85 grams of iron. The second sample requires 380 grams of ammonium nitrate, 110 grams of phospherous, and 90 grams of iron. How many grams of each additive are required for sample 1, and how many grams of each additive are required for sample 2?

28. The following table shows the carbohydrate, fat, and protein content of three food types.

	Grams per 100 grams		
Food Type	Carbohydrate	Fat	Protein
I	13	10	13
II	4	4	3
III	1	0	10

A nutritionist must prepare two diets from these three food groups. The first diet must contain 23 grams of carbohydrate, 18 grams of fat, and 39 grams of protein. The second diet must contain 35 grams of carbohydrate, 28 grams of fat, and 42 grams of protein. How many grams of each food type are required for the first diet, and how many grams of each food type are required for the second diet?

In Exercises 29 to 32, solve the input-output problems.

29. A simplified economy has three industries: manufacturing, transportation, and service. The input-output matrix for this economy is

$$\begin{bmatrix} 0.20 & 0.15 & 0.10 \\ 0.10 & 0.30 & 0.25 \\ 0.20 & 0.10 & 0.10 \end{bmatrix}$$

Find the gross output needed to satisfy the consumer demand of $120 million worth of manufacturing,

$60 million of transportation, and $55 million worth of service.

30. A four-sector economy is comprised of manufacturing, agriculture, service, and transportation. The input-output matrix for this economy is

$$\begin{bmatrix} 0.10 & 0.05 & 0.20 & 0.15 \\ 0.20 & 0.10 & 0.30 & 0.10 \\ 0.05 & 0.30 & 0.20 & 0.40 \\ 0.10 & 0.20 & 0.15 & 0.20 \end{bmatrix}$$

Find the gross output needed to satisfy a consumer demand of $80 million worth of manufacturing, $100 million worth of agriculture, $50 million worth of service, and $80 million of transportation.

31. A conglomerate is composed of three industries; coal, iron, and steel. To produce $1 of coal requires $.05 of coal, $.02 of iron, and $.10 of steel. To produce $1 of iron requires $.20 of coal, $.03 of iron, and $.12 of steel. To produce $1 of steel requires $.15 of coal, $.25 of iron, and $.05 of steel. How much should each industry produce to allow for a consumer demand of $30 million worth of coal, $5 million worth of iron, and $25 million worth of steel?

32. A conglomerate has three divisions: plastics, semiconductors, and computers. For each $1 of output, the plastics division needs $.01 worth of plastics, $.03 worth of semiconductors, and $.10 worth of computers. Each $1 of output from the semiconductor division requires $.08 worth from plastics, $.05 worth from semiconductors, and $.15 worth from computers. For each $1 of output, the computer division needs $.20 worth from plastics, $.20 worth from semiconductors, and $.10 from computers. The conglomerate estimates consumer demand of $100 million from the plastics division, $75 million from the semiconductor division, and $150 million from the computer division. At what level should each division produce to satisfy this demand?

Supplemental Exercises

33. Let

$$A = \begin{bmatrix} 2 & -3 \\ -6 & 9 \end{bmatrix} \quad \text{and} \quad B = \begin{bmatrix} -3 & 15 \\ -2 & 10 \end{bmatrix}.$$

Show that $AB = \mathbf{0}$, the 2×2 zero matrix. This illustrates that for matrices, if $AB = \mathbf{0}$, it is not necessarily so that $A = \mathbf{0}$ or $B = \mathbf{0}$.

34. Show that if a matrix A has an inverse and $AB = \mathbf{0}$, then $B = \mathbf{0}$.

35. Let

$$A = \begin{bmatrix} 2 & -1 \\ -4 & 2 \end{bmatrix}, B = \begin{bmatrix} 3 & 4 \\ 1 & 5 \end{bmatrix}, \text{ and } C = \begin{bmatrix} 4 & 7 \\ 3 & 11 \end{bmatrix}.$$

Show that $AB = AC$ but that $B \neq C$. This illustrates that the cancellation rule of real numbers may not apply to matrices.

36. (Continuation of Exercise 35.) Show that if A is a matrix that has an inverse and $AB = AC$, then $B = C$.

37. Show that if

$$A = \begin{bmatrix} a & b \\ c & d \end{bmatrix} \quad \text{and} \quad ad - bc \neq 0,$$

then

$$A^{-1} = \frac{1}{ad - bc} \begin{bmatrix} d & -b \\ -c & a \end{bmatrix}.$$

38. Use the result of Exercise 37 to show that a square matrix of order 2 has an inverse if and only if $ad - bc \neq 0$.

39. Use the result of Exercise 37 to find the inverse of each matrix:

a. $\begin{bmatrix} 2 & -3 \\ 4 & -5 \end{bmatrix}$,   **b.** $\begin{bmatrix} 5 & 6 \\ 3 & 4 \end{bmatrix}$,   **c.** $\begin{bmatrix} 0 & -1 \\ 4 & 4 \end{bmatrix}$

40. Let

$$A = \begin{bmatrix} 3 & -2 \\ 1 & 1 \end{bmatrix} \quad \text{and} \quad B = \begin{bmatrix} 2 & -1 \\ 2 & 3 \end{bmatrix}.$$

Use Exercise 37 to show that

$$A^{-1} = \frac{1}{5} \begin{bmatrix} 1 & 2 \\ -1 & 3 \end{bmatrix} \quad \text{and} \quad B^{-1} = \frac{1}{8} \begin{bmatrix} 3 & 1 \\ -2 & 2 \end{bmatrix}.$$

Now show that $(AB)^{-1} = B^{-1} \cdot A^{-1}$.

41. Generalize the last result in Exercise 40. That is, show that if A and B are square matrices of order n and each have an inverse matrix, then $(AB)^{-1} = B^{-1} \cdot A^{-1}$. (*Hint:* Begin with the equation $(AB)(AB)^{-1} = I$, where I is the identity matrix. Now multiply each side of the equation by A^{-1} and then B^{-1}.)

11.4

Determinants

Associated with each square matrix A is a number called the *determinant* of A. We will denote the determinant of the matrix A by det(A) or by $|A|$. For the remainder of this chapter, we assume that all matrices are square matrices.

The Determinant of a 2 × 2 Matrix

The **determinant** of the matrix $A = [a_{ij}]$ of order 2 is

$$|A| = \begin{vmatrix} a_{11} & a_{12} \\ a_{21} & a_{22} \end{vmatrix} = a_{11}a_{22} - a_{21}a_{12} .$$

Caution   Be careful not to confuse the notation for a matrix and that for a determinant. The symbol [ ] (brackets) is used for a matrix; the symbol | | (vertical bars) is used for the determinant of the matrix.

An easy way to remember the formula for the determinant of a 2 × 2 matrix is to recognize that the determinant is the difference in the products of the diagonal elements.

$$\begin{vmatrix} a_{11} & a_{12} \\ a_{21} & a_{22} \end{vmatrix} = a_{11}a_{22} - a_{21}a_{12} .$$

EXAMPLE 1   Find the Determinant of a Matrix of Order 2

Find the determinant of the matrix $A = \begin{bmatrix} 5 & 3 \\ 2 & -3 \end{bmatrix}$.

Solution

$$|A| = \begin{vmatrix} 5 & 3 \\ 2 & -3 \end{vmatrix} = 5(-3) - 2(3) = -15 - 6 = -21$$

■ *Try Exercise **2**, page 590.*

To define the determinant of a matrix of order greater than 2, we first need two definitions.

The Minor of a Matrix

The **minor** M_{ij} of the element a_{ij} of a square matrix A of order $n \geq 3$ is the determinant of the matrix of order $n - 1$ obtained by deleting the ith row and jth column of A.

Consider the matrix $A = \begin{bmatrix} 2 & -1 & 5 \\ 4 & 3 & -7 \\ 8 & -7 & 6 \end{bmatrix}$.

The minor M_{23} is the determinant of matrix A after row 2 and column 3 are deleted from A.

$$\begin{vmatrix} 2 & -1 & 5 \\ 4 & 3 & 7 \\ 8 & -7 & 6 \end{vmatrix},$$

thus $M_{23} = \begin{vmatrix} 2 & -1 \\ 8 & -7 \end{vmatrix} = 2(-7) - 8(-1) = -14 + 8 = -6$

The minor M_{31} is the determinant of matrix A after row 3 and column 1 are deleted from A.

$$\begin{vmatrix} 2 & -1 & 5 \\ 4 & 3 & -7 \\ 8 & -7 & 6 \end{vmatrix},$$

thus $M_{31} = \begin{vmatrix} -1 & 5 \\ 3 & -7 \end{vmatrix} = (-1)(-7) - 3(5) = 7 - 15 = -8$

The second definition we need is the *cofactor* of a matrix.

Cofactor of a Matrix

> The **cofactor** C_{ij} of the element a_{ij} of a square matrix A is given by $C_{ij} = (-1)^{i+j}M_{ij}$, where M_{ij} is the minor of a_{ij}.

Remark When $i + j$ is an even integer, $(-1)^{i+j} = 1$. When $i + j$ is an odd integer, $(-1)^{i+j} = -1$. Thus,

$$C_{ij} = \begin{cases} M_{ij}, & i + j = \text{even integer}, \\ -M_{ij}, & i + j = \text{odd integer}. \end{cases}$$

EXAMPLE 2 **Find a Minor and a Cofactor of a 3 × 3 Matrix.**

Given $A = \begin{bmatrix} 4 & 3 & -2 \\ 5 & -2 & 4 \\ 3 & -2 & -6 \end{bmatrix}$, find M_{32} and C_{12}.

Solution

$$M_{32} = \begin{vmatrix} 4 & -2 \\ 5 & 4 \end{vmatrix} = 4(4) - 5(-2) = 16 + 10 = 26$$

$$C_{12} = (-1)^{1+2}M_{12} = -M_{12} = -\begin{vmatrix} 5 & 4 \\ 3 & -6 \end{vmatrix} = -(-30 - 12) = 42$$

■ *Try Exercise **14**, page 590.*

The definitions of minors and cofactors are used to define the determinant of a matrix of order 3 or greater.

Determinants by Expanding by Cofactors

Given the square matrix A of order 3 or greater, the determinant of A is the sum of the products of the elements of any row or column and their cofactors. For the rth row of A, the determinant of A is

$$|A| = a_{r1}C_{r1} + a_{r2}C_{r2} + a_{r3}C_{r3} + \cdots + a_{rn}C_{rn}.$$

For the cth column of A, the determinant of A is

$$|A| = a_{1c}C_{1c} + a_{2c}C_{2c} + a_{3c}C_{3c} + \cdots + a_{nc}C_{nc}.$$

EXAMPLE 3   Evaluate a 3 × 3 Determinant by Cofactors

Evaluate the determinant of the matrix $A = \begin{bmatrix} 2 & 3 & -1 \\ 4 & -2 & 3 \\ 1 & -3 & 4 \end{bmatrix}$ by expanding about   a.   row 2   b.   column 3.

Solution   By the definition, any row or column can be used in the expansion. To illustrate the method, we arbitrarily choose row 2 and then column 3.

a.
$$|A| = 4C_{21} + (-2)C_{22} + 3C_{23} = 4(-M_{21}) + (-2)M_{22} + 3(-M_{23})$$

$$= -4\begin{vmatrix} 3 & -1 \\ -3 & 4 \end{vmatrix} + (-2)\begin{vmatrix} 2 & -1 \\ 1 & 4 \end{vmatrix} + (-3)\begin{vmatrix} 2 & 3 \\ 1 & -3 \end{vmatrix}$$

$$= -4(9) + (-2)(9) + (-3)(-9)$$

$$= -36 - 18 + 27 = -27$$

b.
$$|A| = (-1)C_{13} + 3C_{23} + 4C_{33} = (-1)M_{13} + (-3M_{23}) + 4M_{33}$$

$$= (-1)\begin{vmatrix} 4 & -2 \\ 1 & -3 \end{vmatrix} + (-3)\begin{vmatrix} 2 & 3 \\ 1 & -3 \end{vmatrix} + 4\begin{vmatrix} 2 & 3 \\ 4 & -2 \end{vmatrix}$$

$$= (-1)(-10) + (-3)(-9) + 4(-16)$$

$$= 10 + 27 - 64 = -27$$

■ *Try Exercise* **20,** *page 590.*

Remark   Example 3 illustrates that whatever row or column is used, expanding by cofactors gives the same value for the determinant. When you are evaluating determinants, choose the most convenient row or column, which usually is the row or column containing the most zeros.

Evaluating determinants by expanding by cofactors is very time-consuming for determinants of large orders. For example, a determinant of order 10 has more than 3 million addends, and each addend is the product of ten numbers.

The easiest determinants to evaluate have many zeros in a row or column. It is possible to transform a determinant into one that has many zeros by using elementary row operations.

Effects of Elementary Row Operations on the Determinant of a Matrix

If A is a square matrix of order n, then the following elementary row operations produce the indicated change in the determinant of A.

1. Interchanging any two rows of A changes the sign of $|A|$.
2. Multiplying a row of A by a constant k multiplies the determinant of A by k.
3. Adding a multiple of a row of A to another row does not change the value of the determinant of A.

Remark The properties of determinants just stated remain true if the word "row" is replaced by "column." In that case, we would have elementary *column* operations.

To illustrate these properties, consider the matrix $A = \begin{bmatrix} 2 & 3 \\ 1 & -2 \end{bmatrix}$. The determinant of A is $|A| = 2(-2) - 1(3) = -7$. Now consider each of the elementary row operations.

Interchange the rows of A and evaluate the determinant.

$$\begin{vmatrix} 1 & -2 \\ 2 & 3 \end{vmatrix} = 1(3) - 2(-2) = 3 + 4 = 7 = -|A|$$

Multiply row 2 of A by -3 and evaluate the determinant.

$$\begin{vmatrix} 2 & 3 \\ -3 & 6 \end{vmatrix} = 2(6) - (-3)3 = 12 + 9 = 21 = -3|A|$$

Multiply row 1 of A by -2 and add to row 2. Evaluate the determinant.

$$\begin{vmatrix} 2 & 3 \\ -3 & -8 \end{vmatrix} = 2(-8) - (-3)(3) = -16 + 9 = -7 = |A|.$$

These elementary row operations are used to rewrite a matrix in *diagonal* form. A matrix is in **diagonal** form if all elements below (or above) the main diagonal are zero. The matrices

$$A = \begin{bmatrix} 2 & -2 & 3 & 1 \\ 0 & -2 & 4 & 2 \\ 0 & 0 & 6 & 9 \\ 0 & 0 & 0 & -5 \end{bmatrix} \quad \text{and} \quad B = \begin{bmatrix} 3 & 0 & 0 & 0 \\ 2 & -3 & 0 & 0 \\ 6 & 4 & -2 & 0 \\ 8 & 3 & 4 & 2 \end{bmatrix}$$

are in diagonal form.

Determinant of a Matrix in Diagonal Form

> Let A be a square matrix of order n in diagonal form. The determinant of A is the product of the elements on the main diagonal.
>
> $$|A| = a_{11}a_{22}a_{33} \cdots a_{nn}.$$

For matrix A given above, $|A| = 2(-2)(6)(-5) = 120$.

EXAMPLE 4 **Evaluate a Determinant by Elementary Row Operations**

Evaluate the determinant by rewriting in diagonal form.

$$\begin{vmatrix} 2 & 1 & -1 & 3 \\ 2 & 2 & 0 & 1 \\ 4 & 5 & 4 & -3 \\ 2 & 2 & 7 & -3 \end{vmatrix}$$

Solution   Rewrite the matrix in diagonal form by using elementary row operations.

$$\begin{vmatrix} 2 & 1 & -1 & 3 \\ 2 & 2 & 0 & 1 \\ 4 & 5 & 4 & -3 \\ 2 & 2 & 7 & -3 \end{vmatrix} \begin{matrix} -1R_1 + R_2 \\ -2R_1 + R_3 \\ -1R_1 + R_4 \\ = \end{matrix} \begin{vmatrix} 2 & 1 & -1 & 3 \\ 0 & 1 & 1 & -2 \\ 0 & 3 & 6 & -9 \\ 0 & 1 & 8 & -6 \end{vmatrix}$$

$$\begin{matrix} \text{Factor 3,} \\ \text{from row 3.} \\ = \end{matrix} 3 \begin{vmatrix} 2 & 1 & -1 & 3 \\ 0 & 1 & 1 & -2 \\ 0 & 1 & 2 & -3 \\ 0 & 1 & 8 & -6 \end{vmatrix} \begin{matrix} -1R_2 + R_3 \\ -1R_3 + R_4 \\ = \end{matrix} 3 \begin{vmatrix} 2 & 1 & -1 & 3 \\ 0 & 1 & 1 & -2 \\ 0 & 0 & 1 & -1 \\ 0 & 0 & 7 & -4 \end{vmatrix}$$

$$\begin{matrix} -7R_3 + R_4 \\ = \end{matrix} 3 \begin{vmatrix} 2 & 1 & -1 & 3 \\ 0 & 1 & 1 & -2 \\ 0 & 0 & 1 & -1 \\ 0 & 0 & 0 & 3 \end{vmatrix} = 3(2)(1)(1)(3) = 18$$

■ *Try Exercise **42**, page 591.*

Remark   The last example used only elementary row operations to reduce the matrix to diagonal form. Elementary column operations could also have been used, or a combination of row and column operations could have been used.

A matrix whose determinant is zero is called a **singular matrix**. In some cases it is possible to recognize when the determinant of a matrix is zero.

Conditions for a Zero Determinant

If A is a square matrix, then $|A| = 0$ when any one of the following is true.

1. A row (column) consists entirely of zeros.
2. Two rows (columns) are identical.
3. One row (column) is a constant multiple of a second row (column).

Proof: To prove part (2) of this theorem, let A be the given matrix and $D = |A|$. Now interchange the two identical rows. Then $|A| = -D$. (Why?[4]) Thus,

$$D = -D.$$

Zero is the only real number that is its own additive inverse, and hence $D = |A| = 0$.

The proofs of the other two properties are left as exercises.

The last property of determinants that we will discuss is a product property.

Product Property of Determinants

If A and B are square matrices of order n, then

$$|AB| = |A||B|.$$

Now consider a matrix A with an inverse A^{-1}. Then by the last theorem,

$$|A \cdot A^{-1}| = |A||A^{-1}|.$$

But $A \cdot A^{-1} = I$, the identity matrix, and $|I| = 1$. (Why?[5]) Therefore,

$$1 = |A||A^{-1}|.$$

From the last equation, $|A| \neq 0$. And, in particular,

$$|A^{-1}| = \frac{1}{|A|}.$$

These results are summarized in the following theorem.

Existence of the Inverse of a Square Matrix

If A is a square matrix of order n, then A has a multiplicative inverse if and only if $|A| \neq 0$. Furthermore,

$$|A^{-1}| = \frac{1}{|A|}$$

---

[4] Interchanging two rows of a matrix changes the sign of the determinant of the matrix.

[5] The identity matrix is in diagonal form with 1s on the main diagonal. Thus $|I|$ is the product of 1s or $|I| = 1$.

Remark   We proved only part of this theorem. It remains to show that given $|A| \neq 0$, then A has an inverse. This proof can be found in most texts on linear algebra.

EXERCISE SET 11.4 _____

In Exercises 1 to 8, evaluate the determinants.

1. $\begin{vmatrix} 2 & -1 \\ 3 & 5 \end{vmatrix}$

2. $\begin{vmatrix} 2 & 9 \\ -6 & 2 \end{vmatrix}$

3. $\begin{vmatrix} 5 & 0 \\ 2 & -3 \end{vmatrix}$

4. $\begin{vmatrix} 0 & -8 \\ 3 & 4 \end{vmatrix}$

5. $\begin{vmatrix} 4 & 6 \\ 2 & 3 \end{vmatrix}$

6. $\begin{vmatrix} -3 & 6 \\ 4 & -8 \end{vmatrix}$

7. $\begin{vmatrix} 0 & 9 \\ 0 & -2 \end{vmatrix}$

8. $\begin{vmatrix} -3 & 9 \\ 0 & 0 \end{vmatrix}$

In Exercises 9 to 12, evaluate the indicated minor and cofactor for the determinant

$$\begin{vmatrix} 5 & -2 & -3 \\ 2 & 4 & -1 \\ 4 & -5 & 6 \end{vmatrix}.$$

9. M_{11}, C_{11} 10. M_{21}, C_{21} 11. M_{32}, C_{32} 12. M_{33}, C_{33}

In Exercises 13 to 16, evaluate the indicated minor and cofactor for the determinant

$$\begin{vmatrix} 3 & -2 & 3 \\ 1 & 3 & 0 \\ 6 & -2 & 3 \end{vmatrix}.$$

13. M_{22}, C_{22} 14. M_{13}, C_{13} 15. M_{31}, C_{31} 16. M_{23}, C_{23}

In Exercises 17 to 26, evaluate the determinant by expanding by cofactors.

17. $\begin{vmatrix} 2 & -3 & 1 \\ 2 & 0 & 2 \\ 3 & -2 & 4 \end{vmatrix}$

18. $\begin{vmatrix} 3 & 1 & -2 \\ 2 & -5 & 4 \\ 3 & 2 & 1 \end{vmatrix}$

19. $\begin{vmatrix} -2 & 3 & 2 \\ 1 & 2 & -3 \\ -4 & -2 & 1 \end{vmatrix}$

20. $\begin{vmatrix} 3 & -2 & 0 \\ 2 & -3 & 2 \\ 8 & -2 & 5 \end{vmatrix}$

21. $\begin{vmatrix} 2 & -3 & 10 \\ 0 & 2 & -3 \\ 0 & 0 & 5 \end{vmatrix}$

22. $\begin{vmatrix} 6 & 0 & 0 \\ 2 & -3 & 0 \\ 7 & -8 & 2 \end{vmatrix}$

23. $\begin{vmatrix} 0 & -2 & 4 \\ 1 & 0 & -7 \\ 5 & -6 & 0 \end{vmatrix}$

24. $\begin{vmatrix} 5 & -8 & 0 \\ 2 & 0 & -7 \\ 0 & -2 & -1 \end{vmatrix}$

25. $\begin{vmatrix} 4 & -3 & 3 \\ 2 & 1 & -4 \\ 6 & -2 & -1 \end{vmatrix}$

26. $\begin{vmatrix} -2 & 3 & 9 \\ 4 & -2 & -6 \\ 0 & -8 & -24 \end{vmatrix}$

In Exercises 27 to 40, without expanding, give a reason for each equality.

27. $\begin{vmatrix} 2 & -1 & 3 \\ 0 & 0 & 0 \\ 3 & 4 & 1 \end{vmatrix} = 0$

28. $\begin{vmatrix} 2 & 3 & 0 \\ 1 & -2 & 0 \\ 4 & 1 & 0 \end{vmatrix} = 0$

29. $\begin{vmatrix} 1 & 4 & -1 \\ 2 & 4 & 12 \\ 3 & 1 & 4 \end{vmatrix} = 2\begin{vmatrix} 1 & 4 & -1 \\ 1 & 2 & 6 \\ 3 & 1 & 4 \end{vmatrix}$

30. $\begin{vmatrix} 1 & -3 & 4 \\ 4 & 6 & 1 \\ 0 & -9 & 3 \end{vmatrix} = -3\begin{vmatrix} 1 & 1 & 4 \\ 4 & -2 & 1 \\ 0 & 3 & 3 \end{vmatrix}$

31. $\begin{vmatrix} 1 & 5 & -2 \\ 2 & -1 & 4 \\ 3 & 0 & -2 \end{vmatrix} = \begin{vmatrix} 1 & 5 & -2 \\ 0 & -11 & 8 \\ 3 & 0 & -2 \end{vmatrix}$

32. $\begin{vmatrix} 1 & 1 & -3 \\ 2 & 2 & 5 \\ 1 & -2 & 4 \end{vmatrix} = \begin{vmatrix} 1 & 1 & -3 \\ 2 & 2 & 5 \\ 0 & -3 & 7 \end{vmatrix}$

33. $\begin{vmatrix} 4 & -3 & 2 \\ 6 & 2 & 1 \\ -2 & 2 & 4 \end{vmatrix} = 2\begin{vmatrix} 2 & -3 & 2 \\ 3 & 2 & 1 \\ -1 & 2 & 4 \end{vmatrix}$

34. $\begin{vmatrix} 2 & -1 & 3 \\ 3 & 0 & 1 \\ -4 & 2 & -6 \end{vmatrix} = 0$

35. $\begin{vmatrix} 2 & -4 & 5 \\ 0 & 3 & 4 \\ 0 & 0 & -2 \end{vmatrix} = -12$

36. $\begin{vmatrix} 3 & 0 & 0 \\ 2 & -1 & 0 \\ 3 & 4 & 5 \end{vmatrix} = -15$

37. $\begin{vmatrix} 3 & 5 & -2 \\ 2 & 1 & 0 \\ 9 & -2 & -3 \end{vmatrix} = -\begin{vmatrix} 9 & -2 & -3 \\ 2 & 1 & 0 \\ 3 & 5 & -2 \end{vmatrix}$

38. $\begin{vmatrix} 6 & 0 & -2 \\ 2 & -1 & -3 \\ 1 & 5 & -7 \end{vmatrix} = -\begin{vmatrix} 0 & 6 & -2 \\ -1 & 2 & -3 \\ 5 & 1 & -7 \end{vmatrix}$

39. $a^3\begin{vmatrix} 1 & 1 & 1 \\ a & a & a \\ a^2 & a^2 & a^2 \end{vmatrix} = \begin{vmatrix} a & a & a \\ a^2 & a^2 & a^2 \\ a^3 & a^3 & a^3 \end{vmatrix}$

40. $\begin{vmatrix} 1 & 1 & 1 \\ 2 & 2 & 2 \\ 3 & 3 & 3 \end{vmatrix} = 0$

In Exercises 41 to 50, evaluate the determinant by rewriting the determinant in diagonal form by using elementary row or column operations.

41. $\begin{vmatrix} 2 & 4 & 1 \\ 1 & 2 & -1 \\ 1 & 2 & 2 \end{vmatrix}$

42. $\begin{vmatrix} 3 & -2 & -1 \\ 1 & 2 & 4 \\ 2 & -2 & 3 \end{vmatrix}$

43. $\begin{vmatrix} 1 & 2 & -1 \\ 2 & 3 & 1 \\ 3 & 4 & 3 \end{vmatrix}$

44. $\begin{vmatrix} 1 & 2 & 5 \\ -1 & 1 & -2 \\ 3 & 1 & 10 \end{vmatrix}$

45. $\begin{vmatrix} 0 & -1 & 1 \\ 1 & 0 & -2 \\ 2 & 2 & 0 \end{vmatrix}$

46. $\begin{vmatrix} 2 & -1 & 3 \\ 1 & 1 & 1 \\ 3 & -4 & 5 \end{vmatrix}$

47. $\begin{vmatrix} 1 & 2 & -1 & 2 \\ 1 & -2 & 0 & 3 \\ 3 & 0 & 1 & 5 \\ -2 & -4 & 1 & 6 \end{vmatrix}$

48. $\begin{vmatrix} 1 & -1 & -1 & 2 \\ 0 & 2 & 4 & 6 \\ 1 & 1 & 4 & 12 \\ 1 & -1 & 0 & 8 \end{vmatrix}$

49. $\begin{vmatrix} 1 & 2 & 3 & -1 \\ 6 & 5 & 9 & 8 \\ 2 & 4 & 12 & -1 \\ 1 & 2 & 6 & -1 \end{vmatrix}$

50. $\begin{vmatrix} 1 & 2 & 0 & -2 \\ -1 & 1 & 3 & 5 \\ 2 & 1 & 4 & 0 \\ -2 & 5 & 2 & 6 \end{vmatrix}$

Supplemental Exercises

The area of a triangle with vertices (x_1, y_1), (x_2, y_2), and (x_3, y_3) can be given as the absolute value of the determinant

$$\frac{1}{2}\begin{vmatrix} x_1 & y_1 & 1 \\ x_2 & y_2 & 1 \\ x_3 & y_3 & 1 \end{vmatrix}.$$

Use this formula to find the area of the triangles whose coordinates are given in Exercises 51 to 54.

51. $(2, 3)$, $(-1, 0)$, $(4, 8)$ **52.** $(-3, 4)$, $(1, 5)$, $(5, -2)$

53. $(4, 9)$, $(8, 2)$, $(-3, -2)$ **54.** $(0, 4)$, $(-5, 7)$, $(2, 9)$

55. Given a square matrix of order 3 where one row is a constant multiple of a second row, show that the determinant of the matrix is zero. (*Hint:* Use an elementary row operation and part 2 of the theorem for conditions for a zero determinant.)

56. Given a square matrix of order 3 with a zero as every element in a column, show that the determinant of the matrix is zero. (*Hint:* Expand the determinant by cofactors using the column of zeros.)

57. Show that the determinant $\begin{vmatrix} x & y & 1 \\ x_1 & y_1 & 1 \\ x_2 & y_2 & 1 \end{vmatrix} = 0$ is the equation of a line through the points (x_1, y_1) and (x_2, y_2).

58. Use Exercise 57 to find the equation of the line passing through the points $(2, 3)$ and $(-1, 4)$.

59. Use Exercise 57 to find the equation of the line passing through the points $(-3, 4)$, and $(2, -3)$.

60. Show that $\begin{vmatrix} a_1 & b_1 \\ a_2 & b_2 \end{vmatrix} = \begin{vmatrix} a_1 & b_1 \\ ka_1 + a_2 & kb_1 + b_2 \end{vmatrix}$. What property of determinants does this illustrate?

61. Surveyors use a formula to find the area of a plot of land. *Surveyor's Area Formula* If the vertices (x_1, y_1), (x_2, y_2), (x_3, y_3), ..., (x_n, y_n) of a simple polygon are listed counterclockwise around the perimeter, the area of the polygon is

$$A = \frac{1}{2}\left\{ \begin{vmatrix} x_1 & x_2 \\ y_1 & y_2 \end{vmatrix} + \begin{vmatrix} x_2 & x_3 \\ y_2 & y_3 \end{vmatrix} + \begin{vmatrix} x_3 & x_4 \\ y_3 & y_4 \end{vmatrix} + \cdots + \begin{vmatrix} x_n & x_1 \\ y_n & y_1 \end{vmatrix} \right\}.$$

Use the Surveyor's Area Formula to find the area of the polygon with vertices $(8, -4)$, $(25, 5)$, $(15, 9)$, $(17, 20)$, $(0, 10)$.

11.5

Cramer's Rule

An application of determinants is to solve a system of linear equations. Consider the system

$$\begin{cases} a_{11}x_1 + a_{12}x_2 = b_1 \\ a_{21}x_1 + a_{22}x_2 = b_2 \end{cases}$$

To eliminate x_2 from this system, we first multiply the top equation by a_{22} and the bottom equation by a_{12}. Then subtract.

$$a_{22}a_{11}x_1 + a_{22}a_{12}x_2 = a_{22}b_1$$

$$a_{12}a_{21}x_1 + a_{12}a_{22}x_2 = a_{12}b_2$$

$$(a_{22}a_{11} - a_{12}a_{21})x_1 = a_{22}b_1 - a_{12}b_2$$

This can also be written as

$$\begin{vmatrix} a_{11} & a_{12} \\ a_{21} & a_{22} \end{vmatrix} x_1 = \begin{vmatrix} b_1 & a_{12} \\ b_2 & a_{22} \end{vmatrix}$$

or

$$x_1 = \frac{\begin{vmatrix} b_1 & a_{12} \\ b_2 & a_{22} \end{vmatrix}}{\begin{vmatrix} a_{11} & a_{12} \\ a_{21} & a_{22} \end{vmatrix}}, \quad \begin{vmatrix} a_{11} & a_{12} \\ a_{21} & a_{22} \end{vmatrix} \neq 0$$

In a similar manner we can find x_2. The results are given in Cramer's Rule for a system of two linear equations.

Cramer's Rule for a System of Two Linear Equations

Let

$$\begin{cases} a_{11}x_1 + a_{12}x_2 = b_1 \\ a_{21}x_1 + a_{22}x_2 = b_2 \end{cases}$$

be the system of equations for which the determinant of the coefficient matrix is not zero. The solution of the system of equations is the ordered pair whose coordinates are

$$x_1 = \frac{\begin{vmatrix} b_1 & a_{12} \\ b_2 & a_{22} \end{vmatrix}}{\begin{vmatrix} a_{11} & a_{12} \\ a_{21} & a_{22} \end{vmatrix}} \quad \text{and} \quad x_2 = \frac{\begin{vmatrix} a_{11} & b_1 \\ a_{21} & b_2 \end{vmatrix}}{\begin{vmatrix} a_{11} & a_{12} \\ a_{21} & a_{22} \end{vmatrix}}.$$

Note that the denominator is the determinant of the coefficient matrix of the variables. The numerator of x_1 is formed by replacing column 1 of the coefficient determinant with the constants b_1 and b_2. The determinant for the numerator of x_2 is formed by replacing column 2 of the coefficient determinant by the constants b_1 and b_2.

EXAMPLE 1   Solve a System of Equations by Using Cramer's Rule

Solve the following system of equations using Cramer's Rule:

$$\begin{cases} 5x_1 - 3x_2 = 6 \\ 2x_1 + 4x_2 = -7 \end{cases}$$

Solution

$$x_1 = \frac{\begin{vmatrix} 6 & -3 \\ -7 & 4 \end{vmatrix}}{\begin{vmatrix} 5 & -3 \\ 2 & 4 \end{vmatrix}} = \frac{3}{26}, \qquad x_2 = \frac{\begin{vmatrix} 5 & 6 \\ 2 & -7 \end{vmatrix}}{\begin{vmatrix} 5 & -3 \\ 2 & 4 \end{vmatrix}} = -\frac{47}{26}$$

The solution is $\left(\dfrac{3}{26}, -\dfrac{47}{26} \right)$.

■ *Try Exercise 4, page 595.*

Cramer's Rule can be used for a system of three linear equations in three variables. For example, consider the system of equations

$$\begin{cases} 2x - 3y + z = 2 \\ 4x \quad\quad + 2z = -3 \\ 3x + y - 2z = 1 \end{cases} \qquad (1)$$

To solve this system of equations, we extend the concepts behind the solution for a system of two linear equations. The solution of the system has the form (x, y, z) where

$$x = \frac{D_x}{D}, \qquad y = \frac{D_y}{D}, \qquad z = \frac{D_z}{D}.$$

The determinant D is the determinant of the coefficient matrix. The determinants D_x, D_y, and D_z are the determinants of the matrices formed by replacing the first, second, and third columns, respectively, by the constants. For the system of equations (1),

$$x = \frac{D_x}{D} \qquad y = \frac{D_y}{D} \qquad z = \frac{D_z}{D}$$

where $D = \begin{vmatrix} 2 & -3 & 1 \\ 4 & 0 & 2 \\ 3 & 1 & -2 \end{vmatrix} = -42 \qquad D_x = \begin{vmatrix} 2 & -3 & 1 \\ -3 & 0 & 2 \\ 1 & 1 & -2 \end{vmatrix} = 5,$

$$D_y = \begin{vmatrix} 2 & 2 & 1 \\ 4 & -3 & 2 \\ 3 & 1 & -2 \end{vmatrix} = 49, \qquad D_z = \begin{vmatrix} 2 & -3 & 2 \\ 4 & 0 & -3 \\ 3 & 1 & 1 \end{vmatrix} = 53.$$

Thus

$$x = -\frac{5}{42} \qquad y = -\frac{49}{42} \qquad z = -\frac{53}{42}$$

The solution of the system (1) is

$$\left(-\frac{5}{42}, -\frac{49}{42}, -\frac{53}{42}\right).$$

Cramer's Rule can be extended to a system of n linear equations in n variables.

Cramer's Rule

Let

$$\begin{cases} a_{11}x_1 + a_{12}x_2 + a_{13}x_3 + \cdots + a_{1n}x_n = b_1 \\ a_{21}x_1 + a_{22}x_2 + a_{23}x_3 + \cdots + a_{2n}x_n = b_2 \\ a_{31}x_1 + a_{32}x_2 + a_{33}x_3 + \cdots + a_{3n}x_n = b_3 \\ \qquad \vdots \qquad \vdots \qquad \vdots \qquad \qquad \vdots \qquad \vdots \\ a_{n1}x_1 + a_{n2}x_2 + a_{n3}x_3 + \cdots + a_{nn}x_n = b_n \end{cases}$$

be a system of n equations in n variables. The solution of the system is given by $(x_1, x_2, x_3, \ldots, x_n)$ where

$$x_1 = \frac{D_1}{D}, \quad x_2 = \frac{D_2}{D}, \quad x_3 = \frac{D_3}{D}, \quad \cdots, \quad x_n = \frac{D_n}{D}.$$

where D is the determinant of the coefficient matrix and $D \neq 0$. D_i is the determinant formed by replacing the ith column of the coefficient matrix with the column of constants $b_1, b_2, b_3, \ldots, b_n$.

Because the determinant of the coefficient matrix must be nonzero to use Cramer's Rule, this method is not appropriate for systems of linear equations with no solution or infinitely many solutions. In fact, the only time a system of linear equations will have a unique solution is when the coefficient determinant is not zero, a fact summarized in the following theorem.

Systems of Linear Equations with Unique Solutions

A system of n linear equations in n variables has a unique solution if and only if the determinant of the coefficient matrix is not zero.

Cramer's Rule is also useful when we want to determine only a value of a single variable in a system of equations.

EXAMPLE 2 **Determine the Value of a Single Variable in a System of Linear Equations**

Find x_3 for the system of equations

$$\begin{cases} 4x_1 & + 3x_3 - 2x_4 = 2 \\ 3x_1 + x_2 + 2x_3 - x_4 = 4 \\ x_1 - 6x_2 - 2x_3 + 2x_4 = 0 \\ 2x_1 + 2x_2 \quad - x_4 = -1 \end{cases}$$

Solution Find D and D_3.

$$D = \begin{vmatrix} 4 & 0 & 3 & -2 \\ 3 & 1 & 2 & -1 \\ 1 & -6 & -2 & 2 \\ 2 & 2 & 0 & -1 \end{vmatrix} = 39 \qquad D_3 = \begin{vmatrix} 4 & 0 & 2 & -2 \\ 3 & 1 & 4 & -1 \\ 1 & -6 & 0 & 2 \\ 2 & 2 & -1 & -1 \end{vmatrix} = 96$$

Thus, $x_3 = \dfrac{96}{39} = \dfrac{32}{13}$.

■ *Try Exercise* **24,** *page 596.*

EXERCISE SET 11.5

In Exercises 1 to 20, solve each system of equations by using Cramer's Rule.

1. $\begin{cases} 3x_1 + 4x_2 = 8 \\ 4x_1 - 5x_2 = 1 \end{cases}$

2. $\begin{cases} x_1 - 3x_2 = 9 \\ 2x_1 - 4x_2 = -3 \end{cases}$

3. $\begin{cases} 5x_1 + 4x_2 = -1 \\ 3x_1 - 6x_2 = 5 \end{cases}$

4. $\begin{cases} 2x_1 + 5x_2 = 9 \\ 5x_1 + 7x_2 = 8 \end{cases}$

5. $\begin{cases} 7x_1 + 2x_2 = 0 \\ 2x_1 + x_2 = -3 \end{cases}$

6. $\begin{cases} 3x_1 - 8x_2 = 1 \\ 4x_1 + 5x_2 = -2 \end{cases}$

7. $\begin{cases} 3x_1 - 7x_2 = 0 \\ 2x_1 + 4x_2 = 0 \end{cases}$

8. $\begin{cases} 5x_1 + 4x_2 = -3 \\ 2x_1 - x_2 = 0 \end{cases}$

9. $\begin{cases} 1.2x_1 + 0.3x_2 = 2.1 \\ 0.8x_1 - 1.4x_2 = -1.6 \end{cases}$

10. $\begin{cases} 3.2x_1 - 4.2x_2 = 1.1 \\ 0.7x_1 + 3.2x_2 = -3.4 \end{cases}$

11. $\begin{cases} 3x_1 - 4x_2 + 2x_3 = 1 \\ x_1 - x_2 + 2x_3 = -2 \\ 2x_1 + 2x_2 + 3x_3 = -3 \end{cases}$

12. $\begin{cases} 5x_1 - 2x_2 + 3x_3 = -2 \\ 3x_1 + x_2 - 2x_3 = 3 \\ x_1 - 2x_2 + 3x_3 = -1 \end{cases}$

13. $\begin{cases} x_1 + 4x_2 - 2x_3 = 0 \\ 3x_1 - 2x_2 + 3x_3 = 4 \\ 2x_1 + x_2 - 3x_3 = -1 \end{cases}$

14. $\begin{cases} 4x_1 - x_2 + 2x_3 = 6 \\ x_1 + 3x_2 - x_3 = -1 \\ 2x_1 + 3x_2 - 2x_3 = 5 \end{cases}$

15. $\begin{cases} 2x_2 - 3x_3 = 1 \\ 3x_1 - 5x_2 + x_3 = 0 \\ 4x_1 + 2x_3 = -3 \end{cases}$

16. $\begin{cases} 2x_1 + 5x_2 = 1 \\ x_1 - 3x_3 = -2 \\ 2x_1 - x_2 + 2x_3 = 4 \end{cases}$

17. $\begin{cases} 4x_1 - 5x_2 + x_3 = -2 \\ 3x_1 + x_2 = 4 \\ x_1 - x_2 + 3x_3 = 0 \end{cases}$

18. $\begin{cases} 3x_1 - x_2 + x_3 = 5 \\ x_1 + 3x_3 = -2 \\ 2x_1 + 2x_2 - 5x_3 = 0 \end{cases}$

19. $\begin{cases} 2x_1 + 2x_2 - 3x_3 = 0 \\ x_1 - 3x_2 + 2x_3 = 0 \\ 4x_1 - x_2 + 3x_3 = 0 \end{cases}$

20. $\begin{cases} x_1 + 3x_2 = -2 \\ 2x_1 - 3x_2 + x_3 = 1 \\ 4x_1 + 5x_2 - 2x_3 = 0 \end{cases}$

In Exercises 21 to 26, solve for the indicated variable.

21. $\begin{cases} 2x_1 - 3x_2 + 4x_3 - x_4 = 1 \\ x_1 + 2x_2 + 2x_4 = -1 \\ 3x_1 + x_2 - 2x_4 = 2 \\ x_1 - 3x_2 + 2x_3 - x_4 = 3 \end{cases}$

Solve for x_2.

22. $\begin{cases} 3x_1 + x_2 - 2x_3 + 3x_4 = 4 \\ 2x_1 - 3x_2 + 2x_3 = -2 \\ x_1 + x_2 - 2x_3 + 2x_4 = 3 \\ 2x_1 + 3x_3 - 2x_4 = 4 \end{cases}$

Solve for x_4.

23. $\begin{cases} x_1 - 3x_2 + 2x_3 + 4x_4 = 0 \\ 3x_1 + 5x_2 - 6x_3 + 2x_4 = -2 \\ 2x_1 - x_2 + 9x_3 + 8x_4 = 0 \\ x_1 + x_2 + x_3 - 8x_4 = -3 \end{cases}$

Solve for x_1.

24. $\begin{cases} 2x_1 + 5x_2 - 5x_3 - 3x_4 = -3 \\ x_1 + 7x_2 + 8x_3 - x_4 = 4 \\ 4x_1 + x_3 + x_4 = 3 \\ 3x_1 + 2x_2 - x_3 = 0 \end{cases}$

Solve for x_3.

25. $\begin{cases} 3x_2 - x_3 + 2x_4 = 1 \\ 5x_1 + x_2 + 3x_3 - x_4 = -4 \\ x_1 - 2x_2 + 9x_4 = 5 \\ 2x_1 + 2x_3 = 3 \end{cases}$

Solve for x_4.

26. $\begin{cases} 4x_1 + x_2 - 3x_4 = 4 \\ 5x_1 + 2x_2 - 2x_3 + x_4 = 7 \\ x_1 - 3x_2 + 2x_3 - 2x_4 = -6 \\ 3x_3 + 4x_4 = -7 \end{cases}$

Solve for x_1.

Supplemental Exercises

27. A solution of the system of equations

$$\begin{cases} 2x_1 - 3x_2 + x_3 = 9 \\ x_1 + x_2 - 2x_3 = -3 \\ 4x_1 - x_2 - 3x_3 = 3 \end{cases}$$

is $(1, -2, 1)$. However, this solution can not be found by using Cramer's Rule. Explain.

28. Verify the solution for x_2 given in Cramer's Rule for a System of Two Equations, by solving the system of equations

$$\begin{cases} a_{11}x_1 + a_{12}x_2 = b_1 \\ a_{21}x_1 + a_{22}x_2 = b_2 \end{cases}$$

for x_2 by using the elimination method.

29. For what value of k does the system of equations

$$\begin{cases} kx + 3y = 7 \\ kx - 2y = 5 \end{cases}$$

have a unique solution?

30. For what values of k does the system of equations

$$\begin{cases} kx + 4y = 5 \\ 9x - ky = 2 \end{cases}$$

have a unique solution?

31. For what values of k does the system of equations

$$\begin{cases} x + 2y - 3z = 4 \\ 2x + ky - 4z = 5 \\ x - 2y + z = 6 \end{cases}$$

have a unique solution?

32. For what values of k does the system of equations

$$\begin{cases} kx_1 + x_2 = 1 \\ x_2 - 4x_3 = 1 \\ x_1 + kx_3 = 1 \end{cases}$$

have a unique solution?

33. Find real values for r and s so that $ru + sv = w$, where u, v, and w are complex numbers and $u = 2 + 3i$, $v = 4 - 2i$, and $w = -6 + 15i$.

34. Find real numbers for r and s such that $ru + sv = w$, where $u = 3 - 4i$, $v = 1 + 2i$, and $w = 4 - 22i$.

Chapter 11 Review

11.1   **Gauss-Jordan Elimination Method**

A matrix is a rectangular array of numbers. A matrix with m rows and n columns is of order $m \times n$ or dimension $m \times n$.

For a system of equations, it is possible to form a coefficient matrix, an augmented matrix, and a constant matrix.

Echelon Form

An augmented matrix is in echelon form if all the following conditions are satisfied:

1. The first nonzero number in any row is a 1.

2. Rows are arranged so that the column containing the first nonzero number is to the left of the column containing the first nonzero number of the next row.

3. All rows consisting entirely of zeros appear at the bottom of the matrix.

The Gauss-Jordan elimination method is used to solve a system of equations.

Elementary Row Operations

The elementary row operations for a matrix are:

1. Interchanging two rows

2. Multiplying all the elements in a row by the same nonzero number

3. Replacing a row by the sum of that row and a multiple of any other row

11.2   The Algebra of Matrices

Two matrices A and B are equal if and only if $a_{ij} = b_{ij}$ for every i and j.

The sum of two matrices of the same order is the matrix whose elements are the sum of the corresponding elements of the two matrices.

The $m \times n$ zero matrix is the matrix whose elements are all zeros.

The product of a real number and a matrix is called scalar multiplication.

To multiply two matrices, the number of columns of the first matrix must equal the number of rows of the second matrix.

In general, matrix multiplication is not commutative.

The multiplicative identity matrix is the matrix with 1s on the main diagonal and zeros everywhere else.

11.3   The Inverse of a Matrix

The multiplicative inverse of a square matrix A, denoted by A^{-1}, has the property that

$$A \cdot A^{-1} = A^{-1} \cdot A = I_n ,$$

where I_n is the multiplicative identity matrix.

A singular matrix is one that does not have a multiplicative inverse.

Input-output analysis attempts to determine the necessary output of industries to satisfy each other's demands plus the demands of consumers.

11.4   Determinants

Associated with each square matrix is a number called the determinant of the matrix.

The minor of the element a_{ij} of a square matrix is the determinant of the matrix obtained by deleting the ith row and jth column of A.

The cofactor of the element a_{ij} of a square matrix is $(-1)^{i+j}M_{ij}$, where M_{ij} is the minor of the element.

The value of a determinant can be found by multiplying the elements of any row or column by their respective cofactors and then adding the results. This is called expanding by cofactors.

11.5   Cramer's Rule

Cramer's Rule is a method of solving a system of n equations in n variables.

CHALLENGE EXERCISES

In Exercises 1 to 15, answer true or false. If the answer is false, give an example.

1. If $A = \begin{bmatrix} 2 & 3 \\ 1 & 4 \end{bmatrix}$, then $A^2 = \begin{bmatrix} 4 & 9 \\ 1 & 16 \end{bmatrix}$.

2. Every matrix has an additive inverse.

3. Every square matrix has a multiplicative inverse.

4. Let the matrices A, B, and C be square matrices of order n. If $AB = AC$, then $B = C$.

5. It is possible to find the determinant of every square matrix.

6. If A and B are square matrices of order n, then $\det(A + B) = \det(A) + \det(B)$.

7. Cramer's Rule can be used to solve any system of three equations in three variables.

8. If A and B are matrices of order n, then $AB - BA = 0$.

9. A nonsingular matrix has a multiplicative inverse.

10. If A, B, and C are square matrices of order n, then the product ABC depends on which two matrices are multiplied first. That is, $(AB)C$ produces a different result than $A(BC)$.

11. The Gauss-Jordon method for solving a system of linear equations can be applied only to systems of equations that have the same number of variables as equations.

12. If A is a square matrix of order n, then $\det(2A) = 2\det(A)$.

13. If A and B are matrices, then the product AB is defined when the number of columns of A equals the number of rows of B.

14. If A and B are square matrices of order n and $AB = 0$ (the zero matrix), then $A = 0$ or $B = 0$.

15. If $A = \begin{bmatrix} 3 & 6 \\ -1 & -2 \end{bmatrix}$, then $A^5 = A$.

REVIEW EXERCISES

In Exercises 1 to 18, perform the indicated operations. Let

$$A = \begin{bmatrix} 2 & -1 & 3 \\ 3 & 2 & -1 \end{bmatrix}, \quad B = \begin{bmatrix} 0 & -2 \\ 4 & 2 \\ 1 & -3 \end{bmatrix}, \quad C = \begin{bmatrix} 2 & 6 & 1 \\ 1 & 2 & -1 \\ 2 & 4 & -1 \end{bmatrix},$$

and $D = \begin{bmatrix} -3 & 4 & 2 \\ 4 & -2 & 5 \end{bmatrix}$.

1. $3A$
2. $-2B$
3. $-A + D$
4. $2A - 3D$
5. AB
6. DB
7. BA
8. BD
9. C^2
10. C^3
11. BAC
12. ADB
13. $AB - BA$
14. $DB - BD$
15. $(A - D)C$
16. $AC - DC$
17. C^{-1}
18. $|C|$

In Exercises 19 to 34, solve the system of equations by using the Gauss-Jordan method.

19. $\begin{cases} 2x - 3y = 7 \\ 3x - 4y = 10 \end{cases}$

20. $\begin{cases} 3x + 4y = -9 \\ 2x + 3y = -7 \end{cases}$

21. $\begin{cases} 4x - 5y = 12 \\ 3x + y = 9 \end{cases}$

22. $\begin{cases} 2x - 5y = 10 \\ 5x + 2y = 4 \end{cases}$

23. $\begin{cases} x + 2y + 3z = 5 \\ 3x + 8y + 11z = 17 \\ 2x + 6y + 7z = 12 \end{cases}$

24. $\begin{cases} x - y + 3z = 10 \\ 2x - y + 7z = 24 \\ 3x - 6y + 7z = 21 \end{cases}$

25. $\begin{cases} 2x - y - z = 4 \\ x - 2y - 2z = 5 \\ 3x - 3y - 8z = 19 \end{cases}$

26. $\begin{cases} 3x - 7y + 8z = 10 \\ x - 3y + 2z = 0 \\ 2x - 8y + 7z = 5 \end{cases}$

27. $\begin{cases} 4x - 9y + 6z = 54 \\ 3x - 8y + 8z = 49 \\ x - 3y + 2z = 17 \end{cases}$

28. $\begin{cases} 3x + 8y - 5z = 6 \\ 2x + 9y - z = -8 \\ x - 4y - 2z = 16 \end{cases}$

29. $\begin{cases} x + y + 2z = -5 \\ 2x + 3y + 5z = -13 \\ 2x + 5y + 7z = -19 \end{cases}$

30. $\begin{cases} x - 2y + 3z = 9 \\ 3x - 5y + 8z = 25 \\ x \qquad - z = 5 \end{cases}$

31. $\begin{cases} w + 2x - y + 2z = 1 \\ 3w + 8x + y + 4z = 1 \\ 2w + 7x + 3y + 2z = 0 \\ w + 3x - 2y + 5z = 6 \end{cases}$

32. $\begin{cases} w - 3x - 2y + z = -1 \\ 2w - 5x \qquad + 3z = 1 \\ 3w - 7x + 3y \qquad = -18 \\ 2w - 3x - 5y - 2z = -8 \end{cases}$

33. $\begin{cases} w + 3x + y - 4z = 3 \\ w + 4x + 3y - 6z = 5 \\ 2w + 8x + 7y - 5z = 11 \\ 2w + 5x \qquad - 6z = 4 \end{cases}$

34. $\begin{cases} w + 4x - 2y + 3z = 6 \\ 2w + 9x - y + 5z = 13 \\ w + 7x + 6y + 5z = 9 \\ 3w + 14x \qquad + 7z = 20 \end{cases}$

In Exercises 35 to 46, find the inverse, if it exists, of the given matrix.

35. $\begin{bmatrix} 2 & -2 \\ 3 & -2 \end{bmatrix}$

36. $\begin{bmatrix} 3 & 4 \\ 2 & 3 \end{bmatrix}$

37. $\begin{bmatrix} -2 & 3 \\ 2 & 4 \end{bmatrix}$

38. $\begin{bmatrix} 5 & -4 \\ 3 & 2 \end{bmatrix}$

39. $\begin{bmatrix} 1 & 2 & 1 \\ 2 & 6 & 4 \\ 3 & 8 & 6 \end{bmatrix}$

40. $\begin{bmatrix} 1 & -3 & 2 \\ 3 & -8 & 7 \\ 2 & -3 & 6 \end{bmatrix}$

41. $\begin{bmatrix} 3 & -2 & 7 \\ 2 & -1 & 5 \\ 3 & 0 & 10 \end{bmatrix}$

42. $\begin{bmatrix} 4 & 9 & -11 \\ 3 & 7 & -8 \\ 2 & 6 & -3 \end{bmatrix}$

43. $\begin{bmatrix} 1 & -1 & 2 & 3 \\ 2 & -1 & 6 & 5 \\ 3 & -1 & 9 & 6 \\ 2 & -2 & 4 & 7 \end{bmatrix}$

44. $\begin{bmatrix} 1 & 2 & -2 & 1 \\ 3 & 7 & -3 & 1 \\ 2 & 7 & 4 & 3 \\ 1 & 4 & 2 & 4 \end{bmatrix}$

45. $\begin{bmatrix} 3 & 7 & -1 & 8 \\ 2 & 5 & 0 & 5 \\ 3 & 6 & -4 & 8 \\ 2 & 4 & -4 & 4 \end{bmatrix}$

46. $\begin{bmatrix} 3 & 1 & 5 & -5 \\ 2 & 1 & 4 & -3 \\ 3 & 0 & 4 & -3 \\ 4 & 1 & 8 & 1 \end{bmatrix}$

In Exercises 47 to 50, solve the given system of equation for each set of constants. Use the inverse matrix method.

47. $\begin{cases} 3x + 4y = b_1 \\ 2x + 3y = b_2 \end{cases}$
 a. $b_1 = 2, b_2 = -3$
 b. $b_1 = -2, b_2 = 4$

48. $\begin{cases} 2x - 5y = b_1 \\ 3x - 7y = b_2 \end{cases}$
 a. $b_1 = -3, b_2 = 4$
 b. $b_1 = 2, b_2 = -5$

49. $\begin{cases} 2x + y - z = b_1 \\ 4x + 4y + z = b_2 \\ 2x + 2y - 3z = b_3 \end{cases}$
 a. $b_1 = -1, b_2 = 2, b_3 = 4$
 b. $b_1 = -2, b_2 = 3, b_3 = 0$

50. $\begin{cases} 3x - 2y + z = b_1 \\ 3x - y + 3z = b_2 \\ 6x - 4y + z = b_3 \end{cases}$
 a. $b_1 = 0, b_2 = 3, b_3 = -2$
 b. $b_1 = 1, b_2 = 2, b_3 = -4$

In Exercises 51 to 58, evaluate the determinants by using elementary row or column operations.

51. $\begin{vmatrix} 2 & 6 & 4 \\ 1 & 2 & 1 \\ 3 & 8 & 6 \end{vmatrix}$

52. $\begin{vmatrix} 3 & 0 & 10 \\ 3 & -2 & 7 \\ 2 & -1 & 5 \end{vmatrix}$

53. $\begin{vmatrix} 3 & -8 & 7 \\ 2 & -3 & 6 \\ 1 & -3 & 2 \end{vmatrix}$

54. $\begin{vmatrix} 4 & 9 & -11 \\ 2 & 6 & -3 \\ 3 & 7 & -8 \end{vmatrix}$

55. $\begin{vmatrix} 1 & -1 & 2 & 1 \\ 2 & -1 & 6 & 3 \\ 3 & -1 & 8 & 7 \\ 3 & 0 & 9 & 9 \end{vmatrix}$

56. $\begin{vmatrix} 1 & 2 & -2 & 3 \\ 3 & 7 & -3 & 11 \\ 2 & 3 & -5 & 11 \\ 2 & 6 & 1 & 8 \end{vmatrix}$

57. $\begin{vmatrix} 1 & 2 & -2 & 1 \\ 2 & 5 & -3 & 1 \\ 2 & 0 & -10 & 1 \\ 3 & 8 & -4 & 1 \end{vmatrix}$ **58.** $\begin{vmatrix} 1 & 3 & -2 & 0 \\ 3 & 11 & -4 & 4 \\ 2 & 9 & -8 & 2 \\ 3 & 12 & -10 & 2 \end{vmatrix}$

In Exercises 59 to 64, solve the system of equations by using Cramer's Rule.

59. $\begin{cases} 2x_1 - 3x_2 = 2 \\ 3x_1 + 5x_2 = 2 \end{cases}$ **60.** $\begin{cases} 3x_1 + 4x_2 = -3 \\ 5x_1 - 2x_2 = 2 \end{cases}$

61. $\begin{cases} 2x_1 + x_2 - 3x_3 = 2 \\ 3x_1 + 2x_2 + x_3 = 1 \\ x_1 - 3x_2 + 4x_3 = -2 \end{cases}$ **62.** $\begin{cases} 3x_1 + 2x_2 - x_3 = 0 \\ x_1 + 3x_2 - 2x_3 = 3 \\ 4x_1 - x_2 - 5x_3 = -1 \end{cases}$

63. $\begin{cases} 2x_2 + 5x_3 = 2 \\ 2x_1 - 5x_2 + x_3 = 4 \\ 4x_1 + 3x_2 = 2 \end{cases}$ **64.** $\begin{cases} 2x_1 - 3x_2 - 4x_3 = 2 \\ x_1 - 2x_2 + 2x_3 = -1 \\ 2x_1 + 7x_2 - x_3 = 2 \end{cases}$

In Exercises 65 and 66, solve for the indicated variable.

65. $\begin{cases} x_1 - 3x_2 + x_3 + 2x_4 = 3 \\ 2x_1 + 7x_2 - 3x_3 + x_4 = 2 \\ -x_1 + 4x_2 + 2x_3 - 3x_4 = -1 \\ 3x_1 + x_2 - x_3 - 2x_4 = 0 \end{cases}$

Solve for x_3.

66. $\begin{cases} 2x_1 + 3x_2 - 2x_3 + x_4 = -2 \\ x_1 - x_2 - 3x_3 + 2x_4 = 2 \\ 3x_1 + 3x_2 - 4x_3 - x_4 = 4 \\ 5x_1 - 5x_2 - x_3 + 2x_4 = 7 \end{cases}$

Solve for x_2.

In Exercises 67 and 68, solve the input-output problems.

67. An electronics conglomerate has three divisions, which produce computers, monitors, and disk drives. For each $1 of output, the computer division needs $.05 worth of computers, $.02 worth of monitors, and $.03 of disk drives. For each $1 of output, the monitor division needs $.06 worth of computers, $.04 worth of monitors, and $.03 worth of disk drives. For each $1 of output, the disk drive division requires $.08 worth of computers, $.04 worth of monitors, and $.05 worth of disk drives. If sales level estimates are $30 million for the computer division, $12 million for the monitor division, and $21 million for the disk drive division, at what level should each division produce to satisfy this demand?

68. A manufacturing conglomerate has three divisions, which produce paper, lumber, and prefabricated walls. For each $1 of output, the lumber division needs $.07 worth of lumber, $.03 worth of paper, and $.03 of prefabricated walls. For each $1 of output, the paper division needs $.04 worth of lumber, $.07 worth of paper, and $.03 worth of prefabricated wall. For each $1 of output, the prefabricated walls division requires $.07 worth of lumber, $.04 worth of paper, and $.02 worth of prefabricated walls. If sales estimates are $27 million for the lumber division, $18 million for the paper division, and $10 million for the prefabricated walls division, at what level should each division produce to satisfy this demand?

12 Sequences, Series, and Probability

Did you ever notice that remarkable coincidence? Bernard Shaw is 61 years old, H. G. Wells is 51, G. K. Chesterton is 41, you're 31 and I'm 21 — all the great authors of the world in arithmetic progression.

F. SCOTT FITZGERALD (1896–1940)
—in a letter to Shane Leslie

601

A SNOWFLAKE WITH INFINITE PERIMETER

We started this book with a discussion of infinities. Now that we have reached the last chapter, it seems appropriate to end with a discussion of infinity. We are going to construct a geometric snowflake that has an infinite perimeter.

We begin with an equilateral triangle with each side 1 unit long. The perimeter is three units. Now construct an identical but smaller triangle onto the middle third of each side. The new snowflake has twelve line segments each of length $\frac{1}{3}$ unit. The total perimeter is four ($12 \cdot \frac{1}{3}$) units. We now repeat the procedure again and construct identical smaller triangles on each of the 12 line segments. The result is forty-eight line segments each of length $\frac{1}{9}$. The perimeter of the snowflake is $\frac{48}{9}$.

If we were to continue this procedure again and again, it can be shown[1] that the perimeter of each succeeding snowflake is $\frac{4}{3}$ of the preceding perimeter. Thus, the perimeter continues to grow and grow without any bound. The perimeter becomes infinitely large.

Now let's examine the area of each snowflake. Let A be the area of the original triangle. In the second stage, each new triangle has an area that is $\frac{1}{9}$ of A. Since there are 3 new triangles, the area is now the original area plus the area of the 3 new triangles. The area of the snowflake is

$$A + 3\left(\frac{1}{9}\right)A = \frac{4}{3}A.$$

For the next snowflake, each new triangle has an area equal to $\frac{1}{81}A$. The total area of the 12 new triangles is

$$12\left(\frac{1}{81}A\right) = \frac{4}{27}A.$$

The area of the snowflake now is

$$\frac{4}{3}A + \frac{4}{27}A = \frac{40}{27}A.$$

At each stage after the first, the area added is $\frac{4}{9}$ the preceding area. Continuing in this way, we can show[2] that as we add more and more triangles, the area of the snowflake becomes closer and closer or $\frac{8}{5}A$. That is, the area is finite.

---

[1] The method we would use is called mathematical induction, a topic presented in this chapter.

[2] The result of adding all these triangles is an infinite geometric series, another topic of this chapter.

We seem to have a paradoxical situation, a snowflake with an infinite perimeter but a finite area! Later in this chapter, you can work an exercise that verifies the calculations we have shown.

12.1

Infinite Sequences and Summation Notation

The *ordered* list of numbers 2, 4, 8, 16, 32, ... is called an infinite sequence. The list is ordered simply because order makes a difference. The sequence 2, 8, 4, 16, 32, ..., contains the same numbers, but in a different order. Therefore it is a different infinite sequence.

An infinite sequence can be thought of as a pairing between positive numbers and real numbers. For example, the sequence 1, 4, 9, 16, 25, 36, ..., n^2... pairs a natural number with its square.

$$
\begin{array}{ccccccccc}
1 & 2 & 3 & 4 & 5 & 6 & \dots & n & \dots \\
\downarrow & \downarrow & \downarrow & \downarrow & \downarrow & \downarrow & & \downarrow & \\
1 & 4 & 9 & 16 & 25 & 36 & \dots & n^2 & \dots
\end{array}
$$

This pairing of numbers allows us to define an infinite sequence as a function with domain the natural numbers.

Infinite Sequence

An **infinite sequence** is a function whose domain is the positive integers and whose range is a set of real numbers.

Remark   Although the positive integers do not include zero, it is occasionally convenient to include zero in the domain of an infinite sequence. Also, we will frequently use the word "sequence" instead of the phrase "infinite sequence."

As an example of a sequence, let $f(n) = 2n - 1$. The range of this function is

$$f(1), f(2), f(3), f(4), \dots, \quad f(n), \quad \dots$$

$$1, \quad 3, \quad 5, \quad 7, \dots, \quad 2n - 1, \dots$$

The elements in the range of a sequence are called the **terms** of the sequence. For our example, the terms of the sequence are 1, 3, 5, 7, ..., $2n - 1$, ... The **first term** of the sequence is 1, the **second term** is 3, and so on. The *n*th **term** or **general term** is $2n - 1$.

Rather than use functional notation for sequences, it is customary to use a subscript notation. Thus a_n represents the *n*th term of a sequence.

Using this notation, we would write

$$a_n = 2n - 1.$$

Thus $a_1 = 1$, $a_2 = 3$, $a_3 = 5$, $a_4 = 7$.

EXAMPLE 1   Find the Terms of a Sequence

a. Find the first three terms of the sequence $a_n = \dfrac{1}{n(n + 1)}$;

b. Find the eighth term of the sequence $a_n = \dfrac{2^n}{n^2}$.

Solution

a. $a_1 = \dfrac{1}{1(1 + 1)} = \dfrac{1}{2}$, $a_2 = \dfrac{1}{2(2 + 1)} = \dfrac{1}{6}$, $a_3 = \dfrac{1}{3(3 + 1)} = \dfrac{1}{12}$

b. $a_8 = \dfrac{2^8}{8^2} = \dfrac{256}{64} = 4$

■ *Try Exercise* **6**, *page 608.*

An **alternating sequence** is one in which the signs of the terms *alternate* between positive and negative values. The sequence defined by $a_n = (-1)^{n+1} \cdot 1/n$ is an alternating sequence.

$$a_1 = (-1)^{1+1} \cdot \dfrac{1}{1} = 1 \qquad a_2 = (-1)^{2+1} \cdot \dfrac{1}{2} = -\dfrac{1}{2} \qquad a_3 = (-1)^{3+1} \cdot \dfrac{1}{3} = \dfrac{1}{3}$$

Some of the terms of the sequence are

$$1, -\dfrac{1}{2}, \dfrac{1}{3}, -\dfrac{1}{4}, \dfrac{1}{5}, -\dfrac{1}{6}.$$

A **recursively** defined sequence is one in which each succeeding term of the sequence is defined by using some of the the preceding terms. For example, let $a_1 = 1$, $a_2 = 1$, $a_{n+1} = a_{n-1} + a_n$.

$$a_3 = a_1 + a_2 = 1 + 1 = 2 \quad n = 2$$
$$a_4 = a_2 + a_3 = 1 + 2 = 3 \quad n = 3$$
$$a_5 = a_3 + a_4 = 2 + 3 = 5 \quad n = 4$$
$$a_6 = a_4 + a_5 = 3 + 5 = 8 \quad n = 5$$

This recursive sequence 1, 1, 2, 3, 5, 8, ... is called the Fibonacci sequence, named after Leonardo Fibonacci (1180?–?1250), an Italian mathematician.

EXAMPLE 2   Find Terms of a Sequence Defined Recursively

Let $a_1 = 1$, $a_n = na_{n-1}$. Find a_2, a_3, and a_4.

Solution

$$a_2 = 2a_1 = 2 \cdot 1 = 2 \qquad a_3 = 3a_2 = 3 \cdot 2 = 6 \qquad a_4 = 4a_3 = 4 \cdot 6 = 24$$

■ *Try Exercise 28, page 608.*

It is possible to find a nonrecursive formula for the general term a_n of the sequence defined recursively in Example 2 by

$$a_1 = 1 \qquad a_n = na_{n-1}.$$

Consider the term a_5 of that sequence.

$$a_5 = 5a_4$$
$$= 5 \cdot 4a_3$$
$$= 5 \cdot 4 \cdot 3a_2$$
$$= 5 \cdot 4 \cdot 3 \cdot 2a_1$$
$$= 5 \cdot 4 \cdot 3 \cdot 2 \cdot 1$$

Continuing in this manner for a_n, we have

$$a_n = na_{n-1}$$
$$= n(n - 1)a_{n-2}$$
$$= n(n - 1)(n - 2)a_{n-3}$$
$$\vdots$$
$$= n(n - 1)(n - 2)(n - 3) \cdots 2 \cdot 1$$

The quantity $n \cdot (n - 1) \cdots 3 \cdot 2 \cdot 1$ is called **n factorial** and is written $n!$.

The Factorial of a Number

If n is a positive integer, then $n!$, read "n factorial," is

$$n! = n \cdot (n - 1) \cdots 3 \cdot 2 \cdot 1$$

We also define

$$0! = 1.$$

Remark   It may seem strange to define $0! = 1$, but as we shall see later, it is a reasonable definition.

Examples of factorials include

$$5! = 5 \cdot 4 \cdot 3 \cdot 2 \cdot 1 = 120$$
$$10! = 10 \cdot 9 \cdot 8 \cdot 7 \cdot 6 \cdot 5 \cdot 4 \cdot 3 \cdot 2 \cdot 1 = 3,628,800$$
$$8! = 8 \cdot 7 \cdot 6 \cdot 5 \cdot 4 \cdot 3 \cdot 2 \cdot 1 = 40,320$$

Note that we can write $12!$ as

$$12! = 12 \cdot 11! = 12 \cdot 11 \cdot 10! = 12 \cdot 11 \cdot 10 \cdot 9!$$

In general,

$$n! = n \cdot (n - 1)!$$

EXAMPLE 3   Evaluate a Factorial Expression

Evaluate each factorial expresssion.   a. $\dfrac{8!}{5!}$ b. $6! - 4!$

Solution

a. $\dfrac{8!}{5!} = \dfrac{8 \cdot 7 \cdot 6 \cdot 5!}{5!} = 8 \cdot 7 \cdot 6 = 336$

b. $6! - 4! = (6 \cdot 5 \cdot 4 \cdot 3 \cdot 2 \cdot 1) - (4 \cdot 3 \cdot 2 \cdot 1) = 720 - 24 = 696$

■ *Try Exercise **64**, page 608.*

Another important way of obtaining a sequence is by adding the terms of a given sequence. For example, consider the sequence whose general term is given by $a_n = 1/2^n$. The terms of this sequence are

$$\frac{1}{2}, \frac{1}{4}, \frac{1}{8}, \frac{1}{16}, \frac{1}{32}, \cdots, \frac{1}{2^n}, \cdots$$

From this sequence we can generate a new sequence that is the sum of terms of $1/2^n$.

$$S_1 = \frac{1}{2}$$

$$S_2 = \frac{1}{2} + \frac{1}{4} = \frac{3}{4}$$

$$S_3 = \frac{1}{2} + \frac{1}{4} + \frac{1}{8} = \frac{7}{8}$$

$$S_4 = \frac{1}{2} + \frac{1}{4} + \frac{1}{8} + \frac{1}{16} = \frac{15}{16}$$

and, in general,

$$S_n = \frac{1}{2} + \frac{1}{4} + \frac{1}{8} + \frac{1}{16} + \cdots + \frac{1}{2^n}$$

The term S_n is called the **nth partial sum** of the infinite sequence, and the sequence $S_1, S_2, S_3, \ldots, S_n$ is called the **sequence of partial sums.**

A convenient notation used for partial sums is called **summation notation.** The sum of the first n terms of a sequence a_n is represented by using the Greek letter Σ (sigma).

$$\sum_{i=1}^{n} a_i = a_1 + a_2 + a_3 + \cdots + a_n$$

This sum is called a **series.** The letter i is called the **index of the summation;** n is the **upper limit** of the summation; 1 is the **lower limit** of the summation.

EXAMPLE 4 **Find the Value of a Series**

Evaluate the series   a. $\displaystyle\sum_{i=1}^{4} \frac{i}{i+1}$ b. $\displaystyle\sum_{j=2}^{5} (-1)^j j^2$

Solution

a. $\displaystyle\sum_{i=1}^{4} \frac{i}{i+1} = \frac{1}{2} + \frac{2}{3} + \frac{3}{4} + \frac{4}{5} = \frac{163}{60}$

b. $\displaystyle\sum_{j=2}^{5} (-1)^j j^2 = (-1)^2 2^2 + (-1)^3 3^2 + (-1)^4 4^2 + (-1)^5 5^2$

$$= 4 - 9 + 16 - 25 = -14$$

■ *Try Exercise* **44,** *page 608.*

Remark   Example 4 illustrates that the index of the sum does not have to be i; any letter can be used. Second, the lower limit of the summation does not have to be a 1.

Properties of Summation Notation

If a_n and b_n are sequences and c a real number, then

(1) $\displaystyle\sum_{i=1}^{n} (a_i \pm b_i) = \sum_{i=1}^{n} a_i \pm \sum_{i=1}^{n} b_i$

(2) $\displaystyle\sum_{i=1}^{n} ca_i = c \sum_{i=1}^{n} a_i$

(3) $\displaystyle\sum_{i=1}^{n} c = nc$

The proof of property (1) depends on the commutative and associative properties of real numbers.

$$\sum_{i=1}^{n} (a_i \pm b_i) = (a_1 \pm b_1) + (a_2 \pm b_2) + \cdots + (a_n \pm b_n)$$

$$= (a_1 + a_2 + \cdots + a_n) \pm (b_1 + b_2 + \cdots + b_n)$$

$$= \sum_{i=1}^{n} a_i \pm \sum_{i=1}^{n} b_i$$

Property (2) is proved by using the distributive property and is left as an exercise.

To prove property (3), let $a_n = c$. That is, each a_n is equal to the same constant c. (This is called a **constant sequence.**) Then

$$\sum_{i=1}^{n} a_n = a_1 + a_2 + \cdots + a_n$$

$$= c + c + \cdots + c = nc$$

EXERCISE SET 12.1

In Exercises 1 to 24, find the first three terms and the eighth term of the sequence that has the given nth term.

1. $a_n = n(n - 1)$

2. $a_n = 2n$

3. $a_n = 1 - \dfrac{1}{n}$

4. $a_n = \dfrac{n + 1}{n}$

5. $a_n = \dfrac{(-1)^{n+1}}{n^2}$

6. $a_n = \dfrac{(-1)^{n+1}}{n(n + 1)}$

7. $a_n = \dfrac{(-1)^{2n-1}}{3n}$

8. $a_n = \dfrac{(-1)^n}{2n - 1}$

9. $a_n = \left(\dfrac{2}{3}\right)^n$

10. $a_n = \left(\dfrac{-1}{2}\right)^n$

11. $a_n = 1 + (-1)^n$

12. $a_n = 1 + (-0.1)^n$

13. $a_n = (1.1)^n$

14. $a_n = \dfrac{n}{n^2 + 1}$

15. $a_n = \dfrac{(-1)^{n+1}}{\sqrt{n}}$

16. $a_n = \dfrac{3^{n-1}}{2^n}$

17. $a_n = n!$

18. $a_n = \dfrac{n!}{(n - 1)!}$

19. $a_n = \log n$

20. $a_n = \ln n$ (natural logarithm)

21. a_n is the digit in the nth place in the decimal expansion of $\frac{1}{7}$.

22. a_n is the digit in the nth place in the decimal expansion of $\frac{1}{13}$.

23. $a_n = 3$

24. $a_n = -2$

In Exercises 25 to 34, find the first three terms of the recursively defined sequences.

25. $a_1 = 5$, $a_n = 2a_{n-1}$

26. $a_1 = 2$, $a_n = 3a_{n-1}$

27. $a_1 = 2$, $a_n = na_{n-1}$

28. $a_1 = 1$, $a_n = n^2 a_{n-1}$

29. $a_1 = 2$, $a_n = (a_{n-1})^2$

30. $a_1 = 4$, $a_n = \dfrac{1}{a_{n-1}}$

31. $a_1 = 2$, $a_n = 2na_{n-1}$

32. $a_1 = 2$, $a_n = (-3)na_{n-1}$

33. $a_1 = 3$, $a_n = (a_{n-1})^{1/n}$

34. $a_1 = 2$, $a_n = (a_{n-1})^n$

35. $a_1 = 1$, $a_2 = 3$, $a_n = \dfrac{1}{2}(a_{n-1} + a_{n-2})$. Find a_3, a_4, and a_5.

36. $a_1 = 1$, $a_2 = 4$, $a_n = (a_{n-1})(a_{n-2})$. Find a_3, a_4, and a_5.

In Exercises 37 to 50, evaluate the series.

37. $\displaystyle\sum_{i=1}^{5} i$

38. $\displaystyle\sum_{i=1}^{4} i^2$

39. $\displaystyle\sum_{i=1}^{5} i(i - 1)$

40. $\displaystyle\sum_{i=1}^{7} (2i + 1)$

41. $\displaystyle\sum_{k=1}^{4} \dfrac{1}{k}$

42. $\displaystyle\sum_{k=1}^{6} \dfrac{1}{k(k + 1)}$

43. $\displaystyle\sum_{j=1}^{8} 2j$

44. $\displaystyle\sum_{i=1}^{6} (2i + 1)(2i - 1)$

45. $\displaystyle\sum_{i=3}^{5} (-1)^i 2^i$

46. $\displaystyle\sum_{i=3}^{5} \dfrac{(-1)^i}{2^i}$

47. $\displaystyle\sum_{n=1}^{7} \log\left(\dfrac{n + 1}{n}\right)$

48. $\displaystyle\sum_{n=2}^{8} \ln\left(\dfrac{n}{n + 1}\right)$

49. $\displaystyle\sum_{k=0}^{8} \dfrac{8!}{k!(8 - k)!}$

50. $\displaystyle\sum_{k=0}^{7} \dfrac{1}{k!}$

In Exercises 51 to 58, write the given series in summation notation.

51. $\dfrac{1}{1} + \dfrac{1}{4} + \dfrac{1}{9} + \dfrac{1}{16} + \dfrac{1}{25} + \dfrac{1}{36}$

52. $2 + 4 + 6 + 8 + 10 + 12 + 14$

53. $2 - 4 + 8 - 16 + 32 - 64 + 128$

54. $1 - 8 + 27 - 64 + 125$

55. $7 + 10 + 13 + 16 + 19$

56. $30 + 26 + 22 + 18 + 14 + 10$

57. $\dfrac{1}{2} + \dfrac{1}{4} + \dfrac{1}{8} + \dfrac{1}{16}$

58. $1 - \dfrac{2}{3} + \dfrac{4}{9} - \dfrac{8}{27} + \dfrac{16}{81} - \dfrac{32}{243}$

In Exercises 59 to 66, evaluate the factorial expression.

59. $7! - 6!$

60. $(4!)^2$

61. $\dfrac{9!}{7!}$

62. $\dfrac{10!}{5!}$

63. $\dfrac{8!}{3!5!}$

64. $\dfrac{12!}{4!8!}$

65. $\dfrac{100!}{99!}$

66. $\dfrac{100!}{98!2!}$

Supplemental Exercises

67. Newton's approximation to the square root of a number is given by the recursive sequence

$$a_1 = \dfrac{N}{2} \qquad a_n = \dfrac{1}{2}\left(a_{n-1} + \dfrac{N}{a_{n-1}}\right).$$

Approximate $\sqrt{7}$ by computing a_4. Compare this result with the calculator value of $\sqrt{7} \approx 2.6457513$.

68. Use the formula in Exercise 67 to approximate $\sqrt{10}$ by finding a_5.

69. Let $a_1 = N$ and $a_n = \sqrt{a_{n-1}}$. Using a calculator, find a_{20} when $N = 7$. (*Hint:* Enter seven into your calculator and then press the $\boxed{\sqrt{}}$ key nineteen times. Make a conjecture as to the value of a_{100}.)

70. Using a calculator (or computer), evaluate $\sum\limits_{k=1}^{n} \dfrac{(-1)^{k+1}}{2k-1}$ for $n = 20$. Compare your answer to $\pi/4$. As n increases, the series better approximates $\pi/4$.

71. Let $a_1 = 1$ and $a_n = na_{n-1}$. Evaluate $1 + \sum\limits_{k=1}^{4} \dfrac{1}{a_k}$. Compare your answer to the value of the e, the base of the natural logarithm. As n increases, the series better approximates e.

72. It is not possible to define a sequence by giving a finite number of terms of the sequence. For example, the question "What is the next term in the sequence 2, 4, 6, 8, ...?" does not have a unique answer. Verify this statement by finding a formua for a_n such that the first four terms of the sequence are 2, 4, 6, 8 and the next term is 43.

73. Extend the result of Exercise 72 by finding a formula for a_n that will give the first four terms as 2, 4, 6, 8 and the fifth term x, where x is any real number.

74. Let $a_n = i^n$, where i is the imaginary unit. Find the first eight terms of the sequence defined by a_n. Find a_{237}.

75. Let $a_n = \left[\dfrac{1}{2}(-1 + i\sqrt{3})\right]^n$. Find the first six terms of the sequence defined by a_n. Find a_{99}.

76. By using a calculator, evaluate $\sqrt{2\pi n}\left(\dfrac{n}{e}\right)^n$, where e is the base of the natural logarithms for $n = 10$, 20, and 30. This formula is called Stirling's formula and is used as an approximation for $n!$. For $n > 20$, the error in the approximation is less than 0.1 percent.

77. Prove that $\sum\limits_{i=1}^{n} ca_i = c\sum\limits_{i=1}^{n} a_i$.

12.2
Arithmetic Sequences and Series

In the sequence

$$2, 5, 8, 11, 14, \ldots, 3n - 1, \ldots,$$

notice that the difference between successive terms is always 3. Such a sequence is an *arithmetic sequence* or an *arithmetic progression*. These sequences have the following property: the difference between successive terms is the same constant. This constant is called the *common difference*. For the sequence above, the common difference is 3.

In general, an arithmetic sequence can be defined as follows.

Arithmetic Sequence

Let d be a real number. A sequence a_n is an **arithmetic sequence** if

$$a_{i+1} - a_i = d \quad \text{for all } i.$$

The number d is the **common difference** for the sequence.

Further examples of arithmetic sequences and their common difference include

3, 8, 13, 18, . . . , $5n - 2$, . . .   Common difference $= 5$.

11, 7, 3, -1, . . . , $-4n + 15$, . . .   Common difference $= -4$.

1, 2, 3, 4, . . . , n, . . .   Common difference $= 1$.

Consider an arithmetic sequence in which the first term is a_1 and the common difference is d. By adding the common difference to each successive term of the arithmetic sequence, a formula for the nth term can be found.

$$a_1 = a_1$$
$$a_2 = a_1 + d$$
$$a_3 = a_2 + d = a_1 + d + d = a_1 + 2d$$
$$a_4 = a_3 + d = a_1 + 2d + d = a_1 + 3d$$

Note the relationship between the term number and the multiplier of d. The multiplier is one less than the term number.

Formula for the nth Term of an Arithmetic Sequence

> The **nth term of an arithmetic sequence** with common difference of d is given by
>
> $$a_n = a_1 + (n - 1)d .$$

EXAMPLE 1   Find the nth Term of an Arithmetic Sequence

a. Find the twenty-fifth term of the arithmetic sequence whose first three terms are -12, -6, 0, . . .

b. The fifteenth term of an arithmetic sequence is -3 and the first term is 25. Find the tenth term.

Solution

a. Find the common difference: $d = a_2 - a_1 = -6 - (-12) = 6$. Use the formula $a_n = a_1 + (n - 1)d$ with $n = 25$.

$$a_{25} = -12 + (25 - 1)(6)$$
$$= -12 + 24(6) = -12 + 144 = 132$$

b. Solve the equation $a_n = a_1 + (n - 1)d$ for d, given that $n = 15$, $a_1 = 25$, and $a_{15} = -3$.

$$-3 = 25 + (14)d$$
$$d = -2$$

Now find the tenth term.

$$a_{10} = 25 + (9)(-2) = 7$$

■ *Try Exercise **16**, page 614.*

The **arithmetic mean** of two numbers a and b is $(a + b)/2$. The three numbers a, $(a + b)/2$, b form an arithmetic sequence. In general, given two numbers a and b, it is possible to insert k numbers in such a way that the sequence

$$a, c_1, c_2, \ldots, c_k, b$$

is an arithmetic sequence. This is called *inserting k arithmetic means between a and b*.

EXAMPLE 2   Insert Arithmetic Means

Insert three arithmetic means between 3 and 13.

Solution   After inserting the three terms, the sequence will be

$$a = 3, c_2, c_3, c_4, b = 13$$

The first term of the sequence is 3, the fifth term is 13, and n is 5. Thus,

$$a_n = a_1 + (n - 1)d$$
$$13 = 3 + 4d$$
$$d = \frac{5}{2}$$

The three arithmetic means are

$$c_1 = a + d = 3 + \frac{5}{2} = \frac{11}{2}$$
$$c_2 = a + 2d = 3 + 2\left(\frac{5}{2}\right) = 8$$
$$c_3 = a + 3d = 3 + 3\left(\frac{5}{2}\right) = \frac{21}{2}$$

■ *Try Exercise 36, page 614.*

Consider the arithmetic sequence given by

$$1, 3, 5, \ldots, 2n - 1, \ldots$$

Adding successive terms of this sequence, we generate a sequence of partial sums. The sum of the terms of an arithmetic sequence is called an **arithmetic series**.

$$S_1 = 1$$
$$S_2 = 1 + 3 = 4$$
$$S_3 = 1 + 3 + 5 = 9$$
$$S_4 = 1 + 3 + 5 + 7 = 16$$
$$S_5 = 1 + 3 + 5 + 7 + 9 = 25$$
$$\vdots \qquad \vdots$$
$$S_n = 1 + 3 + \cdots + (2n - 1)$$

The first five terms of this sequence are 1, 4, 9, 16, 25. It appears from this example that the sum of the first n odd integers is n^2. Shortly, we will be able to prove this result by using the following formula.

Formula for nth Partial Sum of an Arithmetic Sequence

> The **nth partial sum of an arithmetic sequence** a_n with common difference d is
>
> $$S_n = \frac{n}{2}(a_1 + a_n).$$

Proof: We write S_n in both forward and reverse order.

$$S_n = a_1 + a_2 + a_3 + \cdots + a_{n-2} + a_{n-1} + a_n$$
$$S_n = a_n + a_{n-1} + a_{n-2} + \cdots + a_3 + a_2 + a_1$$

Add the two partial sums

$$2S_n = (a_1 + a_n) + (a_2 + a_{n-1}) + (a_3 + a_{n-2}) + \cdots$$
$$+ (a_{n-2} + a_3) + (a_{n-1} + a_2) + (a_n + a_1)$$

Consider the term $(a_3 + a_{n-2})$. Using the formula for the nth term of an arithmetic sequence, we have

$$a_3 = a_1 + (3 - 1)d = a_1 + 2d$$
$$a_{n-2} = a_1 + [(n - 2) - 1]d = a_1 + nd - 3d$$

Thus,

$$a_3 + a_{n-2} = (a_1 + 2d) + (a_1 + nd - 3d)$$
$$= a_1 + (a_1 + nd - d) = a_1 + [a_1 + (n - 1)d]$$
$$= a_1 + a_n$$

In a similar manner, we can show that each term of $2S_n$ equals $(a_1 + a_n)$. Since there are n such terms, we have

$$2S_n = n(a_1 + a_n)$$
$$S_n = \frac{n}{2}(a_1 + a_n).$$

There is an alternate form for the formula for the sum of n terms of an arithmetric sequence.

Alternate Formula for the Sum of an Arithmetic Series

> If a_n is an arithmetic sequence, then the **nth partial sum, S_n,** is
>
> $$S_n = \frac{n[2a_1 + (n - 1)d]}{2}$$

The proof of this theorem is left as an exercise.

EXAMPLE 3   Find a Partial Sum of an Arithmetic Sequence

a.   Find the sum of the first 100 positive odd integers

b.   Find the sum of the first 50 terms of the arithmetic sequence whose first three terms are $2, \dfrac{13}{4}$, and $\dfrac{9}{2}$.

Solution   Use the formula $S_n = \dfrac{n}{2}[2a_1 + (n - 1)d]$.

a.   We have $a_1 = 1$, $d = 2$, and $n = 100$. Thus

$$S_n = \frac{100}{2}[2(1) + (100 - 1)2] = 10{,}000$$

b.   We have $a_1 = 2$, $d = \dfrac{5}{4}$, and $n = 50$. Thus

$$S_n = \frac{50}{2}\left[2(2) + (50 - 1)\frac{5}{4}\right] = \frac{6525}{4}.$$

■ *Try Exercise* **22,** *page 614.*

The first n positive integers $1, 2, 3, 4, \ldots, n$ are part of an arithmetic sequence with a common difference of 1, $a_1 = 1$, and $a_n = n$. A formula for the sum of the first n positive integers can be found by using the formula for the nth partial sum of an arithmetic sequence.

$$S_n = \frac{n}{2}(a_1 + a_n)$$

Thus

$$S_n = \frac{n}{2}(1 + n) = \frac{n(n + 1)}{2}.$$

This proves the following theorem.

Sum of the First n Positive Integers

The sum of the first n positive integers is given by

$$S_n = \frac{n(n + 1)}{2}.$$

To find the sum of the first eighty-five positive integers, use $n = 85$.

$$S_{85} = \frac{85(85 + 1)}{2} = 3655$$

EXERCISE SET 12.2

In Exercises 1 to 14, find the ninth, twenty-fourth, and nth term of the arithmetic sequence.

1. 6, 10, 14, . . .

2. 7, 12, 17, . . .

3. 6, 4, 2, . . .

4. 11, 4, −3, . . .

5. −8, −5, −2, . . .

6. −15, −9, −3, . . .

7. 1, 4, 7, . . .

8. −4, 1, 6, . . .

9. a, $a + 2$, $a + 4$, . . .

10. $a − 3$, $a + 1$, $a + 5$, . . .

11. log 7, log 14, log 28, . . .

12. ln 4, ln 16, ln 64, . . .

13. $\log a$, $\log a^2$, $\log a^3$, . . .

14. $\log_2 5$, $\log_2 5a$, $\log_2 5a^2$, . . .

15. The fourth and fifth terms of an arithmetic sequence are 13 and 15. Find the twentieth term.

16. The sixth and eighth terms of an arithmetic sequence are −14 and −20. Find the fifteenth term.

17. The fifth and seventh terms of an arithmetic sequence are −19 and −29. Find the seventeenth term.

18. The fourth and seventh terms of an arithmetic sequence are 22 and 34. Find the twenty-third term.

In Exercises 19 to 32, find the nth partial sum of the arithmetic sequence.

19. $a_n = 3n + 2$; $n = 10$

20. $a_n = 4n − 3$; $n = 12$

21. $a_n = 3 − 5n$; $n = 15$

22. $a_n = 1 − 2n$; $n = 20$

23. $a_n = 6n$; $n = 12$

24. $a_n = 7n$; $n = 14$

25. $a_n = n + 8$; $n = 25$

26. $a_n = n − 4$; $n = 25$

27. $a_n = −n$; $n = 30$

28. $a_n = 4 − n$; $n = 40$

29. $a_n = n + x$; $n = 12$

30. $a_n = 2n − x$; $n = 15$

31. $a_n = nx$; $n = 20$

32. $a_n = −nx$; $n = 14$

33. Show that the sum of the first n positive odd integers is n^2.

34. Show that the sum of the first n positive even integers is $n^2 + n$.

In Exercises 35 to 38, insert k arithmetic means between the given numbers.

35. −1 and 23; $k = 5$

36. 7 and 19; $k = 5$

37. 3 and $\dfrac{1}{2}$; $k = 4$

38. $\dfrac{11}{3}$ and 6; $k = 4$

39. Logs are stacked so that there are twenty-five logs on the bottom row, twenty-four logs on the second row, and so on, decreasing by one log each row. How many logs are stacked on the sixth row? How many logs are in the six rows?

40. The seating section in a theater has twenty-seven seats in the first row, twenty-nine seats in the second row, and so on increasing, by two seats each row for a total of ten rows. How many seats are in the tenth row, and how many seats are there in the section?

41. A contest offers fifteen prizes. The first prize is $5000, and each successive prize is $250 less than the preceding prize. What is the value of the fifteenth prize? What is the total amount of money distributed in prizes?

42. An exercise program calls for walking 15 minutes each day for a week. Each week thereafter, the amount of time spent walking increases by 5 minutes per day. In how many weeks will a person be walking 60 minutes each day?

43. An object dropped from a cliff will fall 16 feet the first second, 48 feet the second second, 80 feet the third second, and so on, increasing by 32 feet each second. What is the total distance the object will fall in 7 seconds?

44. The distance a ball rolls down a ramp each second is given by the arithmetic sequence whose nth term is $2n − 1$ feet. Find the distance the ball rolls during the tenth second and the total distance the ball travels in 10 seconds.

Supplemental Exercises

45. The sum of the interior angles of a triangle is 180°, the sum is 360° for a quadralateral, 540° for a pentagon, and so on. Assuming this pattern continues, find the sum of the interior angles for a dodecagon (twelve sides).

46. If $f(x)$ is a linear function, show that $f(n)$, where n is a positive integer, is an arithmetic sequence.

47. Find the formula for a_1 in terms of a_1 and n for the sequence that is defined recursively by $a_1 = 3$, $a_n = a_{n−1} + 5$.

48. Find a formula for a_1 in terms of a_1 and n for the sequence that is defined recursively by $a_1 = 4$, $a_n = a_{n−1} − 3$.

49. Suppose a_n and b_n are two sequences such that $a_1 = 4$, $a_n = b_{n−1} + 5$ and $b_1 = 2$, $b_n = a_{n−1} + 1$. Show that a_n and b_n are arithmetic sequences. Find a_{100}.

50. Suppose a_n and b_n are two sequences such that $a_1 = 1$, $a_n = b_{n−1} + 7$ and $b_1 = −2$, $b_n = a_{n−1} + 1$. Show that a_n and b_n are arithmetic sequences. Find a_{50}.

51. Prove the Alternate Formula for the Sum of an Arithmetic Series Theorem.

12.3

Geometric Sequences and Series

Arithmetic sequences are characterized by a common *difference* between successive terms. In this section we discuss a *geometric sequence* that is characterized by a common *ratio* between successive terms.

The sequence

$$3, 6, 12, 24, \ldots, 3(2^{n-1}), \ldots$$

is a geometric sequence. Notice that the ratio of any two successive terms is 2.

$$\frac{6}{3} = 2 \qquad \frac{12}{6} = 2 \qquad \frac{24}{12} = 2.$$

Geometric Sequence

> Let r be a constant real number. A sequence is a **geometric sequence** if
>
> $$\frac{a_{i+1}}{a_i} = r \quad \text{for all } i \text{ and } r \neq 0.$$

EXAMPLE 1 **Determine If a Sequence Is a Geometric Sequence**

Which of the following are geometric sequences?

a. $4, -2, 1, \ldots, 4\left(-\frac{1}{2}\right)^{n-1}, \ldots$

b. $1, 4, 9, \ldots, n^2, \ldots$

Solution To determine if the sequence is a geometric sequence, calculate the ratio of successive terms.

a.

$$\frac{a_{i+1}}{a_i} = \frac{4\left(-\dfrac{1}{2}\right)^i}{4\left(-\dfrac{1}{2}\right)^{i-1}} = -\frac{1}{2}.$$ Because the ratio of successive terms is a constant, the sequence is a geometric sequence.

b.

$$\frac{a_{i+1}}{a_i} = \frac{(i+1)^2}{i^2} = \left(1 + \frac{1}{i}\right)^2$$ Because the ratio of successive terms is not a constant, the sequence is not a geometric sequence.

■ *Try Exercise* **6,** *page 621.*

Consider a geometric sequence in which the first term is a_1 and the common ratio is r. By multiplying each successive term of the geometric sequence by the common ratio, we can derive a formula for the nth term.

$$a_1 = a_1$$

$$a_2 = a_1 r$$

$$a_3 = a_2 r = (a_1 r) r = a_1 r^2$$

$$a_4 = a_3 r = (a_1 r^2) r = a_1 r^3$$

Note the relationship between the number of the term and the number that is the exponent on r. The exponent on r is one less than the number of the term. With this observation, we can write a formula for the nth term of a geometric sequence.

nth Term of a Geometric Sequence

> The **nth term of a geometric sequence** with first term a_1 and common ratio r is
>
> $$a_n = a_1 r^{n-1}.$$

EXAMPLE 2   Find the nth Term of a Geometric Sequence

Find the nth term of the geometric sequence whose first three terms are
a. 4, 8/3, 16/9, . . .   b. 5, −10, 20, . . .

Solution

a. $r = \dfrac{8/3}{4} = \dfrac{2}{3}$ and $a_1 = 4$. Thus $a_n = 4\left(\dfrac{2}{3}\right)^{n-1}$.

b. $r = \dfrac{-10}{5} = -2$ and $a_1 = 5$. Thus $a_n = 5(-2)^{n-1}$.

■ *Try Exercise* **18,** *page 621.*

Adding the terms of a geometric sequence, we can define the nth partial sum of a geometric sequence in a manner similar to that of an arithmetic sequence. Consider the geometric sequence 1, 2, 4, 8, . . . , 2^{n-1}, . . .

$$S_1 = 1$$

$$S_2 = 1 + 2 = 3$$

$$S_3 = 1 + 2 + 4 = 7$$

$$S_4 = 1 + 2 + 4 + 8 = 15$$

$$\vdots \qquad \vdots$$

$$S_n = 1 + 2 + \cdots + 2^{n-1}$$

The first four terms of the sequence of partial sums are 1, 3, 7, 15.

To find a general formula for S_n, the nth term of the sequence of partial sums of a geometric sequence, let

$$S_n = a + ar + ar^2 + \cdots + ar^{n-1}.$$

Multiply each side of this equation by r.

$$S_n = a_1 + a_1r + a_1r^2 + \cdots a_1r^{n-2} + a_1r^{n-1}$$

$$rS_n = \qquad a_1r + a_1r^2 + \cdots + \qquad a_1r^{n-1} + a_1r^n$$

Subtract the two equations.

$$S_n - rS_n = a_1 - a_1r^n$$

$$S_n(1 - r) = a_1(1 - r^n)$$

If $r \neq 1$, then

$$S_n = \frac{a_1(1 - r^n)}{1 - r}.$$

This proves the following theorem.

The nth Partial Sum of a Geometric Sequence

> The nth partial sum of a geometric sequence with first term a_1 and common ratio r is
>
> $$S_n = \frac{a_1(1 - r^n)}{1 - r} \quad r \neq 1.$$

Remark   If $r = 1$, then the sequence is given by $a_1, a_1, a_1, a_1, \ldots, a_1, \ldots,$ and the nth partial sum equals na_1. Why?[3]

EXAMPLE 3   Find the nth Partial Sum of a Geometric Sequence

Find the nth partial sum of the given geometric sequences.

a. $5, 15, 45, \ldots, 5(3)^{n-1}, \ldots; n = 4$ b. $\displaystyle\sum_{n=1}^{17} 3\left(\frac{3}{4}\right)^{n-1}$

Solution

a.   We have $a_1 = 5$, $r = 3$, and $n = 4$. Thus

$$S_4 = \frac{5[1 - 3^4]}{1 - 3} = \frac{5(-80)}{-2} = 200$$

---

[3] When $r = 1$, the sequence is the constant sequence a_1. The nth partial sum of a constant sequence is na_1.

b.   When $n = 1$, $a_1 = 3$. The first term is 3. The second term is $\dfrac{9}{4}$, there-

fore the common ratio is $r = \dfrac{3}{4}$. Thus $S_{17} = \dfrac{3[1 - (3/4)^{17}]}{1 - (3/4)} \approx 11.909797$.

■ *Try Exercise **40**, page 621.*

Infinite Geometric Series

Following are two examples of geometric sequences for which $|r| < 1$:

$$3, \frac{3}{4}, \frac{3}{16}, \frac{3}{64}, \frac{3}{256}, \frac{3}{1024}, \cdots \qquad r = \frac{1}{4}$$

$$2, -1, \frac{1}{2}, -\frac{1}{4}, \frac{1}{8}, -\frac{1}{16}, \frac{1}{32}, \cdots \qquad r = -\frac{1}{2}$$

Note that when the absolute value of the common ratio of a geometic sequence is less than 1, the terms of the geometric sequence approach zero as n increases. We write for $|r| < 1$, $|r|^n \to 0$ as $n \to \infty$.

Consider again the geometric sequence

$$3, \frac{3}{4}, \frac{3}{16}, \frac{3}{64}, \frac{3}{256}, \frac{3}{1024}, \cdots$$

The nth partial sums for $n = 3$, 6, 9, and 12 are given in the following table, along with the value of r^n.

n	S_n	r^n
3	3.93750000	0.01562500
6	3.99902344	0.00024414
9	3.99998474	0.00000381
12	3.99999976	0.00000006

Note that as n increases, S_n is closer to 4 and r^n is closer to zero. By finding more values of S_n for larger values of n, we would find that $S_n \to 4$ as $n \to \infty$. As n becomes larger and larger, S_n is the sum of more and more terms of the sequence. The sum of *all* the terms of a sequence is called an **infinite series**. If the sequence is a geometric sequence, we have an **infinite geometric series**.

Sum of an Infinite Geometric Sequence

If a_n is a geometric sequence with $|r| < 1$ and first term a_1, then the value of the infinite geometric series is

$$S = \frac{a_1}{1 - r}.$$

A formal proof of this formula requires topics that typically are studied in calculus. We can, however, give an intuitive argument.

Let
$$S_n = \frac{a_1(1 - r^n)}{1 - r}.$$

When $|r| < 1$, $|r|^n \approx 0$ when n is large. Thus

$$S_n = \frac{a_1(1 - r^n)}{1 - r} \approx \frac{a_1(1 - 0)}{1 - r} = \frac{a_1}{1 - r}.$$

An infinite series is represented by $\sum\limits_{n=1}^{\infty} a_n$. One application of infinite geometric series concerns repeating decimals. Consider the repeating decimal

$$0.\overline{6} = \frac{6}{10} + \frac{6}{100} + \frac{6}{1000} + \frac{6}{10,000} + \cdots.$$

The right-hand side is a geometric series with $a_1 = \frac{6}{10}$ and common ratio $r = \frac{1}{10}$. Thus

$$S = \frac{6/10}{1 - (1/10)} = \frac{6/10}{9/10} = \frac{2}{3}.$$

The repeating decimal $0.\overline{6} = \frac{2}{3}$. Any repeating decimal can be written as a ratio of two integers by using an infinite geometric series.

EXAMPLE 4   Find the Value of an Infinite Geometric Series

a. Evaluate the infinite geometric series $\sum\limits_{n=0}^{\infty} \left(-\frac{2}{3}\right)^n$.

b. Write $0.3\overline{45}$ as the ratio of two integers in lowest terms.

Solution

a. To find the first term, we let $n = 0$. Then $a_1 = \left(-\frac{2}{3}\right)^0 = 1$. The common ratio $r = -\frac{2}{3}$. Thus,

$$S = \frac{1}{1 - (-2/3)} = \frac{1}{(5/3)} = \frac{3}{5}.$$

b. $$0.3\overline{45} = \frac{3}{10} + \left[\frac{45}{1000} + \frac{45}{100,000} + \frac{45}{10,000,000} + \cdots\right]$$

The terms in the brackets form an infinite geometric series. Evaluate that series with $a_1 = \frac{45}{1000}$ and $r = \frac{1}{100}$, and then add the term $\frac{3}{10}$.

$$\frac{45}{1000} + \frac{45}{100,000} + \frac{45}{10,000,000} + \cdots = \frac{45/1000}{1 - (1/100)} = \frac{1}{22}$$

$$0.3\overline{45} = \frac{3}{10} + \frac{1}{22} = \frac{19}{55}$$

■ *Try Exercise **62**, page 622.*

Caution The sum of an infinite geometric series is not defined for $|r| > 1$. For example, the geometric series

$$2 + 4 + 8 + 16 + 32 + \cdots$$

increases without bound. However, applying the formula $S = a_1/(1 - r)$ with $r = 2$ produces $S = -2$ which is not correct.

In an earlier chapter we discussed compound interest by using exponential functions. As an extension of this idea, suppose that A dollars is deposited on December 31 for each of the next 5 years into an account earning i percent annual interest compounded annually. Using the compound interest formula, we can find the total value of all the deposits. The table shows the growth of the investment.

Deposit Number	Value of Each Deposit of $A after 4 Years	
1	$A(1 + i)^4$	Value of first deposit after 4 years
2	$A(1 + r)^3$	Value of second deposit after 3 years
3	$A(1 + r)^2$	Value of third deposit after 2 years
4	$A(1 + r)$	Value of fourth deposit after 1 year
5	A	Value of fifth deposit

The total value of the investment after the last deposit, called the **future value** of the investment, is the sum of the values of each deposit.

$$P_5 = A + A(1 + i) + A(1 + i)^2 + A(1 + i)^3 + A(1 + i)^4$$

This is a geometric series with first term A and common ratio $1 + i$. Thus, using the formula for the nth partial sum of a geometric sequence

$$S = \frac{a_1(1 - r^n)}{1 - r},$$

we have

$$P_5 = \frac{A[1 - (1 + i)^5]}{1 - (1 + i)} = \frac{A[(1 + i)^5 - 1]}{i}.$$

Deposits of equal amounts at equal intervals of time are called **annuities.** When the amounts are deposited at the end of a compounding period (as in our example), we have an **ordinary annuity.**

Future Value of an Ordinary Annuity

Let $r = i/n$ and $m = nt$, where i is the annual interest rate, n is the number of compounding periods per year, and t is the number of years. Then the future value of an ordinary annuity is given by

$$P = \frac{A[(1 + r)^m - 1]}{r}.$$

EXAMPLE 5   **Find the Future Value of an Annuity**

An employee savings plan allows any employee to deposit $25 per month into a savings account earning 6 percent annual interest compounded monthly. Find the future value of this savings plan if an employee makes the deposits for 10 years.

Solution   We are given $A = 25$, $i = 0.06$, $n = 12$, and $t = 10$. Thus,

$$r = \frac{i}{n} = \frac{0.06}{12} = 0.005 \quad \text{and} \quad m = nt = 12(10) = 120.$$

$$P = \frac{25[(1.005)^{120} - 1]}{0.005} \approx 4096.9837.$$

The future value is $4096.98.

■ *Try Exercise 70, page 622.*

EXERCISE SET 12.3

In Exercises 1 to 12, determine which sequences are geometric. For geometric sequences, find the common ratio.

1. $4, 16, 64, \ldots, 4^n, \ldots$

2. $1, 6, 36, \ldots, 6^{n-1}, \ldots$

3. $1, \frac{1}{2}, \frac{1}{3}, \ldots, \frac{1}{n}, \ldots$

4. $\frac{1}{2}, \frac{1}{4}, \frac{1}{8}, \ldots, \frac{1}{2^n}, \ldots$

5. $2^x, 2^{2x}, 2^{3x}, \ldots, 2^{nx}, \ldots$

6. $e^x, -e^{2x}, e^{3x}, \ldots, (-1)^{n-1}e^{nx}, \ldots$

7. $3, 6, 12, \ldots, 3(2^{n-1}), \ldots$

8. $5, -10, 20, \ldots, 5(-2)^{n-1}, \ldots$

9. $x^2, x^4, x^6, \ldots, x^{2n}, \ldots$

10. $3x, 6x^2, 9x^3, \ldots, 3nx^n, \ldots$

11. $\ln 5, \ln 10, \ln 15, \ldots, \ln 5n, \ldots$

12. $\log x, \log x^2, \log x^4, \ldots, \log x^{2^{n-1}}, \ldots$

In Exercises 13 to 32, find the nth term of the geometric sequence.

13. $2, 8, 32, \ldots$

14. $1, 5, 25, \ldots$

15. $-4, 12, -36, \ldots$

16. $-3, 6, -12, \ldots$

17. $6, 4, \frac{8}{3}, \ldots$

18. $8, 6, \frac{9}{2}, \ldots$

19. $-6, 5, -\frac{25}{6}, \ldots$

20. $-2, \frac{4}{3}, -\frac{8}{9}, \ldots$

21. $9, -3, 1, \ldots$

22. $8, -\frac{4}{3}, \frac{2}{9}, \ldots$

23. $1, -x, x^2, \ldots$

24. $2, 2a, 2a^2, \ldots$

25. $c^2, c^5, c^8, \ldots$

26. $-x^2, x^4, -x^6, \ldots$

27. $\frac{3}{100}, \frac{3}{10,000}, \frac{3}{1,000,000}, \ldots$

28. $\frac{7}{10}, \frac{7}{10,000}, \frac{7}{10,000,000}, \ldots$

29. $0.5, 0.05, 0.005, \ldots$

30. $0.4, 0.004, 0.00004, \ldots$

31. $0.45, 0.0045, 0.000045, \ldots$

32. $0.234, 0.000234, 0.000000234, \ldots$

33. Find the third term of a geometric sequence whose first term is 2 and whose fifth term is 162.

34. Find the fourth term of a geometric sequence whose third term is 1 and whose eighth term is $\frac{1}{32}$.

35. Find the second term of a geometric sequence whose third term is 4/3 and whose sixth term is $-32/27$.

36. Find the fifth term of a geometric sequence whose fourth term is 8/9 and whose seventh term is 64/243.

In Exercises 37 to 46, find the sum of the geometric series.

37. $\sum_{n=1}^{5} 3^n$

38. $\sum_{n=1}^{7} 2^n$

39. $\sum_{n=1}^{6} \left(\frac{2}{3}\right)^n$

40. $\sum_{n=1}^{14} \left(\frac{4}{3}\right)^n$

41. $\sum_{n=0}^{8} \left(-\frac{2}{5}\right)^n$

42. $\sum_{n=0}^{7} \left(-\frac{1}{3}\right)^n$

43. $\sum_{n=1}^{10} (-2)^{n-1}$

44. $\sum_{n=0}^{7} 2(5)^n$

45. $\displaystyle\sum_{n=0}^{9} 5(3)^n$ **46.** $\displaystyle\sum_{n=0}^{10} 2(-4)^n$

In Exercises 47 to 56, find the sum of the infinite geometric series.

47. $\displaystyle\sum_{n=1}^{\infty} \left(\frac{1}{3}\right)^n$ **48.** $\displaystyle\sum_{n=1}^{\infty} \left(\frac{3}{4}\right)^n$

49. $\displaystyle\sum_{n=1}^{\infty} \left(-\frac{2}{3}\right)^n$ **50.** $\displaystyle\sum_{n=1}^{\infty} \left(-\frac{3}{5}\right)^n$

51. $\displaystyle\sum_{n=1}^{\infty} \left(\frac{9}{100}\right)^n$ **52.** $\displaystyle\sum_{n=1}^{\infty} \left(\frac{7}{10}\right)^n$

53. $\displaystyle\sum_{n=1}^{\infty} (0.1)^n$ **54.** $\displaystyle\sum_{n=1}^{\infty} (0.5)^n$

55. $\displaystyle\sum_{n=0}^{\infty} (-0.4)^n$ **56.** $\displaystyle\sum_{n=0}^{\infty} (-0.8)^n$

In Exercises 57 to 68, find a rational number as the quotient of two integers for each repeating decimal.

57. $0.\overline{3}$ **58.** $0.\overline{5}$

59. $0.\overline{45}$ **60.** $0.\overline{63}$

61. $0.\overline{123}$ **62.** $0.3\overline{95}$

63. $0.4\overline{22}$ **64.** $0.\overline{355}$

65. $0.25\overline{4}$ **66.** $0.37\overline{2}$

67. $1.20\overline{84}$ **68.** $2.25\overline{90}$

69. Find the future value of an ordinary annuity that calls for depositing $100 at the end of every 6 months for 8 years into an account that earns 9 percent interest compounded semiannually.

70. To save for the replacement of a computer, a business deposits $250 at the end of each month into an account that earns 8 percent annual interest compounded monthly. Find the future value of the ordinary annuity in 4 years.

Supplemental Exercises

71. If the sequence a_n is a geometric sequence, make a conjecture and give a proof as to the sequence $\log a_n$.

72. If the sequence a_n is an arithmetic sequence, make a conjecture and give a proof as to the sequence 2^{a_n}.

73. A remarkable set of numbers, called the Cantor set, after Georg Cantor, is formed in the following way. Begin with the interval of real numbers $[0, 1]$. Remove the middle one-third. Now remove the middle one-third of the remaining interval. Continue indefinitely by removing the middle one-third of each remaining interval.

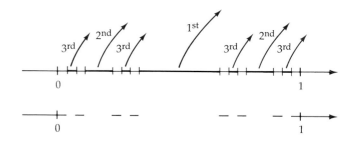

Find the total amount that has been removed. The remaining points form the Cantor set.

74. Consider a square with a side of length 1. Construct another square, inside the first one by connecting the midpoints of the first square. What is the area of the inscribed square? Continue constructing squares in the same way. Find the area of the nth inscribed square.

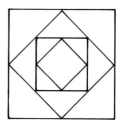

75. The product $P_n = a_1 \cdot a_2 \cdot a_3 \cdots a_n$ is called the nth partial product of a sequence. Find a formula for the nth partial product of the geometric sequence whose nth term is ar^{n-1}.

76. Let $f(x) = ab^x$, a, $b > 0$. Show that if x is restricted to positive integers n, then $f(n)$ is a geometric sequence.

77. A ball is dropped from a height of 5 feet. The ball rebounds 80 percent of the distance after each fall. Use an infinite geometric series to find the total distance the ball traveled.

78. The bob of a pendulum swings through an arc of 30 inches on its first swing. Each successive swing is 90 percent of the length of the previous swing. Find the total distance the bob travels before coming to rest.

79. Some people can trace their ancestry back ten generations, which means two parents, four grandparents, eight great-grandparents, and so on. How many grandparents does this include?

80. Find the perimeter of the snowflake curve given at the beginning of this chapter.

81. Find the area of the snowflake curve given at the beginning of this chapter.

12.4

Mathematical Induction

Consider the sequence

$$\frac{1}{1 \cdot 2}, \frac{1}{2 \cdot 3}, \frac{1}{3 \cdot 4}, \ldots, \frac{1}{n(n+1)}, \ldots$$

and the sequence of partial sums for this sequence.

$$S_1 = \frac{1}{1 \cdot 2} = \frac{1}{2}$$

$$S_2 = \frac{1}{1 \cdot 2} + \frac{1}{2 \cdot 3} = \frac{2}{3}$$

$$S_3 = \frac{1}{1 \cdot 2} + \frac{1}{2 \cdot 3} + \frac{1}{3 \cdot 4} = \frac{3}{4}$$

$$S_4 = \frac{1}{1 \cdot 2} + \frac{1}{2 \cdot 3} + \frac{1}{3 \cdot 4} + \frac{1}{4 \cdot 5} = \frac{4}{5}$$

This pattern suggests the conjecture that

$$S_n = \frac{1}{1 \cdot 2} + \frac{1}{2 \cdot 3} + \frac{1}{3 \cdot 4} + \cdots + \frac{1}{n(n+1)} = \frac{n}{n+1}.$$

How can we be sure that the pattern does not break down when $n = 50$ or maybe $n = 2000$ or some other large number? As we will show, this conjecture is true for all values of n.

As a second example, let's conjecture that the expression $n^2 - n + 41$ is a prime number for all positive integers. To test this conjecture, we will try various values of n.

n	$n^2 - n + 41$	
1	41	Prime
2	43	Prime
3	47	Prime
4	53	Prime
5	61	Prime

The results suggest that the conjecture is true. But again, how can we be sure? In fact, this conjecture is false when $n = 41$. In that case we have

$$n^2 - n + 41 = (41)^2 - 41 + 41 = (41)^2,$$

and $(41)^2$ is not a prime.

These two examples illustrate that just verifying a conjecture for a few values of n does not constitute a proof of the conjecture. To prove theorems about statements involving positive integers, a process called

mathematical induction is used. This process is based on an axiom called the **induction axiom.**

Induction Axiom

Suppose S is a set of positive integers with the following two properties:

1. 1 is an element of S.
2. If the positive integer k is in S, then $k + 1$ is in S.

Then, S contains all the positive integers.

Part 2 of this axiom states that if some positive integer, say 8, is in S, then $8 + 1$ or 9 is in S. But since 9 is in S, part 2 says that $9 + 1$ or 10 is in S, and so on. Part 1 states that 1 is in S. Thus 2 is in S; thus 3 is in S; thus 4 is in S; . . . Therefore all the positive integers are in S.

The induction axiom is used to prove the *Principle of Mathematical Induction.*

Principle of Mathematical Induction

Let P_n be a statement about a positive integer n. If

1. P_1 is true, and
2. The truth of P_k implies the truth of P_{k+1}

then P_n is true for all positive integers.

Remark   Part 2 of the Principle of Mathematical Induction is referred to as the **induction hypothesis.** When applying this step, we assume the statement P_k is true and then try to prove that P_{k+1} is also true.

As an example, we will prove that the first conjecture we made in this section is true for all positive integers. Every induction proof has the two distinct parts stated in the theorem. First we must show the result is true for $n = 1$. Second, we assume that the statement is true for some positive integer k and, using that assumption, prove that the statement is true for $n = k + 1$.

Prove that

$$S_n = \frac{1}{1 \cdot 2} + \frac{1}{2 \cdot 3} + \frac{1}{3 \cdot 4} + \cdots + \frac{1}{n(n + 1)} = \frac{n}{n + 1}$$

for all positive integers n.

Proof: 1. For $n = 1$,

$$S_1 = \frac{1}{1 \cdot 2} = \frac{1}{2} = \frac{1}{1 + 1}.$$

The statement is true for $n = 1$.

2. Assume the statement is true for some positive integer k.

$$S_k = \frac{1}{1 \cdot 2} + \frac{1}{2 \cdot 3} + \frac{1}{3 \cdot 4} + \cdots + \frac{1}{k(k+1)} = \frac{k}{k+1} \quad \text{Induction hypothesis}$$

Now verify the formula is true when $n = k + 1$. That is, verify that

$$S_{k+1} = \frac{k+1}{(k+1)+1} = \frac{k+1}{k+2} \quad \text{This is the goal of the induction proof.}$$

It is helpful when proving a theorem about sums to note that

$$S_{k+1} = S_k + a_{k+1}.$$

Begin by noting that $a_k = \dfrac{1}{k(k+1)}$; thus, $a_{k+1} = \dfrac{1}{(k+1)(k+2)}$.

$$S_{k+1} = S_k \qquad + a_{k+1}$$

$$= \frac{k}{k+1} + \frac{1}{(k+1)(k+2)} \qquad \text{By the induction hypothesis and substituting for } a_{k+1}.$$

$$= \frac{k(k+2)}{(k+1)(k+2)} + \frac{1}{(k+1)(k+2)}$$

$$= \frac{k(k+2)+1}{(k+1)(k+2)}$$

$$= \frac{k^2 + 2k + 1}{(k+1)(k+2)} = \frac{(k+1)^2}{(k+1)(k+2)}$$

$$S_{k+1} = \frac{k+1}{k+2}$$

Because we have verified the two parts of the Principle of Mathematical Induction, we can conclude that the statement is true for all positive integer values.

EXAMPLE 1 Prove by Mathematical Induction

Prove that $1^2 + 2^2 + 3^2 + \cdots + n^2 = \dfrac{n(n+1)(2n+1)}{6}$.

Solution Verify the two parts of the Principle of Mathematical Induction.

1. Let $n = 1$.

$$S_1 = 1^2 = 1 = \frac{1(1+1)(2 \cdot 1 + 1)}{6}.$$

2. Assume the statement is true for some positive integer k.

$$S_k = 1^2 + 2^2 + 3^2 + \cdots + k^2 = \frac{k(k+1)(2k+1)}{6} \quad \text{Induction hypothesis}$$

Verify that the statement is true when $n = k + 1$. Show that

$$S_{k+1} = \frac{(k + 1)(k + 2)(2k + 3)}{6}.$$

Because $a_k = k^2$, $a_{k+1} = (k + 1)^2$.

$$S_{k+1} = \qquad S_k \qquad + a_{k+1}$$

$$= \frac{k(k + 1)(2k + 1)}{6} + (k + 1)^2$$

$$= \frac{k(k + 1)(2k + 1)}{6} + \frac{6(k + 1)^2}{6} = \frac{k(k + 1)(2k + 1) + 6(k + 1)^2}{6}$$

$$= \frac{(k + 1)[k(2k + 1) + 6(k + 1)]}{6} = \frac{(k + 1)(2k^2 + 7k + 6)}{6}$$

$$= \frac{(k + 1)(k + 2)(2k + 3)}{6}$$

By the Principle of Mathematical Induction, the statement is true for all positive integers.

■ *Try Exercise 8, page 628.*

Mathematical induction can also be used to prove statements about sequences, products, and inequalities.

EXAMPLE 2   Prove a Product Formula by Mathematical Induction

Prove that

$$P_n = \left(1 + \frac{1}{1}\right)\left(1 + \frac{1}{2}\right)\left(1 + \frac{1}{3}\right)\cdots\left(1 + \frac{1}{n}\right) = n + 1$$

Solution

1. Verify for $n = 1$.

$$\left(1 + \frac{1}{1}\right) = 2 = 1 + 1$$

2. Assume the statement is true for some positive integer k.

$$P_k = \left(1 + \frac{1}{1}\right)\left(1 + \frac{1}{2}\right)\left(1 + \frac{1}{3}\right)\cdots\left(1 + \frac{1}{k}\right) = k + 1 \quad \text{Induction hypothesis}$$

Verify that the statement is true when

$$n = k + 1.$$

That is, prove

$$P_{k+1} = k + 2.$$

$$P_{k+1} = \left(1 + \frac{1}{1}\right)\left(1 + \frac{1}{2}\right)\left(1 + \frac{1}{3}\right)\cdots\left(1 + \frac{1}{k}\right)\left(1 + \frac{1}{k+1}\right)$$

$$= P_k\left(1 + \frac{1}{k+1}\right) = (k+1)\left(1 + \frac{1}{k+1}\right) = k + 1 + 1$$

$$P_{k+1} = k + 2$$

The statement is true for all positive integers.

■ *Try Exercise **12**, page 628.*

EXAMPLE 3   **Prove an Inequality by Mathematical Induction**

Prove that $1 + 2n \leq 3^n$ for all positive integers.

Solution

1. Let $n = 1$. Then $1 + 2(1) = 3 \leq 3^1$. The statement is true when n is 1.
2. Assume the statement is true for some positive integer k.

$$1 + 2k \leq 3^k \quad \text{Induction hypothesis}$$

Now prove the statement is true for $n = k + 1$. That is, we must prove that $1 + 2(k + 1) \leq 3^{k+1}$.

$$3^{k+1} = 3^k(3)$$

$$\geq (1 + 2k)(3) \quad \text{Why?}[4]$$

$$= 6k + 3$$

$$> 2k + 2 + 1 \quad 6k > 2k, \text{ and } 3 = 2 + 1.$$

$$= 2(k + 1) + 1$$

Thus $1 + 2(k + 1) \leq 3^{k+1}$. By the Principle of Mathematical Induction, $1 + 2k \leq 3^k$ for all positive integers.

■ *Try Exercise **16**, page 629.*

The Principle of Mathematical Induction can be extended to cases where the beginning index is greater than 1.

Extended Principle of Mathematical Induction

Let P_n be a statement about a positive integer n. If

1. P_j is true for some positive integer j, and

2. For $k \geq j$ the truth of P_k implies the truth of P_{k+1} then P_n is true for all positive integers $n \geq j$.

---

[4] By the induction hypothesis, $1 + 2k \leq 3^k$.

EXAMPLE 4 **Prove by the Extended Principle of Mathematical Induction**

For $n \geq 3$, prove that $n^2 > 2n + 1$.

Solution

1. Let $n = 3$. Then $3^2 = 9$; $2(3) + 1 = 7$. Thus, $n^2 > 2n + 1$ for $n = 3$.

2. Assume the statement is true for some positive integer $k \geq 3$.

$$k^2 > 2k + 1 \quad \text{Induction hypothesis}$$

Verify that the statement is true when $n = k + 1$. That is, show that

$$(k + 1)^2 > 2(k + 1) + 1 = 2k + 3.$$

$$(k + 1)^2 = k^2 + 2k + 1$$
$$> (2k + 1) + 2k + 1 \quad \text{Induction hypothesis}$$
$$> 2k + 1 + 1 + 1 \quad \quad 2k > 1$$
$$= 2k + 3$$

Thus, $(k + 1)^2 > 2k + 3$. By the Principle of Mathematical Induction, $n^2 > 2n + 1$ for all $n \geq 3$.

■ *Try Exercise 14, page 629.*

EXERCISE SET 12.4

In Exercises 1 to 12, use mathematical induction to prove each statement.

1. $\displaystyle\sum_{k=1}^{n} 3k - 2 = 1 + 4 + 7 + \cdots + 3n - 2 = \dfrac{n(3n - 1)}{2}$

2. $\displaystyle\sum_{k=1}^{n} 2k = 2 + 4 + 6 + \cdots + 2n = n(n + 1)$

3. $\displaystyle\sum_{k=1}^{n} k^3 = 1 + 8 + 27 + \cdots + n^3 = \dfrac{n^2(n + 1)^2}{4}$

4. $\displaystyle\sum_{k=1}^{n} 2^k = 2 + 4 + 8 + \cdots + 2^n = 2(2^n - 1)$

5. $\displaystyle\sum_{k=1}^{n} 4k - 1 = 3 + 7 + 11 + \cdots + 4n - 1 = n(2n + 1)$

6. $\displaystyle\sum_{k=1}^{n} 3^k = 3 + 9 + 27 + \cdots + 3^n = \dfrac{3(3^n - 1)}{2}$

7. $\displaystyle\sum_{k=1}^{n} (2k - 1)^3 = 1 + 27 + 125 + \cdots + (2n - 1)^3$
$$= n^2(2n^2 - 1)$$

8. $\displaystyle\sum_{k=1}^{n} k(k + 1) = 2 + 6 + 12 + \cdots + \cdots + n(n + 1)$
$$= \dfrac{n(n + 1)(n + 2)}{3}$$

9. $\displaystyle\sum_{k=1}^{n} \dfrac{1}{(2k - 1)(2k + 1)} = \dfrac{1}{1 \cdot 3} + \dfrac{1}{3 \cdot 5} + \dfrac{1}{5 \cdot 7} + \cdots$
$$+ \dfrac{1}{(2n - 1)(2n + 1)} = \dfrac{n}{2n + 1}$$

10. $\displaystyle\sum_{k=1}^{n} \dfrac{1}{2k(2k + 2)} = \dfrac{1}{2 \cdot 4} + \dfrac{1}{4 \cdot 6} + \dfrac{1}{6 \cdot 8} + \cdots$
$$+ \dfrac{1}{2k(2k + 2)} = \dfrac{n}{4(n + 1)}$$

11. $\displaystyle\sum_{k=1}^{n} k^4 = 1 + 16 + 81 + \cdots + n^4$
$$= \dfrac{n(n + 1)(2n + 1)(3n^2 + 3n - 1)}{30}$$

12. $P_n = \left(1 - \dfrac{1}{2}\right)\left(1 - \dfrac{1}{3}\right)\left(1 - \dfrac{1}{4}\right)\cdots\left(1 - \dfrac{1}{n + 1}\right) = \dfrac{1}{n + 1}$

In Exercises 13 to 20, use mathematical induction to prove the inequalities.

13. $\left(\dfrac{3}{2}\right)^n > n + 1, \; n \geq 4$ **14.** $\left(\dfrac{4}{3}\right)^n > n, \; n \geq 7$

15. If $0 < a < 1$, show that $a^{n+1} < a^n$ for all positive integers.

16. If $a > 1$, show that $a^{n+1} > a^n$ for all positive integers.

17. $1 \cdot 2 \cdot 3 \cdot \cdots \cdot n > 2^n, \; n \geq 4$

18. $\dfrac{1}{\sqrt{1}} + \dfrac{1}{\sqrt{2}} + \dfrac{1}{\sqrt{3}} + \cdots + \dfrac{1}{\sqrt{n}} \geq \sqrt{n}$

19. For $a > 0$, show that $(1 + a)^n \geq 1 + na$.

20. $\log_{10} n < n$ for all positive integers. (*Hint:* Because $\log_{10} x$ is an increasing function, $\log_{10}(n + 1) \leq \log_{10}(n + n)$.)

In Exercises 21 to 30, use mathematical induction to prove the statements.

21. 2 is a factor of $n^2 + n$ for all positive integers.

22. 3 is a factor of $n^3 - n$ for all positive integers.

23. 4 is a factor of $5^n - 1$ for all positive integers. (*Hint:* $5^{k+1} - 1 = 5 \cdot 5^k - 5 + 4$.)

24. 5 is a factor of $6^n - 1$ for all positive integers.

25. $(xy)^n = x^n y^n$ for all all positive integers.

26. $\left(\dfrac{x}{y}\right)^n = \dfrac{x^n}{y^n}$ for all positive integers.

27. For $a \neq b$, show that $(a - b)$ is a factor of $a^n - b^n$, where n is a positive integer. (*Hint:*
$$a^{k+1} - b^{k+1} = (a \cdot a^k - ab^k) + (ab^k - b \cdot b^k).)$$

28. For $a \neq -b$, show that $(a + b)$ is a factor of $a^{2n+1} + b^{2n+1}$, where n is a positive integer. (*Hint:*
$$a^{2k+3} + b^{2k+3} = (a^{2k+2} + b^{2k+2})(a + b) - ab(a^{2k+1} + b^{2k+1}).)$$

29. For $r \neq 1$, $\displaystyle\sum_{k=1}^{n} ar^{k-1} = \dfrac{a(1 - r^n)}{1 - r}$

30. $\displaystyle\sum_{k=1}^{n} (ak + b) = \dfrac{n[(n + 1)a + 2b]}{2}$

Supplemental Exercises

In Exercises 31 to 35, use mathematical induction to prove the statements.

31. Using a calculator, find the smallest integer N for which $\log N! > N$. Now prove $\log n! > n$ for all $n > N$.

32. Let a_n be a sequence for which there is a number r and an integer N for which $a_{n+1}/a_n < r$ for $n \geq N$. Show that $a_{N+k} < a_N r^k$ for each positive integer k.

33. For constant positive integers m and n, show that $(x^m)^n = x^{mn}$.

34. Prove that $\displaystyle\sum_{i=0}^{n} \dfrac{1}{i!} \leq 3 - \dfrac{1}{n}$ for all positive integers.

35. Prove $\left(\dfrac{n + 1}{n}\right)^n < n$ for all integers $n \geq 3$.

36. To give a proof by mathematical induction, it is important that both parts of the principle of mathematical induction be verified. For example, consider the formula
$$2 + 4 + 8 + \cdots + 2^n \overset{?}{=} 2^{n+1} + 1$$

a. Show that if we assume the formula is true for some positive integer k, then the formula is true for $k + 1$.

b. Show that the formula is not true for $n = 1$. Now show that the formula is not valid for any n by showing that the left side is always an even number and the right side is always an odd number.

Thus, although the second part of the principle of induction is valid, the first part is not and the formula is not correct.

37. The Tower of Hanoi is a game that consists of three pegs and n disks of distinct diameter arranged on one of the pegs so that the largest disk is on the bottom, then the next largest, and so on. The object of the game is to move all the disks from one peg to a second peg. The rules require that only one disk be moved at a time and that a larger disk may not be placed on a smaller disk. All pegs may be used. Show that it is possible to complete the game in $2^n - 1$ moves.

38. A legend says that in the center of the universe, high priests have the task of moving sixty-four golden disks from one of three diamond needles by using the rules of the game in Exercise 37. When they have completed the transfer, the universe will cease to exist. If one move is made each second, and the priest started 5 billion years ago (the approximate age of the earth), how many more years, assuming the legend to be true, would the universe exist?

39. Use mathematical induction to prove DeMoivre's Theorem: $[r(\cos \theta + i \sin \theta)]^n = r^n(\cos n\theta + i \sin n\theta)$ for all positive integers.

12.5

The Binomial Theorem

In certain situations in mathematics it is necessary to write $(a + b)^n$ as the sum of its terms. Because $(a + b)$ is a binomial, this process is called **expanding the binomial.** For small values of n, it is relatively easy to write the expansion by using multiplication.

Earlier in the text, we found

$$(a + b)^1 = a + b$$

$$(a + b)^2 = a^2 + 2ab + b^2$$

$$(a + b)^3 = a^3 + 3a^2b + 3ab^2 + b^3.$$

Building on these, we can write a few more expansions.

$$(a + b)^4 = a^4 + 4a^3b + 6a^2b^2 + 4ab^3 + b^4$$

$$(a + b)^5 = a^5 + 5a^4b + 10a^3b^2 + 10a^2b^3 + 5ab^4 + b^5$$

We could continue to build on previous expansions and eventually have quite a comprehensive list of binomial expansions. Instead, however, we will look for a theorem that will allow us to expand $(a + b)^n$ directly without multiplying.

Look at the variable parts of each expansion above. Note that for each $n = 1, 2, 3, 4, 5$

1. The first term is a^n. The exponent on a decreases by 1 for each successive term.

2. The exponent on b increases by 1 for each successive term. The last term is b^n.

3. The degree of each term is n.

To find a pattern for the coefficients in each expansion, first note that there are $n + 1$ terms and that the coefficient of the first and last term is 1. To find the remaining coefficients, consider the expansion of $(a + b)^5$.

$$(a + b)^5 = a^5 + 5a^4b + 10a^3b^2 + 10a^2b^3 + 5ab^4 + b^5$$

$$\frac{5}{1} = 5; \quad \frac{5 \cdot 4}{2 \cdot 1} = 10; \quad \frac{5 \cdot 4 \cdot 3}{3 \cdot 2 \cdot 1} = 10; \quad \frac{5 \cdot 4 \cdot 3 \cdot 2}{4 \cdot 3 \cdot 2 \cdot 1} = 5$$

Observe from these patterns that there is a strong relationship to factorials. In fact, we can express each coefficient by using factorial notation.

$$\frac{5!}{1! \, 4!} = 5; \quad \frac{5!}{2! \, 3!} = 10; \quad \frac{5!}{3! \, 2!} = 10; \quad \frac{5!}{4! \, 1!} = 5$$

In each denominator, the first factorial is the exponent of b and the second factorial is the exponent of a.

In general, we will conjecture that the coefficient of the term $a^{n-k}b^k$ in the expansion of $(a + b)^n$ is $\dfrac{n!}{k!\,(n - k)!}$. Each coefficient of a binomial expansion is called a **binomial coefficient** and is denoted by $\dbinom{n}{k}$.

Formula for a Binomial Coefficient

The coefficient of the term whose variable part is $a^{n-k}b^n$ in the expansion of $(a + b)^n$ is

$$\binom{n}{k} = \frac{n!}{k!\,(n - k)!}$$

where k is the exponent of b for that term.

Remark   The first term of the expansion of $(a + b)^n$ can be thought of as $a^n b^0$. In that case, we can calculate the coefficient of that term as

$$\binom{n}{0} = \frac{n!}{0!\,(n - 0)!} = \frac{n!}{1 \cdot n!} = 1.$$

EXAMPLE 1   **Evaluate a Binomial Coefficient**

Evaluate each binomial coefficient:

a. $\dbinom{9}{6}$ b. $\dbinom{10}{10}$

Solution

a. $\dbinom{9}{6} = \dfrac{9!}{6!\,3!} = \dfrac{9 \cdot 8 \cdot 7 \cdot 6!}{6! \cdot 3 \cdot 2 \cdot 1} = 84$

b. $\dbinom{10}{10} = \dfrac{10!}{10!\,0!} = 1.$ Remember $0! = 1$.

■ *Try Exercise* **4,** *page 634.*

We are now ready to state the Binomial Theorem for positive integers.

Binomial Theorem for Positive Integers

If n is a positive integer, then

$$(a + b)^n = \sum_{i=0}^{n} \binom{n}{i} a^{n-i} b^i$$

$$= \binom{n}{0} a^n + \binom{n}{1} a^{n-1}b + \binom{n}{2} a^{n-2}b^2 + \cdots + \binom{n}{n} b^n$$

Proof: The proof of this theorem is by mathematical induction.

1. For $n = 1$, the statement reduces to $a + b = a + b$.
2. Assume the statement is true when $n = k$ and verify the statement for $n = k + 1$.

The induction hypothesis states that

$$(a + b)^k = \binom{k}{0}a^k + \binom{k}{1}a^{k-1}b + \binom{k}{2}a^{k-2}b^2 + \cdots + \binom{k}{i-1}a^{k-i+1}b^{i-1}$$
$$+ \binom{k}{i}a^{k-i}b^i + \cdots + \binom{k}{k-1}ab^{k-1} + \binom{k}{k}b^k.$$

Now multiply each side of this equation by $(a + b)$ and collect like terms. We suggest that you try this for $(a + b)^4(a + b)$ to see how terms will be arranged.

$$(a + b)^{k+1} = \binom{k}{0}a^{k+1} + \left[\binom{k}{1} + \binom{k}{0}\right]a^k b + \left[\binom{k}{2} + \binom{k}{1}\right]a^{k-1}b^2 + \cdots$$
$$+ \left[\binom{k}{i} + \binom{k}{i-1}\right]a^{k-i+1}b^i + \cdots + \binom{k}{k}b^{k+1}$$

The coefficient of the general term $a^{k-i+1}b^i$ shown above can be simplified.

$$\binom{k}{i} + \binom{k}{i-1} = \frac{k!}{i!\,(k-i)!} + \frac{k!}{(i-1)!\,(k-i+1)!}$$

$$= \frac{k!\,(k-i+1)}{i!\,(k-i+1)!} + \frac{k!\,i}{i!\,(k-i+1)!} = \frac{k!\,[(k-i+1)+i]}{i!\,(k-i+1)!}$$

$$= \frac{k!\,(k+1)}{i!\,(k-i+1)!} = \frac{(k+1)!}{i!\,(k-i+1)!} = \binom{k+1}{i}$$

Thus, we have shown that the general term of $(a + b)^{k+1}$ is the same term as given in the theorem, with n replaced by $k + 1$. Thus, the theorem is true for all positive integers.

EXAMPLE 2   Expand the Sum of Two Terms

Expand $(2x^2 + 3)^4$.

Solution

$$(2x^2 + 3)^4 = \binom{4}{0}(2x^2)^4 + \binom{4}{1}(2x^2)^3(3) + \binom{4}{2}(2x^2)^2(3)^2$$
$$+ \binom{4}{3}(2x^2)(3)^3 + \binom{4}{4}(3)^4$$
$$= 16x^8 + 96x^6 + 216x^4 + 216x^2 + 81$$

■ *Try Exercise* **18**, *page 634.*

EXAMPLE 3   Expand a Difference of Two Terms

Expand $(\sqrt{x} - 2y)^5$.

Solution

$$(\sqrt{x} - 2y)^5 = \binom{5}{0}(\sqrt{x})^5 + \binom{5}{1}(\sqrt{x})^4(-2y) + \binom{5}{2}(\sqrt{x})^3(-2y)^2$$
$$+ \binom{5}{3}(\sqrt{x})^2(-2y)^3 + \binom{5}{4}(\sqrt{x})(-2y)^4 + \binom{5}{5}(-2y)^5$$
$$= x^{5/2} - 10x^2y + 40x^{3/2}y^2 - 80xy^3 + 80x^{1/2}y^4 - 32y^5$$

■ *Try Exercise* **20**, *page 634.*

Remark   If one of the terms a or b in $(a + b)^n$ is negative, the terms of the expansion alternate signs.

The Binomial Theorem can also be used to find a specific term in the expansion of $(a + b)^n$.

Formula for the *i*th Term of a Binomial Expansion

> The *i*th term of the expansion of $(a + b)^n$ is given by
> $$\binom{n}{i-1}a^{n-i+1}b^{i-1}.$$

Remark   Note that the exponent on b is *one less* than the term number.

The proof of this theorem is left as an exercise.

EXAMPLE 4   **Find the *i*th Term of a Binomial Expansion**

Find the fourth term in the expansion of $(2x^3 - 3y^2)^5$.

Solution   With $a = 2x^3$ and $b = -3y^2$, and using the last theorem with $i = 4$ and $n = 5$, we have

$$\binom{5}{3}(2x^3)^2(-3y^2)^3 = -1080x^6y^6$$

The fourth term is $-1080x^6y^6$.

■ *Try Exercise* **34,** *page 634.*

A pattern for the coefficients of the terms of an expanded binomial can be found by writing the coefficients in a triangular array, known as **Pascal's Triangle.**

$(a + b)^1$:					1		1			
$(a + b)^2$:				1		2		1		
$(a + b)^3$:			1		3		3		1	
$(a + b)^4$:		1		4		6		4		1
$(a + b)^5$:	1		5		10		10		5	1

Each row begins and ends with the number 1. Any other number in a row is the sum of the two closest numbers above it. For example, $4 + 6 = 10$. Thus each succeeding row can be found from the preceding row. For example, the sixth row is

$$1 \quad 6 \quad 15 \quad 20 \quad 15 \quad 6 \quad 1.$$

This triangle can be used to expand a binomial for small values of n.

EXAMPLE 5   Use Pascal's Triangle to Expand a Binomial

Find the seventh row of Pascal's Triangle and use it to write the expansion of $(a + b)^7$.

Solution   The seventh row of Pascal's Triangle is

$$1 \quad 7 \quad 21 \quad 35 \quad 35 \quad 21 \quad 7 \quad 1.$$

$$(a + b)^7 = a^7 + 7a^6b + 21a^5b^2 + 35a^4b^3 + 35a^3b^4 + 21a^2b^5 + 7ab^6 + b^7$$

■ *Try Exercise 22, page 634.*

EXERCISE SET 12.5

In Exercises 1 to 8, evaluate the binomial coefficients.

1. $\binom{7}{4}$ **2.** $\binom{8}{6}$ **3.** $\binom{9}{2}$ **4.** $\binom{10}{5}$

5. $\binom{12}{9}$ **6.** $\binom{6}{5}$ **7.** $\binom{11}{0}$ **8.** $\binom{14}{14}$

In Exercises 9 to 28, expand the binomial.

9. $(x - y)^6$ **10.** $(a - b)^5$ **11.** $(x + 3)^5$

12. $(x - 5)^4$ **13.** $(2x - 1)^7$ **14.** $(2x + y)^6$

15. $(x + 3y)^6$ **16.** $(x - 4y)^5$ **17.** $(2x - 5y)^4$

18. $(3x + 2y)^4$ **19.** $\left(x + \dfrac{1}{x}\right)^6$ **20.** $(2x - \sqrt{y})^7$

21. $(x^2 - 4)^7$ **22.** $(x - y^3)^6$ **23.** $(2x^2 + y^3)^5$

24. $(2x - y^3)^6$ **25.** $\left(\dfrac{2}{x} - \dfrac{x}{2}\right)^4$ **26.** $\left(\dfrac{a}{b} + \dfrac{b}{a}\right)^3$

27. $(s^{-2} + s^2)^6$ **28.** $(2r^{-1} + s^{-1})^5$

In Exercises 29 to 36, find the indicated term without expanding completely.

29. $(3x - y)^{10}$; eighth term

30. $(x + 2y)^{12}$; fourth term

31. $(x + 4y)^{12}$; third term

32. $(2x - 1)^{14}$; thirteenth term

33. $(\sqrt{x} - \sqrt{y})^9$; fifth term

34. $(x^{-1/2} + x^{1/2})^{10}$; sixth term

35. $\left(\dfrac{a}{b} + \dfrac{b}{a}\right)^{11}$; ninth term

36. $\left(\dfrac{3}{x} - \dfrac{x}{3}\right)^{13}$; seventh term

37. Find the term that contains b^8 in the expansion of $(2a - b)^{10}$.

38. Find the term that contains s^7 in the expansior of $(3r + 2s)^9$.

39. Find the term that contains y^8 in the expansion of $(2x + y^2)^6$.

40. Find the term that contains b^9 in the expansion of $(a - b^3)^8$.

41. Find the middle term of $(3a - b)^{10}$.

42. Find the middle term of $(a + b^2)^8$.

43. Find the two middle terms of $(s^{-1} + s)^9$.

44. Find the two middle terms of $(x^{1/2} - y^{1/2})^7$.

In Exercises 45 to 50, use the Binomial Theorem to simplify the powers of the complex numbers.

45. $(2 - i)^4$ **46.** $(3 + 2i)^3$

47. $(1 + 2i)^5$ **48.** $(1 - 3i)^5$

49. $\left(\dfrac{\sqrt{2}}{2} + i\dfrac{\sqrt{2}}{2}\right)^8$ **50.** $\left(\dfrac{1}{2} + i\dfrac{\sqrt{3}}{2}\right)^6$

Supplemental Exercises

51. Let n be a positive integer. Expand and simplify $\dfrac{(x + h)^n - x^n}{h}$, where x is any real number and $h \neq 0$.

52. Show that $\dbinom{n}{k} = \dbinom{n}{n - k}$ for all positive integers n and k with $0 \leq k \leq n$.

53. Show that $\displaystyle\sum_{k=0}^{n} \dbinom{n}{k} = 2^n$. (*Hint:* Use the Binomial Theorem with $x = 1$, $y = 1$.)

54. Prove that $\dbinom{n}{k} + \dbinom{n}{k + 1} = \dbinom{n + 1}{k + 1}$, n and k integers, $0 \leq k \leq n$.

55. Prove that $\displaystyle\sum_{i=0}^{n} (-1)^i \dbinom{n}{i} = 0$.

56. Approximate $(0.98)^8$ by evaluating the first three terms of $(1 - 0.02)^8$.

57. Approximate $(1.02)^8$ by evaluating the first three terms of $(1 + 0.02)^8$.

There is an extension of the Binomial Theorem called the *Multinomial Theorem*. This theorem is used in determining probabilities. *Multinomial Theorem*   If n, r, and k are positive integers, then the coefficient of $a^r b^k c^{n-r-k}$ in the expansion of $(a + b + c)^n$ is
$$\frac{n!}{r!\,k!\,(n - r - k)!}.$$

In Exercises 58 to 61, use the Multinomial Theorem to find the indicated coefficient.

58. Find the coefficient of $a^2 b^3 c^5$ in the expansion of $(a + b + c)^{10}$.

59. Find the coefficient of $a^5 b^2 c^2$ in the expansion of $(a + b + c)^9$.

60. Find the coefficient of $a^4 b^5$ in the expansion of $(a + b + c)^9$.

61. Find the coefficient of $a^3 c^5$ in the expansion of $(a + b + c)^8$.

12.6
Permutations and Combinations

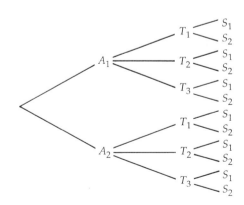

Figure 12.1

An electronics store offers a three-component stereo system for \$250. A buyer must choose one amplifier, one tuner, and one pair of speakers. If the store has two models of amplifiers, three models of tuners, and two speaker models, how many different stereo systems could a consumer purchase?

This problem belongs to a class of problems called "counting problems." The problem is to *count* the number of ways the conditions of the problem can be satisfied. One way to do this is to make a list and then count the items on the list. We will organize the list in a table using A_1 and A_2 for the amplifiers; T_1, T_2, and T_3 for the tuners; and S_1 and S_2 for the speakers.

By counting the possible systems that can be purchased, we find there are twelve different systems. Another way to arrive at this result is to find the product of the number of options available.

$$\begin{matrix} \text{Number of} \\ \text{amplifiers} \end{matrix} \times \begin{matrix} \text{number of} \\ \text{tuners} \end{matrix} \times \begin{matrix} \text{number of} \\ \text{speakers} \end{matrix} = \begin{matrix} \text{number of} \\ \text{systems} \end{matrix}$$

$$2 \quad \times \quad 3 \quad \times \quad 2 \quad = \quad 12$$

In some states, such as California, a standard car license plate begins with a nonzero digit, followed by three letters, followed by three more digits. What is the maximum number of car license plates that could be issued in California? If we begin a list of the possible license plates, it soon becomes apparent that listing them all would be very time consuming and impractical.

$$1AAA000, \quad 1AAA001, \quad 1AAA002, \quad 1AAA003, \ldots$$

Instead, the following counting principle is used. This principle forms the basis for all counting problems.

Fundamental Counting Principle

> Let $T_1, T_2, T_3, \ldots, T_n$ be a sequence of n conditions. Suppose that T_1 can occur in m_1 ways, T_2 can occur in m_2 ways, T_3 can occur in m_3 ways, and so on until finally T_n can occur in m_n ways. Then the number of ways of satisfying the conditions $T_1, T_2, T_3, \ldots, T_n$ in succession is given by the product
>
> $$m_1 m_2 m_3 \cdots m_n.$$

Applying the counting principle to the California license plate problem, we have

Condition	Number of Ways
T_1: a nonzero digit	$m_1 = 9$
T_2: letter of the alphabet	$m_2 = 26$
T_3: letter of the alphabet	$m_3 = 26$
T_4: letter of the alphabet	$m_4 = 26$
T_5: a digit	$m_5 = 10$
T_6: a digit	$m_6 = 10$
T_7: a digit	$m_7 = 10$

$$\begin{matrix} \text{Number of car} \\ \text{license plates} \end{matrix} = 9 \cdot 26 \cdot 26 \cdot 26 \cdot 10 \cdot 10 \cdot 10 = 158,184,000$$

When we formed the product of the options for the stereo system we used, the counting principle.

Condition	Number of Ways
T_1: an amplifier	$m_1 = 2$
T_2: a tuner	$m_2 = 3$
T_3: speakers	$m_3 = 2$

$$\text{Number of systems} = 2 \cdot 3 \cdot 2 = 12$$

EXAMPLE 1 Apply the Fundamental Counting Principle

An automobile dealer offers three midsized cars. A customer selecting one of these cars has the option of three different engines, five different colors, and four different interior packages. How many different selections can the customer make?

Solution

T_1: midsized car	$m_1 = 3$
T_2: engines	$m_2 = 3$
T_3: color	$m_3 = 5$
T_4: interior	$m_4 = 4$

Number of different selections $= 3 \cdot 3 \cdot 5 \cdot 4 = 180$.

■ *Try Exercise **26**, page 641.*

An application of the fundamental counting principle is to determine the number of arrangements of distinct elements in a definite order.

Permutation

> A **permutation** is an arrangement of distinct objects in a definite order.

For example, *abc* and *bca* are two of the possible permutations of the three elements *a*, *b*, *c*.

Consider a race with ten runners. In how many different orders can the runners finish first, second, and third (assuming no ties)?

Any one of the 10 runners could finish first:	$m_1 = 10$
Any one of the remaining 9 runners could be second:	$m_2 = 9$
Any one of the remaining 8 runners could be third:	$m_3 = 8$

By the fundamental counting principle, there are $10 \cdot 9 \cdot 8 = 720$ possible first-, second- and third-place finishes for the ten runners. Using the language of permutations, we would say "There are 720 permutations of 10 objects (the runners) taken 3 (the possible finishes) at a time."

Permutations occur so frequently in counting problems that a formula rather than the counting principle is often used.

Formula for a Permutation of *n* Distinct Objects Taken *r* at a Time

> The number of permutations of *n* distinct objects taken *r* at a time is
>
> $$P(n, r) = \frac{n!}{(n - r)!}.$$

EXAMPLE 2   **Find the Number of Permutations**

In how many ways can a president, vice president, secretary, and treasurer be selected from a committee of fifteen people?

Solution   There are fifteen distinct people to place in four positions. Thus $n = 15$ and $r = 4$.

$$P(15, 4) = \frac{15!}{(15 - 4)!} = \frac{15!}{11!} = \frac{15 \cdot 14 \cdot 13 \cdot 12 \cdot 11!}{11!} = 32{,}760$$

■ *Try Exercise* **18,** *page 641.*

EXAMPLE 3   **Find the Number of Seating Permutations**

Six people attend a movie and all sit in the same row with six seats.
a.   Find the number of ways the group can sit together.
b.   Find the number of ways the group can sit together if two people in the group must sit together.
c.   Find the number of ways the group can sit together if two people in the group refuse to sit together.

Solution

a.   There are six distinct people to place in six distinct positions. Thus $n = 6$ and $r = 6$.

$$P(6, 6) = \frac{6!}{(6 - 6)!} = \frac{6!}{0!} = \frac{6!}{1} = 720$$

b.   Think of the two people who must sit together as a single object and count the number of arrangements of the *five* objects (AB), C, D, E, F. Thus $n = 5$ and $r = 5$.

$$P(5, 5) = \frac{5!}{(5 - 5)!} = \frac{5!}{0!} = \frac{5!}{1} = 120$$

There are also 120 arrangements with A and B reversed (BA), C, D, E, F. Thus the total number of arrangements is $120 + 120 = 240$.

c.   From part a, there are 720 possible seating arrangements. From part b, there are 240 arrangements with two specific people next to each other. Thus there are $720 - 240 = 480$ arrangements where two specific people are not seated together.

■ *Try Exercise* **14,** *page 641.*

To this point we have been counting the number of distinct arrangements of objects. In some cases we may be interested in determining the number of ways of selecting objects without regard to the order of the selection. For example, suppose we want to select a committee of three people from five candidates denoted by *A, B, C, D,* and *E.* One possible committee is *A, C, D*. If we select *D, C, A*, we still have the same committee because the order of the selection is not important. An arrangement of objects for which the order of the selection is not important is a **combination**.

Formula for the Combination of *n* Objects Taken *r* at a Time

The number of combinations of *n* objects taken *r* at a time is

$$C(n, r) = \frac{n!}{r!(n - r)!}$$

Remark Notice that this formula is the same formula as for binomial coefficients.

EXAMPLE 4 Find the Number of Combinations of *n* Objects Taken *r* at a Time

A standard deck of playing cards consists of fifty-two cards. How many five-card hands can be chosen from this deck?

Solution We have *n* = 52 and *r* = 5. Thus

$$C(52, 5) = \frac{52!}{5!(52 - 5)!} = \frac{52!}{5!47!} = \frac{52 \cdot 51 \cdot 50 \cdot 49 \cdot 48 \cdot 47!}{5 \cdot 4 \cdot 3 \cdot 2 \cdot 1 \cdot 47!} = 2{,}598{,}960.$$

■ *Try Exercise **16**, page 641.*

EXAMPLE 5 Find the Number of Combinations of *n* Objects Taken at a Time

A chemist has nine samples of a solution, of which four are type *A* and five are type *B*. If the chemist chooses three of the solutions at random, in how many ways will the chemist have a. exactly one type *A* solution, b. more than one type *A* solution?

Solution

a. The chemist has chosen three solutions, one of which is type *A*. If one is type *A*, then two are type *B*. The number of ways of choosing one type *A* solution from four type *A* solutions is *C*(4, 1).

$$C(4, 1) = \frac{4!}{1!(4 - 1)!} = \frac{4!}{1!3!} = 4$$

The number of ways of choosing two type *B* solutions from five type *B* solutions is *C*(5, 2).

$$C(5, 2) = \frac{5!}{2!(5 - 2)!} = \frac{5!}{2!3!} = 10$$

By the counting principle, there are

$$C(4, 1) \cdot C(5, 2) = 4 \cdot 10 = 40$$

ways to have one type A and two type B solutions.

b.   More than one type A solution means two type A and one type B or three type A and zero type B solutions. We first calculate the number of ways of choosing two type A and one type B.

$$C(4, 2) = \frac{4!}{2!2!} = 6 \quad \text{Number of ways of choosing two type } A \text{ solutions from four type } A \text{ solutions}$$

$$C(5, 1) = \frac{5!}{1!4!} = 5 \quad \text{Number of ways of choosing one type } B \text{ solution from five type } B \text{ solutions}$$

The number of ways to have two type A and one type B is

$$C(4, 2) \cdot C(5, 1) = 6 \cdot 5 = 30 \,.$$

Now we calculate the number of ways of choosing three type A and zero type B solutions.

$$C(4, 3) = \frac{4!}{3!1!} = 4 \quad \text{Number of ways of choosing three type } A \text{ solutions from four type } A \text{ solutions.}$$

$$C(5, 0) = \frac{5!}{0!5!} = 1 \quad \text{Number of ways of choosing zero type } B \text{ solutions from five type } B \text{ solutions}$$

The number of ways to have three type A and zero type B is

$$C(4, 3) \cdot C(5, 0) = 4 \cdot 1 = 4 \,.$$

The number of ways to have more than one type A solution is

$$C(4, 2) \cdot C(5, 1) + C(4, 3) \cdot C(5, 0) = 30 + 4 = 34 \,.$$

■ *Try Exercise* **28,** *page 641.*

The difficult part of counting is determining whether the counting principle, the permutation formula, or the combination formula should be used. Following is a summary of guidelines.

Guidelines for Solving Counting Problems

1. The counting principle will always work but is not always the easiest method to apply.
2. When reading a problem, ask yourself "Is the order of the selection process important?" If the answer is yes, the arrangements are permutations. If the answer is no, the arrangements are combinations.

EXERCISE SET 12.6

In Exercises 1 to 10, evaluate each of the quantities.

1. $P(6, 2)$ **2.** $P(8, 7)$ **3.** $C(8, 4)$ **4.** $C(9, 2)$

5. $P(8, 0)$ **6.** $P(9, 9)$ **7.** $C(7, 7)$ **8.** $C(6, 0)$

9. $C(10, 4)$ **10.** $P(10, 4)$

11. How many different ways can six employees be assigned to six different jobs?

12. First-, second-, and third-place prizes are to be awarded in a dance contest in which twelve contestants are entered. In how many ways can the prizes be awarded?

13. There are five mailboxes outside a post office. How many ways can three letters be deposited into the five boxes?

14. How many different committees of three people can be selected from nine people?

15. A company has more than 676 employees. Explain why there must be at least 2 employees who have the same first and last initials.

16. A car holds six passengers, three in the front seat and three in the back seat. How many different seating arrangements of six people are possible if one person refuses to sit in front and one person refuses to sit in the back seat?

17. A committee of six people is chosen from six senators and eight representatives. How many committees are possible if there are to be three senators and three representatives on the committee?

18. The numbers 1, 2, 3, 4, 5, 6 are to be arranged. How many different arrangements are possible if

a. all the even numbers come first?

b. the arrangements are such that they alternate between even and odd?

19. A true-false examination contains ten questions. How many ways can a person answer the questions on this test by just guessing? Assume that all questions are answered.

20. A twenty-question four-option multiple choice examination is given as a pre-employment test. How many ways could a prospective employee answer the questions on this test by just guessing? Assume that all questions are answered.

21. A state lottery game requires a person to select six different numbers from forty numbers. The order of the selection is not important. How many ways can this be done?

22. A student must answer eight of ten questions on an exam. How many different choices can the student make?

23. A warehouse receives a shipment of ten computers, of which three are defective. Five computers are then randomly selected from the ten and delivered to a store.

a. In how many ways can the store receive no defective computers?

b. In how many ways can the store receive one defective computer?

c. In how many ways can the store receive all three defective computers?

24. A television manufacturer uses a code for the serial number of a television. The first symbol is the letter A, B, or C and represents the location of the manufacturing plant. The next two symbols are 01, 02, . . . , 12 and represents the month the set was manufactured. The next symbol is a 5, 6, 7, 8, or 9 and represents the year the set was manufactured. The last seven symbols are digits. How many serial numbers are possible?

25. Five cards are chosen at random from an ordinary deck of playing cards. In how many ways can the cards be chosen so that
a. all are hearts;   **b.** all are the same suit;   **c.** exactly three are kings;   **d.** two or more are aces?

26. A quality control inspector receives a shipment of ten computer disk drives and randomly selects three of the drives for testing. If two of the disk drives in the shipment are defective, find the number of ways in which the inspector could select at most one defective drive.

27. A basketball team has twelve members. In how many ways can five players be chosen if   **a.** the selection is random;   **b.** the two tallest players are always among the five selected?

28. The numbers 1, 2, 3, 4, 5, 6 are arranged in random order. In how many ways will the numbers 1 and 2 appear next to one another and in the order 1, 2?

29. Seven identical balls are randomly placed in seven available containers in such a way that two balls are in one container. Of the remaining six containers, each receives at most one ball. Find the number of ways this can be accomplished.

30. Seven points lie in a plane in such a way that no three points lie on the same line. How many lines are determined by 7 points?

31. A chess tournament has twelve participants. How many games must be scheduled if every player must play every other player exactly once?

32. Eight couples attend a benefit at which two prizes are given. In how many ways can two names be randomly drawn so that the prizes are not awarded to the same couple?

33. Suppose there are twelve distinct points on a circle. How many different triangles can be formed with vertices at the given points?

34. In how many ways can a student answer a twenty-question true-false test if the student marks ten of the questions true and ten of the questions false?

35. From a group of fifteen people a committee of eight is formed. From the committee a president, secretary, and treasurer are selected. Find the number of ways the two consecutive operations can be carried out.

36. From a group of twenty people a committee of twelve is formed. From the committee of twelve, a subcommittee of four people is chosen. Find the number of ways the two consecutive operations can be carried out.

Supplemental Exercises

37. Generalize Exercise 30. That is, given n points in a plane, no three of which lie on the same line, how many lines are determined by n points?

38. Seven people are asked the month of their birth. In how many ways will   **a.** no two people have a birthday in the same month?   **b.** at least two people have a birthday in the same month?

39. From a penny, nickel, dime, and quarter, how many different sums of money can be formed using one or more of the coins?

40. Five sticks of equal length are broken into a short piece and a long piece. The ten pieces are randomly arranged in five pairs. In how many ways will each pair consist of a long stick and a short stick? This exercise actually has a practical side. When cells are exposed to harmful radiation, some chromosomes break. If two long sides unite or two short sides unite, the cell dies.

41. Four random digits are drawn (repetitions are allowed). Among the four digits, in how many ways will two or more repetitions occur?

42. An aimless tourist, standing on a street corner, tosses a coin. If the result is heads, the tourist walks one block north. If the result is tails, the tourist walks one block south. At the new corner, the coin is tossed again and the same rule applied. If the coin is tossed ten times, in how many ways will the tourist be back at the original corner? This problem is an elementary example of what is called a "random walk." Random walk problems have many applications to physics, chemistry, and economics.

12.7
Introduction to Probability

Many events in the world around us have random character, such as the chances of an accident on a certain freeway, the chances of winning a state lottery, and the chances that the nucleus of an atom will undergo fission. By repeatedly observing such events, it is often possible to recognize certain patterns. **Probability** is the mathematical study of random patterns.

When a weather reporter predicts 30 percent rain, the forecaster is saying that similar weather conditions have led to rain 30 times out of 100. When a fair coin is tossed, we expect heads to occur $\frac{1}{2}$ or 50 percent of the time. The numbers 30 percent (or 0.3) and $\frac{1}{2}$ are the probabilities of the events.

An activity with an observable outcome is called an **experiment**. Examples of experiments include

1. Flipping a coin and observing the side facing upward

2. Observing the incidence of a disease in a certain population

3. Observing the length of time a person waits in a checkout line in a grocery store

The **sample space** of an experiment is the set of *all possible* outcomes of that experiment.

Consider the experiment of tossing one coin three times and recording the upward side of the coin. The sample space would be

$$S = \{HHH, HHT, HTH, THH, HTT, THT, TTH, TTT\}.$$

EXAMPLE 1   **List the Elements of a Sample Space**

Suppose among five batteries, two are defective. Two batteries are randomly drawn from the five and tested for defects. List the elements in the sample space.

Solution   Label the nondefective batteries N_1, N_2, N_3 and the defective batteries D_1, D_2. The sample space is

$$S = \{N_1D_1, N_2D_1, N_3D_1, N_1D_2, N_2D_2, N_3D_2, N_1N_2, N_1N_3, N_2N_3, D_1D_2\}$$

■ *Try Exercise* **6,** *page 648.*

An **event** E is any subset of a sample space. For the sample space defined in Example 1, several of the events we could define are

E_1: There are no defective batteries.

E_2: At least one battery is defective.

E_3: Both batteries are defective.

Because an event is a subset of the sample space, each of these events can be expressed as sets.

$$E_1 = \{N_1N_2, N_1N_3, N_2N_3\}$$
$$E_2 = \{N_1D_1, N_2D_1, N_3D_1, N_1D_2, N_2D_2, N_3D_2, D_1D_2\}$$
$$E_3 = \{D_1D_2\}$$

There are two methods by which elements are drawn from a sample space, with replacement and without replacement. *With replacement* means that after the element is drawn it is then returned to the sample space. The same element could be selected on the next drawing. When drawing elements *without replacement*, the element is not returned to the sample space and therefore is not available for any subsequent drawing.

EXAMPLE 2   **List the Elements of an Event**

Two numbers are drawn from the digits 1, 2, 3, 4 with replacement and without replacement. Express each event as a set.

a. E_1: the second number is greater than or equal to the first number;

b. E_2: both numbers are less than zero.

Solution

a.   With replacement: $E_1 = \{11, 12, 13, 14, 22, 23, 24, 33, 34, 44\}$

Without replacement: $E_1 = \{12, 13, 14, 23, 24, 34\}$

b. $E_3 = \varnothing$ Choosing from the digits 1, 2, 3, 4, this event is impossible. The impossible event is denoted by the empty set.

■ *Try Exercise* **14**, *page 648.*

The probability of an event is defined in terms of the concepts of sample space and event.

Probability of an Event

> Let $n(S)$ and $n(E)$ represent the number of elements in the sample space S and the event E, respectively. The probability of event E, $P(E)$, is
>
> $$P(E) = \frac{n(E)}{n(S)}$$

Remark   Because E is a subset of S, $n(E) \leq n(S)$. Thus $P(E) \leq 1$. If E is an impossible event, then $E = \varnothing$ and $n(E) = 0$. Thus $P(E) = 0$. If E is the event that *always* occurs, then $E = S$ and $n(E) = n(S)$. Thus $P(E) = 1$. Combining these elements, we have for any event E,

$$0 \leq P(E) \leq 1.$$

EXAMPLE 3   Calculate the Probability of an Event

A coin is tossed three times. What is the probability that   a. E_1: two or more heads will appear   b. E_2: at least one tail will appear?

Solution   First, determine the number of elements in the sample space. The sample space for this experiment is

$$S = \{HHH, HHT, HTH, THH, HTT, THT, TTH, TTT\}.$$

Therefore $n(S) = 8$. Now determine the number of elements in each event. Then calculate the probability of the event by using $P(E) = \dfrac{n(E)}{n(S)}$.

a. $E_1 = \{HHH, HHT, HTH, THH\}.$

Thus, $P(E_1) = \dfrac{n(E_1)}{n(S)} = \dfrac{4}{8} = \dfrac{1}{2}.$

b. $E_2 = \{HHT, HTH, THH, HTT, THT, TTH, TTT\}.$

$$P(E_2) = \frac{n(E_2)}{n(S)} = \frac{7}{8}$$

■ *Try Exercise* **22**, *page 648.*

Calculating probabilities by listing and then counting the elements of a sample space is not always practical. Instead, we will use the counting principles developed in the last section to determine the number of elements in the sample space and in an event.

EXAMPLE 4   Use the Counting Principles to Calculate a Probability

A state lottery game allows a person to choose five numbers from the integers 1 to 40. Repetitions of numbers are not allowed. If three or more numbers match the numbers chosen by the lottery, the player wins a prize. Find the probability that a player will match   a. exactly three numbers   b. exactly four numbers.

Solution   The sample space S is the number of ways five numbers can be chosen from forty numbers. This is a combination because the order of the drawing is not important.

$$n(S) = C(40, 5) = \frac{40!}{5!\,35!} = 658{,}008$$

We will call the five numbers chosen by the state lottery "lucky" and the remaining thirty-five numbers "unlucky."

a.   Let E_1 be the event a player has three lucky and therefore two unlucky numbers. The three lucky numbers are chosen from the five lucky numbers. There are $C(5, 3)$ ways to do this. The two unlucky numbers are chosen from the thirty-five unlucky numbers. There are $C(35, 2)$ ways to do this. By the counting principle, the number of ways the event E_1 can occur is

$$n(E_1) = C(5, 3) \cdot C(35, 2) = 5950 \,.$$

$$P(E_1) = \frac{n(E_1)}{n(S)} = \frac{C(5, 3) \cdot C(35, 2)}{C(40, 5)} = \frac{5950}{658{,}008} \approx 0.009042$$

b.   Let E_2 be the event a player has four lucky numbers and one unlucky number. The number of ways a person can select four lucky numbers and one unlucky number is $C(5, 4) \cdot C(35, 1)$.

$$P(E_2) = \frac{n(E_2)}{n(S)} = \frac{C(5, 4) \cdot C(35, 1)}{C(40, 5)} = \frac{175}{658{,}008} \approx 0.000266$$

■ Try Exercise 32, page 648.

The expression "one or the other of two events occurs" is written as the union of the two sets. For example, if an experiment leads to the sample space

$$S = \{1,\ 2,\ 3,\ 4,\ 5,\ 6\}$$

and the events are

Draw a number less than four $E_1 = \{1,\ 2,\ 3\}\,,$

Draw an even number $E_2 = \{2,\ 4,\ 6\}\,,$

then the event $E_1 \cup E_2$ is described by drawing a number less than four *or* an even number.

Thus,

$$E_1 \cup E_2 = \{1,\ 2,\ 3\} \cup \{2,\ 4,\ 6\} = \{1,\ 2,\ 3,\ 4,\ 6\}\,.$$

Two events, E_1 and E_2, that cannot occur at the same time are **mutu-**

ally exclusive events. Using set notation, if $E_1 \cap E_2 = \varnothing$, then E_1 and E_2 are mutually exclusive.

For example, use the same sample space as above, if

$$E_3 = \{1, 3, 5\} \quad \text{Draw an odd number.}$$

Then $E_2 \cap E_3 = \varnothing$ and the events E_2 and E_3 are mutually exclusive. On the other hand,

$$E_1 \cap E_2 = \{2\}$$

so the events E_1 and E_2 are not mutually exclusive.

One of the axioms of probability involves the union of mutually exclusive events.

A Probability Axiom

If E_1 and E_2 are mutually exclusive events, then

$$P(E_1 \cup E_2) = P(E_1) + P(E_2).$$

If the events are not mutually exclusive, the addition rule for probabilities can be used.

Addition Rule for Probabilities

If E_1 and E_2 are two events, then

$$P(E_1 \cup E_2) = P(E_1) + P(E_2) - P(E_1 \cap E_2).$$

The probability axiom and addition rule are useful when calculating probabilities of events connected by the word "or."

Using the calculations of Example 4, we can find the probability that a player will have three or four lucky numbers in the lottery. Because the events E_1 and E_2 as defined in Example 4 are mutually exclusive,

$$P(E_1 \cup E_2) = P(E_1) + P(E_2) = 0.009042 + 0.000266 = 0.009308.$$

As an example of nonmutually exclusive events, draw a card at random from a deck of ordinary playing cards. Find the probability of drawing an ace or a heart.

$$S = \{52 \text{ ordinary playing cards}\}$$

Let $E_1 = \{\text{an ace}\}$ and $E_2 = \{\text{a heart}\}$. Then,

$$P(E_1) = \frac{n(E_1)}{n(S)} = \frac{4}{52} = \frac{1}{13}, \qquad P(E_2) = \frac{n(E_2)}{n(S)} = \frac{13}{52} = \frac{1}{4}.$$

We have $E_1 \cup E_2 = \{\text{an ace } or \text{ a heart}\}$ and $E_1 \cap E_2 = \{\text{ace of hearts}\}$. First, we find $P(E_1 \cap E_2)$.

$$P(E_1 \cap E_2) = \frac{n(E_1 \cap E_2)}{n(S)} = \frac{1}{52}$$

Now we can find $P(E_1 \cup E_2)$.

$$P(E_1 \cup E_2) = P(E_1) + P(E_2) - P(E_1 \cap E_2) = \frac{1}{13} + \frac{1}{4} - \frac{1}{52} = \frac{16}{52} = \frac{4}{13}$$

Two events are **independent** if the outcome of the first event does not influence the outcome of the second event. As an example, consider tossing a fair coin twice. The outcome on the first toss has no bearing on the outcome of the second toss. The two events are independent.

Now consider drawing two cards in succession from a regular deck of playing cards. The probability that the second card drawn will be an ace depends on the card drawn first.

Probability Rule for Independent Events

If E_1 and E_2 are two independent events, then the probability that both E_1 *and* E_2 will occur is

$$P(E_1) \cdot P(E_2).$$

EXAMPLE 5 **Calculate the Probability of Independent Events**

A television survey suggests that 40 percent of a city population watch a nightly news program. If three people are randomly selected from the city, assuming the events are independent what is the probability that all three watch a nightly news program?

Solution If E is the event that one person watches a nightly news program, then we are given that $P(E) = 0.4$. The probability that all three people watch a news program is

$$P(E) \cdot P(E) \cdot P(E) = (0.4)(0.4)(0.4) = 0.064.$$

■ *Try Exercise* **28,** *page 648.*

Following are five guidelines for calculating probabilities.

Guidelines for Calculating a Probability

1. The word "or" usually means to add the probabilities of each event.
2. The word "and" usually means to multiply the probabilities of each event.
3. The phrase "at least n" means n or more. At least 5 would be 5 or more.
4. The phrase "at most n" means n or less. At most 5 would be 5 or less.
5. "Exactly n" means just that. Exactly 5 heads in 7 tosses of a coin would mean 5 heads *and therefore* 2 tails.

EXERCISE SET 12.7

In Exercises 1 to 10, list the elements in the sample space defined by the given experiment.

 1. Two people are selected from two senators and three representatives.

 2. A letter is chosen at random from the word "Tennessee."

 3. A fair coin is tossed and then a random integer between one and four, inclusive, is selected.

 4. A fair coin is tossed four times.

 5. Two identical tennis balls are randomly placed in three tennis ball cans.

 6. Two people are selected from one Republican, one Democrat, and one Independent.

 7. Three cards are randomly chosen from the ace of hearts, ace of spades, ace of clubs, and ace of diamonds.

 8. Three letters addressed to A, B, and C are randomly put into the three envelopes addressed to A, B, and C.

 9. Two vowels are randomly chosen from a, e, i, o, and u.

10. Three computer disks are randomly chosen from one defective disk and three nondefective disks.

In Exercises 11 to 15, use the sample space defined by the experiment of tossing a fair coin four times. Express each event as a subset of the sample space.

11. There are no tails.

12. There are exactly two heads.

13. There are at most two heads.

14. There are more than two heads.

15. There are twelve tails.

In Exercises 16 to 20, use the sample space defined by the experiment of choosing two random numbers, in succession, from the integers 1, 2, 3, 4, 5, 6. The numbers are chosen with replacement. Express each event as a subset of the sample space.

16. The sum of the numbers is 7.

17. The two numbers are the same.

18. The first number is greater than the second number.

19. The second number is a 4.

20. The sum of the two numbers is greater than 1.

In Exercises 21 to 44, calculate the probabilities of the events.

21. From a deck of regular playing cards, one card is chosen at random. What is the probability that   **a.** the card is a king;   **b.** the card is a spade?

22. A single number is chosen from the digits 1, 2, 3, 4, 5, and 6. Find the probability that the number is an even number or divisible by 3.

23. An economist predicts that the probability of an in-

crease in gross national product (GNP) is 0.64 and that the probability of an increase in inflation is 0.55. The economist also predicts that the probability of an increase in GNP and inflation is 0.22. Find the probability of an increase in GNP or an increase in inflation.

24. Four digits are selected from the digits 1, 2, 3, 4 and a number is formed. Find the probability that the number is greater than 3000, assuming digits can be repeated.

25. Four digits are selected from the digits 1, 2, 3, 4 and a number is formed. Find the probability that the number is greater than 3000, assuming digits cannot be repeated.

26. An owner of a construction firm has bid for the contracts of two buildings. If the contractor estimates that the probability of getting the first contract is $\frac{1}{2}$, getting the second contract is $\frac{1}{5}$, and getting both contracts is $\frac{1}{10}$, find the probability that the contractor will get at least one of the two building contracts.

27. A box contains 500 envelopes, of which 50 have $100 in cash, 75 have $50 in cash, and 125 have $25. If an envelope is selected at random from this box, what is the probability it will contain at least $50?

28. A missile radar detection system consists of two radar screens. The probability that any one of the radar screens will detect an incoming missile is 0.95. If a missile enters the detection space of this radar, assuming radar detections are independent events, what is the probability that at least one of the radar screens will detect it?

29. An oil drilling venture involves drilling four wells in different parts of the country. For each well, the probability that it will be profitable is 0.10, and the probability that it will be unprofitable is 0.90. If these events are independent, what is the probability of drilling at least one unprofitable well?

30. A software firm is considering marketing two newly designed spreadsheet programs, A and B. To test the appeal of the programs, the firm installs the programs in four corporations. After 2 weeks, the firm asks the corporations to evaluate each program. If the corporations have no preference, what is the probability that all four will choose product A?

31. A manufacturer of computer disks claims that 1 disk in 1000 will have a defect. Of the next three disks manufactured, what is the probability that all three are defective?

32. A shipment of ten calculators contains two defective calculators. Two calculators are chosen from the shipment. What is the probability that   **a.** both are defective;   **b.** at least one is defective?

33. A manufacturing company uses an acceptance scheme on articles it produces before shipping the articles. Crates of twenty items are readied for shipment, and

three articles are randomly selected and tested for defectives. If no defectives are found the shipment is sent. If any defectives are found, the entire box is tested. What is the probability of sending a shipment that contains three defective articles?

34. A magician claims to be able to read minds. To test this claim, five cards, numbered 1 to 5, are used. A subject selects two cards from the five and concentrates on the numbers. What is the probability that the magician can correctly identify the two numbers by just guessing?

35. Consider a lottery that sells 100 tickets and awards two prizes. If you purchase 5 tickets, what is the probability of winning a prize?

36. A quality control inspector receives a shipment of twenty computer consoles. From the lot of twenty, the inspector randomly chooses three consoles. If two of the consoles in the shipment are defective, what is the probability that the inspector will choose at least one defective console?

37. A jury of twelve people is selected from thirty people, fifteen women and fifteen men. What is the probability that the jury will have six men and six women?

38. Suppose a small town has two plumbers, A and B. On a certain day, three residents select a plumber at random. What is the probability that all three residents chose plumber A?

39. Six persons are arranged in a line. What is the probability that two specific people, say A and B, are standing next to each other?

40. Five random digits are chosen with replacement. What is the probability that the digit 0 does not occur?

41. A fruit grower claims that three-fourths of the current orange crop have suffered frost damage. Find the probability that among four oranges, exactly three have frost damage.

42. Suppose that airplane engines operate independently and that the probability that any one engine will fail is 0.03. A plane can make a safe flight if at least one-half of its engines operate. Is a safe flight more likely to occur in a two-engine or a four-engine plane? Why?

43. Three girls and three boys are randomly placed in six seats. What is the probability that boys and girls will be in alternate seats?

44. A committee of four is chosen from three accountants and five actuaries. Find the probability that the committee consists of two accountants and two actuaries.

Supplemental Exercises

45. A club has nine members. One member starts a rumor by telling a second club member, who repeats the rumor to a third person, and so on. At each stage, the recipient of the rumor is chosen at random from the nine club members. What is the probability that the rumor will be told three times without it returning to the originator?

46. As a test for extrasensory perception (ESP), ten cards, five black and five white, are shuffled and then a person looks at each card. In another room, the ESP subject attempts to guess whether the card is black or white. Additionally, the ESP subject must guess black five times and white five times. If the ESP subject has no extrasensory perception, what is the probability that the subject will correctly name eight of the ten cards?

47. The telephone extensions at a university are four-digit numbers. If two telephone numbers are randomly chosen from the telephone book, what is the probability that the last two digits (and no others) match?

48. Each arrangement of the letters of the word "Tennessee" are placed on a piece of paper and placed in a bowl. One piece of paper is selected at random. What is the probability that the first letter is a T?

Chapter 12 Review ▬▬▬▬▬▬▬▬▬▬▬▬

12.1 Infinite Sequences and Summation Notation

An infinite sequence is a function whose domain is the positive integers and whose range is a set of real numbers.

An alternating sequence is one in which the signs of the terms *alternate* between plus and minus values.

A recursively defined sequence is defined by using the preceding terms of the sequence.

If n is a positive integer, then n factorial, $n!$, is the product of the first n positive integers.

$$n! = n(n - 1)(n - 2) \cdots 3 \cdot 2 \cdot 1.$$

If a_n is a sequence, then $S_n = \sum_{i=1}^{n} a_i$ is the nth partial sum of the sequence.

12.2 Arithmetic Sequences and Series

Given that d is a constant, the sequence a_n is an arithmetic sequence if $a_{i+1} - a_i = d$ for all i. The constant d is called the common difference for the arithmetic sequence.

The nth term of an arithmetic sequence is $a_n = a_1 + (n - 1)d$.

If a_n is an arithmetic sequence, then the nth partial sum of the sequence is given by

$$S_n = \sum_{i=1}^{n} a_i = \frac{n(a_1 + a_n)}{2}$$

The nth partial sum of an arithmetic sequence is called a finite arithmetic series.

12.3 Geometric Sequences and Series

Given that $r \neq 0$ is a constant, the sequence a_n is a geometric sequence if $a_{i+1}/a_i = r$ for all i. The ratio r is called the common ratio for the geometric sequence.

The nth term of the geometric sequence is $a_n = a_1 r^{n-1}$, where a_1 is the first term of the sequence and r is the common ratio.

If a_n is a geometric sequence, then the nth partial sum of the sequence is given by

$$S_n = \sum_{i=1}^{n} a_i = \frac{a_1(1 - r^n)}{1 - r}, \quad r \neq 1.$$

The sum of all the terms of a geometric sequence is an infinite geometric series. If $|r| < 1$, then the sum of an infinite geometric series is given by

$$S = \sum_{i=1}^{\infty} a_i = \frac{a_1}{1 - r}$$

12.4 Mathematical Induction

Principle of Mathematical Induction
Let P_n be a statement about a positive integer n. If

1. P_1 is true, and
2. The truth of P_k implies the truth of P_{k+1}, then P_n is true for all positive integers.

12.5 The Binomial Theorem

Binomial Theorem for Positive Integers
If n is a positive integer, then

$$(a + b)^n = \sum_{i=0}^{n} \binom{n}{i} a^{n-i} b^i.$$

The Bionomial Theorem can be used to find the ith term in the expansion of a binomial. The ith term of $(a + b)^n$ is

$$\binom{n}{i-1} a^{n-i+1} b^{i-1}.$$

12.6 Permutations and Combinations

The fundamental counting principle is used to count the number of ways a sequence of n conditions can occur. A permutation is an arrangement of distinct objects in a definite order. The formula for the permutations of n distinct objects taken r at a time is

$$P(n, r) = \frac{n!}{(n - r)!}.$$

A combination is an arrangement of objects for which the order of the selection is not important. The formula for the number of combinations of n objects taken r at a time is

$$C(n, r) = \frac{n!}{r!(n - r)!}.$$

12.7 Introduction to Probability

Probability is the mathematical study of random events. The sample space of an experiment is the set of all possible outcomes of an experiment. An event is any subset of a sample space.

If S is the sample space of an experiment and E is an event in the sample space, then the probability of an event is given by

$$P(E) = \frac{n(E)}{n(S)}$$

where $n(E)$ and $n(S)$ are the number of elements in E and S, respectively.

CHALLENGE EXERCISES

In Exercises 1 to 15, answer true or false. If the statement is false, give an example.

1. $0! \cdot 4! = 0$

2. $\left(\sum_{i=1}^{3} a_i\right)\left(\sum_{i=1}^{3} b_i\right) = \sum_{i=1}^{3} a_i b_i$

3. $\dfrac{n(n-1)(n-2)\cdots(n-k+1)}{k!} = C(n, k)$

4. No two terms of a sequence can be equal.

5. $1, 8, 27, 64, \ldots, k^3, \ldots$ is a geometric sequence.

6. $a_1 = 2, a_{n+1} = a_n - 3$ defines an arithmetic sequence.

7. $0.\overline{9} = 1$.

8. Adding all the terms of an infinite sequence produces an infinite sum.

9. Because the first step of an induction proof is normally easy, this step can be omitted.

10. In the expansion of $(a + b)^8$, the exponent on a for the fifth term is 5.

11. The counting principle states that if there is n ways to satisfy one condition and m ways to satisfy a second condition, then there are $n + m$ ways to satisfy both conditions.

12. The permutations of n things taken r at a time is given by $n!/r!$.

13. If E is an event in a sample space, then $0 \le P(E) \le 1$ where $P(E)$ is the probability of E.

14. If A and B are mutually exclusive events, then $P(A \cap B) = 1$.

15. If a coin is tossed five times, then the probability of observing HHHHH is the same as the probability of observing HTHHT.

REVIEW EXERCISES

In Exercises 1 to 20, find the third and seventh terms of the sequence defined by a_n.

1. $a_n = n^2$
2. $a_n = n!$
3. $a_n = 3n + 2$
4. $a_n = 1 - 2n$
5. $a_n = 2^{-n}$
6. $a_n = 3^n$
7. $a_n = \dfrac{1}{n!}$
8. $a_n = \dfrac{1}{n}$
9. $a_n = \left(\dfrac{2}{3}\right)^n$
10. $a_n = \left(-\dfrac{4}{3}\right)^n$
11. $a_1 = 2, a_n = 3a_{n-1}$
12. $a_1 = -1, a_n = 2a_{n-1}$
13. $a_1 = 1, a_n = -na_{n-1}$
14. $a_1 = 2, a_n = n^2 a_{n-1}$
15. $a_1 = 4, a_n = a_{n-1} + 2$
16. $a_1 = 3, a_n = a_{n-1} - 3$
17. $a_1 = 1, a_2 = 2, a_n = a_{n-1}a_{n-2}$
18. $a_1 = 1, a_2 = 2, a_n = a_{n-1}/a_{n-2}$
19. $a_1 = -1, a_n = 3na_{n-1}$
20. $a_1 = 2, a_n = -2na_{n-1}$

21–40. Classify the sequences defined in Exercises 1 to 20 as arithmetic, geometric, or neither.

In Exercises 41 to 56, find the indicated sum of the series.

41. $\sum_{n=1}^{9} (2n - 3)$

42. $\sum_{i=1}^{11} (1 - 3i)$

43. $\sum_{k=1}^{8} (4k + 1)$

44. $\sum_{i=1}^{10} (i^2 + 3)$

45. $\sum_{n=1}^{6} 3 \cdot 2^n$

46. $\sum_{i=1}^{5} 2 \cdot 4^{i-1}$

47. $\sum_{k=1}^{9} (-1)^k 3^k$

48. $\sum_{i=1}^{8} (-1)^{i+1} 2^i$

49. $\sum_{i=1}^{10} \left(\dfrac{2}{3}\right)^i$

50. $\sum_{i=1}^{11} \left(\dfrac{3}{2}\right)^i$

51. $\sum_{n=1}^{9} \dfrac{(-1)^{n+1}}{n^2}$

52. $\sum_{k=1}^{5} \dfrac{(-1)^{k+1}}{k!}$

53. $\sum_{n=1}^{\infty} \left(\dfrac{1}{4}\right)^n$

54. $\sum_{i=1}^{\infty} \left(-\dfrac{5}{6}\right)^i$

55. $\sum_{k=1}^{\infty} \left(-\dfrac{4}{5}\right)^k$

56. $\sum_{j=0}^{\infty} \left(\dfrac{1}{5}\right)^j$

In Exercises 57 to 64, prove each statement by induction.

57. $\sum_{i=1}^{n} (5i + 1) = \dfrac{n(5n + 7)}{2}$

58. $\sum_{i=1}^{n} (3 - 4i) = n(1 - 2n)$

59. $\sum_{i=0}^{n} \left(-\dfrac{1}{2}\right)^i = \dfrac{2(1 - (-1/2)^{n+1})}{3}$

60. $\sum_{i=0}^{n} (-1)^i = \dfrac{1 - (-1)^{n+1}}{2}$

61. $n^n \ge n!$

62. $n! > 4^n; n \ge 9$

63. 3 is a factor of $n^3 + 2n$ for all positive integers.

64. Let $a_1 = \sqrt{2}$ and $a_n = (\sqrt{2})^{a_{n-1}}$. Prove that $a_n < 2$ for all positive integers n.

In Exercises 65 to 68, use the Binomial Theorem to expand each binomial.

65. $(4a - b)^5$

66. $(x + 3y)^6$

67. $(\sqrt{a} + 2\sqrt{b})^8$

68. $\left(2x - \dfrac{1}{2x}\right)^7$

69. Find the fifth term in the expansion of $(3x - 4y)^7$.

70. Find the eighth term in the expansion of $(1 - 3x)^9$.

71. A computer password consists of eight letters. How many passwords are possible? Assume there is no difference between lower case and upper case letters.

72. The serial number on an airplane begins with the letter N, followed by six numerals, followed by one letter. How many serial numbers are possible?

73. From a committee of fifteen members, a president, vice president, and treasurer are elected. In how many ways can this be accomplished?

74. The emergency staff for a hospital consists of four supervisors and twelve regular employees. How many shifts of four people can be formed if each shift must contain exactly one supervisor?

75. From twelve people, a committee of five people is formed. In how many ways can this be accomplished if there are two people among the twelve who refuse to serve together on the committee?

76. A shipment of ten calculators contains two defective ones. A quality control inspector randomly chooses four of the calculators for testing. What is the probability that the inspector will choose one defective calculator?

77. A nickel, dime, and quarter are tossed. What is the probability that the nickel and dime will show heads and the quarter will show tails? What is the probability that only one of the coins will show tails?

78. A deck of ten cards contains five red and five black cards. If four cards are drawn from the deck, what is the probability that two are red and two are black?

79. For the 1000 numbers 000 to 999, what is the probability that the middle digit is greater than the other two digits?

80. Two numbers are chosen, with replacement, from the digits 1, 2, 3, 4, 5, 6 and their sum is recorded. Now two more digits are selected and their sum noted. This process continues until the sum is 7 or the original sum is obtained. If the original sum was 9, what is the probability of having another sum of 9 before having a sum of 7? (*Hint:* Assume the events are independent. The probability can be found by summing an infinite geometric series.)

81. Which of the following has the greater probability: Drawing an ace and a ten card (ten, jack, queen, king) from a regular deck of fifty-two playing cards or drawing an ace and a ten card from two decks of regular playing cards?

82. From the digits 1, 2, 3, 4, and 5 two numbers are chosen without replacement. What is the probability that the second number is greater than the first number?

83. A room contains twelve people who are wearing badges numbered 1 to 12. If three people are randomly selected, what is the probability that the person wearing badge 6 will be included?

Using Tables

Logarithmic Tables

If a calculator is not available, then the values of logarithmic functions can be found by using the tables in this book.

Let x be a number written in scientific notation. That is, let $x = a \cdot 10^k$ where $1 \le a < 10$ and k is an integer. Applying some properties of logarithms, we have

$$\log x = \log(a \cdot 10^k)$$
$$= \log a + \log 10^k \quad \text{The Product Property}$$
$$= \log a + k \quad\quad \log_b b^p = p.$$

TABLE A1   Portion of a Common Logarithm Table

x	0	1	2	3
2.0	.3010	.3032	.3054	.3075
2.1	.3222	.3243	.3263	.3284
2.2	.3424	.3444	.3464	.3483
2.3	.3617	.3636	.3655	.3674
2.4	.3802	.3820	.3838	.3856
2.5	.3979	.3997	.4014	.4031
2.6	.4150	.4166	.4183	.4200

This last equation states that given any real number x, the real number $\log x$ is the sum of the logarithm of a number between one and ten, and an integer k. The real number $\log a$ is called the **mantissa** of $\log x$ and k is called the **characteristic** of $\log x$.

The Table of Common Logarithms in Appendix II gives the common logarithms accurate to 4 decimal places of numbers between 1.00 and 9.99 in increments of 0.01. For example, from the section of the Table of Common Logarithms at the left, $\log 2.43 \approx 0.3856$.

EXAMPLE 1   Evaluate Logarithms

Find   a. $\log 54{,}600$ b. $\log 54.6$ c. $\log 0.00546$

Solution

a. $\log 54{,}600 = \log(5.46 \times 10^4)$ Write 54,600 in scientific notation.

$\phantom{\log 54{,}600} = \log 5.46 + \log 10^4$ The Product Property

$\phantom{\log 54{,}600} \approx 0.7372 + 4 = 4.7372$ $\log 10^4 = 4$

b. $\log 54.6 = \log(5.46 \times 10^1)$

$ = \log 5.46 + \log 10^1$

$ \approx 0.7372 + 1 = 1.7372$

c. $\log 0.00546 = \log(5.46 \times 10^{-3})$

$ = \log 5.46 + \log 10^{-3}$

$ \approx 0.7372 + (-3)$

In Example 1c, adding the characteristic to the mantissa would result in a negative logarithm. This form of a logarithm is inconvenient to use with logarithmic tables because these tables contain only positive mantissas. The value of log 0.00546 can be written in many forms.

For example, $\log 0.00546 \approx 0.7372 + (-3)$

$$\log 0.00546 \approx 0.7372 + (7 - 10) = 7.7372 - 10$$

$$\log 0.00546 \approx 0.7372 + (1 - 4) = 1.7372 - 4$$

Of these forms, log 0.00546 = 7.7372 − 10 is the most common.

Caution   When a logarithm has a negative characteristic like $0.7372 + (-3)$, sometimes $0.7372 + (-3)$ is incorrectly written as −3.7372. The correct result is −2.2628.

As a final note, when using a calculator, log 0.00546 will be displayed as −2.26280736. By adding and subtracting 10 to this number, we can write the number with a positive mantissa.

$$\log 0.00546 \approx -2.26280736$$

$$= (-2.26280736 + 10) - 10$$

$$= 7.73719264 - 10$$

Adding and subtracting 10 is a somewhat arbitrary choice. Any integer that will produce a positive mantissa could be used. For example,

$$\log 0.00546 \approx -2.26280736$$

$$= (-2.26280736 + 3) - 3$$

$$= 0.73719264 - 3$$

We now examine the process of finding x given log x. For example, given log $x = 0.3522$, locate 0.3522 in the *body* of the Table of Common Logarithms. Find the number that corresponds to this mantissa. The number is 2.25. Thus log 2.25 ≈ 0.3522 and 2.25 is called the antilogarithm of 0.3522. We write antilog 2.25 ≈ 0.3522.

TABLE A2   Portion of a Common Logarithm Table

x	3	4	5	6
2.0	.3075	.3096	.3118	.3139
2.1	.3284	.3304	.3324	.3345
2.2	.3483	.3502	.3522	.3541
2.3	.3674	.3692	.3711	.3729
2.4	.3856	.3874	.3892	.3909
2.5	.4031	.4048	.4065	.4082
2.6	.4200	.4216	.4232	.4249

EXAMPLE 2   **Evaluate Antilogarithms**

Find   a.   antilog 3.4639   b.   antilog −1.4881

Solution

a.   antilog 3.4639 = antilog(0.4639 + 3)   Write the number as the sum of the mantissa and the characteristic.

$\approx 2.91 \times 10^3$ Use the table to find the number that corresponds to the mantissa. The characteristic is the exponent on 10.

= 2910

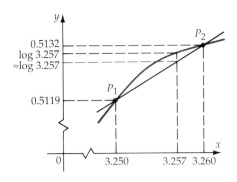

Figure A.1

b. The first step is to write the antilogarithm with a positive mantissa.

$$\text{antilog } -1.4881 = \text{antilog}(-1.4881 + 2 - 2)$$
$$= \text{antilog}(0.5119 - 2)$$
$$\approx 3.25 \times 10^{-2} = 0.0325$$

Linear Interpolation

Many functions can be approximated, over small intervals, by a straight line. Using this technique, we can approximate logarithms and antilogarithms of numbers that are not given in the Table of Common Logarithms.

A portion of the graph of $y = \log x$ is shown in Figure A.1. The points $P_1(3.250, 0.5119)$ and $P_2(3.260, 0.5132)$ are two points on the graph of $y = \log x$. The value of $\log 3.257$ is on the curve. An approximation for this value is on the straight line connecting P_1 and P_2. Because the slopes are the same between any two points on a straight line, we can write

$$\frac{y_2 - y_1}{x_2 - x_1} = \frac{y - y_1}{x - x_1}.$$

Substituting the coordinates of P_1, P_2, and using $x = 3.257$, we have

$$\frac{0.5132 - 0.5119}{3.260 - 3.250} = \frac{y - 0.5119}{3.257 - 3.250}.$$

Solving for y, we have
$$\frac{0.0013}{0.01} = \frac{y - 0.5199}{0.007}$$
$$0.00091 = y - 0.5119$$
$$0.51281 = y$$

$\log 3.257 \approx 0.5128$ rounded to the nearest ten thousandth.

A more convenient method of using linear interpolation is illustrated in the following example.

EXAMPLE 3   Evaluate Logarithms Using Linear Interpolation

Find $\log 0.02903$.

Solution   Arrange the numbers and their corresponding mantissas in a table. Indicate the differences between the numbers and the mantissas as shown using d to represent the unknown difference between the mantissa of 2.900 and 2.910.

$$\log 0.02903 = \log(2.903 \times 10^{-2}) = \log 2.903 + \log 10^{-2}$$

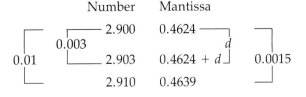

Write a proportion and solve for d.

$$\frac{0.003}{0.01} = \frac{d}{0.0015}$$

$$0.0005 = d$$

Add $d = 0.0005$ to the mantissa 0.4624.

$$\log 0.02903 = \log 2.903 + \log 10^{-2}$$
$$\approx (0.4624 + 0.0005) + (-2)$$
$$= 0.4629 + (-2) = 8.4629 - 10$$

Therefore $\log 0.02903 \approx 8.4629 - 10$.

Linear interpolation can also be used to find an antilogarithm.

EXAMPLE 4   Evaluate Antilogarithms Using Linear Interpolation

Find antilog(9.8465 − 10).

Solution

$$
\begin{array}{ccc}
 & \text{Mantissa} & \text{Antilog} \\
 & 0.8463 & 7.02 \\
0.0007 \left[\; 0.0002 \left[\begin{array}{c} \\ 0.8465 \end{array} \right. \right. & & \left. \begin{array}{c} 7.02 + d \end{array} \right] d \; \right] 0.01 \\
 & 0.8470 & 7.03
\end{array}
$$

Write a proportion and solve for d.

$$\frac{0.0002}{0.0007} = \frac{d}{0.01}$$

$$0.003 \approx d$$

Thus,

$$\text{antilog}(9.8465 - 10) \approx (7.02 + d) \times 10^{-1} = (7.02 + 0.003) \times 10^{-1}$$
$$= 7.023 \times 10^{-1} = 0.7023$$

Using Trigonometric Tables

The values of trigonometric functions can be found by using the Table of Values of Trigonometric Functions found in Appendix III. These tables give the values of the trigonometric functions, accurate to four significant digits, in increments of 10′.

To use the tables of trigonometric functions, first note that the table has degrees listed in both the right- and left-hand columns and the trigonometric functions are listed in both the top and bottom rows of the tables. For angles listed at the left, use the top row of functions. For angles listed at the right, use the bottom row of functions. For example, to determine the value of sin 27°40', locate 27°40' in the left-hand column of the table and read across until you locate the column at the top headed by "sin." Thus, sin 27°40' ≈ 0.4643.

To determine the value of cos 72°20', locate 72°20' in the right-hand column of the table and read across until you locate the column at the bottom headed by "cos." Thus, cos 72°20' ≈ 0.3035.

The Table of Trigonometric Functions only gives the values of the trigonometric functions from 0° to 90°. To find the value of a trigonometric function for other angles, first find the reference angle. Then attach the appropriate sign depending on the quadrant where the terminal side of the angle lies.

EXAMPLE 1   Evaluate Trigonometric Functions Using the Reference Angle

Find   a. tan 153°40'   b. sec(−87°10')   c. sin 3.9910

Solution

a. tan 153°40' = −tan 26°20'   180° − 153°40' = 26°20'

$$≈ −0.4950$$

b. sec(−87°10') = sec 87°10' ≈ 20.23

c. In this case, we are finding the sine of a real number. Find the reference number in the column headed by θ.

$$\sin 3.9910 ≈ −\sin 0.8494 ≈ −0.7509 \quad 3.9910 − π ≈ 0.8494$$

We now examine the process of finding θ given a trigonometric function of θ. For example, given cos θ = 0.5736 with θ an angle in the first quadrant, locate 0.5736 in the body of the table in the cosine column. Thus θ ≈ 55° and cos 55° ≈ 0.5736. Because the trigonometric functions are periodic, it is necessary to specify some bounds on θ.

EXAMPLE 2   Solve a Trigonometric Equation

Solve the equation for 0° ≤ θ < 360°.

a. sin θ = 0.5225   b. tan θ = −1.611

Solution

a. Let θ' be the reference angle so that sin θ' = 0.5225. From the table, θ' ≈ 31°30'. Because sin θ is positive in quadrants I and II,

$$θ ≈ 31°30' \quad \text{or} \quad θ ≈ 180° − 31°30' = 148°30'.$$

b. Let θ' be the reference angle so that tan θ' = 1.611. From the table, θ' ≈ 58°10'. Because tan θ is negative in quadrants II and IV,

$$θ ≈ 180° − 58°10' = 121°50' \quad \text{or} \quad θ ≈ 360° − 58°10' = 301°50'.$$

Using the method of linear interpolation, we can approximate trigonometric function values that are not given in the tables.

EXAMPLE 3   Evaluate Trigonometric Functions

Find   a. $\cos 58°28'$ b. $\tan(-0.5828)$ c. $\csc 231°12'$

Solution

a.

$$10' \left[\; 8' \left[\begin{array}{l} \cos 58°20' \approx 0.5250 \\ \cos 58°28' \approx 0.5250 + d \end{array} \right] d \;\; \right] -0.0025$$
$$\cos 58°30' \approx 0.5225$$

Write a proportion and solve for d.

$$\frac{8}{10} = \frac{d}{-0.0025}$$

$$d = -0.0020$$

$$\cos 58°28' \approx 0.5250 + (-0.0020) = 0.5230.$$

b.   Let $\theta' = 0.5828$ be the reference number. Look for 0.5828 in the θ column.

$$0.0029 \left[\; 0.0010 \left[\begin{array}{l} \tan 0.5818 \approx 0.6577 \\ \tan 0.5828 \approx 0.6577 + d \end{array} \right] d \;\; \right] 0.0042$$
$$\tan 0.5847 \approx 0.6619$$

Write a proportion and solve for d.

$$\frac{0.0010}{0.0029} = \frac{d}{0.0042}$$

$$d \approx 0.0014$$

$$\tan(-0.5828) = -\tan 0.5828 \approx -(0.6577 + 0.0014) = -0.6591.$$

c.   Let $\theta' = 51°12'$ be the reference angle for $231°12'$.

$$10' \left[\; 2' \left[\begin{array}{l} \csc 51°10' \approx 1.284 \\ \csc 51°12' \approx 1.284 + d \end{array} \right] d \;\; \right] -0.0030$$
$$\csc 51°20' \approx 1.281$$

Write and solve a proportion for d.

$$\frac{2}{10} = \frac{d}{-0.0030}$$

$$d \approx -0.0006$$

$$\csc 231°12' = -\csc 51°12' = -(1.284 - 0.0006) = -1.283.$$

We can also use interpolation to find θ given the value of the trigonometric function of θ.

EXAMPLE 4   Solve a Trigonometric Equation

Solve the equation $\cot \theta = 1.044$ where $0° \leq \theta < 360°$.

Solution   Search the body of the trigonometric table to find the two values of θ', where θ' is the reference angle, for which $\cot \theta$ is closest to 1.044.

$$10' \left[\; d \left[\begin{array}{l} \cot 43°40' \approx 1.048 \\ \cot(43°40' + d) \approx 1.044 \end{array} \right. \begin{array}{r} \Big] -0.004 \\ \end{array} \right. \left. \begin{array}{r} \\ \Big] -0.006 \end{array} \right.$$
$$\cot 43°50' \approx 1.042$$

Write a proportion and solve for d.

$$\frac{d}{10} = \frac{-0.004}{-0.0060}$$

$$d \approx 7'$$

Thus $\theta' \approx 43°40' + 7' = 43°47'$. Because $\cot \theta$ is positive in the first and third quadrants,

$$\theta \approx 43°47' \quad \text{or} \quad \theta \approx 180° + 43°47' = 223°47'.$$

EXAMPLE 5   Solve a Trigonometric Equation

Solve the equation $\sin \theta = -0.5672$ where $0 \leq \theta < 2\pi$.

Solution   Find two values of θ', where θ' is the reference number for θ, for which $\sin \theta$ is closest to 0.5672.

$$0.0029 \left[\; d \left[\begin{array}{l} \sin 0.6021 \approx 0.5664 \\ \sin(0.6021 + d) \approx 0.5672 \end{array} \right. \begin{array}{r} \Big] 0.0008 \\ \end{array} \right. \left. \begin{array}{r} \\ \Big] 0.0024 \end{array} \right.$$
$$\sin 0.6050 \approx 0.5688$$

Write a proportion and solve for d.

$$\frac{d}{0.0029} = \frac{0.0008}{0.0024}$$

$$d \approx 0.0010$$

Thus $\theta' \approx 0.6021 + 0.0010 = 0.6031$. Because $\sin \theta$ is negative in the third and fourth quadrants,

$$\theta \approx \pi + 0.6031 \approx 3.7447 \quad \text{or} \quad \theta \approx 2\pi - 0.6031 \approx 5.6801.$$

Appendix *II*

Logarithmic and Exponential Tables

Common Logarithms

x	0	1	2	3	4	5	6	7	8	9
1.0	.0000	.0043	.0086	.0128	.0170	.0212	.0253	.0294	.0334	.0374
1.1	.0141	.0453	.0492	.0531	.0569	.0607	.0645	.0682	.0719	.0755
1.2	.0792	.0828	.0864	.0899	.0934	.0969	.1004	.1038	.1072	.1106
1.3	.1139	.1173	.1206	.1239	.1271	.1303	.1335	.1367	.1399	.1430
1.4	.1461	.1492	.1523	.1553	.1584	.1614	.1644	.1673	.1703	.1732
1.5	.1761	.1790	.1818	.1847	.1875	.1903	.1931	.1959	.1987	.2014
1.6	.2041	.2068	.2095	.2122	.2148	.2175	.2201	.2227	.2253	.2279
1.7	.2304	.2330	.2355	.2380	.2405	.2430	.2455	.2480	.2504	.2529
1.8	.2553	.2577	.2601	.2625	.2648	.2672	.2695	.2718	.2742	.2765
1.9	.2788	.2810	.2833	.2856	.2878	.2900	.2923	.2945	.2967	.2989
2.0	.3010	.3032	.3054	.3075	.3096	.3118	.3139	.3160	.3181	.3201
2.1	.3222	.3243	.3263	.3284	.3304	.3324	.3345	.3365	.3385	.3404
2.2	.3424	.3444	.3464	.3483	.3502	.3522	.3541	.3560	.3579	.3598
2.3	.3617	.3636	.3655	.3674	.3692	.3711	.3729	.3747	.3766	.3784
2.4	.3802	.3820	.3838	.3856	.3874	.3892	.3909	.3927	.3945	.3962
2.5	.3979	.3997	.4014	.4031	.4048	.4065	.4082	.4099	.4116	.4133
2.6	.4150	.4166	.4183	.4200	.4216	.4232	.4249	.4265	.4281	.4298
2.7	.4314	.4330	.4346	.4362	.4378	.4393	.4409	.4425	.4440	.4456
2.8	.4472	.4487	.4502	.4518	.4533	.4548	.4564	.4579	.4594	.4609
2.9	.4624	.4639	.4654	.4669	.4683	.4698	.4713	.4728	.4742	.4757
3.0	.4771	.4786	.4800	.4814	.4829	.4843	.4857	.4871	.4886	.4900
3.1	.4914	.4928	.4942	.4955	.4969	.4983	.4997	.5011	.5024	.5038
3.2	.5051	.5065	.5079	.5092	.5105	.5119	.5132	.5145	.5159	.5172
3.3	.5185	.5198	.5211	.5224	.5237	.5250	.5263	.5276	.5289	.5302
3.4	.5315	.5328	.5340	.5353	.5366	.5378	.5391	.5403	.5416	.5428
3.5	.5441	.5453	.5465	.5478	.5490	.5502	.5514	.5527	.5539	.5551
3.6	.5563	.5575	.5587	.5599	.5611	.5623	.5635	.5647	.5658	.5670
3.7	.5682	.5694	.5705	.5717	.5729	.5740	.5752	.5763	.5775	.5786
3.8	.5798	.5809	.5821	.5832	.5843	.5855	.5866	.5877	.5888	.5899
3.9	.5911	.5922	.5933	.5944	.5955	.5966	.5977	.5988	.5999	.6010
4.0	.6021	.6031	.6042	.6053	.6064	.6075	.6085	.6096	.6107	.6117
4.1	.6128	.6138	.6149	.6160	.6170	.6180	.6191	.6201	.6212	.6222
4.2	.6232	.6243	.6253	.6263	.6274	.6284	.6294	.6304	.6314	.6325
4.3	.6335	.6345	.6355	.6365	.6375	.6385	.6395	.6405	.6415	.6425
4.4	.6435	.6444	.6454	.6464	.6474	.6484	.6493	.6503	.6513	.6522
4.5	.6532	.6542	.6551	.6561	.6571	.6580	.6590	.6599	.6609	.6618
4.6	.6628	.6637	.6646	.6656	.6665	.6675	.6684	.6693	.6702	.6712
4.7	.6721	.6730	.6739	.6749	.6758	.6767	.6776	.6785	.6794	.6803
4.8	.6812	.6821	.6830	.6839	.6848	.6857	.6866	.6875	.6884	.6893
4.9	.6902	.6911	.6920	.6928	.6937	.6946	.6955	.6964	.6972	.6981

A9

Common Logarithms (continued)

x	0	1	2	3	4	5	6	7	8	9
5.0	.6990	.6998	.7007	.7016	.7024	.7033	.7042	.7050	.7059	.7067
5.1	.7076	.7084	.7093	.7101	.7110	.7118	.7126	.7135	.7143	.7152
5.2	.7160	.7168	.7177	.7185	.7193	.7202	.7210	.7218	.7226	.7235
5.3	.7243	.7251	.7259	.7267	.7275	.7284	.7292	.7300	.7308	.7316
5.4	.7324	.7332	.7340	.7348	.7356	.7364	.7372	.7380	.7388	.7396
5.5	.7404	.7412	.7419	.7427	.7435	.7443	.7451	.7459	.7466	.7474
5.6	.7482	.7490	.7497	.7505	.7513	.7520	.7528	.7536	.7543	.7551
5.7	.7559	.7566	.7574	.7582	.7589	.7597	.7604	.7612	.7619	.7627
5.8	.7634	.7642	.7649	.7657	.7664	.7672	.7679	.7686	.7694	.7701
5.9	.7709	.7716	.7723	.7731	.7738	.7745	.7752	.7760	.7767	.7774
6.0	.7782	.7789	.7796	.7803	.7810	.7818	.7825	.7832	.7839	.7846
6.1	.7853	.7860	.7868	.7875	.7882	.7889	.7896	.7903	.7910	.7917
6.2	.7924	.7931	.7938	.7945	.7952	.7959	.7966	.7973	.7980	.7987
6.3	.7993	.8000	.8007	.8014	.8021	.8028	.8035	.8041	.8048	.8055
6.4	.8062	.8069	.8075	.8082	.8089	.8096	.8102	.8109	.8116	.8122
6.5	.8129	.8136	.8142	.8149	.8156	.8162	.8169	.8176	.8182	.8189
6.6	.8195	.8202	.8209	.8215	.8222	.8228	.8235	.8241	.8248	.8254
6.7	.8261	.8267	.8274	.8280	.8287	.8293	.8299	.8306	.8312	.8319
6.8	.8325	.8331	.8338	.8344	.8351	.8357	.8363	.8370	.8376	.8382
6.9	.8388	.8395	.8401	.8407	.8414	.8420	.8426	.8432	.8439	.8445
7.0	.8451	.8457	.8463	.8470	.8476	.8482	.8488	.8494	.8500	.8506
7.1	.8513	.8519	.8525	.8531	.8537	.8543	.8549	.8555	.8561	.8567
7.2	.8573	.8579	.8585	.8591	.8597	.8603	.8609	.8615	.8621	.8627
7.3	.8633	.8639	.8645	.8651	.8657	.8663	.8669	.8675	.8681	.8686
7.4	.8692	.8698	.8704	.8710	.8716	.8722	.8727	.8733	.8739	.8745
7.5	.8751	.8756	.8762	.8768	.8774	.8779	.8785	.8791	.8797	.8802
7.6	.8808	.8814	.8820	.8825	.8831	.8837	.8842	.8848	.8854	.8859
7.7	.8865	.8871	.8876	.8882	.8887	.8893	.8899	.8904	.8910	.8915
7.8	.8921	.8927	.8932	.8938	.8943	.8949	.8954	.8960	.8965	.8971
7.9	.8976	.8982	.8987	.8993	.8998	.9004	.9009	.9015	.9020	.9025
8.0	.9031	.9036	.9042	.9047	.9053	.9058	.9063	.9069	.9074	.9079
8.1	.9085	.9090	.9096	.9101	.9106	.9112	.9117	.9122	.9128	.9133
8.2	.9138	.9143	.9149	.9154	.9159	.9165	.9170	.9175	.9180	.9186
8.3	.9191	.9196	.9201	.9206	.9212	.9217	.9222	.9227	.9232	.9238
8.4	.9243	.9248	.9253	.9258	.9263	.9269	.9274	.9279	.9284	.9289
8.5	.9294	.9299	.9304	.9309	.9315	.9320	.9325	.9330	.9335	.9340
8.6	.9345	.9350	.9355	.9360	.9365	.9370	.9375	.9380	.9385	.9390
8.7	.9395	.9400	.9405	.9410	.9415	.9420	.9425	.9430	.9435	.9440
8.8	.9445	.9450	.9455	.9460	.9465	.9469	.9474	.9479	.9484	.9489
8.9	.9494	.9499	.9504	.9509	.9513	.9518	.9523	.9528	.9533	.9538
9.0	.9542	.9547	.9552	.9557	.9562	.9566	.9571	.9576	.9581	.9586
9.1	.9590	.9595	.9600	.9605	.9609	.9614	.9619	.9624	.9628	.9633
9.2	.9638	.9643	.9647	.9652	.9657	.9661	.9666	.9671	.9675	.9680
9.3	.9685	.9689	.9694	.9699	.9703	.9708	.9713	.9717	.9722	.9727
9.4	.9731	.9736	.9741	.9745	.9750	.9754	.9759	.9763	.9768	.9773
9.5	.9777	.9782	.9786	.9791	.9795	.9800	.9805	.9809	.9814	.9818
9.6	.9823	.9827	.9832	.9836	.9841	.9845	.9850	.9854	.9859	.9863
9.7	.9868	.9872	.9877	.9881	.9886	.9890	.9894	.9899	.9903	.9908
9.8	.9912	.9917	.9921	.9926	.9930	.9934	.9939	.9943	.9948	.9952
9.9	.9956	.9961	.9965	.9969	.9974	.9978	.9983	.9987	.9991	.9996

Natural Exponential Function

x	e^x	e^{-x}	x	e^x	e^{-x}	x	e^x	e^{-x}
.00	1.00000	1.00000	.40	1.49182	.67032	.80	2.22554	.44032
.01	1.01005	.99005	.41	1.50682	.66365	.85	2.33965	.42741
.02	1.02020	.98020	.42	1.52196	.65705	.90	2.45960	.40657
.03	1.03045	.97045	.43	1.53726	.65051	.95	2.58571	.38674
.04	1.04081	.96079	.44	1.55271	.64404	1.00	2.71828	.36788
.05	1.05127	.95123	.45	1.56831	.63763	1.10	3.00416	.33287
.06	1.06184	.94176	.46	1.58407	.63128	1.20	3.32011	.30119
.07	1.07251	.93239	.47	1.59999	.62500	1.30	3.66929	.27253
.08	1.08329	.92312	.48	1.61607	.61878	1.40	4.05519	.24659
.09	1.09417	.91393	.49	1.63232	.61263	1.50	4.48168	.22313
.10	1.10517	.90484	.50	1.64872	.60653	1.60	4.95302	.20189
.11	1.11628	.89583	.51	1.66529	.60050	1.70	5.47394	.18268
.12	1.12750	.88692	.52	1.68203	.59452	1.80	6.04964	.16529
.13	1.13883	.87810	.53	1.69893	.58860	1.90	6.68589	.14956
.14	1.15027	.86936	.54	1.71601	.58275	2.00	7.38905	.13533
.15	1.16183	.86071	.55	1.73325	.57695	2.10	8.16616	.12245
.16	1.17351	.85214	.56	1.75067	.57121	2.20	9.02500	.11080
.17	1.18530	.84366	.57	1.76827	.56553	2.30	9.97417	.10025
.18	1.19722	.83527	.58	1.78604	.55990	2.40	11.02316	.09071
.19	1.20925	.82696	.59	1.80399	.55433	2.50	12.18248	.08208
.20	1.22140	.81873	.60	1.82212	.54881	3.00	20.08551	.04978
.21	1.23368	.81058	.61	1.84043	.54335	3.50	33.11545	.03020
.22	1.24608	.80252	.62	1.85893	.53794	4.00	54.59815	.01832
.23	1.25860	.79453	.63	1.87761	.53259	4.50	90.01713	.01111
.24	1.27125	.78663	.64	1.89648	.52729	5.00	148.41316	.00674
.25	1.28403	.77880	.65	1.91554	.52205	5.50	224.69193	.00409
.26	1.29693	.77105	.66	1.93479	.51685	6.00	403.42879	.00248
.27	1.30996	.76338	.67	1.95424	.51171	6.50	665.14163	.00150
.28	1.32313	.75578	.68	1.97388	.50662	7.00	1096.63316	.00091
.29	1.33643	.74826	.69	1.99372	.50158	7.50	1808.04241	.00055
.30	1.34986	.74082	.70	2.01375	.49659	8.00	2980.95799	.00034
.31	1.36343	.73345	.71	2.03399	.49164	8.50	4914.76884	.00020
.32	1.37713	.72615	.72	2.05443	.48675	9.00	8130.08392	.00012
.33	1.39097	.71892	.73	2.07508	.48191	9.50	13359.72683	.00007
.34	1.40495	.71177	.74	2.09594	.47711	10.00	22026.46579	.00005
.35	1.41907	.70469	.75	2.11700	.47237			
.36	1.43333	.69768	.76	2.13828	.46767			
.37	1.44773	.69073	.77	2.15977	.46301			
.38	1.46228	.68386	.78	2.18147	.45841			
.39	1.47698	.67706	.79	2.20340	.45384			

Natural Logarithms

x	$\ln x$	x	$\ln x$	x	$\ln x$
		4.5	1.5041	9.0	2.1972
0.1	-2.3026	4.6	1.5261	9.1	2.2083
0.2	-1.6094	4.7	1.5476	9.2	2.2192
0.3	-1.2040	4.8	1.5686	9.3	2.2300
0.4	-0.9163	4.9	1.5892	9.4	2.2407
0.5	-0.6931	5.0	1.6094	9.5	2.2513
0.6	-0.5108	5.1	1.6292	9.6	2.2618
0.7	-0.3567	5.2	1.6487	9.7	2.2721
0.8	-0.2231	5.3	1.6677	9.8	2.2824
0.9	-0.1054	5.4	1.6864	9.9	2.2925
1.0	0.0000	5.5	1.7047	10	2.3026
1.1	0.0953	5.6	1.7228	11	2.3979
1.2	0.1823	5.7	1.7405	12	2.4849
1.3	0.2624	5.8	1.7579	13	2.5649
1.4	0.3365	5.9	1.7750	14	2.6391
1.5	0.4055	6.0	1.7918	15	2.7081
1.6	0.4700	6.1	1.8083	16	2.7726
1.7	0.5306	6.2	1.8245	17	2.8332
1.8	0.5878	6.3	1.8405	18	2.8904
1.9	0.6419	6.4	1.8563	19	2.9444
2.0	0.6931	6.5	1.8718	20	2.9957
2.1	0.7419	6.6	1.8871	25	3.2189
2.2	0.7885	6.7	1.9021	30	3.4012
2.3	0.8329	6.8	1.9169	35	3.5553
2.4	0.8755	6.9	1.9315	40	3.6889
2.5	0.9163	7.0	1.9459	45	3.8067
2.6	0.9555	7.1	1.9601	50	3.9120
2.7	0.9933	7.2	1.9741	55	4.0073
2.8	1.0296	7.3	1.9879	60	4.0943
2.9	1.0647	7.4	2.0015	65	4.1744
3.0	1.0986	7.5	2.0149	70	4.2485
3.1	1.1314	7.6	2.0281	75	4.3175
3.2	1.1632	7.7	2.0412	80	4.3820
3.3	1.1939	7.8	2.0541	85	4.4427
3.4	1.2238	7.9	2.0669	90	4.4998
3.5	1.2528	8.0	2.0794	100	4.6052
3.6	1.2809	8.1	2.0919	110	4.7005
3.7	1.3083	8.2	2.1041	120	4.7875
3.8	1.3350	8.3	2.1163	130	4.8676
3.9	1.3610	8.4	2.1282	140	4.9416
4.0	1.3863	8.5	2.1401	150	5.0106
4.1	1.4110	8.6	2.1518	160	5.0752
4.2	1.4351	8.7	2.1633	170	5.1358
4.3	1.4586	8.8	2.1748	180	5.1930
4.4	1.4816	8.9	2.1861	190	5.2470

Appendix ***III*** | **Trigonometric Tables**

Trigonometric Functions in Degrees and Radians

θ (degrees)	θ (radians)	sin θ	cos θ	tan θ	cot θ	sec θ	csc θ	θ (radians)	θ (degrees)
7°00'	.1222	.1219	.9925	.1228	8.144	1.008	8.206	1.4486	83°00'
10	.1251	.1248	.9922	.1257	7.953	1.008	8.016	1.4457	50
20	.1280	.1276	.9918	.1287	7.770	1.008	7.834	1.4428	40
30	.1309	.1305	.9914	.1317	7.596	1.009	7.661	1.4399	30
40	.1338	.1334	.9911	.1346	7.429	1.009	7.496	1.4370	20
50	.1367	.1363	.9907	.1376	7.269	1.009	7.337	1.4341	10
8°00'	.1396	.1392	.9903	.1405	7.115	1.010	7.185	1.4312	82°00'
10	.1425	.1421	.9899	.1435	6.968	1.010	7.040	1.4283	50
20	.1454	.1449	.9894	.1465	6.827	1.011	6.900	1.4254	40
30	.1484	.1478	.9890	.1495	6.691	1.011	6.765	1.4224	30
40	.1513	.1507	.9886	.1524	6.561	1.012	6.636	1.4195	20
50	.1542	.1536	.9881	.1554	6.435	1.012	6.512	1.4166	10
9°00'	.1571	.1564	.9877	.1584	6.314	1.012	6.392	1.4137	81°00'
10	.1600	.1593	.9872	.1614	6.197	1.013	6.277	1.4108	50
20	.1629	.1622	.9868	.1644	6.084	1.013	6.166	1.4079	40
30	.1658	.1650	.9863	.1673	5.976	1.014	6.059	1.4050	30
40	.1687	.1679	.9858	.1703	5.871	1.014	5.955	1.4021	20
50	.1716	.1708	.9853	.1733	5.769	1.015	5.855	1.3992	10
10°00'	.1745	.1736	.9848	.1763	5.671	1.015	5.759	1.3963	80°00'
10	.1774	.1765	.9843	.1793	5.576	1.016	5.665	1.3934	50
20	.1804	.1794	.9838	.1823	5.485	1.016	5.575	1.3904	40
30	.1833	.1822	.9833	.1853	5.396	1.017	5.487	1.3875	30
40	.1862	.1851	.9827	.1883	5.309	1.018	5.403	1.3846	20
50	.1891	.1880	.9822	.1914	5.226	1.018	5.320	1.3817	10
11°00'	.1920	.1908	.9816	.1944	5.145	1.019	5.241	1.3788	79°00'
10	.1949	.1937	.9811	.1974	5.066	1.019	5.164	1.3759	50
20	.1978	.1965	.9805	.2004	4.989	1.020	5.089	1.3730	40
30	.2007	.1994	.9799	.2035	4.915	1.020	5.016	1.3701	30
40	.2036	.2022	.9793	.2065	4.843	1.021	4.945	1.3672	20
50	.2065	.2051	.9787	.2095	4.773	1.022	4.876	1.3643	10
12°00'	.2094	.2079	.9781	.2126	4.705	1.022	4.810	1.3614	78°00'
10	.2123	.2108	.9775	.2156	4.638	1.023	4.745	1.3584	50
20	.2153	.2136	.9769	.2186	4.574	1.024	4.682	1.3555	40
30	.2182	.2164	.9763	.2217	4.511	1.024	4.620	1.3526	30
40	.2211	.2193	.9757	.2247	4.449	1.025	4.560	1.3497	20
50	.2240	.2221	.9750	.2278	4.390	1.026	4.502	1.3468	10
13°00'	.2269	.2250	.9744	.2309	4.331	1.026	4.445	1.3439	77°00'
10	.2298	.2278	.9737	.2339	4.275	1.027	4.390	1.3410	50
20	.2327	.2306	.9730	.2370	4.219	1.028	4.336	1.3381	40
30	.2356	.2334	.9724	.2401	4.165	1.028	4.284	1.3352	30
40	.2385	.2363	.9717	.2432	4.113	1.029	4.232	1.3323	20
50	.2414	.2391	.9710	.2462	4.061	1.030	4.182	1.3294	10
		cos θ	sin θ	cot θ	tan θ	csc θ	sec θ	θ (radians)	θ (degrees)

θ (degrees)	θ (radians)	sin θ	cos θ	tan θ	cot θ	sec θ	csc θ	θ (radians)	θ (degrees)
0°00'	.0000	.0000	1.0000	.0000	—	1.000	—	1.5708	90°00'
10	.0029	.0029	1.0000	.0029	343.8	1.000	343.8	1.5679	50
20	.0058	.0058	1.0000	.0058	171.9	1.000	171.9	1.5650	40
30	.0087	.0087	1.0000	.0087	114.6	1.000	114.6	1.5621	30
40	.0116	.0116	.9999	.0116	85.94	1.000	85.95	1.5592	20
50	.0145	.0145	.9999	.0145	68.75	1.000	68.76	1.5563	10
1°00'	.0175	.0175	.9998	.0175	57.29	1.000	57.30	1.5533	89°00'
10	.0204	.0204	.9998	.0204	49.10	1.000	49.11	1.5504	50
20	.0233	.0233	.9997	.0233	42.96	1.000	42.98	1.5475	40
30	.0262	.0262	.9997	.0262	38.19	1.000	38.20	1.5446	30
40	.0291	.0291	.9996	.0291	34.37	1.000	34.38	1.5417	20
50	.0320	.0320	.9995	.0320	31.24	1.001	31.26	1.5388	10
2°00'	.0349	.0349	.9994	.0349	28.64	1.001	28.65	1.5359	88°00'
10	.0378	.0378	.9993	.0378	26.43	1.001	26.45	1.5330	50
20	.0407	.0407	.9992	.0407	24.54	1.001	24.56	1.5301	40
30	.0436	.0436	.9990	.0437	22.90	1.001	22.93	1.5272	30
40	.0465	.0465	.9989	.0466	21.47	1.001	21.49	1.5243	20
50	.0495	.0494	.9988	.0495	20.21	1.001	20.23	1.5213	10
3°00'	.0524	.0523	.9986	.0524	19.08	1.001	19.11	1.5184	87°00'
10	.0553	.0552	.9985	.0553	18.07	1.002	18.10	1.5155	50
20	.0582	.0581	.9983	.0582	17.17	1.002	17.20	1.5126	40
30	.0611	.0610	.9981	.0612	16.35	1.002	16.38	1.5097	30
40	.0640	.0640	.9980	.0641	15.60	1.002	15.64	1.5068	20
50	.0669	.0669	.9978	.0670	14.92	1.002	14.96	1.5039	10
4°00'	.0698	.0698	.9976	.0699	14.30	1.002	14.34	1.5010	86°00'
10	.0727	.0727	.9974	.0729	13.73	1.003	13.76	1.4981	50
20	.0756	.0756	.9971	.0758	13.20	1.003	13.23	1.4952	40
30	.0785	.0785	.9969	.0787	12.71	1.003	12.75	1.4923	30
40	.0814	.0814	.9967	.0816	12.25	1.003	12.29	1.4893	20
50	.0844	.0843	.9964	.0846	11.83	1.004	11.87	1.4864	10
5°00'	.0873	.0872	.9962	.0875	11.43	1.004	11.47	1.4835	85°00'
10	.0902	.0901	.9959	.0904	11.06	1.004	11.10	1.4806	50
20	.0931	.0929	.9957	.0934	10.71	1.004	10.76	1.4777	40
30	.0960	.0958	.9954	.0963	10.39	1.005	10.43	1.4748	30
40	.0989	.0987	.9951	.0992	10.08	1.005	10.13	1.4719	20
50	.1018	.1016	.9948	.1022	9.788	1.005	9.839	1.4690	10
6°00'	.1047	.1045	.9945	.1051	9.514	1.006	9.567	1.4661	84°00'
10	.1076	.1074	.9942	.1080	9.255	1.006	9.309	1.4632	50
20	.1105	.1103	.9939	.1110	9.010	1.006	9.065	1.4603	40
30	.1134	.1132	.9936	.1139	8.777	1.006	8.834	1.4573	30
40	.1164	.1161	.9932	.1169	8.556	1.007	8.614	1.4544	20
50	.1193	.1190	.9929	.1198	8.345	1.007	8.405	1.4515	10
		cos θ	sin θ	cot θ	tan θ	csc θ	sec θ	θ (radians)	θ (degrees)

Trigonometric Functions in Degrees and Radians (continued)

θ (degrees)	θ (radians)	sin θ	cos θ	tan θ	cot θ	sec θ	csc θ	(radians)	(degrees)
21°00'	.3665	.3584	.9336	.3839	2.605	1.071	2.790	1.2043	69°00'
10	.3694	.3611	.9325	.3872	2.583	1.072	2.769	1.2014	50
20	.3723	.3638	.9315	.3906	2.560	1.074	2.749	1.1985	40
30	.3752	.3665	.9304	.3939	2.539	1.075	2.729	1.1956	30
40	.3782	.3692	.9293	.3973	2.517	1.076	2.709	1.1926	20
50	.3811	.3719	.9283	.4006	2.496	1.077	2.689	1.1897	10
22°00'	.3840	.3746	.9272	.4040	2.475	1.079	2.669	1.1868	68°00'
10	.3869	.3773	.9261	.4074	2.455	1.080	2.650	1.1839	50
20	.3898	.3800	.9250	.4108	2.434	1.081	2.632	1.1810	40
30	.3927	.3827	.9239	.4142	2.414	1.082	2.613	1.1781	30
40	.3956	.3854	.9228	.4176	2.394	1.084	2.595	1.1752	20
50	.3985	.3881	.9216	.4210	2.375	1.085	2.577	1.1723	10
23°00'	.4014	.3907	.9205	.4245	2.356	1.086	2.559	1.1694	67°00'
10	.4043	.3934	.9194	.4279	2.337	1.088	2.542	1.1665	50
20	.4072	.3961	.9182	.4314	2.318	1.089	2.525	1.1636	40
30	.4102	.3987	.9171	.4348	2.300	1.090	2.508	1.1606	30
40	.4131	.4014	.9159	.4383	2.282	1.092	2.491	1.1577	20
50	.4160	.4041	.9147	.4417	2.264	1.093	2.475	1.1548	10
24°00'	.4189	.4067	.9135	.4452	2.246	1.095	2.459	1.1519	66°00'
10	.4218	.4094	.9124	.4487	2.229	1.096	2.443	1.1490	50
20	.4247	.4120	.9112	.4522	2.211	1.097	2.427	1.1461	40
30	.4276	.4147	.9100	.4557	2.194	1.099	2.411	1.1432	30
40	.4305	.4173	.9088	.4592	2.177	1.100	2.396	1.1403	20
50	.4334	.4200	.9075	.4628	2.161	1.102	2.381	1.1374	10
25°00'	.4363	.4226	.9063	.4663	2.145	1.103	2.366	1.1345	65°00'
10	.4392	.4253	.9051	.4699	2.128	1.105	2.352	1.1316	50
20	.4422	.4279	.9038	.4734	2.112	1.106	2.337	1.1286	40
30	.4451	.4305	.9026	.4770	2.097	1.108	2.323	1.1257	30
40	.4480	.4331	.9013	.4806	2.081	1.109	2.309	1.1228	20
50	.4509	.4358	.9001	.4841	2.066	1.111	2.295	1.1199	10
26°00'	.4538	.4384	.8988	.4877	2.050	1.113	2.281	1.1170	64°00'
10	.4567	.4410	.8975	.4913	2.035	1.114	2.268	1.1141	50
20	.4596	.4436	.8962	.4950	2.020	1.116	2.254	1.1112	40
30	.4625	.4462	.8949	.4986	2.006	1.117	2.241	1.1083	30
40	.4654	.4488	.8936	.5022	1.991	1.119	2.228	1.1054	20
50	.4683	.4514	.8923	.5059	1.977	1.121	2.215	1.1025	10
27°00'	.4712	.4540	.8910	.5095	1.963	1.122	2.203	1.0996	63°00'
10	.4741	.4566	.8897	.5132	1.949	1.124	2.190	1.0966	50
20	.4771	.4592	.8884	.5169	1.935	1.126	2.178	1.0937	40
30	.4800	.4617	.8870	.5206	1.921	1.127	2.166	1.0908	30
40	.4829	.4643	.8857	.5243	1.907	1.129	2.154	1.0879	20
50	.4858	.4669	.8843	.5280	1.894	1.131	2.142	1.0850	10
		cos θ	sin θ	cot θ	tan θ	csc θ	sec θ	(radians)	(degrees)

θ (degrees)	θ (radians)	sin θ	cos θ	tan θ	cot θ	sec θ	csc θ	(radians)	(degrees)
14°00'	.2443	.2419	.9703	.2493	4.011	1.031	4.134	1.3265	76°00'
10	.2473	.2447	.9696	.2524	3.962	1.031	4.086	1.3235	50
20	.2502	.2476	.9689	.2555	3.914	1.032	4.039	1.3206	40
30	.2531	.2504	.9681	.2586	3.867	1.033	3.994	1.3177	30
40	.2560	.2532	.9674	.2617	3.821	1.034	3.950	1.3148	20
50	.2589	.2560	.9667	.2648	3.776	1.034	3.906	1.3119	10
15°00'	.2618	.2588	.9659	.2679	3.732	1.035	3.864	1.3090	75°00'
10	.2647	.2616	.9652	.2711	3.689	1.036	3.822	1.3061	50
20	.2676	.2644	.9644	.2742	3.647	1.037	3.782	1.3032	40
30	.2705	.2672	.9636	.2773	3.606	1.038	3.742	1.3003	30
40	.2734	.2700	.9628	.2805	3.566	1.039	3.703	1.2974	20
50	.2763	.2728	.9621	.2836	3.526	1.039	3.665	1.2945	10
16°00'	.2793	.2756	.9613	.2867	3.487	1.040	3.628	1.2915	74°00'
10	.2822	.2784	.9605	.2899	3.450	1.041	3.592	1.2886	50
20	.2851	.2812	.9596	.2931	3.412	1.042	3.556	1.2857	40
30	.2880	.2840	.9588	.2962	3.376	1.043	3.521	1.2828	30
40	.2909	.2868	.9580	.2994	3.340	1.044	3.487	1.2799	20
50	.2938	.2896	.9572	.3026	3.305	1.045	3.453	1.2770	10
17°00'	.2967	.2924	.9563	.3057	3.271	1.046	3.420	1.2741	73°00'
10	.2996	.2952	.9555	.3089	3.237	1.047	3.388	1.2712	50
20	.3025	.2979	.9546	.3121	3.204	1.048	3.356	1.2683	40
30	.3054	.3007	.9537	.3153	3.172	1.049	3.326	1.2654	30
40	.3083	.3035	.9528	.3185	3.140	1.049	3.295	1.2625	20
50	.3113	.3062	.9520	.3217	3.108	1.050	3.265	1.2595	10
18°00'	.3142	.3090	.9511	.3249	3.078	1.051	3.236	1.2566	72°00'
10	.3171	.3118	.9502	.3281	3.047	1.052	3.207	1.2537	50
20	.3200	.3145	.9492	.3314	3.018	1.053	3.179	1.2508	40
30	.3229	.3173	.9483	.3346	2.989	1.054	3.152	1.2479	30
40	.3258	.3201	.9474	.3378	2.960	1.056	3.124	1.2450	20
50	.3287	.3228	.9465	.3411	2.932	1.057	3.098	1.2421	10
19°00'	.3316	.3256	.9455	.3443	2.904	1.058	3.072	1.2392	71°00'
10	.3345	.3283	.9446	.3476	2.877	1.059	3.046	1.2363	50
20	.3374	.3311	.9436	.3508	2.850	1.060	3.021	1.2334	40
30	.3403	.3338	.9426	.3541	2.824	1.061	2.996	1.2305	30
40	.3432	.3365	.9417	.3574	2.798	1.062	2.971	1.2275	20
50	.3462	.3393	.9407	.3607	2.773	1.063	2.947	1.2246	10
20°00'	.3491	.3420	.9397	.3640	2.747	1.064	2.924	1.2217	70°00'
10	.3520	.3448	.9387	.3673	2.723	1.065	2.901	1.2188	50
20	.3549	.3475	.9377	.3706	2.699	1.066	2.878	1.2159	40
30	.3578	.3502	.9367	.3739	2.675	1.068	2.855	1.2130	30
40	.3607	.3529	.9356	.3772	2.651	1.069	2.833	1.2101	20
50	.3636	.3557	.9346	.3805	2.628	1.070	2.812	1.2072	10
		cos θ	sin θ	cot θ	tan θ	csc θ	sec θ	(radians)	(degrees)

Trigonometric Functions in Degrees and Radians (continued)

θ (degrees)	θ (radians)	sin θ	cos θ	tan θ	cot θ	sec θ	csc θ	θ (radians)	θ (degrees)
35°00'	.6109	.5736	.8192	.7002	1.428	1.221	1.743	.9599	55°00'
10	.6138	.5760	.8175	.7046	1.419	1.223	1.736	.9570	50
20	.6167	.5783	.8158	.7089	1.411	1.226	1.729	.9541	40
30	.6196	.5807	.8141	.7133	1.402	1.228	1.722	.9512	30
40	.6225	.5831	.8124	.7177	1.393	1.231	1.715	.9483	20
50	.6254	.5854	.8107	.7221	1.385	1.233	1.708	.9454	10
36°00'	.6283	.5878	.8090	.7265	1.376	1.236	1.701	.9425	54°00'
10	.6312	.5901	.8073	.7310	1.368	1.239	1.695	.9396	50
20	.6341	.5925	.8056	.7355	1.360	1.241	1.688	.9367	40
30	.6370	.5948	.8039	.7400	1.351	1.244	1.681	.9338	30
40	.6400	.5972	.8021	.7445	1.343	1.247	1.675	.9308	20
50	.6429	.5995	.8004	.7490	1.335	1.249	1.668	.9279	10
37°00'	.6458	.6018	.7986	.7536	1.327	1.252	1.662	.9250	53°00'
10	.6487	.6041	.7969	.7581	1.319	1.255	1.655	.9221	50
20	.6516	.6065	.7951	.7627	1.311	1.258	1.649	.9192	40
30	.6545	.6088	.7934	.7673	1.303	1.260	1.643	.9163	30
40	.6574	.6111	.7916	.7720	1.295	1.263	1.636	.9134	20
50	.6603	.6134	.7898	.7766	1.288	1.266	1.630	.9105	10
38°00'	.6632	.6157	.7880	.7813	1.280	1.269	1.624	.9076	52°00'
10	.6661	.6180	.7862	.7860	1.272	1.272	1.618	.9047	50
20	.6690	.6202	.7844	.7907	1.265	1.275	1.612	.9018	40
30	.6720	.6225	.7826	.7954	1.257	1.278	1.606	.8988	30
40	.6749	.6248	.7808	.8002	1.250	1.281	1.601	.8959	20
50	.6778	.6271	.7790	.8050	1.242	1.284	1.595	.8930	10
39°00'	.6807	.6293	.7771	.8098	1.235	1.287	1.589	.8901	51°00'
10	.6836	.6316	.7753	.8146	1.228	1.290	1.583	.8872	50
20	.6865	.6338	.7735	.8195	1.220	1.293	1.578	.8843	40
30	.6894	.6361	.7716	.8243	1.213	1.296	1.572	.8814	30
40	.6923	.6383	.7698	.8292	1.206	1.299	1.567	.8785	20
50	.6952	.6406	.7679	.8342	1.199	1.302	1.561	.8756	10
40°00'	.6981	.6428	.7660	.8391	1.192	1.305	1.556	.8727	50°00'
10	.7010	.6450	.7642	.8441	1.185	1.309	1.550	.8698	50
20	.7039	.6472	.7623	.8491	1.178	1.312	1.545	.8668	40
30	.7069	.6494	.7604	.8541	1.171	1.315	1.540	.8639	30
40	.7098	.6517	.7585	.8591	1.164	1.318	1.535	.8610	20
50	.7127	.6539	.7566	.8642	1.157	1.322	1.529	.8581	10
41°00'	.7156	.6561	.7547	.8693	1.150	1.325	1.524	.8552	49°00'
10	.7185	.6583	.7528	.8744	1.144	1.328	1.519	.8523	50
20	.7214	.6604	.7509	.8796	1.137	1.332	1.514	.8494	40
30	.7243	.6626	.7490	.8847	1.130	1.335	1.509	.8465	30
40	.7272	.6648	.7470	.8899	1.124	1.339	1.504	.8436	20
50	.7301	.6670	.7451	.8952	1.117	1.342	1.499	.8407	10
		cos θ	sin θ	cot θ	tan θ	csc θ	sec θ	θ (radians)	θ (degrees)

θ (degrees)	θ (radians)	sin θ	cos θ	tan θ	cot θ	sec θ	csc θ	θ (radians)	θ (degrees)
28°00'	.4887	.4695	.8829	.5317	1.881	1.133	2.130	1.0821	62°00'
10	.4916	.4720	.8816	.5354	1.868	1.134	2.118	1.0792	50
20	.4945	.4746	.8802	.5392	1.855	1.136	2.107	1.0763	40
30	.4974	.4772	.8788	.5430	1.842	1.138	2.096	1.0734	30
40	.5003	.4797	.8774	.5467	1.829	1.140	2.085	1.0705	20
50	.5032	.4823	.8760	.5505	1.816	1.142	2.074	1.0676	10
29°00'	.5061	.4848	.8746	.5543	1.804	1.143	2.063	1.0647	61°00'
10	.5091	.4874	.8732	.5581	1.792	1.145	2.052	1.0617	50
20	.5120	.4899	.8718	.5619	1.780	1.147	2.041	1.0588	40
30	.5149	.4924	.8704	.5658	1.767	1.149	2.031	1.0559	30
40	.5178	.4950	.8689	.5696	1.756	1.151	2.020	1.0530	20
50	.5207	.4975	.8675	.5735	1.744	1.153	2.010	1.0501	10
30°00'	.5236	.5000	.8660	.5774	1.732	1.155	2.000	1.0472	60°00'
10	.5265	.5025	.8646	.5812	1.720	1.157	1.990	1.0443	50
20	.5294	.5050	.8631	.5851	1.709	1.159	1.980	1.0414	40
30	.5323	.5075	.8616	.5890	1.698	1.161	1.970	1.0385	30
40	.5352	.5100	.8601	.5930	1.686	1.163	1.961	1.0356	20
50	.5381	.5125	.8587	.5969	1.675	1.165	1.951	1.0327	10
31°00'	.5411	.5150	.8572	.6009	1.664	1.167	1.942	1.0297	59°00'
10	.5440	.5175	.8557	.6048	1.653	1.169	1.932	1.0268	50
20	.5469	.5200	.8542	.6088	1.643	1.171	1.923	1.0239	40
30	.5498	.5225	.8526	.6128	1.632	1.173	1.914	1.0210	30
40	.5527	.5250	.8511	.6168	1.621	1.175	1.905	1.0181	20
50	.5556	.5275	.8496	.6208	1.611	1.177	1.896	1.0152	10
32°00'	.5585	.5299	.8480	.6249	1.600	1.179	1.887	1.0123	58°00'
10	.5614	.5324	.8465	.6289	1.590	1.181	1.878	1.0094	50
20	.5643	.5348	.8450	.6330	1.580	1.184	1.870	1.0065	40
30	.5672	.5373	.8434	.6371	1.570	1.186	1.861	1.0036	30
40	.5701	.5398	.8418	.6412	1.560	1.188	1.853	1.0007	20
50	.5730	.5422	.8403	.6453	1.550	1.190	1.844	.9977	10
33°00'	.5760	.5446	.8387	.6494	1.540	1.192	1.836	.9948	57°00'
10	.5789	.5471	.8371	.6536	1.530	1.195	1.828	.9919	50
20	.5818	.5495	.8355	.6577	1.520	1.197	1.820	.9890	40
30	.5847	.5519	.8339	.6619	1.511	1.199	1.812	.9861	30
40	.5876	.5544	.8323	.6661	1.501	1.202	1.804	.9832	20
50	.5905	.5568	.8307	.6703	1.492	1.204	1.796	.9803	10
34°00'	.5934	.5592	.8290	.6745	1.483	1.206	1.788	.9774	56°00'
10	.5963	.5616	.8274	.6787	1.473	1.209	1.781	.9745	50
20	.5992	.5640	.8258	.6830	1.464	1.211	1.773	.9716	40
30	.6021	.5664	.8241	.6873	1.455	1.213	1.766	.9687	30
40	.6050	.5688	.8225	.6916	1.446	1.216	1.758	.9657	20
50	.6080	.5712	.8208	.6959	1.437	1.218	1.751	.9628	10
		cos θ	sin θ	cot θ	tan θ	csc θ	sec θ	θ (radians)	θ (degrees)

Trigonometric Functions in Degrees and Radians (continued)

θ (degrees)	θ (radians)	sin θ	cos θ	tan θ	cot θ	sec θ	csc θ		
42°00′	.7330	.6691	.7431	.9004	1.111	1.346	1.494	.8378	**48°00′**
10	.7359	.6713	.7412	.9057	1.104	1.349	1.490	.8348	50
20	.7389	.6734	.7392	.9110	1.098	1.353	1.485	.8319	40
30	.7418	.6756	.7373	.9163	1.091	1.356	1.480	.8290	30
40	.7447	.6777	.7353	.9217	1.085	1.360	1.476	.8261	20
50	.7476	.6799	.7333	.9271	1.079	1.364	1.471	.8232	10
43°00′	.7505	.6820	.7314	.9325	1.072	1.367	1.466	.8203	**47°00′**
10	.7534	.6841	.7294	.9380	1.066	1.371	1.462	.8174	50
20	.7563	.6862	.7274	.9435	1.060	1.375	1.457	.8145	40
30	.7592	.6884	.7254	.9490	1.054	1.379	1.453	.8116	30
40	.7621	.6905	.7234	.9545	1.048	1.382	1.448	.8087	20
50	.7650	.6926	.7214	.9601	1.042	1.386	1.444	.8058	10
44°00′	.7679	.6947	.7193	.9657	1.036	1.390	1.440	.8029	**46°00′**
10	.7709	.6967	.7173	.9713	1.030	1.394	1.435	.7999	50
20	.7738	.6988	.7153	.9770	1.024	1.398	1.431	.7970	40
30	.7767	.7009	.7133	.9827	1.018	1.402	1.427	.7941	30
40	.7796	.7030	.7112	.9884	1.012	1.406	1.423	.7912	20
50	.7825	.7050	.7092	.9942	1.006	1.410	1.418	.7883	10
45°00′	.7854	.7071	.7071	1.000	1.000	1.414	1.414	.7854	**45°00′**
		cos θ	sin θ	cot θ	tan θ	csc θ	sec θ	θ (radians)	θ (degrees)

Solutions to Selected Exercises

Exercise Set 1.1, page 9

2. **a.** $\dfrac{10}{2}$ and $\dfrac{0}{4}$ are integers.

   b. $2.8\overline{10}$, -4.25, $-\dfrac{1}{4}$, $\dfrac{10}{2}$, $\dfrac{2}{10}$, $0.131313\ldots$, and $\dfrac{0}{4}$ are rational numbers.

   c. π and $0.131131113\ldots$ are irrational numbers.

   d. All are real numbers.

12. $\dfrac{2a}{5} + \dfrac{3a}{7} = \dfrac{2a \cdot 7}{5 \cdot 7} + \dfrac{3a \cdot 5}{7 \cdot 5} = \dfrac{29a}{35}$

24. Symmetric property of equality

26. Commutative property of addition

42. $\{0, 1, 2, 3, 4\} \cap \{1, 3, 6, 10\} = \{1, 3\}$

Exercise Set 1.2, page 16

8. $-\dfrac{3}{2} > -3$

26. $[-2, 1)$

38. $-2 \le x < 3$

70. $2 - \sqrt{3}$, since $2 - \sqrt{3} > 0$

80. $|-108 - 22| = |-130| = 130$

Exercise Set 1.3, page 25

4. $-2^4 = -16$

18. $\left(\dfrac{4 \cdot 5^{-1}}{2^{-3}}\right)^{-2} = \dfrac{5^2}{4^2 \cdot 2^6} = \dfrac{25}{16 \cdot 64} = \dfrac{25}{1024}$

30. $\left(\dfrac{3pq^2}{-2pq^3r^2}\right)^4 = \left(\dfrac{3}{-2qr^2}\right)^4 = \dfrac{81}{16q^4r^8}$

48. $25{,}600 = 2.56 \times 10^4$

Exercise Set 1.4, page 31

24. $(5y^2 - 7y + 3) + (2y^2 + 8y + 1) = 7y^2 + y + 4$

28. $(7s^2 - 4s + 11) - (-2s^2 + 11s - 9)$
$$= 7s^2 - 4s + 11 + 2s^2 - 11s + 9$$
$$= 9s^2 - 15s + 20$$

32.
$$
\begin{array}{r}
3x^2 - 8x - 5 \\
5x - 7 \\
\hline
-21x^2 + 56x + 35 \\
15x^3 - 40x^2 - 25x \quad\quad\quad \\
\hline
15x^3 - 61x^2 + 31x + 35
\end{array}
$$

36. $(5x - 3)(2x + 7) = 10x^2 + 35x - 6x - 21$
$$= 10x^2 + 29x - 21$$

60. $(4x^2 - 3y)(4x^2 - 3y) = (4x^2)^2 - (3y^2)^2 = 16x^4 - 9y^2$

72. $-x^2 - 5x + 4 = -(-5)^2 - 5(-5) + 4$
$$= -25 + 25 + 4 = 4$$

82. $\dfrac{1}{6}n^3 - \dfrac{1}{2}n^2 + \dfrac{1}{3}n = \dfrac{1}{6}(21)^3 - \dfrac{1}{2}(21)^2 + \dfrac{1}{3}(21)$
$$= 1330 \text{ committees}$$

84. **a.** $0.015v^2 + v + 10 = 0.015(30)^2 + 30 + 10$
$$= 53.5 \text{ feet}$$

   b. $0.015(55)^2 + 55 + 10 = 110.375$ feet

Exercise Set 1.5, page 41

6. $6a^3b^2 - 12a^2b + 72ab^3 = 6ab(a^2b - 2a + 12b^2)$

10. $10z^3 - 15z^2 - 4z + 6 = 5z^2(2z - 3) - 2(2z - 3)$
$$= (2z - 3)(5z^2 - 2)$$

18. $x^2 + 9x + 20 = (x + 4)(x + 5)$

22. $8a^2 - 26a + 15 = (4a - 3)(2a - 5)$

26. $8x^2 + 10xy - 25y^2 = (4x^2 - 5y^2)(2x^2 + 5y^2)$

28. $x^4 + 11x^2 + 18 = (x^2 + 9)(x^2 + 2)$

36. $81b^2 - 16c^2 = (9b - 4c)(9b + 4c)$

46. $b^2 - 24b + 144 = (b - 12)^2$

52. $b^3 + 64 = (b + 4)(b^2 - 4b + 16)$

62. $81y^4 - 16 = (9y^2 - 4)(9y^2 + 4)$
$$= (3y - 2)(3y + 2)(9y^2 + 4)$$

72. $4y^2 - 4yz + z^2 - 9 = (2y - z)^2 - 9$
$$= (2y - z - 3)(2y - z + 3)$$

Exercise Set 1.6, page 50

8. $\dfrac{2x^3 - 6x^2 + 5x - 15}{9 - x^2} = \dfrac{2x^2(x - 3) + 5(x - 3)}{9 - x^2}$

$$= \dfrac{(x - 3)(2x^2 + 5)}{(3 - x)(3 + x)} = -\dfrac{2x^2 + 5}{x + 3}$$

16. $\dfrac{x^2 - 16}{x^2 + 7x + 12} \cdot \dfrac{x^2 - 4x - 21}{x^2 - 4x}$

$$= \dfrac{(x - 4)(x + 4)(x + 3)(x - 7)}{(x + 3)(x + 4)x(x - 4)} = \dfrac{x - 7}{x}$$

30. $\dfrac{3y - 1}{3y + 1} - \dfrac{2y - 5}{y - 3} = \dfrac{(3y - 1)(y - 3)}{(3y + 1)(y - 3)} - \dfrac{(2y - 5)(3y + 1)}{(y - 3)(3y + 1)}$

$$= \dfrac{(3y^2 - 10y + 3) - (6y^2 - 13y - 5)}{(3y + 1)(y - 3)}$$

$$= \dfrac{-3y^2 + 3y + 8}{(3y + 1)(y - 3)}$$

42. $\dfrac{3 - \dfrac{2}{a}}{5 + \dfrac{3}{a}} = \dfrac{\left(3 - \dfrac{2}{a}\right)a}{\left(5 + \dfrac{3}{a}\right)a} = \dfrac{3a - 2}{5a + 3}$

60. $\dfrac{e^{-2} - f^{-1}}{ef} = \dfrac{\dfrac{1}{e^2} - \dfrac{1}{f}}{ef} = \dfrac{f - e^2}{e^2f} \div \dfrac{ef}{1}$

$$= \dfrac{f - e^2}{e^2f} \cdot \dfrac{1}{ef} = \dfrac{f - e^2}{e^3f^2}$$

64. a. $\dfrac{v_1 + v_2}{1 + \dfrac{v_1v_2}{c^2}} = \dfrac{1.2 \times 10^8 + 2.4 \times 10^8}{1 + \dfrac{(1.2 \times 10^8)(2.4 \times 10^8)}{(6.7 \times 10^8)^2}} \approx 3.4 \times 10^8$

b. $\dfrac{v_1 + v_2}{1 + \dfrac{v_1 + v_2}{c^2}} = \dfrac{c^2(v_1 + v_2)}{c^2\left(1 + \dfrac{v_1 + v_2}{c^2}\right)} = \dfrac{c^2(v_1 + v_2)}{c^2 + v_1 + v_2}$

Exercise Set 1.7, page 62

2. $49^{1/2} = 7$

6. $16^{3/2} = (2^4)^{3/2} = 2^6 = 64$

38. $\left(\dfrac{r^3s^{-2}}{rs^4}\right)^{1/2} = (r^{3-1}s^{-2-4})^{1/2} = (r^2s^{-6})^{1/2} = rs^{-3} = \dfrac{r}{s^3}$

46. $\sqrt[3]{-64} = \qquad\qquad -4$

52. $(\sqrt[4]{25})^2 = 25^{2/4} = 25^{1/2} = 5$

70. $\sqrt{75} = \sqrt{25 \cdot 3} = 5\sqrt{3}$

86. $\sqrt[4]{25a^2b^2} = \sqrt[4]{5^2a^2b^2} = (5^2a^2b^2)^{1/4} = 5^{1/2}a^{1/2}b^{1/2} = \sqrt{5ab}$

98. $5\sqrt[3]{3} + 2\sqrt[3]{81} = 5\sqrt[3]{3} + 2\sqrt[3]{3^4} = 5\sqrt[3]{3} + 6\sqrt[3]{3} = 11\sqrt[3]{3}$

102. $(\sqrt{7} + 4)(\sqrt{7} - 1) = 7 - \sqrt{7} + 4\sqrt{7} - 4 = 3 + 3\sqrt{7}$

116. $\dfrac{2}{\sqrt[3]{4}} = \dfrac{2}{\sqrt[3]{4}} \cdot \dfrac{\sqrt[3]{2}}{\sqrt[3]{2}} = \dfrac{2\sqrt[3]{2}}{\sqrt[3]{8}} = \dfrac{2\sqrt[3]{2}}{2} = \sqrt[3]{2}$

124. $\dfrac{5}{\sqrt{y} - \sqrt{3}} = \dfrac{5}{09\overline{9}07 - \sqrt{3}} \cdot \dfrac{\sqrt{y} + \sqrt{3}}{\sqrt{y} + \sqrt{3}} = \dfrac{5(\sqrt{y} + \sqrt{3})}{y - 3}$

$$= \dfrac{5\sqrt{y} + 5\sqrt{3}}{y - 3}$$

Exercise Set 1.8, page 69

2. $3 + \sqrt{-25} = 3 + i\sqrt{25} = 3 + 5i$

12. $(1 - 3i) + (6 + 2i) = 7 - i$

16. $(-3 + i) - (-8 + 2i) = -3 + i + 8 - 2i = 5 - i$

24. $(5 - 3i)(-2 - 4i) = -10 - 20i + 6i + 12i^2$
$$= -10 - 14i - 12 = -22 - 14i$$

34. $\dfrac{5 - 7i}{5 + 7i} = \dfrac{5 - 7i}{5 + 7i} \cdot \dfrac{5 - 7i}{5 - 7i} = \dfrac{25 - 35i - 35i + 49i^2}{25 - 49i^2}$

$\qquad = \dfrac{25 - 70i - 49}{25 + 49} = \dfrac{-24 - 70i}{74}$

$\qquad = -\dfrac{12}{37} - \dfrac{35}{37}i$

54. $i^{28} = (i^4)^7 = 1$

68. $\sqrt{-3}\,\sqrt{-121} = i\sqrt{3} \cdot i\sqrt{121} = i\sqrt{3} \cdot 11i$

$\qquad = 11i^2\sqrt{3} = -11\sqrt{3}$

Exercise Set 2.1, page 83

2.
$$-3y + 20 = 2$$
$$-3y = -18$$
$$y = 6$$

12.
$$\dfrac{1}{2}x + 7 - \dfrac{1}{4}x = \dfrac{19}{2}$$
$$4\left(\dfrac{1}{2}x + 7 - \dfrac{1}{4}x\right) = 4\left(\dfrac{19}{2}\right)$$
$$2x + 28 - x = 38$$
$$x = 38 - 28$$
$$x = 10$$

20.
$$5(x + 4)(x - 4) = (x - 3)(5x + 4)$$
$$5(x^2 - 16) = 5x^2 - 11x - 12$$
$$5x^2 - 80 = 5x^2 - 11x - 12$$
$$-80 + 12 = -11x$$
$$-68 = -11x$$
$$\dfrac{68}{11} = x$$

30.
$$2x + \dfrac{1}{3} = \dfrac{6x + 1}{3}$$
$$\dfrac{6x + 1}{3} = \dfrac{6x + 1}{3}$$

This equation is an identity.

38.
$$\dfrac{4}{y + 2} = \dfrac{7}{y - 4}$$
$$4(y - 4) = 7(y + 2)$$
$$4y - 16 = 7y + 14$$
$$4y - 7y = 14 + 16$$
$$-3y = 30$$
$$y = -10$$

58.
$$4x - 7y = -15$$
$$4x = 7y - 15$$
$$x = \dfrac{1}{4}(7y - 15)$$
$$x = \dfrac{7}{4}y - \dfrac{15}{4}$$

66.
$$\dfrac{2}{x - 2} = \dfrac{y}{x - 3}$$
$$2x - 6 = y(x - 2)$$
$$2x - 6 = xy - 2y$$
$$2x - xy = 6 - 2y$$
$$x(2 - y) = 6 - 2y$$
$$x = \dfrac{6 - 2y}{2 - y}$$

Exercise Set 2.2, page 91

2.
$$P = S - Sdt$$
$$Sdt = S - P$$
$$t = \dfrac{S - P}{Sd}$$

4.
$$A = P + Prt$$
$$A = P(1 + rt)$$
$$\dfrac{A}{1 + rt} = P$$

22.

$$P = 2L + 2W$$

$$110 = 2x + 2\left(\frac{1}{2}x + 1\right)$$

$$110 = 2x + x + 2$$

$$36 = x$$

$$L = 36 \text{ meters}$$

$$W = \frac{1}{2}x + 1 = 19 \text{ meters}$$

30.

$$\text{Time} = 7.5$$

$$t_1 + t_2 = 7.5$$

$$t_2 = 7.5 - t_1$$

$$15t_1 = 10t_2$$

$$15t_1 = 10(7.5 - t_1)$$

$$15t_1 = 75 - 10t_1$$

$$25t_1 = 75$$

$$t_1 = 3 \text{ hours}$$

$$D = 3t_1 = 3(15)$$

$$D = 45 \text{ miles}$$

38.

$$\text{Profit} = \text{Income} - \text{Cost}$$

$$2337 = 0.75x - 0.18x$$

$$2337 = .57x$$

$$4100 = x$$

4100 glasses of orange juice were sold.

40. Let x = cost last year.

$$x + 0.04x = 26$$

$$1.04x = 26$$

$$x = \$25$$

42.

5%	x
7%	$7500 - x$

$$.05x + .07(7500 - x) = 405$$

$$.05x + 525 - .07x = 405$$

$$-.02x = -120$$

$$x = 6000$$

$$7500 - x = 1500$$

\$6000 was invested at 5%. \$1500 was invested at 7%.

46.

.40	x
.24	4
.30	$4 + x$

$$.40x + .24(4) = .30(4 + x)$$

$$.40x + 0.96 = 1.2 + .30x$$

$$.10x = 0.24$$

$$x = 2.4$$

2.4 liters of 40% sulfuric acid should be mixed with 4 liters of a 24% sulfuric acid solution.

56.

A	$\frac{1}{3}$	x
B	$\frac{1}{4}$	x

$$12\left(\frac{1}{3}x + \frac{1}{4}x\right) = 12 \cdot 1$$

$$4x + 3x = 12$$

$$7x = 12$$

$$x \approx 1.71$$

It would take 1.71 hours to print the report.

Exercise Set 2.3, page 104

4.

$$12w^2 - 41w + 24 = 0$$

$$(4w - 3)(3w - 8) = 0$$

$$4w - 3 = 0 \quad 3w - 8 = 0$$

$$w = \frac{3}{4} \qquad w = \frac{8}{3}$$

18.

$$y^2 = 225$$

$$y = \pm\sqrt{225}$$

$$y = \pm 15$$

30.

$$x^2 + 8x - 10 = 0$$

$$x^2 + 8x + 16 = 10 + 16$$

$$(x + 4)^2 = 26$$

$$x + 4 = \pm\sqrt{26}$$

$$x = -4 \pm \sqrt{26}$$

$$x = -4 + \sqrt{26} \quad x = -4 - \sqrt{26}$$

38.
$$2x^2 + 10x - 3 = 0$$

$$x^2 + 5x + \frac{25}{4} = \frac{3}{2} + \frac{25}{4}$$

$$\left(x + \frac{5}{2}\right)^2 = \frac{31}{4}$$

$$x + \frac{5}{2} = \pm\sqrt{\frac{31}{4}}$$

$$x = -\frac{5}{2} \pm \frac{\sqrt{31}}{2}$$

$$x = \frac{-5 + \sqrt{31}}{2} \quad x = \frac{-5 - \sqrt{31}}{2}$$

40.
$$4x^2 - 4x + 15 = 0$$

$$x^2 - x + \frac{1}{4} = -\frac{15}{4} + \frac{1}{4}$$

$$\left(x - \frac{1}{2}\right)^2 = -\frac{14}{4}$$

$$x - \frac{1}{2} = \pm\sqrt{\frac{-14}{4}}$$

$$x = \frac{1}{2} \pm \frac{i\sqrt{14}}{2}$$

$$x = \frac{1 + i\sqrt{14}}{2} \quad x = \frac{1 - i\sqrt{14}}{2}$$

46. $a = 1$, $b = -5$, $c = -24$. Thus

$$x = \frac{-(-5) \pm \sqrt{(-5)^2 - 4(1)(-24)}}{2(1)} = \frac{5 \pm 11}{2} = 8 \quad \text{or} \quad -3$$

50.
$$2x^2 + 4x - 1 = 0$$

$$x = \frac{-4 \pm \sqrt{4^2 - 4(2)(-1)}}{4}$$

$$x = \frac{-4 \pm \sqrt{16 + 8}}{4} = \frac{-4 \pm \sqrt{24}}{4}$$

$$x = \frac{-4 \pm 2\sqrt{6}}{4} = \frac{-2 \pm \sqrt{6}}{2}$$

$$x = \frac{-2 + \sqrt{6}}{2} \quad x = \frac{-2 - \sqrt{6}}{2}$$

62.
$$x^2 + 3x - 11 = 0$$

$$b^2 - 4ac = 3^2 - 4(1)(-11)$$

$$= 9 + 44 = 53$$

Two distinct real numbers.

72.
$$x^2 + 4x - 21 = 0, \quad -7, 3$$

$$r_1 + r_2 = -4 \quad r_1 r_2 = -21$$

$$-\frac{b}{a} = -\frac{4}{1} = -4 \quad \frac{c}{a} = \frac{-21}{1} = -21$$

-7 and 3 are roots.

78.
$$60 = LW$$

$$W = \frac{60}{L}$$

$$2L + 2W = 34$$

$$2L + 2\frac{60}{L} = 34$$

$$2L^2 + 120 = 34L$$

$$2L^2 - 34L + 120 = 0$$

$$2(L^2 - 17L + 60) = 0$$

$$2(L - 5)(L - 12) = 0$$

$$L = 5 \quad L = 12$$

$$12W = 60$$

$$W = 5$$

$W = 5$ feet, $L = 12$ feet.

80. Let P = perimeter and A = area.

$$P = 4w + 2l = 400$$

$$2w + l = 200$$

$$A = 4800 = lw$$

$$l = \frac{4800}{w}$$

$$2w + \frac{4800}{w} = 200$$

$$2w^2 + 4800 = 200w$$

$$w^2 - 100w + 2400 = 0$$

$$(w - 60)(w - 40) = 0$$

$$w = 60 \qquad w = 40$$

$$l = \frac{4800}{60} = 80 \quad l = \frac{4800}{40} = 120$$

There are two solutions: 60 ft $\times$ 80 ft or 40 ft $\times$ 120 ft.

Exercise Set 2.4, page 111

6.
$$x^4 - 36x^2 = 0$$
$$x^2(x^2 - 36) = 0$$
$$x^2(x - 6)(x + 6) = 0$$
$$x = 0, \ x = 6, \ x = -6$$

14. $\sqrt{10 - x} = 4$ Check: $\sqrt{10 - (-6)} = 4$
$$10 - x = 16 \qquad\qquad \sqrt{16} = 4$$
$$-x = 6 \qquad\qquad\qquad 4 = 4$$
$$x = -6$$

The solution is -6.

16.
$$x = \sqrt{5 - x} + 5$$
$$(x - 5)^2 = (\sqrt{5 - x})^2$$
$$x^2 - 10x + 25 = 5 - x$$
$$x^2 - 9x + 20 = 0$$
$$(x - 5)(x - 4) = 0$$
$$x = 5 \quad x = 4$$

Check: $5 = \sqrt{5 - 5} + 5$ $4 = \sqrt{5 - 4} + 5$
$$5 = 0 + 5 \qquad\qquad 4 = 1 + 5$$
$$5 = 5 \qquad\qquad\quad 4 = 6 \quad \text{No}$$

The solution is 5.

20.
$$(\sqrt{x + 7} - 2)^2 = (\sqrt{x - 9})^2$$
$$x + 7 - 4\sqrt{x + 7} + 4 = x - 9$$
$$-4\sqrt{x + 7} = -20$$
$$(\sqrt{x + 7})^2 = (5)^2$$
$$x + 7 = 25$$
$$x = 18$$

Check: $\sqrt{18 + 7} - 2 = \sqrt{18 - 9}$
$$\sqrt{25} - 2 = \sqrt{9}$$
$$5 - 2 = 3$$
$$3 = 3$$

The solution is 18.

Exercise Set 2.5, page 121

8.
$$-4(x - 5) \geq 2x + 15$$
$$-4x + 20 \geq 2x + 15$$
$$-6x \geq -5$$
$$x \leq \frac{5}{6}$$

32.
$$(4z + 7)^{1/3} = 2$$
$$[(4z + 7)^{1/3}]^3 = 2^3$$
$$4z + 7 = 8$$
$$4z = 1$$
$$z = \frac{1}{4}$$

Check: $\left[4\left(\dfrac{1}{4}\right) + 7\right]^{1/3} = 2$
$$8^{1/3} = 2$$
$$2 = 2$$

The solution is $\dfrac{1}{4}$.

42. $x^4 - 10x^2 + 9 = 0$. Let $u = x^2$.
$$u^2 - 10u + 9 = 0$$
$$(u - 9)(u - 1) = 0$$
$$u = 9 \qquad u = 1$$
$$x^2 = 9 \qquad x^2 = 1$$
$$x = \pm 3 \quad x = \pm 1$$

The solutions are 3, -3, 1, and -1.

52. $6x^{2/3} - 7x^{1/3} - 20 = 0$. Let $u = x^{1/3}$.
$$6u^2 - 7u - 20 = 0$$
$$(3u + 4)(2u - 5) = 0$$
$$u = -\frac{4}{3} \qquad u = \frac{5}{2}$$
$$x^{1/3} = -\frac{4}{3} \qquad x^{1/3} = \frac{5}{2}$$
$$x = -\frac{64}{27} \qquad x = \frac{125}{8}$$

12. Company A: $19 + .12x$ x is the number of miles driven
Company B: $12 + .21x$
$$19 + .12x < 12 + .21x$$
$$77.7 < x$$

Company A is less expensive if you drive over 78 miles.

16.
$$-16 < 2x + 5 < 9$$
$$-21 < 2x \quad\quad < 4$$
$$-\frac{21}{2} < \quad x \quad\quad < 2$$

22.
$$41 < F \quad\quad\quad < 68$$
$$41 \le \frac{9}{5}C + 32 \le 68$$
$$9 \le \frac{9}{5}C \quad\quad \le 36$$
$$5 \le C \quad\quad\quad \le 20$$

30.
$$x^2 + 5x + 6 = 0$$
$$(x + 2)(x + 3) = 0$$
$$x = -2 \quad x = -3 \quad \text{critical values}$$
The solution is $(-3, -2)$.

Exercise Set 2.6, page 126

10.
$$|2x - 3| = 21$$
$$2x - 3 = 21 \quad \text{or} \quad 2x - 3 = -21$$
$$2x = 24 \quad\quad\quad 2x = -18$$
$$x = 12 \quad\quad\quad x = -9$$
The solution set is $\{12, -9\}$.

30.
$$|2x - 9| < 7$$
$$-7 < 2x - 9 < 7$$
$$2 < 2x \quad\quad < 16$$
$$1 < x \quad\quad < 8$$
The solution set is $(1, 8)$.

34.
$$|2x - 5| \ge 1$$
$$2x - 5 \le -1 \quad \text{or} \quad 2x - 5 \ge 1$$
$$2x \le 4 \quad\quad\quad\quad 2x \ge 6$$
$$x \le 2 \quad\quad\quad\quad x \ge 3$$
The solution set is $(-\infty, 2] \cup [3, \infty)$.

Section 3.1, page 136

2. $(-4, 3)$

$(-2, 0)$ $(0, 2)$

$(-3, -5)$

38.
$$\frac{x - 2}{x + 3} = 0$$
$$x = 2 \quad x = -3 \quad \text{critical values}$$
The solution set is $(-\infty, -3) \cup (2, \infty)$.

46. The denominator $x - 2 = 0$ when $x = 2$. Also
$$\frac{3x + 1}{x - 2} = 4 \quad \text{when } 3x + 1 = 4x - 8.$$
$$x = 2 \quad x = 9 \quad \text{critical values}$$
The solution set is $(2, 9]$.

54.
$$3x^2 - 312x + 5925 = 0$$
$$(x - 25)(x - 79) = 0$$
$$x = 25 \quad x = 79 \quad \text{critical values}$$
The solution set is $[25, 79]$. Thus the monthly revenue is greater than or equal to $5925 when $25 \le x \le $79.

46.
$$|x - c| < d$$
$$-d < x - c < d$$
$$-d + c < x \quad\quad < d + c$$
The solution set is $(-d + c, d + c)$.

48.
$$|x^2 - 2| > 1$$
$$x^2 - 2 = 1 \quad x^2 - 2 = -1$$
$$x^2 = 3 \quad\quad x^2 = 1$$
$$x = \pm\sqrt{3} \quad x = \pm 1 \quad \text{critical values}$$
The solution set is $(-\infty, -\sqrt{3}) \cup (-1, 1) \cup (\sqrt{3}, \infty)$.

12. $d = \sqrt{(-10 - (-5))^2 + (14 - 8)^2}$
$$= \sqrt{(-5)^2 + 6^2} = \sqrt{61}$$

24. $M = \left(\dfrac{x_1 + x_2}{2}, \dfrac{y_1 + y_2}{2} \right)$
$$= \left(\frac{-5 + 6}{2}, \frac{-2 + 10}{2} \right) = \left(\frac{1}{2}, 4 \right)$$

38.

46.

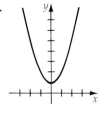

42.

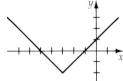

Exercise Set 3.2, page 145

14. symmetrical with respect to x-axis

26. No, because $(-y) = 3(-x) - 2$ is not equivalent to the original equation $y = 3x - 2$.

40. Yes, because $2x = 5 - 2y$ is equivalent to the original equation $2y = 5 - 2x$.

70.

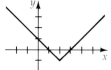

90. $(x - 0)^2 + (y - 0)^2 = 13^2$ because $r = \sqrt{5^2 + 12^2} = 13$.

96.
$$x^2 - 6x + y^2 - 4y = -12$$
$$x^2 - 6x + 9 + y^2 - 4y + 4 = -12 + 9 + 4$$
$$(x - 3)^2 + (y - 2)^2 = 1^2$$
Center $(3, 2)$, radius 1

Exercise Set 3.3, page 152

2. Given $g(x) = 2x^2 + 3$

a. $g(3) = 2(3)^2 + 3 = 18 + 3 = 21$

b. $g(-1) = 2(-1)^2 + 3 = 2 + 3 = 5$

c. $g(0) = 2(0)^2 + 3 = 0 + 3 = 3$

d. $g\left(\dfrac{1}{2}\right) = 2\left(\dfrac{1}{2}\right)^2 + 3 = \dfrac{1}{2} + 3 = \dfrac{7}{2}$

e. $g(c) = 2(c)^2 + 3 = 2c^2 + 3$

f. $g(c + 5) = 2(c + 5)^2 + 3 = 2c^2 + 20c + 50 + 3$
$= 2c^2 + 20c + 53$

14.
$$x^2 - 2y = 2$$
$$-2y = -x^2 + 2$$
$$y = \frac{1}{2}x^2 - 1$$

y is a function of x because each x value will yield one y value.

22. The domain is the set of all real numbers.

34. Even, because $f(-x) = (-x)^2 + 1 = x^2 + 1 = f(x)$.

52. $v(t) = 44{,}000 - 4200t, \quad 0 \le t \le 8$.

54. Let y be the yield and let n be the number of trees. Then

$$y(n) = \begin{cases} 240n & \text{if } n \le 30 \\ [240 - 10(n - 30)]n & \text{if } n > 30 \end{cases}$$

Note n is a natural number.

60. $t = \dfrac{\sqrt{1 + x^2}}{2} + \dfrac{3 - x}{8}$

Exercise Set 3.4, page 162

8. Yes, because every vertical line intersects the graph in one point.

48.

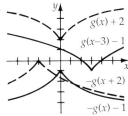

50.

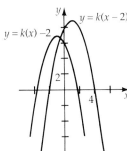

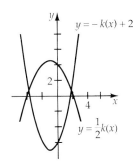

52. Domain is the set of real numbers. Range is the set of real numbers.

70.

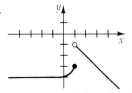

Exercise Set 3.5, page 174

2. $m = \dfrac{1-4}{5-(-2)} = -\dfrac{3}{7}$

14. $m = \dfrac{f(a+h) - f(a)}{a+h-a} = \dfrac{f(a+h) - f(a)}{h}$

16.

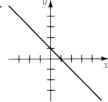

28. $y - 5 = -2(x - 0)$

$\quad\quad y = -2x + 5$

40. $m = \dfrac{-8 - (-6)}{2 - 5} = \dfrac{-2}{-3} = \dfrac{2}{3}$

$$y - (-6) = \frac{2}{3}(x - 5)$$

$$y = \frac{2}{3}x - \frac{10}{3} - 6$$

$$y = \frac{2}{3}x - \frac{28}{3}$$

48. The graph of $x + y = 10$ has $m = -1$.

$$y + 1 = (-1)(x - 2)$$

$$y = -x + 2 - 1$$

$$y = -x + 1$$

52. The graph of $2x - y = 7$ has $m = 2$. Thus we use a slope of $-\frac{1}{2}$.

$$y - 4 = -\frac{1}{2}(x + 3)$$

$$y = -\frac{1}{2}x - \frac{3}{2} + \frac{8}{2}$$

$$y = -\frac{1}{2}x + \frac{5}{2}$$

60. a. Annual depreciation $= \dfrac{40{,}090 - 8500}{15} = 2106$

$$V(n) = 40{,}090 - 2106n$$

b. $V(4) = 40{,}090 - 2106(4) = 40{,}090 - 8424 = \$31{,}666$

Exercise Set 3.6, page 183

2. f

10. $f(x) = x^2 + 6x - 1$

$= x^2 + 6x + 9 + (-1 - 9)$

$= (x + 3)^2 - 10$

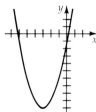

20. $f(x) = -x^2 - 6x$

$= -(x^2 + 6x)$

$= -(x^2 + 6x + 9) + 9$

$= -(x + 3)^2 + 9$

$= -(x - (-3))^2 + 9$

Maximum value of f is 9 when $x = -3$.

30. $P(x) = -3x^2 + 96x - 368$

$P(x) = -3(x^2 - 32x \qquad) - 368$

$P(x) = -3(x^2 - 32x + 256) - 368 + 768$

$P(x) = -3(x - 16)^2 + 400$

a. $x = 16$ will maximize the profit.

b. The maximum profit is $P(16) = 400$.

36. $x = \dfrac{+6}{2(1)} = 3$

$y = f(3) = 3^2 - 6(3) = -9$

Vertex is $(3, -9)$.

46. The y-intercept is $(0, 0)$.

$$-x^2 + 4x = 0$$

$$x(-x + 4) = 0$$

$$x = 0 \quad x = 4$$

The x-intercepts are $(0, 0)$ and $(4, 0)$.

56. $h(t) = -16t^2 + 64t + 80$

$$t = \frac{-b}{2a} = \frac{-64}{2(-16)} = 2$$

$$h(2) = -16(2)^2 + 64(2) + 80$$

$$= -64 + 128 + 80 = 144$$

a. The vertex $(2, 144)$ gives us the maximum height of 144 feet.

b. The vertex of the graph of h is $(2, 144)$ thus the time when it achieves this maximum height is at time $t = 2$ seconds.

c. $-16t^2 + 64t + 80 = 0$

$$-16(t^2 - 4t - 5) = 0$$

$$-16(t + 1)(t - 5) = 0$$

$$t = -1 \quad t - 5 = 0$$

$$\text{no} \qquad t = 5$$

The projectile will have a height of 0 feet at time $t = 5$ seconds.

Exercise Set 3.7, page 189

10. $f(x) + g(x) = \sqrt{x - 4} - x$ domain $\{x \mid x \geq 4\}$

$f(x) - g(x) = \sqrt{x - 4} + x$ domain $\{x \mid x \geq 4\}$

$f(x)g(x) = -x\sqrt{x - 4}$ domain $\{x \mid x \geq 4\}$

$f(x)/g(x) = -\dfrac{\sqrt{x - 4}}{x}$ domain $\{x \mid x \geq 4\}$

14. $(f + g)(x) = (x^2 - 3x + 2) + (2x - 4) = x^2 - x - 2$

$(f + g)(-7) = (-7)^2 - (-7) - 2 = 49 + 7 - 2 = 54$

30. $\dfrac{f(x + \Delta x) - f(x)}{\Delta x} = \dfrac{[4(x + \Delta x) - 5] - (4x - 5)}{\Delta x}$

$= \dfrac{4x + 4(\Delta x) - 5 - 4x + 5}{\Delta x}$

$= \dfrac{4(\Delta x)}{\Delta x} = 4$

38.
$$(g \circ f)(x) = g[f(x)] = g[2x - 7]$$
$$= 3[2x - 7] + 2 = 6x - 19$$
$$(f \circ g)(x) = f[g(x)] = f[3x + 2]$$
$$= 2[3x + 2] - 7 = 6x - 3$$

54. $(h \circ g)(x) = -3x^4 + 30x^3 - 75x^2 + 4$
$(h \circ g)(0) = -3(0)^4 + 30(0)^3 - 75(0)^2 + 4 = 4$

Exercise Set 3.8, page 196

2.
$$(f \circ g)(x) = f[g(x)] = f[2x + 6]$$
$$= \frac{1}{2}[2x + 6] - 3 = x + 3 - 3 = x$$
$$(g \circ f)(x) = g[f(x)] = g\left[\frac{1}{2}x - 3\right]$$
$$= 2\left[\frac{1}{2}x - 3\right] + 6 = x - 6 + 6 = x$$

10.
$$g(x) = \frac{2}{3}x + 4$$
$$y = \frac{2}{3}x + 4$$
$$x = \frac{2}{3}y + 4$$
$$x - 4 = \frac{2}{3}y$$
$$\frac{3}{2}x - 6 = y$$

Thus $g^{-1}(x) = \frac{3}{2}x - 6$.

18.
$$G(x) = \frac{3x}{x - 5}, \quad x \neq 5$$
$$y = \frac{3x}{x - 5}$$
$$x = \frac{3y}{y - 5}$$
$$xy - 5x = 3y$$
$$xy - 3y = 5x$$
$$y = \frac{5x}{x - 3}$$

Thus $G^{-1}(x) = \frac{5x}{x - 3}, \quad x \neq 3$.

32. $f(x) = x^2 + 6x - 6, \; x \geq -3$
domain f is $\{x \,|\, x \geq -3\}$, domain f^{-1} is $\{x \,|\, x \geq -15\}$
range f is $\{y \,|\, y \geq -15\}$, range f^{-1} is $\{y \,|\, y \geq -3\}$
$$y = x^2 + 6x - 6$$
$$x = y^2 + 6y - 6$$
$$x + 6 = y^2 + 6y$$
$$x + 15 = y^2 + 6y + 9 \quad \text{(complete the square)}$$
$$x + 15 = (y + 3)^2$$

Choose the positive root, since range f^{-1} is $\{y \,|\, y \geq -3\}$.
$$\sqrt{x + 15} = y + 3$$
$$-3 + \sqrt{x + 15} = y$$
Thus $f^{-1}(x) = -3 + \sqrt{x + 15}$

36.

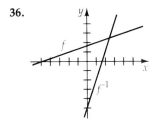

Exercise Set 3.9, page 202

14. $m = kn$

$92 = k \cdot 23$

$\dfrac{92}{23} = k$

$k = 4$

24.

$$r = kv^2$$

$$140 = k \cdot 60^2$$

$$\frac{140}{60 \cdot 60} = k$$

$$\frac{7}{180} = k$$

Thus $r = \dfrac{7}{180} \cdot 65^2 \approx 164.3$ feet

28.

$$I = \frac{k}{d^2}$$

$$50 = \frac{k}{10^2}$$

$$5000 = k$$

Thus $I = \dfrac{5000}{d^2}$

$$= \frac{5000}{15^2}$$

$$= \frac{5000}{225} \approx 22.2 \text{ footcandles}$$

30.

$$i = kpr$$

$$78 = k2600 \cdot (.04)$$

$$k = \frac{78}{2600(0.04)} = 0.75$$

Thus $i = 0.75 \cdot 4400 \cdot (.06) = \$198$.

32.

$$L = kbd^2$$

$$200 = k \cdot 2 \cdot 6^2$$

$$\frac{25}{9} = \frac{200}{2 \cdot 6^2} = k$$

Thus $L = \dfrac{25}{9} \cdot 4 \cdot 4^2 = \dfrac{1600}{9} \approx 177.8$ pounds.

34.

$$L = \frac{kd^4}{h^2}$$

$$6 = \frac{k \cdot 2^4}{10^2}$$

$$k = \frac{6 \cdot 10^2}{2^4} = \frac{600}{16} = 37.5$$

Thus $L = \dfrac{37.5(3^4)}{14^2} \approx 15.5$ tons.

Exercise Set 4.1, page 217

6.

$$2x^2 - x - 5 \overline{\smash{\big)}\ \begin{array}{l} x^2 - 9 \\ 2x^4 - x^3 - 23x^2 + 9x + 45 \end{array}}$$

$$\underline{2x^2 - x^3 - 5x^2}$$

$$-18x^2 + 9x + 45$$

$$\underline{-18x^2 + 9x + 45}$$

$$0$$

12.

$$5 \,\underline{\big|\ \begin{array}{cccc} 5 & 6 & -8 & 1 \\ & 25 & 155 & 735 \end{array}}$$

$$\ \begin{array}{cccc} 5 & 31 & 147 & 736 \end{array}$$

$$\frac{5x^3 + 6x^2 - 8x + 1}{x - 5} = 5x^2 + 31x + 147 + \frac{736}{x - 5}$$

32.

$$3 \,\underline{\big|\ \begin{array}{cccc} 2 & -1 & 3 & -1 \\ & 6 & 15 & 54 \end{array}}$$

$$\ \begin{array}{cccc} 2 & 5 & 18 & 53 \end{array}$$

$$P(c) = P(3) = 53$$

42.

$$-6 \,\underline{\big|\ \begin{array}{cccc} 1 & 4 & -27 & -90 \\ & -6 & 12 & 90 \end{array}}$$

$$\ \begin{array}{cccc} 1 & -2 & -15 & 0 \end{array}$$

A remainder of 0 implies that $x + 6$ is a factor of $P(x)$.

Exercise Set 4.2, page 224

2. Since $a_n = -2$ is negative and $n = 3$ is odd, the graph of P goes up to the far left and down to the far right.

12. a.
$$
\begin{array}{r|rrrr}
1 & 2 & -5 & 4 & 10 \\
  &   & 2 & -3 & 1 \\
\hline
  & 2 & -3 & 1 & 11
\end{array}
\qquad g(1) = 11
$$

b.
$$
\begin{array}{r|rrrr}
-1 & 2 & -5 & 4 & 10 \\
   &   & -2 & 7 & -11 \\
\hline
   & 2 & -7 & 11 & -1
\end{array}
\qquad g(-1) = -1
$$

c.
$$
\begin{array}{r|rrrr}
2 & 2 & -5 & 4 & 10 \\
  &   & 4 & -2 & 4 \\
\hline
  & 2 & -1 & 2 & 14
\end{array}
\qquad g(2) = 14
$$

d.
$$
\begin{array}{r|rrrr}
3 & 2 & -5 & 4 & 10 \\
  &   & 6 & 3 & 21 \\
\hline
  & 2 & 1 & 7 & 31
\end{array}
\qquad g(3) = 31
$$

e.
$$
\begin{array}{r|rrrr}
0 & 2 & -5 & 4 & 10 \\
  &   & 0 & 0 & 0 \\
\hline
  & 2 & -5 & 4 & 10
\end{array}
\qquad g(0) = 10
$$

f.
$$
\begin{array}{r|rrrr}
-5 & 2 & -5 & 4 & 10 \\
   &   & -10 & 75 & -395 \\
\hline
   & 2 & -15 & 79 & -385
\end{array}
\qquad g(-5) = -385
$$

Exercise Set 4.3, page 232

12. $p = \pm1, \pm2, \pm3, \pm5, \pm6, \pm10, \pm15, \pm30$

$q = \pm1$

$\dfrac{p}{q} = \pm1, \pm2, \pm3, \pm5, \pm6, \pm10, \pm15, \pm30$

24.
$$
\begin{array}{r|rrrr}
3 & 1 & 0 & -19 & -28 \\
  &   & 3 & 9 & -30 \\
\hline
  & 1 & 3 & -10 & -58
\end{array}
\qquad
\begin{array}{r|rrrr}
-2 & 1 & 0 & -19 & -28 \\
   &   & -2 & 4 & 30 \\
\hline
   & 1 & -2 & -15 & 2
\end{array}
$$

$$
\begin{array}{r|rrrr}
4 & 1 & 0 & -19 & -28 \\
  &   & 4 & 16 & -12 \\
\hline
  & 1 & 4 & -3 & -40
\end{array}
\qquad
\begin{array}{r|rrrr}
-3 & 1 & 0 & -19 & -28 \\
   &   & -3 & 9 & 30 \\
\hline
   & 1 & -3 & -10 & 2
\end{array}
$$

$$
\begin{array}{r|rrrr}
5 & 1 & 0 & -19 & -28 \\
  &   & 5 & 25 & 30 \\
\hline
  & 1 & 5 & 6 & 2
\end{array}
\qquad
\begin{array}{r|rrrr}
-4 & 1 & 0 & -19 & -28 \\
   &   & -4 & 16 & 12 \\
\hline
   & 1 & -4 & -3 & -16
\end{array}
$$

$$
\begin{array}{r|rrrr}
-5 & 1 & 0 & -19 & -28 \\
   &   & -5 & 25 & -30 \\
\hline
   & 1 & -5 & 6 & -58
\end{array}
$$

5 is an upper bound and -5 is a lower bound.

32. Since P has 2, -3, and 1 as zeros, the x-intercepts of the graph of P are $(2, 0)$, $(-3, 0)$, $(-1, 0)$.

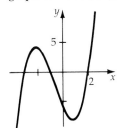

48.
$$
\begin{array}{r|rrrr}
0 & 4 & -1 & -6 & 1 \\
  &   & 0 & 0 & 0 \\
\hline
  & 4 & -1 & -6 & 1
\end{array}
\qquad P(0) = 1
$$

$$
\begin{array}{r|rrrr}
1 & 4 & -1 & -6 & 1 \\
  &   & 4 & 3 & -3 \\
\hline
  & 4 & 3 & -3 & -2
\end{array}
\qquad P(1) = -2
$$

Because $P(0)$ and $P(1)$ have different signs, P must have a real zero between 0 and 1.

36. One positive real zero since the polynomial has one variation in sign.

$$(-x)^3 - 19(-x) - 30 = -x^3 + 19x - 30$$

2 or no negative real zeros because $-x^3 + 19x - 30$ has two variations in sign.

48. One positive and two or no negative real zeros (see Exercise 36).

$$
\begin{array}{r|rrr}
5 & 1 & 0 & -19 & -30 \\
  &   & 5 & 25 & 30 \\
\hline
  & 1 & 5 & 6 & 0
\end{array}
\qquad
\begin{array}{r|rr}
-2 & 1 & 2 \\
   &   & -2 \\
\hline
   & 1 & 0
\end{array}
$$

$$
\begin{array}{r|rrr}
-3 & 1 & 5 & 6 \\
   &   & -3 & -6 \\
\hline
   & 1 & 2 & 0
\end{array}
$$

The zeros of $x^3 - 19x - 30$ are 5, -2, and -3.

Exercise Set 4.4, page 238

2.

$$
\begin{array}{r|rrrr}
5 + 3i & 3 & -29 & 92 & 34 \\
 & & 15 + 9i & -97 + 3i & -34 \\
\hline
 & 3 & -14 + 9i & -5 + 3i & 0
\end{array}
$$

$$
\begin{array}{r|rrr}
5 - 3i & 3 & -14 + 9i & -5 + 3i \\
 & & 15 - 9i & 5 - 3i \\
\hline
 & 3 & 1 & 0
\end{array}
\qquad
-\frac{1}{3}
\begin{array}{r|rr}
 & 3 & 1 \\
 & & -1 \\
\hline
 & 3 & 0
\end{array}
$$

The zeros of $3x^3 - 29x^2 + 92x + 34$ are $5 + 3i$, $5 - 3i$

and $-\dfrac{1}{3}$.

20. $6x^3 - 23x^2 - 4x = x(6x^2 - 23x - 4)$
$ = x(6x + 1)(x - 4)$

40. Since P has real coefficients, use the Conjugate Pair Theorem.

$$
\begin{aligned}
P &= (x - 3 - 2i)(x - 3 + 2i)(x - 7) \\
&= (x^2 - 6x + 13)(x - 7) \\
&= x^3 - 13x^2 + 55x - 91
\end{aligned}
$$

Exercise Set 4.5, page 249

2.

$$x^2 - 4 = 0$$

$$(x - 2)(x + 2) = 0$$

$$x = 2 \quad \text{or} \quad x = -2$$

The vertical asymptotes are $x = 2$ and $x = -2$.

6. The horizontal asymptote is $y = 0$ (x-axis) because the degree of the denominator is larger than the degree of the numerator.

10. Vertical asymptote: $x - 2 = 0$
$ x = 2$

Horizontal asymptote: $y = 0$

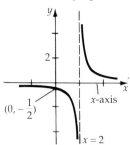

26. Vertical asymptote: $x^2 - 6x + 9 = 0$
$ (x - 3)(x - 3) = 0$
$ x = 3$

Horizontal asymptote is $y = \frac{1}{1} = 1$ (the Theorem on Horizontal Asymptotes) because numerator and denominator both have degree 2.

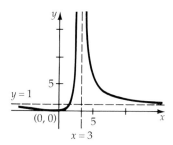

34.

$$
\begin{array}{r}
x + 1 \\
x^2 - 3x + 5 \overline{\big)\, x^3 - 2x^2 + 3x + 4} \\
\underline{x^3 - 3x^2 + 5x} \\
x^2 - 2x + 4 \\
\underline{x^2 - 3x + 5} \\
x - 1
\end{array}
$$

$$F(x) = x + 1 + \frac{x - 1}{x^2 - 3x + 5}$$

Slant asymptote: $y = x + 1$

40.

$$2x + 5 = 0$$

$$2x = -5$$

$$x = -\frac{5}{2}$$

$$
\begin{array}{r}
\frac{1}{2}x \;-\; \frac{13}{4} \\[4pt]
2x + 5 \overline{\big)\, x^2 \;-\; 4x \;-\; 5} \\[4pt]
\underline{x^2 + \frac{5}{2}x} \\[4pt]
-\frac{13}{2}x \;-\; 5 \\[4pt]
\underline{-\frac{13}{2}x \;-\; \frac{65}{4}} \\[4pt]
\frac{45}{4}
\end{array}
$$

$$F(x) = \frac{x^2 - 4x}{2x + 5} = \frac{1}{2}x - \frac{13}{4} + \frac{45/4}{2x + 5}$$

Vertical asymptote: $x = -\dfrac{5}{2}$

Slant asymptote: $y = \dfrac{1}{2}x - \dfrac{13}{4}$

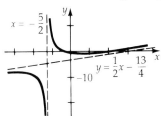

50.

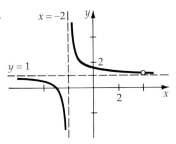

Exercise Set 4.6, page 254

2. $P(1) = (1)^3 + 18(1) - 30 = -11$

$P(2) = (2)^3 + 18(2) - 30 = 14$ different signs

P has a zero between 1 and 2 (zero location theorem).

8. $P(0) = -10$

$P(1) = 9$

$P\!\left(\dfrac{1}{2}\right) = \left(\dfrac{1}{2}\right)^3 + 18\left(\dfrac{1}{2}\right) - 10 = -0.875$

$P\!\left(\dfrac{3}{4}\right) = (.75)^3 + 18(.75) - 10 = 3.921875$

$P\!\left(\dfrac{5}{8}\right) = \left(\dfrac{5}{8}\right)^3 + 18\left(\dfrac{5}{8}\right) - 10 = 1.494140625$

P has a zero between $\dfrac{1}{2}$ and $\dfrac{5}{8}$.

14.

$P(3) = 1$

$P(2.9) = -1.411$

$P(2.99) \approx .751$

$P(2.96) \approx .014$

$P(2.95) \approx -.228$

P has a zero between 2.95 and 2.96.

Exercise Set 5.1, page 266

26.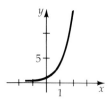

36.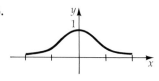

38.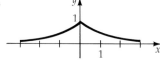

44.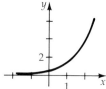

46.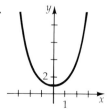

Exercise Set 5.2, page 275

2.
$$\log_{10} 1000 = 3$$
$$1000 = 10^3$$

12.
$$3^5 = 243$$
$$\log_3 243 = 5$$

22.
$$\log_{10} \frac{1}{1000} = n$$
$$10^n = \frac{1}{1000}$$
$$10^n = 10^{-3}$$
$$n = -3$$

32. $\log_b x^2 y^3 = \log_b x^2 + \log_b y^3 = 2 \log_b x + 3 \log_b y$

42. $\log_b 20 = \log_b 2^2 \cdot 5 = 2 \log_b 2 + \log_b 5$
$$= 2(0.3562) + (0.8271) = 1.5395$$

52. $5 \log_3 x - 4 \log_3 y + 2 \log_3 z$
$$= \log_3 x^5 - \log_3 y^4 + \log_3 z^2$$
$$= \log_3 \frac{x^5 z^2}{y^4}$$

62. $\log_5 37 = \dfrac{\log 37}{\log 5} \approx 2.243589$

Exercise Set 5.3, page 280

2.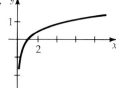

10.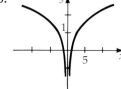

12.

14.

20.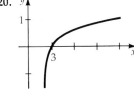

42.

$$I = 100,000 I_0$$

$$R = \log \frac{100,000 I_0}{I_0}$$

$$R = \log 100,000$$

$$R = 5$$

44. Let I = the intensity of the smaller earthquake. Then $1000I$ = the intensity of the larger earthquake.

$$R_1 = \log \frac{I}{I_0} \qquad R_2 = \log \frac{1000I}{I_0} = \log 1000 \frac{I}{I_0}$$

$$R_2 = \log 1000 \frac{I}{I_0} = \log 1000 + \log \frac{I}{I_0}$$

$$= \log 10^3 + R_1 = 3 + R_1$$

The first earthquake has a Richter scale measure that is 3 more than the second earthquake.

Exercise Set 5.4, page 286

2.

$$3^x = 243$$

$$3^x = 3^5$$

$$x = 5$$

12.

$$6^x = 50$$

$$\log(6^x) = \log 50$$

$$x \log 6 = \log 50$$

$$x = \frac{\log 50}{\log 6} \approx 2.18$$

24.

$$\log_3 x + \log_3(x + 6) = 3$$

$$\log_3 x(x + 6) = 3$$

$$3^3 = x(x + 6)$$

$$27 = x^2 + 6x$$

$$x^2 + 6x - 27 = 0$$

$$(x + 9)(x - 3) = 0$$

$$x = 3 \quad \text{or} \quad -9$$

Since $\log_3 x$ is only defined for $x > 0$, the only solution is $x = 3$.

32.

$$\ln x = \frac{1}{2} \ln\left(2x + \frac{5}{2}\right) + \frac{1}{2} \ln 2$$

$$\ln x = \frac{1}{2} \ln 2\left(2x + \frac{5}{2}\right)$$

$$\ln x = \frac{1}{2} \ln (4x + 5)$$

$$\ln x = \ln(4x + 5)^{1/2}$$

$$x = \sqrt{4x + 5}$$

$$x^2 = 4x + 5$$

$$0 = x^2 - 4x - 5$$

$$0 = (x - 5)(x + 1)$$

$$x = 5, -1$$

Check: $\ln 5 = \frac{1}{2} \ln\left(10 + \frac{5}{2}\right) + \frac{1}{2} \ln 2$

$$1.6094 \approx 1.2629 + 0.3466$$

Since $\ln(-1)$ is not defined, -1 is not a solution. Thus the only solution is $x = 5$.

42.

$$\frac{10^x + 10^{-x}}{2} = 8$$

$$10^x(10^x + 10^{-x}) = (16)10^x$$

$$10^{2x} + 1 = 16(10^x)$$

$$10^{2x} - 16(10^x) + 1 = 0$$

Let $u = 10^x$.

$$u^2 - 16u + 1 = 0$$

$$u = \frac{16 \pm \sqrt{16^2 - 4(1)(1)}}{2} = 8 \pm 3\sqrt{7}$$

$$10^x = 8 \pm 3\sqrt{7}$$

$$x \log 10 = \log(8 \pm 3\sqrt{7})$$

$$x = \log(8 \pm 3\sqrt{7}) = \pm 1.20241$$

50. $\text{pH} = -\log(1.26 \times 10^{-12}) = 11.89$ basic solution

52.

$$\text{pH} = 3.1$$

$$-\log[\text{H}_3\text{O}^+] = 3.1$$

$$\text{H}_3\text{O}^+ = 10^{-3.1} \approx 7.9 \times 10^{-4}$$

Exercise Set 5.5, page 295

4. a. $P = 12{,}500$, $r = .08$, $t = 10$, $n = 1$.

$$B = 12{,}500\left(1 + \frac{.08}{1}\right)^{10} = \$26{,}986.56$$

b. $n = 365$

$$B = 12{,}500\left(1 + \frac{.08}{365}\right)^{3650} \approx \$27{,}816.82$$

c. $n = 8760$

$$B = 12{,}500\left(1 + \frac{.08}{8760}\right)^{87600} \approx \$27{,}819.16$$

6. $P = 32{,}000$, $r = .08$, $t = 3$.

$$B = Pe^{rt}$$

$$B = 32{,}000e^{3(.08)} \approx \$40{,}679.97$$

10.

$$N(t) = N_0 e^{kt}$$

$$N(0) = 53{,}700 = N_0 \quad \text{(by definition)}$$

$$N(6) = 58{,}100$$

Use $t = 6$ and $N_0 = 53{,}700$.

$$58{,}100 = 53{,}700e^{k(6)}$$

$$\frac{58{,}100}{53{,}700} = e^{6k}$$

$$\ln\left(\frac{58{,}100}{53{,}700}\right) = 6k \ln e$$

$$\frac{\ln\left(\frac{58{,}100}{53{,}700}\right)}{6} = k$$

$$k \approx 0.0131$$

$$N(t) = 53{,}700e^{0.0131t}$$

12. $P(t) = 15{,}600(e^{0.09t})$

a. $P(10) = 15{,}600(e^{.9}) \approx 38{,}400$

b. We must determine the value of t when the population $N(t) = 15{,}600 \cdot 2 = 31{,}200$.

$$31{,}200 = 15{,}600e^{.09t}$$

$$\frac{31{,}200}{15{,}600} = e^{.09t}$$

$$t = \frac{\ln\left(\frac{31{,}200}{15{,}600}\right)}{.09} \approx 8$$

The population will double by 1992.

14.
$$N(138) = 100e^{138k}$$
$$50 = 100e^{138k}$$
$$.5 = e^{138k}$$
$$\ln .5 = 138k \ln e$$
$$k = \frac{\ln .5}{138} \approx -.005023$$
$$N(t) = 100e^{-.005023t}$$

16.
$$N(t) = N_0(0.5)^{t/138}$$
$$N(730) = N_0(0.5)^{730/138} \approx N_0(0.0256)$$

After 2 years (730 days), only 2.56% of the polonium remains.

18.
$$N(t) = N_0(0.5)^{t/5730}$$
$$N(t) = .65N_0$$
$$.65N_0 = N_0(0.5)^{t/5730}$$
$$t = 5730 \frac{\ln .65}{\ln 0.5} \approx 3560$$

The bone is approximately 3560 years old.

22.
$$N(3.4 \times 10^{-5}) = 10 \log\left(\frac{3.4 \times 10^{-5}}{10^{-16}}\right)$$
$$= 10(\log 3.4 + \log 10^{11})$$
$$= 10 \log 3.4 + 110 \approx 115.3 \text{ decibels}$$

Exercise Set 6.1, page 312

22. $35°42' = 35° + 42'' = 35° + 42''\left(\dfrac{1°}{3600''}\right)$

$\qquad = 35° + 0.012° = 35.012°$

30. $55.44° = 55° + 0.44°\left(\dfrac{60'}{1°}\right) = 55° + 26.4'$

$\qquad = 55° + 26' + 0.4'\left(\dfrac{60''}{1'}\right)$

$\qquad = 55° + 26' + 24'' = 55°26'24''$

42. $585° = 585°\left(\dfrac{\pi}{180°}\right) = \dfrac{13}{4}\pi$

52. $\dfrac{11\pi}{18} = \dfrac{11\pi}{18}\left(\dfrac{180°}{\pi}\right) = 110°$

74. $\theta = \dfrac{s}{r} = \dfrac{4}{7} = 0.57$

$\qquad = 0.57\left(\dfrac{180°}{\pi}\right) \approx 32.7°$

80. $s = r\theta = 5\,(144°)\left(\dfrac{\pi}{180°}\right) \approx 12.57 \text{ meters}$

82. $\theta = \dfrac{3}{8}(2\pi) = \dfrac{3}{4}\pi$

84. $\theta_2 = \dfrac{1.2}{0.8}(240°)\left(\dfrac{\pi}{180}\right) = 2\pi$

86. $\omega = \dfrac{\theta}{t} = \dfrac{2\pi}{86400} = 7.27 \times 10^{-5} \text{ radian}/s$

90. $v = \omega r$

$\qquad = \dfrac{450 \cdot 2\pi \cdot 60 \cdot 15}{12 \cdot 520} \approx 40.16 \text{ mph}$

Exercise Set 6.2, page 319

8. $a = 10$, $b = 6$, $r = \sqrt{10^2 + 6^2} = 2\sqrt{34}$.

$$\sin \theta = \frac{b}{r} = \frac{6}{2\sqrt{34}} = \frac{3\sqrt{34}}{34} \qquad \csc \theta = \frac{\sqrt{34}}{3}$$

$$\cos \theta = \frac{a}{r} = \frac{10}{2\sqrt{34}} = \frac{5\sqrt{34}}{34} \qquad \sec \theta = \frac{\sqrt{34}}{5}$$

$$\tan \theta = \frac{b}{a} = \frac{6}{10} = \frac{3}{5} \qquad \cot \theta = \frac{5}{3}$$

18. Since $\tan \theta = \frac{b}{a} = \frac{4}{3}$, let $b = 4$ and $a = 3$.

$$r = \sqrt{3^2 + 4^2} = 5$$

$$\sec \theta = \frac{r}{a} = \frac{5}{3}$$

48. $\sec \dfrac{3\pi}{8} = 2.6131$

70. $\sin \dfrac{\pi}{3} \cos \dfrac{\pi}{4} - \tan \dfrac{\pi}{4} = \dfrac{\sqrt{3}}{2} \cdot \dfrac{\sqrt{2}}{2} - 1$

$$= \frac{\sqrt{6}}{4} - 1 = \frac{\sqrt{6} - 4}{4}$$

Exercise Set 6.3, page 326

6. $a = -6$, $b = -9$, $r = \sqrt{(-6)^2 + (-9)^2} = \sqrt{117} = 3\sqrt{13}$.

$$\sin \theta = \frac{b}{r} = \frac{-9}{3\sqrt{13}} = -\frac{3}{\sqrt{13}} = -\frac{3\sqrt{13}}{13} \qquad \csc \theta = -\frac{\sqrt{13}}{3}$$

$$\cos \theta = \frac{a}{r} = \frac{-6}{3\sqrt{13}} = -\frac{2}{\sqrt{13}} = -\frac{2\sqrt{13}}{13} \qquad \sec \theta = -\frac{\sqrt{13}}{2}$$

$$\tan \theta = \frac{b}{a} = \frac{-9}{-6} = \frac{3}{2} \qquad \cot \theta = \frac{2}{3}$$

10. The reference angle for $225°$ is $45°$.

$$\sin 225° = -\frac{\sqrt{2}}{2} \qquad \csc 225° = -\sqrt{2}$$

$$\cos 225° = -\frac{\sqrt{2}}{2} \qquad \sec 225° = -\sqrt{2}$$

$$\tan 225° = 1 \qquad \cot 225° = 1$$

26. $\cos \dfrac{7\pi}{4} \tan \dfrac{4\pi}{3} + \cos \dfrac{7\pi}{6} = \dfrac{\sqrt{2}}{2} \cdot (\sqrt{3}) + \left(-\dfrac{\sqrt{3}}{2}\right)$

$$= \frac{\sqrt{6}}{2} - \frac{\sqrt{3}}{2} = \frac{\sqrt{6} - \sqrt{3}}{2}$$

40. $\sec \phi = \dfrac{2\sqrt{3}}{3} = \dfrac{r}{a}$, $r = 2\sqrt{3}$, $a = 3$,

$$b = \pm\sqrt{(2\sqrt{3})^2 - 3^2} = \pm\sqrt{3},$$

$$b = -\sqrt{3} \quad \text{in quadrant IV.}$$

$$\sin \phi = \frac{-\sqrt{3}}{2\sqrt{3}} = -\frac{1}{2}$$

54. $\sin 398° \approx 0.6157$

56. $\sin 740° \approx 0.3420$

Exercise Set 6.4, page 334

48. $\sin \dfrac{37\pi}{4} = \sin\left(\dfrac{5\pi}{4} + \dfrac{32\pi}{4}\right) = \sin\left(\dfrac{5\pi}{4} + 8\pi\right)$

$$= \sin\left(\frac{5\pi}{4} + 4(2\pi)\right) = \sin \frac{5\pi}{4} = -\frac{\sqrt{2}}{2}$$

58.

$$\tan \phi = \frac{y}{x}$$

$$\tan(\phi - 180°) = \frac{-y}{-x} = \frac{y}{x}$$

$$\tan \phi = \tan(\phi - 180°)$$

80. $\cos \phi \sec^2 \phi - \dfrac{\cos \phi}{\cot^2 \phi} = \cos \phi \dfrac{1}{\cos^2 \phi} - \dfrac{\cos \phi}{\dfrac{\cos^2 \phi}{\sin^2 \phi}}$

$$= \dfrac{1}{\cos \phi} - \cos \phi \dfrac{\sin^2 \phi}{\cos^2 \phi}$$

$$= \dfrac{1}{\cos \phi} - \dfrac{\sin^2 \phi}{\cos \phi}$$

$$= \dfrac{1 - \sin^2 \phi}{\cos \phi}$$

$$= \dfrac{\cos^2 \phi}{\cos \phi} = \cos \phi$$

96. $\sin \phi = \dfrac{1}{\csc \phi} = \dfrac{1}{\sqrt{2}} = \dfrac{\sqrt{2}}{2}$

98. $\cos \phi = -\sqrt{1 - \sin^2 \phi}$ since ϕ is in quadrant II.

$$\cos \phi = -\sqrt{1 - \left(\dfrac{1}{2}\right)^2} = -\sqrt{1 - \dfrac{1}{4}} = -\dfrac{\sqrt{3}}{2}$$

$$\tan \phi = \dfrac{\sin \phi}{\cos \phi} = \dfrac{\dfrac{1}{2}}{-\dfrac{\sqrt{3}}{2}} = -\dfrac{1}{\sqrt{3}} = -\dfrac{\sqrt{3}}{3}$$

106. $\tan \phi = \dfrac{\sin \phi}{\cos \phi} = \dfrac{\sin \phi}{\pm\sqrt{1 - \sin^2 \phi}} = \pm\dfrac{\sin \phi}{\sqrt{1 - \sin^2 \phi}}$

Exercise Set 6.5, page 345

20. $f(x) = \dfrac{3}{2} \sin x,\ a = \left|\dfrac{3}{2}\right| = \dfrac{3}{2},\ p = 2\pi.$

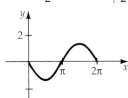

28. $f(x) = 3 \cos \dfrac{\pi x}{3},\ a = 3,\ p = \dfrac{2\pi}{\pi/3} = 6.$

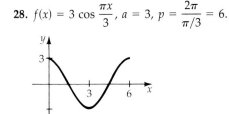

32. $f(x) = \dfrac{1}{2} \sin 2.5x,\ a = \dfrac{1}{2},\ p = \dfrac{2\pi}{2.5} = \dfrac{4\pi}{5}.$

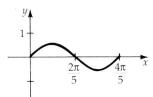

34. $f(x) = -\dfrac{3}{4} \cos 5x,\ a = \left|-\dfrac{3}{4}\right| = \dfrac{3}{4},\ p = \dfrac{2\pi}{5}.$

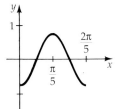

40. $f(x) = -\left|3 \sin \dfrac{2}{3}x\right|$

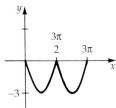

Exercise Set 6.6, page 353

30. $f(x) = -3 \tan 3x$, $p = \dfrac{2\pi}{3}$.

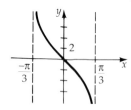

32. $f(x) = \dfrac{1}{2} \cot 2x$, $p = \dfrac{\pi}{2}$.

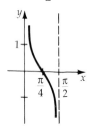

38. $f(x) = 3 \csc \dfrac{\pi x}{2}$, $p = \dfrac{2\pi}{\pi/2} = 4$.

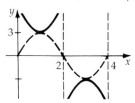

42. $f(x) = \sec \dfrac{x}{2}$, $p = \dfrac{2\pi}{1/2} = 4\pi$.

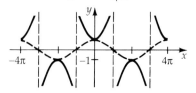

Exercise Set 6.7, page 359

20. $f(x) = \cos\left(2x - \dfrac{\pi}{3}\right)$

$$0 \le 2x - \dfrac{\pi}{3} \le 2\pi$$

$$\dfrac{\pi}{3} \le 2x \qquad \le \dfrac{7\pi}{3}$$

$$\dfrac{\pi}{6} \le x \qquad \le \dfrac{7\pi}{6}$$

Period $= \pi$, phase shift $= \dfrac{\pi}{6}$.

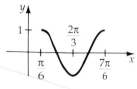

22. $f(x) = \tan(x - \pi)$

$$-\dfrac{\pi}{2} < x - \pi < \dfrac{\pi}{2}$$

$$\dfrac{\pi}{2} < x \qquad < \dfrac{3\pi}{2}$$

Period $= \pi$, phase shift $= \pi$.

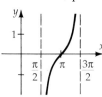

28. $f(x) = \sec\left(2x + \dfrac{\pi}{6}\right)$

$$0 \le 2x + \dfrac{\pi}{6} \le 2\pi$$

$$-\dfrac{\pi}{12} \le x \qquad\qquad \le \dfrac{11\pi}{12}$$

Period $= \pi$, phase shift $= -\dfrac{\pi}{12}$.

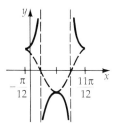

40. $f(x) = -2\cos\left(x + \dfrac{\pi}{3}\right) + 3$, $p = 2\pi$, phase angle $= -\dfrac{\pi}{3}$.

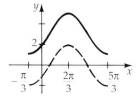

44. $f(x) = \cos\dfrac{x}{3} + 4$, $p = 6\pi$.

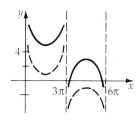

50. $f(x) = \dfrac{2}{3}x - \sin\dfrac{x}{2}$

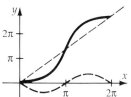

52. $f(x) = -\sin x + \cos x$

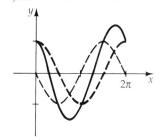

58. $f(x) = x\cos x$

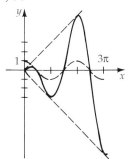

Exercise Set 6.8, page 365

20. Amplitude $= 3$, frequency $= 1/\pi$, period $= \pi$.
Since $2\pi/b = \pi$, we have $b = 2$. Thus

$$y = 3\cos 2t$$

28. Amplitude $= 1.5$.

$$f = \dfrac{1}{2\pi}\sqrt{\dfrac{k}{m}} = \dfrac{1}{2\pi}\sqrt{\dfrac{3}{27}} = \dfrac{1}{2\pi} \cdot \dfrac{1}{3} = \dfrac{1}{6\pi}, \qquad p = 6\pi$$

$$y = 1.5\cos 2\pi ft = 1.5\cos\left[2\pi\left(\dfrac{1}{6\pi}\right)t\right] = 1.5\cos\dfrac{1}{3}t$$

30. Amplitude $= 4$.

$$f = \dfrac{1}{2\pi}\sqrt{\dfrac{g}{l}} = \dfrac{1}{2\pi}\sqrt{\dfrac{32}{20}} = \dfrac{1}{2\pi} \cdot \dfrac{4}{\sqrt{10}} = \dfrac{\sqrt{10}}{5\pi}$$

$$p = \dfrac{5\pi}{\sqrt{10}} = \dfrac{\pi\sqrt{10}}{2}$$

$$y = 4\cos 2\pi ft = 4\cos\left[2\pi\left(\dfrac{\sqrt{10}}{5\pi}\right)t\right] = 4\cos\dfrac{2\sqrt{10}}{}$$

Exercise Set 7.1, page 374

30. $\sin x \cot x \sec x = \sin x \cdot \dfrac{\cos x}{\sin x} \cdot \dfrac{1}{\cos x} = 1$

40. $\sin^4 x - \cos^4 x = (\sin^2 x + \cos^2 x)(\sin^2 x - \cos^2 x)$

$\qquad\qquad = 1(\sin^2 x - \cos^2 x)$

$\qquad\qquad = \sin^2 x - \cos^2 x$

52. $\dfrac{2 \sin x \cot x + \sin x - 4 \cot x - 2}{2 \cot x + 1}$

$\qquad = \dfrac{(\sin x)(2 \cot x + 1) - 2(2 \cot x + 1)}{2 \cot x + 1}$

$\qquad = \dfrac{(2 \cot x + 1)(\sin x - 2)}{2 \cot x + 1}$

$\qquad = \sin x - 2$

62. $\dfrac{\dfrac{1}{\sin x} + \dfrac{1}{\cos x}}{\dfrac{1}{\sin x} - \dfrac{1}{\cos x}} = \dfrac{\dfrac{1}{\sin x} + \dfrac{1}{\cos x}}{\dfrac{1}{\sin x} - \dfrac{1}{\cos x}} \cdot \dfrac{\sin x \cos x}{\sin x \cos x}$

$\qquad = \dfrac{\cos x + \sin x}{\cos x - \sin x}$

$\qquad = \dfrac{\cos x + \sin x}{\cos x - \sin x} \cdot \dfrac{\cos x - \sin x}{\cos x - \sin x}$

$\qquad\qquad = \dfrac{\cos^2 x - \sin^2 x}{\cos^2 x - 2 \sin x \cos x + \sin^2 x}$

$\qquad\qquad = \dfrac{\cos^2 x - \sin^2 x}{1 - 2 \sin x \cos x}$

72. $\dfrac{\dfrac{1}{\tan x} + \cot x}{\dfrac{1}{\tan x} + \tan x} = \dfrac{\dfrac{1}{\tan x} + \cot x}{\dfrac{1}{\tan x} + \tan x} \cdot \dfrac{\tan x}{\tan x}$

$\qquad\qquad = \dfrac{1 + 1}{1 + \tan^2 x}$

$\qquad\qquad = \dfrac{2}{\sec^2 x}$

Exercise Set 7.2, page 382

20. $\sin \dfrac{11\pi}{12} = \sin\left(\dfrac{2\pi}{3} + \dfrac{\pi}{4}\right)$

$\qquad = \sin \dfrac{2\pi}{3} \cos \dfrac{\pi}{4} + \cos \dfrac{2\pi}{3} \sin \dfrac{\pi}{4}$

$\qquad = \dfrac{\sqrt{3}}{2} \cdot \dfrac{\sqrt{2}}{2} + \left(-\dfrac{1}{2} \cdot \dfrac{\sqrt{2}}{2}\right)$

$\qquad = \dfrac{\sqrt{6}}{4} - \dfrac{\sqrt{2}}{4} = \dfrac{\sqrt{6} - \sqrt{2}}{4}$

22. $\cos \dfrac{\pi}{12} = \cos\left(\dfrac{\pi}{3} - \dfrac{\pi}{4}\right)$

$\qquad = \cos \dfrac{\pi}{3} \cos \dfrac{\pi}{4} + \sin \dfrac{\pi}{3} \sin \dfrac{\pi}{4}$

$\qquad = \dfrac{1}{2} \cdot \dfrac{\sqrt{2}}{2} + \dfrac{\sqrt{3}}{2} \cdot \dfrac{\sqrt{2}}{2}$

$\qquad = \dfrac{\sqrt{2}}{4} + \dfrac{\sqrt{6}}{4} = \dfrac{\sqrt{2} + \sqrt{6}}{4}$

24. $\tan \dfrac{7\pi}{12} = \tan\left(\dfrac{\pi}{3} + \dfrac{\pi}{4}\right)$

$\qquad = \dfrac{\tan \pi/3 + \tan \pi/4}{1 - \tan \pi/3 \tan \pi/4}$

$\qquad = \dfrac{\sqrt{3} + 1}{1 - \sqrt{3}(1)} = \dfrac{1 + \sqrt{3}}{1 - \sqrt{3}} \cdot \dfrac{1 + \sqrt{3}}{1 + \sqrt{3}}$

$\qquad = -2 - \sqrt{3}$

62. $\sin \alpha = \dfrac{24}{25}$, $\cos \alpha = -\dfrac{7}{25}$, $\cos \beta = -\dfrac{4}{5}$, $\sin \beta = -\dfrac{3}{5}$

a. $\cos(\beta - \alpha) = \cos \beta \cos \alpha + \sin \beta \sin \alpha$

$\qquad = -\dfrac{4}{5} \cdot \left(-\dfrac{7}{25}\right) + \left(-\dfrac{3}{5}\right) \cdot \dfrac{24}{25}$

$\qquad = \dfrac{28}{125} - \dfrac{72}{125}$

$\qquad = -\dfrac{44}{125}$

b.

$$\sin(\alpha + \beta) = \sin \alpha \cos \beta + \cos \alpha \sin \beta$$

$$= \frac{24}{25} \cdot \left(-\frac{4}{5}\right) + \left(-\frac{7}{25}\right) \cdot \left(-\frac{3}{5}\right)$$

$$= -\frac{96}{125} + \frac{21}{125}$$

$$= -\frac{75}{125} = -\frac{3}{5}$$

70.

$$\cos(\theta + \pi) = \cos \theta \cos \pi - \sin \theta \sin \pi$$

$$= (\cos \theta)(-1) - (\sin \theta)(0)$$

$$= -\cos \theta$$

82.

$$\cos 5x \cos 3x + \sin 5x \sin 3x = \cos(5x - 3x)$$

$$= \cos 2x$$

$$= \cos(x + x)$$

$$= \cos x \cos x - \sin x \sin x$$

$$= \cos^2 x - \sin^2 x$$

86.

$$\sin(\alpha - \beta) - \sin(\alpha + \beta)$$

$$= \sin \alpha \cos \beta - \cos \alpha \sin \beta - \sin \alpha \cos \beta - \cos \alpha \sin \beta$$

$$= -2 \cos \alpha \sin \beta$$

Exercise Set 7.3, page 389

26.

$$\cos \theta = \frac{24}{25} \qquad \sin \theta = -\sqrt{1 - \left(\frac{24}{25}\right)^2} \qquad \tan \theta = \frac{-7/25}{24/25}$$

$$= -\frac{7}{25} \qquad\qquad = -\frac{7}{24}$$

$$\sin 2\theta = 2 \sin \theta \cos \theta \qquad \cos 2\theta = \cos^2 \theta - \sin^2 \theta$$

$$= 2\left(\frac{-7}{25}\right)\left(\frac{24}{25}\right) \qquad = \left(\frac{24}{25}\right)^2 - \left(-\frac{7}{25}\right)^2$$

$$= -\frac{336}{625} \qquad\qquad = \frac{527}{625}$$

$$\tan 2\theta = \frac{2 \tan \theta}{1 - \tan^2 \theta}$$

$$= \frac{2(-7/24)}{1 - (-7/24)^2}$$

$$= \frac{-7/12}{1 - 49/576}$$

$$= -\frac{336}{527}$$

38. $\sin \alpha = -\dfrac{7}{25}$, $\cos \alpha = -\sqrt{1 - \left(-\dfrac{7}{25}\right)^2} = -\dfrac{24}{25}$

$$\sin \frac{\alpha}{2} = \sqrt{\frac{1 - \cos \alpha}{2}} \qquad \cos \frac{\alpha}{2} = -\sqrt{\frac{1 + \cos \alpha}{2}}$$

$$= \sqrt{\frac{1 + 24/25}{2}} \qquad = \sqrt{\frac{1 - 24/25}{2}}$$

$$= \frac{7\sqrt{2}}{10} \qquad\qquad = -\frac{\sqrt{2}}{10}$$

$$\tan \frac{\alpha}{2} = \frac{1 - \cos \alpha}{\sin \alpha}$$

$$= \frac{1 + 24/25}{-7/25} = -7$$

54.

$$\frac{1}{1 - \cos 2x} = \frac{1}{1 - 1 + 2 \sin^2 x}$$

$$= \frac{1}{2 \sin^2 x} = \frac{1}{2} \csc^2 x$$

58.

$$\frac{2 \sin x \cos x}{\cos^2 x - \sin^2 x} = \frac{\sin 2x}{\cos 2x} = \tan 2x$$

62.

$$\sin 4x = 2 \sin 2x \cos 2x$$

$$= 2(2 \sin x \cos x)(\cos^2 x - \sin^2 x)$$

$$= 4 \sin x \cos^3 x - 4 \sin^3 x \cos x$$

68.

$$\sin 3x + \sin x = \sin(2x + x) + \sin x$$

$$= \sin 2x \cos x + \cos 2x \sin x + \sin x$$

$$= (2 \sin x \cos x) \cos x$$

$$\qquad + (1 - 2 \sin^2 x) \sin x + \sin x$$

$$= 2 \sin x \cos^2 x + \sin x - 2 \sin^3 x$$

$$\qquad + \sin x$$

$$= 2 \sin x(1 - \sin^2 x) + 2 \sin x - 2 \sin^3 x$$

$$= 2 \sin x - 2 \sin^3 x + 2 \sin x - 2 \sin^3 x$$

$$= 4 \sin x - 4 \sin^3 x$$

72.

$$\cos^2 \frac{x}{2} = \left[\pm \sqrt{\frac{1 + \cos x}{2}}\right]^2 = \frac{1 + \cos x}{2}$$

$$= \frac{1 + \cos x}{2} \cdot \frac{\sec x}{\sec x} = \frac{\sec x + 1}{2 \sec x}$$

78.

$$\tan^2 \frac{x}{2} = \left(\frac{1 - \cos x}{\sin x}\right)^2$$

$$= \frac{(1 - \cos x)^2}{\sin^2 x}$$

$$= \frac{(1 - \cos x)^2}{1 - \cos^2 x}$$

$$= \frac{(1 - \cos x)^2}{(1 - \cos x)(1 + \cos x)}$$

$$= \frac{1 - \cos x}{1 + \cos x}$$

$$= \frac{\dfrac{1}{\cos x} - \dfrac{\cos x}{\cos x}}{\dfrac{1}{\cos x} + \dfrac{\cos x}{\cos x}}$$

$$= \frac{\sec x - 1}{\sec x + 1}$$

Exercise Set 7.4, page 397

12.

$$\sin 195° \cos 15° = \frac{1}{2}[\sin(195° + 15°) + \sin(195° - 15°)]$$

$$= \frac{1}{2}(\sin 210° + \sin 180°)$$

$$= \frac{1}{2}\left(-\frac{1}{2} + 0\right) = -\frac{1}{4}$$

22.

$$\cos 3\theta + \cos 5\theta = 2\cos\frac{3\theta + 5\theta}{2}\cos\frac{3\theta - 5\theta}{2}$$

$$= 2\cos 4\theta \cos(-\theta)$$

$$= 2\cos 4\theta \cos \theta$$

36.

$$\sin 5x \cos 3x = \frac{1}{2}[\sin(5x + 3x) + \sin(5x - 3x)]$$

$$= \frac{1}{2}(\sin 8x + \sin 2x)$$

$$= \frac{1}{2}(2\sin 4x \cos 4x + 2\sin x \cos x)$$

$$= \sin 4x \cos 4x + \sin x \cos x$$

44.

$$\frac{\cos 5x - \cos 3x}{\sin 5x + \sin 3x} = \frac{-2\sin\dfrac{5x + 3x}{2}\sin\dfrac{5x - 3x}{2}}{2\sin\dfrac{5x + 3x}{2}\cos\dfrac{5x - 3x}{2}}$$

$$= -\frac{\sin 4x \sin x}{\sin 4x \cos x} = -\tan x$$

62. $a = 1, b = \sqrt{3}, k = \sqrt{(\sqrt{3})^2 + (1)^2} = 2$. Thus α is a first quadrant angle.

$$\sin \beta = \left|\frac{\sqrt{3}}{2}\right| = \frac{\sqrt{3}}{2}$$

$$\beta = \frac{\pi}{3}, \qquad \alpha = \frac{\pi}{3}$$

$$y = k\sin(x + \alpha)$$

$$y = 2\sin\left(x + \frac{\pi}{3}\right)$$

70. From Exercise 62, we know

$$y = \sin x + \sqrt{3}\cos x$$

$$y = 2\sin\left(x + \frac{\pi}{3}\right)$$

Exercise Set 7.5, page 409

2.

$$y = \sin^{-1}\frac{\sqrt{2}}{2}$$

$$\sin y = \frac{\sqrt{2}}{2} \quad \text{for} \quad -\frac{\pi}{2} \le y \le \frac{\pi}{2}$$

$$\frac{\pi}{}$$

10.

$$y = \sec^{-1}\frac{2\sqrt{3}}{3}$$

$$\sec y = \frac{2\sqrt{3}}{3} \quad \text{for} \quad 0 \le y \le \pi$$

$$\cos y = \frac{\sqrt{3}}{2}$$

$$y = \frac{\pi}{6}$$

18. $$y = \tan^{-1} 1$$

$$\tan y = 1 \quad \text{for} \quad -\frac{\pi}{2} \le y \le \frac{\pi}{2}$$

$$y = \frac{\pi}{4}$$

30. $y = \sin^{-1}(-0.9650)$
$y \approx -74.8°$

40. $y = \csc^{-1}(-10.9856)$
$y \approx -5.2°$

54. $y = \sin^{-1}\left(\cos\frac{7\pi}{6}\right) = \sin^{-1}\left(-\frac{\sqrt{3}}{2}\right)$

$$y = -\frac{\pi}{3}$$

58. $y = \cot^{-1}\left(\cos\frac{2\pi}{3}\right) = \cot^{-1}\left(-\frac{1}{2}\right) = \tan^{-1}(-2)$

$y \approx -1.1071$

70. Let $x = \cos^{-1} 3/5$. Thus $\cos x = 3/5$ and $\sin x = \sqrt{1-(3/5)^2} = 4/5$.

$$y = \tan\left(\cos^{-1}\frac{3}{5}\right) = \tan x$$

$$y = \frac{\sin x}{\cos x} = \frac{4/5}{3/5} = \frac{4}{3}$$

84. $$\tan^{-1} x = \sin^{-1}\frac{24}{25}$$

$$\tan(\tan^{-1} x) = \tan\left(\sin^{-1}\frac{24}{25}\right)$$

$$x = \tan\left(\sin^{-1}\frac{24}{25}\right)$$

$$x = \frac{24}{7}$$

Exercise Set 7.6, page 415

44. $2\sin^2 x + 5\sin x + 3 = 0$

$$\sin x = \frac{-5 \pm \sqrt{5^2 - 4(2)(3)}}{2(2)} = \frac{-5 \pm 1}{4}$$

$$\sin x = -1 \qquad \sin x = -\frac{3}{2}$$

$$x = 270° \qquad \text{no solution}$$

The solution is 270°.

52. $\sin x + 2\cos x = 1$

$$\sin x = 1 - 2\cos x$$

$$(\sin x)^2 = (1 - 2\cos x)^2$$

$$\sin^2 x = 1 - 4\cos x + 4\cos^2 x$$

94. $$\sin^{-1} x + \cos^{-1}\frac{4}{5} = \frac{\pi}{6}$$

$$\sin^{-1} x = \frac{\pi}{6} - \cos^{-1}\frac{4}{5}$$

$$\sin(\sin^{-1} x) = \sin\left(\frac{\pi}{6} - \cos^{-1}\frac{4}{5}\right)$$

$$x = \sin\frac{\pi}{6}\cos\left(\cos^{-1}\frac{4}{5}\right) - \cos\frac{\pi}{6}\sin\left(\cos^{-1}\frac{4}{5}\right)$$

$$= \frac{1}{2}\cdot\frac{4}{5} - \frac{\sqrt{3}}{2}\cdot\frac{3}{5} = \frac{4-3\sqrt{3}}{10}$$

102. Let $\alpha = \cos^{-1} x$ and $\beta = \cos^{-1}(-x)$. Thus $\cos\alpha = x$ and $\cos\beta = -x$. We have $\sin\alpha = \sqrt{1-x^2}$ and we have $\sin\beta = \sqrt{1-x^2}$ because α is in quadrant I and β is in quadrant II.

$$\cos^{-1} x + \cos^{-1}(-x) = \alpha + \beta$$

$$= \cos^{-1}[\cos(\alpha+\beta)]$$

$$= \cos^{-1}(\cos\alpha\cos\beta - \sin\alpha\sin\beta)$$

$$= \cos^{-1}[x(-x) - \sqrt{1-x^2}\cdot\sqrt{1-x^2}]$$

$$= \cos^{-1}(-x^2 - 1 + x^2)$$

$$= \cos^{-1}(-1) = \pi$$

106.

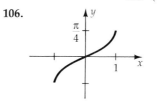

$$1 - \cos^2 x = 1 - 4\cos x + 4\cos^2 x$$

$$0 = 5\cos^2 x - 4\cos x$$

$$0 = \cos x(5\cos x - 4)$$

$$\cos x = 0 \qquad\qquad 5\cos x - 4 = 0$$

$$x = 90°, 270° \qquad\qquad \cos x = \frac{4}{5}$$

$$x \approx 36.9°, 323.1°$$

The solutions are 90°, 270°, 36.9°, and 323.1°.

56. $2\cos^2 x - 5\cos x - 5 = 0$

$$\cos x = \frac{5 \pm \sqrt{(-5)^2 - 4(2)(-5)}}{2(2)} = \frac{5 \pm \sqrt{65}}{4}$$

$\cos x \approx 3.26 \qquad \cos x \approx -0.7656$

no solution $\qquad x \approx 140.0°, \ 220.0°$

The solutions are $140.0°$ and $220.0°$.

66. $\cos 2x = -\dfrac{\sqrt{3}}{2}$

$$2x = \frac{5\pi}{6} + 2k\pi \quad \text{or} \quad 2x = \frac{7\pi}{6} + 2k\pi, \quad k \text{ an integer}$$

$$x = \frac{5\pi}{12} + k\pi \quad \text{or} \quad x = \frac{7\pi}{12} + k\pi$$

Exercise Set 8.1, page 426

14. $B = 90.00° - 45.89° = 44.11°$

$\tan 45.89° = \dfrac{a}{1.228} \qquad\qquad \cos 45.89° = \dfrac{1.228}{c}$

$a = 1.228 \tan 45.89° \qquad\qquad c = \dfrac{1.228}{\cos 45.89°}$

$\qquad \approx 1.267 \qquad\qquad\qquad\qquad \approx 1.764$

28. $A = 70°, \ b = 5.2$

$$\tan 70° = \frac{a}{5.2}$$

$$a = 5.2 \tan 70° \approx 14 \text{ meters}$$

Exercise Set 8.2, page 431

4. $A = 180° - 78° - 28° = 74°$

$\dfrac{b}{\sin B} = \dfrac{c}{\sin C} \qquad\qquad \dfrac{a}{\sin A} = \dfrac{c}{\sin C}$

$\dfrac{b}{\sin 28°} = \dfrac{44}{\sin 78°} \qquad\qquad \dfrac{a}{\sin 74°} = \dfrac{44}{\sin 78°}$

$b = \dfrac{44 \sin 28°}{\sin 78°} \approx 21 \qquad a = \dfrac{44 \sin 74°}{\sin 78°} \approx 43$

14. $\qquad \sin 22.6° = \dfrac{h}{13.8}$

$\qquad\qquad h = 13.8 \sin 22.6 \approx 5.30$

two solutions exist.

32. $\qquad C = 33.8° + 56.2° = 90.0°$

$\qquad c = \sqrt{453^2 + 1520^2} \approx 1590 \text{ yards}$

34. $\quad \alpha = 360° - 335.4° = 24.6° \qquad a = 2(480) = 960$

$\qquad C = 24.6° + 65.4° = 90° \qquad\qquad b = 2(215) = 430$

$\qquad\qquad\qquad\qquad\qquad\qquad\qquad c = \sqrt{a^2 + b^2}$

$\qquad\qquad\qquad\qquad\qquad\qquad\qquad\quad = \sqrt{960^2 + 430^2}$

$\qquad\qquad\qquad\qquad\qquad\qquad\qquad\quad \approx 1050 \text{ mi}$

$$\frac{13.8}{\sin A} = \frac{5.55}{\sin 22.6}$$

$$\sin A = 0.9555$$

$$A \approx 72.9° \quad \text{or} \quad 107.1°$$

$A = 72.9° \qquad\qquad\qquad A = 107.1°$

$C = 180° - 72.9° - 22.6° \quad C = 180° - 107.1° - 22.6°$

$C = 84.5° \qquad\qquad\qquad\quad C = 50.3°$

$\dfrac{c}{\sin 84.5°} = \dfrac{5.55}{\sin 22.6°} \qquad \dfrac{c}{\sin 50.3°} = \dfrac{5.55}{\sin 22.6°}$

$c = \dfrac{5.55 \sin 84.5°}{\sin 22.6°} \approx 14.4 \quad c = \dfrac{5.55 \sin 50.3°}{\sin 22.6°} \approx 11.1$

84. $\quad 2 \sin x \cos x - 2\sqrt{2} \sin x - \sqrt{3} \cos x + \sqrt{6} = 0$

$2 \sin x(\cos x - \sqrt{2}) - \sqrt{3}(\cos x - \sqrt{2}) = 0$

$(\cos x - \sqrt{2})(2 \sin x - \sqrt{3}) = 0$

$\cos x = \sqrt{2} \qquad \sin x = \dfrac{\sqrt{3}}{2}$

no solution $\qquad x = \dfrac{\pi}{3}, \dfrac{2\pi}{3}$

The solutions are $\dfrac{\pi}{3}$ and $\dfrac{2\pi}{3}$.

20. $\alpha = 65°$

$B = 65° + 8° = 73°$

$A = 180° - 50° - 65° = 65°$

$C = 180° - 65° - 73° = 42°$

$$\frac{b}{\sin B} = \frac{c}{\sin C}$$

$$\frac{b}{\sin 73°} = \frac{20}{\sin 42°}$$

$$b = \frac{20 \sin 73°}{\sin 42°}$$

$$b \approx 29 \text{ mi}$$

24. $A = 56°,\ C = 56°,\ B = 35°,$

$$\theta = 56° - 35° = 21°$$

$$= 180° - 35° - 21° = 124°$$

$$\frac{AC}{\sin B} = \frac{BC}{\sin \alpha}$$

$$\frac{AC}{\sin 124°} = \frac{6}{\sin 35°}$$

$$AC \approx 8.7 \text{ m}$$

Exercise Set 8.3, page 438

12. $c^2 = a^2 + b^2 - 2ab \cos C$

$c^2 = 14.2^2 + 9.30^2 - 2(14.2)(9.30)\cos 9.20°$

$c^2 \approx 27.4$

$c \approx 5.24$

18. $\cos A = \dfrac{b^2 + c^2 - a^2}{2bc}$

$\cos A = \dfrac{132^2 + 160^2 - 108^2}{2(132)(160)} \approx 0.7424$

$A \approx 42.1°$

26. $K = \dfrac{1}{2}ac \sin B$

$K = \dfrac{1}{2}(32)(25)\sin 127° \approx 319$

28. $A = 180° - 102° - 27° = 51°$

$K = \dfrac{a^2 \sin B \sin C}{2 \sin A}$

$K = \dfrac{8.5^2 \sin 102° \sin 27°}{2 \sin 51°} \approx 20.6$

36. $s = \dfrac{1}{2}(a + b + c)$

$= \dfrac{1}{2}(10.2 + 13.3 + 15.4) = 19.45$

$K = \sqrt{s(s - a)(s - b)(s - c)}$

$= \sqrt{19.45(19.45 - 10.2)(19.45 - 13.3)(19.45 - 15.4)}$

≈ 66.9

48. $\alpha = 90° - 74° = 16°$

$A = 16° + 90° + 32° = 138°$

$B = 4 \cdot 16 = 64$

$C = 3 \cdot 22 = 66$

$a^2 = b^2 + c^2 - 2bc \cos A$

$a^2 = 64^2 + 66^2 - 2(64)(66) \cos 138°$

$a^2 = 14{,}730$

$a \approx 121 \text{ mi}$

56. $S = \dfrac{1}{2}(324 + 412 + 516) = 626$

$K = \sqrt{626(626 - 324)(626 - 412)(626 - 516)}$

$K \approx 66{,}710$

$\text{cost} = 4.15(66{,}710) = \$276{,}848$

Exercise Set 8.4, page 450

2.
$$|P_1 P_2| = \sqrt{[-5 - (-4)]^2 + [6 - (-7)]^2}$$
$$|P_1 P_2| = \sqrt{1 + 169}$$
$$P_1 P_2 = \sqrt{170}$$

$$\alpha = \tan^{-1}\left|\frac{6 - (-7)}{-5 - (-4)}\right| = \tan^{-1}\frac{13}{1} = 85.6°$$

$\theta = 360° - 85.6°$ θ is in quadrant IV

$\theta = 274.4°$

8.

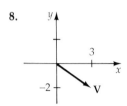

22. $x = -3,\ y = 4$

28. $x = 15 \cos 140° \approx -11.5$
$y = 15 \sin 140° \approx 9.6$

32. $(\mathbf{i} - 4\mathbf{j}) - (-3\mathbf{i} - 2\mathbf{j}) = [1 - (-3)]\mathbf{i} + [-4 - (-2)]\mathbf{j}$
$= 4\mathbf{i} - 2\mathbf{j}$

34. $-4\mathbf{V} = -4(-\mathbf{i} + 2\mathbf{j}) = 4\mathbf{i} - 8\mathbf{j}$

40.
$$\frac{3}{4}\mathbf{U} - \frac{1}{2}\mathbf{V} = \frac{3}{4}(2\mathbf{i} + 3\mathbf{j}) - \frac{1}{2}(-\mathbf{i} + 2\mathbf{j})$$
$$= \left(\frac{3}{2}\mathbf{i} + \frac{9}{4}\mathbf{j}\right) + \left(\frac{1}{2}\mathbf{i} - \mathbf{j}\right)$$
$$= \left(\frac{3}{2} + \frac{1}{2}\right)\mathbf{i} + \left(\frac{9}{4} - 1\right)\mathbf{j}$$
$$= 2\mathbf{i} + \frac{5}{4}\mathbf{j}$$

56.
$$\mathbf{U} \cdot \mathbf{V} = (15\mathbf{i} - 10\mathbf{j}) \cdot (18\mathbf{i} + 16\mathbf{j})$$
$$= (15)(18) + (-10)(16)$$
$$= 270 - 160 = 110$$

62.
$$\cos\theta = \frac{\mathbf{U} \cdot \mathbf{V}}{|\mathbf{U}|\,|\mathbf{V}|}$$
$$\cos\theta = \frac{(-1)(4) + (5)(7)}{\sqrt{(-1)^2 + 5^2}\sqrt{4^2 + 7^2}} = 0.7541$$
$$\theta \approx 41.1°$$

68.
$$\alpha = \tan^{-1}\left|\frac{0.8}{2.6}\right| \approx 17.1°$$
$$\theta = 90° - 17.1° = 72.9°$$

The heading is 72.9°.

70.
$$\alpha = \theta$$
$$\sin\alpha = \frac{120}{800}$$
$$\alpha = 8.6°$$

72.
$$W = \mathbf{F} \cdot \mathbf{S}$$
$$W = |\mathbf{F}|\,|\mathbf{S}|\cos 35°$$
$$W = (800\cos 35°)(45) = 29{,}489 \text{ ft-lb}$$

Exercise Set 8.5, page 459

12.
$$r = \sqrt{1^2 + (\sqrt{3})^2} = 2$$
$$\alpha = \tan^{-1}\left|\frac{\sqrt{3}}{1}\right| = \tan^{-1}\sqrt{3} = 60°$$
$$\theta = 60°$$
$$z = 2(\cos 60° + i\sin 60°)$$

26. $z = 4\left(\cos\frac{5\pi}{3} + i\sin\frac{5\pi}{3}\right) = 4\left(\frac{1}{2} - \frac{\sqrt{3}}{2}i\right) = 2 - 2i\sqrt{3}$

40. $z_1 z_2 = 4(\cos 120° + i\sin 120°) \cdot 6(\cos 315° + i\sin 315°)$
$= 24[\cos(120° + 315°) + i\sin(120° + 315°)]$
$= 24(\cos 75° + i\sin 75°)$

52.
$$\frac{z_1}{z_2} = \frac{10(\cos\pi/3 + i\sin\pi/3)}{5(\cos\pi/4 + i\sin\pi/4)}$$
$$= 2\left[\cos\left(\frac{\pi}{3} - \frac{\pi}{4}\right) + i\sin\left(\frac{\pi}{3} - \frac{\pi}{4}\right)\right]$$
$$= 2\left(\cos\frac{\pi}{12} + i\sin\frac{\pi}{12}\right) \approx 2[0.9659 + 0.2588i]$$
$$= 1.93 + 0.518i$$

70.

$$z = 1 + i\sqrt{3} \qquad \alpha = \tan^{-1}\left|\frac{\sqrt{3}}{1}\right|$$

$$r = \sqrt{1^2 + (\sqrt{3})^2} = 2 \qquad \alpha = 60°$$

$$\theta = 60°$$

$$z = 2(\cos 60° + i \sin 60°)$$

$$(1 + i\sqrt{3})^8 = [2(\cos 60° + i \sin 60°)]^8$$

$$= 2^8[\cos(8 \cdot 60°) + i \sin (8 \cdot 60°)]$$

$$= 256(\cos 480° + i \sin 480°)$$

$$= 256(\cos 120° + i \sin 120°)$$

78. $w = \cos\dfrac{315° + 360° \cdot k}{5} + i \sin\dfrac{315° + 360° \cdot k}{5}$

$$k = 0, 1, 2, 3, 4$$

$k = 0 \quad w_1 = \cos\dfrac{315°}{5} + i \sin\dfrac{315°}{5}$

$k = 1 \quad w_2 = \cos\dfrac{315° + 360°}{5} + i \sin\dfrac{315° + 360°}{5}$

$k = 2 \quad w_3 = \cos\dfrac{315° + 360° \cdot 2}{5} + i \sin\dfrac{315° + 360° \cdot 2}{5}$

$k = 3 \quad w_4 = \cos\dfrac{315° + 360° \cdot 3}{5} + i \sin\dfrac{315° + 360° \cdot 3}{5}$

$k = 4 \quad w_5 = \cos\dfrac{315° + 360° \cdot 4}{5} + i \sin\dfrac{315° + 360° \cdot 4}{5}$

84. $-1 + i = \sqrt{2}(\cos 135° + i \sin 135°)$

$$w = (\sqrt{2})^{1/5}\left(\cos\frac{135° + 360°k}{5} + i \sin\frac{135° + 360°k}{5}\right)$$

$$k = 0, 1, 2, 3, 4$$

$k = 0 \quad w_1 = 2^{1/10}\left(\cos\dfrac{135°}{5} + i \sin\dfrac{135°}{5}\right)$

$k = 1 \quad w_2 = 2^{1/10}\left(\cos\dfrac{135° + 360°}{5} + i \sin\dfrac{135° + 360°}{5}\right)$

$k = 2$

$$w_3 = 2^{1/10}\left(\cos\dfrac{135° + 360° \cdot 2}{5} + i \sin\dfrac{135° + 360° \cdot 2}{5}\right)$$

$k = 3$

$$w_3 = 2^{1/10}\left(\cos\dfrac{135° + 360° \cdot 3}{5} + i \sin\dfrac{135° + 360° \cdot 3}{5}\right)$$

$k = 4$

$$w_4 = 2^{1/10}\left(\cos\dfrac{135° + 360° \cdot 4}{5} + i \sin\dfrac{135° + 360° \cdot 4}{5}\right)$$

Exercise Set 9.1, page 471

4.

$$x^2 = -\frac{1}{4}y$$

$$4p = -\frac{1}{4}$$

$$p = -\frac{1}{16}$$

vertex $(0, 0)$, focus $\left(0, -\dfrac{1}{16}\right)$, directrix $y = \dfrac{1}{16}$

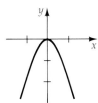

20.

$$x^2 + 5x - 4y - 1 = 0$$

$$x^2 + 5x = 4y + 1$$

$$x^2 + 5x + \frac{25}{4} = 4y + 1 + \frac{25}{4}$$

$$\left(x + \frac{5}{2}\right)^2 = 4\left(y + \frac{29}{16}\right)$$

$$4p = 4$$

$$p = 1$$

vertex $\left(-\dfrac{5}{2}, -\dfrac{29}{16}\right)$, focus $\left(-\dfrac{5}{2}, -\dfrac{13}{16}\right)$, directrix $y = -\dfrac{45}{16}$

28. vertex $(0, 0)$, focus $(5, 0)$, $p = 5$ since focus is $(p, 0)$

$$y^2 = 4px$$
$$y^2 = 4(5)x$$
$$y^2 = 20x$$

Exercise 9.2, page 479

8. Write equation in standard form.

$$\frac{x^2}{12} + \frac{y^2}{25} = 1$$

Thus $a^2 = 25$, $b^2 = 12$, $c^2 = 25 - 12 = 13$.
Center $(0, 0)$, vertices $(0, 5)$, $(0, -5)$ and foci $(0, \sqrt{13})$, $(0, -\sqrt{13})$.

14.
$$9x^2 + 16y^2 + 36x - 16y - 104 = 0$$
$$9x^2 + 36x + 16y^2 - 16y - 104 = 0$$
$$9(x^2 + 4x) + 16(y^2 - y) = 104$$
$$9(x^2 + 4x + 4) + 16\left(y^2 - y + \frac{1}{4}\right) = 104 + 36 + 4$$
$$9(x + 2)^2 + 16\left(y - \frac{1}{2}\right)^2 = 144$$
$$\frac{(x + 2)^2}{16} + \frac{\left(y - \frac{1}{2}\right)^2}{9} = 1$$

center $(-2, \frac{1}{2})$, $a = 4$, $b = 3$, $c = \sqrt{4^2 - 3^2} = \sqrt{7}$
vertices $(2, \frac{1}{2})$ and $(-6, \frac{1}{2})$, foci $(-2 + \sqrt{7}, \frac{1}{2})$ and $(-2 - \sqrt{7}, \frac{1}{2})$

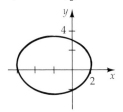

22.
$$2b = 6$$
$$b = 3$$

Since the foci are $(0, 4)$ and $(0, -4)$, $c = 4$, $a^2 = 4^2 + 3^2 = 25$.

$$\frac{x^2}{9} + \frac{y^2}{25} = 1$$

30. vertex $(2, -3)$, focus $(0, -3)$. $h = 2$ and $k = 3$ since the vertex is $(2, -3)$, and $p = -2$ since focus is $(h + p, k)$ and $h = 2$

$$(y - k)^2 = 4p(x - h)$$
$$(y + 3)^2 = 4(-2)(x - 2)$$
$$(y + 3)^2 = -8(x - 2)$$

30.
$$2b = 8$$
$$b = 4$$
$$b^2 = 16$$

The equation of the ellipse is of the form

$$\frac{(x + 4)^2}{a^2} + \frac{(y - 1)^2}{16} = 1$$
$$\frac{(0 + 4)^2}{a^2} + \frac{(4 - 1)^2}{16} = 1$$
$$\frac{16}{a^2} + \frac{9}{16} = 1$$
$$\frac{16}{a^2} = \frac{7}{16}$$
$$a^2 = \frac{256}{7}$$

36. Because the foci are $(0, -3)$ and $(0, 3)$, $c = 3$ and center is $(0, 0)$.

$$e = \frac{c}{a}$$
$$\frac{1}{4} = \frac{3}{a}$$
$$a = 12$$
$$3^2 = 12^2 - b^2$$
$$b^2 = 144 - 9 = 135$$

The equation of the ellipse is $\dfrac{x^2}{135} + \dfrac{y^2}{144} = 1$.

Exercise Set 9.3, page 489

4.
$$a^2 = 25 \qquad b^2 = 36$$
$$a = 5 \qquad b = 6$$
$$c^2 = a^2 + b^2 = 25 + 36 = 61$$
$$c = \sqrt{61}$$

Transverse axis is on y-axis because y^2 term is positive. Center $(0, 0)$, foci $(0, \sqrt{61})$, $(0, -\sqrt{61})$, and asymptotes $y = \dfrac{5}{6}x$ and $y = -\dfrac{5}{6}x$.

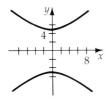

18. $16x^2 - 9y^2 - 32x - 54y + 79 = 0$

$$16(x^2 - 2x + 1) - 9(y^2 + 6y + 9) = -79 - 65 = -144$$

$$\frac{(y + 3)^2}{16} - \frac{(x - 1)^2}{9} = 1$$

Transverse axis is parallel to y-axis because y^2 term is positive. Center is at $(1, -3)$; because $a^2 = 16$ or $a = 4$ vertices are $(1, -3 + 4)$ and $(1, -3 - 4)$, or $(1, 1)$ and $(1, -7)$.

$$c^2 = a^2 + b^2 = 16 + 9 = 25$$
$$c = \sqrt{25} = 5$$

Foci are $(1, -3 + 5)$ and $(1, -3 - 5)$, or $(1, 2)$ and $(1, -8)$. Since $b^2 = 9$ or $b = 3$, asymptotes are $y + 3 = \dfrac{4}{3}(x - 1)$ and $y + 3 = -\dfrac{4}{3}(x - 1)$.

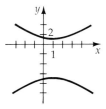

40. Because the vertices are $(2, 3)$ and $(-2, 3)$, $a = 2$ and center is $(0, 3)$.

$$e = \frac{c}{a}$$
$$\frac{5}{2} = \frac{c}{2}$$
$$c = 5$$
$$5^2 = 2^2 + b^2$$
$$b^2 = 25 - 4 = 21$$

The equation of the hyperbola is $\dfrac{x^2}{4} - \dfrac{(y - 3)^2}{21} = 1$.

Exercise Set 9.4, page 501

14.

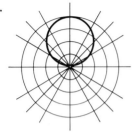

18.

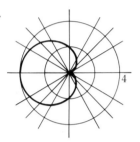

20.

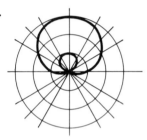

28.

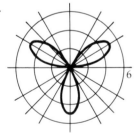

32.

44.

$$x = r \cos \theta \qquad\qquad y = r \sin \theta$$

$$= 2 \cos\left(-\frac{\pi}{3}\right) \qquad = 2 \sin\left(-\frac{\pi}{3}\right)$$

$$= (2)\left(\frac{1}{2}\right) = 1 \qquad = (2)\left(-\frac{\sqrt{3}}{2}\right) = -\sqrt{3}$$

The rectangular coordinates are $(1, -\sqrt{3})$.

48.

$$r = \sqrt{x^2 + y^2} \qquad\qquad \theta = \tan^{-1}\frac{y}{x}$$

$$= \sqrt{(12)^2 + (-5)^2} \qquad = \tan^{-1}\left(-\frac{5}{12}\right)$$

$$= \sqrt{144 + 25} = 13 \qquad \approx 337.4°$$

The polar coordinates are $(13, 337.4°)$.

56. $r = \cot \theta = \dfrac{\cos \theta}{\sin \theta}$. Substitute $\sqrt{x^2 + y^2}$ for r, x/r for $\cos \theta$ and y/r for $\sin \theta$.

$$\sqrt{x^2 + y^2} = \frac{x}{y}$$

$$x^2 + y^2 = \frac{x^2}{y^2}$$

$$y^4 + x^2 y^2 - x^2 = 0$$

58.

$$r = \frac{2}{1 - \sin \theta}$$

$$r - r \sin \theta = 2$$

$$\sqrt{x^2 + y^2} - y = 2$$

$$\sqrt{x^2 + y^2} = 2 - y$$

$$x^2 + y^2 = 4 - 4y + y^2$$

$$x^2 + 4y - 4 = 0$$

Exercise Set 10.1, page 513

6.
$$\begin{cases} 8x + 3y = -7 \\ \quad\quad x = 3y + 15 \end{cases}$$

$$8(3y + 15) + 3y = -7$$

$$24y + 120 + 3y = -7$$

$$y = -\frac{127}{27}$$

$$x = 3\left(-\frac{127}{27}\right) + 15 = \frac{8}{9}$$

The solution is $\left(\dfrac{8}{9}, -\dfrac{127}{27}\right)$.

18.
$$\begin{cases} 3x - 4y = 8 \\ 6x - 8y = 9 \end{cases}$$

$$8y = 6x - 9$$

$$y = \frac{3}{4}x - \frac{9}{8}$$

$$3x - 4\left(\frac{3}{4}x - \frac{9}{8}\right) = 8$$

$$3x - 3x + \frac{9}{2} = 8$$

$$0 = \frac{7}{2}$$

The system of equations has no solution.

20.
$$\begin{cases} 5x + 2y = 2 \\ \quad\quad y = -\dfrac{5}{2}x + 1 \end{cases}$$

$$5x + 2\left(-\frac{5}{2}x + 1\right) = 2$$

$$5x - 5x + 2 = 2$$

$$2 = 2$$

Let $x = c$. Then $y = -\dfrac{5}{2}c + 1$. Thus the solutions are

$\left(c, -\dfrac{5}{2}c + 1\right)$.

24.
$$\begin{cases} \quad 3x - 8y = -6 \\ -5x + 4y = 10 \end{cases}$$

$$\begin{array}{r} 3x - 8y = -6 \\ -10x + 8y = 20 \\ \hline -7x = 14 \\ x = -2 \end{array}$$

$$3(-2) - 8y = -6$$

$$-8y = 0$$

$$y = 0$$

The solution is $(-2, 0)$.

28.
$$\begin{cases} 4x + 5y = 2 \\ 8x - 15y = 9 \end{cases}$$

$$\begin{array}{r} 12x + 15y = 6 \\ 8x - 15y = 9 \\ \hline 20x = 15 \end{array}$$

$$x = \frac{3}{4}$$

$$4\left(\frac{3}{4}\right) + 5y = 2$$

$$y = -\frac{1}{5}$$

The solution is $\left(\dfrac{3}{4}, -\dfrac{1}{5}\right)$.

30.
$$\begin{cases} 4x - 2y = 9 \\ 2x - \ y = 3 \end{cases}$$

$$\begin{array}{r} 4x - 2y = 9 \\ 4x - 2y = 6 \\ \hline 0 = 15 \end{array}$$

The system of equations has no solution.

44. Rate of canoeist with the current: $r + w$
Rate of canoeist against the current: $r - w$

$$r \cdot t = d$$

$$(r + w) \cdot 2 = 12$$

$$(r - w) \cdot 4 = 12$$

$$\begin{array}{r} \begin{cases} r + w = 6 \\ r - w = 3 \end{cases} \\ \hline 2r = 9 \end{array}$$

$$r = 4.5$$

$$4.5 + w = 6$$

$$w = 1.5$$

Rate of canoeist = 4.5 mph
Rate of current = 1.5 mph

Exercise Set 10.2, page 524

12.
$$\begin{cases} 3x + 2y - 5z = 6 & (1) \\ 5x - 4y + 3z = -12 & (2) \\ 4x + 5y - 2z = 15 & (3) \end{cases}$$

$$\begin{cases} 3x + 2y - 5z = 6 \\ 11y - 17z = 33 \quad (4) \\ 4x + 5y - 2z = 15 \end{cases} \qquad \begin{aligned} 15x + 10y - 25z &= 30 \\ -15x + 12y - 9z &= 36 \\ \hline 22y - 34z &= 66 \end{aligned}$$

$$\begin{cases} 3x + 2y - 5z = 6 \\ 11y - 17z = 33 \quad (4) \\ y + 2z = 3 \quad (5) \end{cases} \qquad \begin{aligned} 12x + 8y - 20z &= 24 \\ -12x - 15y + 6z &= -45 \\ \hline -7y - 14z &= -21 \end{aligned}$$

$$\begin{cases} 3x + 2y - 5z = 6 \\ 11y - 17z = 33 \\ z = 0 \quad (6) \end{cases} \qquad \begin{aligned} 11y - 17z &= 33 \\ -11y - 22z &= -33 \\ \hline -39z &= 0 \end{aligned}$$

$$11y - 0 = 33$$
$$y = 3$$
$$3x + 2(3) - 5(0) = 6$$
$$x = 0$$

The solution is $(0, 3, 0)$.

16.
$$\begin{cases} 2x + 3y + 2z = 14 & (1) \\ x - 3y + 4z = 4 & (2) \\ -x + 12y - 6z = 2 & (3) \end{cases}$$

$$\begin{cases} 2x + 3y + 2z = 14 \\ 3y - 2z = 2 \quad (4) \\ -x + 12y - 6z = 2 \end{cases} \qquad \begin{aligned} 2x + 3y + 2z &= 14 \\ -2x + 6y - 8z &= -8 \\ \hline 9y - 6z &= 6 \end{aligned}$$

$$\begin{cases} 2x + 3y + 2z = 14 \\ 3y - 2z = 2 \quad (4) \\ 27y - 10z = 18 \quad (5) \end{cases} \qquad \begin{aligned} 2x + 3y + 2z &= 14 \\ -2x + 24y - 12z &= 4 \\ \hline 27y - 10z &= 18 \end{aligned}$$

$$\begin{cases} 2x + 3y + 2z = 14 \\ 3y - 2z = 2 \\ z = 0 \quad (7) \end{cases} \qquad \begin{aligned} -27y + 18z &= -18 \\ 27y - 10z &= 18 \\ \hline 8z &= 0 \end{aligned}$$

$$3y - 2(0) = 2$$
$$y = \frac{2}{3}$$
$$2x + 3\left(\frac{2}{3}\right) + 2(0) = 14$$
$$x = 6$$

The solution is $\left(6, \dfrac{2}{3}, 0\right)$.

18.
$$\begin{cases} 2x + 3y - 6z = 4 & (1) \\ 3x - 2y - 9z = -7 & (2) \\ 2x + 5y - 6z = 8 & (3) \end{cases}$$

$$\begin{cases} 2x + 3y - 6z = 4 \\ 13y = 8 \quad (4) \\ 2x + 5y - 6z = 8 \end{cases} \qquad \begin{aligned} 6x + 9y - 18z &= 12 \\ -6x + 4y + 18z &= 14 \\ \hline 13y &= 26 \end{aligned}$$

$$\begin{cases} 2x + 3y - 6z = 4 \\ y = 2 \\ y = 2 \quad (5) \end{cases} \qquad \begin{aligned} 2x + 3y - 6z &= 4 \\ -2x - 5y + 6z &= -8 \\ \hline -2y &= -4 \end{aligned}$$

$$\begin{cases} 2x + 3y - 6z = 4 \\ y = 2 \\ 0 = 0 \quad (6) \end{cases} \qquad \begin{aligned} y &= 2 \\ -y &= -2 \\ \hline 0 &= 0 \end{aligned}$$

The equations are dependent. Let $z = c$.
$$2x + 3(2) - 6c = 4$$
$$x = 3c - 1$$

The solutions are $(3c - 1, 2, c)$.

20.
$$\begin{cases} x - 3y + 4z = 9 & (1) \\ 3x - 8y - 2z = 4 & (2) \end{cases}$$

$$\begin{cases} x - 3y + 4z = 9 \\ y - 14z = 23 \quad (3) \end{cases} \qquad \begin{aligned} -3x + 9y - 12z &= -27 \\ 3x - 8y - 2z &= 4 \\ \hline y - 14z &= -23 \end{aligned}$$

$$y = 14z - 23$$
$$x - 3(14z - 23) + 4z = 9$$
$$x = 38z - 60$$

Let $z = c$. The solutions are $(38c - 60, 14c - 23, c)$.

32.
$$\begin{cases} 5x + 2y + 3z = 0 & (1) \\ 3x + y - 2z = 0 & (2) \\ 4x - 7y + 5z = 0 & (3) \end{cases}$$

$$\begin{array}{r} 15x + 6y + 9z = 0 \\ -15x - 5y + 10z = 0 \\ \hline y + 19z = 0 \end{array} \qquad \begin{array}{r} 20x + 8y + 12z = 0 \\ -20x + 35y - 25z = 0 \\ \hline 43y - 13z = 0 \end{array}$$

$$\begin{cases} 5x + 2y + 3z = 0 \\ y + 19z = 0 & (4) \\ 43y - 13z = 0 & (5) \end{cases}$$

$$\begin{cases} 5x + 2y + 3z = 0 \\ y + 19z = 0 \\ z = 0 & (6) \end{cases} \qquad \begin{array}{r} -43y - 817z = 0 \\ 43y - 13z = 0 \\ \hline -830z = 0 \end{array}$$

The only solution is $(0, 0, 0)$.

34.
$$\begin{cases} -2 = a(1)^2 + b(1) + c \\ -4 = a(3)^2 + b(3) + c \\ -2 = a(2)^2 + b(2) + 2 \end{cases} \text{ or } \begin{cases} a + b + c = -2 & (1) \\ 9a + 3b + c = -4 & (2) \\ 4a + 2b + c = -2 & (3) \end{cases}$$

$$\begin{cases} a + b + c = -2 \qquad \text{Eliminate } c \text{ from (2) and (3).} \\ 8a + 2b = -2 \quad (4) \\ 3a + b = 0 \quad (5) \end{cases}$$

$$\begin{cases} a + b + c = -2 \quad \text{Eliminate } b \text{ from (5).} \\ 8a + 2b = -2 \\ -a = -1 \end{cases}$$

$$8(-1) + 2b = -2$$
$$b = 3$$
$$-1 + 3 + c = -2$$
$$c = -4$$

The equation is $y = -x^2 + 3x - 4$.

Exercise Set 10.3, page 529

4.
$$\begin{cases} 2x^2 + 3y^2 = 5 & (1) \\ x^2 - 3y^2 = 4 & (2) \end{cases}$$

$$\begin{array}{r} 2x^2 + 3y^2 = 5 \quad (1) \\ x^2 - 3y^2 = 4 \\ \hline 3x^2 = 9 \end{array}$$

$$x^2 = 3$$
$$x = \pm\sqrt{3}$$
$$(\sqrt{3})^2 - 3y^2 = 4$$
$$y^2 = -\frac{1}{3}$$

$y^2 = -\dfrac{1}{3}$ has no real solutions. The graphs of the equations do not intersect.

10.
$$\begin{cases} y = x^3 - 2x^2 + 5x + 1 \\ y = x^2 + 7x - 5 \end{cases}$$

$$x^3 - 2x^2 + 5x + 1 = x^2 + 7x - 5$$
$$x^3 - 3x^2 - 2x + 6 = 0$$
$$x^2(x - 3) - 2(x - 3) = 0$$
$$(x^2 - 2)(x - 3) = 0$$
$$x = \pm\sqrt{2} \quad \text{or} \quad x = 3$$

$$y = (\sqrt{2})^2 + 7\sqrt{2} - 5 \qquad y = 3^2 + 7(3) - 5$$
$$y = 7\sqrt{2} - 3 \qquad\qquad y = 25$$
$$y = (-\sqrt{2})^2 + 7(-\sqrt{2}) - 5$$
$$y = -7\sqrt{2} - 3$$

The solutions are $(\sqrt{2}, 7\sqrt{2} - 3)$, $(-\sqrt{2}, -7\sqrt{2} - 3)$, and $(3, 25)$.

12.
$$\begin{cases} 2x^2 + 3y^2 = 11 \\ 3x^2 + 2y^2 = 14 \end{cases}$$

$$\begin{array}{r} -4x^2 - 6y^2 = -22 \\ 9x^2 + 6y^2 = 42 \\ \hline 5x^2 = 20 \end{array}$$

$$x^2 = 4$$
$$x = \pm 2$$
$$2(\pm 2)^2 + 3y^2 = 11$$
$$3y^2 = 3$$
$$y = \pm 1$$

The solutions are $(2, 1)$, $(2, -1)$, $(-2, 1)$, and $(-2, -1)$.

18. $\begin{cases} x^2 + y^2 - 4y = 4 & (1) \\ 5x - 2y = 2 & (2) \end{cases}$

Substitute y from (2) into (1).

$$x^2 + \left(\frac{5}{2}x - 1\right)^2 - 4\left(\frac{5}{2}x - 1\right) = 4$$

$$x^2 + \frac{25}{4}x^2 - 5x + 1 - 10x + 4 = 4$$

$$\frac{29}{4}x^2 - 15x + 1 = 0$$

$$29x^2 - 60x + 4 = 0$$

$$(29x - 2)(x - 2) = 0$$

$$x = \frac{2}{29} \qquad x = 2$$

Substitute for x in (2).

$$5\left(\frac{2}{29}\right) - 2y = 2 \qquad 5(2) - 2y = 2$$

$$y = -\frac{24}{29} \qquad\qquad y = 4$$

The solutions are $\left(\frac{2}{29}, -\frac{24}{29}\right)$ and $(2, 4)$.

20. $\begin{cases} (x + 2)^2 + (y - 3)^2 = 10 & (1) \\ (x - 3)^2 + (y + 1)^2 = 13 & (2) \end{cases}$

$$\begin{aligned} x^2 + 4x + 4 + y^2 - 6y + 9 &= 10 \\ \underline{x^2 - 6x + 9 + y^2 + 2y + 1 = 13} \\ 10x - 5 \qquad -8y + 8 &= -3 \end{aligned}$$

$$10x - 8y = -6$$

$$y = \frac{5x + 3}{4}$$

$$(x + 2)^2 + \left(\frac{5x - 9}{4}\right)^2 = 10$$

$$x^2 + 4x + 4 + \frac{25x^2 - 90x + 81}{16} = 10$$

$$16x^2 + 64x + 64 + 25x^2 - 90x + 81 = 160$$

$$41x^2 - 26x - 15 = 0$$

$$(41x + 15)(x - 1) = 0$$

$$x = -\frac{15}{41} \quad \text{or} \quad x = 1$$

$$y = \frac{5}{4}\left(-\frac{15}{41}\right) + \frac{3}{4} \qquad y = \frac{5(1) + 3}{4}$$

$$y = \frac{12}{41} \qquad\qquad y = 2$$

The solutions are $\left(-\frac{15}{41}, \frac{12}{41}\right)$ and $(1, 2)$.

Exercise Set 10.4, page 535

14. $\dfrac{7x + 44}{(x + 4)(x + 6)} = \dfrac{A}{x + 4} + \dfrac{B}{x + 6}$

$$7x + 44 = A(x + 6) + B(x + 4)$$

$$7x + 44 = (A + B)x + (6A + 4B)$$

$$\begin{cases} 7 = A + B \\ 44 = 6A + 4B \end{cases}$$

The solution is $A = 8$, $B = -1$.

$$\dfrac{7x + 44}{x^2 + 10x + 24} = \dfrac{8}{x + 4} + \dfrac{-1}{x + 6}$$

20. $\dfrac{x^3 - 13x - 9}{x^2 - x - 12} = x + 1 + \dfrac{3}{(x - 4)(x + 3)}$

$$\dfrac{3}{(x - 4)(x + 3)} = \dfrac{A}{x - 4} + \dfrac{B}{x + 3}$$

$$3 = A(x + 3) + B(x - 4)$$

$$3 = (A + B)x + 3A - 4B$$

$$\begin{cases} 0 = A + B \\ 3 = 3A - 4B \end{cases}$$

The solutions are $A = \dfrac{3}{7}$, $B = -\dfrac{3}{7}$.

$$\dfrac{x^3 - 13x - 9}{x^2 - x - 12} = x + 1 + \dfrac{3}{7(x - 4)} + \dfrac{-3}{7(x + 3)}$$

22.
$$\frac{x - 18}{x(x - 3)^2} = \frac{A}{x} + \frac{B}{x - 3} + \frac{C}{(x - 3)^2}$$

$$x - 18 = A(x - 3)^2 + Bx(x - 3) + Cx$$

$$x - 18 = Ax^2 - 6Ax + 9A + Bx^2 - 3Bx + Cx$$

$$x - 18 = (A + B)x^2 + (-6A - 3B + C)x + 9A$$

$$\begin{cases} 0 = A + B \\ 1 = -6A - 3B + C \\ -18 = 9A \end{cases}$$

The solution is $A = -2$, $B = 2$, $C = -5$.

$$\frac{x - 18}{x(x - 3)^2} = \frac{-2}{x} + \frac{2}{x - 3} + \frac{-5}{(x - 3)^2}$$

24.
$$\frac{9x^2 - 3x + 49}{(x - 1)(x^2 + 10)} = \frac{A}{x - 1} + \frac{Bx + C}{x^2 + 10}$$

$$9x^2 - 3x + 49 = A(x^2 + 10) + (Bx + C)(x - 1)$$

$$9x^2 - 3x + 49 = (A + B)x^2 + (-B + C)x + (10A - C)$$

$$\begin{cases} 9 = A + B \\ -3 = -B + C \\ 49 = 10A - C \end{cases}$$

The solution is $A = 5$, $B = 4$, $C = 1$.

$$\frac{9x^2 - 3x + 49}{x^3 - x^2 + 10x - 10} = \frac{5}{x - 1} + \frac{4x + 1}{x^2 + 10}$$

30.
$$\frac{2x^3 + 9x + 1}{(x^2 + 7)^2} = \frac{Ax + B}{x^2 + 7} + \frac{Cx + D}{(x^2 + 7)^2}$$

$$2x^3 + 9x + 1 = (Ax + B)(x^2 + 7) + Cx + D$$

$$2x^3 + 9x + 1 = Ax^3 + Bx^2 + (7A + C)x + (7B + D)$$

$$\begin{cases} 2 = A \\ 0 = B \\ 9 = 7A + C \\ 1 = 7B + D \end{cases}$$

The solutions are $A = 2$, $B = 0$, $C = -5$, $D = 1$.

$$\frac{2x^3 + 9x + 1}{x^4 + 14x^2 + 49} = \frac{2}{x^2 + 7} + \frac{-5x + 1}{(x^2 + 7)^2}$$

Exercise Set 10.5, page 541

6.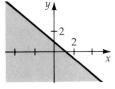

20.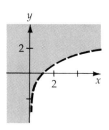

36.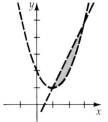

42.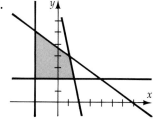

12.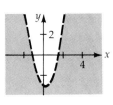

26.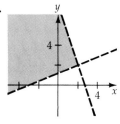

40.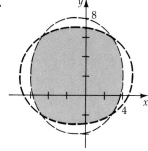

Exercise Set 10.6, page 547

12. $C = 4x + 3y$

	C	
$(0, 8)$	24	
$(2, 4)$	20	minimum
$(5, 2)$	26	
$(11, 0)$	44	
$(20, 0)$	80	
$(20, 20)$	140	
$(0, 20)$	60	

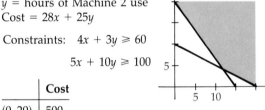

22. x = hours of Machine 1 use
y = hours of Machine 2 use
Cost $= 28x + 25y$

Constraints: $4x + 3y \geq 60$

$5x + 10y \geq 100$

	Cost	
$(0, 20)$	500	
$(12, 4)$	436	minimum
$(20, 0)$	560	

To achieve the minimum cost, use Machine 1 for 12 hours and Machine 2 for 4 hours.

Exercise Set 11.1, page 562

14. $\begin{bmatrix} 1 & -3 & 1 & 8 \\ 2 & -5 & -3 & 2 \\ 1 & 4 & 1 & 1 \end{bmatrix} \xrightarrow[-1R_1 + R_3]{-2R_1 + R_2} \begin{bmatrix} 1 & -3 & 1 & 8 \\ 0 & 1 & -5 & -14 \\ 0 & 7 & 0 & -7 \end{bmatrix}$

$\xrightarrow{-7R_2 + R_3} \begin{bmatrix} 1 & -3 & 1 & 8 \\ 0 & 1 & -5 & -14 \\ 0 & 0 & 35 & 91 \end{bmatrix}$

$\xrightarrow{(1/35)R_3} \begin{bmatrix} 1 & -3 & 1 & 8 \\ 0 & 1 & -5 & -14 \\ 0 & 0 & 1 & \frac{13}{5} \end{bmatrix}$

24. Let x = number of standard models.
Let y = number of deluxe models.
Cost $= 25x + 35y$

Constraints: $x + 3y \leq 24$

$x + y \leq 10$

$2x + y \leq 16$

	Cost	
$(0, 0)$	0	
$(6, 8)$	280	
$(6, 4)$	290	
$(3, 7)$	320	maximum
$(8, 0)$	200	

To maximize profits, produce 3 standard models and 7 deluxe models.

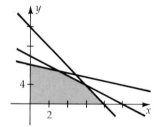

$\begin{cases} x - 3y + z = 8 \\ \quad\quad y - 5z = -14 \\ \quad\quad\quad z = \dfrac{13}{5} \end{cases}$

By back substitution, the solution is $\left(\dfrac{12}{5}, -1, \dfrac{13}{5} \right)$.

18.
$$\begin{bmatrix} 3 & -5 & 2 & 4 \\ 1 & -3 & 2 & 4 \\ 5 & -11 & 6 & 12 \end{bmatrix} \xrightarrow{R_2 \longleftrightarrow R_1} \begin{bmatrix} 1 & -3 & 2 & 4 \\ 3 & -5 & 2 & 4 \\ 5 & -11 & 6 & 12 \end{bmatrix}$$

$$\xrightarrow[-5R_1 + R_3]{-3R_1 + R_2} \begin{bmatrix} 1 & -3 & 2 & 4 \\ 0 & 4 & -4 & -8 \\ 0 & 4 & -4 & -8 \end{bmatrix}$$

$$\xrightarrow{(1/4)R_2} \begin{bmatrix} 1 & -3 & 2 & 4 \\ 0 & 1 & -1 & -2 \\ 0 & 4 & -4 & -8 \end{bmatrix}$$

$$\xrightarrow{-4R_2 + R_3} \begin{bmatrix} 1 & -3 & 2 & 4 \\ 0 & 1 & -1 & -2 \\ 0 & 0 & 0 & 0 \end{bmatrix}$$

$$\begin{cases} x - 3y + 2z = 4 \\ \quad\;\; y - z = -2 \\ \qquad\qquad 0 = 0 \end{cases}$$

$$y - z = -2 \quad \text{or} \quad y = z - 2$$

$$x - 3(z - 2) + 2z = 4$$

$$x - 3z + 6 + 2z = 4$$

$$x = z - 2$$

Let $z = c$. The solutions are $(c - 2, c - 2, c)$.

20.
$$\begin{bmatrix} 2 & 5 & 2 & -1 \\ 1 & 2 & -3 & 5 \\ 5 & 12 & 1 & 10 \end{bmatrix} \xrightarrow{R_2 \longleftrightarrow R_1} \begin{bmatrix} 1 & 2 & -3 & 5 \\ 2 & 5 & 2 & -1 \\ 5 & 12 & 1 & 10 \end{bmatrix}$$

$$\xrightarrow[-5R_1 + R_3]{-2R_1 + R_2} \begin{bmatrix} 1 & 2 & -3 & 5 \\ 0 & 1 & 8 & -11 \\ 0 & 2 & 16 & -15 \end{bmatrix}$$

$$\xrightarrow{-2R_2 + R_3} \begin{bmatrix} 1 & 2 & -3 & 5 \\ 0 & 1 & 8 & -11 \\ 0 & 0 & 0 & 7 \end{bmatrix}$$

$$\begin{cases} x + 2y - 3z = 5 \\ \quad\;\; y + 8z = -11 \\ \qquad\qquad 0 = 7 \end{cases}$$

Because $0 = 7$ is a false equation, the system of equations has no solution.

36.
$$\begin{bmatrix} 1 & -1 & 3 & -5 & 10 \\ 2 & -3 & 4 & 1 & 7 \\ 3 & 1 & -2 & -2 & 6 \end{bmatrix}$$

$$\xrightarrow[-3R_1 + R_3]{-2R_1 + R_2} \begin{bmatrix} 1 & -1 & 3 & -5 & 10 \\ 0 & -1 & -2 & 11 & -13 \\ 0 & 4 & -11 & 13 & -24 \end{bmatrix}$$

$$\xrightarrow{-1R_2} \begin{bmatrix} 1 & -1 & 3 & -5 & 10 \\ 0 & 1 & 2 & -11 & 13 \\ 0 & 4 & -11 & 13 & -24 \end{bmatrix}$$

$$\xrightarrow{-4R_2 + R_3} \begin{bmatrix} 1 & -1 & 3 & -5 & 10 \\ 0 & 1 & 2 & -11 & 13 \\ 0 & 0 & -19 & 57 & -76 \end{bmatrix}$$

$$\xrightarrow{(-1/19)R_3} \begin{bmatrix} 1 & -1 & 3 & -5 & 10 \\ 0 & 1 & 2 & -11 & 13 \\ 0 & 0 & 1 & -3 & 4 \end{bmatrix}$$

$$\begin{cases} t - u + 3v - 5w = 10 \\ \quad\;\; u + 2v - 11w = 13 \\ \qquad\qquad v - 3w = 4 \end{cases}$$

$$v = 3w + 4$$

$$u + 2(3w + 4) - 11w = 13$$

$$u = 5w + 5$$

$$t - (5w + 5) + 3(3w + 4) - 5w = 10$$

$$t = w + 3$$

Let w be any real number c. The solution of the system of equations is $(c + 3, 5c + 5, 3c + 4, c)$.

Exercise Set 11.2, page 572

6. a. $A + B = \begin{bmatrix} 2 & -2 \\ 3 & 4 \\ 1 & 0 \end{bmatrix} + \begin{bmatrix} -1 & 8 \\ 2 & -2 \\ -4 & 3 \end{bmatrix} = \begin{bmatrix} 1 & 6 \\ 5 & 2 \\ -3 & 3 \end{bmatrix}$

b. $A - B = \begin{bmatrix} 2 & -2 \\ 3 & 4 \\ 1 & 0 \end{bmatrix} - \begin{bmatrix} -1 & 8 \\ 2 & -2 \\ -4 & 3 \end{bmatrix} = \begin{bmatrix} 3 & -10 \\ 1 & 6 \\ 5 & -3 \end{bmatrix}$

c. $2B = 2\begin{bmatrix} -1 & 8 \\ 2 & -2 \\ -4 & 3 \end{bmatrix} = \begin{bmatrix} -2 & 16 \\ 4 & -4 \\ -8 & 6 \end{bmatrix}$

d. $2A - 3B = 2\begin{bmatrix} 2 & -2 \\ 3 & 4 \\ 1 & 0 \end{bmatrix} - 3\begin{bmatrix} -1 & 8 \\ 2 & -2 \\ -4 & 3 \end{bmatrix} = \begin{bmatrix} 7 & -28 \\ 0 & 14 \\ 14 & -9 \end{bmatrix}$

16. $AB = \begin{bmatrix} -1 & 2 & 0 \\ 2 & -1 & 1 \\ -2 & 2 & -1 \end{bmatrix}\begin{bmatrix} 2 & -1 & 0 \\ 1 & 5 & -1 \\ 0 & -1 & 3 \end{bmatrix} = \begin{bmatrix} 0 & 11 & -2 \\ 3 & -8 & 4 \\ -2 & 13 & -5 \end{bmatrix}$

$BA = \begin{bmatrix} 2 & -1 & 0 \\ 1 & 5 & -1 \\ 0 & -1 & 3 \end{bmatrix}\begin{bmatrix} -1 & 2 & 0 \\ 2 & -1 & 1 \\ -2 & 2 & -1 \end{bmatrix} = \begin{bmatrix} -4 & 5 & -1 \\ 11 & -5 & 6 \\ -8 & 7 & -4 \end{bmatrix}$

Exercise Set 11.3, page 581

6. $\begin{bmatrix} 1 & 3 & -2 & 1 & 0 & 0 \\ -1 & -5 & 6 & 0 & 1 & 0 \\ 2 & 6 & -3 & 0 & 0 & 1 \end{bmatrix}$

$\xrightarrow[-2R_1 + R_3]{R_1 + R_2} \begin{bmatrix} 1 & 3 & -2 & 1 & 0 & 0 \\ 0 & -2 & 4 & 1 & 1 & 0 \\ 0 & 0 & 1 & -2 & 0 & 1 \end{bmatrix}$

$\xrightarrow{(-1/2)R_2} \begin{bmatrix} 1 & 3 & -2 & 1 & 0 & 0 \\ 0 & 1 & -2 & -\frac{1}{2} & -\frac{1}{2} & 0 \\ 0 & 0 & 1 & -2 & 0 & 1 \end{bmatrix}$

$\xrightarrow[2R_3 + R_1]{2R_3 + R_2} \begin{bmatrix} 1 & 3 & 0 & -3 & 0 & 2 \\ 0 & 1 & 0 & -\frac{9}{2} & -\frac{1}{2} & 2 \\ 0 & 0 & 1 & -2 & 0 & 1 \end{bmatrix}$

$\xrightarrow{-3R_2 + R_1} \begin{bmatrix} 1 & 0 & 0 & \frac{21}{2} & \frac{3}{2} & -4 \\ 0 & 1 & 0 & -\frac{9}{2} & -\frac{1}{2} & 2 \\ 0 & 0 & 1 & -2 & 0 & 1 \end{bmatrix}$

The inverse matrix is $\begin{bmatrix} \frac{21}{2} & \frac{3}{2} & -4 \\ -\frac{9}{2} & -\frac{1}{2} & 2 \\ -2 & 0 & 1 \end{bmatrix}$.

10. $\begin{bmatrix} 1 & -2 & 2 & 1 & 0 & 0 \\ 2 & -3 & 1 & 0 & 1 & 0 \\ 3 & -6 & 6 & 0 & 0 & 1 \end{bmatrix}$

$\xrightarrow[-3R_1 + R_3]{-2R_1 + R_2} \begin{bmatrix} 1 & -2 & 2 & 1 & 0 & 0 \\ 0 & 1 & -3 & -2 & 1 & 0 \\ 0 & 0 & 0 & -3 & 0 & 1 \end{bmatrix}$

Because if there are zeros below the rightmost 1 along the main diagonal, the matrix does not have an inverse.

20. $\begin{bmatrix} 1 & 2 & -1 \\ 2 & 3 & -1 \\ 3 & 6 & -2 \end{bmatrix}\begin{bmatrix} x \\ y \\ z \end{bmatrix} = \begin{bmatrix} 5 \\ 8 \\ 14 \end{bmatrix}$

The inverse of the coefficient matrix is $\begin{bmatrix} 0 & 2 & -1 \\ -1 & -1 & 1 \\ -3 & 0 & 1 \end{bmatrix}$.

Multiplying each side of the equation by the inverse, we have

$\begin{bmatrix} x \\ y \\ z \end{bmatrix} = \begin{bmatrix} 0 & 2 & -1 \\ -1 & -1 & 1 \\ -3 & 0 & 1 \end{bmatrix}\begin{bmatrix} 5 \\ 8 \\ 14 \end{bmatrix} = \begin{bmatrix} 2 \\ 1 \\ -1 \end{bmatrix}$

The solution is $(2, 1, -1)$.

Exercise Set 11.4, page 590

2. $\begin{vmatrix} 2 & 9 \\ -6 & 2 \end{vmatrix} = 2 \cdot 2 - (-6)(9) = 4 + 54 = 58$

14. $M_{13} = \begin{vmatrix} 1 & 3 \\ 6 & -2 \end{vmatrix} = 1(-2) - 6(3) = -2 - 18 = -20$

$C_{13} = (-1)^{1+3} \cdot M_{13} = 1 \cdot M_{13} = 1(-20) = -20$

20. $\begin{vmatrix} 3 & -2 & 0 \\ 2 & -3 & 2 \\ 8 & -2 & 5 \end{vmatrix} = 3C_{11} + (-2)C_{12} + 0 \cdot C_{13}$

$= 3\begin{vmatrix} -3 & 2 \\ -2 & 5 \end{vmatrix} + 2\begin{vmatrix} 2 & 2 \\ 8 & 5 \end{vmatrix} + 0\begin{vmatrix} 2 & -3 \\ 8 & -2 \end{vmatrix}$

$= 3(-15 + 4) + 2(10 - 16)$

$= 3(-11) + 2(-6) = -33 + (-12)$

$= -45$

42. Let $D = \begin{vmatrix} 3 & -2 & -1 \\ 1 & 2 & 4 \\ 2 & -2 & 3 \end{vmatrix}$. Then

$D = -\begin{vmatrix} 1 & 2 & 4 \\ 3 & -2 & -1 \\ 2 & -2 & 3 \end{vmatrix} = -\begin{vmatrix} 1 & 2 & 4 \\ 0 & -8 & -13 \\ 0 & -6 & -5 \end{vmatrix}$

$= 8\begin{vmatrix} 1 & 2 & 4 \\ 0 & 1 & \frac{13}{8} \\ 0 & -6 & -5 \end{vmatrix} = 8\begin{vmatrix} 1 & 2 & 4 \\ 0 & 1 & \frac{13}{8} \\ 0 & 0 & \frac{19}{4} \end{vmatrix}$

$= 8(1)(1)(\frac{19}{4}) = 38$

Exercise Set 11.5, page 595

4. $x_1 = \dfrac{\begin{vmatrix} 9 & 5 \\ 8 & 7 \end{vmatrix}}{\begin{vmatrix} 2 & 5 \\ 5 & 7 \end{vmatrix}} = \dfrac{63 - 40}{14 - 25} = \dfrac{23}{-11} = -\dfrac{23}{11}$

$x_2 = \dfrac{\begin{vmatrix} 2 & 9 \\ 5 & 8 \end{vmatrix}}{\begin{vmatrix} 2 & 5 \\ 5 & 7 \end{vmatrix}} = \dfrac{16 - 45}{14 - 25} = \dfrac{-29}{-11} = \dfrac{29}{11}$

The solution is $\left(-\dfrac{23}{11}, \dfrac{29}{11}\right)$.

24. $x_3 = \dfrac{\begin{vmatrix} 2 & 5 & -3 & -3 \\ 1 & 7 & 4 & -1 \\ 4 & 0 & 3 & 1 \\ 3 & 2 & 0 & 0 \end{vmatrix}}{\begin{vmatrix} 2 & 5 & -5 & -3 \\ 1 & 7 & 8 & -1 \\ 4 & 0 & 1 & 1 \\ 3 & 2 & -1 & 0 \end{vmatrix}} = \dfrac{157}{168} \approx 0.9345$

Exercise Set 12.1, page 608

6. $a_n = \dfrac{(-1)^{n+1}}{n(n+1)}$, $a_1 = \dfrac{(-1)^{1+1}}{1(1+1)} = \dfrac{1}{2}$,

$a_2 = \dfrac{(-1)^{2+1}}{2(2+1)} = -\dfrac{1}{6}$, $a_3 = \dfrac{(-1)^{3+1}}{3(3+1)} = \dfrac{1}{12}$,

$a_8 = \dfrac{(-1)^{8+1}}{8(8+1)} = -\dfrac{1}{72}$

28. $a_1 = 1$, $a_2 = 2^2 \cdot a_1 = 4 \cdot 1 = 4$, $a_3 = 3^2 \cdot a_2 = 9 \cdot 4 = 36$

44. $\displaystyle\sum_{i=1}^{6} (2i+1)(2i-1) = \sum_{i=1}^{6} (4i^2 - 1)$

$= (4 \cdot 1^2 - 1) + (4 \cdot 2^2 - 1)$
$\quad + (4 \cdot 3^2 - 1) + (4 \cdot 4^2 - 1)$
$\quad + (4 \cdot 5^2 - 1) + (4 \cdot 6^2 - 1)$
$= 3 + 15 + 35 + 63 + 99 + 143$
$= 358$

64. $\dfrac{12!}{4!\,8!} = \dfrac{12 \cdot 11 \cdot 10 \cdot 9 \cdot 8!}{4!\,8!} = \dfrac{12 \cdot 11 \cdot 10 \cdot 9}{4 \cdot 3 \cdot 2 \cdot 1} = 495$

Exercise Set 12.2, page 614

16. $a_6 = -14$, $a_8 = -20$, $d = \dfrac{a_8 - a_6}{8 - 6} = \dfrac{-20 - (-14)}{2}$

$= \dfrac{-6}{2} = -3$

$a_n = a_1 + (n-1)d$
$a_6 = a_1 + (6-1)d$
$-14 = a_1 + 5(-3)$
$a_1 = 1$
$a_{15} = 1 + (15-1)(3) = 1 + (14)(-3) = -41$

22. $a_n = 1 - 2n$, $n = 20$

$S_n = \dfrac{n}{2}(a_1 + a_n)$

$a_1 = 1 - 2(1) = -1$

$a_{20} = 1 - 2(20) = -39$

$S_{20} = \dfrac{20}{2}(-1 + (-39))$

$= 10(-40) = -400$

36. $a = 7$, c_1, c_2, c_3, c_4, c_5, $b = 19$

$a_n = a_1 + (n-1)d$
$19 = 7 + 6d$
$d = 2$
$c_1 = a_1 + d = 7 + 2 = 9$
$c_2 = a_1 + 2d = 7 + 4 = 11$
$c_3 = a_1 + 3d = 7 + 6 = 13$
$c_4 = a_1 + 4d = 7 + 8 = 15$
$c_5 = a_1 + 5d = 7 + 10 = 17$

Exercise Set 12.3, page 621

6. $\dfrac{a_{n+1}}{a_n} = \dfrac{(-1)^n e^{(n+1)x}}{(-1)^{n-1} e^{nx}} = -e^{(n+1)x - nx} = -e^x$

Because x is a constant, $-e^x$ is a constant and the sequence is a geometric sequence.

18. $\dfrac{a_2}{a_1} = \dfrac{6}{8} = \dfrac{3}{4} = r$

$a_n = a_1 r^{n-1}$

$a_n = 8\left(\dfrac{3}{4}\right)^{n-1}$

40. $r = \dfrac{4}{3}, \ a_1 = \dfrac{4}{3}, \ n = 14$

$S_n = \dfrac{a_1(1 - r^n)}{1 - r}$

$S_{14} = \dfrac{\dfrac{4}{3}\left(1 - \left(\dfrac{4}{3}\right)^{14}\right)}{1 - \dfrac{4}{3}} \approx 220.49$

62. $0.3\overline{95} = \dfrac{3}{10} + \dfrac{95}{1000} + \dfrac{95}{100{,}000} + \cdots = \dfrac{3}{10} + \dfrac{95/1000}{1 - 1/100}$

$= \dfrac{3}{10} + \dfrac{95}{990} = \dfrac{392}{990} = \dfrac{196}{495}$

70. $P = \dfrac{A[(1 + r)^m - 1]}{r}; \quad A = 250, \ r = .08/12, \ m = 12(4)$

$P = \dfrac{250[(1.00667)^{48} - 1]}{0.00667} \approx 14088.63$

Exercise Set 12.4, page 628

8. $S_n = 2 + 6 + 8 + \cdots + n(n + 1) = \dfrac{n(n + 1)(n + 2)}{3}$

when $n = 1$, $S_1 = 1(1 + 1) = 2$; $\dfrac{n(n + 1)(n + 2)}{3} = 2$

Therefore the statement is true for $n = 1$.
Assume true for $n = k$.

$S_k = 2 + 6 + 8 + \cdots + k(k + 1)$

$= \dfrac{k(k + 1)(k + 2)}{3}$ Induction Hypothesis

Prove the statement is true for $n = k + 1$. That is, prove

$S_{k+1} = \dfrac{(k + 1)(k + 2)(k + 3)}{3}.$

Since $a_k = k(k + 1)$, $a_{k+1} = (k + 1)(k + 2)$

$S_{k + 1} = S_k + a_{k+1} = \dfrac{k(k + 1)(k + 2)}{3} + (k + 1)(k + 2)$

$= \dfrac{k(k + 1)(k + 2) + 3(k + 1)(k + 2)}{3}$

$= \dfrac{(k + 1)[k(k + 2) + 3(k + 2)]}{3}$

$= \dfrac{(k + 1)[k^2 + 2k + 3k + 6]}{3}$

$= \dfrac{(k + 1)(k^2 + 5k + 6)}{3} = \dfrac{(k + 1)(k + 2)(k + 3)}{3}$

12. $P_n = \left(1 - \dfrac{1}{2}\right)\left(1 - \dfrac{1}{3}\right)\cdots\left(1 - \dfrac{1}{n + 1}\right) = \dfrac{1}{n + 1}$

Let $n = 1$, then $P_1 = \left(1 - \dfrac{1}{2}\right) = \dfrac{1}{2}$; $\dfrac{1}{n + 1} = \dfrac{1}{2}$

The statement is true for $n = 1$.
Assume the statement is true for $n = k$.

$P_k = \left(1 - \dfrac{1}{2}\right)\left(1 - \dfrac{1}{3}\right)\cdots\left(1 - \dfrac{1}{k + 1}\right) = \dfrac{1}{k + 1}$

Prove the statement is true for $n = k + 1$. That is, prove

$P_{k+1} = \left(1 - \dfrac{1}{2}\right)\left(1 - \dfrac{1}{3}\right)\cdots\left(1 - \dfrac{1}{k + 1}\right)\left(1 - \dfrac{1}{k + 2}\right)$

$= \dfrac{1}{k + 2}$

Since $a_k = \left(1 - \dfrac{1}{k + 1}\right)$, $a_{k+1} = \left(1 - \dfrac{1}{k + 2}\right)$,

$P_{k+1} = P_k \cdot a_{k+1} = \dfrac{1}{k + 1} \cdot \left(1 - \dfrac{1}{k + 2}\right)$

$= \dfrac{1}{k + 1} \cdot \dfrac{k + 1}{k + 2} = \dfrac{1}{k + 2}$

14. When $n = 7$, $\left(\frac{4}{3}\right)^7 \approx 7.49 > 7$. Thus the statement is true for $n = 7$. Assume statement is true for $n = k$, $k \geq 7$.

$$\left(\frac{4}{3}\right)^k > k \quad \text{Induction Hypothesis}$$

Prove the statement is true for $n = k + 1$. That is, prove

$$\left(\frac{4}{3}\right)^{k+1} > k + 1.$$

By the induction hypothesis

$$\left(\frac{4}{3}\right)^k > k$$

Thus $\left(\frac{4}{3}\right)^{k+1} = \left(\frac{4}{3}\right)\left(\frac{4}{3}\right)^k > \frac{4}{3}k = k + \frac{1}{3}k > k + 1$.

The last inequality is true because $k \geq 7$, thus $\frac{1}{3}k > 1$.

Exercise Set 12.5, page 634

4. $\displaystyle\binom{10}{5} = \frac{10!}{5!\,5!} = \frac{10 \cdot 9 \cdot 8 \cdot 7 \cdot 6 \cdot 5!}{5!\,5!} = \frac{10 \cdot 9 \cdot 8 \cdot 7 \cdot 6}{5 \cdot 4 \cdot 3 \cdot 2 \cdot 1}$

$\qquad = 252$

18. $(3x + 2y)^4 = (3x)^4 + 4(3x)^3(2y) + 6(3x)^2(2y)^2$
$\qquad\qquad + 4(3x)(2y)^3 + (2y)^4$

$\qquad = 81x^4 + 216x^3y + 216x^2y^2 + 96xy^3 + 16y^4$

20. $(2x - \sqrt{y}\,)^7 = \displaystyle\binom{7}{0}(2x)^7 + \binom{7}{1}(2x)^6(-\sqrt{y}\,)$

$\qquad + \displaystyle\binom{7}{2}(2x)^5(-\sqrt{y}\,)^2 + \binom{7}{3}(2x)^4(-\sqrt{y}\,)^3$

$\qquad + \displaystyle\binom{7}{4}(2x)^3(-\sqrt{y}\,)^4 + \binom{7}{5}(2x)^2(-\sqrt{y}\,)^5$

$\qquad + \displaystyle\binom{7}{6}(2x)(-\sqrt{y}\,)^6 + \binom{7}{7}(-\sqrt{y}\,)^7$

16. If $a > 1$, show that $a^{n+1} > a^n$ for all positive integers. Because $a > 1$, $a \cdot a > a \cdot 1$ or $a^2 > a$. Thus the statement is true when $n = 1$.

Assume the statement is true for $n = k$:

$$a^{k+1} > a^k \quad \text{Induction Hypothesis}$$

Prove the statement is true for $n = k + 1$. That is, prove

$$a^{k+2} > a^{k+1}.$$

Because $a^{k+1} > a^k$ and $a > 0$,

$$a(a^{k+1}) > a(a^k)$$
$$a^{k+2} > a^{k+1}$$

$\qquad = 128x^7 - 448x^6\sqrt{y} + 672x^5y - 560x^4y\sqrt{y}$

$\qquad\quad + 280x^3y^2\sqrt{y} - 84x^2y^2\sqrt{y}$

$\qquad\quad + 14xy^3 - y^3\sqrt{y}$

22. The sixth row of Pascal's Triangle is

$$1 \quad 6 \quad 15 \quad 20 \quad 15 \quad 6 \quad 1$$

$(x - y^3)^6 = x^6 - 6x^5y^3 + 15x^4y^6$

$\qquad - 20x^3y^9 + 15x^2y^{12} - 6xy^{15} + y^{18}$

34. $\displaystyle\binom{10}{6-1}(x^{-1/2})^{10-6+1}(x^{1/2})^{6-1} = \binom{10}{5}(x^{-1/2})^5(x^{1/2})^5 = 252$

Exercise Set 12.6, page 641

14. $C(9, 3) = \dfrac{9!}{3!\,6!} = \dfrac{9 \cdot 8 \cdot 7 \cdot 6!}{3!\,6!} = 84$

16. Let A be the person who refuses to sit in the front seat and B the person who refuses to sit in the back seat. There are $4 \cdot 3 \cdot 2 \cdot 1 = 4! = 24$ ways in which the remaining people can be seated.

18. a. Let E be even and O be odd

$$\begin{array}{cccccc} \text{E} & \text{E} & \text{E} & \text{O} & \text{O} & \text{O} \\ 3 & \cdot 2 & \cdot 1 & \cdot 3 & \cdot 2 & \cdot 1 \end{array} = 6 \cdot 6 = 36$$

b. The first number can be even or odd. Thus there are 6 choices for the first number. The next number must be of opposite parity (there are 3 of these) and the parity must alternate. Thus, the number of arrangements is

$$6 \cdot 3 \cdot 2 \cdot 2 \cdot 1 \cdot 1 = 72.$$

26. At most one means one or none. In this case one or no defective drives.

One defective drive: $C(2, 1) \cdot C(8, 2) = 2 \cdot 28 = 56$

No defective drives: $C(8, 3) = 56$

One or no defective $= 56 + 56 = 112$

28. $P(5, 5) = 120$

Exercise Set 12.7, page 648

6. $\{(R, D), (R, I), (D, I)\}$

14. $\{HHHT, HHTH, HTHH, THHH, HHHH\}$

22. Let $E = \{2, 4, 6\}$ and $T = \{3, 6\}$.

$$E \cup T = \{2, 3, 4, 6\}$$

$$P(E \cup T) = \frac{N(E \cup T)}{N(S)} = \frac{4}{6} = \frac{2}{3}$$

28. Let A and B be the radar stations and let D indicate detection and N not detected. Here are the possibilities.

A	B	probability
D	D	(.95) (.95)
D	N	(.95) (.05)
N	D	(.05) (.95)

The probability of detection is the sum of these probabilities.

Probability of detection $= 0.9025 + 0.0475 + 0.0475$

$$= 0.9975$$

32. a. $\dfrac{2}{10} \cdot \dfrac{1}{9} = \dfrac{2}{90} = \dfrac{1}{45}$

b. $\dfrac{C(2, 1)C(8, 1)}{C(10, 2)} + \dfrac{C(2, 2)C(8, 1)}{C(10, 2)} = \dfrac{16}{45} + \dfrac{1}{45} = \dfrac{17}{45}$

Answers to Odd-Numbered Exercises

Exercise Set 1.1, page 9

1. a. 0, 4, $\sqrt{4}$, $\sqrt{9}$ are integers **b.** 0, 4, 1/5, 11/3, $\sqrt{4}$, $\sqrt{9}$, 3.1$\overline{4}$, and 3.14 are rational numbers.
c. $-0.272272227\ldots$ is an irrational number. **d.** all are real numbers. **3.** $-7 - (-15) = -7 + 15 = 8$ **5.** 16 **7.** 17 **9.** -30
11. $-3a/7$ **13.** $-7a/20$ **15.** $-10/21$ **17.** $-18/5$ **19.** $-2a/15$ **21. a.** 26/55 of the pool **b.** 26x/165 of the pool
23. Identity property of multiplication **25.** Associative property of addition **27.** Reflexive property of equality
29. Closure property of addition **31.** Closure property of multiplication **33.** Inverse property of addition
35. Commutative property of addition **37.** Transitive property of equality **39.** Identity property of addition **41.** $\{1, 3\}$
43. $\{1, 3\}$ **45.** $\{0, 2, 4\}$ **47.** $\{0, 1, 2, 3, 4, 5, 11\}$ **49.** $\{1, 3, 5, 6, 10, 11\}$ **51.** $\{0, 1, 2, 3, 4, 6, 8, 10\}$ **53.** $-\sqrt{2} - 7$
55. 8/59 **57.** $2/(2 - \pi)$ **59.** $2 - 1 \neq 1 - 2$ **61.** $(3 - 1) - 5 = -3; 3 - (1 - 5) = 7$
63. a. yes **b.** no **c.** yes **d.** no **65.** false **67.** true **69.** false **71.** true **73.** true
75. a. $0.\overline{72}$ **b.** 0.825 **c.** $0.\overline{285714}$ **d.** $0.1\overline{35}$ **77.** 0.16620626, 0.16662040, 0.16666204, 0.1667,

$$\frac{\sqrt{x + 9} - 3}{x} \text{ is approaching } 0.1\overline{6} = 1/6.$$

79. $A = \{4, 6, 8, 9, 10\}$ **81.** $C = \{53, 59\}$ **83.** All properties except for the identity property of addition and Inverse properties
85. all of the properties
87. a. $T > L$ and L is the largest prime, so T can't be prime. **b.** Since the prime numbers $2, 3, 5, 7, \ldots L$ all are factors of $T - 1$, none can be factors of T, so T can't be a composite number.

Exercise Set 1.2, page 16

1. **3.** **5.** **7.** $5/2 < 4$ **9.** $2/3 > 0.6666$ **11.** $1.75 < 2.23$

13. $\sqrt{5} > 2$ **15.** $\sqrt{12} > 3$ **17.** $0.4 < 4/9$ **19.** $0/2 = -0/5$ **21.** $22/7 > \pi$ **23.** $22/7 > 3.14159$

25. $(3, 5)$ **27.** $(-\infty, 3)$ **29.** $[0, 3)$ **31.** $(-\infty, -3) \cup [2, \infty)$

33. $(3, 4)$ **35.** $(-1, 1]$ **37.** $-4 \leq x \leq 1$ **39.** $1 < x < 5$

41. $x \geq 2.5$ **43.** $x < 2$ **45.** $x \leq 2$ or $x > 3$ **47.** $x < 3$ or $x > 3$

49. $(-\infty, 3]$ **51.** $-1 < x \leq 2$, $(-1, 2]$ **53.** $1 \leq x \leq 4$ **55.** $-2 \leq x < \pi$

57. 4 **59.** 27.4 **61.** -3 **63.** 40 **65.** $\sqrt{2} + 1$ **67.** $-\sqrt{2} + 1$ **69.** $-\sqrt{3} + 2$ **71.** $-\pi/3 + 2$ **73.** $-3\pi + 10$
75. 7 **77.** 8 **79.** 50 **81.** 33 **83.** $\pi + 3$ **85.** $\sqrt{2} + 10$ **87.** $|a - 2|$ **89.** $|m - n|$ **91.** $|z - 5| > 4$ **93.** $(-\infty, 3) \cup (3, \infty)$
95. $(-3, 3)$ **97.** $x^2 + 1$ **99.** $w^2 + \pi$ **101.** false **103.** true **105.** 111/271, $\sqrt{2}/\sqrt{11}$, 351/820 **107.** $I \leq 120$
109. $2 \leq A < 3$ **111.** $|x - 2| < |x - 6|$ **113.** $|x - a| < \delta$ **115.** all real numbers

A65

Exercise Set 1.3, page 25

1. -64 **3.** -64 **5.** 1   **7.** 1   **9.** 243   **11.** $1/27$ **13.** 32   **15.** 64   **17.** $125/63$ **19.** $81/16$ **21.** $4/2025$
23. $32/5$ **25.** 1   **27.** $6x^7y^4$ **29.** $8c^9/125$ **31.** $36x^4/y^4$ **33.** $2y^2/9x$ **35.** $y/3x$ **37.** y^6/x^4 **39.** $(b^2 + a)/ab^2$
41. b/ac^2 **43.** 16   **45.** 1   **47.** 7.34×10^1 **49.** 1.9×10^6 **51.** 1.63×10^{11} **53.** 3.2×10^{-5} **55.** 7×10^{-3}
57. 8.21×10^{-8} **59.** 6500   **61.** 0.0000731   **63.** 80,000,000,000   **65.** 0.000217   **67.** 100,000,000,000   **69.** 3.75
71. three   **73** six   **75.** one   **77.** three   **79.** two   **81.** four   **83.** 2.158924997   **85.** 6645.309867
87. 1.97×10^4 seconds   **89.** \$4873.50   **91.** x^3y^{n+1} **93.** $x^{4n-2}y^4$ **95.** y^{n+1}/x^{5n} **97.** $3^{(3^3)} > (3^3)^3$

Exercise Set 1.4, page 31

1. D   **3.** H   **5.** G   **7.** B   **9.** J   **11. a.** $x^2 + 2x - 7$ **b.** 2   **c.** $1, 2, -7$ **d.** 1   **e.** $x^2, 2x, -7$
13. a. $x^3 - 1$ **b.** 3   **c.** $1, -1$ **d.** 1   **e.** $x^3, -1$
15. a. $2x^4 + 3x^3 + 4x^2 + 5$ **b.** 4   **c.** $2, 3, 4, 5$ **d.** 2   **e.** $2x^4, 3x^3, 4x^2, 5$ **17.** 3   **19.** 5   **21.** 2   **23.** $5x^2 + 11x + 3$
25. $9w^3 + 8w^2 - 2w + 6$ **27.** $-2r^2 + 3r - 12$ **29.** $-3u^2 - 2u + 4$ **31.** $8x^3 + 18x^2 - 67x + 40$
33. $6x^4 - 19x^3 + 26x^2 - 29x + 10$ **35.** $10x^2 + 22x + 4$ **37.** $y^2 + 3y + 2$ **39.** $4z^2 - 19z + 12$ **41.** $a^2 + 3a - 18$
43. $b^2 + 2b - 24$ **45.** $10x^2 - 57xy + 77y^2$ **47.** $18x^2 + 55xy + 25y^2$ **49.** $12w^2 - 40wx + 33x^2$ **51.** $6p^2 - 11pq - 35q^2$
53. $12d^2 + 4d - 8$ **55.** $r^3 + s^3$ **57.** $60c^3 - 49c^2 + 4$ **59.** $9x^2 - 25$ **61.** $9x^4 - 6x^2y + y^2$ **63.** $16w^2 + 8wz + z^2$
65. $x^2 - 4x + 4 + 2xy - 4y + y^2$ **67.** $x^2 + 10x + 25 - y^2$ **69.** 29   **71.** -17 **73.** -1 **75.** 33   **77. a.** 48.46   **b.** 51.44
79 a. 10.994998   **b.** 10.99949998   **81.** 11,175 committees   **83.** 510 ft   **85.** $a^3 + 3a^2b + 3ab^2 + b^3$ **87.** $x^3 - 3x^2 + 3x - 1$
89. $8x^3 - 36x^2y + 54xy^2 - 27y^3$ **91. a.** n **b.** n **c.** n **d.** no degree   **93. a.** $41, 43, 47, 53$ **b.** $n = 41$

Exercise Set 1.5, page 41

1. $5(x + 4)$ **3.** $-3x(5x + 4)$ **5.** $2xy(5x + 3 - 7y)$ **7.** $(x - 3)(2a + 4b)$ **9.** $(3x + 1)(x^2 + 2)$ **11.** $(x - 1)(ax + b)$
13. $(3w + 2)(2w^2 - 5)$ **15.** $(3ax^2 - 2b)(4ax - 5)$ **17.** $(x + 3)(x + 4)$ **19.** $(a - 12)(a + 2)$ **21.** $(6x + 1)(x + 4)$
23. $(17x + 4)(3x - 1)$ **25.** $(3x + 8y)(2x - 5y)$ **27.** $(x^2 + 5)(x^2 + 1)$ **29.** $(6x^2 + 5)(x^2 + 3)$ **31.** $(6x^3 - 7)(3x^3 - 4)$
33. $(x - 3)(x + 3)$ **35.** $(2a - 7)(2a + 7)$ **37.** $(1 - 10x)(1 + 10x)$
39. $(x^2 - 3)(x^2 + 3)$ **41.** $(x + 3)(x + 7)$ **43.** $(x + 5)^2$ **45.** $(a - 7)^2$ **47.** $(2x + 3)^2$ **49.** $(z^2 + 2w^2)^2$ **51.** $(x - 2)(x^2 + 2x + 4)$
53. $(2x - 3y)(4x^2 + 6xy + 9y^2)$ **55.** $(2 - x^2)(4 + 2x^2 + x^4)$ **57.** $(x - 3)(x^2 - 3x + 3)$ **59.** $2(3x - 1)(3x + 1)$
61. $(2x - 1)(2x + 1)(4x^2 + 1)$ **63.** $a(3x - 2y)(4x - 5y)$ **65.** $b(3x + 4)(x - 1)(x + 1)$ **67.** $2b(6x + y)(6x + y)$
69. $(w - 3)(w^2 - 12w + 39)$ **71.** $(x + 3y - 1)(x + 3y + 1)$ **73.** not factorable   **75.** $(a - 1)(a^2 + a + 1)(a + 1)(a^2 - a + 1)$
77. $(2x - 5)^2(3x + 5)$ **79.** $(2x - y)(2x + y + 1)$ **81.** 8   **83.** 64   **85.** $(x^n - 1)(x^n + 1)(x^{2n} + 1)$ **87.** $\pi(R - r)(R + r)$
89. $r^2(4 - \pi)$ **91. a.** I + II + III   **b.** II + III + V   **c.** I + II + III = II + III + V

Exercise Set 1.6, page 50

1. $\dfrac{x + 4}{3}$ **3.** $\dfrac{x - 3}{x - 2}$ **5.** $\dfrac{a^2 - 2a + 4}{a - 2}$ **7.** $-\dfrac{x + 8}{x + 2}$ **9.** $-\dfrac{4y^2 + 7}{y + 7}$ **11.** $-\dfrac{8}{a^3b}$ **13.** $\dfrac{10}{27q^2}$ **15.** $\dfrac{x(3x + 7)}{2x + 3}$ **17.** $\dfrac{x + 3}{2x + 3}$

19. $\dfrac{(2y + 3)(3y - 4)}{(2y - 3)(y + 1)}$ **21.** $\dfrac{1}{a - 8}$ **23.** $\dfrac{3p - 2}{r}$ **25.** $\dfrac{8x(x - 4)}{(x - 5)(x + 3)}$ **27.** $\dfrac{3y - 4}{y + 4}$ **29.** $\dfrac{7z(2z - 5)}{(2z - 3)(z - 5)}$ **31.** $\dfrac{-2x^2 + 14x - 3}{(x - 3)(x + 3)(x + 4)}$

33. $\dfrac{(2x - 1)(x + 5)}{x(x - 5)}$ **35.** $\dfrac{-q^2 + 12q + 5}{(q - 3)(q + 5)}$ **37.** $\dfrac{3x^2 - 7x - 13}{(x - 3)(x + 4)(x - 3)(x - 4)}$ **39.** $\dfrac{(x + 2)(3x - 1)}{x^2}$ **41.** $\dfrac{4x + 1}{x - 1}$ **43.** $\dfrac{x - 2y}{y(y - x)}$

45. $\dfrac{(5x + 9)(x + 3)}{(x + 2)(4x + 3)}$ **47.** $\dfrac{(b + 3)(b - 1)}{(b - 2)(b + 2)}$ **49.** $\dfrac{x - 1}{x}$ **51.** $2 - m^2$ **53.** $\dfrac{-x^2 + 5x + 1}{x^2}$ **55.** $\dfrac{-x - 7}{x^2 + 6x - 3}$ **57.** $\dfrac{2x - 3}{x + 3}$

59. $\dfrac{a + b}{ab(a - b)}$ **61.** $\dfrac{(b - a)(b + a)}{ab(a^2 + b^2)}$ **63. a.** 136.55 mph   **b.** $\dfrac{2v_1v_2}{v_1 + v_2}$ **65.** $\dfrac{2x + 1}{x(x + 1)}$ **67.** $\dfrac{3x^2 - 4}{x(x - 2)(x + 2)}$ **69.** $\dfrac{x^2 + 9x + 25}{(x + 5)^2}$

71. $\dfrac{x(1 - 4xy)}{(1 - 2xy)(1 + 2xy)}$ **73.** 0.5   **75.** 0.6   **77.** $0.\overline{615384}$ **79.** 0.61803399   **81.** 0.2   **83.** $5/4 \neq 3/10$ **85.** $2/3, 4/5, 6/7$

Exercise Set 1.7, page 62

1. 3   **3.** -3 **5.** 8   **7.** -16 **9.** 16   **11.** $1/4$ **13.** $1/9$ **15.** $3/4$ **17.** $8/125$ **19.** 100   **21.** 7   **23.** 5   **25.** $x^{11/10}$
27. $4a^2$ **29.** $3xy^3$ **31.** $a^{1/2}b^{3/10}$ **33.** $a^2 + 7a$
35. $p - q$ **37.** $m^2n^{3/2}$ **39.** 1   **41.** $r^{(m-n)/nm}$ **43.** 2   **45.** -6 **47.** $3/4$ **49.** 2   **51.** 2   **53.** 6   **55.** 49   **57.** $\sqrt{3x}$

59. $5\sqrt[4]{xy}$ **61.** $\sqrt[3]{25w^2}$ **63.** $(17k)^{1/3}$ **65.** $a^{2/5}$ **67.** $\left(\dfrac{7a}{3}\right)^{1/2}$ **69.** $3\sqrt{5}$ **71.** $2\sqrt[3]{3}$ **73.** $-3\sqrt[3]{3}$ **75.** $-2\sqrt[3]{4}$ **77.** $15\sqrt{5}$

79. $10\sqrt[3]{35}$ **81.** $2xy\sqrt{6x}$ **83.** $-2ay^2\sqrt[3]{2y}$ **85.** $\sqrt{3x}$ **87.** $\sqrt[3]{4m^2n}$ **89.** $\sqrt[4]{9x^3y}$ **91.** $7\sqrt{2}$ **93.** $6\sqrt{3}$ **95.** $-3\sqrt{2}$ **97.** 0

99. $3x\sqrt{2xy}$ **101.** $29 + 11\sqrt{5}$ **103.** $2x - 9$ **105.** $50y + 10\sqrt{6yz} + 3z$ **107.** $x + 10\sqrt{x - 3} + 22$ **109.** $2x + 14\sqrt{2x + 5} + 54$
111. $\sqrt{2}$ **113.** $\sqrt{10}/6$ **115.** $3\sqrt[3]{4}/2$ **117.** $2\sqrt[3]{x}/x$ **119.** $\sqrt{5}/3$ **121.** $\dfrac{\sqrt{6xy}}{9y}$ **123.** $\dfrac{3\sqrt{5} - 3\sqrt{x}}{5 - x}$ **125.** $-\dfrac{2\sqrt{7} + 7}{3}$
127. $\dfrac{5}{3\sqrt{5}}$ **129.** $\dfrac{4}{21 - 7\sqrt{5}}$ **131.** $\dfrac{a^2\sqrt{30}}{3}$ **133.** $\dfrac{3\sqrt{2xy}}{2y}$ **135.** $\dfrac{\sqrt{30y}}{2y}$ **137.** $\dfrac{\sqrt[3]{12x^2y}}{x}$ **139.** 1.516 **141.** -4.441
143. 2.09×10^1 **145.** 2.79×10^2 **147.** -6.31 **149. a.** 81% **b.** 66% **151.** $8/5$ **153.** $-19/12$ **155.** $(x - \sqrt{7})(x + \sqrt{7})$
157. $(x + 3\sqrt{2})(x + 3\sqrt{2})$ **159.** $\sqrt{3^2 + 4^2} \neq 3 + 4$ **161.** Step 5

Exercise Set 1.8, page 69

1. $2 + 3i$ **3.** $4 - 11i$ **5.** $8 + i\sqrt{3}$ **7.** $7 + 4i$ **9.** $0 + 9i$ **11.** $5 + 12i$ **13.** $4 - 3i$ **15.** $-2 - 5i$ **17.** $12 - 2i$
19. $-2 + 11i$ **21.** $16 + 16i$ **23.** $23 + 2i$ **25.** $74 + 0i$ **27.** $-117 - i$ **29.** $\dfrac{1}{2} - \dfrac{i}{2}$ **31.** $\dfrac{7}{58} + \dfrac{3}{58}i$ **33.** $\dfrac{5}{13} + \dfrac{12}{13}i$
35. $\dfrac{1}{61} + \dfrac{11}{61}i$ **37.** $1 - 6i$ **39.** $-16 - 30i$ **41.** $-29 - 17i$ **43.** $-2 - 2i$ **45.** $-16 + 0i$ **47.** $0 + 75i$ **49.** $-i$ **51.** i
53. -1 **55.** -1 **57.** $-i$ **59.** i **61.** -1 **63.** $-i$ **65.** -2 **67.** $-8\sqrt{5}$ **69.** 11 **71.** $9 + 40i$ **73.** $\dfrac{1}{2} \pm \dfrac{\sqrt{3}}{2}i$
75. $-1 \pm i$ **77.** $-\dfrac{3}{2} \pm \dfrac{\sqrt{3}}{2}i$ **79.** $-\dfrac{1}{4} \pm \dfrac{\sqrt{23}}{4}i$ **81.** 5 **83.** $\sqrt{29}$ **85.** $\sqrt{65}$ **87.** 3 **95.** no   **97.** $66 + 6\sqrt{5} - 14\sqrt{3}$
99. 1 **101.** 0 **103.** $(x + 3i)(x - 3i)$ **105.** $(2x + 9i)(2x - 9i)$ **107.** 0 **109.** 0

Challenge Exercises, page 73

1. true   **3.** true   **5.** false, $(2 \oplus 4) \oplus 6 \neq 2 \oplus (4 \oplus 6)$ **7.** false, $\sqrt{(-2)^2} \neq -2$ **9.** true

Review Exercises, page 74

1. integer, rational number, real number   **3.** rational number, real number   **5.** $7/2 > 3$ **7.** Distributive property
9. Associative property of multiplication   **11.a.** yes **b.** yes **c.** yes **d.** no
13. both closure properties, both commutative properties, both associative properties, additive identity, additive inverse, distributive properties   **15.** $(-4, 2)$ **17.** $-3 \leq x < 2$
19. 7 **21.** $4 - \pi$ **23.** 17 **25.** -36 **27.** $12x^8y^3$ **29.** 6.2×10^5 **31.** $35{,}000$ **33.** $-a^2 - 2a - 1$
35. $6x^4 + 5x^3 - 13x^2 + 22x - 20$ **37.** $3(x + 5)^2$ **39.** $4(5a^2 - b^2)$ **41.** $4ab^3\sqrt{3b}$ **43.** $6x\sqrt{2y}$ **45.** $\dfrac{3y\sqrt{15y}}{5}$ **47.** $\dfrac{7\sqrt[3]{4x}}{2}$
49. $-3y^2\sqrt[3]{5x^2y}$ **51.** $3 - 8i, 3 + 8i$ **53.** $5 + 2i$ **55.** $25 - 19i$ **57.** 1 **59.** 1

Exercise Set 2.1, page 83

1. 15 **3.** -4 **5.** $9/2$ **7.** $108/23$ **9.** $2/9$ **11.** 12 **13.** 16 **15.** 9 **17.** 75 **19.** $1/2$ **21.** $22/13$ **23.** $95/18$
25. 1200 **27.** 2500 **29.** identity   **31.** conditional equation   **33.** contradiction   **35.** contradiction   **37.** 31 **39.** 2
41. no solution   **43.** no solution   **45.** $7/2$ **47.** 6 **49.** -4 **51.** -12 **53.** 1 **55.** no solution   **57.** $x = -\dfrac{3}{2}y + 3$
59. $x = \dfrac{d - 5c}{2c - 5}, c \neq 5/2$ **61.** $x = \dfrac{ab - ay}{b}, b \neq 0$ **63.** $x = \dfrac{y - y_1 + mx_1}{m}, m \neq 0$ **65.** $x = \dfrac{3 - 4l - 4w}{l + w}, l \neq -w$
67. $x = m + n, m \neq n$ **69.** 5.47 **71.** no solution   **73.** 0.000879 **75.** no   **77.** yes   **79.** $x = \dfrac{c - b}{a}$ **81.** $\dfrac{10\sqrt{7}}{7}$
83. $-\dfrac{7\sqrt{3}}{6}$ **85.** $\dfrac{1}{a - b}$

Exercise Set 2.2, page 91

1. $h = \dfrac{3V}{\pi r^2}$ 3. $t = \dfrac{I}{Pr}$ 5. $m_1 = \dfrac{Fd^2}{Gm_2}$ 7. $v_0 = \dfrac{s + 16t^2}{t}$ 9. $T_w = \dfrac{-Q_w + m_w c_w T_f}{m_w c_w}$ 11. $d = \dfrac{a_n - a_1}{n - 1}$ 13. $r = \dfrac{S - a_1}{S}$

15. $f_1 = \dfrac{w_1 f - w_2 f_2 + w_2 f}{w_1}$ 17. $v_{LC} = \dfrac{f_{LC} v - f_v v}{f_v}$ 19. 100   21. 30 feet by 57 feet   23. 12 cm, 36 cm, 36 cm   25. 872, 873

27. 18, 20   29. 240 meters   31. 2 hours   33. 2 miles per hour   35. 98   37. 850 sunglasses   39. $937.50
41. $7600 invested at 8%, $6400 invested at 6.5%   43. $3750   45. $18\frac{2}{11}$ grams   47. 64 liters   49. 1200 at $14 and 1800 at $25
51. $6\frac{2}{3}$ lb of the $12 coffee and $13\frac{1}{3}$ lb of the $9 coffee   53. 10 grams   55. 7.875 hours   57. $13\frac{1}{3}$ hours
59. $10.05 for book, $0.05 for bookmark   61. 6.25 feet   63. 40 lb   65. $9\frac{3}{7}$ hours   67. 1383.81 feet   69. $\frac{2}{9}$ hour
71. 84 years old

Exercise Set 2.3, page 104

1. 5, −3   3. −24, 3/8   5. 0, 7/3   7. 4/3, −2/5   9. 1/2, −4   11. 8, 2   13. 3, 8/3   15. 7   17. ±9   19. $\pm 2\sqrt{6}$

21. ±2i   23. 11, −1   25. 7/2   27. −1/2   29. $-3 \pm 2\sqrt{2}$ 31. 5, −3   33. 0, −10   35. $-\dfrac{3}{2} \pm \dfrac{\sqrt{13}}{2}$ 37. $\dfrac{-2 \pm \sqrt{6}}{2}$

39. $\dfrac{4 \pm \sqrt{13}}{3}$ 41. $\dfrac{-3 \pm \sqrt{29}}{10}$ 43. $\dfrac{-3 \pm 2\sqrt{6}}{3}$ 45. −3, 5   47. $\dfrac{-1 \pm \sqrt{5}}{2}$ 49. $\dfrac{-2 \pm \sqrt{2}}{2}$ 51. $\dfrac{5 \pm i\sqrt{11}}{6}$ 53. $\dfrac{-1 \pm \sqrt{21}}{10}$

55. $\dfrac{-3 \pm \sqrt{41}}{4}$ 57. $-\dfrac{\sqrt{2}}{2}, -\sqrt{2}$ 59. $\dfrac{3 \pm i\sqrt{11}}{2}$ 61. 81, 2 distinct real numbers   63. −116, 2 distinct complex numbers

65. 0, 1 real number   67. ±24   69. 49/4   71. yes   73. yes   75. yes   77. 10 feet by 3.5 feet   79. 100 feet by 150 feet

81. 10, 12   83. $\dfrac{5 + \sqrt{29}}{2}$ 85. 35 mph for the first part and 45 mph for the last part   87. 12 hours   89. ±1.32i   91. 1.17, −3.95

93. 3.35, −0.62   95. $t = \dfrac{v_0 \pm \sqrt{v_0^2 - 4g(s - s_0)}}{2g}$ 97. $y = \dfrac{2 \pm \sqrt{4 + 3x}}{x}$ 99. $x = \dfrac{-y \pm yi\sqrt{47}}{6}$ 101. $y = \dfrac{-1 \pm \sqrt{4x + 33}}{2}$

105. a. $w = l(-1 + \sqrt{5})/2$ b. 62.42 feet   107. 22

Exercise Set 2.4, page 111

1. 0, ±5   3. 2, ±1   5. 0, ±3   7. 0, −5, 8   9. 0, ±4   11. 2, $-1 \pm i\sqrt{3}$ 13. 40   15. 3   17. 7   19. 7   21. 9
23. 5/2   25. 1, −6   27. 4   29. 1, 5   31. 2   33. 23, −31   35. 2, −1/8   37. 0, 1/256   39. −1, −59/3   41. $\pm\sqrt{7}, \pm\sqrt{2}$
43. ±2, $\pm\sqrt{6}/2$ 45. $\sqrt[3]{2}, -\sqrt[3]{3}$ 47. $-\sqrt[3]{36}/3, \sqrt[3]{98}/7$ 49. 1, 16   51. −1/27, 64   53. $\pm\sqrt{15}/3$ 55. ±1   57. 1/2, −1/5

59. 256/81, 16   61. ±0.62, ±1.62   63. ±0.34, ±2.98   65. $x = \pm\sqrt{9 - y^2}$ 67. $x = y + 2\sqrt{yz} + z$ 69. $x = \dfrac{7 - 2y^2}{2y}$ 71. 9, 36

73. 3 inches   75. 10.5 mm   77. 87 feet   79. a. 8.93 in   b. $5\sqrt{3}$ in   81. $s = \left(\dfrac{-275 + 5\sqrt{3025 + 176T}}{2}\right)^2$

Exercise Set 2.5, page 121

1. $x < 4$ 3. $x < -6$ 5. $x \le -3$ 7. $x \ge -13/8$ 9. $x < 2$ 11. If you write more than 57 checks a month.
13. at least 34 sales   15. $-3/4 < x \le 4$ 17. $11/3 \ge x \ge 1/3$ 19. $11/4 > x \ge -3/8$ 21. $20 \le C \le 40$
23. {12, 14, 16}, {14, 16, 18}   25. $(-\infty, -7) \cup (0, \infty)$ 27. [−4, 4]   29. (−5, −2)   31. $(-\infty, -4] \cup [7, \infty)$ 33. [−1/2, 4/3]
35. $(-\infty, -5/4] \cup [3/2, \infty)$ 37. (−4, 1)   39. $(-\infty, -8) \cup [5, \infty)$ 41. $(-\infty, -7/2) \cup [0, \infty)$ 43. $(-\infty, -1) \cup (2, 4)$
45. $(-\infty, 5) \cup [12, \infty)$ 47. $(-2/3, 0) \cup (5/2, \infty)$ 49. (−∞, 5)   51. [4, ∞)   53. (0, 210)   55. [−9, ∞)   57. [−3, 3]
59. $(-\infty, -4] \cup [4, \infty)$ 61. $(-\infty, -3] \cup [5, \infty)$ 63. (−∞, ∞)   65. $(-\infty, 3) \cup (3, 6) \cup (6, \infty)$ 67. (−3, ∞)   69. $[-\sqrt{3}, 0] \cup [\sqrt{3}, \infty)$
71. (−∞, ∞)   73. $[-2, 0] \cup [5, -\infty)$ 75. $(-\infty, -2\sqrt{6}] \cup [2\sqrt{6}, \infty)$ 77. $(-\infty, -2\sqrt{14}] \cup [2\sqrt{14}, \infty)$ 79. $1 < t < 3$
81. a. $x > 4$ b. $0 < x < 1$ c. $x > \sqrt{2}$ 83. $a > 1$

Exercise Set 2.6, page 126

1. {−4, 4}   3. {7, 3}   5. {−5, −7}   7. {−34, 6}   9. {8, −3}   11. {2, −8}   13. {20, −12}   15. no solution   17. {12, −18}
19. $\left\{\dfrac{a + b}{2}, \dfrac{a - b}{2}\right\}$ 21. {a + δ, a − δ}   23. (−4, 4)   25. (−8, 10)   27. $(-\infty, -33) \cup (27, \infty)$ 29. $(-\infty, -3/2) \cup (5/2, \infty)$
31. $(-\infty, -8] \cup [2, \infty)$ 33. [−4/3, 8]   35. $(-\infty, -4] \cup [28/5, \infty)$

37. $(-\infty, \infty)$ **39.** $\{4\}$ **41.** no solution   **43.** $(-\infty, \infty)$ **45.** $(3 - b, 3 + b)$ **47.** $(-\sqrt{2}, \sqrt{2})$ **49.** $(-4, -2) \cup (2, 4)$
51. $(-\infty, -5] \cup [-4, -3] \cup [-2, \infty)$ **53.** $(-\infty, -\sqrt{26}] \cup [-\sqrt{17}, \sqrt{17}] \cup [\sqrt{26}, \infty)$ **55.** false, not true for $x < 0$.   **57.** true
59. true   **61.** $\{x \mid x \geq -4\}$ **63.** $\{x \mid x \leq -7\}$ **65.** $\{x \mid x \geq -7/2\}$ **67.** $\{x \mid x = 3, -5\}$ **69.** $(-5, -1) \cup (1, 5)$ **71.** $(-7, -3] \cup [3, 7)$
73. $(a - \delta, a) \cup (a, a + \delta)$ **75.** $(2, 4) \cup (8, 10)$ **77.** $(-\infty, -20/3] \cup [28/3, \infty)$ **79.** $(1/2, \infty)$ **81. a.** $|x - 3| < 8$ **b.** $|x - j| < k$
83. a. $|s - 4.25| \leq 0.01$ **b.** $4.24 \leq s \leq 4.26$

Challenge Exercises, page 128

1. false, $(-3)^2 = 9$ **3.** true   **5.** false, $100 > 1$ but $1/100 \not> 1/1$ **7.** false, $\sqrt{1} + \sqrt{1} = 2$, but $1 + 1 \neq 4$
9. false, $2(x^2 - 16) = 0$ has roots -4 and 4.

Review Exercises, page 128

1. $3/2$ **3.** $1/2$ **5.** $-38/15$ **7.** $0, 5/3$ **9.** $\pm 2\sqrt{3}/3, \pm\sqrt{10}/2$ **11.** $-5, 3$ **13.** 4 **15.** -4 **17.** $-2, -4$ **19.** $5, 1$
21. $2, -3$ **23.** $-2, -1$ **25.** $(-\infty, 2]$ **27.** $[-5, 2]$ **29.** $[145/9, 35]$ **31.** $(-\infty, 0] \cup [3, 4]$ **33.** $(-\infty, -3) \cup (4, \infty)$

35. $(-\infty, 5/2] \cup (3, \infty)$ **37.** $(2/3, 2)$ **39.** $(-2, 0) \cup (0, 2)$ **41.** $(1, 2) \cup (2, 3)$ **43.** $h = \dfrac{V}{\pi r^2}$ **45.** $b_1 = \dfrac{2A - hb_2}{h}$ **47.** $m = \dfrac{e}{c^2}$

49. 80 **51.** 24 nautical miles   **53.** \$1750 in the 4% account, \$3750 in the 6% account   **55.** \$864   **57.** 18 hours   **59.** 13 feet

Exercise Set 3.1, page 136

1.   **3.**   **5.** 13 **7.** $4\sqrt{6}$ **9.** $2\sqrt{35}$ **11.** $7\sqrt{5}$ **13.** $\sqrt{1261}$ **15.** $\sqrt{89}$ **17.** $\sqrt{38 - 12\sqrt{6}}$

19. $2\sqrt{a^2 + b^2}$ **21.** $-x\sqrt{10}$ **23.** $(3, 2)$ **25.** $(6, 4)$ **27.** $(5/2, 6)$
29. $(-0.875, 3.91)$ **31.** yes   **33.** yes   **35.** no   **37.**   **39.**   **41.**

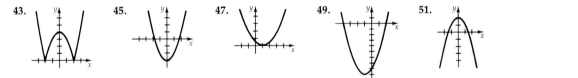

43.   **45.**   **47.**   **49.**   **51.**

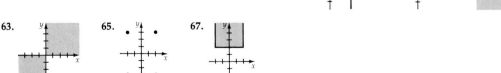

53. $(12, 0), (-4, 0)$ **55. a.** yes   **b.** yes   **c.** yes   **d.** yes   **57.**   **59.**   **61.**

63.   **65.**   **67.**

69. $(13, 5)$ **71.** $(7, -6)$ **73.** yes   **75.** $x^2 - 6x + y^2 - 8y = 0$ **77.** $9x^2 + 25y^2 = 225$

Exercise Set 3.2, page 145

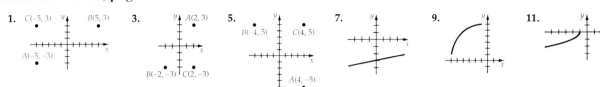

1. $C(-5, 3)$, $B(5, 3)$, $A(-5, -3)$ **3.** $A(2, 3)$, $B(-2, -3)$, $C(2, -3)$ **5.** $B(-4, 5)$, $C(4, 5)$, $A(4, -5)$ **7.**   **9.**   **11.**

13. a. no **b.** yes   **15. a.** no **b.** no   **17. a.** yes **b.** yes   **19. a.** yes **b.** yes   **21. a.** yes **b.** yes   **23. a.** no **b.** no
25. no   **27.** yes   **29.** yes   **31.** yes   **33.** no   **35.** no   **37.** yes   **39.** yes   **41.** no   **43.** yes   **45.** yes   **47.** yes
49. $(0, 12/5)$, $(6, 0)$ **51.** $(0, \sqrt{5})$, $(0, -\sqrt{5})$; $(5, 0)$ **53.** $(0, 4)$, $(0, -4)$; $(-4, 0)$ **55.** $(0, \pm 2)$; $(\pm 2, 0)$ **57.** $(0, \pm 4)$; $(\pm 4, 0)$

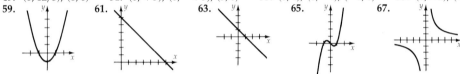

59.   **61.**   **63.**   **65.**   **67.**

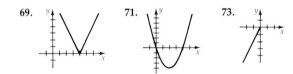

69.   **71.**   **73.**

75. center $(0, 0)$, radius 6   **77.** center $(0, 0)$, radius 10   **79.** center $(1, 3)$, radius 7   **81.** center $(-2, -5)$, radius 5
83. center $(8, 0)$, radius 1/2   **85.** $(x - 4)^2 + (y - 1)^2 = 2^2$ **87.** $(x - 1/2)^2 + (y - 1/4)^2 = (\sqrt{5})^2$ **89.** $(x - 0)^2 + (y - 0)^2 = 5^2$
91. $(x - 1)^2 + (y - 3)^2 = 5^2$ **93.** $(x - 4)^2 + (y - 1)^2 = 2^2$ **95.** center $(3, 0)$, radius 2   **97.** center $(2, 5)$, radius 3
99. center $(7, -4)$, radius 3   **101.** center $(-1/2, 0)$, radius 4   **103.** center $(1/2, -1/3)$, radius 1/6
105.   **107.**   **109.**   **111.** $(x + 1)^2 + (y - 7)^2 = 5^2$ **113.** $(x - 7)^2 + (y - 11)^2 = 11^2$

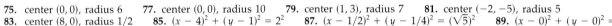

115. $(x + 3)^2 + (y - 3)^2 = 3^2$

Exercise Set 3.3, page 152

1. a. 5 **b.** -4 **c.** -1 **d.** 1 **e.** $3k - 1$ **f.** $3k + 5$ **3. a.** $\sqrt{5}$ **b.** 3 **c.** 3 **d.** $\sqrt{21}$ **e.** $\sqrt{r^2 + 2r + 6}$ **f.** $\sqrt{c^2 + 5}$
5. a. 1/2 **b.** 1/2 **c.** 5/3 **d.** $1/\pi$ **e.** $1/(c^2 + 4)$ **f.** $1/|2 + h|$
7. a. -4 **b.** $-2\sqrt{3}$ **c.** $-\sqrt{14}$ **d.** $-\sqrt{39}/2$ **e.** $-\sqrt{16 - a^2}$ **f.** $-\sqrt{16 - a^2}$ **9. a.** 1 **b.** 1 **c.** -1 **d.** -1 **e.** 1 **f.** -1 **11.** yes
13. no   **15.** no   **17.** yes   **19.** no   **21.** all real numbers   **23.** all real numbers   **25.** $\{x \mid x \neq -2\}$ **27.** $\{x \mid x \geq -7\}$
29. $\{x \mid -2 \leq x \leq 2\}$ **31.** $\{x \mid x > -4\}$ **33.** even   **35.** odd   **37.** even   **39.** even   **41.** even   **43.** even   **45.** even
47. neither   **49. a.** $25 - l = w$ **b.** $A = 25l - l^2$ **51.** $v(t) = 80,000 - 6500t$, $0 \leq t \leq 10$
53. $C(x) = [19.95 - 0.05(x - 50)]x$, $50 < x < 200$ **55. a.** $C(x) = 2000.00 + 22.80x$ **b.** $R(x) = 37.00x$ **c.** $P(x) = 14.20x - 2000.00$
57. $h(r) = 15 - 5r$ **59.** $d = \sqrt{(3t)^2 + (50)^2}$, $0 \leq t \leq 60$ **61.** 0.56, 0.66, 0.77, 0.88, 0.99   **63.** 275, 375, 385, 390, 394   **65.** 4   **67.** 2
69. 0   **71. a.** 36 **b.** 13 **c.** 12 **d.** 30 **e.** $13k - 2$ **f.** $8k - 11$ **73.** $4\sqrt{21}$ **75.** 9   **77.** 17

Exercise Set 3.4, page 162

1. yes   **3.** yes   **5.** yes   **7.** yes   **9.** no   **11.** yes   **13.** no   **15.** yes   **17.** decreasing on $(-\infty, 0]$; increasing on $[0, \infty)$
19. increasing on $(-\infty, \infty)$ **21.** increasing on $(-\infty, -3]$; increasing on $[-3, 0]$;   **23.** constant on $(-\infty, 0]$; increasing on $[0, \infty)$
    decreasing on $[0, 3]$; increasing on $[3, \infty)$
25. decreasing on $(-\infty, 0]$; constant on $[0, 1]$;   **27.** no   **29.** yes   **31.** no   **33.** no   **35.** no   **37.** even   **39.** neither
    increasing on $[1, \infty)$

41. even   **43.** neither   **45.** neither   **47.**    **49.**

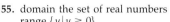

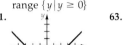

51. domain the set of real numbers   **53.** domain the set of real numbers   **55.** domain the set of real numbers
range $\{y \mid y \geq -1\}$ range the set of real numbers   range $\{y \mid y \geq 0\}$

57. domain the set of real numbers   **59.** domain the set of real numbers   **61.**   **63.**
range $\{y \mid y \geq 0\}$ range $\{y \mid y \geq -2\}$

65.   **67.**   **69.**   **71.**   **73.**

75. a. $f(x) = \dfrac{2}{(x+1)^2 + 1} + 1$ **b.** $f(x) = -\dfrac{2}{(x-2)^2 + 1}$ **77.** $h(x) = 3x - 3$ **79.** $h(x) = -x^2 + 5x - 4$ **81.** $h(x) = -x^3 - x^2 + 6x$

Exercise Set 3.5, page 174

1. $-3/2$ **3.** $-1/2$ **5.** the line does not have slope   **7.** 6   **9.** 9/19   **11.** $\dfrac{f(3+h) - f(3)}{h}$ **13.** $\dfrac{f(h) - f(0)}{h}$

15.   **17.**   **19.**

21.   **23.**   **25.**

27. $y = x + 3$ **29.** $y = 3x - 1$ **31.** $y = \dfrac{3}{4}x + \dfrac{1}{2}$ **33.** $y = (0)x + 4 = 4$ **35.** $y = 2x + 1$ **37.** $y = -4x - 10$

39. $y = -\dfrac{3}{4}x + \dfrac{13}{4}$ **41.** $y = \dfrac{12}{5}x - \dfrac{29}{5}$ **43.** $y = \dfrac{4}{5}x - \dfrac{2}{5}$ **45.** $y = -\dfrac{25}{7}x + \dfrac{543}{70}$ **47.** $y = -\dfrac{3}{4}x + \dfrac{15}{4}$ **49.** $y = 2x + 10$

51. $y = x + 1$ **53.** $y = -\dfrac{5}{3}x - \dfrac{7}{3}$ **55.** $\dfrac{f(x_1 + h) - f(x_1)}{h}$ **57.** $F = \dfrac{9}{5}C + 32$ **59. a.** $V(n) = 8280 - 690n$ **b.** $6210

61. a. $275   **b.** $283   **c.** $355   **d.** $8

63. a. $C(t) = 19{,}500.00 + 6.75t$ **b.** $R(t) = 55.00t$ **c.** $P(t) = 48.25t - 19{,}500.00$ **d.** approximately 405 days   **67.** $y = -2x + 11$

69. $y = -\dfrac{13}{18}x + \dfrac{1}{18}$ **71.** $5x + 3y = 15$ **73.** $2x + 4y = 1$ **75.** $3x + y = 17$

Exercise Set 3.6, page 183

1. d   **3.** b   **5.** g   **7.** c   **9.** $f(x) = (x + 2)^2 - 3$ **11.** $f(x) = (x - 4)^2 - 11$

13. $f(x) = \left(x - \left(-\dfrac{3}{2}\right)\right)^2 - \dfrac{5}{4}$ **15.** $f(x) = -(x - 2)^2 + 6$ **17.** $f(x) = -3\left(x - \dfrac{1}{2}\right)^2 + \dfrac{31}{4}$

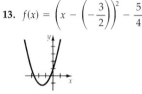

19. -16, minimum   **21.** 11, maximum   **23.** $-1/8$, minimum   **25.** -11, minimum   **27.** 35, maximum

29. a. 27 feet   **b.** $22\frac{5}{16}$ feet   **c.** 20.1 feet from the center   **31.** 300 tennis rackets

33. a. $w = \dfrac{600 - 3x}{2}$ **b.** $A = 300x - \dfrac{3}{2}x^2$ **c.** $w = 100$ ft, $x = 150$ ft   **35.** $(5, -25)$ **37.** $(0, -10)$ **39.** $(3, 10)$ **41.** $(3/4, 47/8)$

43. $(1/8, 17/16)$ **45.** y-intercept $(0, 0)$; x-intercepts $(0, 0)$ and $(-6, 0)$ **47.** y-intercept $(0, 9)$; x-intercept $(-3, 0)$

49. y-intercept $(0, 4)$; x-intercepts $(-4, 0)$, $(1, 0)$ **51.** y-intercept $(0, 1)$; no x-intercepts   **53.** y-intercept $(0, 4)$; no x-intercepts

55. a. 256 feet   **b.** $t = 4$ seconds   **c.** $t = 0$ seconds and $t = 8$ seconds   **57.** $f(x) = \dfrac{3}{4}x^2 - 3x + 4$

59. a. $w = 16 - x$ **b.** $A = 16x - x^2$ **63.** increases the height of each point on the graph by c units   **65.** 4, 4

Exercise Set 3.7, page 189

1. $f(x) + g(x) = x^2 - x - 12$,   domain all real numbers
$f(x) - g(x) = x^2 - 3x - 18$,   domain all real numbers
$f(x) \cdot g(x) = x^3 + x^2 - 21x - 45$,   domain all real numbers
$\dfrac{f(x)}{g(x)} = x - 5$,   domain $\{x \,|\, x \neq -3\}$

3. $f(x) + g(x) = 3x + 12$,   domain all real numbers
$f(x) - g(x) = x + 4$,   domain all real numbers
$f(x) \cdot g(x) = 2x^2 + 16x + 32$,   domain all real numbers
$\dfrac{f(x)}{g(x)} = 2$,   domain $\{x \,|\, x \neq -4\}$

5. $f(x) + g(x) = x^3 + 2x^2 + 8x$,   domain all real numbers
$f(x) - g(x) = x^3 + 2x^2 + 6x$,   domain all real numbers
$f(x) \cdot g(x) = x^4 + 2x^3 + 7x^2$,   domain all real numbers
$\dfrac{f(x)}{g(x)} = x^2 + 2x + 7$,   domain $\{x \,|\, x \neq 0\}$

7. $f(x) + g(x) = 4x^2 + 7x - 12$,   domain all real numbers
$f(x) - g(x) = x - 2$,   domain all real numbers
$f(x) \cdot g(x) = 4x^4 + 14x^3 - 12x^2 - 41x + 35$,   domain all real numbers
$\dfrac{f(x)}{g(x)} = 1 + \dfrac{x - 2}{2x^2 + 3x - 5}$,   domain $\{x \,|\, x \neq 1, x \neq -5/2\}$

9. $f(x) + g(x) = \sqrt{x - 3} + x$,   domain $\{x \,|\, x \geq 3\}$
$f(x) - g(x) = \sqrt{x - 3} - x$,   domain $\{x \,|\, x \geq 3\}$
$f(x) \cdot g(x) = x\sqrt{x - 3}$,   domain $\{x \,|\, x \geq 3\}$
$\dfrac{f(x)}{g(x)} = \dfrac{\sqrt{x - 3}}{x}$,   domain $\{x \,|\, x \geq 3\}$

11. $f(x) + g(x) = \sqrt{4 - x^2} + 2 + x$,   domain $\{x \,|\, -2 \leq x \leq 2\}$
$f(x) - g(x) = \sqrt{4 - x^2} - 2 - x$,   domain $\{x \,|\, -2 \leq x \leq 2\}$
$f(x) \cdot g(x) = (\sqrt{4 - x^2})(2 + x)$,   domain $\{x \,|\, -2 \leq x \leq 2\}$
$\dfrac{f(x)}{g(x)} = \dfrac{\sqrt{4 - x^2}}{2 + x}$,   domain $\{x \,|\, -2 < x \leq 2\}$

13. 18   **15.** $-9/4$ **17.** 30   **19.** 12   **21.** 300   **23.** $-384/125$ **25.** $-5/2$ **27.** $-1/4$ **29.** 2   **31.** $2x + h$

33. $4x + 2h + 4$ **35.** $-8x - 4h$ **37.** $(g \circ f)(x) = 6x + 3$, $(f \circ g)(x) = 6x - 16$

39. $(g \circ f)(x) = x^2 + 4x + 1$ **41.** $(g \circ f)(x) = -5x^3 - 10x$ **43.** $(g \circ f)(x) = \dfrac{1 - 5x}{x + 1}$
$(f \circ g)(x) = x^2 + 8x + 11$ $(f \circ g)(x) = -125x^3 - 10x$ $(f \circ g)(x) = \dfrac{2}{3x - 4}$

45. $(g \circ f)(x) = \dfrac{\sqrt{1 - x^2}}{x}$ **47.** $(g \circ f)(x) = -\dfrac{2|5 - x|}{3}$ **49.** 66   **51.** 51   **53.** -4 **55.** 41   **57.** $-3848/625$
$(f \circ g)(x) = \dfrac{1}{x - 1}$ $(f \circ g)(x) = \dfrac{3|x|}{|5x + 2|}$

59. $6 + 2\sqrt{3}$ **61.** $16c^2 + 4c - 6$ **63.** $9k^4 + 36k^3 + 45k^2 + 18k - 4$ **65.** 25.39   **67.** 5.474×10^{-6}

75. a. $(s \circ m)(x) = 87 + 49{,}300/x$ **b.** \$89   **77. a.** 2   **b.** 3/2   **c.** 5/3   **d.** 8/5

Exercise Set 3.8, page 196

9. $f^{-1}(x) = \dfrac{x-1}{4}$ **11.** $F^{-1}(x) = \dfrac{1-x}{6}$ **13.** $j^{-1}(t) = \dfrac{t-1}{2}$ **15.** $f^{-1}(v) = \sqrt[3]{1-v}$ **17.** $f^{-1}(x) = -\dfrac{4x}{x+3}$, $x \neq -3$

19. $M^{-1}(t) = \dfrac{5}{1-t}$, $t \neq 1$ **21.** $r^{-1}(t) = -\sqrt{\dfrac{1}{t}}$, $t > 0$ **23.** $J^{-1}(x) = \sqrt{x-4}$, $x \geq 4$

25. $f^{-1}(x) = \sqrt{x-3}$, domain $\{x \mid x \geq 3\}$, range $\{y \mid y \geq 0\}$ **27.** $f^{-1}(x) = x^2$, domain $\{x \mid x \geq 0\}$, range $\{y \mid y \geq 0\}$
f has domain $\{x \mid x \geq 0\}$, range $\{y \mid y \geq 3\}$ f has domain $\{x \mid x \geq 0\}$, range $\{y \mid y \geq 0\}$

29. $f^{-1}(x) = \sqrt{9-x^2}$, domain $\{x \mid 0 \leq x \leq 3\}$, range $\{y \mid 0 \leq y \leq 3\}$ **31.** $f^{-1}(x) = 2 + \sqrt{x+3}$, domain $\{x \mid x \geq -3\}$, range $\{y \mid y \geq 2\}$
f has domain $\{x \mid 0 \leq x \leq 3\}$, range $\{y \mid 0 \leq y \leq 3\}$ f has domain $\{x \mid x \geq 2\}$, range $\{y \mid y \geq -3\}$

33. $f^{-1}(x) = -4 - \sqrt{x+25}$, domain $\{x \mid x \geq -25\}$, range $\{y \mid y \leq -4\}$ **35.**
f has domain $\{x \mid x \leq -4\}$, range $\{y \mid y \geq -25\}$

37.    **39.**   **41.**

43.   **45.**   **47.**

49. $f^{-1}(x) = \dfrac{x-b}{a}$, $a \neq 0$ **51.** $f^{-1}(x) = \dfrac{x+1}{1-x}$, $x \neq 1$ **53.** no   **53.** yes   **57.** yes   **59.** no   **61.** 5   **63.** 4

Exercise Set 3.9, page 202

1. $d = kt$ **3.** $y = k/x$ **5.** $m = knp$ **7.** $V = klwh$ **9.** $A = ks^2$ **11.** $F = km_1m_2/d^2$ **13.** $y = kx$, $k = 4/3$
15. $r = kt^2$, $k = 1/81$ **17.** $T = krs^2$, $k = 7/25$ **19.** $V = klwh$, $k = 1$ **21.** 1.02 liters   **23.** 437.5 lbs/ft^2
25. **a.** approximately 3.3 seconds **b.** approximately 3.7 ft   **27.** 112 db   **29.** $330
31. **a.** 9 times larger **b.** 3 times larger **c.** 27 times larger   **33.** 6 times larger   **35.** approximately 3.2 mi/sec
37. approximately 410 vibrations per second   **39.** approximately 295 lbs   **41.** approximately 142,000,000 mi

Challenge Exercises, page 207

1. false, consider $f(x) = x^2$ and find $f(3)$ and $f(-3)$ **3.** false, let $f(x) = x^2$ and $g(x) = 2x$ and consider $(f \circ g)(0)$ and $(g \circ f)(0)$
5. false let $f(x) = 3x$ **7.** true   **9.** true

Review Exercises, page 208

1. 10   **3.** 5   **5.** $\sqrt{181}$ **7.** $(-1/2, 10)$ **9.**   **11.** **a.** no **b.** yes **c.** no   **13.** **a.** no **b.** no **c.** yes

15. **a.** yes **b.** yes **c.** yes   **17.** **a.** yes **b.** yes **c.** yes   **19.** center $(3, -4)$, radius 9   **21.** $(x-2)^2 + (y+3)^2 = 5^2$
23. **a.** 2 **b.** 10 **c.** $3t^2 + 4t - 5$ **d.** $3x^2 + 6xh + 3h^2 + 4x + 4h - 5$ **e.** $9t^2 + 12t - 5$ **f.** $27t^2 + 12t - 5$
25. **a.** 5 **b.** -11 **c.** $x^2 - 12x + 32$ **d.** $x^2 + 4x - 8$ **27.** $8x + 4h - 3$ **29.** the set of all real numbers   **31.** $\{x \mid -5 \leq x \leq 5\}$

33.

increasing on $[3, \infty)$
decreasing on $(-\infty, 3]$

35.

increasing on $[-2, 2]$
constant on $(-\infty, -2] \cup [2, \infty)$

37.

increasing $(-\infty, \infty)$

39.

a. domain is set of all real numbers
range $\{y \,|\, y \le 4\}$
b. even

41.

a. domain is set of all real numbers
range $\{y \,|\, y \ge 4\}$
b. even

43.

a. domain is set of all real numbers
range is set of all real numbers
b. odd

45. $y = -2x + 1$ **47.** $y = \dfrac{3}{4}x + \dfrac{19}{2}$

49. $(1, 8)$ **51.** $(5, 161)$ **53.** $(0, 1050)$ **55.** $f(x) = (x + 3)^2 + 1$ **57.** $f(x) = -(x + 4)^2 + 19$

59. $f(x) = -3(x - 2/3)^2 - 11/3$ **61.** $F(x) = (x + 2)^2 - 11$ **63.** $P(x) = 3(x - 0)^2 - 4$ **65.** $W(x) = -4(x + 3/4)^2 + 33/4$

67. $f(x) + g(x) = x^2 + x - 6,$ domain is set of all real numbers
$f(x) - g(x) = x^2 - x - 12,$ domain is set of all real numbers
$f(x) \cdot g(x) = x^3 + 3x^2 - 9x - 27,$ domain is set of all real numbers
$\dfrac{f(x)}{g(x)} = x - 3,$ domain $\{x \,|\, x \ne -3\}$

69. yes   **71.** yes   **73.** $f^{-1}(x) = \dfrac{x + 4}{3}$ **75.** $h^{-1}(x) = -2x - 4$ **77.** 25, 25   **79.** $t \approx 3.7$ seconds

Exercise Set 4.1, page 217

1. $5x^2 - 9x + 10 - \dfrac{10}{x + 3}$ **3.** $x^3 + 5x^2 - 9x - 45$ **5.** $x^2 - \dfrac{100}{3} + \dfrac{100x + 10}{3(3x^2 + x + 1)}$ **7.** $4x^2 + 1 + \dfrac{11}{5x^2 - 2}$

9. $x + 4 + \dfrac{6x - 3}{x^2 + x - 4}$ **11.** $4x^2 + 3x^2 + 12 + \dfrac{17}{x - 2}$ **13.** $4x^2 - 4x + 2 + \dfrac{1}{x + 1}$ **15.** $x^4 + 4x^3 + 6x^2 + 24x + 101 + \dfrac{403}{x - 4}$

17. $x^4 + x^3 + x^2 + x + 1$ **19.** $8x^2 + 6$ **21.** $x^7 + 2x^6 + 5x^5 + 10x^4 + 21x^3 + 42x^2 + 85x + 170 + \dfrac{344}{x - 2}$

23. $x^5 - 3x^4 + 9x^3 - 27x^2 + 81x - 242 + \dfrac{716}{x + 3}$ **25.** $3x - 3.1 + \dfrac{4.07}{x - 0.3}$ **27.** $2x^2 - 11x - 17 + \dfrac{3}{x}$ **29.** $1 + \dfrac{6}{x + 2}$ **31.** 25

33. 45   **35.** -2230 **37.** -80 **39.** -187 **41.** yes   **43.** yes   **45.** yes   **47.** yes   **49.** yes   **51.** yes   **65.** 13

Exercise Set 4.2, page 224

1. up to far left, up to far right   **3.** down to far left, up to far right   **5.** down to far left, down to far right
7. down to far left, up to far right   **9.** up to far left, down to far right   **11. a.** -7 **b.** 33 **c.** -2 **d.** 58 **e.** 9 **f.** 169
13. a. -77 **b.** 97 **c.** 15 **d.** 5 **e.** -18 **f.** 38   **15. a.** 22 **b.** 74 **c.** -286 **d.** 176 **e.** -748 **f.** -1592
17. a. 126 **b.** 86 **c.** 14 **d.** 126 **e.** 134 **f.** 110   **19. a.** -8 **b.** 82 **c.** 168 **d.** 782 **e.** 6482/81 **f.** 23,192   **21.** $-7/2, -2, 3$
23. $0, 2/5, 1$ **25.** $-5, -7/3, 11/2$ **27.** $-3, 0, 2$ **29.** 1
31.   **33.**   **35.**   **37.**   **39.**

41.   **43.**   **45.**   **53.** 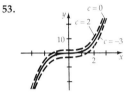   **55.**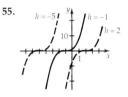

57. 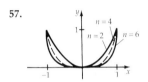   **59.**

Exercise Set 4.3, page 232

1. 3 (multiplicity two), -5 (multiplicity one)   **3.** 0 (multiplicity two), $-5/3$ (multiplicity two)
5. 2 (multiplicity one), -2 (multiplicity one),   **7.** 5 (multiplicity two), -2 (multiplicity two)
-3 (multiplicity two)
9. -3 (multiplicity one), 3 (multiplicity one),   **11.** $\pm 1, \pm 2, \pm 4, \pm 8$ **13.** $\pm 1, \pm 2, \pm 3, \pm 4, \pm 6, \pm 12, \pm 1/2, \pm 3/2$
-1 (multiplicity one), 1 (multiplicity one)
15. $\pm 1, \pm 2, \pm 4, \pm 1/2, \pm 1/3, \pm 2/3, \pm 4/3, \pm 1/6$ **17.** $\pm 1, \pm 3, \pm 9, \pm 1/2, \pm 3/2, \pm 9/2$ **19.** $\pm 1, \pm 7, \pm 1/2, \pm 7/2, \pm 1/4, \pm 7/4$
21. $\pm 1, \pm 2, \pm 4, \pm 8, \pm 16, \pm 32$ **23.** upper bound 2, lower bound -5 **25.** upper bound 4, lower bound -4
27. upper bound 1, lower bound -4 **29.** upper bound 1, lower bound -5 **31.** upper bound 4, lower bound -2
33. upper bound 2, lower bound -1 **35.** one positive, two or zero negative   **37.** two or zero positive, one negative
39. one positive, three or one negative   **41.** one positive, two or zero negative   **43.** three or one positive, one negative
45. one positive, no negative   **47.** $2, -1, -4$ **49.** $3, -4, 1/2$ **51.** $1/2, -1/3, -2$ **53.** $1, -1, -9/2$
55. $1/2, 4, \sqrt{3}, -\sqrt{3}$ **57.** $2, -1$ **59.** $0, -2, 1 + \sqrt{2}, 1 - \sqrt{2}$ **61.** $-1, 2$ **63.** $-3/2, 1, 8$ **71.** yes   **73.** yes

Exercise Set 4.4, page 238

1. $1 - i, 1/2$ **3.** $i, -3$ **5.** $-i\sqrt{2}, 1, \sqrt{5}, -\sqrt{5}$ **7.** $1 + \frac{1}{2}i, 1 - \frac{1}{2}i$ **9.** $1 - 3i, 1 + 2i, 1 - 2i$ **11.** $-i, 3, -1$
13. $2, -3, 2i, -2i$ **15.** $1/2, -3, 1 + 5i, 1 - 5i$ **17.** $1, 3 + 2i, 3 - 2i$ **19.** $x(x - 2)(x + 1)$ **21.** $x(x^2 + 9)$
23. $(x^2 + 6)(x + 2)(x - 2)$ **25.** $(x^2 + 2)(x^2 + 1)$ **27.** $(x + 1)(x - 3)(x^2 + 4)$ **29.** $x^3 - 3x^2 - 10x + 24$
31. $x^3 - 3x^2 + 4x - 12$ **33.** $x^4 - 10x^3 + 63x^2 - 214x + 290$ **35.** $x^5 - 22x^4 + 212x^3 - 1012x^2 + 2251x - 1830$
37. $4x^3 - 19x^2 + 224x - 159$ **39.** $x^3 + 13x + 116$ **41.** $x^4 - 18x^3 + 131x^2 - 458x + 650$ **43.** $3x^3 - 12x^2 + 3x + 18$
45. $-2x^4 + 4x^3 + 36x^2 - 140x + 150$ **49.** $P(x) = (x - 2)^3(x^2 + 9)$ **51.** $P(x) = \frac{1}{2}x^5 - 4x^4 + \frac{25}{2}x^3 - 19x^2 + 14x - 4$

Exercise Set 4.5, page 249

1. vertical asymptotes $x = 0$, $x = -3$ **3.** vertical asymptote $x = 4/3$, $x = -1/2$ **5.** horizontal asymptote $y = 4$

7. horizontal asymptote $y = 30$ **9.** vertical asymptote $x = -4$ **11.** vertical asymptote $x = 3$
horizontal asymptote $y = 0$ horizontal asymptote $y = 0$

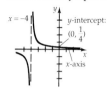

13. vertical asymptote $x = 0$ **15.** vertical asymptote $x = -4$ **17.** vertical asymptote $x = 2$
horizontal asymptote $y = 0$ horizontal asymptote $y = 1$ horizontal asymptote $y = -1$

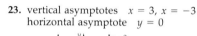

19. vertical asymptotes $x = 3$, $x = -3$ **21.** vertical asymptote $x = -3$, $x = 1$ **23.** vertical asymptotes $x = 3$, $x = -3$
horizontal asymptote $y = 0$ horizontal asymptote $y = 0$ horizontal asymptote $y = 0$

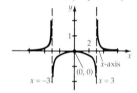

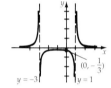

25. vertical asymptote $x = -2$ **27.** vertical asymptote none   **29.** vertical asymptotes $x = 3$, $x = -3$
horizontal asymptote $y = 1$ horizontal asymptote $y = 0$ horizontal asymptote $y = 2$

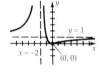

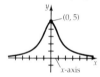

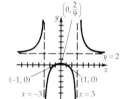

31. vertical asymptotes $x = -1 + \sqrt{2}$, $x = -1 - \sqrt{2}$ **33.** $y = 3x - 7$ **35.** $y = x$ **37.** vertical asymptote $x = 0$
horizontal asymptote $y = 1$ slant asymptote $y = x$

39. vertical asymptote $x = -3$ **41.** vertical asymptote $x = 4$ **43.** vertical asymptote $x = -2$
slant asymptote $y = x - 6$ slant asymptote $y = 2x + 13$ slant asymptote $y = x - 3$

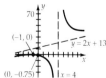

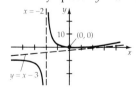

45. vertical asymptotes $x = 2$, $x = -2$ **47.**

slant asymptote $y = x$

49.

51.

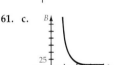

53.

55.   **57. a.** \$1333.33 **b.** \$8000 **c.**   **59. a.** 611 **b.** 1777 **c.** $y = 2000$ **d.** 1905

61. c.   **63.** $(-2, 2)$ **65.** $(0, 1), (-4, 1)$

Exercise Set 4.6, page 254

7. 2.0625   **9.** 1.1875   **11.** -1.5625 **13.** 0.855   **15.** 1.275   **17.** 2.125   **19.** 1.45   **21.** 1.25

Challenge Exercises, page 257

1. false, consider $x^3 - x^2 - ix^2 - 9x + 9 + 9i$ which has the zero $1 + i$ **3.** true   **5.** false, consider $f(x) = \dfrac{x}{x^2 + 1}$ **7.** true

9. true   **11.** true   **13.** true

Review Exercises, page 257

1. $x + 4 + \dfrac{-5x - 29}{x^2 + x + 3}$ **3.** $-x + \dfrac{3x^2 - 12x - 3}{x^3 + x}$ **5.** $3x^2 - 5x - 1$ **7.** $4x^2 + x + 8 + \dfrac{22}{x - 3}$ **9.** $3x^2 - 6x + 7 - \dfrac{13}{x + 2}$

11. $3x^2 + 5x - 11$ **13.** 77   **15.** 33   **21.**   **23.**   **25.**

27. $\pm 1, \pm 2, \pm 3, \pm 6$ **29.** $\pm 1, \pm 2, \pm 3, \pm 4, \pm 6, \pm 12, \pm 1/3, \pm 2/3, \pm 4/3, \pm 1/5,$ **31.** ± 1

$\pm 2/5, \pm 3/5, \pm 4/5, \pm 6/5, \pm 12/5, \pm 1/15, \pm 2/15, \pm 4/15$

33. no positive and three or one negative   **35.** one positive and one negative   **37.** $1, -2, -5$ **39.** $-2, -1/2, -4/3$ **41.** 1

43. $2x^3 - 3x^2 - 23x + 12$ **45.** $x^4 - 3x^3 + 27x^2 - 75x + 50$ **47.** vertical asymptote $x = -2$ **49.** vertical asymptote $x = -1$

horizontal asymptote $y = 3$ slant asymptote $y = 2x + 3$

51.   **53.**   **55.**   **57.**

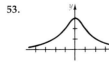

Exercise Set 5.1, page 266

1. 4.72880   **3.** 442.335   **5.** 2.17458   **7.** 164.022   **9.** 5.65223   **11.** 0.969476   **13.** 70.4503   **15.** 19.8130   **17.** 14.0940
19. 15.1543   **21.** 3353.33   **23.** 8103.08

25. 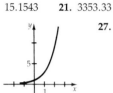   **27.**   **29.**   **31.**   **33.**

35. 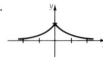   **37.**   **39.**   **41.** 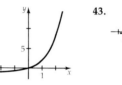   **43.**

45.   **47.**   **49.** 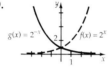   **51.**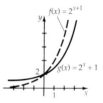

53.

Value of n	Value of $\left(1 + \dfrac{1}{n}\right)^n$
5	2.48832
50	2.691588029
500	2.715568520
5000	2.718010041
50,000	2.718254646
500,000	2.718280469
5,000,000	2.718281828

55. a. -2 **b.** 4   **c.** does not exist
h is not an exponential function because b is not a positive constant.

57.   **61. a.** 20,000   **b.** 40,000   **c.** 320,000   **63.** e^π **65.**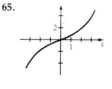

Exercise Set 5.2, page 275

1. $10^2 = 100$ **3.** $5^3 = 125$ **5.** $3^4 = 81$ **7.** $b^t = r$ **9.** $3^{-3} = \frac{1}{27}$ **11.** $\log_2 16 = 4$ **13.** $\log_7 343 = 3$ **15.** $\log_{10} 10,000 = 4$
17. $\log_b j = k$ **19.** $\log_b b = 1$ **21.** 6   **23.** 5   **25.** 3   **27.** -2 **29.** 0   **31.** $\log_b x + \log_b y + \log_b z$ **33.** $\log_3 x - 4\log_3 z$
35. $\frac{1}{2}\log_b x - 3\log_b y$ **37.** $\log_b x + \frac{2}{3}\log_b y - \frac{1}{3}\log_b z$ **39.** $\frac{1}{2}\log_7(x + z^2) - \log_7(x^2 - y)$ **41.** 0.9208   **43.** 1.1292   **45.** -0.4709

47. 1.7479   **49.** 1.3562   **51.** $\log_{10} x^2(x + 5)$ **53.** $\log_b \sqrt{\dfrac{(x - y)^3(x + y)}{z}}$ **55.** $\log_8(x + y)$ **57.** $\ln x^2(x - 3)^4$

59. $\ln \dfrac{xz}{y}$ **61.** 1.5395   **63.** 0.86719   **65.** -1.7308 **67.** -2.3219 **69.** 0.87357   **71.** 3.06   **73.** 3190

75. 0.00334   **77.** 7.40   **79.** 2.00   **81.** 0.300   **83.** $[1, 10^{1000}]$ **85.** $[e^e, e^{e^3}]$ **87.** $(0, 1)$

91. reflexive property of equality, definition of $\log_b x = n$

Exercise Set 5.3, page 280

1.   3.   5.   7. 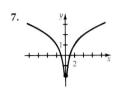   9.

11. 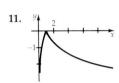   13. 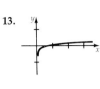   15.   17. 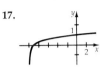   19.

21. 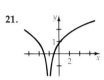   23.   25.   27.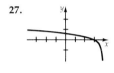

29. $\{x \mid x > 0, x \neq 1\}$ **31.** $\{x \mid x < -3 \text{ or } x > 3\}$ **33.** $\{x \mid x \neq 0\}$ **35.** $\{y \mid y \neq 0\}$ **37.** all real numbers   **39.** all real numbers
41. **a.** 15   **b.** 54   **c.** $6,565,990,000   **43.** 8.6   **45. a.** 3   **b.** 1.386   **c.** 3.296

47. The domain of F is $\{x \mid x \neq 0\}$, but the domain of G is $\{x \mid x > 0\}$.   **49.** 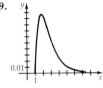   **51.** domain $\{x \mid x \geq 1\}$, range $\{y \mid y > 0\}$

53. domain $\{x \mid -1 < x < 1\}$,   range $\{y \mid y \geq 100\}$ **55.** domain $\{x \mid x > 1\}$, range all real numbers   **57.** 8.8

Exercise Set 5.4, page 286

1. 6   **3.** 3   **5.** $-3/2$ **7.** $-6/5$ **9.** 3   **11.** 2.64   **13.** -4.36 **15.** 0.163   **17.** -0.25 **19.** 2.3   **21.** 5.0852
23. $2 + 2\sqrt{2}$ **25.** 199/95   **27.** 3   **29.** 10^{10} **31.** 25   **33.** 2   **35.** 3, -3 **37.** no solution   **39.** 5   **41.** $\log(20 + \sqrt{401})$
43. $\frac{1}{2}\log\frac{3}{2}$ **45.** $\ln(15 \pm 4\sqrt{14})$ **47.** $\ln(1 + \sqrt{65}) - \ln 8$ **49.** 2.2   **51.** 3.2×10^{-5}
53. **a.** 8500, 10,285   **b.** in 6 years   **55. a.** 60°F   **b.** in 27 minutes
57. The second step, because $\log 0.5 < 0$, thus the inequality sign must be reversed.

59. no solutions   **61.** 2 solutions   **63.** 2 solutions   **65.** 4 solutions   **67.** $x = \dfrac{y}{y - 1}$

   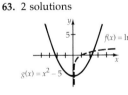 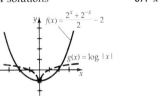

69. $e^{0.336} \approx 1.4$ **71.** $\left(0, \dfrac{1}{\ln 2}\right)$

Exercise Set 5.5, page 295

1. **a.** $9724.05   **b.** $11,256.80   **3. a.** $48,885.72   **b.** $49,282.20   **c.** $49,283.29   **5.** $24,730.82   **7. a.** 2200   **b.** 17,600
9. $N(t) = 22,600(e^{.01368t})$ **11. a.** 18,400   **b.** 1994   **13.** $N(t) = 100e^{-.000418t}$ **15.** 39.1%   **17.** 2161 years old
19. **a.** 0.056   **b.** 42°   **c.** 54 minutes   **d.** never   **21.** 10 times more powerful   **23.** 3.01 decibels
25. **a.** 211 hours   **b.** 1386 hours   **27.** 3.1 years   **29. a.** 21.7, 0.87   **b.** 1086, .88   **c.** 72,382, .92

31. a. 0.023639 **b.** 6.8 billion **c.** 80 billion   **33. a.** 16,000 **b.** 10,500 (nearest 500)   **35.** $\dfrac{1}{k} \ln\left[\dfrac{P(t)\,(m - P_0)}{P_0(m - P(t))}\right]$

37. a. 286,500 gallons (nearest 500) **b.** 250,500 gallons (nearest 500) **c.** 34.5 hours   **39.** 15.8 times stronger   **41.** 6.93%
43. a. 96 **b.** 3385 **c.** 13,395 **d.** 39,751

Challenge Exercises, page 299

1. true   **3.** true   **5.** false, $h(x) = \left(\frac{1}{2}\right)^x$ is a decreasing function   **7.** true   **9.** true
11. false, $\log(1 + 1) = \log 2$, but $\log 1 + \log 1 = 0$ **13.** true

Review Exercises, page 300

1. 2   **3.** 3   **5.** −2   **7.** −3   **9.** 1000   **11.** 7   **13.** 15.6729   **15.** 5.47395   **17.** 13.6458

19. 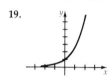   **21.** 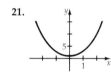   **23.** 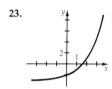   **25.**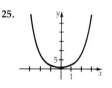

27. 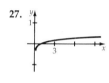   **29.**   **31.**   **33.** $4^3 = 64$ **35.** $\sqrt{2}^4 = 4$ **37.** $\log_5 125 = 3$ **39.** $\log_{10} 1 = 0$

41. $\log_b x + 3 \log_b y - \log_b z$ **43.** $\ln x + 3 \ln y$ **45.** $\log x^2 \sqrt[3]{x + 1}$ **47.** $\ln\left(\dfrac{\sqrt{2xy}}{z^3}\right)$ **49.** 2.86754   **51.** −0.117233

53. 1.00×10^{248} **55.** 1.41×10^{22} **57.** $\dfrac{\ln 30}{\ln 4}$ **59.** 4   **61.** 4   **63.** $\dfrac{\ln 3}{2 \ln 4}$ **65.** 10^{1000} **67.** 1,000,005   **69.** 81   **71.** 4

73. 4.2   **75. a.** $20,323.79 **b.** $20,339.99   **77.** $4438.10   **79.** $N(t) = e^{0.3219t}$ **81.** $N(t) = 2.899e^{0.3219t}$

Exercise Set 6.1, page 312

1. 75°   **3.** 3.7°   **5.** 12.45°   **7.** 43°51′   **9.** 79°4′25″   **11.** 113°   **13.** 7.66°   **15.** 56.2°   **17.** 134.22°   **19.** 78.133°
21. 16.733°   **23.** 47.333°   **25.** 165.615°   **27.** 95.469°   **29.** 36°36′   **31.** 66°43′12″   **33.** 6°11′24″   **35.** 132°34′48″   **37.** $\pi/12$
39. $7\pi/4$ **41.** $7\pi/2$ **43.** 0.51   **45.** 2.90   **47.** 10.65   **49.** 30°   **51.** 67.5°   **53.** 660°   **55.** 68.75°   **57.** 36.67°   **59.** 250.96°
61. $\pi/3$ **63.** $\pi/12$ **65.** $\pi/2 - 1.22$ **67.** $\pi/4$ **69.** $\pi/8$ **71.** $\pi - 1.76$ **73.** 4, 229.2°   **75.** 2.38, 136.6°   **77.** 6.28 in
79. 18.33 cm   **81.** 3π **83.** $5\pi/12$ **85.** $\pi/30$ rad/s   **87.** 100π rad/s   **89.** 69.14 rad/s   **91.** 53.55 mph   **93.** 30.79 ft
95. 486.95 ft²   **97.** 1199.9 cm²   **99.** 840,000 mi   **101 a.** 6206 revolutions **b.** 5378 revolutions   **103.** 157 cm/s
105. a. 69.1 mi **b.** 1.15 mi **c.** 0.0192 mi

Exercise Set 6.2, page 319

1. $\sin \theta = 3\sqrt{10}/10$ $\csc \theta = \sqrt{10}/3$ **3.** $\sin \theta = \sqrt{2}/2$ $\csc \theta = \sqrt{2}$ **5.** $\sin \theta = \sqrt{2}/2$ $\csc \theta = \sqrt{2}$
$\cos \theta = \sqrt{10}/10$ $\sec \theta = \sqrt{10}$ $\cos \theta = \sqrt{2}/2$ $\sec \theta = \sqrt{2}$ $\cos \theta = \sqrt{2}/2$ $\sec \theta = \sqrt{2}$
$\tan \theta = 3$ $\cot \theta = 1/3$ $\tan \theta = 1$ $\cot \theta = 1$ $\tan \theta = 1$ $\cot \theta = 1$
7. $\sin \theta = 2\sqrt{13}/13$ $\csc \theta = \sqrt{13}/2$ **9.** $\sin \theta = \sqrt{5}/5$ $\csc \theta = \sqrt{5}$ **11.** $\sin \theta = \sqrt{2}/2$ $\csc \theta = \sqrt{2}$
$\cos \theta = 3\sqrt{13}/13$ $\sec \theta = \sqrt{13}/3$ $\cos \theta = 2\sqrt{5}/5$ $\sec \theta = \sqrt{5}/2$ $\cos \theta = \sqrt{2}/2$ $\sec \theta = \sqrt{2}$
$\tan \theta = 2/3$ $\cot \theta = 3/2$ $\tan \theta = 1/2$ $\cot \theta = 2$ $\tan \theta = 1$ $\cot \theta = 1$
13. 3/4   **15.** 4/5   **17.** 3/4   **19.** 12/13   **21.** 13/5   **23.** $\sqrt{10}/3$ **25.** 1   **27.** 2   **29.** $\sqrt{3}$ **31.** $\sqrt{2}/2$ **33.** 1/2
35. $\sqrt{3}$ **37.** 0.2079   **39.** 0.6249   **41.** 1.1190   **43.** 0.8221   **45.** 1.0053   **47.** 0.4816   **49.** 1.0729   **51.** 0.3153   **53.** 1.2331
55. $\sqrt{2}$ **57.** 0   **59.** −3/4   **61.** $\sqrt{3}/6$ **63.** 5/4   **65.** $\sqrt{3} - \sqrt{6}$ **67.** $\sqrt{3}$ **69.** $\dfrac{3\sqrt{2} + 2\sqrt{3}}{6}$ **71.** $\dfrac{3 - \sqrt{3}}{3}$
73. $2\sqrt{2} - \sqrt{3}$ **75.** 30°, $\pi/6$ **77.** 30°, $\pi/6$ **79.** 45°, $\pi/4$ **81.** 45°, $\pi/4$ **83.** 30°, $\pi/6$ **85.** 60°, $\pi/3$ **87.** 0.5932
89. 1.0051   **91.** 0.6327   **101.** −3/4   **103.** 34/9   **105.** $\sqrt{2} - 1$ **107.** 0   **109.** true   **111.** true

Exercise Set 6.3, page 326

1. $\sin \theta = 3\sqrt{13}/13$ $\csc \theta = \sqrt{13}/3$
$\cos \theta = 2\sqrt{13}/13$ $\sec \theta = \sqrt{13}/2$
$\tan \theta = 3/2$ $\cot \theta = 2/3$
3. $\sin \theta = 3\sqrt{13}/13$ $\csc \theta = \sqrt{13}/3$
$\cos \theta = -2\sqrt{13}/13$ $\sec \theta = -\sqrt{13}/2$
$\tan \theta = -3/2$ $\cot \theta = -2/3$
5. $\sin \theta = -5\sqrt{89}/89$ $\csc \theta = -\sqrt{89}/5$
$\cos \theta = -8\sqrt{89}/89$ $\sec \theta = -\sqrt{89}/8$
$\tan \theta = 5/8$ $\cot \theta = 8/5$
7. $\sin \theta = 0$ $\csc \theta$ is undefined
$\cos \theta = -1$ $\sec \theta = -1$
$\tan \theta = 0$ $\cot \theta$ is undefined
9. $\sin 330° = -1/2$ $\csc 330° = -2$
$\cos 330° = \sqrt{3}/2$ $\sec 330° = 2\sqrt{3}/3$
$\tan 330° = -\sqrt{3}/3$ $\cot 330° = -\sqrt{3}$
11. $\sin 210° = -1/2$ $\csc 210° = -2$
$\cos 210° = -\sqrt{3}/2$ $\sec 210° = -2\sqrt{3}/3$
$\tan 210° = \sqrt{3}/3$ $\cot 210° = \sqrt{3}$
13. $\sin \pi/3 = \sqrt{3}/2$ $\csc \pi/3 = 2\sqrt{3}/3$
$\cos \pi/3 = 1/2$ $\sec \pi/3 = 2$
$\tan \pi/3 = \sqrt{3}$ $\cot \pi/3 = \sqrt{3}/3$
15. $\sin 11\pi/6 = -1/2$ $\csc 11\pi/6 = -2$
$\cos 11\pi/6 = \sqrt{3}/2$ $\sec 11\pi/6 = 2\sqrt{3}/2$
$\tan 11\pi/6 = -\sqrt{3}/3$ $\cot 11\pi/6 = -\sqrt{3}$
17. $1/2$ **19.** 0 **21.** 1 **23.** $\dfrac{3\sqrt{2} + 2\sqrt{3}}{6}$ **25.** $-3/2$ **27.** 1 **29.** quadrant I   **31.** quadrant IV
33. quadrant III   **35.** $\tan \phi = \sqrt{3}/3$ **37.** $\sin \phi = -1/2$ **39.** $\cot \phi = -1$ **41.** $\tan \phi = -\sqrt{3}/3$ **43.** $\csc \phi = 2\sqrt{3}/3$
45. $\cot \phi = -\sqrt{3}/3$ **47.** $\cot \phi = -\sqrt{3}/3$ **49.** 0.7986 **51.** -0.4384 **53.** 0.1405 **55.** -0.7880 **57.** 0.5878 **59.** 0.8090
61. -0.8296 **63.** -0.2594 **65.** $30°, 150°$ **67.** $150°, 210°$ **69.** $225°, 315°$ **71.** $240°, 300°$ **73.** $3\pi/4, 7\pi/4$ **75.** $5\pi/6, 11\pi/6$
77. $\pi/3, 2\pi/3$ **79.** $\pi/6, 7\pi/6$ **95.** 0.5625 **97.** -1.2137 **103. a.** 0.479425532 **b.** 0.866021271

Exercise Set 6.4, page 334

1. $\pi/3$ **3.** $7\pi/4$ **5.** $2\pi/3$ **7.** $\pi/4$ **9.** $7\pi/4$ **11.** 2.4 **13.** $30°$ **15.** $330°$ **17.** $126°$ **19.** $(1/2, \sqrt{3}/2)$
21. $(-\sqrt{2}/2, \sqrt{2}/2)$ **23.** $(-\sqrt{3}/2, -1/2)$ **25.** $(1/2, -\sqrt{3}/2)$ **27.** $(\sqrt{3}/2, -1/2)$ **29.** $(0, -1)$ **31.** $(1/2, -\sqrt{3}/2)$
33. $(-1/2, -\sqrt{3}/2)$ **35.** 0.9391 **37.** -3.3805 **39.** -1.1528 **41.** -0.2679 **43.** 0.8090 **45.** 48.0889 **47.** $\sqrt{2}/2$ **49.** $\sqrt{3}/2$
51. $-\sqrt{3}/2$ **53.** $1/2$ **55.** $-\sqrt{3}/2$ **97.** $-\sqrt{3}/2$ **99.** $-1/2$ **101.** -1 **103.** $\pm\sqrt{\sec^2\phi - 1}$ **105.** $\pm\sqrt{1 + \tan^2\phi}$
107. $\pm\sqrt{1 + \cot^2\phi}$ **109. a.** OF   **b.** OB   **c.** EF

Exercise Set 6.5, page 345

1. $2, 2\pi$ **3.** $1, \pi$ **5.** $1/2, \pi$ **7.** $2, 4\pi$ **9.** $1/2, 2\pi$ **11.** $1, 8\pi$ **13.** $2, 6\pi$ **15.** $3, 3\pi$

17.   **19.**   **21.**   **23.**   **25.**

27.   **29.**   **31.**   **33.**   **35.**

37.   **39.**   **41.**

43. $y = \cos 2x$ **45.** $y = 3 \sin \dfrac{2}{3}x$ **47.** $y = -2 \cos \pi x$

49.   **51.**   **53.**   **55.**

57. even function   **59.** $y = 2 \sin \frac{2}{3}x$ **61.** $y = 0.5 \sin \frac{8}{5}x$ **63.** $y = 4 \sin \pi x$ **65.** $y = 3 \cos 4x$ **67.** $y = 1.8 \cos \frac{4}{3}x$

69. $y = 3 \cos \frac{4\pi}{5}x$ **71.** $f(t) = 60 \cos 20t$

period $= \pi/5$

Exercise Set 6.6, page 353

1. $\pi/2 + k\pi$, k an integer   **3.** $\pi/2 + k\pi$, k an integer   **5.** 2π **7.** π **9.** 2π **11.** $2\pi/3$ **13.** $\pi/3$ **15.** 8π **17.** 1

19. 4   **21.**   **23.**   **25.**   **27.**

29.   **31.**   **33.**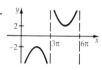

35.   **37.**   **39.**

41. 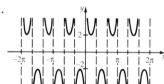   **43.** 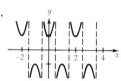   **45.** 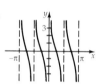   **47.**

49. $f(x) = \frac{3}{2}x$ **51.** $f(x) = \csc \frac{2}{3}x$ **53.** $f(x) = \sec \frac{3}{4}x$ **55.** $y = \tan 3x$ **57.** $y = \sec \frac{8}{3}x$ **59.** $y = \cot \frac{\pi}{2}x$ **61.** $y = \csc \frac{4\pi}{3}x$

63. 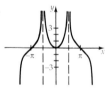   **65.** 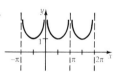   **67.**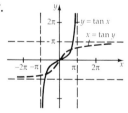

Exercise Set 6.7, page 359

1. $2, \pi/2, 2\pi$ **3.** $1, \pi/8, \pi$ **5.** $4, -\pi/4, 3\pi$ **7.** $5/4, 2\pi/3, 2\pi/3$ **9.** $\pi/8, \pi/2$ **11.** $-3\pi, 6\pi$ **13.** $\pi/16, \pi$ **15.** $-12\pi, 4\pi$

17.    **19.** 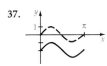   **21.** 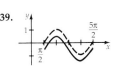   **23.**   **25.**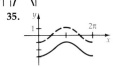

27.   **29.** 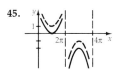   **31.** 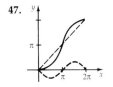   **33.** $f(x) = \sin x + 1$ **35.**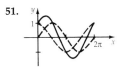

37. 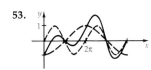   **39.** 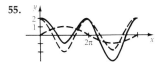   **41.** 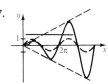   **43.** 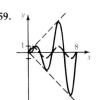   **45.**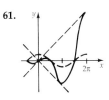

47.    **49.**   **51.** 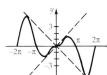   **53.**

55. 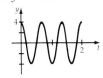   **57.**   **59.**   **61.**

3. $f(x) = \sin\left(2x - \dfrac{\pi}{3}\right)$ **65.** $f(x) = \csc\left(\dfrac{x}{2} - \pi\right)$ **67.** $f(x) = \sec\left(x - \dfrac{\pi}{2}\right)$ **69.** $f(x) = 2\sin\left(2x - \dfrac{2\pi}{3}\right)$ **71.** $f(x) = \tan\left(\dfrac{x}{2} - \dfrac{\pi}{4}\right)$

73. $f(x) = \sec\left(\dfrac{x}{2} - \dfrac{3\pi}{8}\right)$ **75.** 1 **77.** $\cos^2 x + 2$ **79.**   **81.**   **83.** $y = 3\cos\dfrac{\pi}{6}t + 9$ $y = 12$

Exercise Set 6.8, page 365

1. $2, \pi, 1/\pi$ **3.** $3, 3\pi, 1/3\pi$ **5.** $4, 2, 1/2$ **7.** $3/4, 4, 1/4$ **9.** $y = 4\cos 3\pi t$ **11.** $y = \dfrac{3}{2}\cos\dfrac{4\pi}{3}t$

13. $y = 2 \sin 2t$ **15.** $y = \sin \pi t$ **17.** $y = 2 \sin 2\pi t$

  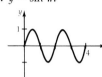

19. $y = \dfrac{1}{2} \cos 4t$ **21.** $y = 3 \cos 2t$ **23.** $y = \dfrac{1}{2} \cos \dfrac{2\pi}{3} t$ **25.** $y = 4 \cos 4t$ **27.** 4π, $1/4\pi$, 2   **29.** $y = \cos \dfrac{4\sqrt{3}}{3} t$

$y = 2 \cos \dfrac{1}{2} t$ $\pi\sqrt{3}/2$, $2\sqrt{3}/3\pi$

31. $\sqrt{10}\,\pi$, $\sqrt{10}/10\pi$, 6   **33.** $y = \dfrac{1}{2} \cos 2\sqrt{2}\,t$,   **35.** period $= 2/3$ **37.** Period increases by a factor of $\sqrt{2}$

$y = 6 \cos \dfrac{\sqrt{10}}{5} t$ $f = \sqrt{2}/\pi$ $y = \dfrac{1}{2} \cos 3\pi t$

Challenge Exercises, page 367

1. false, initial site must be along the positive x-axis   **3.** true   **5.** false, $1 + \tan^2 \theta = \sec^2 \theta$ is an identity   **7.** false, the period is 2π
9. false, $\sin 45° + \cos(90 - 45°) \neq 1$ **11.** false, $\sin^2 30 \neq (\sin 30)^2$ **13.** false, 1 rad $\approx 57°$

Review Exercises, page 368

1. a.   **b.**   **3. a.** $114°48'$ **b.** $-38°22'48''$ **5. a.** $7\pi/4$ **b.** 1.70

7. 3.67 in   **9.** $14.32°$ **11.** 3.49 m²   **13.** 53.78 rad/s   **15.** $\sin \theta = -3\sqrt{10}/10$ $\csc \theta = -\sqrt{10}/3$ **17.** $\sqrt{5}/2$ **19.** $3\sqrt{5}/5$
$\cos \theta = \sqrt{10}/10$ $\sec \theta = \sqrt{10}$
$\tan \theta = -3$ $\cot \theta = -1/3$

21. a. -1 **b.** 0.5365   **23. a.** -3.2361 **b.** 0   **25. a.** $-2\sqrt{3}/3$ **b.** $-\sqrt{3}$ **27 a.** $\sqrt{2}$ **b.** -1 **29.** $-\dfrac{4 + \sqrt{6}}{4}$ **31.** $\dfrac{\sqrt{3} - 2\sqrt{2}}{2}$
33. $-1/2$ **35.** $-\sqrt{2}/2$ **37.** 1.955   **39.** 0.035   **41.** 2.6   **43.** 0.85

51.   **53.**   **55.**   **57.** 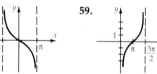   **59.**

61.   **63.**   **65.** 2.5, $\pi/25$, $25/\pi$ **67.** π, $1/\pi$, 0.5, $y = 0.5 \cos 2t$

Exercise Set 7.1, page 374

1. $\cos^2 x - 5 \cos x + 6$ **3.** $\sin^4 x - 17 \sin^2 x + 60$ **5.** $12 \sin^2 x + 7 \sin x - 10$ **7.** $\cos x \sin x + 7 \cos x - 2 \sin x - 14$
9. $3 \sin^2 x - 2 \sin x \cos x - 8 \cos^2 x$ **11.** $(\sin x - 1)^2$ **13.** $(4 \tan x - 3)(2 \tan x - 5)$ **15.** $(2 \sin x - 1)(2 \sin x + 1)$

17. $(\sin x - 3)(\sin^2 x + 3 \sin x + 9)$ **19.** $(\cos x + 1)(\sin x + 1)$ **21.** $\dfrac{\cos x + 3 \sin x}{\sin x \cos x}$ **23.** $\dfrac{2 \sin^2 x - \cos^2 x}{\sin x \cos x}$ **25.** $\dfrac{9 \cos x + 2}{\cos x(3 \cos x + 1)}$

27. $\dfrac{\sin x + 1}{\sin x - 1}$ **83.** $\pm\sqrt{1 - \sin^2 x}$ **85.** $\pm\dfrac{\sqrt{1 - \sin^2 x}}{1 - \sin^2 x}$

Exercise Set 7.2, page 383

1. $\dfrac{\sqrt{6} + \sqrt{2}}{4}$ 3. $\dfrac{\sqrt{6} + \sqrt{2}}{4}$ 5. 0   7. $\dfrac{-\sqrt{6} + \sqrt{2}}{4}$ 9. $-\dfrac{\sqrt{6} + \sqrt{2}}{4}$ 11. $2 + \sqrt{3}$ 13. $\dfrac{\sqrt{6} - \sqrt{2}}{4}$ 15. $-\dfrac{\sqrt{6} + \sqrt{2}}{4}$

17. $2 + \sqrt{3}$ 19. $\dfrac{\sqrt{2} + \sqrt{6}}{4}$ 21. $-\dfrac{\sqrt{6} + \sqrt{2}}{4}$ 23. $-2 - \sqrt{3}$ 25. 0   27. $\sqrt{3}/2$ 29. $1/2$ 31. $1/2$ 33. $\sqrt{3}$

35. $\cos 10°$ 37. $-\sin 40°$ 39. $\sin 5\pi/12$ 41. $\cos 5\pi/12$ 43. $\tan 7\pi/12$ 45. $\sin 5x$ 47. $\cos x$ 49. $\sin 4x$ 51. $\cos 2x$
53. $\sin x$ 55. $\tan 7x$ 57. $-\sin 1$ 59. $\cos 3$ 61. a. $-63/65$ b. $-56/65$ 63. a. $63/65$ b. $56/65$ c. $33/56$
65. a. $-77/85$ b. $-84/85$ c. $-13/84$ 67. a. $-33/65$ b. $-16/65$ c. $63/16$

Exercise Set 7.3, page 389

1. $\sin 4\alpha$ 3. $\cos 4\beta$ 5. $\cos 6\alpha$ 7. $\tan 6\alpha$ 9. $\dfrac{\sqrt{2 - \sqrt{3}}}{2}$ 11. $\sqrt{2} + 1$ 13. $-\dfrac{\sqrt{2 + \sqrt{2}}}{2}$ 15. $\dfrac{\sqrt{2 - \sqrt{2}}}{2}$

17. $\dfrac{\sqrt{2 - \sqrt{2}}}{2}$ 19. $\dfrac{\sqrt{2 - \sqrt{3}}}{2}$ 21. $-2 - \sqrt{3}$ 23. $\dfrac{\sqrt{2 + \sqrt{3}}}{2}$ 25. $\sin 2\theta = -24/25$, $\cos 2\theta = 7/25$, $\tan 2\theta = -24/7$
27. $\sin 2\theta = -240/289$, $\cos 2\theta = 161/289$, $\tan 2\theta = -240/161$ 29. $\sin 2\theta = -336/625$, $\cos 2\theta = -527/625$, $\tan 2\theta = 336/527$
31. $\sin \theta = 240/289$, $\cos 2\theta = -161/289$, $\tan 2\theta = 240/161$ 33. $\sin 2\theta = -720/1681$, $\cos 2\theta = 1519/1681$, $\tan 2\theta = -720/1519$
35. $\sin 2\theta = 240/289$, $\cos 2\theta = -161/289$, $\tan 2\theta = -240/161$ 37. $\sin \alpha/2 = \sqrt{26}/26$, $\cos \alpha/2 = 5\sqrt{26}/26$, $\tan \alpha/2 = 1/5$
39. $\sin \alpha/2 = 5\sqrt{34}/34$, $\cos \alpha/2 = -3\sqrt{34}/34$, $\tan \alpha/2 = -5/3$ 41. $\sin \alpha/2 = \sqrt{5}/5$, $\cos \alpha/2 = 2\sqrt{5}/5$, $\tan \alpha/2 = 1/2$
43. $\sin \alpha/2 = \sqrt{2}/10$, $\cos \alpha/2 = -7\sqrt{2}/10$, $\tan \alpha/2 = -1/7$ 45. $\sin \alpha/2 = \sqrt{17}/17$, $\cos \alpha/2 = 4\sqrt{17}/17$, $\tan \alpha/2 = 1/4$
47. $\sin \alpha/2 = 5\sqrt{34}/34$, $\cos \alpha/2 = -3\sqrt{34}/34$, $\tan \alpha/2 = -5/3$

Exercise Set 7.4, page 397

1. $\sin 3x - \sin x$ 3. $\dfrac{1}{2}[\sin 8x - \sin 4x]$ 5. $\sin 8x + \sin 2x$ 7. $\dfrac{1}{2}[\cos 4x - \cos 6x]$ 9. $1/4$ 11. $-\sqrt{2}/4$ 13. $-1/4$

15. $\dfrac{\sqrt{3} - 2}{4}$ 17. $2 \sin 3\theta \cos \theta$ 19. $2 \cos 2\theta \cos \theta$ 21. $-2 \sin 4\theta \sin 2\theta$ 23. $2 \cos 2\theta \cos \theta$ 25. $\sqrt{6}/2$ 27. $-\sqrt{6}/2$

29. $\sqrt{2}/2$ 31. $\sqrt{6}/2$ 49. $y = \sqrt{2} \sin(x - 135°)$ 51. $y = \sin(x - 60°)$ 53. $y = \dfrac{\sqrt{2}}{2} \sin(x - 45°)$ 55. $y = 17 \sin(x + 61.9°)$

57. $y = 8.5 \sin(x - 20.6°)$ 59. $y = \sqrt{2} \sin(x + 3\pi/4)$ 61. $y = \sin(x + \pi/6)$ 63. $y = 9.8 \sin(x + 1.99)$

65. $y = 5\sqrt{2} \sin(x + 3\pi/4)$ 67.   69.   71.

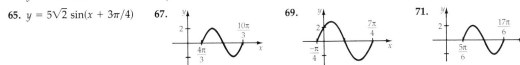

73.   75.   91. $\sqrt{2}$, 90°, 4π 93. 2, $\pi/12$, π 95. 2, 1/3, 2   99. 90°

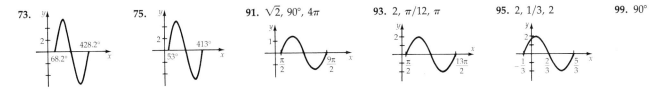

Exercise Set 7.5, page 409

1. $\pi/2$ 3. $5\pi/6$ 5. $-\pi/4$ 7. $\pi/3$ 9. $\pi/3$ 11. $-\pi/4$ 13. $-60°$ 15. 120°   17. 30°   19. 30°   21. 0.4729
23. 1.8087   25. -0.9620 27. 0.7943   29. $-16.1°$ 31. 56.3°   33. $-63.9°$ 35. $-45.7°$ 37. 106.7°   39. 84.0°   41. 1/2
43. 2   45. 3/5   47. 1   49. 1/2   51. $\pi/6$ 53. $\pi/4$ 55. not defined   57. 0.4636   59. $-\pi/6$ 61. $-\pi/3$ 63. 2.1735
65. -0.7430 67. 0.6675   69. 24/25   71. 13/5   73. $\pi/3$ 75. $\pi/6$ 77. $-\pi/4$ 79. 0   81. 24/25   83. 12/13   85. 2

87. 0   89. $\dfrac{2 + \sqrt{15}}{6}$ 91. $\dfrac{7\sqrt{3} - 3\sqrt{7}}{7 + 3\sqrt{21}}$ 93. $7\sqrt{2}/10$ 95. $\dfrac{\sqrt{6} - \sqrt{2}}{4}$ 97. $\sqrt{1 - x^2}$ 99. $\dfrac{\sqrt{x^2 - 1}}{x}$

105.   **107.**   **109.**

111.   **113.**   **115.**

121. $y = \dfrac{1}{3}\tan 5x$ **123.** $y = 3 + \cos\left(x - \dfrac{\pi}{3}\right)$ **125.**   **127.**   **129.** 0.1014

Exercise Set 7.6, page 415

1. $\pi/4, 7\pi/4$ **3.** $\pi/3, 4\pi/3$ **5.** $\pi/4, \pi/2, 3\pi/4, 3\pi/2$ **7.** $\pi/2, 3\pi/2$ **9.** $\pi/6, \pi/4, 3\pi/4, 11\pi/6$ **11.** $\pi/4, 3\pi/4$
13. $\pi/6, \pi/2, 5\pi/6$ **15.** $\pi/6, 5\pi/6, 7\pi/6, 11\pi/6$ **17.** $0, \pi/4, 3\pi/4, \pi, 5\pi/4, 7\pi/4$ **19.** $\pi/6, 5\pi/6, 4\pi/3, 5\pi/3$ **21.** $0, \pi/2, \pi, 3\pi/2$
23. 41.4°, 318.6°   **25.** no solution   **27.** 68.0°, 292°   **29.** no solution   **31.** 12.8°, 167.2°   **33.** 15.5°, 164.5°
35. 0°, 33.7°, 180°, 213.7°   **37.** no solution   **39.** no solution   **41.** 0, 120°, 240°   **43.** 70.5°, 289.5°   **45.** 68.2°, 116.6°, 248.2°, 296.6°
47. 19.5°, 90°, 160.5°, 270°   **49.** 60°, 90°, 300°   **51.** 53.1°, 180°   **53.** 72.4°, 220.2°   **55.** 50.1°, 129.9°, 205.7°, 334.3°   **57.** no solution
59. 22.5°, 157.5°   **61.** $\pi/8 + k\pi/2$ or $5\pi/8 + k\pi/2$ where k is an integer   **63.** $\pi/10 + 2k\pi/5$ where k is an integer
65. $0 + 2k\pi, \pi/3 + 2\pi k, \pi + 2k\pi, 5\pi/3 + 2k\pi$ where k is an integer   **67.** $\pi/2 + k\pi, 5\pi/6 + k\pi$ where k is an integer
69. $0° + 2k\pi$ where k is an integer   **71.** $0, \pi$ **73.** $0, \pi/6, 5\pi/6, \pi, 7\pi/6, 11\pi/6$ **75.** $0, \pi/2, 3\pi/2$ **77.** $0, \pi/3, 2\pi/3, \pi, 4\pi/3, 5\pi/3$
79. $4\pi/3, 5\pi/3$ **81.** $0, \pi/4, 3\pi/4, \pi, 5\pi/4, 7\pi/4$ **83.** $\pi/6, 5\pi/6, \pi$ **85.** $\pi/6, \pi/2$ **87.** $-\pi/3, 0$

89. $0, \pi/4, \pi/2, 3\pi/4, \pi, 5\pi/4, 3\pi/2, 7\pi/4$ **91.** $0, \pi/2, \pi, 3\pi/2$ **93.** $0, \pi/2, 5\pi/6, 7\pi/6, 3\pi/2, 11\pi/6$ **95.** $A = \dfrac{1}{2}r^2\theta$

97. $A = \dfrac{1}{2}r^2(\theta - \sin\theta)$

Challenge Exercises, page 419

1. false, $\dfrac{\tan 45°}{\tan 60°} \neq \dfrac{45°}{60°}$ **3.** false, $\sin^{-1}x \neq \dfrac{1}{\sin x}$ **5.** false, $\sin(30° + 60°) \neq \sin 30° + \sin 60°$
7. false, $\tan 45° = \tan 225°$ but $45° \neq 225°$ **9.** false, $\cos(\cos^{-1}(2)) \neq 2$ **11.** false, $\sin(180° - \theta) = \sin\theta$

Review Exercises, page 419

1. $\dfrac{\sqrt{6} - \sqrt{2}}{4}$ **3.** $\dfrac{\sqrt{6} - \sqrt{2}}{4}$ **5.** $-\dfrac{\sqrt{6} + \sqrt{2}}{4}$ **7.** $\dfrac{\sqrt{2 - \sqrt{2}}}{2}$ **9.** $\sqrt{2} + 1$ **11. a.** 0   **b.** $\sqrt{3}$ **c.** 1/2

13. a. $\sqrt{3}/2$ **b.** $-\sqrt{3}$ **c.** $-\dfrac{\sqrt{2 - \sqrt{3}}}{2}$ **15.** $\sin 6x$ **17.** $\sin 3x$ **19.** $\cos\beta$ **21.** 0.2740   **23.** $-\sqrt{3}/2$ **25.** $-2\sin 3\theta \cos 2\theta$
27. $2\sin 4\theta \sin 2\theta$ **47.** 13/5   **49.** 3/2   **51.** 4/5   **53.** 30°, 150°, 240°, 300°
55. $\pi/2 + 2k\pi, 3.8713 + 2k\pi, 5.553 + 2k\pi$ where k is an integer   **57.** $\pi/12, 5\pi/12, 13\pi/12, 17\pi/12$ **59.**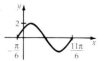

61. 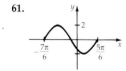   **63.**   **65.**

Exercise Set 8.1, page 426

1. 51°   **3.** 24°   **5.** 1570   **7.** 3.47   **9.** $B = 34°$, $b = 81$, $c = 140$ **11.** $B = 66.2°$, $a = 55.1$, $c = 137$
13. $A = 66.53°$, $b = 497.1$, $a = 1145$ **15.** $A = 68.61°$, $B = 21.39°$, $c = 34.74$ **17.** 14
19. $\sin A = 0.4472$ $\cot A = 0.8944$ $\tan A = 0.5$ **21.** 0.6   **23.** 0.625   **25.** $B = 48°$, $b = 11$ in, $c = 15$ in
$\csc A = 2.2361$ $\sec A = 1.1180$ $\cot A = 2$
27. $A = 41°$, $B = 49°$, $a = 4.0$ ft   **29.** 17 ft   **31.** 2230 ft   **33.** 2.94 mi   **35.** 38 mi   **37.** 5.2 m   **39.** 59 ft   **41.** 6.8°   **43.** 70 s

Exercise Set 8.2, page 431

1. $C = 77°$, $b = 16$, $c = 17$ **3.** $B = 38°$, $a = 18$, $c = 10$ **5.** $C = 32.6°$, $c = 21.6$, $a = 39.8$ **7.** $B = 47.7°$, $a = 57.4$, $b = 76.3$
9. $C = 59°$, $B = 84°$, $b = 46$ or   **11.** no solution   **13.** no solution   **15.** $C = 19.8°$, $B = 145.5°$, $b = 10.7$ **17.** no solution
$C = 121°$, $B = 22°$, $b = 17$ $C = 160.3°$, $B = 4.9°$, $b = 1.6$
19. 8.1 mi   **21.** 4.2 mi   **23.** 193 yd   **25.** 94 ft   **27.** 33.5 ft   **29.** 1170 mi

Exercise Set 8.3, page 438

1. 12.5   **3.** 150.5   **5.** 28.9   **7.** 9.5   **9.** 10.5   **11.** 40.1   **13.** 90.7   **15.** 38.7°   **17.** 89.6°   **19.** 47.9°   **21.** 116.7°
23. 80.3°   **25.** 139   **27.** 52.9   **29.** 58.8   **31.** 299   **33.** 36.3   **35.** 7.3   **37.** 709 mi   **39.** 347 mi   **41.** 40 cm
43. 9.17 in, 24.9 in   **45.** 55.4 cm   **47.** 2764 ft   **49.** 47,520 m^2 **51.** 203 m^2 **53.** 162 in^2 **55.** $41,479   **57.** 6.23 acres
59. 12.5°   **61.** 72.7°   **63.** 52.0 cm   **69.** 137 in^3

Exercise Set 8.4, page 450

1. $\sqrt{10}$, 161.6°   **3.** 9, 0°   **5.** 30, 180°   **7, 9.**

11. 5, 126.9°   **13.** 44.7, 296.6°   **15.** 4.5, 296.6°   **17.** 45.7, 336.8°   **19.** $x = -8$, $y = 5$ **21.** $x = -2$, $y = -6$ **23.** $x = -3$, $y = 4$
25. $x = 6$, $y = -4$ **27.** $x = 3.9$, $y = 4.6$ **29.** $6\mathbf{i} + \mathbf{j}$ **31.** $-5\mathbf{i} - 2\mathbf{j}$ **33.** $6\mathbf{i} + 9\mathbf{j}$ **35.** $5\mathbf{i} + 4\mathbf{j}$ **37.** $-6\mathbf{i} + \frac{17}{2}\mathbf{j}$ **39.** $\frac{5}{6}\mathbf{i} + 3\mathbf{j}$
41. $-3\mathbf{i} + 4\mathbf{j}$ **43.** $\mathbf{i} - 3\mathbf{j}$ **45.** $3.6\mathbf{i} + 3.5\mathbf{j}$ **47.** $6.1\mathbf{i} - 5.1\mathbf{j}$ **49.** -10 **51.** -10 **53.** 24   **55.** 186   **57.** 70.3°   **59.** 63.4°
61. 47.7°   **63.** 177.3°   **65.** 186 lb, 11°   **67.** 378 mph, 120°   **69.** 293 lb   **71.** 1192 ft-lb   **77.** $3\mathbf{i} + \mathbf{j}$ **79.** $\frac{21\sqrt{10}}{5}\mathbf{i} - \frac{7\sqrt{10}}{5}\mathbf{j}$
81. $-\sqrt{2}\mathbf{i} + \sqrt{2}\mathbf{j}$

Exercise Set 8.5, page 459

1–7.   **9.** $z = \sqrt{2}(\cos 315° + i\sin 315°)$ **11.** $2(\cos 330° + i\sin 330°)$ **13.** $3(\cos 90° + i\sin 90°)$

15. $5(\cos 180° + i\sin 180°)$ **17.** $\sqrt{2} + i\sqrt{2}$ **19.** $\frac{\sqrt{2}}{2} - \frac{i\sqrt{2}}{2}$ **21.** $-4.60 + 3.86i$ **23.** 8   **25.** $-\sqrt{3} + i$ **27.** $-3i$

29. $-4\sqrt{2} + 4i\sqrt{2}$ **31.** $\frac{9\sqrt{3}}{2} - \frac{9}{2}i$ **33.** $-0.832 + 1.819i$ **35.** $\frac{3}{2} - \frac{3\sqrt{3}}{2}$ **37.** $-4\sqrt{3} - 4i$ **39.** $6(\cos 225° + i\sin 225°)$

41. $12(\cos 335° + i\sin 335°)$ **43.** $10(\cos 184° + i\sin 184°)$ **45.** $15\left(\cos \frac{\pi}{4} + i\sin \frac{\pi}{4}\right)$ **47.** $15(\cos 165° + i\sin 165°)$

49. $-4 - 4i\sqrt{3}$ **51.** $-\frac{3\sqrt{3}}{2} + \frac{3i}{2}$ **53.** $0 + 3i$ **55.** $-\frac{5}{2} + \frac{5i\sqrt{3}}{2}$ **57.** $-\frac{1}{2} - \frac{i\sqrt{3}}{2}$ **59.** $-\frac{3}{8} + \frac{i\sqrt{3}}{8}$
61. $256(\cos 240° + i\sin 240°)$ **63.** $32(\cos 120° + i\sin 120°)$ **65.** $32(\cos 45° + i\sin 45°)$ **67.** $64(\cos 0° + i\sin 0°)$
69. $32(\cos 270° + i\sin 270°)$ **71.** $1024\sqrt{2}(\cos 315° + i\sin 315°)$ **73.** $\cos 270° + i\sin 270°$ **75.** $1.148 + 2.772i$, $-1.148 - 2.772i$

77. 1.879 + 0.684i **79.** 0.951 + 0.309i **81.** 1 + 0i **83.** 1.070 + 0.213i **85.** −0.276 + 1.563i **87.** $2\sqrt{2} + 2i\sqrt{6}$
        0.347 + 1.970i 0 + i −0.5 + 0.866i −0.213 + 1.070i −1.216 − 1.020i $-2\sqrt{2} - 2i\sqrt{6}$
       −1.532 + 1.286i −0.951 + 0.309i −0.5 − 0.866i −1.070 − 0.213i 1.492 − 0.543i
       −1.879 − 0.684i −0.588 − 0.809i 0.213 − 1.070i
       −0.347 − 1.970i 0.588 − 0.809i
        1.532 − 1.286i

89. cos 67.5° + i sin 67.5°   **91.** 3(cos 0° + i sin 0°)   **93.** 1.19(cos 75° + i sin 75°)
        cos 157.5° + i sin 157.5°   3(cos 120° + i sin 120°)   1.19(cos 165° + i sin 165°)
        cos 247.5° + i sin 247.5°   3(cos 240° + i sin 240°)   1.19(cos 255° + i sin 255°)
        cos 333.7° + i sin 333.5°   1.19(cos 345° + i sin 345°)

Challenge Exercises, page 462

1. false, cannot solve a triangle given the angle opposite one of the given sides   **3.** true   **5.** true   **7.** true   **9.** true
11. false, if $\mathbf{v} = i + j$ and $\mathbf{w} = i - j$, then $\mathbf{v} \cdot \mathbf{w} = 1 - 1 = 0$ **13.** false, $z^2 = r^2(\cos 2\theta + i \sin 2\theta)$ **15.** false, $i = \cos \pi/2 + i \sin \pi/2$

Review Exercises, page 462

1. $B = 53°, a = 11, c = 18$ **3.** $B = 48°, C = 95°, A = 37°$ **5.** $c = 13, A = 55°, B = 90°$ **7.** No triangle is formed
9. $C = 45°, a = 29, b = 35$ **11.** 357   **13.** 917   **15.** 792   **17.** 167   **19.** 4.47, 333°   **21.** 4.47, 153°   **23.** 3.61, 124°

25. 5.10, 11°   **27.** $x = -8, y = 5$ **29.** $x = 10, y = 6$ **31.** $-7i - 3j$ **33.** $-5i - \dfrac{5}{2}j$ **35.** −9   **37.** −9   **39.** 86°   **41.** 125°

43. 3.61, 304°   **45.** $2\sqrt{2}(\cos 315° + i \sin 315°)$ **47.** 3.54 − 3.54i **49.** −7.07 − 7.07i **51.** −6.01 − 13.74i

53. 3(cos 110° + i sin 110°)   **55.** $\sqrt{2}(\cos 285° + i \sin 285°)$ **57.** −0.5 + 0.866i **59.** 0 + 32,768i
61. 1.68(cos 22.5° + i sin 22.5°)   **63.** 1.07(cos 45° + i sin 45°)
        1.68(cos 112.5 + i sin 112.5°)   1.07(cos 117° + i sin 117°)
        1.68(cos 202.5° + i sin 202.5°)   1.07(cos 189° + i sin 189°)
        1.68(cos 292.5° + i sin 292.5°)   1.07(cos 261° + i sin 261°)
                                                   1.07(cos 333° + i sin 333°)

Exercise Set 9.1, page 471

1. vertex: $(0, 0)$ **3.** vertex: $(0, 0)$ **5.** vertex: $(2, -3)$
   focus: $(0, -1)$ focus: $(1/12, 0)$ focus: $(2, -1)$
   directrix: $y = 1$ directrix: $x = -1/12$ directrix: $y = -5$

7. vertex: $(2, -4)$ **9.** vertex: $(-4, 1)$ **11.** vertex: $(2, 2)$
   focus: $(1, -4)$ focus: $(-7/2, 1)$ focus: $(2, 5/2)$
   directrix: $x = 3$ directrix: $x = -9/2$ directrix: $y = 3/2$

13. vertex: $(-4, -10)$ **15.** vertex: $(-7/4, 3/2)$ **17.** vertex: $(-5, -3)$
   focus: $(-4, -39/4)$ focus: $(-2, 3/2)$ focus: $(-9/2, -3)$
   directrix: $y = -41/4$ directrix: $x = -3/2$ directrix: $x = -11/2$

19. vertex: $(-3/2, 13/12)$
focus: $(-3/2, 1/3)$
directrix: $y = 11/6$

21. vertex: $(2, -5/4)$
focus: $(2, -3/4)$
directrix: $y = -7/4$

23. vertex: $(9/2, -1)$
focus: $(35/8, -1)$
directrix: $x = 37/8$

25. vertex: $(1, 1/9)$
focus: $(1, 31/36)$
directrix: $y = -23/36$

27. $x^2 = -16y$

29. $(x + 1)^2 = 4(y - 2)$ **31.** $(x - 3)^2 = 4(y + 4)$ **33.** $(x + 4)^2 = 4(y - 1)$ **35.** 4   **37.** $4|p|$ **39.**

43.

45. $x^2 + y^2 - 8x - 8y - 2xy = 0$

Exercise Set 9.2, page 479

1. vertices: $(0, 5)$, $(0, -5)$
foci: $(0, 3)$, $(0, -3)$

3. vertices: $(3, 0)$, $(-3, 0)$
foci: $(\sqrt{5}, 0)$, $(-\sqrt{5}, 0)$

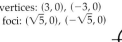

5. vertices $(2, 0)$, $(-2, 0)$
foci: $(1, 0)$, $(-1, 0)$

7. vertices: $(0, 5)$, $(0, -5)$
foci: $(0, 3)$, $(0, -3)$

9. vertices $(0, 4)$, $(0, -4)$
foci: $(0, \sqrt{39}/2)$, $(0, -\sqrt{39}/2)$

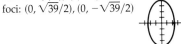

11. vertices: $(3, 6)$, $(3, 2)$
foci: $(3, 4 + \sqrt{3})$, $(3, 4 - \sqrt{3})$

13. vertices: $(-1, -3)$, $(5, -3)$
foci: $(0, -3)$, $(4, -3)$

15. vertices: $(2, 4)$, $(2, -4)$
foci $(2, \sqrt{7})$, $(2, -\sqrt{7})$

17. vertices: $(-1, 6)$, $(-1, -4)$
foci: $(-1, 4)$, $(-1, -2)$

19. vertices: $(11/2, -1)$ $(1/2, -1)$
foci: $(3 + \sqrt{17}/2, -1)$, $(3 - \sqrt{17}/2, -1)$

21. $\dfrac{x^2}{25} + \dfrac{y^2}{9} = 1$ **23.** $\dfrac{x^2}{36} + \dfrac{y^2}{16} = 1$ **25.** $\dfrac{x^2}{36} + \dfrac{y^2}{81/8} = 1$ **27.** $\dfrac{(x+2)^2}{16} + \dfrac{(y-4)^2}{7} = 1$ **29.** $\dfrac{(x-2)^2}{25/24} + \dfrac{(y-4)^2}{25} = 1$

31. $\dfrac{(x-5)^2}{16} + \dfrac{(y-1)^2}{25} = 1$ **33.** $\dfrac{x^2}{25} + \dfrac{y^2}{21} = 1$ **35.** $\dfrac{x^2}{20} + \dfrac{y^2}{36} = 1$ **37.** $\dfrac{(x-1)^2}{25} + \dfrac{(y-3)^2}{21} = 1$ **39.** $\dfrac{x^2}{80} + \dfrac{y^2}{144} = 1$

41. The information does not describe an ellipse.   **43.** $\dfrac{x^2}{36} + \dfrac{y^2}{27} = 1$ **45.** $\dfrac{(x-1)^2}{16} + \dfrac{(y-2)^2}{12} = 1$ **47.** $9/2$ **51.** $x = \pm\dfrac{9\sqrt{5}}{5}$

Exercise Set 9.3, page 489

1. center: $(0, 0)$
vertices: $(\pm 4, 0)$
foci: $(\pm\sqrt{41}, 0)$
asymptotes: $y = \pm 5x/4$

3. center: $(0, 0)$
vertices: $(0, \pm 2)$
foci: $(0, \pm\sqrt{29})$
asymptotes: $y = \pm 2x/5$

5. center: $(3, -4)$
vertices: $(7, -4)$, $(-1, -4)$
foci: $(8, -4)$, $(-2, -4)$
asymptotes: $y + 4 = \pm 3(x - 3)/4$

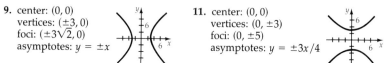

7. center: $(1, -2)$
vertices: $(1, 0)$, $(1, -4)$
foci: $(1, -2 \pm 2\sqrt{5})$
asymptotes: $y + 2 = \pm 1(x - 1)/2$

9. center: $(0, 0)$
vertices: $(\pm 3, 0)$
foci: $(\pm 3\sqrt{2}, 0)$
asymptotes: $y = \pm x$

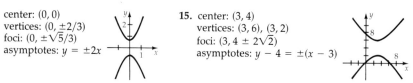

11. center: $(0, 0)$
vertices: $(0, \pm 3)$
foci: $(0, \pm 5)$
asymptotes: $y = \pm 3x/4$

13. center: $(0, 0)$
vertices: $(0, \pm 2/3)$
foci: $(0, \pm\sqrt{5}/3)$
asymptotes: $y = \pm 2x$

15. center: $(3, 4)$
vertices: $(3, 6)$, $(3, 2)$
foci: $(3, 4 \pm 2\sqrt{2})$
asymptotes: $y - 4 = \pm(x - 3)$

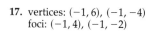

17. center: $(-2, -1)$
vertices: $(-2, 2)$, $(-2, -4)$
foci: $(-2, 2 \pm \sqrt{13})$
asymptotes: $y + 1 = \pm 3(x + 2)/2$

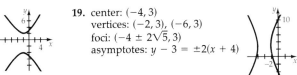

19. center: $(-4, 3)$
vertices: $(-2, 3)$, $(-6, 3)$
foci: $(-4 \pm 2\sqrt{5}, 3)$
asymptotes: $y - 3 = \pm 2(x + 4)$

21. center: $(2, -2)$
vertices: $(2, 1)$, $(2, -5)$
foci: $(2, 3)$, $(2, -7)$
asymptote: $y + 2 = \pm 3(x - 2)/4$

23. center: $(-1, -1)$
vertices: $(-1/2, -1)$, $(-3/2, -1)$
foci: $((-6 \pm \sqrt{13})/6, -1)$
asymptotes: $y + 1 = \pm 2(x + 1)/3$

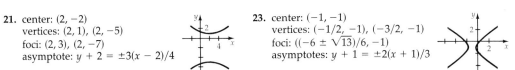

25. $\dfrac{x^2}{9} - \dfrac{y^2}{7} = 1$ **27.** $\dfrac{y^2}{20} - \dfrac{x^2}{5} = 1$ **29.** $\dfrac{y^2}{9} - \dfrac{x^2}{36/7} = 1$ **31.** $\dfrac{y^2}{16} - \dfrac{x^2}{64} = 1$ **33.** $\dfrac{(x-4)^2}{4} - \dfrac{(y-3)^2}{5} = 1$

35. $\dfrac{41(x-4)^2}{144} - \dfrac{41(y+2)^2}{225} = 1$ **37.** $\dfrac{(y-2)^2}{3} - \dfrac{(x-7)^2}{12} = 1$ **39.** $\dfrac{(y-7)^2}{1} - \dfrac{(x-1)^2}{3} = 1$ **41.** $\dfrac{x^2}{4} - \dfrac{y^2}{12} = 1$

43. $\dfrac{7(x-4)^2}{36} - \dfrac{(y-1)^2}{4} = 1$ and $\dfrac{7(y-1)^2}{36} - \dfrac{(x-4)^2}{4} = 1$

45. ellipse   **47.** parabola   **49.** parabola   **51.** ellipse

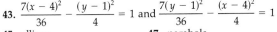

53. $\dfrac{(x - \sqrt{1215}/2)^2}{324} + \dfrac{y^2}{81/4} = 1$ **55.** $\dfrac{x^2}{2500} - \dfrac{y^2}{7500} = 1$ **57.** $\dfrac{x^2}{1} - \dfrac{y^2}{3} = 1$ **59.** $\dfrac{y^2}{9} - \dfrac{x^2}{7} = 1$ **61.** $y + 2 = \pm 9/5$ **67.**

Exercise 9.4, page 501

1–7.

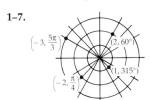

9.

11.

13.

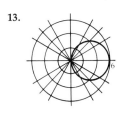

15.

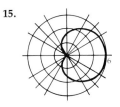

17.

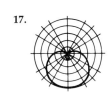

19.   **21.**   **23.**

25. 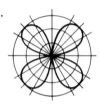 **27.**   **29.**

31. 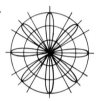 **33.**   **35.**

37.   **39.**   **41.** $(2, -60°)$

43. $(3/2, -3\sqrt{3}/2)$ **45.** $(0, 0)$ **47.** $(5, 53.1°)$ **49.** $x^2 + y^2 - 3x = 0$ **51.** $x = 3$ **53.** $x^2 + y^2 = 16$ **55.** $x^4 - y^2 + x^2 y^2 = 0$
57. $y^2 + 4x - 4 = 0$ **59.** $y = 2x + 6$ **61.** $r = 2\csc\theta$ **63.** $r = 2$ **65.** $r\cos^2\theta = 8\sin\theta$ **67.** $r^2\cos 2\theta = 25$

69.   **71.**   **73.**   **75.**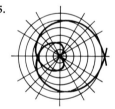

Challenge Exercises, page 503

1. false, a parabola has no asymptotes
3. false, by keeping foci fixed and varying asymptotes, we can make conjugate axis any size needed.
5. false, parabolas have no asymptotes   **7.** false, a parabola can be a function   **9.** true

Review Exercises, page 504

1. vertices: $(\pm 2, 0)$
foci: $(\pm 2\sqrt{2}, 0)$
asymptotes: $y = \pm x$

3. vertices: $(-1, -1)$, $(7, -1)$
foci: $(3 \pm 2\sqrt{3}, -1)$

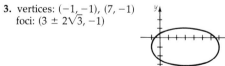

5. vertex: $(-2, 1)$
foci: $(-29/16, 1)$

7. vertices: $(-2, -2)$, $(-2, 4)$
foci: $(-2, 1 \pm \sqrt{5})$

9. vertices: $(-5, 2/3)$, $(7, 2/3)$
foci: $(1 \pm 2\sqrt{13}, 2/3)$
asymptotes: $y - 2/3 = \pm 2(x - 1)/3$

11. vertex: $(-7/2, -1)$
focus: $(-7/2, -3)$

13. $\dfrac{(x-2)^2}{25} + \dfrac{(y-3)^2}{16} = 1$ **15.** $\dfrac{(x+2)^2}{16} - \dfrac{(y-2)^2}{20} = 1$ **17.** $x^2 = 3(y+2)/2$ or $(y+2)^2 = 12x$ **19.** $\dfrac{x^2}{36} - \dfrac{y^2}{4/9} = 1$

21. $(y-3)^2 = -8x$ **23.** $\dfrac{(x-1)^2}{25} + \dfrac{(y-1)^2}{9} = 1$ **25.**

27.

29.

31.

33.

35. $r \sin^2 \theta = 16 \cos \theta$ **37.** $3r \cos \theta - 2r \sin \theta = 6$ **39.** $y^2 = 8x + 16$

41. $x^4 + y^4 + 2x^2 y^2 - x^2 + y^2 = 0$

Exercise Set 10.1, page 513

1. $(2, -4)$ **3.** $(-6/5, 27/5)$ **5.** $(3, 4)$ **7.** $(1, -1)$ **9.** $(3, -4)$ **11.** $(2, 5)$ **13.** $(-1, -1)$ **15.** $(62/25, 34/25)$ **17.** no solution
19. $(c, -4c/3 + 2)$ **21.** $(2, -4)$ **23.** $(0, 3)$ **25.** $(3c/5, c)$ **27.** $(-1/2, 2/3)$ **29.** no solution   **31.** $(-6, 3)$ **33.** $(2, -3/2)$
35. $(2\sqrt{3}, 3)$ **37.** $(38/17\pi, 3/17)$ **39.** $(\sqrt{2}, \sqrt{3})$ **41.** 120 mph, 30 mph   **43.** 25 mph, 5 mph   **45.** \$12 for iron, \$16 for lead
47. 12 nickels, 7 dimes   **49.** 86   **51.** \$14,000 at 6%, \$11,000 at 6.5%   **53.** 8 gm of 40% gold, 12 gm of 60% gold
55. 20 ml of 13% solution, 30 ml of 18% solution   **57.** $x = -58/17$, $y = 52/17$ **59.** $x = -2$, $y = -1$ **61.** $x = 153/26$, $y = 151/26$
63. $x = 2 + 3i$, $y = 1 - 2i$ **65.** $x = 3 - 5i$, $y = 4i$

Exercise Set 10.2, page 524

1. $(2, -1, 3)$ **3.** $(2, 0, -3)$ **5.** $(2, -3, 1)$ **7.** $(-5, 1, -1)$ **9.** $(3, -5, 0)$ **11.** $(0, 2, 3)$ **13.** $(5c - 25, 48 - 9c, c)$

21. no solution   **23.** $\left(\dfrac{25 + 4c}{29}, \dfrac{55 - 26c}{29}, c \right)$ **25.** $(0, 0, 0)$ **27.** $(5c/14, 4c/7, c)$ **29.** $(-11c, -6c, c)$ **31.** $(0, 0, 0)$

33. $y = 2x^2 - x - 3$ **35.** $x^2 + y^2 - 4x + 2y - 20 = 0$ **37.** center $(-7, -2)$, radius 13   **39.** 5 dimes, 10 nickels, 4 quarters
41. 685   **43.** $(3, 5, 2, -3)$ **45.** $(1, -2, -1, 3)$ **47.** $(14a - 7b - 8, -6a + 2b + 5, a, b)$ **49.** $A = -13/2$ **51.** $A \neq -3$, $A \neq 1$
53. $A = -3$ **55.** $3x - 5y - 2z = -2$

Exercise Set 10.3, page 529

1. $(5, 18)$ **3.** $(4, 2), (-4, 2), (4, -2), (-4, -2)$ **5.** $(4, 6), (6, 4)$ **7.** $(19/29, -11/29), (1, 1)$ **9.** $(-2, 9), (1, -3), (-1, 1)$
11. $(2, 1/2), (-2, 1/2), (2, -1/2), (-2, -1/2)$ **13.** $(12/5, 1/5), (2, 1)$ **15.** $(26/5, -3/5), (1, -2)$ **17.** $(39/10, -7/10), (3, 2)$
19. $\left(\dfrac{-3 + \sqrt{3}}{2}, \dfrac{1 + \sqrt{3}}{2} \right), \left(\dfrac{-3 - \sqrt{3}}{2}, \dfrac{1 - \sqrt{3}}{2} \right)$ **21.** $(19/13, 22/13), (1, 4)$ **23.** no solution   **25.** $(1, 5)$ **27.** $(-1, 1), (1, -1)$
29. $(1, -2), (-1, 2)$

Exercise Set 10.4, page 535

1. $A = -3, B = 4$ **3.** $A = -2/5, B = 1/5$ **5.** $A = 1, B = -1, C = 4$ **7.** $A = 1, B = 3, C = 2$ **9.** $A = 1, B = 0, C = 1, D = 0$
11. $\dfrac{3}{x} + \dfrac{5}{x + 4}$ **13.** $\dfrac{7}{x - 9} + \dfrac{-4}{x + 2}$ **15.** $\dfrac{5}{2x + 3} + \dfrac{3}{2x + 5}$ **17.** $\dfrac{20}{11(3x + 5)} + \dfrac{-3}{11(x - 2)}$ **19.** $x + 3 + \dfrac{1}{x - 2} + \dfrac{-1}{x + 2}$
21. $\dfrac{1}{x} + \dfrac{2}{x + 7} + \dfrac{-28}{(x + 7)^2}$ **23.** $\dfrac{2}{x} + \dfrac{3x - 1}{x^2 - 3x + 1}$ **25.** $\dfrac{2}{x + 3} + \dfrac{-1}{(x + 3)^2} + \dfrac{4}{x^2 + 1}$ **27.** $\dfrac{3}{x - 4} + \dfrac{5}{(x - 4)^2}$ **29.** $\dfrac{3x - 1}{x^2 + 10} + \dfrac{4x}{(x^2 + 10)^2}$
31. $\dfrac{1}{2k(k - x)} + \dfrac{1}{2k(k + x)}$ **33.** $x + \dfrac{1}{x} + \dfrac{-2}{x - 1}$ **35.** $2x - 2 + \dfrac{3}{x^2 - x - 1}$ **37.** $\dfrac{1}{5(x + 2)} + \dfrac{4}{5(x - 3)}$
39. $\dfrac{1}{x} + \dfrac{2}{x^2} + \dfrac{3}{x^4} + \dfrac{-2}{x - 2}$ **41.** $\dfrac{4}{3(x - 1)} + \dfrac{2x + 7}{3(x^2 + x + 1)}$

Exercise Set 10.5, page 541

1.   **3.**   **5.**   **7.**

9.   **11.**   **13.**   **15.**

17. 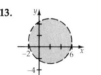   **19.**   **21.**   **23.**

25.   **27.** no solution   **29.**   **31.**

33.   **35.**   **37.**   **39.**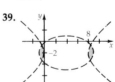

41. 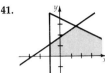 **43.**   **45.**   **47.**

49.   **51.**   **53.**   **55.**

57. a.   **b.**   If x is a negative number, then the inequality is reversed when multiplying both sides of the inequality by a negative number.

Exercise 10.6, page 547

1. minimum at $(0, 8)$: 16   **3.** minimum at $(6, 5)$: 71   **5.** minimum at $(0, 10/3)$: 20   **7.** maximum at $(0, 12)$: 72
9. minimum at $(0, 32)$: 32   **11.** maximum at $(0, 8)$: 56   **13.** minimum at $(2, 6)$: 18   **15.** maximum at $(3, 4)$: 25
17. minimum at $(2, 3)$: 12   **19.** maximum at $(100, 400)$: 3400   **21.** 20 acres of wheat and 40 acres of barley
23. 0 Starter sets and 18 Pro sets   **25.** 24 ounces of group B and 0 ounces of group A
27. 2 4-cylinder engines and 7 6-cylinder engines

Challenge Exercises, page 550

1. false, $\begin{cases} x + y = 1 \\ x + y = 2 \end{cases}$ has no solution.   **3.** false, a homogeneous system is one where one constant term in each equation is zero.
5. true   **7.** true   **9.** false, it is inconsistent.

Review Exercises, page 551

1. $(-18/7, -15/28)$ **3.** $(-3, -1)$ **5.** $(3, 1)$ **7.** $((5 - 3c)/2, c)$ **9.** $(1/2, 3, -1)$ **11.** $((7c - 3)/11, (16c - 43)/11, c)$
13. $(2, (3c + 2)/2, c)$ **15.** $(14c/11, -2c/11, c)$ **17.** $((c + 1)/2, 3c - 1/4, c)$ **19.** $(2, -3)$ **21.** no solution   **23.** $(1/5, 18/5), (1, 2)$

25. $(2, 0), (18/17, -64/17)$ **27.** $(2, 1), (-2, -1)$ **29.** $(2, -3), (-2, 3)$ **31.** $\dfrac{3}{x - 2} + \dfrac{4}{x + 1}$ **33.** $\dfrac{6x - 2}{5(x^2 + 1)} + \dfrac{-6}{5(x + 2)}$

35. $\dfrac{2}{x} + \dfrac{4}{x - 1} + \dfrac{5}{x + 1}$ **37.**   **39.**   **41.**

43.   **45.**   **47.**   **49.**   **51.**

53. 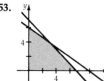   **55.**   **57.**   **59.**

61. maximum at $(4, 5)$: 18   **63.** minimum at $(0, 8)$: 8   **65.** minimum at $(2, 5)$: 27

67. $y = \dfrac{11}{6}x^2 - \dfrac{5}{2}x + \dfrac{2}{3}$ **69.** $z = -2x + 3y + 3$ **71.** wind: 28 mph, plane: 143 mph

73. $(0, 0, 0),\ (1, 1, 1),\ (1, -1, -1),\ (-1, -1, 1),\ (-1, 1, -1)$

Exercise Set 11.1, page 562

1. $\begin{bmatrix} 2 & -3 & 1 & 1 \\ 3 & -2 & 3 & 0 \\ 1 & 0 & 5 & 4 \end{bmatrix} \begin{bmatrix} 2 & -3 & 1 \\ 3 & -2 & 3 \\ 1 & 0 & 5 \end{bmatrix} \begin{bmatrix} 1 \\ 0 \\ 4 \end{bmatrix}$

3. $\begin{bmatrix} 2 & -3 & -4 & 1 & 2 \\ 0 & 2 & 1 & 0 & 2 \\ 1 & -1 & 2 & 0 & 4 \\ 3 & -3 & -2 & 0 & 1 \end{bmatrix} \begin{bmatrix} 2 & -3 & -4 & 1 \\ 0 & 2 & 1 & 0 \\ 1 & -1 & 2 & 0 \\ 3 & -3 & 0 & -2 \end{bmatrix} \begin{bmatrix} 2 \\ 2 \\ 4 \\ 1 \end{bmatrix}$

5. $\begin{bmatrix} 1 & -1 & 2 & 2 \\ 0 & 1 & -1 & -6 \\ 0 & 0 & 1 & -27/2 \end{bmatrix}$

7. $\begin{bmatrix} 1 & -2 & -1 & 3 \\ 0 & 1 & -1 & 1 \\ 0 & 0 & -1 & -\frac{13}{6} \end{bmatrix}$

9. $\begin{bmatrix} 1 & -2 & 3 & -4 \\ 0 & 1 & 2 & -1/2 \\ 0 & 0 & 1 & -2 \\ 0 & 0 & 0 & 0 \end{bmatrix}$

11. $\begin{bmatrix} 1 & -3 & 4 & 2 & 1 \\ 0 & 1 & -1 & -2 & -1 \\ 0 & 0 & 0 & 1 & 3 \end{bmatrix}$

13. $(2, -1, 1)$ **15.** $(1, -2, -1)$

17. $(2 - 2c, 2c + 1/2, c)$ **19.** $(1/2, 1/2, 3/2)$ **21.** $(16c, 6c, c)$ **23.** $(7c + 6, -11c - 8, c)$ **25.** $(c + 2, c, c)$ **27.** no solution

29. $(2, -2, 3, 4)$ **31.** $(21/10, -8/5, 2/5, -5/2)$ **33.** $(3, -3/2, 1, -1)$ **35.** $(27c/2 + 39, 5c/2 + 10, -4c - 10, c)$

37. $(c_1 - 12c_2/7 + 6/7, c_1 - 9c_2/7 + 1/7, c_1, c_2)$ **39.** $(1, 0, -2, 1, 2)$ **41.** $(77c/3 + 151/3, -25/c - 50/3, 14c/3 + 34/3, -3c - 7, c)$

43. $(-3c_1 + 6c_2 + 1, 2, -c_1 + 9c_2 - 8, c_1, c_2)$ **45.** all values $> a$ except $a \neq 1$ and $a \neq -6$ **47.** when $a = -6$ **49.** $z = 2x + 3y - 2$

51. $f(x) = \dfrac{1}{2}x^3 + \dfrac{1}{2}x^2$

Exercise Set 11.2, page 572

1. a. $\begin{bmatrix} 1 & 2 \\ 5 & 4 \end{bmatrix}$ **b.** $\begin{bmatrix} 3 & -4 \\ 1 & 2 \end{bmatrix}$ **c.** $\begin{bmatrix} -2 & 6 \\ 4 & 2 \end{bmatrix}$ **d.** $\begin{bmatrix} 7 & -11 \\ 0 & 3 \end{bmatrix}$

3. a. $\begin{bmatrix} -3 & 0 & 5 \\ 3 & 5 & -5 \end{bmatrix}$ **b.** $\begin{bmatrix} 3 & -2 & 1 \\ -1 & -5 & 1 \end{bmatrix}$ **c.** $\begin{bmatrix} -6 & 2 & 4 \\ 4 & 10 & -6 \end{bmatrix}$ **d.** $\begin{bmatrix} 9 & -5 & 0 \\ -4 & -15 & 5 \end{bmatrix}$

5. a. $\begin{bmatrix} 1 & 5 \\ 3 & -5 \\ 2 & -4 \end{bmatrix}$ **b.** $\begin{bmatrix} -7 & 3 \\ 1 & -1 \\ -4 & 4 \end{bmatrix}$ **c.** $\begin{bmatrix} 8 & 2 \\ 2 & -4 \\ 6 & -8 \end{bmatrix}$ **d.** $\begin{bmatrix} -18 & 5 \\ 1 & 0 \\ -11 & 12 \end{bmatrix}$

7. a. $\begin{bmatrix} -1 & 1 & -1 \\ 2 & 2 & 1 \\ -1 & 2 & 5 \end{bmatrix}$ **b.** $\begin{bmatrix} -3 & 5 & -1 \\ -2 & -4 & 3 \\ -7 & 4 & 1 \end{bmatrix}$ **c.** $\begin{bmatrix} 2 & -4 & 0 \\ 4 & 6 & -2 \\ 6 & -2 & 4 \end{bmatrix}$ **d.** $\begin{bmatrix} -7 & 12 & -2 \\ -6 & -11 & 7 \\ -17 & 9 & 0 \end{bmatrix}$

9. $\begin{bmatrix} -10 & 17 \\ 6 & -8 \end{bmatrix}, \begin{bmatrix} 0 & 22 \\ 1 & -18 \end{bmatrix}$

11. $\begin{bmatrix} 10 & 6 \\ 14 & -7 \end{bmatrix}, \begin{bmatrix} 14 & -1 \\ 0 & -11 \end{bmatrix}$

13. $\begin{bmatrix} 0 & -4 & 5 \\ 6 & 0 & 3 \\ -3 & -2 & 1 \end{bmatrix}, \begin{bmatrix} 5 & -13 \\ 5 & -4 \end{bmatrix}$

15. $\begin{bmatrix} 9 & -2 & -6 \\ 0 & -1 & 2 \\ 4 & -2 & -4 \end{bmatrix} \begin{bmatrix} 4 & -2 & 6 \\ 2 & -3 & 4 \\ 4 & -4 & 3 \end{bmatrix}$

17. $[0, 8]$

19. The product is not possible   **21.** $\begin{bmatrix} 0 & 0 \\ 0 & 0 \end{bmatrix}$ **23.** The product is not possible   **25.** $\begin{bmatrix} 1/3 & -5/3 \\ -1/3 & 4/3 \\ 1/3 & -4/3 \end{bmatrix}$ **27.** $\begin{bmatrix} -1 & 1 \\ 3 & 2 \\ 7 & -2 \end{bmatrix}$

29. $\begin{bmatrix} 1 & -3 \\ 1 & -2 \end{bmatrix}$ **31.** $\begin{bmatrix} 7 & -1 & 1 \\ 1 & 2 & 0 \\ 5 & -1 & 4 \end{bmatrix}$ **33.** $\begin{cases} 8x - 8y = 11 \\ 4x + 3y = 1 \end{cases}$ **35.** $\begin{cases} x - 3y - 2z = 6 \\ 3x + y = 2 \\ 2x - 4y + 5z = 1 \end{cases}$ **37.** $\begin{cases} 2x_1 - x_2 + 2x_4 = 5 \\ 4x_1 + x_2 + 2x_3 - 3x_4 = 6 \\ 6x_1 + x_3 - 2x_4 = 10 \\ 5x_1 + 2x_2 - x_3 - 4x_4 = 8 \end{cases}$

39. $\begin{bmatrix} 6 + 9i & 3 - 6i \\ 3 + 3i & 6 - 3i \end{bmatrix}$ **41.** $\begin{bmatrix} 2 + 2i & -6 + 4i \\ -4 + 6i & 2 + 8i \end{bmatrix}$ **43.** $\begin{bmatrix} 3 + 2i & 3 + i \\ 4 + 3i & 6 - 2i \end{bmatrix}$ **45.** $\begin{bmatrix} 12 - 3i & -3 + 3i \\ 10 + i & 6 - i \end{bmatrix}$ **47.** $\begin{bmatrix} -2 + 11i & 8 - 6i \\ 2 + 6i & 6 - 5i \end{bmatrix}$

Exercise Set 11.3, page 581

1. $\begin{bmatrix} -5 & -3 \\ -2 & -1 \end{bmatrix}$ **3.** $\begin{bmatrix} 5 & -2 \\ -1 & 1/2 \end{bmatrix}$ **5.** $\begin{bmatrix} -16 & -2 & 7 \\ 7 & 1 & -3 \\ -3 & 0 & 1 \end{bmatrix}$ **7.** $\begin{bmatrix} 15 & -1 & -4 \\ -11/2 & 1/2 & 3/2 \\ 3 & 0 & -1 \end{bmatrix}$ **9.** $\begin{bmatrix} 7/2 & -2 & -2 \\ -5/2 & 1 & 2 \\ -1 & 0 & 1 \end{bmatrix}$

11. $\begin{bmatrix} 19/2 & -1/2 & -3/2 & 3/2 \\ 7/4 & 1/4 & -1/4 & 3/4 \\ -7/2 & 1/2 & 1/2 & -1/2 \\ 1/4 & -1/4 & 1/4 & 1/4 \end{bmatrix}$ **13.** $\begin{bmatrix} 2 & 3/5 & -7/5 & 4/5 \\ 4 & -7/5 & -2/5 & 4/5 \\ -6 & 14/5 & -1/5 & -3/5 \\ 3 & -8/5 & 2/5 & 1/5 \end{bmatrix}$ **15.** $(2, 1)$ **17.** $(7/4, -25/8)$ **19.** $(1, -1, 2)$ **21.** $(23, -12, 3)$

23. $(0, 4, -6, -2)$ **25.** 95 adults, 25 children   **27.** Sample 1: 500 g of additive 1, 200 g of additive 2, 300 g of additive 3
Sample 2: 400 g of additive 1, 400 g of additive 2, 200 g of additive 3
29. \$194.40 million manufacturing, \$156.65 million transportation, \$121.70 million services
31. \$39.65 million coal, \$14.15 million iron, \$32.15 million steel   **39. a.** $\begin{bmatrix} -5/2 & 3/2 \\ -2 & 1 \end{bmatrix}$ **b.** $\begin{bmatrix} 2 & -3 \\ -3/2 & 5/2 \end{bmatrix}$ **c.** $\begin{bmatrix} 4 & 1 \\ -1 & 0 \end{bmatrix}$

Exercise Set 11.4, page 590

1. 13   **3.** -15 **5.** 0   **7.** 0   **9.** 19, 19   **11.** 1, -1 **13.** $-9, -9$ **15.** $-9, -9$ **17.** 10   **19.** 53   **21.** 20   **23.** 46
25. 0   **27.** row 2 consists of zeros. Therefore the determinant is zero.   **29.** 2 was factored from row 2
31. Row 1 was multiplied by -2 and added to row 2.   **33.** 2 was factored from column 1.
35. The matrix is in diagonal form. The value of the determinant is the product of the terms on the main diagonal.
37. Row 1 and Row 3 were interchanged. Therefore, the sign of the determinant was changed.
39. Each row of the determinant was multiplied by a.   **41.** 0   **43.** 0   **45.** 6   **47.** -90 **49.** 21   **51.** 9/2   **53.** $46\frac{1}{2}$
59. $7x + 5y = -1$ **61.** 263.5

Exercise Set 11.5, page 595

1. $x_1 = 44/31, x_2 = 29/31$ **3.** $x_1 = 1/3, x_2 = -2/3$ **5.** $x_1 = 2, x_2 = -7$ **7.** $x_1 = 0, x_2 = 0$ **9.** $x_1 = 41/32, x_2 = 15/8$
11. $x_1 = 21/17, x_2 = -3/17, x_3 = -29/17$ **13.** $x_1 = 32/49, x_2 = 13/49, x_3 = 42/49$ **15.** $x_1 = -29/64, x_2 = -25/64, x_3 = -19/32$
17. $x_1 = 50/53, x_2 = 62/53, x_3 = 4/53$ **19.** $x_1 = 0, x_2 = 0, x_3 = 0$ **21.** $x_2 = -35/19$ **23.** $x_1 = -121/131$ **25.** $x_4 = 4/3$
27. The determinant of the coefficient matrix is zero. Thus, Cramer's Rule cannot be used. The system of equations has infinitely many solutions.   **29.** all values of k except $k = 0$ **31.** all values of k except $k = 2$ **33.** $r = 3, s = -3$

Challenge Exercises, page

1. false, $A^2 = A \cdot A = \begin{bmatrix} 7 & 18 \\ 6 & 19 \end{bmatrix}$ **3.** false, a singular matrix does not have a multiplicative inverse   **5.** true
7. false, if the determinant of the coefficient matrix is zero, Cramer's Rule cannot be used to solve the system of equations.
9. false, the matrix must not only be nonsingular but also be square to have a multiplicative inverse.
11. false, if the system of equations has a solution, the Gauss-Jordan method can be used to solve the system. The solutions will be given in terms of one or more of the variables.   **13.** true   **15.** true

Review Exercises, page 598

1. $\begin{bmatrix} 6 & -3 & 9 \\ 9 & 6 & -3 \end{bmatrix}$ **3.** $\begin{bmatrix} -5 & 5 & -1 \\ 1 & -4 & 6 \end{bmatrix}$ **5.** $\begin{bmatrix} -1 & -15 \\ 7 & 1 \end{bmatrix}$ **7.** $\begin{bmatrix} -6 & -4 & 2 \\ 14 & 0 & 10 \\ -7 & -7 & 6 \end{bmatrix}$ **9.** $\begin{bmatrix} 12 & 28 & -5 \\ 2 & 6 & 0 \\ 6 & 16 & -1 \end{bmatrix}$ **11.** $\begin{bmatrix} -12 & -36 & -4 \\ 48 & 124 & 4 \\ -9 & -32 & -6 \end{bmatrix}$

13. not possible   **15.** $\begin{bmatrix} 7 & 24 & 9 \\ -10 & -22 & 1 \end{bmatrix}$ **17.** $\begin{bmatrix} -1 & -5 & 4 \\ 1/2 & 2 & -3/2 \\ 0 & -2 & 1 \end{bmatrix}$ **19.** $(2, -1)$ **21.** $(3, 0)$ **23.** $(3, 1, 0)$ **25.** $(1, 0, -2)$

27. $(3, -4, 1)$ **29.** $(-c - 2, -c - 3, c)$ **31.** $(1, -2, 2, 3)$ **33.** $(-37c + 2, 16c, -7c + 1, c)$ **35.** $\begin{bmatrix} -1 & 1 \\ -3/2 & 1 \end{bmatrix}$ **37.** $\begin{bmatrix} -2/7 & 3/14 \\ 1/7 & 1/7 \end{bmatrix}$

39. $\begin{bmatrix} 2 & -2 & 1 \\ 0 & 3/2 & -1 \\ -1 & -1 & 1 \end{bmatrix}$ **41.** $\begin{bmatrix} -10 & 20 & -3 \\ -5 & 9 & -1 \\ 3 & -6 & 1 \end{bmatrix}$ **43.** $\begin{bmatrix} -1 & -7 & 4 & 2 \\ -6 & -3 & 2 & 3 \\ 1 & 2 & -1 & -1 \\ -2 & 0 & 0 & 1 \end{bmatrix}$ **45.** The matrix does not have an inverse.

47. a. $(-18, 13)$ **b.** $(-22, 16)$ **49. a.** $(-18/7, 23/7, -6/7)$ **b.** $(-31/14, 20/7, 3/7)$ **51.** -2 **53.** -1 **55.** 0 **57.** 0
59. $x_1 = 16/19, x_2 = -2/19$ **61.** $x_1 = 13/44, x_2 = 1/4, x_3 = -17/44$ **63.** $x_1 = 54/69, x_2 = -26/69, x_3 = 38/69$ **65.** $x_3 = 115/126$
67. \$51.75 million computer division, \$14.13 million disk drive division, \$23.64 million moniter division.

Exercise Set 12.1, page 608

1. $0, 2, 6, a_8 = 56$ **3.** $0, 1/2, 2/3, a_8 = 7/8$ **5.** $1, -1/4, 1/9, a_8 = -1/64$ **7.** $-1/3, -1/6, -1/9, a_8 = -1/24$
9. $2/3, 4/9, 8/27, a_8 = 256/6561$ **11.** $0, 2, 0, a_8 = 2$ **13.** $1.1, 1.21, 1.331, a_8 = 2.14358881$ **15.** $1, -\sqrt{2}/2, \sqrt{3}/3, a_8 = -\sqrt{2}/4$
17. $1, 2, 6, a_8 = 40320$ **19.** $0, 0.3010, 0.4771, a_8 = 0.9031$ **21.** $1, 4, 2, a_8 = 4$ **23.** $3, 3, 3, a_8 = 3$ **25.** $5, 10, 20$ **27.** $2, 4, 12$
29. $2, 4, 16$ **31.** $2, 8, 48$ **33.** $3, \sqrt{3}, \sqrt[6]{3}$ **35.** $2, 5/2, 9/4$ **37.** 15 **39.** 40 **41.** $25/12$ **43.** 72 **45.** -24 **47.** 0.9031

49. 256 **51.** $\sum_{i=1}^{6} \frac{1}{i^2}$ **53.** $\sum_{i=1}^{7} 2^i (-1)^{i+1}$ **55.** $\sum_{i=0}^{4} 7 + 3i$ **57.** $\sum_{i=1}^{4} \frac{1}{2^i}$ **59.** 4320 **61.** 72 **63.** 56 **65.** 100 **67.** 2.6457520

69. $a_{20} = 1.0000037, a_{100} \approx 1$ **71.** 2.70833 **73.** $a_n = (x - 2n)\left[\dfrac{n(n-1)(n-2)(n-3)(n-4)}{n!}\right] + 2n$

75. $\frac{1}{2}(-1 + i\sqrt{3}), \frac{1}{2}(-1 - i\sqrt{3}), 1, \frac{1}{2}(-1 + i\sqrt{3}), \frac{1}{2}(-1 - i\sqrt{3}), 1, a_{99} = 1$

Exercise Set 12.2, page 614

1. $a_9 = 38, a_{24} = 98, a_n = 4n + 2$ **3.** $a_9 = -10, a_{24} = -40, a_n = 8 - 2n$ **5.** $a_9 = 16, a_{24} = 61, a_n = 3n - 11$
7. $a_9 = 25, a_{24} = 70, a_n = 3n - 2$ **9.** $a_9 = a + 16, a_{24} = a + 46, a_n = a + 2n - 2$
11. $a_9 = \log 7 + 8 \log 2, a_{24} = \log 7 + 23 \log 2, a_n = \log 7 + (n - 1) \log 2$ **13.** $a_9 = 9 \log a, a_{24} = 24 \log a, a_n = n \log a$
15. 45 **17.** -79 **19.** 185 **21.** -555 **23.** 468 **25.** 525 **27.** -465 **29.** $78 + 12x$ **31.** $210x$ **35.** $3, 7, 11, 15, 19$
37. $5/2, 2, 3/2, 1$ **39.** 20 on 6th row, 135 in the 6 rows   **41.** \$48,750   **43.** 784 ft   **45.** $1800°$ **47.** $a_n = 5n - 2$ **49.** $a_{100} = 301$

Exercise Set 12.3, page 620

1. yes, $r = 4$ **3.** no   **5.** yes, $r = 2^x$ **7.** yes, $r = 2$ **9.** yes, $r = x^2$ **11.** no   **13.** 2^{2n-1} **15.** $4(-3)^{n-1}$ **17.** $6\left(\dfrac{2}{3}\right)^{n-1}$

19. $-6\left(-\dfrac{5}{6}\right)^{n-1}$ **21.** $\left(-\dfrac{1}{3}\right)^{n-3}$ **23.** $(-x)^{n-1}$ **25.** c^{3n-1} **27.** $3\left(\dfrac{1}{100}\right)^n$ **29.** $5(0.1)^n$ **31.** $45(0.01)^n$ **33.** 18 **35.** -2

37. 363 **39.** $\dfrac{1330}{729}$ **41.** $\dfrac{279{,}091}{390{,}625}$ **43.** -341 **45.** $147{,}620$ **47.** $1/2$ **49.** $-2/5$ **51.** $9/91$ **53.** $1/9$ **55.** $5/7$ **57.** $1/3$
59. $5/11$ **61.** $41/333$ **63.** $422/999$ **65.** $229/900$ **67.** $1994/1650$ **69.** \$2271.93
71. Since $\log r$ is a constant, the sequence $\log a_n$ is an arithmetic sequence.   **73.** The total amount removed is 1.   **75.** $a_1^n r^{(n-1)n/2}$
77. 45 ft   **79.** 2046 **81.** $8A/5$

Exercise Set 12.5, page 634

1. 35 **3.** 36 **5.** 220 **7.** 1 **9.** $x^6 - 6x^5y + 15x^4y^2 - 20x^3y^3 + 15x^2y^4 - 6xy^5 + y^6$ **11.** $x^5 + 15x^4 + 90x^3 + 270x^2 + 405x + 243$
13. $128x^7 - 448x^6 + 672x^5 - 560x^4 + 280x^3 - 84x^2 + 14x - 1$ **15.** $x^6 + 18x^5y + 135x^4y^2 + 540x^3y^3 + 1215x^2y^4 + 1458xy^5 + 729y^6$

17. $16x^4 - 160x^3y + 600x^2y^2 - 1000xy^3 + 625y^4$ **19.** $x^6 + 6x^4 + 15x^2 + 20 + \dfrac{15}{x^2} + \dfrac{6}{x^4} + \dfrac{1}{x^6}$

21. $x^{14} - 28x^{12} + 336x^{10} - 2240x^8 + 8960x^6 - 21{,}504x^4 + 28{,}672x^2 - 1638$ **23.** $32x^{10} + 80x^8y^3 + 80x^6y^6 + 40x^4y^9 + 10x^2y^{12} + y^{15}$

25. $\dfrac{16}{x^4} - \dfrac{16}{x^2} + 6 - x^2 + \dfrac{x^4}{16}$ **27.** $s^{-12} + 6s^{-8} + 15s^{-4} + 20 + 15s^4 + 6s^8 + s^{12}$ **29.** $-3240x^3y^7$ **31.** $1056x^{10}y^2$ **33.** $126x^2y^2\sqrt{x}$

35. $165b^5/a^5$ **37.** $180a^2b^8$ **39.** $60x^2y^8$ **41.** $61{,}236a^5b^5$ **43.** $126s^{-1}$, $126s$ **45.** $-7 - 24i$ **47.** $41 - 38i$ **49.** 1

51. $nx^{n-1} + \dfrac{n(n-1)x^{n-2}h}{2} + \dfrac{n(n-1)(n-2)x^{n-3}h^2}{6} + \cdots + h^{n-1}$ **57.** 1.1712 **59.** 756 **61.** 56

Exercise Set 12.6, page 641

1. 30 **3.** 70 **5.** 1 **7.** 1 **9.** 210 **11.** 720 **13.** 125
15. There are 676 ways to arrange 26 letters taken two at a time. Now if there are more than 676 employees then at least two employees will have the same first and last initials. **17.** 76 **19.** 1024 **21.** 3,838,380 **23. a.** 21 **b.** 105 **c.** 21
25. a. 1287 **b.** 5148 **c.** 4512 **d.** 108,336 **27. a.** 792 **b.** 120 **29.** 42 **31.** 66 **33.** 220 **35.** 2,162,160 **37.** $n(n-1)/2$
39. 15 different sums **41.** 4960

Exercise Set 12.7, page 648

1. $\{S_1R_1, S_1R_2, S_1R_3, S_2R_1, S_2R_2, S_2R_3, R_1R_2, R_1R_3, R_2R_3, S_1S_2\}$ **3.** $\{H1, H2, H3, H4, T1, T2, T3, T4\}$
5. Let A, B, C represent cans and 1.2 represent the balls
$\{A00, A01, A02, A12, B00, B01, B02, B12, C00, C01, C02, C12\}$
7. $\{HSC, HSD, HCD, SCD\}$ **9.** $\{ae, ai, ao, au, ei, eo, eu, io, iu, ou\}$ **11.** $\{HHHH\}$
13. $\{TTTT, HTTT, THTT, TTHT, TTTH, TTHH, THTH, HTHT, THHT, HTTH, HHTT\}$ **15.** $\varnothing$ **17.** $\{(1,1), (2,2), (3,3), (4,4), (5,5), (6,6)\}$

19. $\{(1,4), (2,4), (3,4), (4,4), (5,4), (6,4)\}$ **21. a.** $1/13$ **b.** $1/4$ **23.** 0.97 **25.** $1/2$ **27.** $1/4$ **29.** 0.9999 **31.** $\dfrac{1}{1{,}000{,}000{,}000}$

33. $34/57$ **35.** $97/990$ **37.** $\dfrac{11{,}011}{38{,}019}$ **39.** $1/3$ **41.** $27/64$ **43.** $1/10$ **45.** $\left(\dfrac{7}{8}\right)^2$ **47.** $\dfrac{81}{10{,}000}$

Challenge Exercises, page 652

1. false. $0! \cdot 4! = 1 \cdot 4 \cdot 3 \cdot 2 \cdot 1 = 24$ **3.** true **5.** false, $\dfrac{(k+1)^3}{k^3} = (1 + 1/k)^3$ is not a constant **7.** true

9. false, see Exercise 36, Section 4 **11.** false, there are $m - n$ ways **13.** true **15.** true

Review Exercises, page 652

1. $a_3 = 9$, $a_7 = 49$ **3.** $a_3 = 11$, $a_7 = 23$ **5.** $a_3 = 1/8$, $a_7 = 1/128$ **7.** $a_3 = 1/6$, $a_7 = 1/5040$ **9.** $a_3 = 8/27$, $a_7 = 128/2187$
11. $a_3 = 18$, $a_7 = 1458$ **13.** $a_3 = 6$, $a_7 = 5040$ **15.** $a_3 = 8$, $a_7 = 16$ **17.** $a_3 = 2$, $a_7 = 256$ **19.** $a_3 = -54$, $a_7 = -3{,}675{,}160$
21. neither **23.** arithmetic **25.** geometric **27.** neither **29.** geometric **31.** geometric **33.** neither **35.** arithmetic
37. neither **39.** neither **41.** 63 **43.** 152 **45.** 378 **47.** $-14{,}763$ **49.** 1.9653 **51.** 0.8280 **53.** $1/3$ **55.** $-4/9$
65. $1024a^5 - 1280a^4b + 640a^3b^2 - 160a^2b^3 + 20ab^4 - b^5$
67. $a^4 + 16a^{3/2}b^{1/2} + 112a^3b + 448a^{5/2}b^{3/2} + 1120a^2b^2 + 1792a^{3/2}b^{5/2} + 1792ab^3 + 1024a^{1/2}b^{7/2} + 256b^4$ **69.** $241{,}920x^3y^4$ **71.** 26^8 **73.** 2730
75. 672 **77.** $1/8$, $3/8$ **79.** 0.285 **81.** Drawing an ace and 10 card from one deck **83.** $1/4$

Index